Measuring Arthropod Biodiversity

Jean Carlos Santos • Geraldo Wilson Fernandes
Editors

Measuring Arthropod Biodiversity

A Handbook of Sampling Methods

Springer

Editors
Jean Carlos Santos
Department of Ecology
Universidade Federal de Sergipe
São Cristóvão, Sergipe, Brazil

Geraldo Wilson Fernandes
Department of Genetics, Ecology
and Evolution
Instituto de Ciências Biológicas
Universidade Federal de Minas Gerais
Belo Horizonte, Minas Gerais, Brazil

ISBN 978-3-030-53228-4 ISBN 978-3-030-53226-0 (eBook)
https://doi.org/10.1007/978-3-030-53226-0

This Springer imprint is published by the registered company Springer Nature Switzerland AG
The registered company address is: Gewerbestrasse 11, 6330 Cham, Switzerland

Preface

In 1988, E. O. Wilson, in his book *Biodiversity*, stated: "The biological diversity most threatened is also the least explored, and there is no prospect at the moment that the scientific task will be completed before a large fraction of the species vanish." He still goes further to state: "the magnitude and control of biological diversity is not just a central problem of evolutionary biology; it is one of the key problems of science as a whole. At present, there is no way of knowing whether there are 5, 10, or 30 million species on Earth. There is no theory that can predict what this number might turn out to be. [...] unless an effort is made to understand all of diversity, we will fall far short of understanding life in these important respects, and due to the accelerating extinction of species, much of our opportunity will slip away forever." Given the unprecedented levels of human disturbances on the planet, and following the reasoning of E. O. Wilson, it is vital to assess insect biodiversity and predict anthropic pressures on them to provide sound science-based data to build solid and long-lasting policy measures for their conservation and sustainable ecosystem services.

Convinced of E. O. Wilson's brilliant words, 93 authors from different institutions have collaborated to bring forward this book which we believe will aid in the task of identifying this central question of "how many species exist on Earth?" Thus, in this book, we have attempted to gather key state-of-the-art techniques and methodologies appropriate to sampling arthropods in the field according to the emerging ecological trends. Arthropoda is the most diverse taxon on the Earth. This group's biodiversity represents one of the most marvelous and astonishing facets of organic evolution in our planet. Unfortunately, this phylum is also proportionally little known. There is a wide range of sampling or survey methods for investigating different arthropod groups, where each method has its own merits and limitations. In the first part of this book, we focus on chapters discussing sampling methods and techniques for insects, after which, in the second part, we concentrate on arachnids and, finally in the third part, on guilds and functional groups. The book provides an important review for zoologists, entomologists, arachnologists, ecologists, students, researchers, and anyone else interested in arthropod sciences. It helps to choose the most appropriate survey methodology according to each arthropod taxon. We hope that this book will contribute to the advancement of knowledge regarding field assessments and conservation strategy of arthropod species. We expect also that a future edition of this book will include methods for additional arthropod taxa, including other freshwater and marine groups, e.g., shrimps and crabs, as well as other neglected groups, e.g., centipedes and millipedes.

São Cristóvão, Sergipe, Brazil Jean C. Santos
Belo Horizonte, Minas Gerais, Brazil G. Wilson Fernandes

Contents

Contributors

Riuler Corrêa Acosta Programa de Pós-Graduação em Biologia Animal, Universidade Federal do Rio Grande do Sul, Porto Alegre, RS, Brazil

Larissa Simões Corrêa de Albuquerque Departamento de Zoologia, Universidade Federal de Pernambuco, Cidade Universitária, Recife, PE, Brazil

Elis Paula de Almeida Batista Laboratory of Innovation Technologies in Vector Control, Department of Parasitology, Biological Sciences Institute, Federal University of Minas Gerais, Belo Horizonte, Minas Gerais, Brazil

Wanessa Rejane de Almeida Department of Ecology, Universidade Federal de Sergipe, São Cristóvão, Sergipe, Brazil

Walter Santos de Araújo Departamento de Biologia Geral, Centro de Ciências Biológicas e da Saúde, Universidade Estadual de Montes Claros, Montes Claros, Minas Gerais, Brazil

Maria Fernanda Barberena-Aerias School of Natural Sciences and Technology, Universidad Ana G. Méndez, Gurabo, Puerto Rico

Bruno Corrêa Barbosa Laboratório de Ecologia Comportamental e Bioacústica, Universidade Federal de Juiz de Fora, Juiz de Fora, Brazil

Edwin Bedoya-Roqueme Programa de Pós-graduação, Recursos Naturais do Cerrado, RENAC, Universidade Estadual de Goiás, Anápolis, GO, Brazil

Research Group on Marine and Coastal Biodiversity BIODIMARC, Study Group on Arachnology PALPATORES, University of Cordoba, Córdoba, Colombia

Márcio Perez Bolfarini Departamento de Ecologia e Biologia Evolutiva, Universidade Federal de São Carlos, São Carlos, SP, Brazil

Brian V. Brown Entomology Section, Natural History Museum of Los Angeles County, Los Angeles, CA, USA

Bianca Caitano Programa de Pós-Graduação em Ecologia e Conservação da Biodiversidade, Universidade Estadual Santa Cruz, Ilhéus, Bahia, Brazil

Marcos Callisto Laboratório de Ecologia de Bentos, Instituto de Ciências Biológicas, Universidade Federal de Minas Gerais, Belo Horizonte, Minas Gerais, Brazil

Lucas Denadai de Campos Instituto de Biociências, Universidade de São Paulo, São Paulo, SP, Brazil

Junia Y. O. Carreira Depto. Biologia Animal, Instituto de Biologia, Unicamp, Campinas, São Paulo, Brazil

Renan Carrenho Museu de Zoologia, Universidae de São Paulo, São Paulo, SP, Brazil

Gerasimos Cassis Evolution and Ecology Research Centre, School of Biological, Earth and Environmental Sciences, University of New South Wales, Sydney, NSW, Australia

Diego M. P. Castro Laboratório de Ecologia de Bentos, Instituto de Ciências Biológicas, Universidade Federal de Minas Gerais, Belo Horizonte, Minas Gerais, Brazil

Rodrigo Roucourt Cezário LESTES Lab, Federal University of São Carlos, São Carlos, SP, Brazil

University of São Paulo—USP, Ribeirão Preto, SP, Brazil

Jung-Tai Chao Taiwan Forestry Research Institute, Taipei, Taiwan , ROC

Reginaldo Constantino Departamento de Zoologia, Universidade de Brasília, Brasilia, Brazil

Adolfo Cordero-Rivera ECO-EVO Lab, University of Vigo, Pontevedra, Spain

Fábio Correia Costa Departamento de Zoologia, Universidade Federal de Pernambuco, Cidade Universitária, Recife, PE, Brazil

Janaina Camara Siqueira da Cunha Department of Entomology, Texas A&M University, College Station, TX, USA

Gerome Darla-West Department of Biology, York University, Toronto, ON, Canada

Wesley DaRocha Laboratório de Mirmecologia, Centro de Pesquisa do Cacau/CEPLAC, Itabuna, Bahia, Brazil

Jacques Delabie Laboratório de Mirmecologia, Centro de Pesquisa do Cacau/CEPLAC, Itabuna, Bahia, Brazil

Departamento de Ciências Agrárias e Ambientais, Universidade Estadual Santa Cruz, Ilhéus, Bahia, Brazil

Pavel Dodonov Instituto de Biologia, Universidade Federal da Bahia, Rua Barão de Jeremoabo, Salvador, Bahia, Brazil

José F. Dzul-Cauich Red de Interacciones Multitróficas, Instituto de Ecología A.C., Xalapa, Veracruz, Mexico

Álvaro Eduardo Eiras Laboratory of Innovation Technologies in Vector Control, Department of Parasitology, Biological Sciences Institute, Federal University of Minas Gerais, Belo Horizonte, Minas Gerais, Brazil

Geraldo Wilson Fernandes Department of Genetics, Ecology and Evolution, Instituto de Ciências Biológicas, Universidade Federal de Minas Gerais, Belo Horizonte, Minas Gerais, Brazil

Marcos Fianco Programa de Pós-Graduação em Ciências Biológicas (Entomologia), Universidade Federal do Paraná, Curitiba, PR, Brazil

Paloma Pena Firme LESTES Lab, Federal University of São Carlos, São Carlos, SP, Brazil

André V. L. Freitas Depto. Biologia Animal and Museu de Zoologia, Instituto de Biologia, Unicamp, Campinas, São Paulo, Brazil

Grizelle González USDA-FS, International Institute of Tropical Forestry, Río Piedras, Puerto Rico

Patrícia E. Gueratto Depto. Biologia Animal, Instituto de Biologia, Unicamp, Campinas, São Paulo, Brazil

Rhainer Guillermo LESTES Lab, Federal University of São Carlos, São Carlos, SP, Brazil

Mónica Hernández-López Red de Interacciones Multitróficas, Instituto de Ecología A.C., Xalapa, Veracruz, Mexico

Vicente Hernández-Ortiz Red de Interacciones Multitróficas, Instituto de Ecología A.C., Xalapa, Veracruz, Mexico

Wei Huang Department of Environmental Sciences, Nanjing Forestry University, College of Biology and the Environment, University of Puerto Rico, Río Piedras, Puerto Rico

Luciana Iannuzzi Departamento de Zoologia, Universidade Federal de Pernambuco, Cidade Universitária, Recife, PE, Brazil

Cristiano A. Iserhard Programa de Pós-Graduação em Biologia Animal, Depto. de Ecologia, Zoologia e Genética, Instituto de Biologia, Universidade Federal de Pelotas, Pelotas, Rio Grande do Sul, Brazil

Benoit Jahyny Laboratório de Mirmecologia do Sertão, Colegiado de Ciências Biológicas, Campus de Ciências Agrárias, Universidade Federal do Vale do São Francisco, Petrolina, Pernambuco, Brazil

Leandro Juen Laboratory of Ecology and Conservation, Federal University of Pará, Belém, PA, Brazil

Natalia Kirichenko Sukachev Institute of Forest, Siberian Branch of Russian Academy of Sciences, Krasnoyarsk, Russia

Siberian Federal University, Krasnoyarsk, Russia

Elmo Koch Laboratório de Mirmecologia, Centro de Pesquisa do Cacau/CEPLAC, Itabuna, Bahia, Brazil

Ricardo Koroiva Departamento de Sistemática e Ecologia, Universidade Federal da Paraíba, João Pessoa, PB, Brazil

Arturo Goldaracena Lafuente Earth and Life Institute—ELIB, Université catholique de Louvain, Louvain La Neuve, Belgium

Fernando Willyan Trevisan Leivas Departamento de Biodiversidade, Universidade Federal do Paraná, Setor Palotina, Jardim Dallas, Palotina, PR, Brazil

Maurice Leponce Biodiversity Monitoring and Assessment, Royal Belgian Institute of Natural Sciences (RBINS), Brussels, Belgium

Evolutionary Biology and Ecology, Université Libre de Bruxelles (ULB), Brussels, Belgium

Marcos Gonçalves Lhano Centro de Ciências da Natureza, Universidade Federal de São Carlos, São Carlos, SP, Brazil

Carolina Nunes Liberal Departamento de Sistemática e Ecologia, Universidade Federal da Paraíba, João Pessoa, PB, Brazil

Elison. Fabricio B. Lima Universidade Federal do Piauí, Campus Amílcar Ferreira Sobral—CAFS, Floriano, PI, Brazil

Marden S. Linares Laboratório de Ecologia de Bentos, Instituto de Ciências Biológicas, Universidade Federal de Minas Gerais, Belo Horizonte, Minas Gerais, Brazil

Nancy Lo-Man-Hung Instituto de Biociências, Universidade de São Paulo, São Paulo, SP, Brazil

Carlos Lopez-Vaamonde INRAE, UR0633 Zoologie Forestière, Orléans, France

Institut de Recherche sur la Biologie de l'Insecte UMR 7261, CNRS—Université de Tours, Tours, France

Giselle M. Lourenço Departamento de Genética, Ecologia e Evolução, Instituto de Ciências Biológicas, Universidade Federal de Minas Gerais, Belo Horizonte, MG, Brazil

Tatiane Tagliatti Maciel Laboratório de Ecologia Comportamental e Bioacústica, Universidade Federal de Juiz de Fora, Juiz de Fora, Brazil

Artur Campos Dália Maia Departamento de Sistemática e Ecologia, Universidade Federal da Paraíba, João Pessoa, PB, Brazil

Jonathan Majer School of Molecular and Life Sciences, Curtin University, Perth, WA, Australia

School of Biological Sciences, University of Western Australia, Perth, WA, Australia

Clea Mariano Departamento de Ciências Biológicas, Universidade Estadual Santa Cruz, Ilhéus, Bahia, Brazil

Rita Marullo Dipartimento di Agraria, Università degli Studi Mediterranea di Reggio Calabria, Reggio Calabria, Italy

Felipe F. F. Moreira Laboratório de Biodiversidade Entomológica, Fundação Oswaldo Cruz, Instituto Oswaldo Cruz, Rio de Janeiro, RJ, Brazil

Riccardo Mugnai Laboratório BIOCICLOS, Centro de Ciências Agrárias e Ambientais, Universidade Federal do Maranhão, Chapadinha, Maranhão, Brazil

Issei Ohshima Department of Life and Environmental Sciences, Kyoto Prefectural University, Kyoto, Japan

Elliott Centeno de Oliveira Universidade Federal de Pelotas, Capão do Leão, RS, Brazil

Claudia M. Ospina-Sánchez USDA-FS, International Institute of Tropical Forestry, Río Piedras, Puerto Rico

Laurence Packer Department of Biology, York University, Toronto, ON, Canada

Thais Giovannini Pellegrini Departamento de Entomologia, Universidade Federal de Lavras, Campus Universitário da Universidade Federal de Lavras, Lavras, MG, Brazil

Gabrielle C. Pestana LESTES Lab, Federal University of São Carlos, São Carlos, SP, Brazil

Victor Mateus Prasniewski Programa de Pós-Graduação em Ecologia e Conservação da Biodiversidade, Universidade Federal de Mato Grosso, Cuiabá, MT, Brazil

Fábio Prezoto Laboratório de Ecologia Comportamental e Bioacústica, Universidade Federal de Juiz de Fora, Juiz de Fora, Brazil

Marcelo Carvalho de Resende Laboratory of Innovation Technologies in Vector Control, Department of Parasitology, Biological Sciences Institute, Federal University of Minas Gerais, Belo Horizonte, Minas Gerais, Brazil

Department of Entomology, Ministry of Health of Brazil, National Health Foundation, Belo Horizonte, Minas Gerais, Brazil

Aline Richter Programa de Pós-Graduação em Biologia Animal, Depto. de Ecologia, Zoologia e Genética, Instituto de Biologia, Universidade Federal de Pelotas, Pelotas, Rio Grande do Sul, Brazil

Augusto H. B. Rosa Depto. Biologia Animal, Instituto de Biologia, Unicamp, Campinas, São Paulo, Brazil

Renato Portela Salomão Instituto Nacional de Pesquisas da Amazônia, Manaus, AM, Brazil

Jean Carlos Santos Department of Ecology, Universidade Federal de Sergipe, São Cristóvão, Sergipe, Brazil

Jessie P. Santos Depto. Biologia Animal, Instituto de Biologia, Unicamp, Campinas, São Paulo, Brazil

Carlos Eduardo Sarmiento Laboratorio de Sistemática y Biología Comparada de Insectos, Universidad Nacional de Colombia, Edificio 425, Bogotá, Colombia

Timothy Schowalter Entomology Department, Louisiana State University, Baton Rouge, LA, USA

Cristiano F. Schwertner Departamento de Ecologia e Biologia Evolutiva, Instituto de Ciências Ambientais, Químicas e Farmacêuticas, Universidade Federal de São Paulo (UNIFESP), Diadema, SP, Brazil

Museu de Zoologia, Universidae de São Paulo, São Paulo, SP, Brazil

Ricardo R. Siewert Departamento de Zoologia, Universidade Federal do Paraná, Curitiba, Paraná, Brazil

Estevão Alves Silva Instituto Federal de Educação, Ciência e Tecnologia Goiano, Campus Urutaí, Urutaí, GO, Brazil

Universidade do Estado de Mato Grosso, Nova Xavantina, MT, Brazil

Pedro Guilherme Barrios de Souza Dias Departamento de Entomologia do Museu Nacional, Universidade Federal do Rio de Janeiro, Museu Nacional, Rio de Janeiro, RJ, Brazil

Milton de Souza Mendonça Jr Departamento de Ecologia, Instituto de Biociências, Universidade Federal do Rio Grande do Sul, Porto Alegre, Rio Grande do Sul, Brazil

Department of Ecology and Evolutionary Biology, UCLA, Los Angeles, CA, USA

Thamyrys Bezerra de Souza Instituto Nacional da Mata Atlântica, Santa Teresa, ES, Brazil

Carlos Frankl Sperber Laboratório de Orthopterologia, Departamento de Biologia Geral, Universidade Federal de Viçosa, Viçosa, MG, Brazil

Neucir Szinwelski Centro de Ciências Biológicas e da Saúde, Universidade Estadual do Oeste do Paraná, Cascavel, PR, Brazil

Everton Tizo-Pedroso Centro de Ensino e Aprendizagem em Rede, Laboratório de Ecologia Comportamental de Aracnídeos, Programa de Pós-Graduação em Recursos Naturais do Cerrado, Universidade Estadual de Goiás, Anápolis, GO, Brazil

Ana Lúcia Tourinho NEBAM—Núcleo de Estudos da Biodiversidade da Amazônia Matogrossense, Instituto de Ciências Naturais, Humanas e Sociais, Universidade Federal do Mato Grosso, Sinop, MT, Brazil

Maria Virgínia Urso-Guimarães Departamento de Biologia, Laboratório de Sistemática de Diptera, Universidade Federal de São Carlos, Sorocaba, São Paulo, Brazil

Natállia Vicente Laboratório de Orthopterologia, Departamento de Biologia Geral, Universidade Federal de Viçosa, Viçosa, MG, Brazil

Diogo S. Vilela LESTES Lab, Federal University of São Carlos, São Carlos, SP, Brazil

University of São Paulo—USP, Ribeirão Preto, SP, Brazil

Edison Zefa Departamento de Zoologia e Genética, Universidade Federal de Pelotas, Capão do Leão, RS, Brazil

Part I

Chapter 1
Arthropods: Why It Is So Crucial to Know Their Biodiversity?

Jean Carlos Santos, Wanessa Rejane de Almeida, and Geraldo Wilson Fernandes

1.1 Arthropod Biodiversity

The phylum Arthropoda is the taxa with the highest evolutionary success of all of Earth's diversity (Pisani 2009; Minelli et al. 2013) and Insecta, the largest and most diverse taxa within the Arthropoda (Grimaldi and Engel 2005). Divergence of this phylum is estimated to have originated about 520 million years ago (Giribet and Edgecombe 2019). Arthropods constitute a monophyletic group that is well supported by morphological characteristics, such as an articulated chitinous exoskeleton divided into repetitive appendages and separated by arthrodial membrane and by molecular phylogenetics (Giribet and Edgecombe 2019). All extant arthropods are placed in two large monophyletic groups: Chelicerata (e.g., spiders, mites, sea spiders, scorpions, and horseshoe crabs) and Mandibulata [myriapods (e.g., centipedes and millipedes), insects, and crustaceans (e.g., shrimps and crabs)] (Giribet and Edgecombe 2019) (Fig. 1.1).

For decades, researchers have tried to unravel how many species (described and mostly undescribed) of arthropods are there on Earth (Hamilton et al. 2010; May 2010). There is a range of estimates in the scientific literature (Hamilton et al. 2010, 2013; May 2010; Stork et al. 2015); however some predictions consider only insects (Gaston 1991), and these estimates too differ wildly (see Stork 2018). Erwin (1982) was the first study to estimate the total numbers of arthropod species. This visionary paper estimated 30 million species of arthropods in the tropics alone (Erwin 1982). But as to the wide range of estimates, Hamilton et al. (2010), for instance, predict merely 3.7 million and 2.5 million tropical arthropod species. Using probabilistic

J. C. Santos (✉) · W. R. de Almeida
Department of Ecology, Universidade Federal de Sergipe, São Cristóvão, Sergipe, Brazil

G. W. Fernandes
Department of Genetics, Ecology and Evolution, Instituto de Ciências Biológicas, Universidade Federal de Minas Gerais, Belo Horizonte, Minas Gerais, Brazil

J. C. Santos, G. W. Fernandes (eds.), *Measuring Arthropod Biodiversity*,
https://doi.org/10.1007/978-3-030-53226-0_1

Fig. 1.1 Some examples of arthropods (Arthropoda): (**a**) crab (Malacostraca: Decapoda), (**b**) millipede (Myriapoda: Diplopoda), (**c**, **d**) spider (Arachnida: Araneae), (**e**) fly (Insecta: Diptera), (**f**) bee (Insecta: Hymenoptera), (**g**) beetle (Insecta: Coleoptera), and (**h**) butterfly (Insecta: Lepidoptera)

models, Hamilton et al. (2013) estimated global arthropod species to be 6.1 million, with a range between 3.6 and 11.4 million. The most current estimate (Stork 2018) predicts 7.0 and 5.5 million species for arthropods and insects, respectively. According to Catalogue of Life from Zhang (2013, 2020), the estimated number of species known to taxonomists is 1,257,040 species for Arthropoda and 1,053,578 species for Insecta. This represents, respectively, 82% and 69% of all animal diversity (1,525,728 spp.). Considering a global estimate of 10.9 million species globally (Mora et al. 2011), much of this diversity can be attributed to arthropods.

Arthropods in addition to their diversity are also ubiquitous, abundant, and functionally important components of terrestrial and aquatic ecosystems, occupying varied habitats and niches along a range of temporal and spatial scales (Price et al. 2011). In terms of biomass, arthropods are the single most important group of animals, representing a range of 0.2 to 1.0 gigatons of carbon (Gt C) for terrestrial and marine arthropods. This represents 46% of all animal biomass on the planet (Bar-On et al. 2018). Arthropods are considered key components because of their critical ecological functions. In terms of interactions, arthropods' relationships with other organisms and with the environment are extraordinary (Fig. 1.2), e.g., ant-plant interactions (Del-Claro et al. 2018) and pollination mutualisms between insects and plants (New 2017) evidenced by the fact that a majority of plant species are pollinated by insects. Wardhaugh (2015) suggests that over 350,000 described arthropod species, which represents 30% of arthropod species, may regularly visit flowers to feed, mate, or obtain other resources. Another example are the interactions of arthropods and soil (e.g., meso- and macrofauna), crucial in maintaining soil fertility (Culliney 2013; Menta and Remelli 2020), including agricultural systems (Roy et al. 2018). Arthropod species may also represent an important component of litter/soil communities (~85% of all soil species) (Culliney 2013). Furthermore, arthropod richness in tropical rainforests can be huge (Basset et al. 2012). Arthropods can belong to different ecological and functional groups, e.g., decomposers, dispersers, herbivores, parasitoids, predators, and pollinators (Gullan and Cranston 2014; Del Claro and Guillermo 2019).

According to Noriega et al. (2018), arthropods (especially insects) are fundamental to regulation across many ecosystem services due to their ecological functions, e.g., provisioning, supporting, regulating, and cultural services. Arthropods provide several economic services such as food chain supplementation and have uses as medical and industrial products for humans (Ratcliffe et al. 2011; Macadam and Stockan 2015; Noriega et al. 2018). Beyond that, in natural and agroecosystems, in urban areas, and in terrestrial and aquatic ecosystems, arthropods perform many important ecosystem services, such as climate regulation, decomposition, pollination, matter and nutrient cycling, energy flow, pathogen and pest suppression, population regulation, seed dispersal, soil fertility regulation, biodiversity protection, environmental and ecological bioindicators, tourism and recreational services, and scientific, educational, cultural (art, music, and literature), and technological tools, among many other services (Gerlach et al. 2013; Macadam and Stockan 2015; Novais et al. 2016; Schowalter et al. 2018; Samways 2019; Cardoso et al. 2020). On the other hand, arthropod species can also have major impacts on

Fig. 1.2 Arthropods are megadiverse organisms in species and relationships with other organisms. Arthropods are parasites of many species such as (**a**) ticks, an ectoparasite of frogs and other vertebrates, (**b**) parasitic mites on a grasshopper, and (**c**) some plant-parasitic insects can induce galls on plants causing an abnormal growth of cells and tissues. (**d**) Herbivorous insects can cause several damages in the reproductive and vegetative plant parts. (**e**) Insects are excellent pollinators and they have many mutualistic interactions with plants. (**f**) Ants interact with many organisms, including aphids; in this type of interaction, ants consume aphid exudates, and in return, aphids gain protection against natural enemies. Arthropods are important predators in ecosystems, such as (**g**) this spider preying on a butterfly and (**h**) this hemipteran preying on an insect

human activities, e.g., economy, agriculture, health, etc. For instance, arthropod pests (e.g., insects and mites) destroy between 18 and 26% of annual crop production worldwide (Culliney 2014), resulting in economic losses of $ 470 billion annually. Several emerging infectious diseases (e.g., malaria, leishmaniasis, trypanosomiasis, borreliosis) are transmitted by arthropod vectors (e.g., louse, mites, mosquitoes, flies, fleas, and ticks), which are responsible for morbidity, mortality, and economic loss worldwide (Goddard 2008; Bernard et al. 2015). Otherwise, only ca. 1% of all known insect species cause crop losses of 20–80% globally, and fewer than 1% of mosquito species transmit diseases.

The number of studies indicating that arthropods are particularly vulnerable to the biodiversity loss crisis is augmenting. Human impacts are the cause of current arthropod declines, potentially leading them to species extinction. The arthropod declines and extinction are driven by many factors, such as habitat loss and fragmentation, invasive species, pollution, climate warming, co-extinction of dependent species, and overexploitation (Sánchez-Bayo and Wyckhuys 2019; Cardoso et al. 2020). Given their huge ecological importance, the loss of arthropods in the current biodiversity crisis is of particular concern, because arthropods can be highly sensitive to anthropogenic disturbance in comparison to vertebrates or plants (Thomas et al. 2004). Beyond that, many arthropod species are highly specialized in other organisms, e.g., animal parasites (Goater et al. 2014) and plant herbivores (Schoonhoven et al. 2005). If pessimist predictions of loss are realized, anthropogenic activities may have a profound impact on arthropod diversity and on the loss of ecosystem services they provide (Cardoso et al. 2020), reduction in arthropod abundances (Hallmann et al. 2017; Lister and Garcia 2018), restructuring of food webs (Lister and Garcia 2018), and overall loss of diversity (Sánchez-Bayo and Wyckhuys 2019; Eisenhauer et al. 2019; Cardoso et al. 2020). All these shall lead to devastating effects on ecosystem functioning and deep impacts on ecosystem service provisioning. For instance, nearly 90% of flowering plant species, including almost 75% of the world's major crops, are pollinated by insects (Klein et al. 2007). The value of insect pollination services is now being calculated for many countries around the world, but it is estimated to exceed $200 billion annually (Novais et al. 2016; Rader et al. 2016). For these reasons, we urgently need to accurately understand the diversity of arthropods before many species are extinct (Costello et al. 2013). Given all this, information on their abundance, richness, and spatial and temporal distribution is severely limited and needs to be rectified.

1.2 How to Assess Arthropod Biodiversity

Despite the estimations of species and their importance, our knowledge of arthropod species diversity is still considered to be neglected and poor for most parts of the world, especially for tropical ecosystems (Stork 2018). This is an important limiting factor in the advancement of knowledge on global biological diversity. For that reason, collecting and trapping arthropods are perhaps the most basic and necessary ways to access this biodiversity. Therefore, through the sampling methods and techniques in this book, researchers can discover how many and which species are present in several types of ecosystems; they can monitor natural populations or pest or beneficial arthropod populations in agroecosystems, conduct rapid surveys, and discover new species (Nageleisen and Bouget 2009; Samways et al. 2010; Larsen 2016; McCravy 2018). It is absolutely impossible to attempt to collect all species from a group as large as Arthropoda using only one sampling technique. This is due to the life span, biology, life histories, and ecology of the different arthropod taxa and species. Therefore, we must consider sampling specific arthropod groups through several methods, e.g., a method for sampling plant-inhabiting arthropod in target plants of tropical forest understory (Lopes et al. 2019).

Our intention in editing this book is to provide the readers, e.g., zoologists, entomologists, arachnologists, ecologists, students, and researchers, essential information on how to sample arthropod species or a particular taxon in the field, through different sets of methods and techniques. As you will see in our chapters, there are many kinds of methods, techniques, and traps for collecting arthropods. Some of these are simple apparatus (e.g., hand net), and others can be more elaborate (e.g., malaise trap). Some of the methods and techniques described in the chapters are excellent for catching several groups of arthropods and some other meant to only catch one particular taxon. The techniques described can also vary according to habitat complexity or temporal and spatial scales. The idea of this book is to guide you to make the most appropriate choice between the different types of techniques and also according to your taxonomic or functional target group, thus avoiding a mass collection or overkill of nontarget group and species. This also will increase your efficiency in the field and save time and money.

The book addresses researchers from the fields of biology, zoology, and ecology. Each chapter is organized to evaluate and describe the main sampling methods: an introductory overview on the group, field methods, materials and supplies, sampling protocol, effort needed, and benefits and limitations of the techniques. In addition, some chapters also contain additional information, such as specimen preparation and conservation, species identification, data collection and management (treatment, statistical analysis, interpretation), and ecological/conservation implications of arthropod communities. All chapters have been peer-reviewed. To provide a broad overview of the sampling methods, the book is structured in three main parts.

Part I provides an overview of the sampling methods for insect groups, which is subdivided into 12 chapters, highlighting the 4 largest insect groups: (1) Hymenoptera (ants, Chap. 2; bees, Chap. 3; and social wasps, Chap. 4); (2) Lepidoptera (butterflies, Chap. 5); (3) Coleoptera (beetles, Chap. 6); and (4) Diptera (flies and mosquitoes, Chap. 7, and fruit flies, Chap. 8). These "big four" insect groups constitute the most part of all living species on Earth. In addition, we have several interesting groups of insects: dragonflies and damselflies (Odonata, Chap. 9); termites (Blattodea, Chap. 10); grasshoppers, crickets, and katydids (Orthoptera, Chap. 11); hemipterans (Hemiptera, Chap. 12); and thrips (Thysanoptera, Chap. 13).

Part II introduces the sampling methods for arachnids, especially for pseudoscorpions (Pseudoscorpiones, Chap. 14) and spiders and other arachnids (Arachnida, Chap. 15).

The book ends with the Part III. The first two chapters show the sampling methods of two important herbivore guilds. Chapter 16 provides information for gall-inducing arthropods and Chap. 17 for leaf-mining insects. Next, we divided the chapters into functional groups. Chapter 18 deals with canopy insects, Chap. 19 deals with soil and litter fauna, and Chap. 20 deals with aquatic insects. Finally, Chap. 21 introduces a special topic on vector insect diversity.

We hope you enjoy this book!

Acknowledgments We are grateful to Springer for believing in our vision on the need for this publication and their willingness to support us. We would like to express our special gratitude to all the authors. We are also grateful to the reviewers for all the comments (L. Viana, K. Del-Claro, and H. M. T. Silingardi), which improved the quality of this chapter. JCS. thanks the Brazilian National Research Council (CNPq) grant process #312752/2018-0, while GWF thanks CNPq and Fapemig.

References

Bar-On YM, Phillips R, Milo R (2018) The biomass distribution on earth. Proc Natl Acad Sci U S A 115:6506–6511. https://doi.org/10.1073/pnas.1711842115

Basset Y, Cizek L, Cuenoud P et al (2012) Arthropod diversity in a tropical forest. Science 338:1481–1484. https://doi.org/10.1126/science.1226727

Bernard Q, Jaulhac B, Boulanger N (2015) Skin and arthropods: an effective interaction used by pathogens in vector-borne diseases. Eur J Dermatol 25:18–22. https://doi.org/10.1684/ejd.2015.2550

Cardoso P, Barton PS, Birkhofer K et al (2020) Scientists' warning to humanity on insect extinctions. Biol Conserv 242:1–12. https://doi.org/10.1016/j.biocon.2020.108426

Costello MJ, May RM, Stork NE (2013) Can we name earth's species before they go extinct? Science 339:413–416. https://doi.org/10.1126/science.1230318

Culliney TW (2013) Role of arthropods in maintaining soil fertility. Agriculture 3:629–659. https://doi.org/10.3390/agriculture3040629

Culliney TW (2014) Crop losses to arthropods. In: Pimentel D, Peshin R (eds) Integrated pest management. Springer, Dordrecht, pp 201–225. https://doi.org/10.1007/978-94-007-7796-5_8

Del Claro K, Guillermo R (eds) (2019) Aquatic insects. Behavior and ecology. Springer, Cham. https://doi.org/10.1007/978-3-030-16327-3

Del-Claro K, Lange D, Torezan-Silingardi HM et al (2018) The complex ant–plant relationship within tropical ecological networks. In: Dáttilo W, Rico-Gray V (eds) Ecological networks in the tropics. Springer, Cham, pp 59–71. https://doi.org/10.1007/978-3-319-68228-0_5

Eisenhauer N, Bonn A, Guerra CA (2019) Recognizing the quiet extinction of invertebrates. Nat Commun 10:50. https://doi.org/10.1038/s41467-018-07916-1

Erwin TL (1982) Tropical forests: their richness in Coleoptera and other arthropod species. Coleopt Bull 36:74–75. http://www.jstor.org/stable/4007977

Gaston KJ (1991) The magnitude of global insect species richness. Conserv Biol 5:183–196

Gerlach J, Samways M, Pryke J (2013) Terrestrial invertebrates as bioindicators: an overview of available taxonomic groups. J Insect Conserv 17:831–850. https://doi.org/10.1007/s10841-013-9565-9

Giribet G, Edgecombe GD (2019) The phylogeny and evolutionary history of arthropods. Curr Biol 29:592–602. https://doi.org/10.1016/j.cub.2019.04.057

Goater TM, Goater CP, Esch GW (2014) Parasitism: the diversity and ecology of animal parasites, 2nd edn. Cambridge University Press, Cambridge

Goddard J (ed) (2008) Infectious diseases and arthropods, 2nd edn. Humana Press, Totowa, NJ. https://doi.org/10.1007/978-1-60327-400-5

Grimaldi DA, Engel MS (2005) Evolution of the insects. Cambridge University Press, Cambridge

Gullan PJ, Cranston PS (2014) The insects: an outline of entomology, 5th edn. Wiley-Blackwell, Oxford

Hallmann CA, Sorg M, Jongejans E et al (2017) More than 75 percent decline over 27 years in total flying insect biomass in protected areas. PLoS One 12:e0185809. https://doi.org/10.1371/journal.pone.0185809

Hamilton AJ, Basset Y, Benke KK et al (2010) Quantifying uncertainty in estimation of tropical arthropod species richness. Am Nat 176:90–95. https://doi.org/10.1086/652998

Hamilton AJ, Novotný V, Waters EK et al (2013) Estimating global arthropod species richness: refining probabilistic models using probability bounds analysis. Oecologia 171:357–365. https://doi.org/10.1007/s00442-012-2434-5

Klein AM, Vaissière BE, Cane J et al (2007) Importance of pollinators in changing landscapes for world crops. Proc R Soc B 274:303–313. https://doi.org/10.1098/rspb.2006.3721

Larsen TH (ed) (2016) Core standardized methods for rapid biological field assessment. Conservation International, Arlington

Lister BC, Garcia A (2018) Climate-driven declines in arthropod abundance restructure a rainforest food web. Proc Natl Acad Sci U S A 115:10397–10406. https://doi.org/10.1073/pnas.1722477115

Lopes MC, Lamarre GPA, Baraloto C et al (2019) The Amazonas-trap: a new method for sampling plant-inhabiting arthropod communities in tropical forest understory. Entomol Exp Appl 167:534–543. https://doi.org/10.1111/eea.12797

Macadam CR, Stockan JA (2015) More than just fish food: ecosystem services provided by freshwater insects. Ecol Entomol 40:113–123. https://doi.org/10.1111/een.12245

May R (2010) Tropical arthropod species, more or less? Science 329:41–42

McCravy KW (2018) A review of sampling and monitoring methods for beneficial arthropods in agroecosystems. Insects 9:170. https://doi.org/10.3390/insects9040170

Menta C, Remelli S (2020) Soil health and arthropods: from complex system to worthwhile investigation. Insects 11:54. https://doi.org/10.3390/insects11010054

Minelli A, Boxshall G, Fusco G (2013) An introduction to the biology and evolution of arthropods. In: Minelli A, Boxshall G, Fusco G (eds) Arthropod biology and evolution. Springer, Heidelberg, pp 1–15. https://doi.org/10.1007/978-3-642-36160-9_1

Mora C, Tittensor DP, Adl S et al (2011) How many species are there on earth and in the ocean? PLoS Biol 9:e1001127. https://doi.org/10.1371/journal.pbio.1001127

Nageleisen LM, Bouget C (2009) Forest insect studies: methods and techniques. Key considerations for standardisation. In: An overview of the reflections of the "entomological Forest inventories" working group (Inv. Ent. For.). Les Dossiers Forestiers no 19. Office National des Forêts, Paris

New TR (2017) Mutualisms and insect conservation. Springer, Cham. https://doi.org/10.1007/978-3-319-58292-4

Noriega JA, Hortal J, Azcárate FM et al (2018) Research trends in ecosystem services provided by insects. Basic Appl Ecol 26:8–23. https://doi.org/10.1016/j.baae.2017.09.006

Novais SMA, Nunes CA, Santos NB et al (2016) Effects of a possible pollinator crisis on food crop production in Brazil. PLoS One 11:e0167292. https://doi.org/10.1371/journal.pone.0167292

Pisani D (2009) Arthropods (Arthropoda). In: Hedges SB, Kumar S (eds) The timetree of life. Oxford University Press, Oxford, pp 251–254

Price P, Denno R, Eubanks M et al (2011) Insect ecology: behavior, populations and communities. Cambridge University Press, Cambridge. https://doi.org/10.1017/CBO9780511975387

Rader R, Bartomeus I, Garibaldi LA et al (2016) Non-bee insects are important contributors to global crop pollination. Proc Natl Acad Sci U S A 113:146–151. https://doi.org/10.1073/pnas.1517092112

Ratcliffe NA, Mello CB, Garcia ES et al (2011) Insect natural products and processes: new treatments for human disease. Insect Biochem Mol Biol 41:747–769. https://doi.org/10.1016/j.ibmb.2011.05.007

Roy S, Roy MM, Jaiswal AK et al (2018) Soil arthropods in maintaining soil health: thrust areas for sugarcane production systems. Sugar Tech 20:376–391. https://doi.org/10.1007/s12355-018-0591-5

Samways MJ (2019) Insect conservation—a global synthesis. CABI, Wallingford

Samways MJ, McGeoch MA, New TR (2010) Insect conservation: a handbook of approaches and methods. Techniques in ecology and conservation series. Oxford University Press, Oxford

Sánchez-Bayo F, Wyckhuys KAG (2019) Worldwide decline of the entomofauna: a review of its drivers. Biol Conserv 232:8–27. https://doi.org/10.1016/j.biocon.2019.01.020

Schoonhoven LM, van Loon JJA, Dicke M (2005) Insect plant biology. Oxford University Press, Oxford

Schowalter TD, Noriega JA, Tscharntke T (2018) Insect effects on ecosystem services—introduction. Basic Appl Ecol 26:1–7. https://doi.org/10.1016/j.baae.2017.09.011

Stork NE (2018) How many species of insects and other terrestrial arthropods are thereon earth? Annu Rev Entomol 63:31–45. https://doi.org/10.1146/annurev-ento-020117-043348

Stork NE, McBroom J, Gely C et al (2015) New approaches narrow global species estimates for beetles, insects, and terrestrial arthropods. Proc Natl Acad Sci U S A 112:7519–7523. https://doi.org/10.1073/pnas.1502408112

Thomas JA, Telfer MG, Roy DB et al (2004) Comparative losses of British butterflies, birds, and plants and the global extinction crisis. Science 303:1879–1881. https://doi.org/10.1126/science.1095046

Wardhaugh CW (2015) How many species of arthropods visit flowers? Arthropod-Plant Interactions 9:547–565. https://doi.org/10.1007/s11829-015-9398-4

Zhang Z-Q (2013) Animal biodiversity: an outline of higher-level classification and survey of taxonomic richness (addenda 2013). Zootaxa 3703:1–82

Zhang Z-Q (2020) Animal biodiversity: an outline of higher-level classification and survey of taxonomic richness (Addenda 2013) (version 2013). In: Roskov Y, Ower G, Orrell T et al (eds) Species 2000 & ITIS Catalogue of Life: 2020-02-24. http://www.catalogueoflife.org/col. Accessed 16 Apr 2020

Chapter 2
Sampling and Analysis Methods for Ant Diversity Assessment

Jacques Delabie, Elmo Koch, Pavel Dodonov, Bianca Caitano, Wesley DaRocha, Benoit Jahyny, Maurice Leponce, Jonathan Majer, and Clea Mariano

The entomologist, in his routine work of inventory, may be seen to act as a generalist predator, finding a broad spectrum of prey, that is, typically the common species. Alternatively, he/she may be seen too use another strategy, i.e. to sit and wait, typical of many specialists, which requires long periods of waiting.

(Espadaler and López-Soria 1991)

2.1 Introduction

Currently over 15,000 ant species are described out of an estimated total of 21,000 species on Earth (antwiki.org consulted on 10/17/2019). Among the arthropods that inhabit the ground or canopy of tropical forests, ants are ubiquitous and among the most abundant organisms, accounting for over 80% of canopy macroinvertebrate biomass (Majer 1992). Furthermore, it is estimated that ants, together with termites, represent one third of soil fauna biomass in a rainforest and four times that of all vertebrates combined (Beck 1967 cited in Fittkau and Klinge 1973). About 50% of ants in tropical forests may be associated with the soil, where they dominate and

J. Delabie (✉)
Laboratório de Mirmecologia, Centro de Pesquisa do Cacau/CEPLAC, Itabuna, Bahia, Brazil

Departamento de Ciências Agrárias e Ambientais, Universidade Estadual Santa Cruz, Ilhéus, Bahia, Brazil

E. Koch · W. DaRocha
Laboratório de Mirmecologia, Centro de Pesquisa do Cacau/CEPLAC, Itabuna, Bahia, Brazil

P. Dodonov
Instituto de Biologia, Universidade Federal da Bahia, Rua Barão de Jeremoabo, Salvador, Bahia, Brazil

B. Caitano
Programa de Pós-Graduação em Ecologia e Conservação da Biodiversidade, Universidade Estadual Santa Cruz, Ilhéus, Bahia, Brazil

J. C. Santos, G. W. Fernandes (eds.), *Measuring Arthropod Biodiversity*,
https://doi.org/10.1007/978-3-030-53226-0_2

contribute high generic and specific richness (Ryder Wilkie et al. 2010). Due to their diversity, their high behavioral plasticity, and the high population density of these organisms in the local communities, ants play an important role in the dynamics of the environment. They contribute to the ecological balance of environments, and they are important in the energy and biomass flow and the overall structuring of communities of these terrestrial ecosystems (Folgarait 1998), especially in tropical regions. In terms of the biomass and omnipresence of these organisms and also their multiple effects on other species, ants constitute one of the dominant arthropod families in tropical forests. They have a significant impact on all trophic levels as a result of their diet and their various types of associations with other animals, plants, or fungi. In addition, numerous ant species rely on plant-derived food resources such as fruits and seeds. Along with earthworms (Annelida) and termites (Arthropoda, Insecta, Isoptera), they are part of the select group of organisms collectively known as "ecosystem engineers" because they contribute to a majority of the ecological processes that structure environments (Jones et al. 1994; Folgarait 1998).

Social insects are notable for their nesting strategies, sometimes constructing complex structures or exploiting resources already available in the environment. An ant nest (or ant hill) can be defined as a physical structure built and occupied by these insects, a chamber or cavity produced by a plant housing them (domatia), or some other structure that is re-employed by the ants. For example, a nest can be constituted from a single cavity excavated by the workers or a dry leaf of a bromeliad epiphyte that houses the whole population of the colony to a very complex structure of several meters deep and more than 100 square meters of apparent area, which shelters more than 1,000,000 inhabitants (Forti et al. 2011; Antonialli-Junior et al. 2015). Nest contact with outside environment can be direct, as in the case of litter nests of species of the genera *Neoponera* Emery 1901, *Odontomachus* Latreille 1804, and *Wasmannia* Forel 1893, or it can take place through a system of galleries or passages that may end as a chimney as in the case of *Ectatomma tuberculatum* (Olivier 1792) and *Acromyrmex balzani* (Emery 1890), for example

B. Jahyny
Laboratório de Mirmecologia do Sertão, Colegiado de Ciências Biológicas, Campus de Ciências Agrárias, Universidade Federal do Vale do São Francisco, Petrolina, Pernambuco, Brazil

M. Leponce
Biodiversity Monitoring and Assessment, Royal Belgian Institute of Natural Sciences (RBINS), Brussels, Belgium

Evolutionary Biology and Ecology, Université Libre de Bruxelles (ULB), Brussels, Belgium

J. Majer
School of Molecular and Life Sciences, Curtin University, Perth, WA, Australia

School of Biological Sciences, University of Western Australia, Perth, WA, Australia

C. Mariano
Departamento de Ciências Biológicas, Universidade Estadual Santa Cruz, Ilhéus, Bahia, Brazil

(Antonialli-Junior et al. 2015), or as a single or a set of orifices, such as in *Atta* spp. The diversity of such structures is illustrated on Figs. 2.1 and 2.2. Unlike the other social hymenopterans, immature ants are not reared in cells, and that leads to a high degree of interaction between colony individuals of all ages (Sudd and Franks 1987). The success in the use of space by colonies can be exemplified not only by the abundance and diversity of ants but also by the presence of other animals, mainly other arthropods, which may use the ant nest as a shelter or as a base for their own dispersal, as do commensals, inquilines, or parasites (Hölldobler and Wilson 1990; Lopes et al. 2015; Castaño-Meneses et al. 2015, 2017).

In the last 20 years, much new taxonomic, genetic, or ecological information has contributed to the general knowledge concerning ants around the world. New frontiers of biodiversity are being progressively explored, for example, with respect to the occurrence of numerous cryptic species (sensu Bickford et al. 2007) among South American Ponerinae, to the genetic diversity of populations, and to their functional diversity or the richness of their interactions with other organisms (Delabie 2001; Chomicki and Renner 2017). All this has been made possible thanks to huge efforts in the field using diverse collecting methods and also to new insights gained from museum studies and biometric, genetic, biogeographic, and ecological analyses (Agosti et al. 2000).

A last question that is important to be considered relates to rareness. As pointed out by Espadaler and López-Soria (1991), the rarity of a species may be understood in several senses, with the most used arguments being that a rare species may have a small population size, a restricted geographical distribution, or a very specific habitat. However myrmecologists now agree that a rare ant is frequently an ant which has been inadequately sampled; many species are apparently very hard to be found since they are extremely specialized in their habitat or their behavior although, when adequately sampled, they can appear to be quite common (Espadaler and López-Soria 1991; Delabie et al. 2000b; Delabie and Reis 2000).

2.2 Sampling for Ant Taxonomy, Morphology, Genetics, or Ecology

2.2.1 General Considerations on Sampling

Much has already been said about ant sampling techniques (see the comprehensive work of Bestelmeyer et al. 2000, which is accessible online), but here we attempt to present a critical approach to the different available methods for sampling ants, their limitations, and their complementarity. We take into account the current demand for studies on ants in general and their communities in particular, which have exponentially increased during the last decades. In this section, we focus on sampling methods for ant inventories; entire ant colony sampling for a range of purposes, such as behavioral, cytogenetic, or population studies' and ant community studies (i.e.,

Fig. 2.1 Overview of ground nest types. (**a**, **b**) Giant mounds of *Atta vollenweideri*, a leaf-cutting ant. (**c**) Numerous species nest in the leaf litter. (**d**) Mounds of needles and branch fragments in *Formica* species. (**e**) Earth mounds of *Forelius* ants. (**f**) Muddy nest of *Pseudolasius* along a trunk and sheltering ant-tended hemipterans. (**g**) Underground nest of *Megaponera* ants. (**h**) Underground nest of primitive attine ant. All pictures by Maurice Leponce

Fig. 2.2 Overview of aboveground nest types on or inside plants. (**a**) Nest of *Anonychomyrma* in *Myrmecodia* ant plant. (**b**) Ant garden. (**c**) *Dolichoderus* nest on leaf. (**d**) Orifice of an *Azteca* nest established inside a *Cecropia* trunk. (**e**) *Cataulacus* nest inside dead wood. (**f**) *Oecophylla* nest joining leaves. (**g**) *Allomerus* nest in leaf pockets (domatia) of *Hirtella* plants. (**h**) Temporary nest of *Eciton* army ants inside a hollow trunk. (**i, j**) Carton nests of *Crematogaster* and *Azteca* ants, respectively. All pictures by Maurice Leponce

inventorying the ants that dwell in the litter, soil, vegetation, etc., in a given area). Other methods related to specific studies, for example, ant feeding habits (Fowler et al. 1991; Brandão et al. 2012), ant nest densities (Baccaro and Ferraz 2013), seed dispersal (Lengyel et al. 2009), and other kinds of interactions with plants (Corbara et al. 2019) or animals (Delabie 2001; Stadler and Dixon 2008), will not be exhaustively discussed. Methods presented here are valid for both tropical and temperate regions except for ants that are thermophilous and found in lower richness and abundance in cold regions or during cold seasons. In temperate regions there are also much less arboreal species (Floren et al. 2014), and trees are generally shorter.

Sampling depends on the stratum of interest (Table 2.1). Thus, different methods, or a combination of them (e.g., ALL protocol, see Agosti et al. 2000), may be selected for soil-, litter-, or vegetation-associated ants. Also, the environment imposes certain restrictions on the methods used, for example, it would make no sense to use standard ground pitfall traps in mangroves or other flooded habitats.

Regardless of the method used, some considerations must always be taken into account. The first of these is compatibility of data between the sampling methods. Be they pitfall traps, Winkler sack samples, or whatever, such sampling units must be treated independently from each another; otherwise the inferential statistics used may not be valid (Hurlbert 1984; Colegrave and Ruxton 2018). Violation of independence may occur in two non-mutually exclusive general forms: pseudoreplication and spatial (or temporal) autocorrelation. Pseudoreplication may occur when the samples related to one treatment are all clumped in one location. Thus, although there may be several samples, they are not independent from one another, and the effective sampling size is reduced. One example of this is when comparing the ant communities of two types of vegetation such as one single forest and one single savanna area. Or there may be several forest and savanna areas, but if all savanna areas are sampled first, the observed differences among the two areas may be caused by a number of time-associated factors that are not necessarily related to vegetation type. We strongly recommend consulting Hurlbert (1984) for a detailed discussion of pseudoreplication and how to avoid it, randomizing the location of sampling units is often the best choice. There are cases where pseudoreplication is justified (Oksanen 2001), but such cases are the exception, not the norm. If some clumping of sampling units is unavoidable, mixed-effect models or restricted randomization tests may be used.

Spatial autocorrelation occurs when the distance among the sampling units is insufficient to ensure independence among them (Zuur et al. 2009). For example, if two sampling units are located within foraging distance from a single ant nest, they will not be independent. Ideally, the minimum distance should be determined based on empirical data on ant movement and foraging distance. In practice, this information is seldom available, and minimum distances are generally used. For this reason, we generally recommend maintaining an interval of 50 m minimum between two successive samples of Winkler sacks or pitfall traps (Delabie 1999). Nevertheless, we recognize that the surface of the area to be sampled must be large enough to support a sufficient amount of independent units. In many instances, including those using the ALL protocol (Agosti et al. 2000), the intervals between two successive

Table 2.1 Comparison of various methods for sampling ants

Method	Capture pattern	Most captured ants	Behavior which drives the capture	Allows community ecology studies?	Allows taxonomic studies?	Allows cytogenetic studies on immatures?	Allows other studies with entire colonies?	Effort/time consumed for 1 sample	Ease of sampling replicates	References
Active nest search	Resident	All castes	Nesting	With adequate replicates, yes, but difficult	Yes, the best	Yes	Yes	Quite many	More or less	Sosa-Calvo et al. (2015)
Arboreal baiting	Attraction	Workers	Mass recruitment	With adequate replicates, yes	Eventually	No	No	Little	Yes	Camarota et al. (2018)
Arboreal bait lines	Attraction	Workers	Foraging	With adequate replicates, yes	Eventually	No	No	Little	Yes	Leponce et al. (2019)
Arboreal pitfall trap (baited)	Attraction and interception	Workers	Foraging	With adequate replicates, yes	Eventually	No	No	Little	Yes	Oliveira-Santos et al. (2009), Koch et al. (2016) and Vasconcelos et al. (2018)
Arboreal pitfall trap (non-baited)	Interception	Workers	Foraging	With adequate replicates, yes	Eventually	No	No	Little	Yes	Powell et al. (2011)

(continued)

Table 2.1 (continued)

Method	Capture pattern	Most captured ants	Behavior which drives the capture	Allows community ecology studies?	Allows taxonomic studies?	Allows cytogenetic studies on immatures?	Allows other studies with entire colonies?	Effort/time consumed for 1 sample	Ease of sampling replicates	References
Beating	Resident	Workers	Foraging	With adequate replicates, yes	Eventually	No	No	Little	Yes	Majer and Delabie (1994)
Berlese-Tullgren funnel	Resident	All castes	Foraging and nesting	With adequate replicates, yes	Yes	No	Yes	Little	Yes	Bestelmeyer et al. (2000)
Branch clipping	Resident	Workers	Foraging	With adequate replicates, yes	Eventually	No	No	Little	Yes	Dejean et al. (2010, 2015)
Canopy hamper method	Resident	All castes	Nesting	With adequate replicates, yes	Yes	Yes	Yes	Quite many	No	DaRocha et al. (2015)
Ground and underground pitfall (baited)	Attraction and interception	Workers	Foraging	With adequate replicates, yes	Yes	No	No	Little	Yes	Wong and Guénard (2017)
Ground bait	Attraction	Workers	Mass recruitment	With adequate replicates, yes	Eventually	No	No	Little	Yes	Lopes and Vasconcelos (2008)

Ground pitfall (non-baited)	Interception	Workers	Foraging	With adequate replicates, yes	Yes	No	No	Little	Yes	Bestelmeyer et al. (2000) and Gomes et al. (2018)
Insecticide knockdown	Resident	Workers	Foraging	With adequate replicates, yes	Eventually	No	No	Huge	Yes	Floren and Linsenmair (1997)
Light trap	Attraction	Males and gynes (exclusively alates and crepuscular)	Mating flight	No	Yes	No	No	Quite many	No	Nascimento et al. (2004) and Steiner and Haüser (2010)
Malaise trap	Interception	Males and gynes (exclusively alates)	Mating flight	No	Yes	No	No	Quite many	No	Delabie and Reis (2000) and van Achterberg et al. (2010)
Nest traps	Resident	All castes	Nesting	With adequate replicates, yes, but difficult	Yes, the best	Yes	Yes	Quite many	More or less	Cobb et al. (2006) and Davidson et al. (2006)

(continued)

Table 2.1 (continued)

Method	Capture pattern	Most captured ants	Behavior which drives the capture	Allows community ecology studies?	Allows taxonomic studies?	Allows cytogenetic studies on immatures?	Allows other studies with entire colonies?	Effort/time consumed for 1 sample	Ease of sampling replicates	References
Pool collection	Attraction for water?	Males and gynes (exclusively alates at daybreak)	Mating flight	No	Yes	No	No	One sample per day	No	Espadaler and López-Soria (1991)
Sticky traps	Resident	Workers	Foraging	With adequate replicates, yes	No	No	No	Little	Yes	Majer (1990)
Sweeping	Resident	Workers	Foraging	With adequate replicates, yes	Eventually	No	No	Little	Yes	Majer and Delabie (1994)
Monolith extraction of soil (TSBF traps)	Resident	All castes	Foraging and nesting	Very difficult	Yes	No	No	Quite many	More or less	Martins et al. (2020)
Winkler sacks	Resident	All castes	Foraging and nesting	With adequate replicates, yes	Yes	No	No	Little	Yes	Bestelmeyer et al. (2000) and Guénard and Lucky (2011)

samples may have to be reduced to allow the collection of a reasonable number of samples.

2.3 A Primer on Main Sampling Methods for Ants

The various methods that are discussed here are comparatively summarized in Table 2.1. Several of these are illustrated on Figs. 2.3, 2.4 and 2.5.

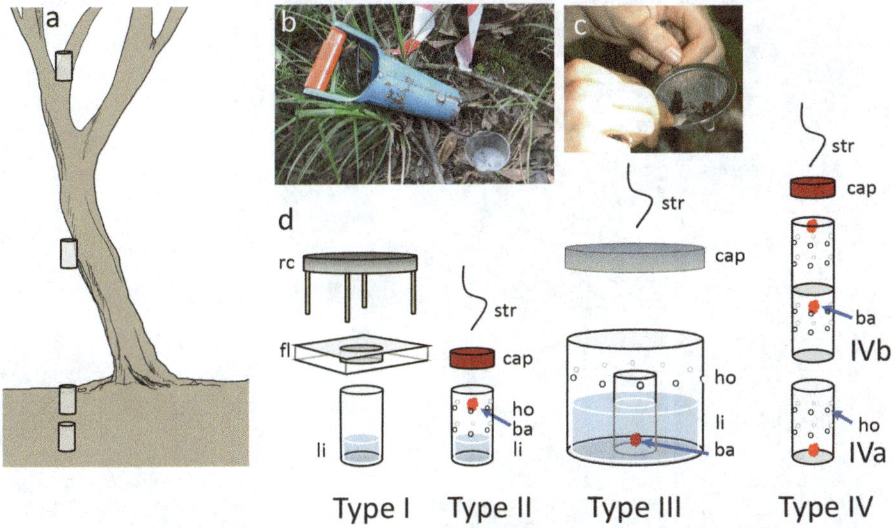

Fig. 2.3 Exploded view of basic designs of pitfall traps used to collect arboreal, epigeic, or hypogeic ants. Relative proportions are indicative and vary according to studies. (**a**) A single trap design (especially types II and III, see below) can be used at all different forest strata. (**b**) The simplest pitfall trap design, a drinking cup filled with a liquid (li) and put in a hole dug with an auger. A flagging tape allows researchers to detect more easily the trap on the forest floor. (**c**) A tea strainer can be used to separate ants from the liquid. (**d**) Four basic designs of pitfall traps, the first three are wet pitfall traps (containing a liquid), and the last one relates to dry pitfall traps. Type I is unbaited (although the liquid used might be attractive, see text) and was already shown in (**b**). To prevent rain water to completely fill the trap, a rain cover (rc) can be added. In inundated forests, a horizontal floater (fl) can maintain the trap at the water surface. Types II and III contain a bait which is separated from the liquid by fixing it above the solution or by putting it within an inner smaller container, respectively. Type IV is similar to type II but without liquid. The bait is placed at the bottom (IVa) or top of the container (IVb). Multiple segments can be superimposed to form a subterranean probe (IVb) inserted in a perforated outer sleeve implanted in the ground. When placed into the soil, type I–IV traps must be closed by a cap, and holes must be drilled laterally to allow the entry of hypogeic ants. To capture epigeic ants, the cap and holes must be removed. To capture arboreal ants, these traps must touch the trunk or branches, and the cap and holes are facultative. Strings (str) or handles allow researchers to hang the traps in trees or to retrieve the device when buried in the soil. All pictures by Maurice Leponce

Fig. 2.4 Overview of main methods used for sampling arboreal ants (except pitfall, see Fig. 2.1). (**a**) Vertical profile of a tropical rainforest showing the strata where ants are found, from top to bottom: canopy, understory, leaf litter, organic soil, and mineral soil. (**b**) Ants on foliage, flowers, and fruits can be collected by vegetation beating with a stick and a tray. (**c**) Ant collection from the

2.3.1 *Berlese-Tullgren Funnel*

Berlese-Tullgren funnels (Fig. 2.4e) are used to extract soil or litter arthropods. In contrast with Winkler sacks, it is a method of active extraction. A source of light and heat (generally a light bulb) dries the medium inside the apparatus, forcing the ants to attempt to escape from heat and desiccation and fall into the funnel, ending with a collecting jar filled with a preservative. Even the more delicate ant species can be extracted by this means (Lynch 1981), as well as other interesting arthropods such as springtails. This type of trap generally requires a source of electrical energy continuously on for 1 to 3 days. Although light is considered more efficient in this trap, this can be a problem, since incandescent bulbs (the source generally preferred for producing both heat and light) are becoming scarcer as they are being replaced on the market by LED bulbs which do not diffuse heat at all. An alternative to using bulbs could be to use natural sunlight, with the sampled medium being exposed to incoming radiation. However, provision has to be made to prevent the ants from escaping from the funnels.

2.3.2 *Beating of the Vegetation*

Ants foraging on foliage, flowers, and fruits can be sampled by holding a beating tray under branches that are then struck with a stick (Majer and Delabie 1994; Vasconcelos et al. 2000) (Fig. 2.4b). Ants are collected from the tray with an aspirator forceps of fine brush that has been dipped in alcohol (Perfecto and Snelling 1995). Another possibility is to fit a collecting jar to the tray, but this increases the sorting time of the sample at the laboratory because ants are mixed with debris and other arthropods (Basset et al. 1997). The effectiveness of this method is low during cold weather (low ant activity) or when the foliage is moist.

Fig. 2.4 (continued) gondola of a canopy crane. (**d**) Canopy hamper protocol. A cloth collector attached to the tree by a rope and a pulley allows researcher to bring back to the ground samples from microhabitats (epiphytes, suspended soil, etc.) where ants nest or forage. (**e**) Branch clipping by a tree climber in the canopy. (**f**) Arboreal bait line protocol. Baits are placed every 5 m along the tree trunk allowing to study the vertical stratification of numerically dominant ant colonies. The line is installed with a slingshot and forms a loop allowing an easy collection of ants at baits. (**g**) Canopy fogging, a cloud of insecticide (usually pyrethrum) is sprayed in the tree crowns. Falling ants and other arthropods are caught in collectors. (**h**) Sticky trap. A band coated with sticky resin glues ants foraging on the tree trunk. All pictures by Maurice Leponce, except (**b**), Roger Le Guen; (**c**), Henri-Pierre Aberlenc; and (**d**), Wesley Duarte DaRocha

Fig. 2.5 Overview of main methods used for sampling ground-dwelling and flying ants (except pitfalls see Fig. 2.1). (**a**) Light trap. (**b**) Malaise traps. (**c**) Collection of leaf litter within a quadrat, sifted before to be put inside a Winkler extractor. (**d**) Three Winkler extractors containing a meshed bag with the sifted leaf litter (shown outside the extractor, in the center). Ants attempt to escape from the leaf litter debris and fall in a collector at the bottom of the apparatus. (**e**) Berlese-Tullgren funnels relying on a heat source (bulb) to force ants to escape from the organic matter placed in the apparatus. (**f**) Visual search of ants within soil samples. (**g**) Mist netting of ant workers foraging on the vegetation. All pictures by Maurice Leponce

2.3.3 *Sweeping of the Vegetation*

Sweeping vegetation with an insect net is also sometimes used to collect ants (Lynch 1981; Majer and Delabie 1994) (Fig. 2.5g). This is particularly well suited for catching foraging workers on shrubs or low vegetation (up to about 3.50 m in height).

2.3.4 *Sticky Traps*

Sticky bands have been used since the eighteenth century to prevent Hemiptera-tending ants or caterpillars climbing on to the trees in production orchards (Samways 1981; Delabie 1989). The viscous material consists of tar, pitch, resin, and mineral oil compound or may be a synthetic resin. Sticky traps (Fig. 2.4h) consisting of a band coated with a sticky resin can be used to sample arboreal ants (Majer 1990). Two parallel bands, correctly installed, allow evaluation of ant activity, upward and downward, on a trunk, simply by comparing the number of ant workers caught on the upper and lower bands. The main problem with this kind of trap is that the pilosity of the ant body retains the sticky agent and is very difficult to clean, compromising the value of the specimen for preservation in a museum collection.

2.3.5 *Branch Clipping*

Branch clipping is a technique used to collect arboreal-dwelling ants in canopy trees. Branches are cut by a tree climber or a scientist accessing the canopy with a balloon (Dejean et al. 2010, 2015) (Fig. 2.4e). Since these branches are marked as part of their territories, arboreal ants remain on them for more than 1 h and can be collected with an aspirator. This technique is also useful for collecting ant colonies living inside branches (e.g., for *Cephalotes* Latreille, 1802, or *Pseudomyrmex* Lund, 1831).

2.3.6 *Insecticide Knockdown*

A fumigator or spraying machine can be used to fumigate or spray a knockdown agent, usually pyrethrum, in the tree canopy (Fig. 2.4g). Knockdown has to be conducted when there is little wind drift, generally during early morning. Ants and other invertebrates that are knocked down fall into a collector. In general, the height of the application is limited by the power of the apparatus that is used for fumigation. It is unlikely that common applicators will be able to apply any insecticide beyond 10 m in height, so it is not an easy method to be used in forests. This can, however, be overcome by accessing the upper canopy with a "cherry picker" or aerial work platform. Canopy knockdown allows the estimation of the biomass of arboreal ants (as well as many other organisms living on trees), in addition to allowing the sampling of ants that live in very discrete colonies and which are rarely caught by other methods. Pyrethrum is often used for canopy spraying and is broken down by sunlight. Re-fogging experiments show little change in species composition after 7 months (Floren and Linsenmair 1997), so the same trees can be used for subsequent resampling. Insecticide knockdown is, however, often not allowed in protected areas (Chen et al. 2012) and is time-consuming at the lab due to the presence of numerous individuals of dominant ants.

2.3.7 *Pitfall Trapping*

Pitfall traps can be used to monitor ants at different strata of the forest, from underground to canopy, provided that a tailored trap design is used (Fowler and Delabie 1995; Berghoff et al. 2002; Frizzo et al. 2011; Chen et al. 2012; Schmidt and Solar 2010; Schmidt et al. 2013; Rabello et al. 2015; Lasmar et al. 2017; Wong and Guénard 2017). Pitfall traps are cheap and easy to operate (Souza et al. 2012). They come in two main categories: wet and dry pitfall traps (Southwood 1978) (Fig. 2.3). Wet pitfall traps contain a liquid that kills the ants. They can be made of single (Fig. 2.3d, types I–II) or double container (type III) and contain (types II–III) or do not contain (type I) a bait. The liquid used usually contains a preservative agent and is in some cases an attractant. Dry traps allow ants to be collected alive (type IV), but they must be emptied regularly or the catch maybe eaten by predators.

Liquids commonly used in wet pitfall traps can be water and detergent (to reduce the surface tension) (Frizzo et al. 2011; Souza et al. 2012), water/detergent/salt (Lasmar et al. 2017; Ribas et al. 2003), water/salt/glycerol (Schmidt et al. 2013), ethylene glycol (Chen et al. 2016), ethanol/ethylene glycol (Andersen et al. 2006), or an attractant such as diluted beer (Powell et al. 2011). Where liquid consists of water with a few drops of detergent, the traps can only be run for a short time (about a day) because the captured insects decompose quickly.

Wet pitfall traps follow three basic designs as shown in Fig. 2.3. Type I, consisting of a pit partly filled with liquid, is the most commonly used trap for collecting epigeic ants (as recommended in the ALL protocol, Agosti et al. 2000). A rain cover can be added (rc, Fig. 2.3d) to prevent flooding and/or obstruction by falling leaves and debris (Nunes et al. 2011). In swamps or inundated forests, floating pitfall traps can be used to collect ants from floating vegetation (Chen et al. 2012). Type I traps are sometimes also used to passively sample arboreal ants (Kaspari 2000; Lasmar et al. 2017). Some type I traps used to collect arboreal ants are an adaption of traps used to collect arboreal spiders (see the trunk trap in Chen et al. 2012). Type I pitfall traps at ground level do not require baits to obtain large catches (Vasconcelos et al. 2018). In very hot and dry environments, such as in the Chaco, water itself may act as an attractant (Delsinne et al. 2008). To collect arboreal ants with type I pitfall traps, it is recommended to use an attracting solution (e.g., sugar-rich as honey diluted in water or nitrogen-rich as diluted human urine) (Powell et al. 2011; Vasconcelos et al. 2018).

Baited wet pitfall traps (types II and III, Fig. 2.3d) are generally preferred for collecting hypogeic and/or arboreal ants (Ribas et al. 2003; Andersen et al. 2006; Wong and Guénard 2017). Baits used in arboreal pitfall traps are generally a source of either proteins and lipids (e.g., canned fish: Ribas et al. 2003; Oliveira-Santos et al. 2009) or carbohydrates (e.g., honey: Frizzo et al. 2011) or a mixture of the two (Ribas et al. 2003). Baits used in hypogeic pitfall traps are also of the same nature, although more diverse food sources (e.g., vegetable oil, cookies, live prey) might allow the capture of a wider diversity of ants or more specialized species (Berghoff et al. 2002, 2003; Wong and Guénard 2017).

In dry pitfall traps (Fig 2.3d, type IV), ants are collected alive, but this runs the risk of aggression or predation between the arthropods that fall in the same container. This method is similar to a classical baiting method, but the bait is protected inside a perforated container. It is especially used to collect hypogeic ants (Morini et al. 2004), sometimes at multiple depths simultaneously using segmented traps (Ryder Wilkie et al. 2007). In some cases, for instance, when monitoring hypogeic army ant movements, the pitfall trap is filled with excavated soil mixed with baits (e.g., oil palm, cookies, or tuna, Berghoff et al. 2002, 2003). A piece of wet sponge can be placed at the bottom of the trap to allow the ants to shelter while keeping them alive, even if there is still the risk of predation. Such traps have also been used to eradicate arboreal ants in manipulation experiments (Klimes et al. 2011).

Pitfall trap catches are heavily influenced by ant activity, and their efficiency is influenced by climatic conditions (Delsinne et al. 2008; Nunes et al. 2011; Rabello et al. 2015), time of exposure (Gomes et al. 2018), and other factors, including the disturbance created by the installation of the traps itself, which may result in an initially higher capture rate (see Lasmar et al. 2017). More details on how to use epigeic and hypogeic pitfall traps can be found in the reviews of Bestelmeyer et al. (2000) and Wong and Guénard (2017), respectively.

2.3.8 Soil Samples

The monolith extraction, or TSBF (tropical soil biology and fertility) method (Anderson and Ingram 1993), is a standardized method of studying soil organisms where soil monoliths (0.25 × 0.25 m and 0.10 m depth) are sampled with an apparatus for study and analysis. It allows the sampling of a constant volume of soil, allowing further comparison between root biomass, earthworm diversity, and a number of other organisms, taken in different types of soils, habitats, climates, or biogeographic regions. In the case of ants, this trap allows catching of only a very limited number of individuals, but in many instances, it is the only method available to catch very specialized ants living underground that are rarely found otherwise (Castro et al. 2018; Martins et al. 2020). Inspection of soil samples is a method included in the ALL standardized protocol for collecting ants (Agosti and Alonso 2000) (Fig. 2.5f).

2.3.9 *Winkler Sacks*

Thanks to the Winkler apparatus (Fig. 2.5e), the study of litter ant communities has expanded in recent years (Besuchet et al. 1987; Agosti et al. 2000) and is especially successful in agroforest and forest areas (Delabie et al. 2007; Groc et al. 2009; Silva and Brandão 2014; Silva et al. 2017; Koch et al. 2019). Its use has been disseminated quite successfully because it is the most efficient trap for catching ants from the soil and litter layers, replacing the Berlese-Tullgren funnel which was formerly the most widely used apparatus for this purpose and whose use requires electric power. The latter also has a proven lower extraction efficiency than the Winkler apparatus (Bestelmeyer et al. 2000). The methodology for collecting the ants of leaf litter we most currently use is adapted from the ALL protocol (Agosti and Alonso 2000; Leponce et al. 2004). It consists of collecting 50 consecutive samples of 1 square meter of sifted litter each in the area where the study is being carried out. These samples are distributed at intervals of 50 m, representing a total sampled area of at least 12.5 hectares. This method takes place in two stages. In the field, the first square meter of the ground to be sampled is marked with a series of tags or with a foldable and portable aluminum quadrat. While protecting the hands with gloves, all the litter, fallen leaves, twigs, and decomposing materials are picked up, starting from the periphery and moving to the middle of the quadrat. This material is then put into a sifter where it is shaken vigorously. The sifted material is then carried in an identified transport bag to the laboratory. At the laboratory the material is placed in a 4 mm mesh lacelike Winkler sack. This is then hung in the calico-framed Winkler apparatus. The original Winkler apparatus could accommodate four sacks, but the "mini-Winkler" that we use is specifically designed for extraction of 1 square meter of leaf litter and accommodates one sack. In general, the sifted material stays in the trap for 1 to 4 days according to the protocol. During this period, the disturbed ants move quickly and randomly in the hung sack and eventually fall into a plastic container with a preservative liquid. More details on this methodology are given in Bestelmeyer et al. (2000). Guénard and Lucky (2011) recommend the regular disturbance of the substrate to increase the effectiveness of the trap.

2.3.10 *Canopy Hamper Method*

This method is adapted from DaRocha et al. (2015, 2016) and can only be performed by people trained to climb trees, which are often very tall. The tree canopy is accessed by using the single rope climbing technique (Perry 1978). To collect tree epiphytes associated with or without suspended soil and their associated fauna (in this case, called phorophytes) or any piece of wood containing an ant nest, a cloth collector with a 70 × 100 cm opening, 90 cm in height, tied to a rectangular iron frame (Fig. 2.4d) can be used. We term this a "canopy hamper." The canopy hamper is connected to a rope that passes through a pulley fixed to the highest point of the canopy, which allows its suspension from one of the main branches of the tree so that it can be positioned immediately below the epiphyte or the piece of branch to be collected. The system is designed to slow down the fall of the object when it is dislodged with a large chisel (160 cm long and with a blade 20 cm wide). The canopy hamper was developed to collect any object with its suspended soil and associated fauna, so that it could be studied when brought to the ground. Several structural measurements of the object can be then taken, and a preliminary collection can be carried out in the field, in order to separate the soil fragments and the most agile animals. Ideally the material collected can be put in large plastic bags or boxes to be transferred to the laboratory where ant nest or individuals and associated fauna can be adequately separated. If placed in plastic, care must be taken to ensure that the sample does not sweat or overheat, such as by placing it in a cool box or portable refrigerator.

Our experience has shown that after completing the collection of the whole epiphytes living on a single tree (DaRocha et al. 2015) and having analyzed data from several phorophytes, we believe that six epiphytes should ideally be collected on each tree (DaRocha et al. 2016). This sample number optimizes crown sampling, while minimizing any negative impacts of epiphyte removal on the phorophyte biota.

This method also samples ant species that normally inhabit the leaf litter and which are probably attracted to bromeliad microhabitats as nesting sites due to favorable conditions created by the aggregation of suspended soil, leaves and roots, and small branches within the rosette. Therefore, it is a method that can be applied for analysis of microhabitat types or conditions and species of host trees on ant communities inhabiting tropical forest canopies. Another form of sampling the ants of the canopy, but which is unfortunately rarely practicable due to technical and financial issues, is to use a crane, as described in Ribeiro et al. (2013) (Fig. 2.4c).

2.3.11 Malaise Traps

Malaise traps (Fig. 2.5b) have generally been successfully used for sampling a range of flying insects, for example, Diptera, many families of Hymenoptera, and a few groups of other insects (Southwood 1978). However, they have seldom been recommended for ant studies since few Formicidae are generally collected by this means, except winged sexual forms, most notably males, during the mating season. Nevertheless, malaise traps allow the study of reproductive phenologies of ant species (together with black-light traps, Kaspari et al. 2001) and allow the sampling of cryptic species that are otherwise never or rarely collected, such as males of *Acanthostichus* Mayr, 1887, *Cylindromyrmex* Mayr, 1870, *Fulakora* Mann, 1919, or *Thaumatomyrmex* Mayr, 1887 (see Delabie et al. 2000b; Delabie and Reis 2000).

The use of this kind of trap for an inventory project must be regarded as complementary to the other methods used, but the ant species sampled often belong to rare or imperfectly sampled genera that may not be found by other methods. Interestingly, many arboreal ant workers are markedly curious and are caught when foraging on the trap. For this reason, workers of *Azteca* spp. are very commonly sampled in this kind of trap, as are workers of *Pseudomyrmex* spp. We observed that the best way to inventory this genus in a forest area is by examining malaise trap material where both alates and workers of *Pseudomyrmex* spp. are especially abundant and representative of the community (Reis and Delabie 1999).

2.3.12 Light Traps

Although exceptions exist [i.e., wingless females, ergatoid males, gamergates (Peeters and Crewe 1984)], ant reproduction and dispersion occur through mating (or nuptial) flights (Hölldobler and Wilson 1990), where both sexes encounter while flying (most of the cases) or sometimes only males fly looking for a wingless female. This kind of flight may occur all year but sometimes with a low intensity in some species where only the male flies in search of a female (e.g., in the case of *Dinoponera* Roger, 1861, *Thaumatomyrmex*, or Dorylinae, see Delabie et al. 2000a, b) or are mostly seasonal and are dependent on climatic factors, mainly temperature and rainfall (Kaspari et al. 2001; Nascimento et al. 2004, 2011). In general, the alates (winged individuals) are strongly attracted by light since the flights are crepuscular or at daybreak, and they can be trapped by light traps (Fig. 2.5a). For some groups of ants, the light trap is practically the only way to verify their occurrence in a determined locality for an inventory. For example, in a single Brazilian locality, we know that the Dorylinae genus *Neivamyrmex* Borgmeier, 1940, is represented by nine species thanks to light traps, since workers of these ants living in soil galleries are seldom found by other means (Cardoso et al. 2011).

2.3.13 Arboreal Bait Lines

In this method, a thin nylon rope is installed over a top branch in the tree crown using a powerful slingshot. Baits are installed every 5 m along the rope, from 2 m above ground up to the top branch (Fig. 2.4f). The rope forms a loop allowing easy installation and recovery of the baits, which must be in contact with the tree trunk. Baits contain a mixture of tuna in vegetable oil and honey, a combination that represents a food rich in proteins, lipids, and carbohydrates. The mixture is wrapped inside a paper towel and tied to the rope. Baits are left for 4 h in the tree before being inspected.

This method was developed to collect territorially dominant arboreal ants and as a rapid assessment method to detect "ant mosaics" (i.e., segregation of ant foraging areas in tree canopies, Dejean et al. 2007; Leponce et al. 2019). The method is less effective for collecting nondominant ants that are less abundant on baits after 4 h. Aggression tests between neighboring ant colonies allow delineating of their distribution on trees.

2.3.14 Collecting Rare Ants in Pools

In 1991, Espadaler and López-Soria alerted myrmecologists about a singular phenomenon: during mating flights, especially early in the morning, winged Mediterranean ants are frequently attracted by swimming pools where they are trapped on the water surface where they can float for hours. Furthermore they observed that a large amount of these alates belonged to rarely collected species, such as, for example, males of *Leptanilla* sp. (Espadaler and López-Soria 1991). Similar observations made in several countries by one of us (JHCD) confirmed that all types of tanks, pools, or bodies of water, even the tiny waves of the seaside, provoke a similar attractiveness to winged ants. Many males or females of species present in the surroundings and not detected otherwise by more classical traps could only be caught in this way. So this collecting method although rather anecdotic can be considered only as an interesting complement for a taxonomic survey or an inventory of local diversity.

2.3.15 Timing and Frequency of Sampling

When and how often to sample ant assemblages are a critical issues if a complete census of the species present is the objective. This is particularly important in cool temperate climates where ant activity is low in winter (Albrecht and Gotelli 2001; Herbers 1989) and in Mediterranean regions where ant activity mirrors variations in temperature (Majer and Greenslade 2012). Although not so variable in tropical regions, there is also a degree of variation in activity there (Levings 1983; Marques and Del-Claro 2006; Jacquemin et al. 2016), most notably in areas subject to inundation (Pereira et al. 2013). In addition to the number of species that may be sampled in different seasons, there may also be variations in assemblage composition through time, with certain species variously favoring hot, cool, moist, or sunny conditions. Therefore, to maximize the capture of all species in a given area, consideration should be given to sampling during contrasting seasons, such as wet and dry seasons or cool and hot seasons, depending on the prevailing climate type where the study is carried out. Ideally, four separate seasons should be sampled. Logistical and financial restrictions may not enable this, so two seasons might suffice, or, if this is not possible, a transitional period between two contrasting seasons could maximize the catch of species.

The time of day for sampling is also important. In semiarid environments like caatinga in Northeastern Brazil, most ant species forage from sunset to sunrise only, and they make their nest under the ground or at the base of plants, never in the leaf litter. Therefore, the sampling method should be adapted for these specific conditions.

2.4 Combining Protocols: Opportunities and Benefits, Weak Points, and Strengths

Any sampling method can give a biased image of true species frequencies, but combining the most efficient and complementary methods allows us to obtain a more representative image of the whole species assemblage (see Fig. 2.2c, d in Leponce et al. 2010 based on the combination of 27 collection methods presented in Delabie et al. 2000a, 2007). The ALL standardized protocol (Agosti et al. 2000) is based on this principle, combining the Winkler, pitfall, deadwood, and soil methods. In a shaded cocoa plantation, these four methods captured over 80% of the hypogeic and leaf litter species that had been found using a total of 17 methods, including several bait types (calculated on the basis of 54 samples, Delabie et al. 2000a). Recently, Souza et al. (2012) compared in terms of cost and time consumption the use of sardine baits, pitfall, and Winkler sacks for monitoring Amazonian ant communities and concluded that pitfall traps are enough to answer most of the questions about ant assemblages and their relations with environmental variables. Other studies that compared two different methods of collecting leaf litter-dwelling ants have been published in several environments (e.g., see Orsolon-Souza et al. 2011; Lopes and Vasconcelos 2008).

2.5 Ant Nest Sampling and Rearing for Cytogenetic and Behavioral Studies and Associated Fauna

Ants are able to occupy different strata of the environment, and their nesting strategies can be qualified as subterranean, superficial (terrestrial, hypogeic or epigeic), or arboreal. Their constructions may be discrete or conspicuous, and each stratum has its distinct nest structures, which in turn require different collection methods.

Complete colonies of ants are used to study behavior, cytogenetics, natural history, or fauna associated with ants (mostly arthropods, some other invertebrates, few cases of vertebrates), among other topics. For some studies, not only the specimen collection but the maintenance of nests in the laboratory is also necessary, with colonies being conserved in artificial nests with the minimum of interferences of the observer to the ant's activities.

Prior to collecting any ant nest, some tools are necessary including shovel, digger, pickaxe, spoon or small trowel, labels, machete, adhesive tape, measuring tape, talcum powder (non-odorous), pocket book and pencil for annotation, brush, flexible clamps, and plastic bags.

Some nests are easy to find because they correspond to structures built by the ants or the nest entrance is evident if it is represented by one or more hole(s), a small chimney, or mounds of loose earth and organic refuse. When the nest is more difficult to find, it can be actively located by finding a foraging ant, following its movements, and waiting for it to find prey or something else that it will bring back to the nest. Alternatively, for a more rapid method, baits, such as fresh killed prey, sardine in oil, tuna fish, dried liver, cereal flakes, cassava, orange peel, and honey-coated apple fragments, can be used (Dominique Fresneau, personal communication, Leal and Oliveira 1998, 2000; Schultz 1993; Weber 1966, 1972; Wheeler 1907), and the ants are followed to the nest. Some grain of small size that is easily carried by workers and whose color contrasts with the substrate can be also offered, as this allows easy following of the ant forager on its way back to the nest (Sosa-Calvo et al. 2015). It works well for larger ants (more than 3 mm) but requires time and patience. The nests of ants forming columns of foragers are easier to locate by following the flow of what they bring back to the nest. Baccaro and Ferraz (2013) used sardine baits for sampling epigeic ant assemblages and were able to determine their nest densities using the software DISTANCE, which is based on the probability that an ant finds the bait according the distance from its colony.

Once the nest entrance is located, mark it with colored ribbon; this procedure is necessary for population studies, nest aggregations, exploitation of food resources, etc. Then register it with GPS. It is strongly recommended that each ant nest be given a serial collection number that will be useful for further observations. This should be maintained in a field pocket book, together with other information that may be necessary about the colony situation, population (including immatures), etc. A few adult individuals from each collected nest should be retained in alcohol as a reference. Furthermore, in the laboratory, a complete series of the different individuals of the colony should be dry mounted when available. Representatives of workers of all classes of size if there is dimorphism or polymorphism (including soldiers), sexual castes, as well as individuals that may have an unusual morphology such as intercastes should also be preserved. All these should be labeled with the serial collection number reported on the field pocket book.

2.5.1 Tracking Underground Nests

Records should be kept of the entry of the nest by means of pictures (always with a scale) and/or drawings. The entrance of the nest should then be characterized by the measurements of the diameter of the chimneys and entrance holes and the measurement of mounds of earth (weight and diameter) (see Cardoso et al. 2011 and Silva et al. 2007, for fungus-farming ants). The ground up to 5 m around the nest should be superficially cleaned, depending on the species (size of individuals and colony population suggest the area to be cleaned), prior to nest excavation and collection of biological material. This is to avoid the possible sampling of foreign organisms at the nest.

Excavations should be started up to 1 m from the nest entrance. Underground nests should be carefully excavated, and each chamber that is found should be characterized and measured. Suggestions for characterization are as follows: access to the nest, chimney, orifice(s) (number), gallery (width, length, number), chamber(s) (dimensions, depth, distance to the exit), number of occupants [immature, workers (soldiers, if any)], presence of the reproductive gyne, guests, etc.

Some authors suggest strategies for improving subterranean ant nest collection and characterization. For fungus-growing species, side trenches of approximately 90 cm depth and at about 30 cm from the entrance hole can be made, so that it retains the soil supporting the nest. The excavation has to be done laterally until the nest chamber(s) is found (Silva et al. 2007). In order to characterize the architecture of the nests, talcum should be sprayed into galleries and chambers (Silva et al. 2007). Further details on methods of excavation of nests of fungus-growing ants can be found in Cardoso et al. (2011) and Sosa-Calvo et al. (2015). These methods can be adapted for characterization of subterranean nests of a range of ant species.

2.5.2 Tracking Nests on the Ground Surface

Nests can be structured or hidden in the litter, in decaying fallen trunks, in dry branches, in abandoned termitaries, in dried fruits such as dry cocoa pods and coconuts, in shells of dead gastropods, etc. Most often, such nests are composed of a single chamber, and the colony is organized in this space. Dry fruits or branches should be collected entirely, individually placed in a numbered and closed bag, and preferably brought to the laboratory. They should be cooled in the refrigerator before opening and sorted in a container illuminated and coated with sticky or slippery substances such as petroleum jelly, heavy mineral oil, Fluon©, or talcum powder to prevent the escape of ants. The use of a magnifying glass or stereomicroscope is strongly suggested. It is possible to observe the occurrence of delimited chambers or spaces in dry branches or decaying trunks in which the population of the colony exists, the refuse, plus the mutualistic Hemiptera and other guests that may be present. All of this must be registered.

2.5.3 Associated Fauna

As well as other invertebrates that may be present, ants could also be guests of other ant species. All animals that are easily observed with the naked eye should be collected with forceps and placed in 70% alcohol. Arthropoda like Arachnida (Acari,

Araneae, Opiliones, Pseudoscorpiones), Insecta [Zygentoma, Collembola, Blattodea, Coleoptera (adults and juveniles)], and Myriapoda and other animals such as Annelida and Gastropoda (Mollusca) are relatively easy to collect.

After a preliminary sorting, the material that constitutes the nest (leaves, dry twigs, soil, etc.) should be sampled in devices as Berlese-Tullgren funnels. For collecting mites and springtails, the "floor" of the nest should be placed in a mini Berlese-Tullgren funnel made with 500 mL disposable cups and gauze, with a saline solution as trapping agent (Palacios-Vargas et al. 2013). After screening, the tiny animals should be taxonomically categorized at least up to the level of order, counted (if an abundance/dominance analysis is required), labeled, and fixed. Mites and springtails should be mounted on a microscope slide in Hoyer's solution for identification.

The study of the fauna associated to ants may refer to the species occurring in the nest (Castaño-Meneses et al. 2019) as well as to the description of the possible interactions that these organisms perform with their host ants. For this purpose, it is necessary not only to collect all the inquilines but also to determine in which sector of the nests they were found. This is easily noted when working with nests of subterranean species or in those that use plant material fallen on the ground or decomposing trunks.

2.5.4 Special Needs for Cytogenetics

For ant cytogenetic studies, the first task is to obtain live brood available at the adequate developmental phase (see below), and at the same time, adult vouchers must be kept for identification. For this purpose it may be necessary to raise a complete colony for a period of time in the laboratory to obtain individuals with adequate development.

At least two individuals or a complete series of individuals per colony should be mounted on pins and conserved as vouchers for further identification and reference, labeled with the serial number of collection as already mentioned. The number of nests to be collected for cytogenetic purposes varies according to the aim of the study. Cytogenetic evaluation of a species that has a wide geographic distribution should obviously involve the collection of nests in different geographical locations, or, in some cases, more than one nest per location can be required for characterizing the karyotypes of the populations. If the study is exploratory, samples from a single nest may be sufficient for the description of the chromosomes of the species. During the colony collection, it is not necessary to separate immatures from imagoes in the field. This procedure should more appropriately be done later in the laboratory, where it will be easier to visualize the pre-pupae to be used for cytogenetic studies and possibly other organisms living on their body (e.g., mites) or inside the ant (mites and parasitoids).

Slides for cytogenetic studies are prepared from fixed cells of the brain ganglion and/or larval testicles at the pre-pupae (pharate) phase (protocol of Imai et al. 1988).

Since this material is not present in the colony throughout the year, it may be necessary to maintain the colony alive in the laboratory for a while. At least ten slides per colony should be prepared (when only a single colony is collected), and this number is sufficient for the application of chromosomal banding techniques for their cytogenetic characterization. In addition to the adults mounted and kept for identification and reference, it is suggested that a series of adults be fixed in absolute alcohol for further use in molecular studies, complementary to cytogenetics.

2.5.5 Laboratory Care of Entire Colonies

To maintain an ant nest in the laboratory, crucial factors are feeding, maintaining moisture and aeration, as well as choosing the artificial nest model and material. In addition, the conditions offered in the laboratory should be appropriate for social life and depend on the aims of the experiment; nests should allow easy viewing and control of their content. Species with different lifestyles vary in relation to specific needs, and thus there is no universal protocol for the structure of an artificial nest. The following exhaustive work is strongly recommended by Czechowski and Pisarski (1992). These authors present different alternatives for the maintenance of ants in the laboratory, from nests whose first aim is the maintenance of the colony and/or exposure to models proposed for behavior studies, with materials such as plexiglass, plastic boxes, or plaster. A kit represented by a compartmentalized area in connection with an open space (piles of drawers measuring 11.5 × 11.5 × 4.0 cm each, with a hole of 2 cm in diameter and covered with transparent plexiglass) allows observation of the activity of the ants and the compartmentalization of the nest (details in Czechowski and Pisarski 1992). To mimic the natural conditions, a very useful substrate is plaster, which allows adaptations related to the ant size, arrangement of the chambers, and degree of maturity of the colony. The pattern of these nests can be determined by microscope slides, plastic petri plates, or bottle caps, whose size will be determined from the species to be allocated in the nest (Czechowski and Pisarski 1992). For fungus ants, descriptions of collection and maintenance of nests are found in Mueller and Wcislo (1998), Solomon et al. (2004), Cardoso et al. (2011), and Sosa-Calvo et al. (2015).

An aspect to keep in mind is the goal of setting up an artificial nest, whether it must imitate the natural conditions of the nest environment without any observation of the ants or if observations regarding behavior or other experiments are expected. Czechowski and Pisarski (1992) present detailed information for the installation and maintenance of ant nests built for several purposes such as these. Pseudomyrmecinae, Myrmicinae, and Formicinae are easily collected with bamboo artificial nests of 1.4 cm diameter and 19 cm in length, arranged at different heights (Cobb et al. 2006). Cylindrical glass tubes can be adapted for colony maintenance following Bhaktar and Whitcomb (1975). These should have about 30 cm in length and should be subdivided into compartments (with rubber or plaster) whose entry diameters vary according to the species to be created. To confine the ants in this space, Fluon© can be applied, and ideally the tubes must be covered with red cellophane paper. Humidity within the artificial nests can be controlled by keeping them upright over different concentrations of saline solution in a connected chamber. Thin gauze is installed to close the end of each tube allowing for changes in moisture (Bhaktar and Whitcomb 1975). For maintaining the ants inside the artificial nest, Ning et al. (2019) suggest a mixture of talc and ethanol in a concentration of 20%, used as an antifouling solution. And finally artificial diets made up of carbohydrates and proteins have to be offered to workers. For carbohydrate supply, the option is sugar solution with 10–15% concentration, which must be prepared for immediate use because of the rapid fermentation. Alternatively, live honeydew-producing aphids or honey could be used. For protein supply, beetle larvae, termites, larvae, and pupae of other species of ants, raw or cooked meat, and cooked egg yolks. Several artificial diets for rearing ants at the laboratory can be found in Bhatkar and Whitcomb (1970) and Porter (1989). For ants that have a specific predatory diet, often with the requirement that the prey has to be given alive, the prey must be found in quantity in the field or raised in parallel in the laboratory. It must be realized that ant keeping takes a lot of time, so technical support may be needed.

2.6 Ant Specimen Preparation and Conservation for Museology and Identification

There are detailed procedures for the preparation of ant specimens to be studied. We suggest consulting the many tips given by Lattke (2000). If these are followed, samples will be kept and sorted so that a more reliable identification will be possible, as well as enabling a finer analysis of the data. Lattke (2000) has made very careful and detailed recommendations for all these steps, and they are easily followed in the literature and on the Internet; we will limit ourselves here to a few tips.

It is important to adopt a uniform sample count between the field, the sorting in the laboratory, and the vouchers that will be kept in a collection. The IBISCA project and some other researchers have solved this by applying a barcode system that follows all these steps (Oliver et al. 2000; and see Fig. 16 in Leponce et al. 2016). In the absence of the necessary equipment (a notebook with a bar code reader and a printer), a field notebook, where the details of the collection (to which is assigned a unique sequential number) are noted, should be adequate. Be careful not to duplicate the codes used to identify or discriminatesamples. A complicated numeration alternating a long sequence of numbers and letters is often a source of error in one of the stages that leads to the assembly of the data for the final analysis.

Samples of captured ants killed in the traps should be kept in alcohol after cleaning the sample. Alcohol diluted at 70% has an established reputation for the preliminary fixation and preservation of this type of material. However, since the cuticle of the ants is not particularly permeable, to obtain specimens without deformation may require alcohol of 90% or 95% strength. Ethanol absorbs the resident water of the ant's body and does not persist for a long time with its relative dehydration. In humid tropical environments, care should especially be taken to preserve samples in high alcohol content fluid; if not there is a risk of losing some of the material by putrefaction. Currently, it is often recommended to keep the biological material in absolute alcohol in addition to keeping it in the freezer for possible future research in molecular biology.

After sorting the samples, some of the specimens will be mounted and tagged for future identification. The remainder should be kept in absolute alcohol (if available), stored in a dry place away from the light or in the freezer for further study, as already suggested. We recommend preserving at least three specimens of each morphospecies in case of ant community analysis; each of them should be individually (or the three recombined on the same pin) mounted on a triangle fixed on an entomological pin; all the ants should be glued with head facing to the left, as this will make comparison easier once all the images for the taxonomic study are oriented in the same way. Each pin should be individually labeled with the sample number given in the field. Other information can be added on separate labels, such as the number of individuals of this morphospecies. The three specimens per morphospecies (if available) will allow the taxonomist who will identify the material to verify the lack of mistakes during the sorting work in the laboratory (sometimes made by inexperienced people). One problem that can arise is common presence in the same sample from a Winkler sack, two or more minute species that are not obviously different but are in fact from different genera (e.g., *Strumigenys* Smith, 1860, *Pheidole* Westwood, 1839, or *Solenopsis* Westwood, 1840).

When the taxonomist gives a name to a specimen, he or she will write or print a label with the ant name. Possibly he or she will not repeat the same to each pin with the same species but will separate all the members of the same morphospecies in a tray in a collection drawer (see Lattke 2000). The label with the name of the ant cannot be removed from the pin that supports the specimen identified (called "voucher") under any circumstances.

After identification, all the series of ants with their identifications will be kept in a reference collection (generally a drawer or a series of them), with each pin perfectly labeled with all the information about the collection place, date, sample number, collector, and other information that could be interesting to note (see details in Lattke 2000). The drawer(s) will be labeled with general information too. The reference collection must be kept up to the period of publication. Afterward, it is possible to absorb the voucher collection into a general ant collection for classical museology. Recommendations for storage and manipulation of an ant collection can be found in Lattke (2000).

Good data organization is at the basis of any scientific study. In general, most researchers have their data organized into spreadsheets. To use tools that make computing and data analysis more efficient, such as programming languages like R or Python, we need to structure the data according to the way the software can read them. Spreadsheet programs are very useful graphical interfaces for designing data tables and handling basic data quality control functions. The formatting, orientation, and insertion of the data may vary according to the requirement or software used for the analysis. In order to effectivelymanage and analyze data, we must first configure our data so that the computer can understand it. The main rules of using spreadsheet programs for data are as follows:

1. Put all variables in columns what you are measuring, such as "height" or "temperature."
2. Place each observation on its own line, each different information = new input/line.
3. Never combine multiple information in a single cell, even if two pieces of information seem to correspond to one single thing, such as pitfall trap identity, it may best be separated into two columns, such as sampling site and pitfall trap identity within the sampling site.
4. Never modify the original raw data.
5. Export the cleaned data to a text-based format like CSV (comma-separated values) format. This ensures that anyone can use the data using a format that is required by most data repositories.

Certain programs or analyses may require the data to be organized in other specific ways. In fact, it is estimated that about 80% of the time spent performing a given data analysis is in data cleaning and preparation (Dasu and Johnson 2003). For details on how to better perform the data organization in a proper and efficient manner, Wickham (2014) and Broman and Woo (2018) are suggested.

2.7 Cautions for Data Analysis in Ecological Studies on Ant Populations or Communities

Considering the nature of the data collected, classical statistical methods, such as linear regression or analysis of variance (ANOVA), may not be feasible for analyzing data on ants. Such techniques require that data be normally distributed, which is not the case for a number of species or for presence/absence data. Data on continuous variables such as biomass could have a normal distribution, but such data are hard to collect and rarely assessed. Hereafter, we provide some suggestions of analyses to be used for the different kinds of data on ants. We also indicate some software and R packages that may be used for these analyses.

2.7.1 Univariate Data

It may be of interest to assess how environmental factors affect a given species of ant, for example, an invasive or threatened species or the most common species collected in a study. For solitary insects and other animals, abundance data may be used, but these data are meaningless for social insects as individuals of one species in a single sample unit will have probably come from the same nest and therefore are not independent. The alternative is to deal with presence/absence data and model the probability of a species occurring in a sample as a function of environmental variable via logistic regression or generalized linear models (GLM).

Unlike linear regression, a GLM assumes that data may come from a variety of statistical distributions or families, which have to be specified in the model. For presence-absence, the binomial family is used. The model may contain one or more explanatory variables, which may be continuous, binary, or categorical, and either significance tests or information criteria (such as the Akaike information criterion – AIC) may be used to determine which variables are significant (Burnham and Anderson 2002).

In some cases, there is just one sample corresponding to each value of the explanatory variables. For example, one may be interested in knowing whether the probability of occurrence of an invasive ant species is affected by the distance to the nearest forest edge. In this case, the presence (1) or absence (0) of a species could be related to the distance from edge with a simple binomial GLM. This approach has several advantages, as, for example, presence/absence data is not subject to overdispersion (Zuur et al. 2009). Difficulties may arise when there is a perfect separation between the presences and absences. In our example, the study species may be present in all samples up to a certain distance from the edge and absent from all more distant samples. A possible solution is the use of randomization tests to assess how likely a perfect separation would be under the null hypothesis.

In other cases there may be more than a single sample for each value of the explanatory variable(s). For example, one may want to assess whether a species of ant is affected by the size of the forest fragment. In this case, several samples may be distributed in each fragment, with the frequency of the ant species calculated (number of occurrences/number of samples). If the samples are independent, frequency is also likely to follow a binomial distribution, and a GLM may be used. Perfect separation is very unlikely to occur, but there may be issues with overdispersion, in which the variance in the data is greater than would be expected for a binomial distribution. If this is the case, an overdispersion parameter may be included (Zuur et al. 2009). If nonlinear patterns are expected, generalized additive models (GAM) (Zuur et al. 2009) would be recommended.

If all the explanatory variables are categorical (e.g., comparisons between preserved and impacted areas), contingency tables are useful. They may be analyzed with tests such as Pearson's chi-square test or Fisher's exact test.

If for some reason the samples are not independent, for example, if there are several study sites and several samples in each of them (e.g., Caitano et al. 2018), generalized linear mixed models (GLMM) may be used. In addition to the explanatory variables of interest, these models contain a random factor stating that the frequency or probability of occurrence of the different species differs between the sites.

There are also ways of including spatial autocorrelation in the analyses, but, in our experience, they often lead to convergence problems and are not straightforward to interpret. Modeling spatial autocorrelation for non-normal data is not trivial, and few softwares or packages include these options. Thus, we recommend that, whenever possible, spatial independence among the sampling units is ensured at the study design phase.

The above considerations apply to univariate community data, such as number of species. In this latter case, the GLM, GAM, and GLMM would employ a Poisson or negative binomial distribution (Zuur et al. 2009).

2.7.2 Co-occurrence Analysis

The analysis of spatial and temporal coexistence patterns of species is of fundamental importance for many ecological questions. It is possible to identify nonrandom patterns of co-occurrence among species from a presence/absence matrix of data. Co-occurrence analyses based on randomized matrices are commonly used; however the computer algorithms used can sometimes produce type I and type II errors. Here, we strongly suggest the use of probabilistic models (see Veech 2013). These models allow estimating the probability that two species may co-occur less frequently (Plt) or more frequently (Pgt) than expected if distributed independently of each other. Thus, associations can be classified as random, negative, and positive. In addition, it is possible to verify how each pair of species is associated. This analysis can be performed in the R environment with the "co-occur" package (Griffith et al. 2016).

2.7.3 Data on Multiple Species

Different to what we have discussed so far, it may be also of interest to analyze how multiple responses (e.g., different species) are simultaneously related to one or more predictor variables. Generally, for ecological studies, it can be investigated if and how the assemblage structure varies according to one or more factors. In such cases, it could be appropriate to use methods specific for multivariate data, which can be quantitative or qualitative, ordered, categorical, or continuous.

Many multivariate methods quantify the differences between the observations, considering the distance between them in a given multivariate space. The measures developed and/or used to measure the distance between objects/multivariate samples are basically divided into three categories: (1) metric distances (e.g., Euclidian, Manhattan, chi-square, and Jaccard), (2) semi-metric distances (Bray-Curtis and Sorensen), and (3) nonmetric distances, the latter type not usually being used by ecologists (for more details on the differences between distances, see Gotelli and Ellison 2011). The most used distance measure is Euclidian distance, which can be interpreted as the distance corresponding to a straight line between two points. Although widely used, Euclidian distance is not recommended when there are variables (e.g., environments) with many zeros, due to a problem known as "double absence". There may be a paradox that two sites with no species in common exhibit a smaller Euclidean distance than two others that share at least some species (for more details see Gotelli and Ellison 2011). The use of the Euclidian, Manhattan, and Bray-Curtis distances is recommended for continuous numerical data, the latter being also recommended as a solution to the "double absence" problem. However, it is common for multivariate data to find situations where it is necessary to measure the similarity or distance between points/objects that are described by presence and absence data; in this case it is recommended to use the distances of Jaccard and Sorensen.

Multivariate techniques are divided in different ways in the literature. One of the most recent and used is the classification of Johnson and Wichern (1992), which separates the applications of multivariate statistical analysis into three groups: **(1) classification and grouping techniques** (principal component analysis, factorial analysis, analysis of canonical correlations, discriminant analysis, cluster analysis, ordering methods), **(2) analysis of the covariance structure** (principal component analysis, structural equation models, canonical correlation analysis), and **(3) inferential analysis on means and linear models** (Hotelling T^2 statistic, multivariate analysis of variance, multivariate linear regression). For ant studies, classification and grouping analysis (group 1) are generally the most used, together with some analysis belonging to group 2; such studies generally seek to verify/illustrate possible differences in ant composition/assemblage according to different environments or to evaluate how certain environmental variables such as temperature, precipitation, and others affect the composition of communities. Hereafter, we present briefly the commonest techniques and multivariate analyses used in ecological studies of ants.

Cluster Analysis This constitutes a multivariate numerical methodology, which encompasses a variety of techniques and algorithms with the common objective of finding and separating objects in similar groups, that is, to divide/classify a given set of observations into a number of homogeneous groups, according to some criterion of homogeneity (Regazzi 2001). To apply these analyses to a set of data, it is very important to choose a coefficient that quantifies how similar two objects are. This coefficient refers to the estimation of a measure of similarity, or dissimilarity, among individuals, or populations, to be grouped. The basic procedure of all clustering methods is similar, starting with the calculation of a matrix of distances between the variables and finishing with a dendrogram (tree graph) in which it is possible to verify the successive fusions of the objects/individuals (Regazzi 2001).

Ordination Techniques These comprise a set of multivariate methods with the purpose of ordering observations or multivariate objects in multidimensional space, reducing them to a smaller number of variables for additional analysis (creation of main axes), for example, identifying response from species or species groups to gradients or environmental disturbances. These analyses can also be used to organize series of data based on species assemblages and environmental variables. There are four broad types of ordination.

Principal component analysis, PCA. This is an ordination method that reduces multivariate data into a smaller number of variables by creating linear combinations of the original variables.

Factor analysis. This is an ordination method that rearranges each original observation/object as a linear combination of a smaller number of parameters, known as "common factors."

Correspondence analysis. This is an ordination method used mainly to relate species distribution to environmental variables. In general the model or hypothesis tested is that the distributions of species abundances are unimodal and approximately normal along the environmental gradient.

Nonmetric multidimensional scaling, NMDS. This is an ordination method that allows preserving the classification order of distances or dissimilarities of the observations. Considered a robust method, it allows the use of any measure of distance or dissimilarity.

Hypothesis tests with multivariate data *Multivariate analysis of variance, MANOVA*. This represents a generalization of the analysis of variance (ANOVA) for univariate data sets. It has immediate application to experimental designs, making it possible to test the hypothesis that the treatments show divergence between itself in relation to the multiple characters evaluated. This analysis allows conclusions about the differences between treatments, while simultaneously considering the whole set of characters observed on each experimental unit. A more rigorous alternative to analyze the statistical significance of multivariate data is to conduct a permutational multivariate analysis of variance—PERMANOVA (Anderson 2001) that uses a geometric partitioning of multivariate variation in the space through dissimilarity to an ANOVA design, with *p*-values obtained using permutation techniques. This analysis is usual even for data sets that are overdispersed or with non-normal behavior; furthermore it is not restricted by distributional assumptions and has recently been extended to accommodate the heterogeneity of intragroup dispersions.

Multivariate multiple regressions. This is a suitable method when the variable response is continuous and one or more predictor variables are continuous. In ecology two methods are commonly used: redundancy analysis, RDA, and canonical correspondence analysis, CCA. The RDA is a variation of multiple regressions to multivariate data, thus assuming a causal relationship between dependent and independent variables. The second method, CCA, focuses on generating a unimodal axis in relation to the response variables and a linear axis in relation to the predictor variables. Significance of the predictor variables can be assessed via randomization tests.

2.7.4 Analysis of Sampling Sufficiency

In many studies, it is interesting (or desirable) to evaluate the sample effort performed. Among the different methods suitable for this, the most used are as follows: (1) comparing the observed species richness to the estimated richness, (2) performing and interpreting a graph of the number of species recorded in relation to the sample area (sample sufficiency curve), or (3) species accumulation graph, also known as collector curve. When a graph of the number of species is produced as a function of the sample area, a curve is generally obtained which rises rapidly and grows more slowly in the subsequent samplings, when only rare species are included. Sampling is considered sufficient when the stability of these curves is reached and new taxa are no more recorded. This kind of curve reaches an asymptote (Ugland et al. 2003), but it is rarely attained in studies on tropical diversity.

In species accumulation curves, the total number of species is calculated for each new sampling unit, and sample adequacy is reached when the curve stabilizes, and the addition of new species does not significantly change the number of species observed. A disadvantage of this is that the shape of the curve may vary according to the order of entry of the samples (Magurran 2004; Schilling and Batista 2008); therefore, this method is more suitable for sequential samplings, such as transect samplings. A possible solution to deal with this bias is to use the species rarefaction curve, which allows establishing an average of species for each unit of sample effort and number of individuals or samples as a function of a random resampling of the species set (Gotelli and Colwell 2001).

2.7.5 Free Software for Data Analysis

Although the free software Past (for Windows and Mac; Hammer et al. 2001) performs binomial GLM, we recommend using R (R Core Team 2020) due to its greater flexibility. The GLM function is part of the base distribution. It demands the specification of the statistical distribution, via the family = binomial argument. In the case of proportional data, the number of samples is specified via the weight's argument.

For generalized additive models, we recommend the R package *mgcv* (Wood 2011).

Generalized linear mixed models may be adjusted with the function *glmer* in the R package lme4 (Bates et al. 2015). The function *glmmPQL* includes options of spatial autocorrelation; however, no information criteria values are available. Generalized additive mixed models may be adjusted with the packages *mgcv* and *gamm4* (Wood and Scheipl 2020).

For model selection tools, including the calculation of different information criteria, we recommend the R package *bbmle* (Bolker 2020).

Contingency table analysis is available from software such as Past (for Windows and Mac) and BioEstat (for Windows). These functions are also available in R, for example, *chisq.test* and *fisher.test*.

Multivariate analyses can be performed in several packages in R, for example, vegan (Oksanen et al. 2012). However, Past is excellent for multivariate analyses, including the main ordination types and many options of distance indices.

For obtaining and comparing different richness estimators, diversity indices, similarity measures, and performance simulation, an option is the software

EstimateS v. 9.1.0 (Colwell 2013), a free software application for Windows and Macintosh operating systems.

Acknowledgments The authors thank the editors for their kind invitation to participate to this volume. Thanks are deserved to Isabelle Bachy (RBINS) who kindly helped to improve the layout of the figures. JHCD and CSFM acknowledge their research grants from CNPq.

References

Agosti D, Alonso LE (2000) The ALL protocol: a standard protocol for the collection of ground-dwelling ants. In: Agosti D, Majer JD, Alonso L et al (eds) Ants. Standard methods for measuring and monitoring biodiversity. Smithsonian Institution Press, Washington, pp 204–206

Agosti D, Majer JD, de Alonso LE et al (eds) (2000) Ants: standard methods for measuring and monitoring biodiversity. Smithsonian Institution, Washington

Albrecht M, Gotelli NJ (2001) Spatial and temporal niche partitioning in grassland ants. Oecologia 126(1):134–141. https://doi.org/10.1007/s004420000494

Andersen AN, Hertog T, Woinarski JCZ (2006) Long-term fire exclusion and ant community structure in an Australian tropical savanna: congruence with vegetation succession. J Biogeogr 33:823–832

Anderson MJ (2001) A new method for non-parametric multivariate analysis of variance. Austral Ecol 26:32–46. https://doi.org/10.1111/j.1442-9993.2001.01070.pp.x

Anderson JM, Ingram JSI (eds) (1993) Tropical soil biology and fertility: a handbook of methods, 2nd edn. CAB International, Wallingford

Antonialli-Junior WF, Giannotti E, Pereira MC et al (2015) Biologia da nidificação e arquitetura de ninhos de formigas poneromorfas do Brasil. In: Delabie JHC, Feitosa RM, Serrão JE et al (eds) As formigas poneromorfas do Brasil. Editus, Ilhéus, pp 285–294

Baccaro FB, Ferraz G (2013) Estimating density of ant nests using distance sampling. Insectes Soc 60(1):103–110. https://doi.org/10.1007/s00040-012-0274-2

Bates D, Maechler M, Bolker B, Walker S (2015) Fitting linear mixed-effects models using lme4. J Stat Softw 67(1):1–48. https://doi.org/10.18637/jss.v067.i01

Basset Y, Springate ND, Aberlenc HP et al (1997) A review of methods for sampling arthropods in tree canopies. In: Stork NE, Adis J, Didham RK (eds) Canopy arthropods. Chapman and Hall, London, pp 27–52

Beck L (1967) Die Bodenfauna des neotropischen Regenwaldes. In: Lent H (ed) Atas do Simpósio sobre a Biota Amazônica. Zoologia, vol 5. Conselho Nacional de Pesquisas, Rio de Janeiro, pp 97–101

Berghoff SM, Weissflog A, Linsenmair KE et al (2002) Foraging of a hypogaeic army ant: a long neglected majority. Insect Soc 49:133–141. https://doi.org/10.1007/s00040-002-8292-0

Berghoff SM, Maschwitz U, Linsenmair KE (2003) Hypogaeic and epigaeic ant diversity on Borneo: evaluation of baited sieve buckets as a study method. Trop Zool 16:153–163. https://doi.org/10.1080/03946975.2003.10531192

Bestelmeyer BT, Agosti D, Alonso LE et al (2000) Field techniques for the study of ground-dwelling ants. In: Agosti D, Majer J, Alonso E et al (eds) Ants. Standard methods for measuring and monitoring biodiversity. Smithsonian Institution Press, Washington, pp 122–144

Besuchet C, Burckhardt DH, Lobl I (1987) The "Winkler/Moczarski" eclector as an efficient extractor for fungus and litter Coleoptera. Coleopt Bull 41:392–394

Bhaktar AP, Whitcomb WH (1975) Rearing arboreal ants in glass tubing. Fla Entomol 58:59–63

Bhatkar AP, Whitcomb WH (1970) Artificial diet for rearing various species of ants. Fla Entomol 53:229–232

Bickford D, Lohman DJ, Sodhi NS et al (2007) Cryptic species as a window on diversity and conservation. Trends Ecol Evol 22:148–155. https://doi.org/10.1016/j.tree.2006.11.004
Bolker B (2020) bbmle: Tools for general maximum likelihood estimation. http://CRAN.R-project.org/package=bbmle
Brandão CRF, Silva RR, Delabie JHC (2012) Neotropical ants (Hymenoptera) functional groups: nutritional and applied implications. In: Panizzi AR, Parra JRP (eds) Bioecology and insect nutrition. CRC, Boca Raton, pp 213–236
Broman KW, Woo KH (2018) Data organization in spreadsheets. Am Stat 72:2–10. https://doi.org/10.1080/00031305.2017.1375989
Burnham KP, Anderson DR (eds) (2002) Model selection and multimodel inference: A practical information-theoretic approach. 2nd edn. Springer-Verlag, New York
Caitano B, Dodonov P, Delabie JHC (2018) Edge, area and anthropization effects on mangrove-dwelling ant communities. Acta Oecol 91:1–6. https://doi.org/10.1016/j.actao.2018.05.004
Camarota F, Vasconcelos HL, Koch EBA et al (2018) Discovery and defense define the social foraging strategy of neotropical arboreal ants. Behav Ecol Sociobiol 72:2–11. 10.1007/s00265-018-2519-1
Cardoso DC, Cristiano MP, Tavares MG (2011) Methodological remarks on rearing basal Attini ants in the laboratory for biological and evolutionary studies: overview of the genus *Mycetophylax*. Insectes Soc 58:427–430. https://doi.org/10.1007/s00040-011-0160-3
Castaño-Meneses G, Palacios-Vargas JG, Carmo AFR (2015) Colêmbolos e outros inquilinos de formigueiros de poneromorfas. In: Delabie JHC, Feitosa RM, Serrão JE et al (eds) As formigas Poneromorfas do Brasil. Editus, Ilhéus, pp 389–401
Castaño-Meneses G, Palacios-Vargas JG, Delabie JHC et al (2017) Springtails (Collembola) associated with nests of fungus-growing ants (Formicidae: Myrmicinae: Attini) in southern Bahia, Brazil. Fla Entomol 100(4):740–742. https://doi.org/10.1653/024.100.0421
Castaño-Meneses G, Santos RJ, Santos JRM et al (2019) Invertebrates associated to Ponerine ants nests in two cocoa farming systems in the southeast of the state of Bahia, Brazil. Trop Ecol 60(1):52–61. https://doi.org/10.1007/s42965-019-00006-3
Castro D, Fernández F, Meneses AD et al (2018) A preliminary checklist of soil ants (Hymenoptera: Formicidae) of Colombian Amazon. Biodiv Data J 6:e29278. https://doi.org/10.3897/BDJ.6.e29278
Chen X, MacGown JA, Adams BJ et al (2012) First record of *Pyramica epinotalis* (Hymenoptera: Formicidae) for the United States. Psyche 2012:1–7. https://doi.org/10.1155/2012/850893
Chen X, Adams B, Sabo A et al (2016) Ant assemblages and co-occurrence patterns in cypress-tupelo swamp. Wetlands 36:1–13. https://doi.org/10.1007/s13157-016-0795-y
Chomicki G, Renner SS (2017) The interactions of ants with their biotic environment. Proc R Soc B 284:20170013. https://doi.org/10.1098/rspb.2017.0013
Cobb M, Watkins K, Silva EV et al (2006) An exploratory study on the use of bamboo pieces for trapping entire colonies of arboreal ants (Hymenoptera: Formicidae). Sociobiology 47:215–223
Colegrave N, Ruxton GD (2018) Using biological insight and pragmatism when thinking about pseudoreplication. Trends Ecol Evol 33:28–35. https://doi.org/10.1016/j.tree.2017.10.007
Colwell RK (2013) EstimateS: statistical estimation of species richness and shared species from samples. Version 9. User's Guide and application published. http://purl.oclc.org/estimates
Corbara B, Leroy C, Orivel J et al (2019) Relaciones entre las hormigas y las plantas en los trópicos del Nuevo Mundo. In: Fernández F, Guerrero R, Delsinne T (eds) Hormigas de Colombia. Universidad Nacional de Colombia, Bogotá, pp 203–254
Czechowski W, Pisarski B (1992) Laboratory methods for rearing ants (Hymenoptera, Formicoidea). Memorabilia Zool 45:1–32
DaRocha WD, Ribeiro SP, Neves FS et al (2015) How does bromeliad distribution structure the arboreal ant assemblage (Hymenoptera: Formicidae) on a single tree in a Brazilian Atlantic forest agroecosystem? Myrmecol News 21:83–92

DaRocha WD, Neves FS, Dáttilo W et al (2016) Epiphytic bromeliads as key components for maintenance of ant diversity and ant-bromeliad interactions in agroforestry system canopies. Forest Ecol Manag 372:128–136. https://doi.org/10.1016/j.foreco.2016.04.011

Dasu T, Johnson T (2003) Exploratory data mining and data cleaning. Wiley, Hoboken

Davidson DW, Arias JA, Mann J (2006) An experimental study of bamboo ants in Western Amazonia. Insect Soc 53:108–114. https://doi.org/10.1007/s00040-005-0843-8

Dejean A, Corbara B, Orivel J et al (2007) Rainforest canopy ants: the implications of territoriality and predatory behavior. Funct Ecosyst Commun 1:105–120

Dejean A, Fisher BL, Corbara B et al (2010) Spatial distribution of dominant arboreal ants in a Malagasy coastal rainforest: gaps and presence of an invasive species. PLoS One 5:e9319. https://doi.org/10.1371/journal.pone.0009319

Dejean A, Ryder S, Bolton B et al (2015) How territoriality and host-tree taxa determine the structure of ant mosaics. The Science of Nature 102(33):5–6. https://doi.org/10.1007/s00114-015-1282-7

Delabie JHC (1989) Avaliação preliminar de técnica alternativa de controle da formiga "pixixica" *Wasmannia auropunctata* em cacauais, Agrotropica, 1(1):75–78

Delabie JHC (1999) Comunidades de formigas (Hymenoptera; Formicidae): métodos de estudo e estudos de casos na Mata Atlântica. Resumos, XII Encontro de Zoologia do Nordeste, Feira de Santana–BA, pp 58–68

Delabie JHC (2001) Trophobiosis between Formicidae and Hemiptera (Sternorrhyncha and Auchenorrhyncha): an overview. Neotrop Entomol 30:501–516

Delabie JHC, Reis YT (2000) Sympatry and mating flight synchrony of three species of *Cylindromyrmex* (Hymenoptera, Formicidae) in southern Bahia, Brazil, and the importance of malaise trap for rare ants' inventory. Rev Bras Entomol 44:109–110

Delabie JHC, Fisher BL, Majer JD et al (2000a) Sampling effort and choice of methods. In: Agosti D, Majer JD, Alonso L et al (eds) Ants. Standard methods for measuring and monitoring biodiversity. Smithsonian Institution Press, Washington, pp 145–154

Delabie JHC, Fresneau D, Pezon A (2000b) Notes on the ecology of *Thaumatomyrmex* spp. (Hymenoptera: Formicidae: Ponerinae) in Southeast Bahia, Brazil. Sociobiology 36(3):571–584

Delabie JHC, Jahyny B, Nascimento IC et al (2007) Contribution of cocoa plantations to the conservation of native ants (Insecta: Hymenoptera: Formicidae) with a special emphasis on the Atlantic Forest fauna of southern Bahia, Brazil. Biodivers Conserv 16:2359–2384. https://doi.org/10.1007/s10531-007-9190-6

Delsinne T, Leponce M, Theunis L et al (2008) Rainfall influences ant sampling in dry forests. Biotropica 40(5):590–596. https://doi.org/10.1111/J.1744-7429.2008.00414.x

Espadaler X, López-Soria L (1991) Rareness of certain Mediterranean ant species: fact or artefact? Insect Soc 38:365–377

Fittkau EJ, Klinge H (1973) On biomass and trophic structure of the central Amazonian rain forest ecosystem. Biotropica 5(1):2–14

Floren A, Linsenmair KE (1997) Diversity and recolonization dynamics of selected arthropod groups on different tree species in a lowland rainforest in Sabah, Malaysia with special reference to Formicidae. In: Stork NE, Adis J, Didham RK (eds) Canopy arthropods. Chapman and Hall, London, pp 344–381

Floren A, Westzel W, Staab M (2014) The contribution of canopy species to overall ant diversity (Hymenoptera: Formicidae) in temperate and tropical ecosystems. Myrmecol News 19:65–74

Folgarait PJ (1998) Ant biodiversity and its relationship to ecosystem functioning: a review. Biodivers Conserv 7:1221–1244

Forti LC, Moreira AA, Andrade APP, Castellani MA, Caldato N (2011) Nidificação e arquitetura de ninhos de formigas-cortadeiras. In: Della Lucia TMC (ed) Formigas-cortadeiras, da Bioecologia ao Manejo. Editora UFV, Viçosa, pp 102–125

Fowler HG, Delabie JHC (1995) Resource partitioning among epigaeic and hypogaeic ants (Hymenoptera: Formicidae) of a Brazilian cocoa plantation. Ecol Austral 5:117–124

Fowler HG, Forti LC, Brandão CRF et al (1991) Ecologia nutricional de formigas. In: Panizzi AP, JRP P (eds) Ecologia Nutricional de Insetos e suas Implicações no Manejo de Pragas. Manole, São Paulo, pp 131–223

Frizzo TLM, Campos RI, Vasconcelos HL (2011) Contrasting effects of fire on arboreal and ground-dwelling ant communities of a Neotropical savanna. Biotropica 44:254–261. https://doi.org/10.1111/j.1744-7429.2011.00797.x

Gomes CB, Souza JLP, Franklin E (2018) A comparison between time of exposure, number of pitfall traps and the sampling cost to capture ground-dwelling poneromorph ants (Hymenoptera: Formicidae). Sociobiology 65(2):138–148. https://doi.org/10.13102/sociobiology.v65i2.1207

Gotelli NJ, Colwell RK (2001) Quantifying biodiversity: procedures and pitfalls in the measurement and comparison of species richness. Ecol Lett 4:379–391. https://doi.org/10.1046/j.1461-0248.2001.00230.x

Gotelli NJ, Ellison AM (2011) Princípios de Estatística em Ecologia. Artmed, Porto Alegre

Griffith DM, Veech JA, Marsh CJ (2016) Cooccur: probabilistic species co-occurrence analysis in R. J Stat Softw 69(2):1–17. https://doi.org/10.18637/jss.v069.c02

Groc S, Orivel J, Dejean A et al (2009) Baseline study of the leaf-litter ant fauna in a French Guianese Forest. Insect Conserv Divers 2:183–193

Guénard B, Lucky A (2011) Shuffling leaf litter samples produces more accurate and precise snapshots of terrestrial arthropod community composition. Environ Entomol 40(6):1523–1529. https://doi.org/10.1603/en11104

Hammer Ø, Harper DAT, Ryan PD (2001) PAST: Paleontological statistics software package for education and data analysis. Palaeontol Electron 4(1):1–9

Herbers JM (1989) Community structure in north temperate ants: temporal and spatial variation. Oecologia 81(2):201–211

Hölldobler B, Wilson EO (1990) The ants. Harvard University Press, Cambridge

Hurlbert SH (1984) Pseudoreplication and the design of ecological field experiments. Ecol Monogr 54:187–211

Imai HT, Crosland MWJ, Crozier RH (1988) Modes of spontaneous chromosome mutation and karyotype evolution in ants with reference to the minimum interaction hypothesis. Jpn J Genet 63:113–125

Jacquemin J, Roisin Y, Leponce M (2016) Spatio-temporal variation in ant (Hymenoptera: Formicidae) communities in leaf-litter and soil layers in a premontane tropical forest. Myrmecol News 22:129–139

Johnson AF, Wichern DW (1992) Applied multivariate statistical analysis, 3rd edn. Prentice-Hall International, Englewood Cliffs, NJ

Jones CG, Lawton JH, Shachak M (1994) Organisms as ecosystem engineers. Oikos 69:373–386

Kaspari M (2000) Do imported fire ants impact canopy arthropods? Evidence from simple arboreal pitfall traps. Southwest Nat 45:118–122. https://doi.org/10.2307/3672451

Kaspari M, Pickering J, Longino JT et al (2001) The phenology of a Neotropical ant assemblage: evidence for continuous and overlapping reproduction. Behav Ecol Sociobiol 50:382–390. https://doi.org/10.1007/s002650100378

Klimes P, Janda M, Ibalim S et al (2011) Experimental suppression of ants foraging on rainforest vegetation in New Guinea: testing methods for a whole-forest manipulation of insect communities. Ecol Entomol 36:94–103. https://doi.org/10.1111/j.1365-2311.2010.01250.x

Koch EBA, Camarota FC, Vasconcelos HL (2016) Plant ontogeny as a conditionality factor in the protective effect of ants on a Neotropical tree. Biotropica 48:198–205. https://doi.org/10.1111/btp.12264

Koch EBA, Santos JRM, Nascimento IC, Delabie JHC (2019) Comparative evaluation of taxonomic and functional diversities of leaf-litter ants of the Brazilian Atlantic Forest. Turk J Zool 43:437–456. https://doi.org/10.3906/zoo-1811-7

Lasmar CJ, Queiroz ACM, Rabello AM et al (2017) Testing the effect of pitfall-trap installation on ant sampling. Insect Soc 64:445–451. https://doi.org/10.1007/s00040-017-0558-7

Lattke JE (2000) Specimen processing, building and curating an ant collection. In: Agosti D, Majer JD, Alonso L et al (eds) Ants: standard methods for measuring and monitoring biodiversity. Smithsonian Institution, Washington, pp 155–171

Leal IR, Oliveira PS (1998) Interactions between fungus-growing ants (Attini), fruits and seeds in cerrado vegetation in Southeast Brazil. Biotropica 30:170–178

Leal IR, Oliveira PS (2000) Foraging ecology of attine ants in a Neotropical savanna: seasonal use of fungal substrate in the cerrado vegetation of Brazil. Insect Soc 47:376–382. https://doi.org/10.1007/PL00001734

Lengyel S, Gove AD, Latimer AM et al (2009) Ants sow the seeds of global diversification in flowering plants. PLoS One 4(5):e5480. https://doi.org/10.1371/journal.pone.0005480

Leponce M, Theunis L, Delabie JHC et al (2004) Scale dependence of diversity measures in a leaf-litter ant assemblage. Ecography 27:253–267. https://doi.org/10.1111/j.0906-7590.2004.03715.x

Leponce M, Meyer C, Häuser CL et al (2010) Challenges and solutions for planning and implementing large-scale biotic inventories. In: Eymann J, Degreef J, Häuser C et al (eds) Manual on field recording techniques and protocols for All Taxa Biodiversity Inventories and Monitoring, vol 8. Abc Taxa, London, pp 18–48

Leponce M, Novotny V, Pascal O et al (2016) Land module of our planet reviewed - Papua New Guinea: aims, methods and first taxonomical results. In: Robillard T, Legendre F, Villemant C, Leponce M (eds) Insects of Mount Wilhelm, Papua New Guinea. Mémoires du Muséum National d'Histoire Naturelle, vol 209, pp 13–48

Leponce M, Delabie JHC, Orivel J et al (2019) Tree-dwelling ant survey (Hymenoptera, Formicidae) in Mitaraka, French Guiana, in Touroult J. (ed.), "Our Planet Reviewed" 2015 large-scale biotic survey in Mitaraka. French Guiana. Zoosystema 41(10):163–179. https://doi.org/10.5252/zoosystema2019v41a10

Levings SC (1983) Seasonal, annual, and among-site variation in the ground ant community of a deciduous tropical forest: some causes of patchy species distributions. Ecol Monogr 53(4):435–455. https://doi.org/10.2307/1942647

Lopes CT, Vasconcelos HL (2008) Evaluation of three methods for sampling ground-dwelling ants in the Brazilian Cerrado. Neotrop Entomol 37(4):399–405. https://doi.org/10.1590/s1519-566x2008000400007

Lopes JMS, Oliveira AR, Delabie JHC et al (2015) A new species of myrmecophile mite of the genus *Oplitis* (Acari: Mesostigmata: Oplitidae) from Brazil. Int J Acarol 41(8):676–680. https://doi.org/10.1080/01647954.2015.1096960

Lynch JF (1981) Seasonal, successional, and vertical segregation in a Maryland ant community. Oikos 37:183–198

Magurran AE (2004) Measuring biological diversity. Blackwell, Oxford

Majer JD (1990) The abundance and diversity of arboreal ants in northern Australia. Biotropica 22:191–199

Majer JD (1992) Comparison of the arboreal ant mosaic in Ghana, Brasil, Papua New Guinea and Australia - its structure and influence on ant diversity. In: LaSalle J, Gauld I (eds) Hymenoptera and biodiversity. CAB International, Wallingford, pp 115–141

Majer JD, Delabie JHC (1994) Comparison of the ant communities of annually inundated and terra firme forests at Trombetas in the Brazilian Amazon. Insect Soc 41:343–359. https://doi.org/10.1007/BF01240639

Majer JD, Greenslade P (2012) Soil and litter invertebrates. In: Specht RL, Rundel PW, Westman WE et al (eds) Mediterranean-type ecosystems: a data source book, vol 19. Springer, Dordrecht, pp 197–226

Marques GDV, Del-Claro K (2006) The ant fauna in a cerrado area: the influence of vegetation structure and seasonality (Hymenoptera: Formicidae). Sociobiology 47(1):235–252

Martins MFO, Thomazini MJ, Baretta D et al (2020) Accessing the subterranean ant fauna (Hymenoptera: Formicidae) in native and modified subtropical landscapes in the Neotropics. Biota Neotropica 20:e20190782. https://doi.org/10.1590/1676-0611-bn-2019-0782

Morini M, Yashima M, Zene Y et al (2004) Observations on *Acanthostichus quadratus* (Hymenoptera: Formicidae: Cerapachyinae) visiting underground baits and fruits of the *Syagrus romanzoffiana*, in an area of the Atlantic Forest, Brazil. Sociobiology 43:573–578

Mueller UG, Wcislo WT (1998) Nesting biology of the fungus-growing ant *Cyphomyrmex longiscapus* Weber (Attini, Formicidae). Insect Soc 45:181–189. https://doi.org/10.1007/s000400050078

Nascimento IC, Delabie JHC, Ferreira PSF, Della Lucia TMC (2004) Mating flight seasonality in the genus *Labidus* at Minas Gerais, in the Brazilian Atlantic forest biome, and *Labidus nero*, junior synonym of *Labidus mars*. Sociobiology 44(3):615–622

Nascimento IC, Delabie JHC, Della Lucia TMC (2011) Phenology of mating flight in Ecitoninae (Hymenoptera: Formicidae) in a Brazilian Atlantic Forest location. Ann Soc Entomol Fr 47(1–2):112–118. https://doi.org/10.1080/00379271.2011.10697702

Ning D, Yang F, Xiao Q et al (2019) A simple and efficient method for preventing ant escape (Hymenoptera: Formicidae). Myrmecol News 29:57–65. https://doi.org/10.25849/myrmecol.news_029:057

Nunes FA, Segundo GBM, Vasconcelos YB et al (2011) Ground-foraging ants (Hymenoptera: Formicidae) and rainfall effect on pitfall trapping in a deciduous thorn woodland (Caatinga), northeastern Brazil. Rev Bio Trop 59:1637–1650. https://doi.org/10.15517/rbt.v59i4.3426

Oksanen L (2001) Logic of experiments in ecology: is pseudoreplication a pseudoissue? Oikos 94:27–38. https://doi.org/10.1034/j.1600-0706.2001.11311.x

Oksanen J, Blanchet F, Kindt R et al (2012) Vegan: Community Ecology Package, R Package Version 2.0–10. https://cran.r-project.org/web/packages/vegan

Oliveira-Santos LGR, Loyola RD, Vargas AB (2009) Armadilhas de dossel: uma técnica para amostrar formigas no estrato vertical de florestas. Neotrop Entomol 38:691–694

Oliver I, Pik A, Britton D, Dangerfield M, Colwell RK, Beattie AJ (2000) Virtual biodiversity assessment systems. Bioscience 50(5):441–450. https://doi.org/10.1641/0006-3568(2000)050[0441:VBAS]2.0.CO;2

Orsolon-Souza G, Esbérard CEL, Mayhé-Nunes AJ et al (2011) Comparison between Winkler's extractor and pitfall traps to estimate leaf litter ants' richness (Formicidae) at a rainforest site in Southeast Brazil. Braz J Biol 71(4):873–880. https://doi.org/10.1590/S1519-69842011000500008

Palacios-Vargas JG, Mejía-Recamier BE, Zeppelini D (eds) (2013) Técnicas atuais para estudo de micro e mesoartrópodes de solo. Eduepb, Campina Grande, Paraíba, Brazil

Peeters C, Crewe R (1984) Insemination controls the reproductive division of labour in a ponerine ant. Naturwissenschaften 71:50–51. https://doi.org/10.1007/BF00365989

Pereira MC, Delabie JHC, Súarez YR et al (2013) Spatial connectivity of aquatic macrophytes and flood cycle influence species richness of an ant community of a Brazilian floodplain. Sociobiology 60(1):41–49

Perfecto I, Snelling R (1995) Biodiversity and the transformation of a tropical agroecosystem: ants in coffee plantations. Ecol Appl 5:1084–1097

Perry DR (1978) A method of access into the crowns of emergent and canopy trees. Biotropica 10:155–157

Porter SD (1989) Effects of diet on the growth of laboratory fire ant colonies (Hymenoptera: Formicidae). J Kans Entomol Soc 62:288–291

Powell S, Costa AN, Lopes CT et al (2011) Canopy connectivity and the availability of diverse nesting resources affect species coexistence in arboreal ants. J Anim Ecol 80:352–360. https://doi.org/10.1111/j.1365-2656.2010.01779.x

R Core Team (2020) R: A language and environment for statistical computing. R Foundation for Statistical Computing, Vienna, Austria. https://www.R-project.org/

Rabello AM, Queiroz ACM, Lasmar CJ et al (2015) When is the best period to sample ants in tropical areas impacted by mining and in rehabilitation process? Insect Soc 62:227–236. https://doi.org/10.1007/s00040-015-0398-2

Regazzi AJ (2001) Análise multivariada, notas de aula INF 766. Universidade Federal de Viçosa, Centro de Ciências Exatas e Tecnológicas. Departamento de Informática, Apostila de disciplina, Viçosa

Reis YT, Delabie JHC (1999) Estudo de uma comunidade de Pseudomyrmecinae usando armadilhas de Malaise numa reserva de Mata Atlântica em Porto Seguro, Bahia. Naturalia 24:119–121

Ribas CR, Schoereder JH, Pic M et al (2003) Tree heterogeneity, resource availability, and larger scale processes regulating arboreal ant species richness. Austral Ecol 28:305–314. https://doi.org/10.1046/j.1442-9993.2003.01290.x

Ribeiro SP, Espírito Santo NB, Delabie JHC et al (2013) Competition, resources and the ant (Hymenoptera: Formicidae) mosaic: a comparison of upper and lower canopy. Myrmecol News 18:113–120

Ryder Wilkie KT, Mertl AL, Traniello JF (2007) Biodiversity below ground: probing the subterranean ant fauna of Amazonia. Naturwissenschaften 94:725–731. https://doi.org/10.1007/s00114-007-0250-2

Ryder Wilkie KT, Mertl AL, Traniello JFA (2010) Species diversity and distribution patterns of the ants of Amazonian Ecuador. PLOS One 5(10):e13146. https://doi.org/10.1371/journal.pone.0013146

Samways MJ (1981) Comparison of ant community structure (Hymenoptera: Formicidae) in citrus orchards under chemical and biological control of red scale, *Aonidiella aurantii* (Maskell) (Hemiptera: Diaspididae). B Entomol Res 71:663–670

Schilling AC, Batista JLF (2008) Curva de acumulação de espécies e suficiência amostral em florestas tropicais. Braz J Bot 31:179–187

Schmidt FA, Solar RRC (2010) Hypogaeic pitfall traps: methodological advances and remarks to improve the sampling of a hidden ant fauna. Insectes Soc 57:261–266. https://doi.org/10.1007/s00040-010-0078-1

Schmidt FA, Ribas CR, Schoereder JH (2013) How predictable is the response of ant assemblages to natural forest recovery? Implications for their use as bioindicators. Ecol Indic 24:158–166. https://doi.org/10.1016/j.ecolind.2012.05.031

Schultz TR (1993) Stalking the wild attine. Notes from Underground 8:7–10

Silva RR, Brandão CRF (2014) Ecosystem-wide morphological structure of leaf litter ant communities along a tropical latitudinal gradient. PLOS One 9:e93049. https://doi.org/10.1371/journal.pone.0093049

Silva MR Jr, Miranda MMA, Luz HP et al (2007) Nidificação de *Cyphomyrmex transversus* Emery (Formicidae: Myrmicinae) em áreas de pastagens no município de Itapetinga, BA. O Biológico 69(2):261–265

Silva FS, Souza RM, Solar RR et al (2017) Ant diversity in Brazilian tropical dry forests across multiple vegetation domains. Environ Res Lett 12:035002. https://doi.org/10.1088/1748-9326/aa5f2a/meta

Solomon SE, Mueller UG, Schultz TR et al (2004) Nesting biology of the fungus growing ants *Mycetarotes* Emery (Attini, Formicidae). Insect Soc 51:333–338. https://doi.org/10.1007/s00040-004-0742-4

Sosa-Calvo J, Jesovnik A, Okonski E et al (2015) Locating, collecting, and maintaining colonies of fungus-farming ants (Hymenoptera: Formicidae: Myrmicinae: Attini). Sociobiology 62(2):300–320. https://doi.org/10.13102/sociobiology.v62i2.300-320

Southwood TRE (1978) Ecological methods: with particular reference to the study of insect populations, 2nd edn. Chapman & Hall, London

Souza JLP, Baccaro FB, Landeiro VL et al (2012) Trade-offs between complementarity and redundancy in the use of different sampling techniques for ground-dwelling ant assemblages. Appl Soil Ecol 56:63–73. https://doi.org/10.1016/j.apsoil.2012.01.004

Stadler B, Dixon T (2008) Mutualism: ants and their insect partners. Cambridge University Press, Cambridge

Steiner A, Haüser CL (2010) Recording insects by light traps. In: Eymann J, Degreef J, Häuser C et al (eds) Manual on field recording techniques and protocols for all taxa biodiversity inventories and monitoring, vol 8. Abc Taxa, London, pp 400–422
Sudd JH, Franks NR (1987) The behavioural ecology of ants. Blackie & Son, Glasgow
Ugland KI, Gray JS, Ellingsen KE (2003) The species–accumulation curve and estimation of species richness. J Animal Ecol 72:888–897. https://doi.org/10.1046/j.1365-2656.2003.00748.x
van Achterberg C, Grootaert P, Shaw MR (2010) Flight interception traps for arthropods. In: Eymann J, Degreef J, Häuser C et al (eds) Manual on field recording techniques and protocols for all taxa biodiversity inventories and monitoring, vol 8. Abc Taxa, London, pp 423–462
Vasconcelos HL, Vilhena JMS, Caliri GJA (2000) Responses of ants to selective logging of a central Amazonian forest. J Appl Ecol 37:508–514. https://doi.org/10.1046/j.1365-2664.2000.00512.x
Vasconcelos HL, Maravalhas JB, Feitosa RM et al (2018) Neotropical savanna ants show a reversed latitudinal gradient of species richness, with climatic drivers reflecting the forest origin of the fauna. J Biogeogr 45:248–258. https://doi.org/10.1111/jbi.13113
Veech JA (2013) A probabilistic model for analyzing species co-occurrence. Glob Ecol Biogeogr 22:252–260. https://doi.org/10.1111/j.1466-8238.2012.00789.x
Weber NA (1966) Fungus-growing ants. Science 153:587–604
Weber NA (1972) Gardening ants: the Attines. Memoirs of the American Philosophical Society 92. American Philosophical Society, Philadelphia
Wheeler WM (1907) Fungus growing-ants of North America. Bull Am Mus Nat Hist 23:1865–1937
Wickham H (2014) Tidy data. Journal of statistical software 59:1–23. https://doi.org/10.18637/jss.v059.i10
Wong MKL, Guénard B (2017) Subterranean ants: summary and perspectives on field sampling methods, with notes on diversity and ecology (hymenoptera: Formicidae). Myrmecol News 25:1–16. https://doi.org/10.25849/myrmecol.news_025:001
Wood SN (2011) Fast stable restricted maximum likelihood and marginal likelihood estimation of semiparametric generalized linear models. J R Stat Soc B 73(1):3–36. https://doi.org/10.1111/j.1467-9868.2010.00749.x
Wood S, Scheipl F (2020) gamm4: Generalized Additive Mixed Models using 'mgcv' and 'lme4'. R package version 0.2-6. https://CRAN.R-project.org/package=gamm4
Zuur A, Ieno EN, Walker N, Saveliev AA et al (2009) Mixed effects models and extensions in ecology with R. Springer, Dordrecht

Chapter 3
Bees: How and Why to Sample Them

Laurence Packer and Gerome Darla-West

3.1 Introduction

The enormous scientific, conservation, economic, and policy interest in bees have resulted in explosive growth in fieldwork aimed at assessing bee diversity and abundance. For example, a Google Scholar search for the term "pollinator sampling" yielded 14 citations for 2019 already by mid-April, yet there were no matches for the entire year of 2000. Indeed, 14 citations were exceeded only by searching for the years 2000 through 2008 combined. Similarly, for "bee sampling", there were 18 citations that clearly related to field sampling of bees for ecological research but none in 2000, and 18 is exceeded only by summed returns from 2000 through 2004. Despite this exponential growth in data acquisition and analysis, there is little consistency in sampling methodology in current bee population studies. Whenever and wherever methods are compared (e.g. Westphal et al. 2008, Nielsen et al. 2011, Bashir et al. 2013, and many others, only some of which are cited below), species richness and other biodiversity metric differences are found to be present and large. These are all index sampling techniques and always biased (i.e. they are an unknown reflection of the absolute or relative abundance of the bees present at a sampling location). This does not invalidate their use; rather, it requires cautious interpretation of studies presenting, using, and comparing techniques using both judgement of the capture patterns and their context and understanding of life histories.

Investigations of surveying techniques often suffer from the problems of limited geographic scope. In most studies, one or only a small number of locations were used. An exception is Westphal et al. (2008) who used a range of sampling methods at a continental scale. Given that each bee species has its own life history, floral preferences, and behaviours and that different habitats and even locations in the same field always contain different suites, ratios, and abundances of species, it is not

L. Packer (✉) · G. Darla-West
Department of Biology, York University, Toronto, ON, Canada

J. C. Santos, G. W. Fernandes (eds.), *Measuring Arthropod Biodiversity*,
https://doi.org/10.1007/978-3-030-53226-0_3

surprising to find that results from different locations diverge. Any large-scale survey of bees in an area that has not been sampled extensively previously should, if possible, undergo some preliminary trials of techniques (Hall 2018). Numbers and species of bees can be substantially, and at times completely, altered simply by changing the timing of surveys within a day, month, or year and according to the number and placement of traps or whether netting or observation is used. Results also differ among years even with identical sampling methods and timelines. In general, the most useful techniques will be ones that capture large, representative, and repeatable numbers of the target populations across a region; are relatively inexpensive, minimally impacted by variation among observers or collectors; and minimize sample processing time and variability in results associated with weather, location, and site conditions. As the reader will see, this is a tough set of conditions to meet, yet we must do the best we can and err on the side of larger sample sizes to capture and illuminate population changes in face of time and treatment effects amidst an ocean of naturally occurring statistical variance and bias. We must remember that we are measuring populations of short-lived insects not antelope, and variance in sampled abundance on different dates within and on the same (Julian or calendar) dates among years is great. The situation is magnified when we consider sampling for social bees for which dozens of individuals might be sampled in a field survey which, in reality, may represent no more than the output from a single queen.

In this review we first outline the pros and cons of lethal versus non-lethal methods for sampling bees and suggest times when the latter is necessary. Next, we survey the different methods that have been used to sample bees. Because more studies have demonstrated the benefits of coloured bowls over, or in combination with, other methods and there is seemingly greater variability in approaches using them, we discuss this method in more detail than others. Next, we discuss the reasons researchers might want to use different trapping methods depending on the research question posed. While it might seem sensible to discuss "why sample bees" before outlining the methods available, we found it impossible to describe the reasons for sampling bees without reference to the methods. We discuss the pros and cons of each technique. Finally, we suggest what the most important methodological research needs of the bee conservation and ecology community might be in regard to bee sampling.

We do not cover the specialized methods that might apply to a small taxonomic subsample of the bees in an area—such as the use of bait lures for orchid bees (e.g. Janzen et al. 1982). Note though the great success of application of this method for long-term orchid bee sampling (Roubik 2001; LeBuhn et al. 2012) and finding new taxa (e.g. Nemésio 2010). Vertebrate lures, in the form of flesh or fish skin, have been shown to attract some bees, but the taxonomic representation with such methods is low, and the pleasures of sorting through trap samples substantially diminished for anyone without anoia. The number of bee species known to be attracted to carrion, faeces (Baumgartner and Roubik 1989), or tears (Bänziger 2018) is both too small and geographically restricted for consideration in a review of sampling methods, although there remains the possibility of discovering new species with them and we would be thrilled to see any publication with a title such as "Panda tears as a bee sampling method".

Despite the use of commercial yellow and blue sticky cards (a coloured surface covered in a sticky compound) for sampling insects elsewhere in the world, it is little mentioned as a survey technique for bees but has been used to effectively capture bees in New Zealand (Larsen et al. 2014). Notably, the New Zealand study applied their own Tangle-Trap© layer to specially crafted large cards which may have greater ability to attract and hold bees than sticky cards used for pest detection; of possible note is the very reduced bee fauna found on New Zealand (~40 vs. ~1600 species in Australia). Further investigations are warranted as such traps may have particular usefulness when hung from bushes and trees.

Bees also are by-catch found within other insect surveys. Pitfall traps and light traps used to collect nocturnal insects and coloured pheromone-based Universal Moth Traps all have been shown to capture bees, sometimes in significant numbers. In all cases that we are aware of, these captures likely have their root in the colour of the traps (usually white, yellow, or blue) or direct attraction to bright light and thus function to some extent like the vane and bowl traps presented below. Traps such as these are often widely deployed as surveys of agricultural pests and present opportunities for opportunistic sampling (Hatten et al. 2013; Hung et al. 2015; Spears et al. 2016).

We also do not discuss the methods used to analyse bee sampling data nor recent developments in automated DNA barcoding of malaise trap samples (e.g. Yu et al. 2012), eDNA samples (Shokralla et al. 2012; Critescu and Hebert 2018), or genetic estimates of population sizes for which the reviews of Lozier and Zayed (2017) and Lopez-Uribe et al. (2017) should be consulted. Molecular techniques will likely play a larger role in the future.

3.2 Lethal Versus Non-lethal Sampling

While it is sometimes possible to learn how to identify bee species in a restricted area by sight without killing specimens, unless for a restricted taxonomic group (e.g. bumble bees, Fowler et al. 2016), this takes years of experience (and usually still involves lethal subsampling to verify the sight identifications). This approach always fails to completely assess species as there are complexes that are impossible to separate by eye (indeed, some cannot be identified even with a microscope (e.g. Packer et al. 2016)); new species move into an area, and as is the case with many of the smaller bees, identification is difficult (Gibbs 2017). Field identification issues even apply to some of the common bumble bees (e.g. Fowler et al. 2016). Bees can also be uncooperative, individuals may be hidden by vegetation, sit at such an angle as their field characteristics cannot be completely seen or move too quickly. Research questions that require counting the numbers of individuals of each species will also suffer at least some slight bias from double counting individuals. The primary issue is the considerable interindividual (researcher/technician/volunteer) ability to both see and accurately identify the study organisms to finer taxonomic categories. Studies using visual surveys primarily partition the bees only into large, multispecies groups based on rough approximations of size or colour (Osborne et al. 2008; Broussard et al. 2011; Oberhauser and LeBuhn 2012). An intermediate approach is to capture and photograph the live specimen in a plastic bag or chamber (Anonymous 2015). Sample sizes, though, suffer because of greatly increased handling time, but identification accuracy is higher; issues of observer bias in netting are relevant and discussed later in this chapter, and identification of small bees or taxonomically complex groups also remains an issue.

Mark-release-recapture methods have been used to assess bee population sizes (e.g. Bischoff 2003; Yamomoto et al. 2014). Larsson and Franzén (2008) found good agreement between population sizes estimated with this method and those obtained from transect surveys with visual identification. However, it is simply logistically impossible to apply such methods to a whole community of bees as the accuracy of the estimates depends strongly on capturing, marking, and recapturing a significant proportion of the population over a short enough period of time in which few emergences and deaths have occurred, a high bar to achieve for most researchers (Bischoff 2003).

Removing bees via lethal collection creates the possibility of impacting subsequent generations and this can be a concern in terms of conservation, moral values, and the validity of further years of sampling if the sampling itself greatly impacts the subsequent year's counts. There is only one published evaluation of the effects of repeated sampling: that study demonstrating no impact (Gezon et al. 2015). Multi-year studies have not generally indicated a consistent decline in numbers of specimens in the year after initial sampling. For example, Grixti and Packer (2006) surveyed bees intensively over 2 years. Of the 11 species for which >100 individuals were collected in the first year, 6 were sampled less frequently in the second, while 5 were found more abundantly. In contrast, in the longest-running lethal trapping study, Onuferko et al. (2018) found that two of the three groups of bees they assessed declined over the course of the 10 years of research. However, they concluded that succession of the study site was the cause of the declines which were most marked for taxa that preferred to nest in patches of bare ground.

Numerous reports exist that fluorescent blue vane traps can capture large numbers of bees. Captures often favour larger species and may be particularly efficient at capturing queen *Bombus*; consequently, deploying these traps needs to be done with caution and perhaps scaled back during early spring when queens are flying, at least until there is an understanding of the population impact, which has not been studied (Kimoto et al. 2012; Gibbs et al. 2017).

There are situations where non-destructive sampling is required—such as very small habitat islands or wherever endangered species are found—as well as cases in which lethal sampling may have a large negative impact or be illegal. The senior author has surveyed bees in two locations in Toronto by excising an antenna from each individual, storing it in ethanol, and then releasing the bee. The antennae were then submitted for DNA barcoding. Almost all antennae yielded DNA barcodes; nearest neighbour identifications averaged over 99.5% similarity, and there was 100% agreement between DNA-based and morphological identifications. While this sounds as if it could be damaging to populations (or cruel), female bees were observed flying around with one antenna missing 2 weeks after initial sampling, and males were observed to resume chasing females on flowers immediately upon release. Nonetheless, there are drawbacks with this approach: (1) not all bee species have DNA barcodes available, (2) some species have indistinguishable barcodes (Gibbs 2017), (3) handling time for antennectomy is somewhat higher than it is for simply putting a collected bee into a killing tube (this latter issue was relatively minor with an average of 46 compared to 27 s, though the difference likely decreases with experience and confidence), and (4) DNA barcoding is likely prohibitively expensive for large-scale surveys at present (although the relative costs of this method to the less obvious costs involved with the accrual of identification expertise remain to be compared and, as just noted, there remain taxa that are currently indistinguishable without molecular methods). Single antennae have also been used to obtain microsatellite data for population genetic research (Oi et al. 2013).

Photographic documentation of free-flying bees can be effective for the more easily identifiable members of a fauna, and it is not surprising that it has been implemented for bumble bees which are abundant, often readily identifiable by colour pattern, and popular with both researchers and the general public. The application of this methodology to citizen science programmes has proven to successfully generate extremely important data. For example, MacPhail et al. (2019) found that a combination of citizen science and museum specimen data demonstrated that a species thought to be at a relatively low risk of extinction was actually critically endangered. On the negative side of the equation, many bees fly away before they can be photographed, are not positioned in the photograph such that their identification marks can be distinguished, and/or are too far away from the camera for a good photograph. Camera surveys (particularly by inexperienced photographers) are highly biased favouring large, easy to identify bees that sit relatively still on flowers that are easy to access. A middle ground for photographic surveys is to capture bees with a net or other live trapping technique and place them in a vial, plastic bag, or squeeze box (Anonymous 2015) and photograph them from multiple angles for identification work prior to releasing. A great deal of time is needed per specimen for these techniques, requiring an enormous effort to achieve similar statistical power as those of lethal surveys.

Trap nests (covered below) could be used as a non-destructive means of sampling, although it typically involves rearing bees in a central location away from the exact site where the traps were placed and we know of no studies that have returned the emerged bees to their point of origin. Consequently, trap nesting is discussed among destructive sampling methods below.

3.3 Sampling Techniques

3.3.1 Netting

Two forms of surveys use nets: spot netting (Fig. 3.1a), where the net is aimed at an individual (or individuals) that the sampler has seen, and sweep netting, whereby a net is methodically swung (usually through flowering vegetation) and many insects (as well as other organisms, plant parts, etc.) are collected.

Netting surveys proportionally capture larger and more visible bees than bowl traps, particularly when conducted by novice netters (Roulston et al. 2007). Nets also undersample the fastest species which more readily elude capture. Of all the methods discussed herein, spot netting is the most biased and least replicable simply because of the different styles and aptitudes of users (or even the same individual at different hours of the same day). Anyone sampling an area at the same time as other collectors has experienced the enormous interindividual differences in specimens and species collected.

Sweeping of vegetation has sometimes been used to sample bees (e.g. MacKay and Knerer 1979; Grixti and Packer 2006), and while it suffers undercounts of rapidly flying species compared to spot netting as well as increased practitioner fatigue, it has the advantage of constraining netting in space and time and requiring little skill. Papers that contrast netting with bowl collecting document differences in the suite of species captured, total numbers captured, and numbers standardized by effort (e.g. Grundel et al. 2011; Bashir et al. 2013; Popic et al. 2013; Spafford and Lortie 2013; Rhoades et al. 2017; Table 3.1). These papers should be consulted as they will indicate general patterns in capture rates by genus and sometimes species.

Another drawback of both types of netting is that they can only be used where nets can reach. They are extremely difficult to deploy among the flowers of tall trees (e.g. from eucalyptus flowers in Australia), in dense vegetation, or where thorny plants are present. Netting is also problematic under conditions of low light intensities, and crepuscular and nocturnal bees are easier to catch in the traps we now discuss or light traps briefly mentioned above.

3.3.2 *Malaise and Flight Intercept Traps*

Malaise traps (Fig. 3.1b) have been used for more than 80 years to intercept flying insects (Malaise 1937) and can accumulate large numbers of bees (Darling and Packer 1988; Deyrup et al. 2002). They are rarely deployed as the sole or main method to capture bees (see Ngo et al. 2013 and Veiga 2013 for exceptions) as malaise traps are expensive and there can be high trap loss in the presence of large wildlife, less well-behaved humans, or wildfire (Veiga 2013). And although they can be deployed for long periods of time, obtaining sufficient sample sizes requires maintaining multiple traps. Studies that compared malaise samples to bowl traps indicated that they largely caught the same pool of species, but, as is always the case, the ranking of the abundance of species differed (Campbell and Hanula 2007; Geroff et al. 2014). McCravy et al. (2016) found that five malaise traps, as they deployed them, likely sampled less than 1/3 of the species richness present (predicted based upon Chao1 estimates), suggesting that more complete sampling would require an enormous investment in equipment, not to mention the considerable length of time it takes to separate and identify the bees of interest from the often overwhelming amount of insect by-catch. However, as the authors pointed out, the short sampling period involved (1 week) meant that the same number of

Fig. 3.1 Insect sampling methods. (**a**) Eli Wyman spot netting a fast flying bee from above *Nolana* flowers (photograph by Laurence Packer). (**b**) A malaise trap—note the tautness of the material, which is essential for it to be less visible to flying insects (photograph by Gerome Darla-West). (**c**) A pan trap, note that traps of different colour can be seen beneath the top one (photograph by Sam Droege). (**d**) Deep cup traps—dug into the ground somewhat to avoid displacement by wind, they can be left in the field with an appropriate fluid for weeks (see text for details) (photograph Laurence Packer). (**e**) A pan trap deployed in a flowering tree with ballast in a smaller cup inside the larger one, to prevent tipping (see text for details) (photograph Elsa Youngsteadt). (**f**) Two vane traps raised above the ground surface (they can also be dug partly into the ground) (photograph by Gerome Darla-West). (**g**) The trap nest design used by Scott MacIvor in his research (photograph Caroline Tiegs). (**h**) A tube taken from (**i**) and partly opened to permit study of the brood cell contents inside (photograph by Madison Marshall)

Table 3.1 Capture proportions in netting and bowl trapping studies in the Mid-Atlantic states, USA; ~ 1/4 million specimen captures, all bee genera used that had >50 total captures/genus; 2079 netting and 6840 bowl trapping events

Genus	Netting (%)
Agapostemon	6
Calliopsis	7
Melitoma	11
Peponapis	19
Augochlorella	20
Osmia	20
Ptilothrix	22
Lasioglossum	24
Eucera	25
Nomada	30
Hoplitis	31
Ceratina	38
Andrena	42
Melissodes	47
Halictus	50
Habropoda	52
Sphecodes	60
Stelis	67
Anthidium	71
Anthophora	73
Apis	75
Augochloropsis	76
Augochlora	81
Pseudopanurgus	85
Megachile	86
Epeolus	86
Colletes	88
Triepeolus	90
Panurginus	90
Hylaeus	91
Bombus	91
Coelioxys	91
Florilegus	91
Svastra	92
Perdita	92
Xylocopa	93
Anthidiellum	98
Lithurgus	98
Heriades	99
Dieunomia	99
All bees	**44**

Results presented as percentage of captures/genus (standardized by collection event) that were from netting; 50% would indicate that you had an equal chance of catching a member of that genus using bowl traps or swinging a net. Recall, though, that this does not account for the fact that you may be able to put out many more transects of bowls in a day than successful netting events. Nor does it account for the fact that most of the netting events here were by individuals with years of experience traps would undoubtedly have given more complete sampling if used over a longer time. Bashir et al. (2013) found that 15 bowls caught slightly more bees than did one malaise trap. Campbell and Hanula (2007) indicated that adding colour panels to malaise traps increased captures, and Darling and Packer (1988) and Russo et al. (2011) suggested that fluid-filled coloured trays that can function both as bowl traps and by treating the malaise trap also as a flight intercept trap (see below) increased catches of insects overall.

Trap type, mesh size, and location choices also impact the numbers of bees captured. Finer meshes or coarser fibres are relatively easier for bees to detect and avoid (Darling and Packer 1988), but larger mesh sizes permit the possible escape of the smallest bees. Trap costs are usually high enough that the total number of malaise traps deployed in any project will be low and consequently more geographically restricted than more extensive surveys using netting, bowl, and vane traps. However, they can be particularly useful in infrequently visited remote sites or in dense vegetation. Two large-scale surveys of insects in Colombia and Thailand (comparatively understudied areas for bees) have yielded numerous new species using malaise traps (e.g. Gonzalez 2004; Pauly 2009; Silva and Packer 2016), and malaise traps may be particularly effective at catching nocturnal and crepuscular bees, which, for obvious reasons, are rarely collected using active sampling methods or those that rely upon visual attraction. Notable may be the type series of *Mexalictus verdazulus* Dumesh comprising over 410 specimens, all captured from a single malaise trap; this is the species with the largest number of known specimens in this rarely collected genus of mostly early morning bees; two other species of which were also collected primarily from malaise traps (Dumesh 2013).

Flight intercept traps (FITs) are similar in some ways to malaise traps in that, as their name suggests, they intercept an insect in flight. Usually lacking a roof, they work best for taxa that drop after hitting an obstruction and thus are not as well suited to collecting bees as they are to some other insects, such as beetles, although some undescribed species have been discovered as a result of their use. An advantage of these traps over malaise traps is that they are smaller, more easily deployed and need less space in the field. Ulyshen et al. (2010) found that FITs placed in the canopy of a deciduous forest in Georgia collected more bees than those placed nearer the ground. In most cases these traps provide opportunities to secondarily sample bees when they are deployed in the capture of other insects, particularly if yellow-, blue-, or white-coloured pans are used to collect the catch.

3.3.3 Moericke Pan-Bowl Traps

Given the bright and novel colours of flowers that usually strongly contrast with the green and/or brown backgrounds of nature, it is not surprising that colour can be used to attract bees to traps. Such traps come in the form of pans (Fig. 3.1c), bowls, bottles, deep cups (Fig. 3.1d), tubes, and tubs. For purposes of simplicity, we will talk about them in aggregation as bowl traps, given that this shape and type now are most often deployed although, as discussed further below, long-term trap deployment works better with deeper cup traps housing propylene glycol than more standard bowl designs.

Bowl traps have been found to successfully capture bees in all the parts of the world where bees are present (Ortiz-Sanchez and Aguirre-Holds 1993; Westphal et al. 2008; Gollan et al. 2011; Popic et al. 2013; Wang et al. 2017) even under local circumstances where bees might only be dispersing through the habitat, such as in salt pans (T. Griswold, pers. comm.), drought conditions in South Africa where bloom was absent (Mayer and Kuhlmann 2004), or large parking lots (S. Droege pers. comm.).

3.3.3.1 The Role of Colour

Bees have a complex visual physiology with many nuances. For example, bees can differentiate ultraviolet (UV) and even iridescent patterns in flowers (Whitney et al. 2009). As a group, flowers exhibit unending diversity of shape, size, colour, pattern, surface structure, and pigmentation with clear behavioural responses of bees to those visual signals (van der Kooi et al. 2019). Yet there are few publications investigating such factors in trap design. Colour has been the primary reported variable in papers looking at bowl trap capture rates, with those studies almost always simply reporting conventional names such as "white", "blue", and "yellow" with no spectral diagrams.

Honey- and bumble bee vision is trichromatic, with peak sensitivities at around 350, 440, and 520–570 nm (Peitsch et al. 1992; Skorupski et al. 2007). These are colours that we perceive (or fail to perceive) as ultraviolet, blue, and green to yellow-green, respectively. Sensitivity decreases to zero over 650 nm, a wavelength we perceive as red. Bowl traps of different colours are universally found to be differentially attractive to bees (e.g. Ortiz-Sanchez and Aguirre-Holds 1993; Campbell and Hanula 2007; Gollan et al. 2011; Vrdoljak and Samways 2011). Heneberg and Bogusch (2014) found that oligolectic bees were more colour-constrained than generalists, not surprisingly preferring bowls of a similar colour to that of their preferred floral hosts, with intraspecific sexual dimorphism in colour preferences. Sircom et al. (2018) found that eusocial bees (they treated only honey and bumble bees as eusocial) had different colour preferences than did other bees.

Researchers normally deploy a combination of blue, white, and yellow bowls to capture as broad a representation of a region's bees as possible. Red is rarely used in bowl trap surveys as most bees do not see that colour (S. Droege, pers. comm. performed two large colour trials in the Eastern United States and found that red bowls caught only a handful of bees, a similar number to that captured with clear bowls). However, that may not be the case in regions where bees are attracted to red or orange flowers. For example, the lead author caught hundreds of a *Macrotera* species in a few hours in red and dark blue traps but few in white and yellow ones, this was a species that commonly visits red and orange cactus flowers.

Sircom et al. (2018) studied the reflectance spectrum of unpainted white bowls as well as two types of yellow- and two types of blue-painted traps, including one fluorescent version of each. The fluorescent and non-fluorescent paints had similar spectra with both yellow paints having high reflectance (over 60%) at all wavelengths over 490 nm but with the fluorescent version having more reflectance than non-fluorescent at wavelengths below 540 nm. In contrast, the blue paints had very low reflectance (<10%) at relevant wavelengths above 430 nm. Both blue shades had higher reflectance than either of the yellow ones below 485 nm and substantially more reflectance (>40%) in the range to which studied bumble bees are most sensitive (420–430 nm) than the yellows (<20%). Perhaps unexpectedly, the plain white bowls had higher reflectance than any of the others below 490 nm and higher reflectance than the blue paints throughout the spectrum. Noted here is the common conflation of fluorescence with UV reflectance. Fluorescent paints take incoming UV light and shift it into the visible spectrum; thus fluorescent yellow will appear brighter yellow to our eyes but will not reflect the UV portion of the spectrum. We are unaware of anyone experimenting with UV reflecting paint for bowl traps although such paints are available in the proper spectral range at least in white (https://www.twilightcoatings.com/collections/coatings-for-horticulture).

Wilson et al. (2016) found that the addition of nectar guides in the form of black lines running to the centre of the bowls in one location in Utah, USA, significantly increased the capture rate, while unpublished studies by K. Graninger in the Washington DC region, USA, documented significantly decreased rates. S. Droege (pers. comm.) found that adding dots and black spots to the sides and bottoms of bowls had similar negative to neutral impacts on capture rates.

In sum, colour of bowl trap strongly influences both the number and species composition of a trap's catch. Switching trap colour across sites or over time will bias the results, at times, severely. Consequently, study designs need to reflect on the long-term repeatability of practitioner's choices of trap and trap colour as well as the continued commercial availability of the colours in the future. It is unclear how subtle, or not so subtle, shifts in reflectance colour's hue, value, or saturation levels impact capture as traps fade through exposure to the sun (even within a single field season) or bowls are replaced by those of another manufacturer.

3.3.3.2 The Role of Size and Shape

Researchers have successfully used traps ranging in size from buckets to 0.75 oz. spit cups. Associations of capture success with size are little studied. At a single site in late summer in Utah, USA, Wilson et al. (2016) compared fluorescent yellow traps of three sizes (20, 8, and 3.5 oz), the largest catching significantly more bees than the other two. However, per volume, the smallest traps caught 3X as many bees as the largest ones. In an unpublished study, S. Droege (pers. comm.) looked at capture rates across seven trap sizes (0.75, 1, 2, 3.25, 4, 6, and 12 oz) from sites from Maine to Tennessee in the United States and found no significant differences in capture rates with size, but a great increase when pro-rated by volume. However, it is worth noting that very small traps can be difficult to relocate. Given the need to take a trap fluid into the field, in most cases, it would be easiest to use smaller traps sampling over a greater area with the same volume of liquid. Additionally, sets of smaller traps can be held in one hand and set on the ground without having to put the remaining traps and jug of water down. These small differences scale quickly to large logistical rewards when deploying hundreds of traps a day.

We know of no study that has compared capture rates of bowls across different bowl shapes. However, it is possible that the low incline of the sides of the standard eating bowls often used by researchers enables some bees to struggle out (anecdotally, we have found dead bees around the lips of traps in situations where there had been no rainfall that could have partially flushed the specimens from the trap). The first author has found that some of the taxa typically undersampled with standard bowls turn up more frequently in almost vertical-sided coloured drinking cups with propylene glycol as the trapping fluid. Capture rates with white Styrofoam© cups were only 20% of that with similar-sized white smooth plastic cups when deployed in a paired experimental design (S. Droege, pers. comm.).

One variant of bowl traps is the use of painted funnels placed over collection buckets of glycol. These have been successfully deployed for years to capture bees in Southwestern United States deserts where rainfall levels are low; success in wetter environments is more limited (yellow plastic automotive funnels drilled and wired with baling wire into the top opening of one quart paint cans; S. Buchmann, K. Wright pers. comm.).

3.3.3.3 The Role of Preservatives

Soapy water using soap that is odour-free (or as odour-free as possible) is the most commonly used liquid placed in bowl traps. The soap breaks the water's surface tension, such that the bees are more likely to sink as well as hastening the bee's death. Traps without soap capture almost no bees. Additionally, an unpublished paired trap study (S. Droege, pers. comm.), comparing citrus-based dish detergents to those with low or no odour, such as Blue Dawn© or lab detergent, found significantly fewer bees in the citrus-based traps. Adding salt to the water solution is required in instances where osmosis will result in undesirable swelling, particularly of the metasomas of bees, when traps are run for multiple days especially in warm weather, and strong salt solution may be used as a preservative in larger traps left out for longer periods or in instances when rain is likely. It is notable that bees have been successfully processed and identified from traps using only soapy water after two weeks in summer conditions despite the strong smell of decomposition.

Propylene glycol is a useful fluid for long-term passive trap use, largely because of its very low evaporation rates, non-toxicity, and preservative nature (Thomas 2008). Ethylene glycol—standard antifreeze—is also used but is poisonous to vertebrates and difficult to find in unadulterated forms. Propylene glycol is a common food additive and can be obtained through chemical companies, veterinary supply houses, plumbing supply houses (often dyed blue), and food grade from apothecary and food supply distributors or in dilute form as RV or non-toxic antifreeze (note, these mixes are often of unknown dilution, containing other additives, and usually coloured red). Thomas et al. (2001) found that a solution of 10% propylene glycol in water worked well when traps were serviced weekly (albeit for surveying fruit flies) and a 50:50 mixture worked well in rainy areas with periodic topping off with 100% (S. Droege, pers. comm.). In dry environments the water fraction will often quickly evaporate, and pure to nearly pure glycol needs to be used if longer periods between checks are expected. In all cases a small amount of non-citrus detergent is needed to cut down surface tension (water and propylene glycol have roughly the same surface tension, although the senior author has found pure propylene glycol to work perfectly well without soap in desert habitats). Other researchers have used saturated salt or sodium benzoate as preservatives in water, though these solutions are uncommonly used.

The effect of different trapping fluids has been little studied, but significant differences in capture rates were found in a New Zealand study (Larsen et al. 2014) as well as one from Canada (Deville and Wheeler 1998). Notably, both studies used commercial engine coolant mixes as well as commercial detergents both containing multiple compounds that could also impact capture rates beyond the primary fluid. As mentioned previously, citrus-based dish detergents added to water were found to significantly impact capture rates compared to laboratory detergents.

The darker the pan trap colour, the hotter the fluid gets and the greater the rate of evaporation. But even traps that have lost all their glycol have yielded identifiable material, albeit not in the best condition, even in windy places where the dried insects have been blown around inside the trap. For example, during a field trip in 2015, the senior author placed numerous bowl traps with propylene glycol along roadsides in northern Chile and left them out for months. Even a trap left out 6 months, run over by a vehicle, with all glycol gone, yielded 50 bees of 5 species, including one extremely rare one.

Following Frank Parker's discovery of large numbers of bees flying around vegetation where bus passengers made restroom stops, he developed a technique of spraying urine over vegetation as an attractant for bees. Upon learning this both authors have compared bowl trap catches with soapy water compared to soapy urine. Fortunately, the latter collected somewhat fewer bees than the former, and further use of bodily fluids seems unwarranted.

3.3.3.4 The Role of Height

Floral height, flight patterns of bees, vegetation density, size and sex of bee, and whether an individual is foraging, nest site searching, or dispersing through a habitat all likely influence the numbers and species composition of bees captured in traps at different heights. Canopy and tree or shrub feeding species are intuitively less likely to be caught in ground-level traps (Gonçalves and Brandão 2008; Gonçalves et al. 2012). Thick ground-level vegetation density also decreases capture rates (Geroff et al. 2014; see also Sheffield et al. 2013). An unpublished study by B. Goggins (S. Droege, pers. comm.) found that in fields with thick vegetation, bowls placed on the ground surrounded by this undisturbed vegetation captured significantly fewer bees when paired with traps in either 0.5-m-wide mown strips or bowls elevated to nestle even with the tops of the vegetation (~0.5 m). Given similar widespread anecdotal evidence, it is clear that moving bowl traps to the canopy level of thick vegetation or creating openings or paths for bowl placement is important. If you can't see the bowls, then bees can't either.

No standards have emerged on how to suspend traps within trees or above vegetation; sometimes elevated traps have a greater number of bees compared to ground traps, other times, the opposite. For some examples, in Ghana, bowl traps raised high into jungle canopy openings were successful in collecting large numbers of mostly meliponine bees (Nuttman et al. 2011). Dave Roubik (pers. comm.) found that bowl traps elevated in Panama along the superstructure of a canopy crane did not catch any of the orchid and stingless bees that were commonly found foraging in the canopy in that region's tropical forests. Suspended open white buckets with clear vanes above them caught large numbers of bees in the canopy of a US, Georgia, bottomland forest, throughout the season, even when flowering had ceased (Ulyshen et al. 2010). Bowl traps suspended from trees in the mid-story of oak-hickory forests in western North Carolina, USA, also showed higher capture rates in the elevated traps than ground traps when sampled throughout the year (Campbell et al. 2018). However, when blue vane traps were used rather than bowl type traps, in the interior of deciduous forests in Massachusetts, USA, and deployed continuously from early spring before trees flowered through the fall, the reverse was found to be the case with more bees caught on the forest floor (J. Milam 2018, unpublished report). Painted open cups suspended in deciduous forests and apple trees in New York, USA, in the spring captured similar numbers of bees in the canopy when compared to the ground (K. Urban-Mead, pers. comm.).

Declining catch rates also have been observed when bowl traps were elevated significantly above the surrounding vegetation. When traps are placed below blooming trees, they can capture almost nothing, as in most closed canopy tropical forests (Gonçalves and Brandão 2008; Gonçalves et al. 2012) and, similarly, in closed canopy deciduous forests after spring canopy closure (Ulyshen et al. 2010; S. Droege, pers. comm.); or traps will capture bee species associated with the blooming herbaceous cover rather than the canopy (Cane et al. 2000). However, K. Urban-Mead (pers. comm.) noted that species composition in New York, USA, deciduous forests was similar between traps in canopy, subcanopy, and on the ground. Finally, there is some evidence that the species composition of bees caught in blooming *Acer* and *Cercis* trees, in North Carolina, USA, in traps is similar to those collected directly from those same flowers using a net (E. Youngsteadt, pers. comm.).

Physically, individual traps (buckets, bowls, cups) can be hung from their rims either singly or in sets immediately below each other (Ulyshen et al. 2010; K. Urban-Mead, pers. comm.), from a platform of stiff construction wire (Campbell et al. 2018), or from both ends of a string that is draped through the branches (E. Youngsteadt, pers. comm.). E. Youngsteadt (pers. comm.) created a stable trap that works when suspended from above by placing a coloured soufflé cup inside a snuggly fitting drinking cup that had damp sand lining the bottom of the drinking cup for trap stability in high wind (Fig. 3.1e). There appears to be a general trend that elevated traps, no matter what the type, capture, on average, larger bees.

3.3.3.5 Context

The number and species composition of flowers have biasing impacts on abundance and species composition of bees captured in bowl trap setups. Published and casual observations indicate that locally abundant and attractive floral resources appear to decrease captures in bowls, with capture rates potentially increasing during periods of drought, following cutting of blooming plants or following a decrease or cessation of seasonal flowering (Cane et al. 2000; Baum and Wallen 2011). Nonetheless, bowl use in unusually densely flowering periods in some deserts and bottomland forests (S. Droege, pers. comm.) has still yielded large numbers of interesting species (e.g. Praz and Packer 2014). Much of the impact of capture rates and degree of bloom may simply have to do with bowls either being placed or not being placed at the level of the flowers.

Narrowly oligolectic bees are often caught in low numbers in bowls (Heneberg and Bogusch 2014), though there are notable exceptions to this [e.g. *Andrena erigeniae* Robertson, *A. violae* Robertson, and *Ptilothrix bombiformis* (Cresson) in eastern North America]. It may be that narrowly specific floral cues are missing from existing traps, or again, it may be that bowls are not placed at the flowering level of the plants, which for oligolectic species in North America tends to be high because they are associated most often with perennials and shrubs (Fowler 2016).

One issue that seems not to have been assessed is the difference in contrast between a bowl and its background, which will vary among localities as well as over time at the same site. Sheffield et al. (2013) controlled for this by placing traps within a grey plastic base that was substantially wider than the bowl itself. This kind of design also decreases the impact of vegetation growth, though it does increase cost, weight, and awkwardness when setting out traps and takes additional time to deploy and retrieve. Its impact on capture rates remains untested.

Bowl traps are usually deployed as part of a multicolour transect with all captures from a transect pooled to create a single sample. Tracking captures by individual bowl or colour is time-consuming and, given the variability and low numbers of bees/bowl, perhaps not greatly informative. Bowls placed too close together will compete with one another for captures; one study in Maryland, USA, using only white bowls, documented that bowls should be placed 3–5 m apart to prevent competition effects (Droege et al. 2010). Trap distances much farther than 10 m can lead to difficulties in re-finding the traps unless they are individually flagged. Shapiro et al. (2014) performed statistical analyses that suggested that a single deployment of 30 bowl traps is likely sufficient to sample the richness of bees along an individual transect. Information in LeBuhn et al. (2012) presents estimates of capture variability for bowl traps and other bee-monitoring techniques, permitting calculations of the required sample sizes under different statistical tests.

3.3.4 Vane Traps

Vane traps have an orthogonal pair of coloured vertical vanes placed over a funnel leading to a collection bucket (Fig. 3.1f). They were first recognized as being useful for sampling bees by Stephen and Rao (2005) in an agricultural setting and subsequently widely deployed in more natural habitats (e.g. Kimoto et al. 2012). The container may remain empty for short-term non-lethal capture or used with dry insecticide (such as dichlorvos strips) or partially filled with soapy water (with or without salt, formalin, and sodium benzoate) or other preservatives such as propylene glycol. The bees are likely attracted to the brightly coloured vanes and funnel, perhaps also to the container beneath, which in some designs is bright yellow. Blue and yellow vanes have been used, with blue substantially outperforming yellow. For example, Hall (2018) found that blue vanes caught six times as many bees as yellow ones (Hall's figure 1b suggests his yellow may have been closer to orange). However, the blue vanes available are generally somewhat translucent, whereas the yellow is commonly entirely opaque (although Hall's were translucent), and so colour per se in most studies of vane traps is likely confounded with differences in hue, value, and saturation. Joshi et al. (2015) found that blue vane traps captured more bees than yellow as well as all three standard colours of bowls in apple orchards on a per trap basis. Although the vanes were placed approximately 0.5 m higher than the bowls (which were also placed above the ground surface), there is likely confounding of trap type with proximity to the apple blossom. These authors also presented data on the reflectance patterns of their traps and showed that there were differences in the spectra between traps at the end of a field season and traps at the beginning: after use, the blue vane traps reflected somewhat less in the blue range of the spectrum and the yellow reflected somewhat more (see their Figs. 8a, b). In a study of fruit tree crop pollination, Gibbs et al. (2017) found that blue vane traps placed in the canopy primarily collected species that were not observed visiting the fruit tree flowers. As noted previously, numbers of captures can be changed by simply changing the number and placement of traps.

Unfortunately, with the more commonly used design, the blue funnel which screws into the trap base deteriorates after a few months in bright sunshine (L. Packer, unpublished observations) limiting the duration that the traps can be used. There is a need for testing a variety of paints, colours, and opacities in vane traps as well as improving their structural integrity for long-term deployment.

When partially filled with propylene glycol, these traps can be left outside for months at a time especially in arid areas. Partial embedding into the ground is needed (half of the bucket depth seems sufficient) to prevent displacement of the trap by wind. This permits extensive sampling in areas where bees are so rare that netting and traps that require more frequent visits are an inefficient use of a melittologist's time. For example, Packer and Graham (2020) found two specimens of a new species of cuckoo bee in a vane trap that had been left in an extremely arid part of Chile for over 3 months. At sites where rainfall might dilute the preservative and/or result in overflow, a clear transparent lid can be concocted to reduce the impact of precipitation. The senior author has had trouble with long-term deployment of vane traps, partially dug into the ground, as a result of rodents gaining access to the inside of the container through the space between the vanes and the funnel and meeting their demise. Bees are still extractable from this situation, although one would likely give the task of processing them to someone else. This problem was eliminated by pushing two heated nails at right angles through the funnel effectively halving the space through which the rodents could gain access (Postlethwaite, pers. comm.).

3.3.5 Trap Nesting

Trap nests are constructions that replicate the kinds of narrow tunnels that cavity-nesting bees most commonly use (MacIvor 2017). They have been employed with a wide range of substrates, dimensions, housing, and height. The most common tube types are either naturally occurring hollow- or soft-centred plant materials, such as *Phragmites*, *Sambucus*, or *Rubus* stems, or cardboard tubes or paper drinking straws. Care has to be taken to ensure that the design does not permit natural enemies to gain access to the entire broods of bees: the senior author's first attempt at trap-nesting bees (with drinking straws in a tin can) resulted in all tubes being occupied by *Osmia bicornis* (L.) but no surviving brood due to 100% parasitism, likely by *Monodontomerus obscurus* Westwood (see also Krunić et al. 2005). Trap nest designs have become extremely popular as a way of making people's homes more "bee-friendly", but such "bee hotels" (MacIvor and Packer 2015) are often of remarkably poor design (some likely acting as sinks rather than sources for populations), and even well-designed ones (Fig. 3.1g, h) will catch many wasps (not in itself a bad thing) and large numbers of non-native species (which probably is) (MacIvor and Packer 2015). Given the popularity of "bee hotels" among the general public, obtaining concrete data on the efficacy of different designs in augmenting native bee populations in different habitats should be considered a high research priority.

While the proportion of bees that use hollow cylinders as a nesting site is taxonomically and numerically restricted, a comparison of multiple methods of sampling bees found that the ecological patterns obtained with such smaller datasets generally paralleled those of more extensive ones and that trap nests can find bees that were undetected with other methods—in this case netting along transects, observation plots, and bowls (Westphal et al. 2008). A distinct advantage of trap nesting is that it permits far more detailed ecological and behavioural analyses than any of the other methods: rates of mortality, fecundity, and parasitism are all obtainable (e.g. Jayasingh and Freeman 1980). It is also relatively easy to place recording equipment at the entrance(s) to determine activity patterns of nesting females; within-nest behaviours can be monitored with designs that incorporate a transparent side to the nest; pollen is more readily identified from trap nests than from ground nests (MacIvor et al. 2014); natural enemy associations and behaviours can be documented (MacIvor and Packer 2015); and sound recordings can be made (Peebles 2011).

Trap nests can be more readily placed at different heights in a forested area than bowls and other fluid-containing trapping methods. Stangler et al. (2016) found that traps 10 m from the ground caught significantly fewer bees than those at 2 m or those at 20 m. Additionally, the two more elevated sets, while capturing fewer species, had higher diversity indices than those close to the ground. Unsurprisingly, different species were sampled with different frequencies at different heights.

There are a few bees that nest under stones and these can be "trap-nested" by placing suitable objects in the environment for them to nest under. Sheffield et al. (2015) placed terracotta saucers into the environment where *Osmia inermis* (Zetterstedt) was common and found that 10% of the artificial nest sites were occupied. Similar approaches could be taken to sample bees that nest in empty snail shells (e.g. *Osmia conjuncta* (Cresson) (Richards et al. 2011).

Emergence traps placed over patches of ground are something of an analogue of trap nests for sampling of the difficult-to-detect ground-nesting bees (Sardiñas and Kremen 2014; Pane and Harmon-Threatt 2017). They do not have the behavioural/ecological data collection advantages of the trap nests noted above, but they do not influence the make-up of the local bee community by adding nest sites suitable for only a portion of the local fauna. Capture rates are relatively low, thus requiring higher trap numbers and/or greater sampling intensity and to get phenological data they need to be checked frequently; sampling strategies such as working in uneven terrain or heavy vegetation, moving traps within a year to maximize trap captures, and reducing trap costs while increasing sampling footprint per trap and trap placement within a study site would be useful avenues of research (Pane and Harmon-Threatt 2017). Such traps will be susceptible to similar sources of damage as are malaise traps.

3.4 Additional Considerations

Sampling continuously throughout the bee flying season evens out the vagaries of intra-annual phenology and punctuated sampling. Several of the trap designs (e.g. vane, larger bowl/cup traps, and malaise traps, all usually filled with propylene glycol) are amenable to continuous trapping. Colour fading, dust accumulation, and dilution or overaccumulation of specimens require periodic repainting, replacement, and specimen removal for the trap to remain effective (Joshi et al. 2015). Similarly, trap nests can be left up for the entire year. Here the decreased availability of unused holes may cause a bias against later emerging species; this can be avoided by replacing nest tubes as they are completed, though this is more time-consuming.

Punctuated sampling is the norm in most studies, and given the rather short activity periods of many solitary bees, extended activity of most eusocial bees, the between year shifts in emergence dates, and the vagaries of weather, this temporal sampling strategy introduces a variance and bias component to the resulting counts that are generally not acknowledged. Given that most bees are protandrous, reporting the sex of all individuals sampled could indicate whether 1 year's sampling was early (male biased) or late (more female biased) in the season for each species.

Given that the amount of mandibular and wing wear increases monotonically (though not necessarily linearly, Foster and Cartar 2011) as a bee ages, comparisons of such relative age measures among years may prove informative. An even better approach would be to survey bees at least twice a week and then subsample the data to investigate cut-offs in sampling frequency for given sampling accumulation goals. This would also permit sampling frequency impacts on the statistical power to detect changes in population and biodiversity statistical parameters.

Bee surveys are often sponsored or undertaken by those interested in the conservation of insects; thus the amount of non-target by-catch of lethal techniques will be of concern. Other flower-using insects (e.g. wasps, flies, beetles) are attracted to and expire in coloured bowl and vane traps, while malaise traps capture an even larger range of flying insects. Studies optimized for bees often have opportunities to look at parallel results for other taxonomic groups, though identification can be problematic. Furthermore, splitting the by-catch among other insect researchers and taxonomists often results in range expansions and/or the discovery of new taxa (Spears and Ramirez 2015).

A disadvantage of passive trapping methods is that they generally do not permit assessment of ecological relationships, especially floral ones, and although a full scopa of pollen may still provide identifiable host plant material, no passive method is likely to prevent cross contamination of pollen among specimens other than via dissection of the gut.

3.5 Why Sample Bees?

All research questions related to the world's most important pollinators require some form of sampling, and fundamental to all of biology is the taxonomic research that results in species descriptions and identification guides. Although this is not something that is normally considered in treatments of sampling methods, it is clearly an essential component of bee biodiversity research, and so we treat taxonomic reasons for sampling before the ecological ones.

3.5.1 Sampling for Taxonomy, Systematics and Biogeography

Each of these research areas requires catching as much of the diversity and richness of a regional bee fauna as possible and, especially for biogeography, from across as many locations as possible. A good general approach here is to combine spot netting with various passive sampling techniques. As most bees do not start flying until some hours after sunrise and cease activity well before sunset, passive traps can be put out before, and brought back in after, the best times for netting, with longer-term deployment possible if the researcher remains in the region for several days or can return to it later.

Passive trapping is a good way to obtain research material from difficult-to-access localities, and propylene glycol allows traps to be in operation for weeks to months especially when deployed in deeper traps, such as those from painted drinking cups or vane traps. This makes good use of fieldtrip time as traps can be placed out over large areas at the beginning of a trip and brought back in on the way back or even on a subsequent fieldtrip (Packer et al. 2017).

Molecular approaches to taxonomy require material preserved in a way that permits DNA amplification. Currently, a greater range of collection and preservation techniques are acceptable compared to those of early allozyme techniques (e.g. Packer et al. 2005) which generally required the use of liquid nitrogen in the field. More recently, strong ethanol (usually 96%) has been used for molecular grade specimen storage and works best if the pickled materials can be stored in a cold and dark place. However, genomic methods are increasingly efficient at the extraction of ultra-conserved elements and other suites of loci even from old pinned specimens and/or other materials containing degraded nucleic acids (Zhang et al. 2019). For those wishing to perform transcriptomic research, DNeasy can preserve fresh net-caught material appropriately. Nuclear and mitochondrial DNA have been successfully sequenced from specimens collected into propylene glycol in traps left out for weeks in bright sunshine and high daytime peak temperatures (Praz and Packer 2014).

3.5.2 Sampling for Conservation Biology

Lethal sampling for conservation is something of an oxymoron. However, in most cases, it is far more efficient than alternative non-lethal techniques and much more likely to result in appropriately high sample sizes necessary for surveying insect groups, increase detection of rare species, and permit the identification and DNA analysis of difficult to identify species. Citizen scientist initiatives using non-destructive sampling via photography are increasingly used to document the relative abundance and, more convincingly, the distributional range of bumble bees (e.g. MacPhail et al. 2019). These initiatives are extremely useful given that the personnel involved would otherwise not sample bees at all. However, given the difficulty associated with species-level identification of most bee species in most parts of the world, photographic approaches would not work well even in relatively depauperate faunas, except for a small proportion of the taxa.

3.5.3 Sampling for Biodiversity Statistics/ Ecological Assessment

The considerable growth in interest in the ecosystem services provided by wild bees (e.g. Grab et al. 2019) has resulted in exponential growth in the survey efforts aimed at assessing wild bee biodiversity. Unfortunately, as outlined above, the diversity of approaches that have been used makes comparison among studies often difficult. Consistency in deployment of techniques across the temporal or spatial extent of a study or in comparison to other studies is critical for replicable ecological research. Small variations in the type, colour, or timing of placement of traps can have significant impacts on the numbers, size, species composition, etc. of the bees collected. Even minor deviations from consistency can create unwanted and usually uncorrectable biases. Researchers should be aware that if they want people to replicate their studies in the future, they need to clearly document all aspects of the types, dimensions, materials, colour signatures, and placement of their traps. A problem arises when specific trap types become unavailable due to changes in manufacturing procedures or cancellation of a product line. Exactly this happened for Packer when Solo stopped making the "bright blue" coloured 12 ounce dinner bowls which had been much more successful at collecting bees in the Atacama Desert than their nearest other blue shade, which is darker. This can be overcome by painting bowls using purchased pigments that are likely to be available more consistently over time (Anonymous 2015) or to not degrade during long-term storage. Similarly, traps that are used for long periods in bright sunshine will suffer some bleaching or loss of gloss, and, although this has been formally assessed only for vane traps (Joshi et al. 2015), it is abundantly obvious under extended field conditions with coloured bowls (whether painted or not). This can happen within a season after repeated or long-term use. Traps should be replaced or repainted regularly with varying frequencies depending on the amount of exposure to the sun or abrasion by wind. All replacements would need to be documented in detail.

Standardization is possible with all described methods. Because visual identification is a learned skill, it is perhaps the least amenable to formal standardization; although there can be little doubt that the development of instantaneous recognition skills is possible, we are aware of no studies where multiple observers have simultaneously assessed the same individual bees in the field to permit validation (for a more detailed and wider ranging assessment of taxonomic validation, see Packer et al. 2018) nor tracked how observer skill changes with time. Photographic methods can provide accurate identifications and have the added advantage of being verifiable through inspection by additional personnel (e.g. bumblebeewatch https://www.bumblebeewatch.org/ and MacPhail et al. 2019), though most bees spotted while in the field cannot be photographed, photos are often unidentifiable, and pictures are skewed towards large, distinctive species. Experts can disagree over an identification even with specimens in hand. For example, Stribling et al. (2008) found 21% identification disagreement among experts with microscopic identifications of freshwater arthropods and that only 1/3 of the disagreements were resolved through further discussion among identifications by the experts. Visual and photographic techniques are usually only appropriate for a small proportion of any local fauna.

Sweep netting can be standardized in terms of the type and size of the net, the extent of sampling in terms of area, and/or the duration of net deployment. However, it is not so easy to standardize the style with which the net is used with variables such as speed of movement, force with which the net hits the vegetation, and the number of sweeps per unit area or unit of time. Gravel (2010) found that differences in sweep net catches among individuals could cause similarity indices to vary substantially such that one collector's sample clustered with the wrong year of data collection. Arm length, and perhaps even brightness or colour pattern of clothing, may have an impact on what an individual catches.

Passive sampling techniques are more readily standardized. Bowl, vane, and malaise traps as well as trap nests can each be made to the same design, deployed in the same number, and set in similar positions for similar durations. Care must be taken with malaise traps, even of identical mesh and colour, as the degree of tautness of the mesh impacts the number and composition of the captures. As mentioned in previous sections, vegetation density has an impact on captures, so in long-term sites, vegetation control or bowl platforms may be necessary (Sheffield et al. 2013). Subtle differences in the way different people put out identical traps may also influence the samples obtained, but this has not been investigated.

3.6 Comparisons among Methods

There is a burgeoning literature of comparisons of sampling techniques (Prado et al. 2017). Most suffer from two problems. 1. The true population size and composition of the bees in the study are almost always unknown. 2. Corrections cannot or are not made to account for sampling effort. Given that the actual number of individuals and bee species are almost always unknowable, it will always be unclear how well a bee-surveying technique acts as an index. Comparisons are useful, but raw totals can easily be changed by simply altering the amount of trapping effort for an individual technique (e.g. time used to net or observe, number of traps deployed, number of locations, number of days or hours within a day). Proportional comparisons are more useful, and such comparisons can and would be indicative of patterns of capture probability across species and genera. If sampling effort is documented, then such studies are also a good indication of the time and money required per individual capture of a target group. Comparative trials are an important part of starting bee surveys in previously unexplored biogeographic regions.

For example, S. Droege (pers. comm.) recently looked at capture proportions in netting and bowl trapping studies in the Mid-Atlantic states, USA. Using a pool of roughly 1/4 million records, he looked at all bee genera with 50 or more total captures across 2079 netting and 6840 bowl trapping events (Table 3.1). An event was defined as a collection of bees netted or captured in a bowl transect from a date and location. Captures of bees for a genus for either netting or bowls were roughly corrected for effort by dividing captures by the total number of collection events for netting or bowls. These results could have been different had the number of bowls put out in an individual collection event or the amount of time spent netting been different; as such, this only represents a comparison of capture rates of given standard procedures. Major differences in capture rates, however, were clear, as illustrated in Table 3.1, with no clear bee taxonomic, size, or social structure associations apparent. With these results in hand, one might choose different survey techniques for *Agapostemon* vs. *Bombus*, yet it is unclear, despite extremely large sample sizes, that netting or bowl trapping is a better index to the actual proportions of bees in the Mid-Atlantic simply because we have no way of knowing what that true value might be. What is apparent, though, is that the best strategy is to use both techniques if the objective is to capture the greatest number of species.

3.7 Future Research Needs

Here we suggest what may be the most urgent areas for additional research effort:

1. Comparisons among methods in a standardized fashion in all the world's major biomes, to illuminate relative capture probabilities for different groups (both taxonomic and functional guild) of bees in different major habitats.
2. Assessment of impacts of variation of trap colours on capture rates, including shades of colours (e.g. shades of blue, shades of yellow), fluorescent vs. non-fluorescent colours, metallic, iridescent, and UV.
3. Direct comparisons among pairs or groups of observers/collectors in skill-based techniques such as spot netting, photography, and in-field observation-based identifications.
4. Assessments of lethal sampling frequency and intensity on subsequent year's populations of bees and the impact of such sampling on the calculation of population change.
5. Assessment of within and between season deterioration of trap colour and visibility—whether due to vegetation changes in the sampled habitat or the visual properties of the traps themselves.
6. Creation of a repository for pictures of traps and their colour spectra both before and after periods of deployment.
7. Impacts of supplied trap nests on bee density, species composition, assessment of bee health, and bee population change.

Acknowledgements We would like to acknowledge all our students, technical assistants, and colleagues who contributed and published information on monitoring, trap design, and sampling. Additionally, we also appreciate the tolerance and support of our families at home. We are particularly grateful to S. Droege for permission to cite many examples of his unpublished research findings and for discussions about the subject. We are grateful to Sam Droege, Madison Marshall, Caroline Tiegs, and Elsa Youngsteadt for permission to use their photographs in Fig. 3.1, to Scott MacIvor for arranging for us to receive some images, and to Liam Graham for amalgamating the photographs into one figure. The senior author's research has been made possible by funding from a wide range of sources, in particular from the Natural Sciences and Engineering Research Council of Canada and the National Geographic Society.

References

Anonymous (2015) The very handy bee manual. http://bio2.elmira.edu/fieldbio/beemanual.pdf. Accessed 23 Sept 2019

Bänziger H (2018) Congregations of tear drinking bees at human eyes: foraging strategies for an invaluable resource by *Lisotrigona* in Thailand (Apidae, Meliponini). Nat Hist Bull Siam Soc 62:161–193

Bashir MA, Saeed S, Sajjad A (2013) Monitoring Hymenoptera and Diptera pollinators in a subtropical forest of southern Punjab, Pakistan. Pak J Agric Sci 50:359–366

Baum KA, Wallen KE (2011) Potential bias in pan trapping as a function of floral abundance. J Kans Entomol Soc 84:155–159. https://doi.org/10.2317/JKES100629.1

Baumgartner DL, Roubik DW (1989) Ecology of necrophilous and filth-gathering stingless bees (Apidae: Meliponinae) of Peru. J Kans Entomol Soc 62:11–22

Bischoff I (2003) Population dynamics of the solitary digger bee *Andrena vaga* Panzer (Hymenoptera, Andrenidae) studied using mark-recapture and nest counts. Pop Ecol 45:197–204. https://doi.org/10.1007/s10144-003-0156-6

Broussard M, Rao S, Stephen WP et al (2011) Native bees, honeybees, and pollination in Oregon cranberries. HortScience 46:885–888. https://doi.org/10.21273/HORTSCI.46.6.885

Campbell JW, Hanula JL (2007) Efficiency of Malaise traps and colored pan traps for collecting flower visiting insects from three forested ecosystems. J Insect Conserv 11:339–408. https://doi.org/10.1007/s10841-006-9055-4

Campbell JW, Vigueira PA, Viguiera CC et al (2018) The effects of repeated prescribed fire and thinning on bees, wasps, and other flower visitors in the understory and midstory of a temperate forest in North Carolina. For Sci 64:299–306. https://doi.org/10.1093/forsci/fxx008

Cane JH, Minckley RL, Kervin LJ (2000) Sampling bees (Hymenoptera: Apiformes) for pollinator community studies: pitfalls of pan-trapping. J Kans Entomol Soc 73:225–231

Critescu ME, Hebert PDN (2018) Uses and misuses of environmental DNA in biodiversity science and conservation. Annu Rev Ecol Evol Syst 49:209–230. https://doi.org/10.1146/annurev-ecolsys-110617-062306

Darling DC, Packer L (1988) Effectiveness of malaise traps in collecting Hymenoptera: the influence of trap design, mesh size, and location. Can Entomol 120:787–796. https://doi.org/10.4039/Ent120787-8

Deville N, Wheeler TA (1998) The effect of different preserving fluids on insect catches in yellow pan traps. Proc Entomol Soc Ont 129:31–37

Deyrup M, Edirisinghe J, Norden B (2002) The diversity and floral hosts of bees at the Archbold Biological Station, Florida (Hymenoptera: Apoidea). Ins Mundi 16:87–120

Droege S, Tepedino VJ, LeBuhn G et al (2010) Spatial patterns of bee captures in North American bowl trapping surveys. Insect Conserv Divers 3:15–23. https://doi.org/10.1111/j.1752-4598.2009.00074.x

Dumesh S (2013) Revision of the rare Mesoamerican bee genus *Mexalictus* (Hymenoptera: Halictidae) with the description of 21 new species. Zootaxa 2013:1–80. https://doi.org/10.11646/zootaxa.3708.1.1

Foster DJ, Cartar RV (2011) What causes wing wear in foraging bumble bees? J Exp Biol 214:1896–1901. https://doi.org/10.1242/jeb.051730

Fowler J (2016) Specialist bees of the northeast: host plants and habitat conservation. Northeast Nat 23:305–320. https://doi.org/10.1656/045.023.0210

Fowler RE, Rotheray EL, Goulson D (2016) Floral abundance and resource quality influence pollinator choice. Insect Conserv Div 9:481–494. https://doi.org/10.1111/icad.12197

Geroff RK, Gibbs J, McCravy KW (2014) Assessing bee (Hymenoptera: Apoidea) diversity of an Illinois restored tallgrass prairie: methodology and conservation considerations. J Insect Conserv 18:951–964. https://doi.org/10.1007/s10841-014-9703-z

Gezon ZJ, Wyman ES, Ascher JS et al (2015) The effect of repeated, lethal sampling on wild bee abundance and diversity. Methods Ecol Evol 6:1044–1054. https://doi.org/10.1111/2041-210X.12375

Gibbs J (2017) DNA barcoding a nightmare taxon: assessing barcode index numbers and barcode gaps for sweat bees. Genome 61:21–31

Gibbs J, Joshi NK, Wilson JK et al (2017) Does passive sampling accurately reflect the bee (Apoidea: Anthophila) communities pollinating apple and sour cherry orchards? Environ Entomol 46:579–588. https://doi.org/10.1093/ee/nvx069

Gollan JR, Ashcroft MB, Batley M (2011) Comparison of yellow and white pan traps in surveys of bee fauna in New South Wales, Australia (Hymenoptera: Apoidea: Anthophila). Aust J Entomol 50:174–178. https://doi.org/10.1111/j.1440-6055.2010.00797.x

Gonçalves RB, Brandão RCF (2008) Diversidade de abelhas (Hymenoptera, Apidae) ao longo de um gradiente latitudinal na Mata Atlântica. Biota Neotrop 8:51–61. https://doi.org/10.1590/S1676-06032008000400004

Gonçalves RB, Santos EF, Scott-Santos CF (2012) Bees (Hymenoptera: Apoidea: Apidae s.l.) captured with Malaise and pan traps along an altitudinal gradient in the Parque Estadual da Serra Do Mar, Ubatuba, São Paulo, Brazil. Check List 8:53–56

Gonzalez VH (2004) A new species of *Acamptopoeum* from Colombia (Hymenoptera: Andrenidae: Panurginae). Caldasia 26:239–243

Grab H, Branstetter MG, Amon N et al (2019) Agriculturally dominated landscapes reduce bee phylogenetic diversity and pollination services. Science 363:282–284. https://doi.org/10.1126/science.aat6016

Gravel A-ID (2010) Bee community comparison in northwestern Patagonia (Argentina). MSc thesis, York University

Grixti JC, Packer L (2006) Changes in the bee fauna (Hymenoptera: Apoidea) of an old field site in southern Ontario revisited after 34 years. Can Entomol 138:147–164

Grundel R, Frohnapple KJ, Jean RP et al (2011) Effectiveness of bowl trapping and netting for inventory of a bee community. Environ Entomol 40:374–380. https://doi.org/10.1603/EN09278

Hall M (2018) Blue and yellow vane traps differ in their sampling effectiveness for wild bees in both open and wooded habitats. Agric For Entomol 20:487–496. https://doi.org/10.1111/afe.12281

Hatten TD, Looney C, Strange JP et al (2013) Bumble bee fauna of Palouse prairie: survey of native bee pollinators in a fragmented ecosystem. J Insect Sci 13:26. https://doi.org/10.1673/031.013.2601

Heneberg P, Bogusch P (2014) To enrich or not to enrich? Are there any benefits of using multiple colors of pan traps when sampling aculeate Hymenoptera? J Insect Conserv 18:1123–1136. https://doi.org/10.1007/s10841-014-9723-8

Hung KLJ, Ascher JS, Gibbs J et al (2015) Effects of fragmentation on a distinctive coastal sage scrub bee fauna revealed through incidental captures by pitfall traps. J Insect Conserv 19:175–179. https://doi.org/10.1007/s10841-015-9763-8

Janzen DH, DeVries PJ, Higgins ML et al (1982) Seasonal and site variation in Costa Rican Euglossine bees at chemical baits in lowland deciduous and evergreen forests. Ecology 63:66–74

Jayasingh DB, Freeman BE (1980) The comparative population dynamics of eight solitary bees and wasps (Aculeata; Apocrita; Hymenoptera) trap-nested in Jamaica. Biotropica 12:214–219

Joshi NK, Leslie T, Rajotte EG et al (2015) Comparative trapping efficiency to characterize bee abundance, diversity, and community composition in apple orchards. Ann Entomol Soc Am 108:785–799. https://doi.org/10.1093/aesa/sav057

Kimoto C, Debano SJ, Thorp RW et al (2012) Investigating temporal patterns of a native bee community in a remnant north American bunchgrass prairie using blue vane traps. J Insect Sci 12:108. https://doi.org/10.1673/031.012.10801

Krunić M, Stanisavljević L, Pinzauti M et al (2005) The accompanying fauna of *Osmia cornuta* and *Osmia rufa* and effective measures of protection. Bull Insectol 58:141–152

Larsen NJ, Minor MA, Cruickshank RH et al (2014) Optimising methods for collecting Hymenoptera, including parasitoids and Halictidae bees, in New Zealand apple orchards. J Asia Pac Entomol 17:375–381. https://doi.org/10.1016/j.aspen.2014.03.004

Larsson M, Franzén M (2008) Estimating the population size of specialized solitary bees. Ecol Entomol 33:232–238. https://doi.org/10.1111/j.1365-2311.2007.00956.x

Lebuhn G, Droege S, Connor EF et al (2012) Detecting insect pollinator declines on regional and global scales. Conserv Biol 27:113–120. https://doi.org/10.1111/j.1523-1739.2012.01962.x

Lopez-Uribe MM, Soro A, Jha S (2017) Conservation genetics of bees: advances in the application of molecular tools to guide bee pollinator conservation. Conserv Genet 18:501–506. https://doi.org/10.1007/s10592-017-0975-1

Lozier JD, Zayed A (2017) Bee conservation in the age of genomics. Conserv Genet 18:713–729. https://doi.org/10.1007/s10592-016-0893-7

MacIvor JS (2017) Cavity-nest boxes for solitary bees: a century of design and research. Apidologie 48:311–327. https://doi.org/10.1007/s13592-016-0477-z

MacIvor JS, Packer L (2015) "Bee hotels" as tools for native pollinator conservation: a premature verdict? PLoS One 10(3):e0122126. https://doi.org/10.1371/journal.pone.0122126

MacIvor JS, Cabral JM, Packer L (2014) Pollen specialization by solitary bees in an urban landscape. Urban Ecosyst 17:139–147. https://doi.org/10.1007/s11252-013-0321-4

MacKay PA, Knerer G (1979) Seasonal occurrence and abundance in a community of wild bees from an old field habitat in southern Ontario. Can Entomol 111:367–376

MacPhail VJ, Richardson LL, Colla SR (2019) Incorporating citizen science, museum specimens, and field work into the assessment of extinction risk of the American bumble bee (*Bombus pensylvanicus* De Geer 1773) in Canada. J Insect Conserv 2019:1–15. https://doi.org/10.1007/s10841-019-00152-y

Malaise R (1937) A new insect trap. Entomol Tids 38:148–160

Mayer C, Kuhlmann M (2004) Synchrony of pollinators and plants in the winter rainfall area of South Africa—observations from a drought year. Trans R Soc S Afr 59:55–57. https://doi.org/10.1080/00359190409519162

McCravy KW, Geroff RK, Gibbs J (2016) Malaise trap sampling efficiency for bees (Hymenoptera: Apoidea) in a restored tallgrass prairie. Fla Entomol 99:321–323. https://doi.org/10.1653/024.099.0230

Nemésio A (2010) Eulaema (Apeulaema) felipei sp. n. (Hymenoptera: Apidae: Euglossina): a new forest-dependent orchid bee found at the brink of extinction in northeastern Brazil. Zootaxa 2424:51–62

Ngo HT, Gibbs J, Griswold T et al (2013) Evaluating bee (Hymenoptera: Apoidea) diversity using Malaise traps in coffee landscapes of Costa Rica. Can Entomol 145:436–453

Nielsen A, Steffan-Dewenter I, Westphal C et al (2011) Assessing bee species richness in two Mediterranean communities: importance of habitat type and sampling techniques. Ecol Res 26:969–983. https://doi.org/10.1007/s11284-011-0852-1

Nuttman CV, Otieno M, Kwapong PK et al (2011) The utility of aerial pan-trapping for assessing insect pollinators across vertical strata. J Kans Entomol Soc 84:260–270. https://doi.org/10.2317/JKES110319.1

Oberhauser K, LeBuhn G (2012) Insects and plants: engaging undergraduates in authentic research through citizen science. Front Ecol Environ 10:318–320. https://doi.org/10.1890/110274

Oi CA, López-Uribe MM, Cervini M et al (2013) Non-lethal method of DNA sampling in Euglossine bees supported by mark-recapture experiments and microsatellite genotyping. J Insect Conserv 17:1071–1079. https://doi.org/10.1007/s10841-013-9582-8

Onuferko TM, Skandalis DA, Coredo RL et al (2018) Rapid initial recovery and long-term persistence of a bee community in a former landfill. Insect Conserv Divers 11:88–99. https://doi.org/10.1111/icad.12261

Ortiz-Sanchez FJ, Aguirre-Holds A (1993) Efecto del color sobre las capturas de abejas mediante trampas Moericke en el sur de Espana (Hymenoptera, Apoidea). Graellsia 49:63–71
Osborne JL, Martin AP, Shortall CR et al (2008) Quantifying and comparing bumblebee nest densities in gardens and countryside habitats. J Appl Ecol 45:784–792. https://doi.org/10.1111/j.1365-2664.2007.01359.x
Packer L, Graham L (2020) Four new species of Isepeolini (Hymenoptera: Apidae) from northern Chile. BMC Zool 5:3. https://doi.org/10.1186/s40850-020-00052-8
Packer L, Zayed A, Grixti JC et al (2005) Conservation genetics of potentially endangered mutualisms: reduced levels of genetic variation in specialist versus generalist bees. Conserv Biol 19:195–202
Packer L, Ali E, Dumesh S et al (2016) The identification of pollinators: where are we and where should we go? In: Gemmill-Herren B (ed) Biotic pollination services to agriculture: sustaining and enhancing a key ecosystem service. Earthscan, London, pp 57–73
Packer L, Litman J, Praz CJ (2017) Systematic position of a remarkable new fideliine bee from northern Chile (Hymenoptera: Apoidea: Megachilidae). Syst Entomol 42:473–488
Packer L, Monckton SK, Onuferko TM, Ferrari RR (2018) Validating taxonomic identifications in entomological research. Insect Conservation and Diversity 11, 1–12. https://doi.org/10.1111/icad.12284
Pane AM, Harmon-Threatt AN (2017) An assessment of the efficacy and peak catch rates of emergence tents for measuring bee nesting. Appl Plant Sci 5. https://doi.org/10.3732/apps.1700007
Pauly A (2009) Classification des Nomiinae de la Région Orientale, de Nouvelle-Guinée et des îles de l'Océan Pacifique (Hymenoptera: Apoidea: Halictidae). Bull Inst R Sci Nat Belg Entomol 79:151–229
Peebles S (2011) Pollination wunder station: extending our senses to the biosphere. Musicworks CD 111
Peitsch D, Fietz A, Hertel H et al (1992) The spectral input systems of hymenopteran insects and their receptor-based colour vision. J Comp Physiol A 170:23–40
Popic TJ, Davila YC, Wardle GM (2013) Evaluation of common methods for sampling invertebrate pollinator assemblages: net sampling out-performs pan traps. PLoS One 8:e66665. https://doi.org/10.1371/journal.pone.0066665
Prado SG, Ngo HT, Florez JA et al (2017) Sampling bees in tropical forests an agroecosystems. J Insect Conserv 21:753–770. https://doi.org/10.1007/s10841-017-0018-8
Praz CJ, Packer L (2014) Phylogenetic position of the bee genera Ancyla and Tarsalia (Hymenoptera: Apidae): a remarkable base compositional bias and an early Paleogene geodispersal from North America to the Old World. Mol Phylogenet Evol 81:258–270. https://doi.org/10.1016/j.ympev.2014.09.003
Rhoades P, Griswold T, Waits L et al (2017) Sampling technique affects detection of factors influencing wild bee communities. J Insect Conserv 21:703–714. https://doi.org/10.1007/s10841-017-0013-0
Richards MH, Rutgers-Kelly A, Gibbs J et al (2011) Bee diversity in naturalizing patches of Carolinian grasslands in southern Ontario. Can Entomol 143:279–299. https://doi.org/10.4039/n11-010
Roubik DW (2001) Ups and downs in pollinator populations when is there a decline? Conserv Ecol 5:2. https://www.jstor.org/stable/26271795
Roulston TH, Smith SA, Brewster AL (2007) A comparison of pan trap and intensive net sampling techniques for documenting a bee (Hymenoptera: Apiformes) fauna. J Kans Entomol Soc 80:179–181
Russo L, Stehouwer R, Herberling JM et al (2011) The composite insect trap: an innovative combination trap for biologically diverse sampling. PLoS One 6:1–7. https://doi.org/10.1371/journal.pone.0021079

Sardiñas HS, Kremen C (2014) Evaluating nesting microhabitat for ground-nesting bees using emergence traps. Basic Appl Ecol 15:161–168. https://doi.org/10.1016/j.baae.2014.02.004

Shapiro LH, Tepedino VJ, Minckley RL (2014) Bowling for bees: optimal sample number for "bee bowl" sampling transects. J Insect Conserv 18:1105–1113. https://doi.org/10.1007/s10841-014-9720-y

Sheffield CS, Kevan PK, Pindar A et al (2013) Bee (Hymenoptera: Apoidea) diversity within apple orchards and old fields in the Annapolis Valley, Nova Scotia, Canada. Can Entomol 145:94–114. https://doi.org/10.4039/tce.2012.89

Sheffield CS, Wilkes MA, Cutler CC et al (2015) An artificial nesting substrate for *Osmia* species that nest under stones, with focus on *Osmia inermis* (Hymenoptera: Megachilidae). Ins Conserv Div 8:189–192. https://doi.org/10.1111/icad.12095

Shokralla S, Spall JL, Gibson JF et al (2012) Next-generation sequencing technologies for environmental DNA research. Mol Ecol 21:1794–1805. https://doi.org/10.1111/j.1365-294X.2012.05538.x

Silva N, Packer L (2016) A new species of *Systropha* from Thailand (Hymenoptera: Halictidae: Rophitinae). J Melittol 61:1–9. https://doi.org/10.17161/jom.v0i61.4721

Sircom J, Jothi GA, Pinksen J (2018) Monitoring bee populations: are eusocial bees attracted to different colours of pan trap than other bees? J Insect Conserv 22:433–441. https://doi.org/10.1007/s10841-018-0071-y

Skorupski P, Döring TF, Chittka L (2007) Photoreceptor spectral sensitivity in island and mainland populations of the bumblebee *Bombus terrestris*. J Comp Physiol A 193:485–494. https://doi.org/10.1007/s00359-006-0206-6

Spafford RD, Lortie CJ (2013) Sweeping beauty: is grassland arthropod community composition effectively estimated by sweep netting? Ecol Evol 3:3347–3358. https://doi.org/10.1002/ece3.688

Spears LR, Ramirez RA (2015) Learning to love leftovers: using by-catch to expand our knowledge in entomology. Am Entomol 61:168–173. https://doi.org/10.1093/ae/tmv046

Spears RA, Looney C, Ikerd H et al (2016) Pheromone lure and trap color affects bycatch in agricultural landscapes of Utah. Environ Entomol 45:1009. https://doi.org/10.1093/ee/nvw085

Stangler ES, Hanson PE, Steffan-Dewenter I (2016) Vertical diversity patterns and biotic interactions of trap-nesting bees along a fragmentation gradient of small secondary rainforest remnants. Apidologie 47:527–538. https://doi.org/10.1007/s13592-015-0397-3

Stephen WP, Rao S (2005) Unscented color traps for non-*Apis* bees (Hymenoptera: Apiformes). J Kans Entomol Soc 78:373–380

Stribling JB, Pavlik KL, Holdsworth SM et al (2008) Data quality, performance, and uncertainty in taxonomic identification for biological assessments. J N Am Benthol Soc 27:906–919

Thomas DB (2008) Nontoxic antifreeze for insect traps. Entomol News 119:361–365. https://doi.org/10.3157/0013-872X-119.4.361

Thomas DB, Holler TC, Heath RR et al (2001) Trap-lure combinations for surveillance of *Anastrepha* fruit flies (Diptera: Tephritidae). Fla Entomol 84:344–351

Ulyshen MD, Soon V, Hanula JL (2010) On the vertical distribution of bees in a temperate deciduous forest. Ins Conserv Div 3:222–228. https://doi.org/10.1111/j.1752-4598.2010.00092.x

van der Kooi CJ, Dyer AG, Kevan PG et al (2019) Functional significance of the optical properties of flowers for visual signalling. Ann Bot 123:263–276. https://doi.org/10.1093/aob/mcy119

Veiga N (2013) Impact of fire on bee communities in Argentina. M.Sc thesis, York University

Vrdoljak SM, Samways MJ (2011) Optimising coloured pan traps to survey flower visiting insects. J Insect Conserv 16:345–354. https://doi.org/10.1007/s10841-011-9420-9

Wang M, Lu X, Ding S et al (2017) Pollinator diversity in different habitats of the agricultural landscape in the middle and lower reaches of the Yellow River based on the three-color pan trap method. Acta Ecol Sin 37:148–155. https://doi.org/10.1016/j.chnaes.2017.06.007

Westphal C, Bommarco R, Carre G et al (2008) Measuring bee diversity in different European habitats and biogeographical regions. Ecol Mon 78:653–671. https://doi.org/10.1890/07-1292.1

Whitney HM, Kolle M, Andrew P et al (2009) Floral iridescence, produced by diffractive optics, acts as a cue for animal pollinators. Science 323:130–133

Wilson SW, Jahner JP, Starley L et al (2016) Sampling bee communities using pan traps: alternative methods increase sample size. J Insect Conserv 20:919–922. https://doi.org/10.1007/s10841-016-9914-6

Yamomoto M, Junqueira CN, Angélica A et al (2014) Estimating crop pollinator population using mark–recapture method. Apidologie 45:205–214. https://doi.org/10.1007/s13592-013-0238-1

Yu DW, Ji Y, Emerson BC et al (2012) Biodiversity soup: metabarcoding of arthropods for rapid biodiversity assessment and biomonitoring. Methods Ecol Evol 3:613–623

Zhang YM, Williams JL, Lucky A (2019) Understanding UCEs: a comprehensive primer on using ultraconserved elements for arthropod phylogenomics. Insect Syst Divers 3:1–12. https://doi.org/10.1093/isd/ixz016

Chapter 4
Social Wasp Sampling Methods

Fábio Prezoto, Tatiane Tagliatti Maciel, Bruno Corrêa Barbosa, and Carlos Eduardo Sarmiento

4.1 Introduction

The term "wasp" encompasses a variety of insects of the Hymenoptera order, especially the Vespidae family (Carpenter and Marques 2001). However, social behavior is only observed in three subfamilies: Stenogastrinae, Vespinae, and Polistinae, with the latter occurring in Brazil where they are popularly known as marimbondos or cabas (Fig. 4.1a, b) (Prezoto et al. 2016).

Vespidae are approximately 4.5–24 mm in length with great variation in color but are commonly found with yellow or all black spots (Fig. 4.1c) and have the following main taxonomic characteristics: median emarginated (reniform) inner margin of the eyes; antennas with 12 segments in females and 13 in males, which are often more curved at the apex; first third of the metasome partially fused with the first sternum overlapping it; anterior wings folded longitudinally when at rest; and the length of the first discoid cell of the anterior wings always greater than the basal (Goulet and Huber 1993; Carpenter and Marques 2001).

Vespidae has approximately 5000 described species, with the richest being Eumeninae consisting of 3579 and Polistinae with 958 species (Pickett and Carpenter 2010). Despite having widely distributed species, the greatest species richness is found in the tropical region (Carpenter 1993; Pickett and Carpenter 2010).

Despite being the least known group among social insects, social wasps play an important role in ecosystems by contributing to pollination (Clemente et al. 2012)

F. Prezoto (✉) · T. T. Maciel · B. C. Barbosa
Laboratório de Ecologia Comportamental e Bioacústica, Universidade Federal de Juiz de Fora, Juiz de Fora, Brazil
e-mail: fabio.prezoto@ufjf.edu.br

C. E. Sarmiento
Laboratorio de Sistemática y Biología Comparada de Insectos, Universidad Nacional de Colombia, Edificio 425, Bogotá, Colombia

J. C. Santos, G. W. Fernandes (eds.), *Measuring Arthropod Biodiversity*,
https://doi.org/10.1007/978-3-030-53226-0_4

Fig. 4.1 (**a**, **b**) Polistinae representatives of the *Mischocyttarus and Polybia* genera best represented in South America. (**c**) Typical colorings of social wasps. (**d**) Social wasp visiting flower. (**e**) Individuals sharing parts of captured prey

and mainly performing a valuable service as a predator for defoliating insect species considered to be crop and urban garden pests (Prezoto et al. 2019) (Fig. 4.1d).

Wasp activity rhythm can be day or night, independent or swarming (Hunt 2007). Independent foundation wasps start their colonies from one or a few associated queens and have open nests, i.e., without protective wrapping; the swarming foundations swarm with several queens accompanied by workers who will start a new colony and in most cases build a closed nest with a protective enclosure (Wenzel 1998; Hunt 2007).

The working hours of most social wasp species are regulated by the sunlight intensity, meaning that daytime wasps start foraging in the early hours of the morning with sunrise and end activity in the late afternoon (Barbosa et al. 2014); night wasps, like the *Apoica*genus, do the opposite; they start their activities at dusk around 6 p.m., lasting around 12 h and ending around 6 a.m., with activity peaks in the early evening and near dawn (Nascimento and Tannure-Nascimento 2005).

These insects perform foraging to collect material for building their nests and obtaining food resources. During these activities wasps look for water, nest-building material, carbohydrates primarily for feeding adults, and protein for nutrition of immature wasps. Carbohydrates can be obtained from nectar, honey, insect exudates, ripe fruit juices, and other sugary products, while the protein is obtained from live prey, mainly Lepidoptera caterpillars, and carcasses (Prezoto et al. 2019).

4.2 Studies on Neotropical Social Wasps

In 1978, British entomologist and ecologist Owain Westmacott Richards produced the complete work on social wasps entitled "The social wasps of the Americas, excluding the Vespinae," which features an extensive review of Neotropical species with details on their distribution, morphology, nesting, and biology (Richards 1978). The first records of social wasps in Brazil occurred during expeditions focused on the taxonomy and distribution of species, and since then the work in the country has grown more and more (Barbosa et al. 2016).

The recent research areas are diversity studies and foraging activity (Barbosa et al. 2016; Prezoto et al. 2019). The first study that aimed to evaluate the diversity of social wasps in Brazil dates from 1982 (Rodrigues and Machado 1982), after which there was a big change in the essence of publications. Researchers began to focus their efforts on sampling wasp diversity in a specific location and over the years developed sampling methodologies and began using ecological indices for data analysis.

In this sense, Mechi (2005) was the first author to apply a diversity index to analyze the social wasp fauna in the Jataí Ecological Station in the interior of São Paulo, which at the time was the Shannon-Wiener index (H′). Only 1 year later, Souza and Prezoto (2006) returned to using ecological indexes, which were a milestone since authors began to apply at least one ecological index in their studies from that date.

Several methods are used for sampling due to the great diversity of social wasp species and the different landscapes in which these studies are performed. The main collection methods can be divided into two groups: active collection, when the collector actively participates in the capture process using tweezers, entomological nets, or other apparatus, and passive or standby collection, when specimen capture is used. Thus, some of the main methods used to collect social wasps will be presented in this chapter.

Social wasps can be interesting models for analyses on the evolution of life history traits and ecological patterns related to environmental factors at the colony or individual level. In this regard, sampling strategies along gradients or comparative situations must be carefully prepared. For example, morphological and physiological responses of the wasps to altitudinal challenges may become very notorious due to their flight habitus to forage. Individuals of the same species along altitudinal gradients can be obtained using net active collection and Malaise or Shuey traps (see below). If simultaneous sampling is not possible, take into consideration that cohort replacement may become an important factor, as individuals from different cohorts may show structural differences related to their own history. Researchers may consider that currently collected foragers were fed as larvae several days or weeks ago when food provision, either prey or nectar, was very different. Thus, sampling efforts in these cases should be as temporally close as possible, and if some information is available about the extension of the life cycle, this should be important for defining maximum time lapse (Rodríguez-Jimenez and Sarmiento 2008).

4.3 Active Sampling Methods

In active methods, the collector must be in the field for the duration of the collections.

4.3.1 Active Searching

This method is very used and recommended for studies on the diversity of social wasps because it allows the collector to visually search and select target species (Barbosa et al. 2016). In this method the insects are captured during their flight through an entomological net, which must be made of a light, strong, and durable material and can present varied sizes. Normally a hoop of approximately 35 cm in diameter is attached to a cable of approximately 100 cm. The procedure also enables obtaining behavioral records of the collected species when they are little known or nonexistent in the literature.

Researchers determine a transect which is traversed by the collector(s), usually during the hours of greatest social wasp activity from 10 a.m. to 3 p.m. (Elisei et al. 2013). The collections should always be standardized regarding the diameter of the hoop net, similarity in the collectors' ability, and the sample effort employed in each location.

Some disadvantages of this method need to be highlighted. The first issue would be the large field effort that this method requires, which makes it difficult to sample for long periods in certain areas. Another point that should be taken into account is the physical limitation to reach high areas such as the canopy or hard to reach places where vegetation above 5 m in height is impossible to reach. The third point is that some species have cryptic behaviors or are very small, which make visualization and capture difficult. Finally, the success and quality of the collections are not only directly affected by the collector's ability to perform the method and their ability to recognize target species but also by the weather conditions on the collection day, since the abundance of these insects is very low on cold and rainy days, for example.

4.3.2 Searching for Nests

This is a variation of the active search that can also be performed in pre-established transects or quadrants; however, in this case, the researcher inspects trees, plants, and building cavities to find active social wasp colonies. Buildings should be carefully inspected, especially roof gutters and poorly visited sections of the structure. This method prevents the diversity of a place from being overestimated or species being mistakenly considered constant, since the number of nests rather than individuals is what is accounted for, thus making the swarming and independent species have the same weight. It is worth mentioning that searching nests is a bit dangerous if you are alone in the field and if you have problems with being bitten by the wasps.

4.3.3 Colony Collection

As with other social insects, Vespidae exhibit several unique traits which are only visible when collecting the nest and the individuals with it, meaning when the colony is collected. Caste composition, development cycle, parasites, parasitoids, commensals, and of course nest architecture are a few of the many issues that can be studied when collecting the entire colony (Somavilla et al. 2015b; Barbosa 2019). As a consequence, the procedures described herein are intended to capture the entire colony and also to reduce the likelihood of an accident. Keeping in mind that sometimes the colonies can be in places which are extremely difficult to access or that imply in destroying the nest, it is better to only collect a sample of adults in these cases and leave the colony aside.

Thus, colony collection must be fully planned in considering all the requirements during and after the process (Sarmiento 2006). These phases will be followed in the next description.

Once the colony is spotted during collection, take into consideration all the actions which are required to remove it. Closely study the place, access, and obstruction objects such as branches, twigs, leafs, etc., and simulate the order of actions to remove these objects. Check for the objects that support the nest, as extra supporting structures (i.e., secondary twigs, leafs) can only be detected after careful study in many cases. Check for pathways. Leave a mark and wait until night to do the collection. *Apoica* is perhaps the only genus that should be collected during the day.

Due to the architectural variation and/or nest location, different strategies for the collection of colonies should be employed. Herein we will describe some of the more common cases. The only recommendations applicable to any nest are probably use red lights during the collection event at night, avoid extra noise and careless movements, and ask two or three people to assist you, especially for large nests, as these lights are not easily detected by the wasps. Use a hat, and use long sleeve shirts tucked into your waist, in addition to a T-shirt. Long pants are also desirable. Sometimes wasps may start walking up in your body, and you may have some unpleasant surprises if they get trapped in your clothes. Use protective suits for honey harvesting in cases of very large nests and very defensive species.

If you do not have freezing facilities in the area, plan your trip to collect the wasps on the last day, and thus you can avoid the wasps rotting. If a freezer is available, you may kill the adults by putting the nest in this low temperature for at least 2 days. The downside of this strategy is that you will not be able to see parasitoids.

Nests Hanging on a Branch These are observed in genera such as *Polybia* and *Epipona*, among others. Carefully remove twigs and leaves which can obstruct quickly wrapping the colony with a large plastic bag. This process can take hours but having a complete colony and an unbroken nest is worth the effort. Clean the single support branch to use it as a handle for the coming steps. Once you have finished the removal of these objects, leave the nest alone to pacify the wasps; in the meantime, prepare a large plastic bag, and put a cotton ball inside; move to the nest location, and quickly and smoothly wrap the nest bottom-up, and close the bag's opening around the support branch with a cord. Next, cut the branch with a hand saw or garden scissors. The supporting branch should be large enough to handle and store the nest inside a plastic box, and the handle will also serve to hang the nest in the definitive storage box at the museum collection. Once you have guaranteed that the colony is safely wrapped in the plastic bag, you can add ethyl acetate or ether into the bag toward the cotton ball. The cotton will reduce the effect of liquids over the bag or the nest.

Nests Under Leaves These are observed in genera such as *Angiopolybia*, *Apoica*, *Mischocyttarus*, *Polistes*, and some *Polybia*. Follow a similar process to that described for nests hanging on branches, but use scissors to cut out extra leaves and facilitate wrapping. Sometimes the nest is at the main vein of the leaf, and in that case, use it as a handle.

Nests Attached to Walls or Trunks These characteristics are very common in genera such as *Synoeca*, *Metapolybia*, and some *Parachartergus*, among others. Sometimes you can block the entrance of the nest with a piece of cotton; if so, do it quickly, and then add ethyl acetate or ether to the cotton, and also inject some through the walls with a syringe. If the nest is attached to a trunk, wrap it with a sheet of plastic tightly tied at the upper and lower parts. Add a ball of cotton before you wrap the nest. Make sure that either ethyl acetate or ether have acted on the adult wasps, remove the plastic, and proceed to transfer adults with tweezers to vials with ethanol; then, use a knife or a cutter to carefully remove the envelope and the combs. Label each one and store them in individual bags. Photograph the nest at several stages of the process such that you can accurately reconstruct the nest once in the lab or collection. If the nest is attached under a roof or similar horizontal surface, you may need to use extra hands to surround it with the plastic bag; inject the ethyl acetate or the ether with the syringe toward the cotton ball, and wait for wasps to stop moving; then, separate the nest from the wall with a cutter or a thin knife, close the bag's opening, and tie the bag tightly.

After collection, open the bag in an open space, and leave it as is for a while to facilitate evaporation of the killing substances. Remove adults and other insects as quickly as possible, and transfer these to vials with ethanol, and remove all larvae if possible. Once you have extracted all the information, it is highly recommended to store the nest in a freezer for several days in order to kill undesirable hosts such as mites of psocopterans. Thoroughly dry the nest after freezing; a handmade oven or exposure to sunlight in a safe area can do the job. Prepare the definitive storing box and all procedures to include the nest into the collection.

4.3.4 Spot Search

This is usually performed on food resources such as inflorescence and fallen fruit, but transects or plots can be used so that insects are carefully captured during entomological foraging and their presence on the resource is recorded.

In addition to a fauna survey, this method is widely used in studies on insect and plant interaction networks, aiming to understand aspects, such as organization, dynamics, and stability of the network (Santos et al. 2010; Clemente et al. 2012), and describe the behaviors presented during foraging (Barbosa et al. 2016; Brügger et al. 2017).

4.3.5 Attractive Liquid

As proposed by Noll and Gomes (2009), this method uses a spray with an attractive solution based on crystal sugar (200 g/L) and salt (25 g/L). The solution should be sprayed at specific points along a previously selected transect, and about 500 mL of solution should be sprayed at each point. After spraying the attractive solution, each point should be observed for 5 min, and the attracted wasps should be collected with an entomological net.

It is important that the sprayer should be cleaned with water and sodium hypochlorite after each collection day to prevent fungal proliferation, which may compromise performance and efficiency in subsequent collections.

4.3.6 Light Trap

There are different types of traps which use light for attracting and collecting insects. Traps for social wasps are generally mounted with a 300 × 200 cm white sheet stretched and illuminated by a 250 W mixed mercury vapor lamp with a field stay of approximately 12 h (Somavilla et al. 2014; Somavilla et al. 2015a). The attracted insects are perched on the sheet and easily collected in jars, tweezers, or containers. The collector must remain close to the trap or make periodic checks. The sheet should be installed by touching the floor, so if insects fall after hitting the fabric they are still trapped inside the trap.

Although little used, the light trap is very effective for capturing social wasp species with nocturnal habit, as is the case of the *Apoica*genus which is hardly captured by other methods. In contrast, a negative point of the method is that light sources require power, in which case it may be necessary to charge batteries or another power source, which makes sampling at remote locations difficult.

4.4 Passive Sampling Methods

Passive methods facilitate sampling over long periods, as field effort is less compared to active methods. In passive collection, traps are responsible for capturing and storing insects until the collector returns to the area; however, it is not possible to observe the behavior of the collected species with this type of approach.

4.4.1 Attractive Trap

The principle of this method lies in the attractiveness of social wasps by the volatilization of the substance used as bait (Saraiva et al. 2017) (Fig. 4.2a, b). Traps are easily made from polyethylene terephthalate (PET) bottles with side openings in the shape of a 2 cm equilateral triangle in the middle portion, filled with fruit juice and hanging from undergrowth and canopy trees (Maciel et al. in press; Brügger et al. 2019).

The demand for carbohydrates and proteins during the foraging activity in social wasps meets the nutritional needs of the colony, so the attractive substances used in the traps can vary from sardines to fruit juices. However, it is known that protein baits record a limited group of species (Barbosa et al. 2016), so it is suggested to use fruit juice (preferably guava or mango) to maximize the performance of attractive traps. It is also important to highlight that industrialized juices have great advantages. Maciel et al. (in press) observed that industrialized juices were more specific regarding the capture of social wasps in the field in relation to fresh juices, which in addition to facilitating screening of the material also has less impact on the local fauna. Moreover, field logistics are facilitated by the use of processed juices as fresh juices need to be prepared in advance.

The use of attractive traps to capture social wasps began in Brazil with the pioneering work by Lorenzato (1985), followed by Santos (1996) and Souza and Prezoto (2006), and standardized and optimized by Maciel et al. (in press). The advantages of this methodology are its practicality, low cost, and optimization of the collection time, since the researcher can make do of collections using other methodologies while the traps are in the field. Its disadvantage is that the captured wasps cannot remain in the traps for a long time because the natural process of decaying the bait also deteriorates the trapped insects, so it is necessary to inspect these traps shortly after installation. Still, this method is the second most used in social wasp diversity studies in the Neotropical region (Barbosa et al. 2016; Maciel et al. 2016).

4.4.2 Shuey Trap

Some attractive traps (such as the Shuey trap, also called Van Someren-Rydon) directed at other groups may also attract these insects due to the general feeding habits of social wasps (Fig. 4.2c). This was used by Arab et al. (2010), who recorded social wasps during a study on Lepidoptera. This trap consists of a tubular structure made of sheer, tulle, or even protective screen with one end closed and the other with a bait-coupled plate, usually a mixture of fermented banana and sugarcane juice.

A disadvantage of this method is leaving the animal alive; they can escape or kill (destroying) the other insects that were also captured if the focus of the study is not just the wasps. The same disadvantage of being alive can also be an advantage, since being alive it takes longer to lose the quality of the insect statues.

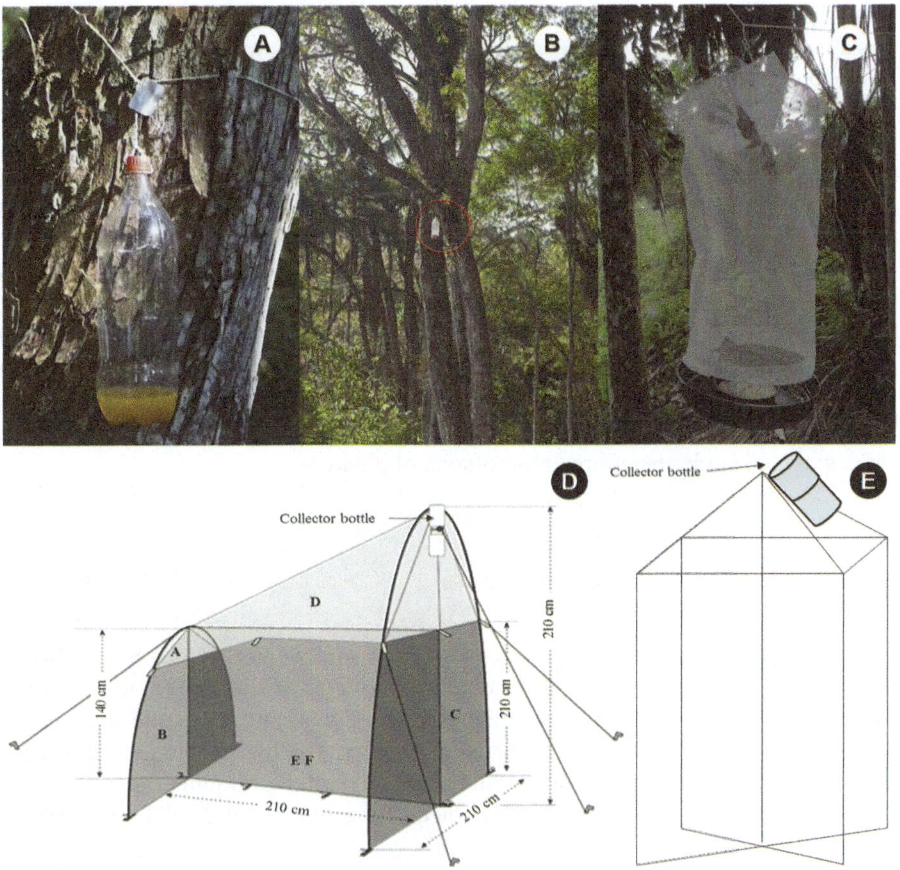

Fig. 4.2 (**a**) Attractive trap. (**b**) Attractive canopy-mounted trap. (**c**) Shuey trap. (**d**) Malaise trap. (**e**) Model of Malaise trap for the canopy, the structure is held by a set of two cross poles both at the top and at the bottom of the mesh. The collecting bottle is at the top. The trap is hung by the top

4.4.3 Malaise Trap

The Malaise trap is also known as the flight intercept trap and was originally described by René Malaise in 1937, and today all tent-type traps are known as the Malaise traps in his honor (Fig. 4.2d, e). Silveira (2002) was the first to use this method on Neotropical social wasps for studies of social wasps in Brazil. It is important that the fabric, organza, or nylon on the walls is darker than the ceiling so that the insects are induced to fly upward into the light and are then captured by the collection flask, which in turn must contain acetate, acetone, ether, or alcohol.

Trapping is set near streams or on open roads and trails where insects with stronger flying ability prefer to fly. It should preferably be mounted transverse to the wind for open areas, while the collection bottle should be oriented as brightly as possible in closed areas, as they are naturally darker. This method can also be used suspended ("suspendible Malaise trap").

This is a considerably efficient method for collecting social wasps and recording a large number of species including exclusive ones (Barbosa et al. 2016); however, it is important to note that the Malaise trap should not be the only form of sampling, as it may underestimate the true richness of species, mainly of the *Polistes* and *Mischocyttarus*genera in certain areas (Silveira 2002).

Regarding logistics, the biggest advantage of this method is that the traps can remain in the field indefinitely (Rafael et al. 2012; Souza et al. 2015). However, it is important to highlight that even though it is a well-used methodology for social wasp diversity work, the Malaise traps have a great impact on local insect fauna because they do not have selectivity, meaning that they collect a wide variety of insects. Therefore, the use of this trap is only suggested for larger studies, where the objective is to study the entire entomofauna of an area.

4.4.4 Möericke ("Pan Trap")

This is the least used trap in the work of social wasp inventories (dos Santos Souza et al. 2016). It is a generalist trap that collects insects attracted by color. It consists of shallow plastic or colored metal trays or plates containing water and some detergent, placed at ground level or at different heights. The disadvantage of this technique is that the trap may overflow with excessive rainfall or liquid evaporation may occur in hot places, and there is still a need for daily removal of the material to prevent deterioration of the collected insects.

4.5 Method Consortium

Although the methods are efficient in themselves, it is suggested to use more than one sampling method in social wasp diversity work whenever possible, preferably an active and a passive one. A consortium of methods was more frequent from the works done in 2006. Based on 98 works published on social wasp fauna in Brazil between 1982 and 2018, it was possible to create a graph showing the performance of the methods in relation to the genera sampled in the Neotropical region (Fig. 4.3).

The attractive trap was able to capture 20 of them of the 21 genera recorded in the Neotropical region, with the passive method showing the best result, followed by the Malaise trap with 10 genera. For active methods, active searching captured 19 genera. This emphasizes that the use of more than one method can optimize capturing species from an area.

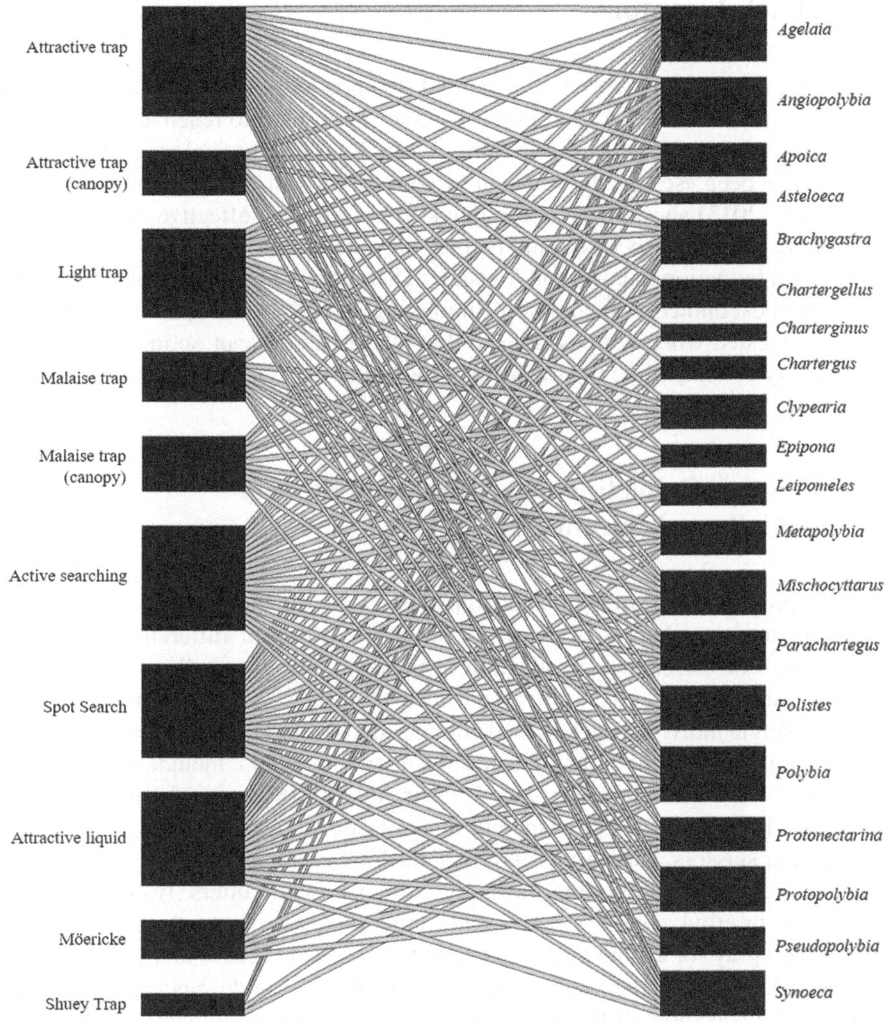

Fig. 4.3 Relationship between methods and method performance in relation to genera sampled in the Neotropical region based on published studies on social wasp fauna in Brazil between 1982 and 2018

It is also important to highlight that species richness and diversity in an area are defined by the complexity of the environment due to its spatial or temporal factors, i.e., the region's productivity and resource richness, predation intensity in the area, heterogeneity environmental adversities, and climatic variations (Silveira et al. 2012; Maciel et al. 2016). Thus, the importance of not only knowing the area to be studied is clear but also work carried out in similar areas, since a consortium of methods and their efficiency will always be limited by the wasp richness of the studied environment.

4.6 Quick Inventory

Although long diversity studies are essential for knowledge of the fauna, they sometimes become unfeasible. For example, monitoring in hard to reach areas is almost impossible on a weekly or even monthly basis. Rapid biological evaluations have alternatively been used in entomofauna surveys (Silva et al. 2013). For social wasps, Silva et al. (2013) showed that these quick inventories are effective tools for sampling more abundant species, thus helping to obtain data in less time as well as providing a reduction in costs and logistics. Therefore, a consortium of methodologies is the best choice in such cases. Employing an active and a passive methodology is suggested wherever possible, so that the researcher can optimize their time and actively collect, while the traps are in the field.

4.7 Prior to Collecting, Assembly, and Storage of Collected Specimens

Take the proper actions to have an adequate space for the incoming specimens. It is important to set a shelf with plenty of space for boxes of different sizes where samples will be stored at the collection or museum. Set your collection catalog to include nests as a new entry, as this is a record which is a bit more complex than that based on single individuals. Collected colonies may include separate data for adults, larvae, parasites, inquilines, and the nest itself, among others. Include the following elements for the collection trip: large plastic boxes to transport the nests, a good sample of different sizes of transparent thick plastic bags, a roll of cord, a portable saw, regular scissors, gardening scissors, a flashlight with an option for red light or a red-colored film to cover the flashlight, strong knives, cutters, tweezers, cotton balls, ether or ethyl acetate, syringe, and rubber gloves. Some authors recommend insecticide spray for killing the wasps; however the issue here is that you may need to clean the specimens thoroughly after the collection. Cyanide vapor is also another option, but this is more difficult to handle and dangerous for humans. CO_2 is a very nice option as it does not leave residues, but you will need to carry a tank. Lastly, 96% alcohol is also widely used to fix insects.

Insects should be packed in suitable containers and properly labeled. It is recommended to place the net collector tissue inside a plastic bag with a cotton pad soaked in acetate, ether, or alcohol to kill all insects to prevent their escape, especially in the active search methods.

Mounting guidelines must be followed for dry storage. In the case of Hymenoptera, the entomological pin should be inserted a little to the right side of the mesoscutum to preserve the best possible side. The insect should be soft-bodied at the assembly time and preferably on a Styrofoam plate that is at least 2 cm thick. The insect should be pressed until it touches the ventral face over the Styrofoam so that the legs are properly arranged. The front legs should be close to the body and facing forward, while the middle and hind legs should be facing back. Pins can be used to position the legs and, if necessary, propped up until the insect has completely dried. Immature forms should always be stored in 96% alcohol.

4.8 The Disappearance of Social Wasps

Concern about biodiversity loss has become a globally relevant issue over the past decade, with high rates of species decline that may in the long run progress to extinction (Pimm 2000; Barnosky et al. 2011). However, until recently, most scientific and government attention was focused on vertebrates, particularly mammals and birds (Manne et al. 1999; Ceballos and Ehrlich 2002), while insects were neglected in conservation and biodiversity studies (Janzen and Hallwachs 2019; Sánchez-Bayo and Wyckhuys 2019).

A topic discussed worldwide is the problem of bees disappearing, and so pollinating insects have attracted the attention of researchers. In parallel with bees, this concern has focused on other insect groups, and the results are unfortunate in every way (e.g., Detoni et al. 2018; Raw 2018; Sánchez-Bayo and Wyckhuys 2019). Along these same lines, the situation is no different for social wasps. Several Neotropical researchers have reported the "disappearance of wasps" in the most varied environments. Many of the abundant species and genera of wasps that were once easily found now require an effort to be found, and when they are abundant, they are far less than recorded decades ago.

Special attention should be paid to the social wasp group, which has been largely neglected because of its aggressive behavior but which can offer much as a study model for understanding social evolution as well as environmentally efficient service providers.

References

Arab A, Cabrini I, de Andrade CFS (2010) Diversity of polistinae wasps (Hymenoptera, Vespidae) in fragments of Atlantic rain Forest with different levels of regeneration in southeastern Brazil. Sociobiology 56:515–526. https://doi.org/10.13102/sociobiology.v62i3.738

Barbosa BC (2019) Ecologia comportamental e de interações: vespas sociais e suas colônias. (Doutorado em Ecologia Aplicada ao Manejo e Conservação de Recursos Naturais) Universidade Federal de Juiz de Fora, Juiz de Fora, Minas Gerais, Brazil, 63 p

Barbosa BC, Paschoalini MF, Prezoto F (2014) Temporal activity patterns and foraging behavior by social wasps (Hymenoptera, Polistinae) on fruits of *Mangifera indica* L.(Anacardiaceae). Sociobiology 61:239–242. https://doi.org/10.13102/sociobiology.v61i2.239-242

Barbosa BC, Detoni M, Maciel TT, Prezoto F (2016) Studies of social wasp diversity in Brazil: over 30 years of research, advancements and priorities. Sociobiology 63:858–880. https://doi.org/10.13102/sociobiology.v63i3.1031

Barnosky A, Matzke N, Tomiya S et al (2011) Has the Earth's sixth mass extinction already arrived?. Nature 471:51–57. https://doi.org/10.1038/nature09678

Brügger BP, Castro BMDC, Prezoto F, Serrão JE, Zanuncio JC (2017) Feeding by the social wasp *Polybia scutellaris* (Hymenoptera: Vespidae) on *Syzygium jambos* (Myrtaceae) fruits in Minas Gerais, Brazil. Fla Entomol 100:172–174. https://doi.org/10.1653/024.100.0125

Brügger BP, Prezoto F, Souza LSA, Zanuncio AJV, Soares MA, Wilcken CF, Zanuncio JC (2019) Use of fruit juice as a method for the collection of social wasps. Fla Entomol 102:592–595. https://doi.org/10.1653/024.102.0315

Carpenter JM (1993) Biogeographic patterns in the Vespidae (Hymenoptera): two views of Africa and South America. In: Goldblatt P (ed) Biological relationships between Africa and South America. New Haven and London: Yale University Press, pp.139–155

Carpenter JM, Marques OM (2001) Contribuição ao estudo dos vespídeos do Brasil. Universidade Federal da Bahia, Departamento de Fitotecnia, Bahia, CD-ROM, 147 p

Ceballos G, Ehrlich PR (2002) Mammal population losses and the extinction crisis. Science 296(5569):904–907

Clemente MA, Lange D, Del-Claro K, Prezoto F, Campos NR, Barbosa BC (2012) Flower-visiting social wasps and plants interaction: network pattern and environmental complexity. Psyche J Entomol 2012:1–10. https://doi.org/10.1155/2012/478431

Detoni M, Barbosa BC, Maciel TT, Santos SJL, Prezoto F (2018) Long-and short-term changes in social wasp community structure in an urban area. Sociobiology 65:305–311. https://doi.org/10.13102/sociobiology.v65i2.2597

dos Santos Souza CA, Vale ACG, Barbosa BC (2016) Checklist de Vespas Sociais (Vespidae: Polistinae) em Duas Áreas Verdes Urbanas no Município de Barra Mansa, Rio de Janeiro, Brazil. EntomoBrasilis 9(3):169–174. https://doi.org/10.12741/ebrasilis.v9i3.620

Elisei T, Nunes J, Junior CR, Junior AF, Prezoto F (2013) What is the ideal weather for social wasp *Polistes versicolor* (Olivier) go to forage? EntomoBrasilis 6:214–216. https://doi.org/10.12741/ebrasilis.v6i3.342

Goulet H, Huber JT (eds.) (1993) Hymenoptera of the world: an identification guide to families. Ottawa: Research Branch Agriculture Canada, 668 p

Hunt JH (2007) The evolution of social wasps. Oxford University Press, Oxford

Janzen DH, Hallwachs W (2019) Perspective: where might be many tropical insects? Biol Conserv 233:102–108. https://doi.org/10.1016/j.biocon.2019.02.030

Lorenzato D (1985) Ocorrência e flutuação populacional de abelhas e vespas em pomares de macieiras Malus domestica Bork e pessegueiros Prunus persica Zucc. no alto vale do Rio do Peixe, Santa Catarina e eficiência de atrativos alimentares sobre esses hymenopteros. Agronomia Sulriograndense 21:87–109

Maciel TT, Barbosa BC, Prezoto F (2016) Armadilhas Atrativas como Ferramenta de Amostragem de Vespas Sociais (Hymenoptera: Vespidae): Uma Meta-Análise. EntomoBrasilis 9:150–157. https://doi.org/10.12741/ebrasilis.v9i3.644

Maciel TT, Barbosa BC, Prezoto F (in press) Advances in the use of attractive traps in collecting neotropical social wasp. Entomologica Americana

Manne L, Brooks T, Pimm S (1999) Relative risk of extinction of passerine birds on continents and islands. Nature 399:258–261. https://doi.org/10.1038/20436

Mechi MR (2005) Comunidade de vespas Aculeata (Hym.) e suas fontes florais. In: Pivello VR, Varanda EM (eds) O Cerrado Pé-de-Gigante: Ecologia e conservação Parque Estadual Vassununga. Secretaria do Meio Ambiente, São Paulo

Nascimento FS, Tannure-Nascimento IC (2005) Foraging patterns in a nocturnal swarm-founding wasp, *Apoica flavissima* van der Vecht (Hymenoptera: Vespidae). Neotrop Entomol 34:177–181. https://doi.org/10.1590/S1519-566X2005000200005

Noll F, Gomes B (2009) An improved bait method for collecting Hymenoptera, especially social wasps (Vespidae: Polistinae). Neotrop Entomol 38:477–481. https://doi.org/10.1590/S1519-566X2009000400006

Pickett KM, Carpenter JM (2010) Simultaneous analysis and the origin of eusociality in the Vespidae (Insecta: Hymenoptera). Arthropod Systematics & Phylogeny 68(1):3–33

Pimm SL (2000) Biodiversity is us. Oikos 90(1):3–6

Prezoto F, Barbosa BC, Maciel TT, Detoni M (2016) Agroecossistemas e o serviço ecológico dos insetos na sustentabilidade. In: Resende LO, Prezoto F, Barbosa BC, Gonçalves EL (eds) Sustentabilidade: Tópicos da Zona da Mata Mineira. Real Consultoria em Negócios Ltda, Juiz de Fora, pp 19–30

Prezoto F, Maciel TT, Detoni M, Mayorquin AZ, Barbosa BC (2019) Pest control potential of social wasps in small farms and urban gardens. Insects 10:1–10. https://doi.org/10.3390/insects10070192

Rafael JA, Melo GAR, Carvalho CJB, Casari SA, Constantino R (2012) Insetos do Brasil: Diversidade e Taxonomia. Ribeirão Preto: Holos, 796 pp

Raw A (2018) A founder-controlled, social wasp assemblage, and a recent severe fall in numbers. Sociobiology 65:325–329. https://doi.org/10.13102/sociobiology.v65i2.1433

Richards OW (1978) The social wasps of the Americas excluding the Vespinae. British Museum (Natural History), London

Rodrigues VM, Machado VLL (1982) Vespídeos sociais: Espécies do Horto Florestal "Navarro de Andrade" de Rio Claro, SP. Naturalia 7:173–175

Rodríguez-Jimenez A, Sarmiento CE (2008) Altitudinal distribution and body resource allocation in a High Mountain social wasp (Hymenoptera: Vespidae). Neotrop Entomol 37:1–7. https://doi.org/10.1590/S1519-566X2008000100001

Sánchez-Bayo F, Wyckhuys KA (2019) Worldwide decline of the entomofauna: a review of its drivers. Biol Conserv 232:8–27. https://doi.org/10.1016/j.biocon.2019.01.020

Santos BB (1996) Ocorrência de vespideos sociais (Hymenoptera, Vespidae) em pomares em Goiânia, Goiás, Brasil. Revista do Setor de Ciências Agrárias 15:43–46

Santos GMM, Aguiar CML, Mello MA (2010) Flower-visiting guild associated with the Caatinga flora: trophic interaction networks formed by social bees and social wasps with plants. Apidologie 41:466–475. https://doi.org/10.1051/apido/2009081

Saraiva NB, Prezoto F, Fonseca M, Das G, Blassioli-Moraes MC, Borges M, Laumann RA, Auad AM (2017) The social wasp *Polybia fastidiosuscula* Saussure (Hymenoptera: Vespidae) uses herbivore-induced maize plant volatiles to locate its prey. J Appl Entomol 141:620–629. https://doi.org/10.1111/jen.12378

Sarmiento CE (2006) Métodos generales de recolección. In: Fernández F, Sharkey MJ (eds) Introducción a los Hymenoptera de la Región Neotropical. Sociedad Colombiana de Entomología y Universidad Nacional de Colombia, Bogotá

Silva NJJ, Morais TA, Santos-Prezoto HH, Prezoto F (2013) Inventário Rápido de Vespas Sociais em Três Ambientes com Diferentes Vegetações. EntomoBrasilis 6:146–149. https://doi.org/10.12741/ebrasilis.v6i2.303

Silveira OT (2002) Surveying neotropical social wasps: an evaluation of methods in the "Ferreira Penna" research station (ECFPn), in Caxiuanã, PA, Brazil (Hym., Vespidae, Polistinae). Pap Avulsos Zool 42:299–323

Silveira OT, Silva SDS, Pereira JLG, Tavares IDS (2012) Local-scale spatial variation in diversity of social wasps in an Amazonian rain forest in Caxiuanã, Pará, Brazil (Hymenoptera, Vespidae, Polistinae). Revista Brasileira de Entomologia 56(3):329–346

Somavilla A, Marques DWA, Barbosa EAS, Pinto JS, Oliveira ML (2014) Vespas Sociais (Vespidae: Polistinae) de uma Área de Floresta Ombrófila Densa Amazônica no Estado do Maranhão, Brasil. EntomoBrasilis 7:183–187. https://doi.org/10.12741/ebrasilis.v7i3.404

Somavilla A, Andena SR, Oliveira ML (2015a) Social wasps (Hymenoptera: Vespidae: Polistinae) of Jaú National Park, Amazonas, Brazil. EntomoBrasilis 8:45–50. https://doi.org/10.12741/ebrasilis.v8i1.447

Somavilla A, Schoeninger K, Carvalho AF, Menezes RS, Del Lama MA, Costa MA, Oliveira ML (2015b) Record of parasitoids in nests of social wasps (Hymenoptera: Vespidae: Polistinae). Sociobiology 62:92–98. https://doi.org/10.13102/sociobiology.v62i1.92-98

Souza MM, Prezoto F (2006) Diversity of social wasps (Hymenoptera: Vespidae) in semideciduous forest and cerrado (Savanna) regions in Brazil. Sociobiology 47:35–147

Souza MM, Perillo LN, Barbosa BC, Prezoto F (2015) Use of flight interception traps of Malaise type and Attractive Traps for social wasps record (Vespidae: Polistinae). Sociobiology 62(3):450–456

Wenzel JW (1998) A generic key to the nests of hornets, yellow jackets, and paper wasps worldwide (Vespidae: Vespinae, Polistinae). Am Mus Novit 3224:1–39

Chapter 5
Sampling Methods for Butterflies (Lepidoptera)

André V. L. Freitas, Jessie P. Santos, Augusto H. B. Rosa, Cristiano A. Iserhard, Aline Richter, Ricardo R. Siewert, Patrícia E. Gueratto, Junia Y. O. Carreira, and Giselle M. Lourenço

5.1 Introduction

The order Lepidoptera is the second-largest insect group in number of species (including moths and butterflies), encompassing about 16% of all living insects (Aguiar et al. 2009). Butterflies (Lepidoptera: Papilionoidea) are diurnal insects usually with bright colors and easy to recognize, and their charismatic appearance made it one of the most well-known groups of insects (Brown and Freitas 1999; Boggs et al. 2003). The butterflies are holometabolous—in other words, they undergo complete metamorphosis, with four distinct stages of development during their life cycle: egg, larva, pupa, and adult. Based on adult food, butterflies can be roughly divided into two groups: (1) fruit-feeding butterflies, which feed on decomposing fruits and other organic matter, and (2) nectar-feeding butterflies, which visit flowers to feed on nectar (DeVries 1987; Brown and Freitas 1999; Freitas et al. 2014) (Fig. 5.1). In addition, a large number of butterfly species complement their diet by drinking salt-rich water from river beaches, puddles, and wet sand.

A. V. L. Freitas (✉)
Depto. Biologia Animal and Museu de Zoologia, Instituto de Biologia, Unicamp, Campinas, São Paulo, Brazil
e-mail: baku@unicamp.br

J. P. Santos · A. H. B. Rosa · P. E. Gueratto · J. Y. O. Carreira
Depto. Biologia Animal, Instituto de Biologia, Unicamp, Campinas, São Paulo, Brazil

C. A. Iserhard · A. Richter
Programa de Pós-Graduação em Biologia Animal, Depto. de Ecologia, Zoologia e Genética, Instituto de Biologia, Universidade Federal de Pelotas, Pelotas, Rio Grande do Sul, Brazil

R. R. Siewert
Departamento de Zoologia, Universidade Federal do Paraná, Curitiba, Paraná, Brazil

G. M. Lourenço
Departamento de Genética, Ecologia e Evolução, Instituto de Ciências Biológicas, Universidade Federal de Minas Gerais, Belo Horizonte, MG, Brazil

J. C. Santos, G. W. Fernandes (eds.), *Measuring Arthropod Biodiversity*,
https://doi.org/10.1007/978-3-030-53226-0_5

Fig. 5.1 Selected examples of nectar-feeding (**a–g**) and fruit-feeding (**h–l**) butterflies. (**a**) *Enantia clarissa* (Pieridae). (**b**) *Heliconius erato phyllis* (Nymphalidae). (**c**) *Actinote genitrix genitrix* (Nymphalidae). (**d**) *Pseudoscada erruca* (Nymphalidae). (**e**) *Melinaea ludovica paraya* (Nymphalidae). (**f**) *Mimoniades versicolor* (Hesperiidae). (**g**) *Hypanartia lethe lethe* (Nymphalidae). (**h**) *Hamadryas laodamia* (Nymphalidae). (**i**) *Myscelia orsis* (Nymphalidae). (**j**) *Nica flavilla flavilla* (Nymphalidae). (**k**) *Pareuptychia ocirrhoe interjecta* (Nymphalidae). (**l**) *Yphthimoides patricia* (Nymphalidae)

The butterflies consist of 18,768 described worldwide species and are divided into seven families: Hesperiidae (4113 spp.), Papilionidae (570 spp.), Pieridae (1164 spp.), Lycaenidae (5201 spp.), Riodinidae (1532 spp.), Hedylidae (36 spp.), and Nymphalidae (6152 spp.) (van Nieukerken et al. 2011; Mitter et al. 2017). Butterfly richness is not equally distributed in the biogeographical regions of the world. The Neotropical region is the richest one, followed by Oriental/Indo-Malayan, Afrotropical, Palearctic, Australian, and Nearctic (Heppner 1991). In this way, it is possible to note that the distribution of the butterfly richness follows a clear latitudinal gradient, with species numbers increasing toward the tropics

(Slansky 1973; Shields 1989; Brown 2004; Larsen 2005). Until now, Pelham (2019) listed 846 species for the United States and Canada, while Lamas (2004) estimated 7784 species in the Neotropics, suggesting that in the Americas roughly 90% of butterfly species have tropical distribution (Bonebrake et al. 2010). Similarly, Tolman and Lewington (2008) listed 440 species for Europe and Britain, compared to the 4000 species found in the Afrotropics (Larsen 2005), again roughly 90% of butterfly species in the western half of the Old World are tropical (Bonebrake et al. 2010).

Butterflies provide many vital and economically important services within terrestrial ecosystems (e.g., nutrient recycling, food resources, and pollination) (Merckx et al. 2013). Indeed, the study of butterfly biology has a long history and has contributed greatly to our understanding of ecology, evolution, biogeography, and conservation (Bates 1862; Wallace 1865; Müller 1879; Fisher et al. 1943; Brown et al. 1974; Gilbert and Singer 1975; Gilbert 1991; Vane-Wright and Ackery 1984; Brown and Freitas 2000; Boggs et al. 2003; Mallet 2004; Bonebrake et al. 2010; Merckx et al. 2013; Chazot et al. 2019).

Taking into account this huge biodiversity, the objective of the present chapter is to provide an overview of the main sampling methods used in studies of butterfly ecology and systematics. The use of standardized sampling methods and protocols allows comparison of data from different locations with the goal of understanding patterns of distribution and abundance of butterflies. For practical reasons, the family Hedylidae will not be treated in the present chapter (their general behavior is more similar to those of nocturnal moths, whose collecting methods will not be discussed here). Also, the chapter is focused on adult butterflies; for immature stages, there are a plethora of different methods that will not be addressed here.

Essentially, sampling is here divided into two main categories: (1) active sampling, when collecting effort and decisions are dependent on the sampler, and (2) passive sampling, usually using a kind of trap (baited or not), where collecting effort and decisions are not dependent on the sampler.

5.2 Active Sampling

5.2.1 Insect Nets

This is the most widely used method of butterfly sampling: the researcher is in charge for sampling effort, and success is entirely dependent on the ability to catch the butterflies. This technique can be used for all families of diurnal butterflies and is the best sampling method for nectar-feeding species. This is a traditional method to study butterfly populations and communities, and it has been largely used for many decades in the formulation of species lists in America and Europe, in monitoring programs, and in studies of ecology, natural history, and conservation (Ebert

1969; Brown 1972; Pollard 1977; Pollard and Yates 1993; Caldas and Robbins 2003; Pozo et al. 2008; Francini 2010; Iserhard et al. 2013; Iserhard et al. 2017).

Studies that require a standardized protocol is generally based on the delimitation of linear transects, which are sampled during a determined period of time (Brown 1972; Pollard 1977; Van Swaay et al. 2015). In some early studies, a standardized and "complete" sampling day (an "entire day count of butterflies") was defined as a survey of at least 4 h of sampling effort (Ebert 1969). In spite of this, in cloudy days (more than 50% of cloud cover), samplings should be intensified to compensate the low butterfly activity (Brown 1972); thus, a suggestion is to double the sampling effort (Caldas and Robbins 2003).

Concerning short-term inventories, a minimum of 3 h per transect is suggested to maximize the sampling effort and to record a large number of species in a given area. The daily sampling period in tropical areas should extend from 07:00 to 18:00 h, an extended period compared to temperate areas (Walpole and Sheldon 1999; Caldas and Robbins 2003). Moreover, the daily activity of butterfly species varies during the day (Marchiori and Romanowski 2006; Wikström et al. 2008); therefore, morning and afternoon samplings are mandatory during the same field expedition or sampling occasion, or at each field expedition, sampling occurs at inverted times of the day (e.g., morning in an expedition and afternoon in another) (Iserhard et al. 2017).

To make sure that most species present in the area or transect were recorded and to generate a comprehensive and reliable species list, at least 14 h of sampling is recommended, an effort suggested as enough to adequately sample the local assemblages (Brown 1972). The 14 h of sampling effort should be made on consecutive days, independent of the number of collectors (Brown 1972, 1997; Iserhard et al. 2013), over a period of up to 15 days (ideally, within 1 week).

During field expeditions, collectors employing butterfly nets (hoop about 40 cm in diameter and handle about 1.20 to 1.80 m) (Fig. 5.2a) should have basic and minimum experience in catching butterflies with this technique (a previous training period of few days is suggested for untrained collectors; see Caldas and Robbins 2003). The transects should be walked in a slow and constant pace (about 1 km/h) searching for flying butterflies along (approximately 5 m ahead) and in the immediate surroundings of the transect (approximately 2.5 m of each side of the transect) (Pollard 1977; Van Swaay et al. 2015), to satisfactorily cover a wide spectrum of flight heights. When available, the researcher may use digital cameras to take pictures for later identification in the laboratory.

Butterflies should be ideally identified on the wing, with captures restricted to those specimens that are difficult to identify. The visual identification procedure is welcome if the collector makes sure that conspicuous individuals in an accessible and visible position allow reliable identification without catching. However, for the purposes of the present chapter, the discussed methods imply that collecting is necessary. In this case, at least one individual of each species should be collected, prepared, and deposited on a scientific collection. After collecting, butterflies should be killed with a minimum manipulation to avoid injuries to the collected individual. Usual killing techniques include (1) a weak thoracic compression using two fingers,

Fig. 5.2 Sampling methods for butterflies: (**a**) A group of people collecting with insect nets. (**b**) A Van Someren-Rydon trap. (**c**) An owl butterfly (*Caligo illioneus*, Nymphalidae) feeding on fermented banana bait. (**d**) Malaise trap installed near a hilltop. (**e**) Two *Phanus* sp. (Hesperiidae) feeding on toilet paper soaked with saliva (Ahrenholz technique). (**f**) Several ithomiines feeding on rotting *Heliotropium indicum* shrubs

(2) a killing jar (common killing agents include ether, chloroform, ethyl acetate, or cyanide compounds), and (3) an ammonia injection with a syringe directly in the thorax (more commonly used on large moths but also useful for some large butterflies). Butterflies should always be manipulated with forceps to not damage wings or remove scales. After that, each individual butterfly should be stored in paper envelopes (glassine is the recommended material), prepared properly, and deposited into an appropriate and curated repository.

This method has several advantages: (1) this is a low-cost method; and (2) all butterfly families are sampled (including nectar-feeding and fruit-feeding) giving a complete picture of entire butterfly assemblages. Conversely, there are several disadvantages: (1) canopy habitats are usually undersampled; (2) collecting success is dependent of the ability of the collector, meaning that the standardized sampling effort and replicability are always hard to achieve; (3) it is impossible to simultaneously sample a variety of habitats (except if different teams are sampling simultaneously, but see last item); and (4) sampling generally does not include crepuscular species (early in the morning and late at afternoon).

This method has been widely used worldwide through many decades and is basically the best (if not the only) way to make inventories in temperate regions.

5.3 Passive Methods

5.3.1 Bait Traps: Van Someren-Rydon

In the last decades, there has been a considerable increase in the number of published studies employing bait traps to capture butterflies, including local and regional butterfly inventories, field guides, studies on population, and community ecology among others (Bonebrake et al. 2010; Freitas et al. 2014; DeVries et al. 2016). Bait traps were originally conceived to capture fruit-feeding butterflies (see introductory section), but other insect groups are occasionally captured, as well (Freitas et al. 2014). Bait traps may attract harmful arthropods that are interested in either the bait or the captured butterflies or both. These are ants, wasps, and spiders, which besides causing damage to the trapped butterflies, can cause injury to the researchers (Van Swaay et al. 2015).

The costs for trap construction and bait preparation can vary among regions and availability of the needed materials. Van Swaay et al. (2015) estimate a cost of US$10 to US$20 per trap, while cheaper options can be built modifying the standard trap (Shuey 1997; Sáfián et al. 2010; DeVries et al. 2016). There are several different models of bait traps, the most used being the Van Someren-Rydon traps (VSR) or variations of this (several other types of bait traps have been proposed; see Freitas et al. 2014). The VSR trap consists in a cylinder made of netting (dark colors are preferred for not scaring the butterflies) and wire closed on top and open at the bottom, attached to a base on which the bait is placed (Fig. 5.2b, c) (Freitas et al.

2014; Van Swaay et al. 2015; DeVries et al. 2016). There are no rules for size or diameter of the cylinder, but usually 100 to 130 cm tall and an opening larger than 25 cm are recommended (allowing the capture of larger owl butterflies). An internal inverted cone is recommended (approximately 20 cm wide at the opening), especially in traps situated in the canopy, to avoid butterflies to escape (Shuey 1997; Uehara-Prado et al. 2005). Due to the behavior of butterflies to fly upward, a transparent plastic may be used to close the top of the trap, as well as to protect the butterflies and baits from rain (Freitas et al. 2003; DeVries et al. 2016). Several types of baits can be used depending of the objectives of the study; a good revision of this and other issues related to sampling with bait traps is available in Freitas et al. (2014). It is worth mentioning that decayed fish (or shrimp) is a major bait to attract Riodinidae and Lycaenidae, and this method has been widely used in recent years (Hall and Willmott 2000; Busby et al. 2017; Checa et al. 2019).

Traps should be hung on tree branches and shrubs or sometimes suspended by supports such as tripods (e.g., in grassland habitats) (DeVries et al. 2016). Since ants usually reach the traps by walking, nylon threads are recommended (alternatively, the suspending threads can be smeared with greasy or equivalent substances to prevent ant access to the traps).

In most ecological studies, three to ten traps are combined, configuring a sampling unit (SU). This is recommended as a way to minimize the effect of the placement of the trap in distinct microhabitats (Shuey 1997). The number of traps per SU is variable, but a minimum of five is recommended (DeVries et al. 2016). The distance among traps vary, but traps placed less than 100 m from each other cannot be considered independent; SUs, for instance, should be distant at least 100 m to each other to ensure that they are independent (Ribeiro and Freitas 2012; Van Swaay et al. 2015; DeVries et al. 2016).

Once the traps are installed, the revisions and bait replacement should occur each 24–48 h (periods longer than 48 h are not recommended), for maintaining a standardized attractiveness of the bait and preventing most captured butterflies from dying in the traps. In places or seasons where species abundances are very high, traps should be checked more frequently (DeVries et al. 2016). Revisions are also important to check trap integrity, since the netting can be easily damaged by small mammals interested in the bait, such as monkeys, marmosets, coatis, and opossums (allowing trap repairs or replacement, if needed). The previous bait should not be discarded on the ground near the traps as this may influence the sampling. The procedure for removing butterflies from traps and killing is the same as described in the previous section. For collected specimens, special care should be given to labeling the envelopes, which should include not only the usual data but also the number and position of each trap where the butterfly has been captured (e.g., number of each trap in the SU and forest strata), all data that will be essential for upcoming ecological analyses. After each sampling occasion, the traps should be removed and cleaned.

Sampling with bait traps has a series of practical advantages (see a recent revision on Freitas et al. 2014), but by far, the most important is the comparability of independent samples, which allow simultaneous sampling with standardized efforts at different sites. The main benefit in the use of bait traps is that as a passive method, the collecting success is not dependent of the experience of the collector (DeVries et al. 2016). Further, this sampling methodology allows replicability at any environment, the sites can be sampled simultaneously, and baits can be standardized according to the group of interest in the study. Finally, butterflies can be marked and released unharmed, allowing the use of mark-recapture techniques (unless the research is using traps with insecticides, as mentioned in Laaksonen et al. 2006).

In short, sampling with bait traps is the best standardized method for ecological studies in tropical regions where fruit-feeding butterflies are an important portion of the butterfly assemblages.

5.3.2 *Malaise Traps*

Malaise traps are a non-attractant flight interception trap developed by the Swedish entomologist René Malaise (Malaise 1937). This kind of trap is very effective for taxonomic, population, and community studies (Matthews and Matthews 1971; Campbell and Hanula 2007; Souza et al. 2015; Perillo et al. 2017). This method is generally focused on the collection of flying insects, such as Hymenoptera, Diptera, and Coleoptera (Gressitt and Gressitt 1962; Brown 2005; Souza et al. 2015). However, other insect groups (e.g., Lepidoptera) are frequently sampled as bycatches, and their information can be used for several purposes (Matthews and Matthews 1971; Rosa et al. 2019). A good example of butterfly sampling using malaise trap is in the study of Covell and Freytag (1979), in which they sampled 43.5% of all butterfly species of the North American state of Kentucky.

Collecting with malaise traps can be easily standardized using the available commercial models, which have also the advantage of being easy to be relocated. The most commonly used commercial model is Townes (1972) (Fig. 5.2d). The capture principle and operation of this trap are based on the interception of the flight using an open tent with a fabric barrier preferably of a dark color. The ceiling of the tent has a sloped, light-colored cover to direct the insects that are attracted by the sunlight to the uppermost collection container with a fixing/killing substance (Campos et al. 2000; Rafael 2002). For a complete size check and materials for making this trap, see Rafael (2002), van Achterberg et al. (2010), and Souza et al. (2015).

Malaise traps can be used in various types of terrestrial habitats and can be fixed in the soil or suspended at various heights (suspendible malaise trap); however, the place where they are placed interferes significantly in the result of collections (Darling and Packer 1988; Rafael 2002). It should be set up where insect flight tends to be concentrated by local circumstances of topography, vegetation density, wind direction, presence of water, and other characteristics (Gressitt and Gressitt 1962). To increase the number of insects collected, it is recommended to mount the trap crosswise to the wind or transversely to natural paths (over streams) or artificial paths (trails, roads) where insects with stronger flights usually fly. In shady areas, such as dense forests, the collecting vial should be directed through the direction of higher luminosity, such as clearings (Rafael 2002). The best scenario is that the traps are kept in the field during all seasons of the year. Traps should be set from 0.5 to 1 km away from each other and monitored (i.e., collecting flasks should be replaced) every 7 or 10 days (depending on the environmental condition) (Souza et al. 2015; Schmidt et al. 2019).

The best fixing/killing substances generally applied are ethanol 70%, 80%, or 96%. While the last two dilutions are interesting since they preserve the DNA for molecular studies, the 70% ethanol implies that the material should be quickly transferred into 80% to 96% ethanol to get the DNA whole and useful for future molecular studies (van Achterberg et al. 2010).

Collecting butterflies (and Lepidoptera in general) with malaise traps presents more cons than pros:

1. Many large insects such as beetles and grasshoppers can break the butterfly specimens inside the container.
2. The alcohol causes butterfly colors to fade with time.
3. It is not particularly effective in capturing large numbers of butterflies (Owen 1971).
4. The time to screen the material on the laboratory is usually excessive (Souza et al. 2015).
5. The high price of each trap (Souza et al. 2015).
6. Climatic conditions such as sun, rain, and wind can cause deterioration of some netting materials, e.g., nylon (Covell and Freytag 1979).
7. Traps can be vandalized in isolated localities (Covell and Freytag 1979).
8. Lack of selectivity, because this technique kills large numbers of insects indiscriminately (Owen 1975; Covell and Freytag 1979).

However, there are some advantages:

1. Sampling is done without any kind of bait (Owen 1971).
2. Records obtained through its use give a good indication of flying activity and abundance from day to day and from month to month (Owen 1971).
3. Possibility of exposition for continuous periods of time and dispense of the presence of the collector for 1 week or more (Campos et al. 2000; Souza et al. 2015).
4. It is lightweight and easy to carry (Rafael 2002).
5. It is possible to recover butterflies specimens from ethanol (Rosa et al. 2019; Schmidt et al. 2019).

5.3.3 The Ahrenholz Technique

The Ahrenholz technique was proposed for attracting skipper butterflies (Hesperiidae) in Neotropical forests (Lamas et al. 1993), although several other butterfly groups (Pieridae, Nymphalidae, and Riodinidae) also visit the baits. Skipper butterflies (Hesperiidae) currently comprise more than 4000 species distributed worldwide and are divided into nine recognized subfamilies, with most of their richness (>70%) concentrated throughout the Neotropics (Li et al. 2019). In South America, skippers are commonly found in forest sites, where hundreds of species can occur. They are also common in grasslands sites, usually feeding on flowering shrubs. The majority of adult skippers is not attracted to the usual traps used for butterflies, including bait traps and colored rags. They are usually collected with entomological nets; however, their swift and erratic flights make their sampling difficult. Additionally, several skippers exhibit a crepuscular behavior and are relatively rarely sampled by diurnal lepidopterists (Austin 2008). This could explain in part why it is common to omit the skippers from butterfly surveys. In this way, the Ahrenholz technique appeared as an attempt to optimize and potentially standardize the sampling of skipper butterflies, using the entomological net to catch these attracted individuals.

The Ahrenholz technique consists in placing small pieces of toilet or tissue paper soaked in human saliva on the upper side of leaves or cleared patches of ground (Lamas et al. 1993), aiming to resemble the whitish patches of bird droppings (Fig. 5.2e). When necessary, the papers can be remoistened with saliva. In the Neotropics, many skippers are known to be attracted to droppings of birds (Downes 1973; Austin et al. 1993), probably because the excreta were suggested as an important source of sodium and amino acids for butterflies (Adler and Pearson 1982). Austin et al. (1993) attested that the attraction of skippers to the swarms of army ant is for feeding on the droppings of birds that follow these swarms. So, an association of the ant-bird-skipper occurs, and these dynamics are still waiting to be better studied in the future.

There are several studies proving the efficacy of the Ahrenholz technique (Austin et al. 1993; Austin 2008; DeVries et al. 2008, 2009; Garcia-Salik et al. 2014), and the results suggest that while not all skippers are attracted to the white baits, the technique is extremely efficient in sampling some skippers considered rare, including the ones showing crepuscular behavior. It is worth mentioning that in places where army ants are absent, the Ahrenholz technique also works, but not as efficiently (O. H. H. Mielke and A. V. L. Freitas, personal communication). This is may be due to a peculiar behavior for some Brazilian skippers: many forest species were observed congregating in the neighborhood of swarms of army ants (Formicidae: Ecitoninae) (Zikán 1929). These ants (*Eciton* and *Labidus*) are exclusively carnivorous, hunting mainly arthropods and exposing them to other predators. Johnson (1954) studied some species of birds foraging at these swarms of army ants and obtaining the majority of their food from arthropods fleeing before the swarm.

However, for an efficient skipper species inventory in Neotropical areas, multiple sampling techniques, including active search and collecting with the use of entomological nets, are necessary.

5.3.4 Collecting with Pyrrolizidine Alkaloids

The Neotropical tribe Ithomiini (Nymphalidae: Danainae) includes almost 400 species of aposematic butterflies that are the main models of most Neotropical mimicry rings (Brown 1987; Willmott and Freitas 2006) (Fig. 5.1d, e). Adults of all species in this tribe and also in the related tribe Danaini (especially males) actively seek four sources of pyrrolizidine alkaloids (PAs), which are used for chemical protection (conferring unpalatability) and as precursors of their sexual pheromones (Brown 1987) (some groups of diurnal moths are also attracted to PA sources for the same reason). Accordingly, several known sources of PAs have been used as baits to attract and collect ithomiines (Beebe 1955; Pliske 1975; Brown 1987). The most used bait is decaying shrubs of *Heliotropium indicum*(Boraginaceae) (but other PA sources also can attract Danaini and Ithomiini butterflies, see Beebe 1955 and Brehm et al. 2007); entire bushes should be uprooted, soaked with water, and left to decay in an open place. After 3 to 5 days, they start to release PAs and volatiles and should be hung on low shrubs inside the forest. These baits not only attracted males of several species of ithomiines but also keep them landed on the bait for several hours without flying upon approach (Fig. 5.2f), allowing a calm collecting of butterflies with forceps (Freitas et al. 2003). Although this method is extremely efficient to sample danaines and ithomiines and is inexpensive, it is hard to standardize, and comparative sampling is hard to achieve. Also, not all ithomiines are equally attracted to PA sources, resulting in a somewhat biased taxonomic sampling. Even though, a recent study using *Heliotropium* as baits on Van Someren-Rydon traps showed that using PA sources are possible and can give valuable comparative data (Araujo 2006).

5.4 Preparation, Maintenance, and Identification

After collecting, specimens need to be kept in good conditions until they can be prepared. Although the ideal is to mount them soon after they are collected, especially Hesperiidae and other species with robust bodies, it is not always possible due to logistics in the field. In such cases, specimens should be kept in glassine envelopes, preferably refrigerated or in boxes with ethyl acetate to prevent them from rotting. It should be noted that adults should not be preserved in alcohol or any other liquid material as it may damage the scales and other delicate parts of their bodies.

5.4.1 Preparation

Relaxing Box In many cases, butterflies in envelopes after collection are dry, making their bodies rigid and brittle. Thus, before mounting, the specimens need to be rehydrated with the aid of a relaxing box (a simple airtight container that contains a wet sponge or damp sand or paper), making them malleable so that they can be pinned.

Pins Butterflies should be mounted on entomological stainless steel pins for use in scientific collections. The specimens should be vertically pinned in the mesothorax so that they lie at an angle of 90° to the longitudinal axis of the body. The thickness of the pin will depend on the size of the thorax of the butterfly, but usually it is used from 2 to 3 mm.

Mounting After being pinned, butterflies need to have their wings extended. For this, spreading boards are used. The individuals are placed in a central groove of the board, and, with the aid of strips of paper and pins, the wings are extended. It is important to note that the wings are not pinned. The posterior margin of the forewings should form a 90° angle with the body and the hind wings slightly below the forewings.

Labeling Labeling is a critical step in the preparation of scientific collections. For label data, it is important to include country, state, county, location, coordinates, elevation, collection date, and collector. A second label of identification must be added, containing the name of the species, author, and date of the original description, besides the name of the specialist that identified the material and the year of the identification.

5.4.2 Maintenance

After mounting, butterflies should be dried out and deposited in a collection (usually in drawers that have unit trays). Drawers are usually kept in cabinets. It is important to keep the humidity and temperature at low levels, so that there is no proliferation of fungi and infestations of other insects that could damage the specimens. Some products are placed in the drawers to prevent such infestations, such as naphthalene and camphor. The collection should still be in a dark place or protected from direct light.

5.4.3 *Identification*

Identification is done through published studies, specimens deposited in a reference collection, or with the help of specialists. In several cases, a detailed study of the morphology is needed.

Wing Study The wings are removed and cleared by an immersion in 70% ethanol, then in sodium hypochlorite (NaOCl) solution until completely cleared, and then again in 70% ethanol to neutralize the action of the NaOCl.

Dissection In some cases, the morphology of the genitalia is mandatory for a correct species identification. The specimens should have their abdomen detached and soaked in a heated 10% potassium hydroxide solution (KOH) for 5 to 10 min and posteriorly dissected to remove the genitalia, so it can be illustrated or photographed. The dissected parts are stored in vials with glycerin.

Books and Articles There are a series of books and articles with images aiming to help the identification of butterflies. Several studies with images and identification keys are used, highlighting the ones of Seitz (1907–1924), Lewis (1973), Evans (1951, 1952, 1953, 1955), and Hayward (1948, 1950) (for Hesperiidae); D'Abrera (1981, 1984, 1987, 1988, 1994, 1995), Brown (1992), and Tyler et al. (1994) (for Papilionidae); Uehara-Prado et al. (2004) and Santos et al. (2011) (for fruit-feeding); and Palo (2017).

Nowadays one of the most accessed webpages for identifying butterflies is the Butterflies of America (www.butterfliesofamerica.com) (Warren et al. 2013), which has 160,500 images (including holotypes) and lists over 8300 butterfly species.

5.5 Data Collection and Management

There are several ways to record data from butterfly surveys. Given the numerous possibilities, the current section presents general recommendations for organizing and structuring the data obtained from butterfly samples, with emphasis on community studies.

Quantitative (number of individuals) and qualitative (presence or absence) observation data and measurement of individuals are easy to record in notebooks when the study involves field expeditions or laboratory experiments. Although templates save time on recording data, they may also limit the records to the pre-established information fields, thus not considering additional information that may be of interest or relevant in the future. Therefore, taking notes allows the observer to record data, including unexpected observations. If data are handwritten, they will need to be digitized. This will facilitate the restructuring process (if necessary) as well as allow the creation of multiple copies, consequently preventing the loss of collected data.

The organization of data in programs or applications with a spreadsheet editor is the best mode to record data or to prepare them for future analysis. It is strongly recommended to record all information on only one data sheet, with columns that refer to the different measured variables and the counting or individual data as rows. In other words, on the data sheet, each row must represent one surveyed individual, and each column must contain all the relevant information on this individual (species name, taxonomic group, genus, date, transect, area, coordinates, sampling methods, weather notes, and other information). Once the data are finished, there are some tools, such as the "pivot table" feature, which allow the mounting of sheets or tables with several structural options. For example, each column in data can turn into fields of pivot table that can be used to construct many different types of tables. This format makes easier and faster the data organization for several future analyses.

5.5.1 Data Analysis

The choice of appropriate statistical analyses will depend on the hypothesis to be tested, the nature of the collected data, as well as the sampling design. A satisfactory and simple description of biodiversity can be achieved with presence/absence data, whereas the use of abundance data can provide a more detailed characterization of the community. Since the species communities are complex and not closed systems, diversity parameters are subject to variation on space and time dimensions, and the perception of these effects also depends on the observed scale. Considering that species and individuals are not always detected along the sampling occasion (Boulinier et al. 1998), in the last decade, many models that incorporate the imperfect detection were developed (Kéry and Royle 2016). Despite the great value, their models are also complex and computationally time-consuming. This chapter is not intended to address all the possible analyses to explore the diversity in its fullness, but some basic and frequently analyses used to characterize and compare local butterfly communities in biodiversity studies are suggested.

Abundance The number of individuals per species allows exploring more options of analysis and better discussion of the results. More than comparing the total number of individuals among communities and sites, abundance data is a basic assumption to calculate most diversity metrics and similarity/dissimilarity indexes in species composition analyses among habitats. To compare the total abundance among sites or communities, there are many options of statistical tests, such as general linear models or generalized linear mixed models (GLM/GLMM), analysis of variance (ANOVA), and so on. Abundance data also can be explored concerning its variation in time with circular analysis. The most suitable method will depend on your sampling design and the distribution of data.

Species Richness In general, the species richness of a given assemblage is difficult to estimate because it depends on the number of individuals (or samples). Since the qualitative data does not take into account abundance, this kind of data can be used to calculate most of the species richness estimators (sample-based data). There are different types of species richness estimators, and they are frequently calculated to predict the total species richness of a community of interest (Magurran 2004; Chao and Chiu 2016). The sample-based species accumulation curve (or collector's curve) is the plot of the accumulated species richness found according to the sampling effort employed. Increasing the number of sampling occasions lead to more species records; consequently the accumulation curve will present an upward trend but will reach an asymptote if the sampled species number reaches the total species richness.

For comparing species richness among communities using quantitative data, the individual-based rarefaction is useful to compare the species richness in two or more samples with different number of individuals. The abundance of the larger sample is rarefied to the total abundance of the smaller one aiming the comparison between the species richness of both samples with the same number of individuals (Chao and Chiu 2016). Some authors suggest considering the extrapolation to equal-sized samples instead of interpolation when the difference in abundance among samples is high (Colwell et al. 2012; Hsieh et al. 2016). Another approach for comparing species richness among communities is the coverage-based rarefaction and extrapolation methods. This method compares samples with equal completeness instead of its size, preserving then a replication principle required for describing diversity (see Chao and Jost 2012 for further details). Recent studies have suggested combining these approaches with the employment of Hill numbers (Chao et al. 2014).

Diversity Classical studies have used a plethora of diversity indexes and measures that incorporate species richness and evenness of species abundances (Magurran 2004; Gotelli and Chao 2013). Recent approaches have included the Hill numbers to generate diversity profiles, which measure the diversity as the effective number of species (Chao et al. 2014). The Hill numbers (equivalent number of equally abundant species) are plotted with a q parameter that controls the sensitivity of species relative abundance. The higher the q value, the more weight is given to the most common species. Thus, when $q = 0$, the diversity is equivalent to the species richness; when $q = 1$, the Hill number is analogous to the Shannon index; and finally, when $q = 2$, the diversity value is equivalent to the Simpson index (Gotelli and Chao 2013; Chao et al. 2014). Through the inspection of diversity profiles of each sample, it is possible to compare, directly, the diversity in the different q orders.

Species Composition Comparisons of species composition among samples employ distance matrices based on different similarity or dissimilarity indexes. Some indexes consider presence and absence of species, while others are more informative, giving weight to each species according to their abundance. Therefore, the input of data for species composition analyses can be both qualitative and quantitative, depending on the type of similarity or dissimilarity measure. The choice of a distance measure is a fundamental decision, since it may influence the results and consequently their interpretation. From the calculated distance matrices, patterns can be explored through different types of cluster or multivariate ordination analyses. Clusters group samples or communities according to their similarity or dissimilarity values, resulting in a dendrogram. The confidence between relations on the dendrogram can be evaluated by a cophenetic correlation coefficient. Ordination analyses aim to organize the information of *n* dimensions (each species in communities) in two main axes of variation, visually grouping the points of samples or communities most similar in species composition to each other in the ordination space. The consistency of the groupings can be verified with an ANOSIM (analysis of similarities, Clarke 1993) or a PERMANOVA (permutational analysis of variance, Anderson 2001). These statistical tests alone demonstrate the difference in the composition of the communities, but since their outputs are only numerical, associating them to ordination analyses improves the visual representation of results.

5.6 Implications for Conservation

Studies involving butterflies provide information that can support conservation actions. Because they have an intimate relationship with the immediate surroundings (Ribeiro et al. 2012), butterflies are excellent indicators of subtle changes in habitats that often may not be detected by studying other organisms. Fruit-feeding butterflies, for example, are known to respond to many types and intensities of disturbances (natural and anthropic), being highly sensitive to fragmentation and changes in forest cover (DeVries et al. 1997; Shahabuddin and Terborgh 1999; Fermon et al. 2003, 2005; Barlow et al. 2007; Uehara-Prado et al. 2007; Ribeiro et al. 2012; Ribeiro and Freitas 2012; Sant'Anna et al. 2014; Filgueiras et al. 2016; Thomas 2016; Lourenço et al. 2019).

In view of this, understanding the differences in butterfly assemblages among habitats is one way to understand the sensitivity of ecological systems to environmental changes and to predict shifts as future climate changes. Thus, conducting well-designed and standardized studies is essential to increase conservation efforts. In addition, because they are large and often colorful, butterflies have great popular appeal, which strengthens the society-conservation link, essential for the effective implementation of conservation programs, since effective conservation programs ensure the protection of endangered species as well as allow the recording of new species and expansion of occurrence of species already described.

Acknowledgments The authors thank the editors, JC Santos and GW Fernandes for the invitation to write this chapter. Two anonymous referees revised the submitted version. Bob Robbins carefully read the last version, made valuable suggestions, and revised English language. AVLF thanks the Conselho Nacional de Desenvolvimento Científico e Tecnológico, CNPq (303834/2015-3), SISBIOTABrasil/CNPq (563332/2010-7), the National Science Foundation (DEB-1256742), and the BIOTA-FAPESP Program (2011/50225-3). JPS thanks the National Science Foundation (DEB-1256742). AHBR, JYOC, and PEG thank Coordenação de Aperfeiçoamento de Pessoal de Nível Superior-Brazil (CAPES) – finance code 001. AR thanks CAPES (88882.345582/2015-01) for graduate fellowship. RRS would like to thank Olaf Hermann Hendrik Mielke, for helpful comments and for the requested bibliography, and CNPq (151873/2018-6) for research fellowship. JYOC thanks CNPq (141443/2018-9) for a research scholarship. GML thanks CNPq (155895/2014-1) and CAPES (88881.133074/2016-01) for a research scholarship. This work was funded in part by USAID and the US National Academy of Sciences (NAS) under the PEER program (sponsor grant award number: AID-OAA-A-11-00012). Butterfly species of this study are registered in the National System for the Management of Genetic Heritage and Associated Traditional Knowledge SISGEN (AF2F72B).

References

Adler PH, Pearson DL (1982) Why do male butterflies visit mud puddles? Can J Zool 60:322–325. https://doi.org/10.1139/z82-043

Aguiar AP, Santos BF, Couri MS et al (2009) Insecta. In: Rocha RM, Boeger WAP (eds) Estado da Arte e Perspectivas para a Zoologia no Brasil. Resultados dos Simpósios do XXVII Congresso Brasileiro de Zoologia. Editora UFPR, Curitiba, pp 131–155

Anderson MJ (2001) A new method for non–parametric multivariate analysis of variance. Austral Ecol 26:32–46. https://doi.org/10.1111/j.1442-9993.2001.01070.pp.x

Araujo IS (2006) Estrutura e influência da sazonalidade na comunidade de borboletas da subfamília ithomiina (Lepidoptera: Nymphalidae) na estação científica Ferreira Penna, Melgaço, Pará. MSc Dissertation, Universidade Federal do Pará

Austin GT (2008) Hesperiidae of Rondônia, Brazil: taxonomic comments on "night" skippers, with descriptions of new genera and species (Lepidoptera: Eudaminae). Insecta Mundi 29:1–36

Austin GT, Brock JP, Mielke OHH (1993) Ants, birds, and skippers. Trop Lepid Res 4:1–11

Barlow J, Overal WL, Araujo IS et al (2007) The value of primary, secondary and plantation forests for fruit-feeding butterflies in the Brazilian Amazon. J Appl Ecol 44:1001–1012. https://doi.org/10.1111/j.1365-2664.2007.01347.x

Bates HW (1862) Contributions to an insect fauna of the Amazon valley. Lepidoptera: Heliconidea. Trans Linn Soc Lond 23:495–566. https://doi.org/10.1111/j.1096-3642.1860.tb00146.x

Beebe W (1955) Two little-known selective insect attractants. Fortschr Zool 40:27–32

Boggs CL, Watt WB, Ehrlich PR (2003) Butterflies: ecology and evolution taking flight. University of Chicago Press, Chicago

Bonebrake TC, Ponisio LC, Boggs CL et al (2010) More than just indicators: a review of tropical butterfly ecology and conservation. Biol Conserv 143:1831–1841. https://doi.org/10.1016/j.biocon.2010.04.044

Boulinier T, Nichols JD, Sauer JR et al (1998) Estimating species richness: the importance of heterogeneity in species detectability. Ecology 79:1018–1028. https://doi.org/10.1890/0012-9658(1998)079[1018:ESRTIO]2.0.CO;2

Brehm G, Hartmann T, Willmott K (2007) Pyrrolizidine alkaloids and pharmacophagous Lepidoptera visitors of *Prestonia amabilis* (Apocynaceae) in a montane rainforest in Ecuador. Ann Missouri Bot Gard 94:463–473. https://doi.org/10.3417/0026-6493(2007)94[463:PAAPLV]2.0.CO;2

Brown JH (2004) Why are there so many species in the tropics? J Biogeogr 41:8–22. https://doi.org/10.1111/jbi.12228

Brown BV (2005) Malaise trap catches and the crisis in Neotropical Dipterology. Am Entomol 51:180–183. https://doi.org/10.1093/ae/51.3.180

Brown KS Jr (1972) Maximizing daily butterfly counts. J Lepid Soc 26:183–196

Brown KS Jr (1987) Chemistry at the solanaceae/ithomiinae interface. Ann Missouri Bot Gard 74:359–397. https://doi.org/10.2307/2399406

Brown KS Jr (1992) Borboletas da Serra do Japi: diversidade, habitats, recursos alimentares e variação temporal. In: Morellato LPC (ed) História natural da Serra do Japi: Ecologia e preservação de uma área florestal no sudeste do Brasil. Editora da UNICAMP/ FAPESP, Campinas, pp 142–186

Brown KS Jr, Freitas AVL (1999) Lepidoptera. In: Joly CA, Bicudo CEM (orgs) (eds) Biodiversidade do Estado de São Paulo, Brasil: síntese do conhecimento ao final do século XX. FAPESP, São Paulo, pp 225–243

Brown KS Jr (1997) Diversity, disturbance, and sustainable use of Neotropical forests: insects as indicators for conservation monitoring. J Insect Conserv 1:25–42. https://doi.org/10.1023/A:1018422807610

Brown KS Jr, Freitas AVL (2000) Atlantic forest butterflies: indicators for landscape conservation. Biotropica 32:934–956. https://doi.org/10.1111/j.1744-7429.2000.tb00631.x

Brown KS Jr, Sheppard PM, Turner JRG (1974) Quaternary refugia in tropical America: evidence from race formation in Heliconius butterflies. Proc R Soc Lond [Biol] 187:369–378. https://doi.org/10.1098/rspb.1974.0082

Busby RC, Faynel C, Moser A et al (2017) Sympatric diversification in the upper Amazon: a revision of the Eumaeine genus *Paraspiculatus* (Lepidoptera: Lycaenidae). Smithson Contrib Zool 649:1–66. https://doi.org/10.5479/si.1943-6696.649

Caldas A, Robbins RK (2003) Modified Pollard transects for assessing tropical butterfly abundance and diversity. Biol Conserv 110:211–219. https://doi.org/10.1016/S0006-3207(02)00190-8

Campbell JW, Hanula JL (2007) Efficiency of Malaise traps and colored pan traps for collecting flower visiting insects from three forested ecosystems. J Insect Conserv 11:399–408. https://doi.org/10.1007/s10841-006-9055-4

Campos WG, Pereira D, Schoereder JH (2000) Comparison of the efficiency of flight-interception trap models for sampling Hymenoptera and other insects. An Soc Entomol Bras 29:381–389. https://doi.org/10.1590/S0301-80592000000300001

Chao A, Chiu CH (2016) Species richness: estimation and comparison. Wiley StatsRef: Statistics Reference Online, pp 1–26. https://doi.org/10.1002/9781118445112.stat03432.pub2

Chao A, Jost L (2012) Coverage–based rarefaction and extrapolation: standardizing samples by completeness rather than size. Ecology 93:2533–2547. https://doi.org/10.1890/11-1952.1

Chao A, Gotelli NJ, Hsieh TC et al (2014) Rarefaction and extrapolation with hill numbers: a framework for sampling and estimation in species diversity studies. Ecol Monogr 84:45–67. https://doi.org/10.1890/13-0133.1

Chazot N, Wahlberg N, Freitas AVL et al (2019) Priors and posteriors in Bayesian timing of divergence analyses: the age of butterflies revisited. Syst Biol 68:797–813. https://doi.org/10.1093/sysbio/syz002

Checa MF, Donoso DA, Rodruiguez J et al (2019) Combining sampling techniques aids monitoring of tropical butterflies. Insect Conserv Divers 12:362–372. https://doi.org/10.1111/icad.12328

Clarke KR (1993) Non–parametric multivariate analyses of changes in community structure. Austral Ecol 18:117–143. https://doi.org/10.1111/j.1442-9993.1993.tb00438.x

Colwell RK, Chao A, Gotelli NJ et al (2012) Models and estimators linking individual-based and sample-based rarefaction, extrapolation, and comparison of assemblages. J Plant Ecol 5:3–21. https://doi.org/10.1093/jpe/rtr044

Covell CV Jr, Freytag PH (1979) The malaise trap as a means of sampling butterfly populations in Kentucky. J Lepid Soc 33:153–161

D'Abrera B (1981) Butterflies of the Neotropical region. Part I. Papilionidae & Pieridae. Hill House, Victoria

D'Abrera B (1984) Butterflies of the Neotropical region. Part II. Danaidae, Ithomidae, Heliconidae & Morphidae. Hill House, Victoria

D'Abrera B (1987) Butterflies of the Neotropical region. Part IV. Nymphalidae (partim). Hill House, Victoria

D'Abrera B (1988) Butterflies of the Neotropical region. Part V. Nymphalidae (conc.) & Satyridae. Hill House, Victoria

D'Abrera B (1994) Butterflies of the Neotropical region. Part VI. Riodinidae. Hill House, Victoria
D'Abrera B (1995) Butterflies of the Neotropical region. Part VII. Lycaenidae. Hill House, Victoria
Darling DC, Packer L (1988) Effectiveness of Malaise traps in collecting Hymenoptera: the influence of trap design, mesh size, and location. Can Entomol 120:787–796. https://doi.org/10.4039/Ent120787-8
DeVries PJ (1987) The butterflies of Costa Rica and their natural history. Princeton University Press, Princeton, NJ
DeVries PJ, Murray D, Lande R (1997) Species diversity in vertical, horizontal, and temporal dimensions of a fruit-feeding butterfly community in an Ecuadorian rainforest. Biol J Linn Soc 62:343–364. https://doi.org/10.1006/bijl.1997.0155
DeVries PJ, Austin GT, Martin NH (2008) Diel activity and reproductive isolation in a diverse assemblage of Neotropical skippers (Lepidoptera: Hesperiidae). Biol J Linn Soc 94:723–736. https://doi.org/10.1111/j.1095-8312.2008.01037.x
DeVries PJ, Austin GT, Martin NH (2009) Estimating species diversity in a guild of Neotropical skippers (Lepidoptera: Hesperiidae) with artificial lures is a sampling problem. Insect Conserv Diver 2:125–134. https://doi.org/10.1111/j.1752-4598.2009.00047.x
DeVries PJ, Hamm CA, Fordyce JA (2016) A standardized sampling protocol for fruit-feeding butterflies (Nymphalidae). In: Larsen TH (ed) Core standardized methods for rapid biological field assessment. Conservation International, Arlington, pp 140–148
Downes JA (1973) Lepidoptera feeding at puddle-margins, dung and carrion. J Lepid Soc 27:89–99
Ebert H (1969) On the frequency of butterflies in eastern Brazil, with a list of the butterfly fauna of Poços de Caldas, Minas Gerais. J Lepid Soc 23:1–52
Evans WH (1951) A catalogue of the American Hesperiidae indicating the classification and nomenclature adopted in the British Museum (Natural History). Part 1. Introduction and Group A Pyrrhopyginae. British Museum (Natural History), London
Evans WH (1952) A catalogue of the American Hesperiidae indicating the classification and nomenclature adopted in the British Museum (Natural History). Part II. Groups B, C, D. Pyrginae section 1. British Museum (Natural History), London
Evans WH (1953) A catalogue of the American Hesperiidae indicating the classification and nomenclature adopted in the British Museum (Natural History). Part III. Groups E, F, G. Pyrginae section 2. British Museum (Natural History), London
Evans WH (1955) A catalogue of the American Hesperiidae indicating the classification and nomenclature adopted in the British Museum (Natural History). Part IV. Groups H to P. Hesperiinae and Megathyminae. British Museum (Natural History), London
Fermon H, Waltert M, Muhlenberg M (2003) Movement and vertical stratification of fruit-feeding butterflies in a managed west African rainforest. J Insect Conserv 7:7–19. https://doi.org/10.1023/A:1024755819790
Fermon H, Waltert M, Vane-Wright RI et al (2005) Forest use and vertical stratification in fruit-feeding butterflies of Sulawesi, Indonesia: impacts for conservation. Biodivers Conserv 14:333–350. https://doi.org/10.1007/s10531-004-5354-9
Filgueiras BKC, Melo DHA, Leal IR et al (2016) Fruit-feeding butterflies in edge-dominated habitats: community structure, species persistence and cascade effect. J Insect Conserv 20:539–548. https://doi.org/10.1007/s10841-016-9888-4
Fisher RA, Corbet SA, Williams CB (1943) The relation between number of species and the number of individuals in a random sample of an animal population. J Anim Ecol 12:42–58. https://doi.org/10.2307/1411
Francini RB (2010) História natural das borboletas do Vale do Rio Quilombo. In: Santos SP (ed) E-book. Privately published. https://archive.org/details/HistoriaNaturalDasBorboletasDoValeDoRioQuilomboSantosSp/page/n16. Accessed 25 Oct 2019. https://doi.org/10.13140/2.1.3862.9441
Freitas AVL, Francini RB, Brown KS Jr (2003) Insetos como indicadores ambientais. In: Cullen L Jr, Rudran R, Valladares-Pádua C (eds) Métodos de estudo em Biologia da Conservação e Manejo da Vida Silvestre. Fundação Boticário e Editora da UFPR, Curitiba, pp 125–151

Freitas AVL, Iserhard CA, Santos JP et al (2014) Studies with butterfly bait traps: an overview. Rev Colomb Entomol 40:203–212
Garcia-Salik LM, Carneiro E, Dolibaina DR, Dias FMS et al (2014) Borboletas da Estação Ecológica do Caiuá, Diamante do Norte, Paraná, Brasil (Lepidoptera: Hesperioidea & Papilionoidea). SHILAP Rev Lepidopt 42:265–280
Gilbert LE (1991) Biodiversity of a central American *Heliconius* community: pattern, process and problems. In: Price PW, Lewinsohn TM, Fernandes GW, Benson WW (eds) Plant-animal interactions. Evolutionary ecology in tropical and temperate regions. Wiley, New York, pp 403–427
Gilbert LE, Singer MC (1975) Butterfly ecology. Annu Rev Ecol Syst 6:365–395. https://doi.org/10.1146/annurev.es.06.110175.002053
Gotelli NJ, Chao A (2013) Measuring and estimating species richness, species diversity, and biotic similarity from sampling data. In: Levin SA (ed) Encyclopedia of biodiversity, 2nd edn. Academic Press, Waltham, pp 195–211. https://doi.org/10.1016/B978-0-12-384719-5.00424-X
Gressitt JL, Gressitt MK (1962) An improved Malaise trap. Pac Insects 4:87–90
Hall JPW, Willmott KR (2000) Patterns of feeding behavior in adult male riodinid butterflies and their relationship to morphology and ecology. Biol J Linn Soc 69:1–23. https://doi.org/10.1111/j.1095-8312.2000.tb01666.x
Hayward KJ (1948) Insecta, Lepidoptera (Rhopalocera), familia Hesperiidarum, subfamiliae Pyrrhopyginarum et Pyrginarum. In: Descole HR (ed) Genera et species animalium argentinorum, vol 1. Guillermo Kraft, Buenos Aires
Hayward KJ (1950) Insecta, Lepidoptera (Rhopalocera), familia Hesperiidarum, subfamiliae Hesperiinarum. In: Descole HR (ed) Genera et species animalium argentinorum, vol 2. Guillermo Kraft, Buenos Aires
Heppner JB (1991) Faunal regions and the diversity of Lepidoptera. In: Tropical Lepidoptera, vol 2. Association for Tropical Lepidoptera, Gainesville. https://doi.org/10.1093/ae/38.4.252
Hsieh TC, Ma KH, Chao A (2016) iNEXT: an R package for rarefaction and extrapolation of species diversity (hill numbers). Methods Ecol Evol 7:1451–1456. https://doi.org/10.1111/2041-210X.12613
Iserhard CA, Brown KS Jr, Freitas AVL (2013) Maximized sampling of butterflies to detect temporal changes in tropical communities. J Insect Conserv 17:615–622. https://doi.org/10.1007/s10841-013-9546-z
Iserhard CA, Romanowski HP, Richter A et al (2017) Monitoring temporal variation to assess changes in the structure of subtropical Atlantic forest butterfly communities. Environ Entomol 46:804–813. https://doi.org/10.1093/ee/nvx115
Johnson RA (1954) The behavior of birds attending army ant raids on Barro Colorado Island, Panama Canal zone. Proc Linn Soc N Y 65:41–70
Kéry M, Royle JA (2016) Applied hierarchical modeling in ecology: analysis of distribution, abundance and species richness in R and BUGS, vol 1. Elsevier, Amsterdam. https://doi.org/10.1016/j.rmb.2017.03.028
Laaksonen J, Laaksonen T, Itämies J et al (2006) A new efficient bait-trap model for Lepidoptera surveys—the "Oulu" model. Entomol Fenn 17:153–160. https://doi.org/10.33338/ef.84301
Lamas G (2004) Atlas of Neotropical Lepidoptera (checklist part 4A. Hesperioidea–Papilionoidea). Association of Tropical Lepidoptera, Gainesville
Lamas G, Mielke OHH, Robbins RK (1993) The Ahrenholz technique for attracting tropical skippers (Hesperiidae). J Lepid Soc 47:80–82
Larsen TB (2005) Butterflies of West Africa. Apollo Books, Stenstrup
Lewis HL (1973) Butterflies of the world. Follett, Chicago
Li W, Cong Q, Shen J et al (2019) Genomes of skipper butterflies reveal extensive convergence of wing patterns. PNAS 116:6232–6237. https://doi.org/10.1073/pnas.1821304116
Lourenço GM, Soares GR, Santos TP et al (2019) Equal but different: natural ecotones are dissimilar to anthropogenic edges. PLoS One 14:1–18. https://doi.org/10.1371/journal.pone.0213008
Magurran AE (2004) Measuring biological diversity. Blackwell, Oxford
Malaise R (1937) A new insect trap. Entomol Tidskr 58:148–160

Mallet J (2004) Poulton, Wallace, and Jordan: how discoveries in Papilio butterflies led to a new species concept 100 years ago. Syst Biodivers 1:441–452. https://doi.org/10.1017/S1477200003001300

Marchiori MO, Romanowski HP (2006) Species composition and diel variation of a butterfly taxocene (Lepidoptera: Papilionoidea and Hesperioidea) in a Restinga forest at Itapuã State Park, Rio Grande do Sul, Brazil. Rev Bras Zool 23:443–454. https://doi.org/10.1590/S0101-81752006000200019

Matthews RW, Matthews JR (1971) The Malaise trap: its utility and potential for sampling insect populations. Great Lakes Entomol 4:117–122

Merckx T, Huertas B, Basset Y et al (2013) A global perspective on conserving butterflies and moths and their habitats. Key Topics in Conserv Biol 2:237–257. https://doi.org/10.1002/9781118520178.ch14

Mitter C, Davis DR, Cummings MP (2017) Phylogeny and evolution of Lepidoptera. Annu Rev Entomol 62:265–283. https://doi.org/10.1146/annurev-ento-031616-035125

Müller F (1879) Ituna and Thyridia; a remarkable case of mimicry in butterflies. Trans Entomol Soc London 1879:xx–xxix

Owen DF (1971) Species diversity in butterflies in a tropical garden. Biol Conserv 3:191–198. https://doi.org/10.1016/0006-3207(71)90164-9

Owen DF (1975) Estimating the abundance and diversity of butterflies. Biol Conserv 8:173–183. https://doi.org/10.1016/0006-3207(75)90061-0

Palo H Jr (2017) Butterflies of Brazil / Borboletas do Brasil, vol 1,2,3. Vento Verde, São Paulo

Pelham JP (2019) A catalogue of the butterflies of the United States and Canada. http://www.butterfliesofamerica.com/US-Can-Cat.htm. Accessed 10 Jun 2019

Perillo LN, Neves FS, Antonini Y et al (2017) Compositional changes in bee and wasp communities along Neotropical mountain altitudinal gradient. PLoS One 12:1–14. https://doi.org/10.1371/journal.pone.0182054

Pliske TE (1975) Attraction of Lepidoptera to plants containing Pyrrolizidine alkaloids. Environ Entomol 4:455–473. https://doi.org/10.1093/ee/4.3.455

Pollard E (1977) A method for assessing changes in the abundance of butterflies. Biol Conserv 12:115–153. https://doi.org/10.1016/0006-3207(77)90065-9

Pollard E, Yates TJ (1993) Monitoring butterflies for ecology and conservation. Chapman & Hall, London

Pozo C, Luis-Martinez A, Llorente-Bousquets J et al (2008) Seasonality and phenology of the butterflies (Lepidoptera: Papilionoidea and Hesperioidea) of Mexico's Calakmul region. Fla Entomol 91:407–422. https://doi.org/10.1653/0015-4040(2008)91[407:SAPOTB]2.0.CO;2

Rafael JA (2002) A amostragem. Protocolo e técnicas de captura de Diptera. In: Costa C et al (eds) Proyecto de red iberoamericana de biogeografia y entomologia sistemática: PRIBES. Sociedade Entomológica Aragonesa, Zaragosa, pp 301–304

Ribeiro DB, Freitas AVL (2012) The effect of reduced-impact logging on fruit-feeding butterflies in Central Amazon, Brazil. J Insect Conserv 16:733–744. https://doi.org/10.1007/s10841-012-9458-3

Ribeiro DB, Batista R, Prado PI, Brown KS Jr et al (2012) The importance of small scales to the fruit-feeding butterfly assemblages in a fragmented landscape. Biodivers Conserv 21:811–827. https://doi.org/10.1007/s10531-011-0222-x

Rosa AHB, Perillo LN, Neves FS et al (2019) Butterflies collected using malaise traps as useful for ecology and conservation. J Threat Taxa 11:14235–14237. https://doi.org/10.11609/jott.4919.11.9.14235-14237

Sáfián S, Csontos G, Winkler D (2010) Butterfly community recovery in degraded rainforest habitats in the upper Guinean Forest zone (Kakum forest, Ghana). J Insect Conserv 15:351–359. https://doi.org/10.1007/s10841-010-9343-x

Sant'anna CLB, Ribeiro DB, Garcia LC et al (2014) Fruit-feeding butterfly communities are influenced by restoration age in tropical forests. Restor Ecol 22:480–485. https://doi.org/10.1111/rec.12091

Santos JP, Iserhard CA, Teixeira MO et al (2011) Fruit-feeding butterflies guide of subtropical Atlantic Forest and Araucaria moist Forest in state of Rio Grande do Sul, Brazil. Biota Neotrop 11:253–274. https://doi.org/10.1590/S1676-06032011000300022

Schmidt O, Schmidt S, Häuser C et al (2019) Using Malaise traps for collecting Lepidoptera (Insecta), with notes on the preparation of Macrolepidoptera from ethanol. Biodivers Data J 7:e32192. https://doi.org/10.3897/BDJ.7.e32192

Seitz A (ed) (1907–1924) Die Gross-schmetterlinge der Erde: Amerikanische Tagfälter, vol 5. Alfred Kernen, Stuttgart

Shahabuddin G, Terborgh JW (1999) Frugivorous butterflies in Venezuelan forest fragments: abundance, diversity and the effects of isolation. J Trop Ecol 15:703–722. https://doi.org/10.1017/S0266467499001121

Shields O (1989) World numbers of butterflies. J Lepid Soc 43:178–183

Shuey JA (1997) Bait trap for quantitative sampling of butterflies. Trop Lepid Res 8:1–4

Slansky F Jr (1973) Latitudinal gradients in species diversity of the New World swallowtail butterflies. J Res Lepid 11:201–217

Souza MM, Perillo LN, Barbosa BC et al (2015) Use of flight interception traps of Malaise type and attractive traps for social wasps record (Vespidae: Polistinae). Sociobiology 62:450–456. https://doi.org/10.13102/sociobiology.v62i3.708

Thomas JA (2016) Butterfly communities under threat. Science 353:216–218. https://doi.org/10.1126/science.aaf8838

Tolman T, Lewington R (2008) Collins butterfly guide: the most complete field guide to the butterflies of Britain and Europe. Harper Collins, London

Townes HK (1972) A light weight Malaise trap. Entomol News 83:239–247

Tyler HA, Brown KS Jr, Wilson KH (1994) Swallowtail butterflies of the Americas: a study in biological dynamics, ecological diversity, biosystematics, and conservation. Scientific Publishers, Gainesville

Uehara-Prado M, Freitas AVL, Francini RB et al (2004) Guia das borboletas frugívoras da Reserva Estadual do Morro Grande e região de Caucaia do Alto, Cotia (São Paulo). Biota Neotrop 4:1–25. https://doi.org/10.1590/S1676-06032004000100007

Uehara-Prado M, Brown KS Jr, Freitas AVL (2005) Biological traits of frugivorous butterflies in a fragmented and a continuous landscape in the south Brazilian Atlantic Forest. J Lepid Soc 59:96–106

Uehara-Prado M, Brown KS Jr, Freitas AVL (2007) Species richness, composition and abundance of fruit-feeding butterflies in the Brazilian Atlantic Forest: comparison between a fragmented and a continuous landscape. Glob Ecol Biogeogr 16:43–54. https://doi.org/10.1111/j.1466-8238.2006.00267.x

Van Achterberg C, Grootaert P, Shaw MR (2010) Flight interception traps for arthropods. In: Eymann J (ed) Manual on field recording techniques and protocols for all taxa biodiversity inventories and monitoring. Belgian National Focal Point to the Global Taxonomy Initiative, Belgium, pp 423–462

Van Nieukerken EJ, Kaila L, Kitching IJ et al (2011) Order Lepidoptera Linnaeus, 1758. Zootaxa 3148:212–221. https://doi.org/10.11646/zootaxa.3148.1.41

Van Swaay C, Regan E, Ling M et al (2015) Guidelines for Standardised Global Butterfly Monitoring. Group on Earth Observations, Biodiversity Observation Network, Leipzig, Germany. GEO BON Technical Series 1. https://doi.org/10.13140/RG.2.1.1610.8885

Vane-Wright RI, Ackery PR (eds) (1984) The biology of butterflies. Academic Press, London

Wallace AR (1865) On the phenomena of variation and geographical distribution as illustrated by the Papilionidae of the Malayan region. Trans Linn Soc Lond 25:1–71. https://doi.org/10.1111/j.1096-3642.1865.tb00178.x

Walpole MJ, Sheldon IR (1999) Sampling butterflies in tropical rainforest: an evaluation of a transect walk method. Biol Conserv 87:85–91. https://doi.org/10.1016/S0006-3207(98)00037-8

Warren AD, Davis KJ, Stangeland EM et al (2013) Illustrated lists of American Butterflies. http://www.butterfliesofamerica.com. Accessed 21 Apr 2019

Wikström L, Milberg P, Bergman C (2008) Monitoring of butterflies in semi-natural grasslands: diurnal variation and weather effects. J Insect Conserv 13:203–211. https://doi.org/10.1007/s10841-008-9144-7
Willmott KR, Freitas AVL (2006) Higher-level phylogeny of the Ithomiinae (Lepidoptera: Nymphalidae): classification, patterns of larval hostplant colonization and diversification. Cladistics 22:297–368. https://doi.org/10.1111/j.1096-0031.2006.00108.x
Zikán JF (1929) Myrmekophilie bei Hesperiden? Ent Rund 46:27–28

Chapter 6
Sampling Methods for Beetles (Coleoptera)

Luciana Iannuzzi, Carolina Nunes Liberal, Thamyrys Bezerra de Souza, Thais Giovannini Pellegrini, Janaina Camara Siqueira da Cunha, Ricardo Koroiva, Larissa Simões Corrêa de Albuquerque, Fábio Correia Costa, Renato Portela Salomão, Artur Campos Dália Maia, and Fernando Willyan Trevisan Leivas

6.1 Introduction

Coleoptera is a monophyletic taxon (Beutel et al. 2019) with more than 380,000 living species distributed worldwide (Nielsen and Mound 1999; Ślipiński et al. 2011; Lawrence et al. 2013). In addition, according to estimates of species richness, there are ca. 1.5 million of beetle species in the world (Stork et al. 2015). They are the most diverse and speciose animal order, comprising more than half of all known living metazoans (Mora et al. 2011; Stork 2018) (Figs. 6.1 and 6.2). Hypotheses regarding the diversification of beetles are related to a shared set of unique ecological, behavioral, and morphological traits (Crowson 1981). From a morphological

L. Iannuzzi (✉) · L. S. C. de Albuquerque · F. C. Costa
Departamento de Zoologia, Universidade Federal de Pernambuco, Cidade Universitária, Recife, PE, Brazil
e-mail: luciana.iannuzzi@ufpe.br

C. N. Liberal · R. Koroiva · A. C. D. Maia
Departamento de Sistemática e Ecologia, Universidade Federal da Paraíba, João Pessoa, PB, Brazil

T. B. de Souza
Instituto Nacional da Mata Atlântica, Santa Teresa, ES, Brazil

T. G. Pellegrini
Departamento de Entomologia, Universidade Federal de Lavras, Campus Universitário da Universidade Federal de Lavras, Lavras, MG, Brazil

J. C. S. da Cunha
Department of Entomology, Texas A&M University, College Station, TX, USA

R. P. Salomão
Instituto Nacional de Pesquisas da Amazônia, Manaus, AM, Brazil

F. W. T. Leivas
Departamento de Biodiversidade, Universidade Federal do Paraná, Setor Palotina, Jardim Dallas, Palotina, PR, Brazil

J. C. Santos, G. W. Fernandes (eds.), *Measuring Arthropod Biodiversity*,
https://doi.org/10.1007/978-3-030-53226-0_6

Fig. 6.1 Coleoptera diversity. (**a**) Melolonthidae, (**b**) Meloidae, (**c**, **d**) Curculionidae, (**e**–**h**) Chrysomelidae, (**i**, **j**) Carabidae. Photographs (**a**–**i**) are from Jean C. Santos. Photograph (**j**) is from Filipe Aléssio

Fig. 6.2 Coleoptera diversity. (**a**) Melolonthidae, (**b**) Tenebrionidae, (**c**) Geotrupidae, (**d**) Curculionidae, (**e**) Meloidae, (**f**) Buprestidae, (**g**) Chrysomelidae, (**h**) Carabidae, (**i**, **j**) Cerambycidae. Photographs (**b**, **c**) and (**j**) are from Filipe Aléssio. Photographs (d) and (**g**) are from Jean C. Santos. Photograph (**h**) is from Rodrigo L. Ferreira

approach, the elytra (sclerotized forewings) protect their membranous wings (hindwings) and body against habitat adversities and predation, also aiding in the control of water loss (Hammond 1979). As for biological and ecological aspects, the occupation of different niches by adults and larvae resulted in a significant reduction of competition for food. Furthermore, coadaptations with different plant and animal groups are recognized as an important element of the radiation of Coleoptera.

In no other group of metazoans is the diversification driven by the association with angiosperms such an evident trend. Although recent interpretation of the fossil record and a time-calibrated phylogeny show that true beetles might have been around for more than 295 Ma, it was not until the Early Cretaceous period some 170 to 210 million years later that the species richness of these insects skyrocketed (Zhang et al. 2018). The radiation of beetles coincides with the rise to dominance of flowering plants in terrestrial ecosystems (Wang et al. 2017), which is assumed to have provided multiple new ecological niches through both antagonistic and mutualistic interactions. Together, angiosperms and holometabolous insects, mainly beetles, have led the Cretaceous Terrestrial Revolution and reshaped the evolutionary history of life on Earth (Benton 2010).

Beetles have a terrestrial origin and extant species vastly occupy land habitats. To a lesser extent, they have also transitioned to freshwater habits (Dijkstra et al. 2014), and even entire lineages of some taxa live at least one stage of their ontogenetic development in water (Hunt et al. 2007). There have actually been multiple shifts from terrestrial to aquatic habits throughout the evolutionary history of the Coleoptera (Hunt et al. 2007). Such appears to be the case of Adephaga, whose phylogenies point toward two origins of aquatic habit (Maddison et al. 2009). Among the water scavenger beetles (Hydrophiloidea), multigene analyses highlight an aquatic ancestry, with several invasions of terrestrial habitats (Bernhard et al. 2006). They are commonly found in streams and rivers, where the semiaquatic species live in the sandy soil interstice. The majority of beetle species, though, inhabit terrestrial environments, living on soil and plants, under the bark of trees, and in caves. Others are inquilines in colonies of social insects (termites and ants) or vertebrate nests. Some are specialized ectoparasites of mammals, as the example of the mammal-nest beetles (Platypsyllinae, Leiodidae), which are ectoparasites of rodents (Lawrence and Newton 1982, Lawrence and Newton 1995; Newton 1998; Peck 2007).

Beetles have diversified feeding habits, but phytophagous species are by far the best represented (Bouchard et al. 2017). Most phytophagous species feed on leaves, wood, or roots, although there are some specialized on pollen (polliniferous) and seeds (spermophagous) (Blanche and Cunningham 2005; Jonason et al. 2013; Bouchard et al. 2017). All representatives of the Curculionidae, Chrysomelidae, and Cerambycidae, three of the five most speciose families of Coleoptera, are phytophagous, and more than 99% of them feed on angiosperms. The high specialization includes species that are endophytic, such as leaf miners or gall inductors (Curculionidae and Mordellidae) (Tooker and Hanks 2004). Predator beetles are arguably the least specialized of the trophic guilds, preying on invertebrates and small vertebrates. Dung beetles, for example, prey Attini ants (Silveira et al. 2006;

Aquino et al. 2018) and millipedes (Larsen et al. 2009; Silva et al. 2018); ladybird beetles (Coccinellidae) are primarily carnivorous, feeding on small arthropods (Hodek et al. 2012); many species of ground beetles (Carabidae) are classified as generalist soil-dwelling predators (Lövei and Sunderland 1996). Detritivorous (or saprophagous) species feed on decaying organic matter, such as dead animals (necrophages), dung (coprophages), and dead plants (plant saprophages) (Marinoni 2001). Among the dung beetles (Scarabaeinae, Scarabaeidae), most species are predominantly coprophagous, although some feed on carrion and fruits or might exhibit the particular strategy of predating or scavenging other arthropods (Hanski and Cambefort 1991; Halffter and Halffter 2009; Scholtz et al. 2009).

Through the activity of organisms, ecological communities provide a large variety of ecosystem services, paramount for the sustenance of humankind and life as whole on our planet. Beetles are key contributors in three such services: pest control, pollination, and improvement of agricultural land through soil aeration and nutrient cycling (Losey and Vaughan 2006; Nichols et al. 2008; Noriega et al. 2018). Pest control is provided mostly by predator beetles that feed on other invertebrates that cause economic losses in agriculture (Scheller 1984; Eitzinger and Traugott 2011; Fonseca et al. 2017). Carabidae, Cantharidae, and Coccinellidae are highlighted as the most important families involved in pest control. For instance, the larvae of Carabidae and Cantharidae predate collembolans from croplands (Eitzinger and Traugott 2011), whereas adult Carabidae and Coccinellidae predate aphids (Scheller 1984; Fonseca et al. 2017). Seed-feeding by some species of Carabidae also reduces the growth of weeds in arable fields (Jonason et al. 2013). Specialized pollen-feeding behavior documented across most Polyphaga superfamilies is also associated with pollination services, and some fruit crops almost exclusively rely on beetle pollinators. This is the case of the atemoya, custard apple, and soursop (*Annona* spp.; Blanche and Cunningham 2005; Costa et al. 2017), as well as the African oil palm (*Elaeis guineensis* Jacq., Tandon et al. 2001; Auffray et al. 2017). Dung beetles (Scarabaeidae, Scarabaeinae) are one of the best-studied groups regarding ecosystem services. Due to their dung burial activities, these insects are responsible for the improvement of soil quality, secondary seed dispersal, and soil aeration (Nichols et al. 2008). Furthermore, because of the competition for food, dung beetles control populations of cattle parasites, such as those of the horned fly (*Haematobia irritans* L.) (Nichols et al. 2008).

It is also important to consider economic losses resulting from beetle activity. There are species that stand out as plant pests and are particularly destructive to agricultural crops, gardens, and natural ecosystems. Some leaf beetles [Chrysomelidae; e.g., *Psylliodes chrysocephala* (L.)] and weevils [Curculionidae; e.g., *Ceutorhynchus obstrictus* (Marsham)] are considered cabbage pests of crops in Europe and the USA (Alford 2003). In addition, there are many species of sap beetles (Nitidulidae; e.g., *Meligethes aeneus* Fab.) that cause severe yield losses in European oilseed crops due to herbivory (Alford 2003). Bark beetles (Curculionidae, Scolytinae) are responsible for high levels of tree mortality in pine forests (Schowalter 2012). Spermophagous beetles are common in granaries and several species are pests (Stejskal et al. 2015). Stored-grain pests are adapted for living under severe water stress and can be particularly destructive. The main grain-feeding beetle pests are recorded among Tenebrionidae (e.g., mealworms, *Tenebrio molitor* L.; flour beetles, *Tribolium* spp.), Bostrichidae, Cucujidae, Curculionidae, Dermestidae, Ptinidae, and Mycetophagidae. It is also important to mention the beetles of medical importance. Some of them can cause irritation injuries to the skin, eyes, and ears; ca. 40 spp. are involved with human accidents (Bellas 1989). Irritation happens when insects are handled or come into accidental contact with the victim, as their glands secrete vesicant toxins (Krinsky 2019). Meloidae, Carabidae, Cantharidae, and Staphylinidae are the main families involved with human accidents (Gullan and Cranston 2005; Pardal et al. 2016).

Given the large taxonomic diversity and complex life histories of beetles, choosing an appropriate sampling method is often a hard task. When the objectives are broad surveys or detailed faunistic inventories, one should take into consideration biological/ecological characteristics of different guilds, functional groups, among other particularities. Inventories are fundamental to assess biodiversity and are an important tool for the development of conservation policies and monitoring programs (Kim 1993). Taking into consideration the biodiversity of insects, especially Coleoptera, a range of questions are answered from surveys using various methodologies (Kim 1993; Hyvärinen et al. 2006). Taxonomic and ecological surveys are the two main approaches used to characterize biodiversity, environmental heterogeneity, types of habitats, biotic components, and the impacts of agriculture and human disturbance on the environment (Braithwaite 1991; Vane-Wright et al. 1991; Kim 1993; Kotze and O'Hara 2003; Rosário et al. 2014). DNA-based inventories are relatively new tools that give an innovative perspective on taxonomy (Lambert et al. 2005; Zhang et al. 2018). They provide a universal database that has been rapidly growing over the past few years with the development of state-of-the-art technologies, helping to answer taxonomic and systematic questions. However, they also demand specific sampling techniques (Wilson 2000; Gould 2002; Blaxter 2003; Zhang et al. 2018).

Our objective is to provide information regarding the appropriate collection methods for beetles, according to their habits and habitats. In this chapter, information regarding the life histories of beetles will be discussed. Next, the main methods used for specimen collection will be reported. With this information, researchers will be able to adequately select which trapping methods to use for optimal sampling, cutting costs and time in their faunistic surveys.

Thus, for a reliable inventory of beetles, it is necessary to explore a plethora of possibilities. There is not a unique or most efficient technique for sampling the entirety of the diversity of beetles inhabiting any given area. Traditionally, nonetheless, a set of trapping methods is used for obtaining reliable and well-represented inventories (Almeida et al. 1998; Rafael 2002; Lamarre et al. 2012). Or if a specific group of beetles is targeted, the use of selective methods and sampling techniques is recommended for optimal efficiency and cost-effectiveness (Lamarre et al. 2012). In both cases, previous knowledge about the biology and ecology of beetles is not only recommended but required (Rafael 2002; Gibb and Oseto 2020).

6.2 Methods for Collecting Beetles

Insect collection methods may be divided into two categories: active and passive collections (Table 6.1; Figs. 6.3, 6.4, 6.5, 6.6 and 6.7). In the **active collecting**, a collector actively finds and collects the insects and can catch specimens by hand or with the aid of equipment, such as forceps, brush, aspirators, hand net (aerial, sweeping, and aquatic nets), or beating sheets (Gibb and Oseto 2020). The advantages of active collections are a directed effort and the specific habitat exploitation (Almeida et al. 1998). However, a good collection depends on the experience of the collector and has the shortcomings of limited time and poor standardization. Moreover, without adequate care, some beetles can cause injury or discomfort to the collector. In the **passive collecting**, the collection is carried out with the use of traps (with or without attractions) (Almeida et al. 1998; Gibb and Oseto 2020), and the choice of trap type will depend on the focus group beetle habit (Russo et al. 2011). Passive methods are advantageous because they are easily standardizable and replicable. These methods generally do not depend on the great experience of the collector. Yet, most traps do not need to be constantly inspected and can remain active for a long period of time. However, the performance of a trap depends on its construction, location, time of day or year, weather, temperature, and kind of attractant (Gibb and Oseto 2020). In addition, they are sometimes expensive and difficult to transport and explore a limited area. There are countless types of traps used to passively collect the beetles, such as malaise, pitfall, light, bottle, and Surber, among others.

The active and passive methods may be used simultaneously so that the methods complement each other and increase the chances of success of the inventory (Hyvärinen et al. 2006; Russo et al. 2011; Lamarre et al. 2012). In the impossibility of applying several methods, the most appropriate one must be chosen according to the beetle's habit. For this, the advantages and shortcomings of the sampling methods need to be known in detail (Hyvärinen et al. 2006).

6.2.1 Terrestrial Beetles

6.2.1.1 Flying Beetles

The evolution of flight enabled the extreme diversification of insects (Ikeda et al. 2012). Flight increases the possibilities of niche exploration, to find food and protection against predators (Grimaldi and Engel 2005), escape from unfavorable environmental conditions, and facilitate outbreeding (Crowson 1981). Long distance dispersal propensity is common among flying beetle taxa (Bouchard et al. 2017). The hindwings (metathoracic) of beetles are used exclusively for flight, whereas the forewings (prothoracic) are never involved. The hindwings can be functional in all species of one family of Coleoptera. In others, the expression of this character can

Table 6.1 Sampling methods for beetles according to their habitats (terrestrial, aquatic and semiaquatic, and subterranean)

Taxa			Terrestrial									
Suborder	Superfamily	Family	Flying	Edaphic	Nests	Fungi and myxomycetes	Ectoparasites	Endogean	Forensic	Immature	Aquatic and semiaquatic	Subterranean
Myxophaga	Sphaeriusoidea	Torridincolidae									15, 17	
		Hydroscaphidae									15, 17	8, 11, 17, 19, 27
Adephaga		Carabidae	2, 5, 18	16, 21	7, 14, 15, 20, 24, 30		15			15, 25		5, 7, 10, 13, 15, 21, 22, 23, 26, 28, 29, 30
		Trachypachidae		21						15, 25		
		Gyrinidae									11, 15	11, 15
		Haliplidae								4, 8, 11	8, 11	3, 8, 11, 27
		Meruidae									15	
Adephaga		Noteridae									8, 27	3, 8, 11, 15, 27
		Aspidytidae									15	
		Dytiscidae								4, 8, 11	8, 11, 17, 19, 27	3, 8, 11, 15, 27

Taxa			Terrestrial									
Suborder	Superfamily	Family	Flying	Edaphic	Nests	Fungi and myxomycetes	Ectoparasites	Endogean	Forensic	Immature	Aquatic and semiaquatic	Subterranean
Polyphaga	Histeroidea	Histeridae	10, 18, 9		1, 7, 12, 13, 14, 15, 17, 20, 21, 30	15			1, 2, 15			15, 21, 22, 28, 29, 30
	Hydrophiloidea	Hydrophilidae								4, 8, 11	8, 11, 17, 27	3, 8, 11, 15, 26
		Sphaeritidae		7, 12, 21								
	Staphylinoidea	Agyrtidae		15								13, 15, 21, 22, 26, 28
		Leiodidae		21, 30	1, 7, 12, 15, 24, 30	15	15					7, 10, 13, 15, 21, 22, 23, 26, 28, 29, 30
		Ptiliidae	2	7, 12, 21	7, 15, 24, 30	15						7, 15, 21, 22, 28, 30
	Staphylinoidea	Staphylinidae	2, 12, 189	7, 21	7, 12, 14, 15, 20, 21, 24, 30	15			1, 2, 15	15, 25		4, 7, 10, 13, 15, 21, 22, 23, 26, 28, 29, 30
		Silphidae		21					1, 2, 15	15		
	Scarabaeoidea	Cetoniidae			1, 7, 15, 2, 24, 30					15		15, 22, 28, 29, 30
		Geotrupidae	18	21						15, 25		15, 21, 22, 28
		Hybosoridae	17	7, 12, 21	7, 12, 15, 24, 30					15, 25		15, 21, 22, 28, 29, 30
		Lucanidae	17, 18	21						15		
		Melolonthidae	17		20					15, 25		
		Ochodaeidae	2, 17, 18									
		Glaresidae	17									
		Passalidae								15		15, 22, 28

(continued)

Table 6.1 (continued)

Taxa			Terrestrial									
Suborder	Superfamily	Family	Flying	Edaphic	Nests	Fungi and myxomycetes	Ectoparasites	Endogean	Forensic	Immature	Aquatic and semiaquatic	Subterranean
	Scarabaeoidea	Scarabaeidae	10, 17, 18	7, 21	7, 12, 14, 15, 21, 24, 30	15	15		1, 2, 15	15, 25		7, 15, 21, 22, 28
		Trogidae	17	21	15					15		15, 21, 22, 28
	Scirtoidea	Eucinetidae		7						15		
		Clambidae		7, 21								7, 15, 21, 22, 28
	Dascilloidea	Rhipiceridae			15		15			15		
		Dascillidae								15, 25		
	Buprestoidea	Buprestidae	5, 18	21						15		15, 21, 22, 28
	Byrrhoidea	Elmidae								27	9, 17, 27	3, 8, 11, 15, 27
		Byrrhidae		7								10, 13, 15, 22, 23, 26, 28
		Chelonariidae	5, 17									
		Heteroceridae	17									
		Callirhipidae	17									
		Ptilodactylidae	17	7						3		7, 15, 22, 28
	Elateroidea	Cantharidae	2, 17, 18									15, 22, 28
		Cerophytidae	2, 17									
		Elateridae	2, 5, 17, 18	21			15			15		15, 21, 22, 26, 28
		Eucnemidae	2, 18									15, 22, 28
		Lycidae	18									15, 22, 28
		Phengodidae	17, 18									15, 22, 28
		Lampyridae	2, 18									15, 22, 28
		Throscidae	2, 17									15, 22, 28
		Jurasaidae	18	15								

Taxa			Terrestrial									
Suborder	Superfamily	Family	Flying	Edaphic	Nests	Fungi and myxomycetes	Ectoparasites	Endogean	Forensic	Immature	Aquatic and semiaquatic	Subterranean
	Derodontoidea	Jacobsoniidae		7								7, 15, 22, 28
	Bostrichoidea	Bostrichidae	18									15, 22, 28
		Dermestidae		21	15				1, 2, 15	15		15, 21, 22,2 8
		Ptinidae	18	21	7, 15, 24, 30					15		7, 15, 21, 22, 28
	Cleroidea	Cleridae	18	16					1, 2, 15	15		
	Cucujoidea	Cucujidae					15			15		15,22,28
		Cerylonidae			15, 17, 24, 30							7, 15, 22, 28, 30
		Erotylidae	5, 18	21								15, 21, 22, 28
		Coccinellidae	18		15					2, 15, 23		15, 22, 28
		Cryptophagidae	2, 12	21	15					15		15, 22, 26, 28
		Endomychidae	12, 17, 18									15, 22, 28
		Nitidulidae	18, 9	21	1, 7, 15, 24, 30				1, 2, 15	15		7, 15, 22, 28, 30
		Latridiidae			7, 14, 15, 24, 30					15		10, 13, 15, 23, 26
		Phalacridae	18									15, 22, 28
		Sphindidae				15				15		
		Bothrideridae					15			15		9, 15, 26
		Passandridae		21	20		15					15

(continued)

Table 6.1 (continued)

Taxa			Terrestrial									
Suborder	Superfamily	Family	Flying	Edaphic	Nests	Fungi and myxomycetes	Ectoparasites	Endogean	Forensic	Immature	Aquatic and semiaquatic	Subterranean
	Tenebrionoidea	Anthicidae	10									15, 22, 28
		Meloidae	18	21	15		15			15		15, 22, 28
		Ripiphoridae	10, 17, 18				15			15		15, 22, 28
		Tenebrionidae	5, 18	21	7, 14, 15,24, 30					15		6, 15, 21, 22, 28, 30
		Oedemeridae	18	21								15, 21, 22, 28
		Stenotrachelidae	10, 18									
		Synchroidae	17, 18									
		Pythidae	17									
		Mordellidae	5, 9, 18									
	Chrysomeloidea	Cerambycidae	2, 5, 18, 9	21						2, 15, 23		15, 21, 22, 28
		Chrysomelidae	5, 18	21						2, 15, 23		15, 21, 22, 28
	Curculionoidea	Brentidae	5, 18		15					15		
		Curculionidae	5, 17, 18	21						2, 15, 23		15, 21, 22, 28
		Curculionidae-Scolytinae		21		15						

(1) Adhesive, (2) aerial sweep net, (3) aquatic leaf litter packs, (4) aquatic net, (5) baiting, (6) beating sheet, (7) Berlese funnel, (8) bottle trap, (9) Carvalho-47, (10) dig and search, (11) D-frame net, (12) flight intercept, (13) flotation, (14) fogging, (15) hand-collecting, (16) Kempson extractor, (17) light, (18) malaise, (19) mark and recapture, (20) nest excavation, (21) pitfall trap, (22) quadrats sampling, (23) shake-bucket, (24) sifting debris, (25) soil samples, (26) subterranean sampling devices, (27) Surber sampler, (28) timed searches, (29) visual inspection, and (30) Winkler extractor. The numbers in parentheses correspond to each capture method. Family and subfamily classification follows the present in Bouchard et al. (2011), Bouchard and Bousquet (2020), and Cherman and Morón (2014)

Fig. 6.3 Flying beetle sampling methods. (**a**) Beating sheet, (**b**) light sheet, (**c**) malaise trap, (**d**) flight interception trap, (**e**) suspended flight interception trap, (**f**) light trap, (**g**) Carvalho-47 trap, (**h**) pantyhose. Photograph (**a**) is from Sheila P. C. Fernandes. Photographs (**c, d, g**) are from João C. Regueira. Photograph (**f**) is from Guilherme M. Viana

Fig. 6.4 Edaphic and immature beetle sampling methods. (**a**) Winkler extractor, (**b**) Berlese-Tullgren funnel extractor, (**c**, **d**) pitfall trap, (**e**, **f**) hand-collecting. Photograph (**e**) is from Tamara G. Carvalho

vary within the family, as in the case of some Carabidae, and some species have fully developed wings (macropterous species), whereas others are brachypterous (AS 1984). The evolutionary loss of flight is attributed to the adaptive selection toward an increased investment in organs essential for survival and reproduction, in exchange for the energetic cost associated with the maintenance of flight capacity (Roff 1990; Ikeda et al. 2012). Flightlessness in Coleoptera is generally thought to have evolved due to changes in the environment or habitat isolation (Ikeda et al.

Fig. 6.5 Beetles from animal nest sampling methods. (**a**) Hand-collecting by general visual inspection, (**b**) hand-collecting by emigration trail visual inspection, (**c**) flotation, (**d**) termites' nest excavation

2008). Approximately 10% of extant beetle species are wingless (Roff 1990), although the overall proportion of flightless taxa should be higher. Flightless beetles appear to be more vulnerable to extinction (Bouchard et al. 2017). However, the loss of the ability to fly may have contributed to beetle diversification, as it potentially decreases the rate of gene flow and leads to differentiation among populations, repeatedly contributing to their speciation over time (Ikeda et al. 2012).

Methods for Collecting Flying Beetles

Active Collecting

Aerial Sweep Net Collections can be targeted by swinging the net rapidly and collecting whatever is at sight or scanning by collecting at random. The aerial net consists of a long cable, with fixed or adjustable length, attached to a wire hoop (rim) lined with a net bag (translucent and fine mesh, green or white, 30 to 38 cm in diameter) (Covell 2009; Almeida et al. 2012; Gibb and Oseto 2020). The collector

Fig. 6.6 Aquatic and semiaquatic beetle sampling methods. (**a**) D-frame net, (**b**) Hand-collecting, (**c**) light trap (Pennsylvania), (**d**) Surber sampler. Photographs (**a**) and (**c**) are from Lucas R. C. Lima

should sweep the net rapidly to capture the specimen, forcing it into the very bottom of the net bag (Covell 2009).

Results are very dependent on the skill and experience of the collector. The aerial sweep net is used primarily for the collection of large-bodied flying beetles (Gibb and Oseto 2020), such as many Throscidae, Lampyridae, and Elateridae (Viviani et al. 2010), Cantharidae (Hernández and Caballero 2016), Ochodaeidae (Arnett et al. 2002), Carabidae (Cincidelinae) (White 1983), Cerambycidae (Monné et al. 2009), Cerophytidae, and Eucnemidae. This method is also used for collection of grab insects in flight or perched on vegetation (Rafael 2002; Gibb and Oseto 2020). For the collection of microcoleopterans (Staphylinidae, Ptiliidae, and

Fig. 6.7 Subterranean beetle sampling methods. (**a**) Active search or hand searching, (**b**) use of brush for active search, (**c**) use of a sucker for active search, (**d**) and (**e**) quadrat sampling. Photographs (**a**–**c**) and (**e**) are from Marconi Souza-Silva. Photograph (**d**) is from Rodrigo L. Ferreira

Cryptophagidae), the aerial sweep net might be attached to a moving car (Peck and Cook 1992; Gibb and Oseto 2020).

Beating Sheet (Fig. 6.3a) This method is used for collecting beetles on the leaves, trunks, branches, and flowers of different plants, mainly shrubs and small trees, but also occasionally large trees (Aguilar Julio 2010). In addition to inventories, this method is also used to harness data about insect-plant interactions and the life history of phytophagous beetle taxa.

It is constructed from a square rigid screen, preferably white, stretched between two rigid structures (wood, PVC pipe) disposed in "X" formation. The screen is held under the plant, which is then shaken. Insects over the plant fall on the screen and are handpicked or collected with the aid of forceps or aspirators (smaller species).

Light Sheet (Black Light) (Fig. 6.3b) The simplest form of black light is composed of a light bulb (black light and/or mixed mercury vapor) powered by an AC or DC energy source (Covell 2009). A white sheet is stretched vertically next to the light source. The sheet serves as a landing support for the phototactic beetles. The attracted specimens are handpicked or removed from the sheet with the aid of a forceps or aspirator (Covell 2009). Despite keeping the collected specimens in excellent condition, it requires the collector to remain close to the trap for the full duration of the collection event.

Passive Collecting

Malaise Trap (Fig. 6.3c) The malaise trap model was proposed in 1937 (Malaise 1937; Townes 1962) and since then has been extensively used with adaptations related to size (Dugdale and Hutcheson 1997), collecting jar position (Achterberg 2009) and color (Rafael 2002; Sheikh et al. 2016). It is a passive trap based on flight interception, commonly used as the standard method for sampling flying insects (Marinoni and Dutra 1997; Sheikh et al. 2016). It consists of an open tentlike structure made of thin and light fabric, preferably dark in color, with a central barrier to intercept the flying insects. Usually the cover of the tent is light-colored so that intercepted insects are directed upward and then collected as they try to escape (Rafael 2002; Covell 2009; Sheikh et al. 2016). At the top of the trap, there may be a collecting bottle filled with a killing/preserving liquid (usually ≥70% ethanol) or adhesive sheets impregnated with pyrethroids for dry collection (Cresswell 1995; Covell 2009; Almeida et al. 2012; Sheikh et al. 2016).

The efficiency of the malaise trap largely depends on its correct setup and placement (Darling and Packer 1988; Rafael 2002). It is recommended that the trap is mounted over clearings and natural trails and that the collecting bottle is directed northward. If the trap is set in an open area, then the collecting bottle faces the most luminous point (Rafael 2002).

The duration of the collection with a malaise trap depends on the purpose of the research. For broad inventories, the trap should be activated continuously, preferably for a long period of time (1 year or more). However, the collecting bottle should be inspected weekly or fortnightly (Rafael 2002; Sheikh et al. 2016). Under higher temperatures and aridity conditions, frequent inspection of collecting bottles and the use of less volatile preserving liquids are recommended (Iannuzzi et al. 2003).

Malaise traps will sample a vast assortment of actively flying beetle taxa, including Cerambycidae, Carabidae, Curculionidae, Buprestidae (Iannuzzi et al. 2003; Skvarla and Dowling 2017), Cantharidae (Hernández and Caballero 2016), Chrysomelidae (Linzmeier and Ribeiro-Costa 2013), Mordellidae (Jackman and Nelson 1995), Ripiphoridae, Synchroidae, Oedemeridae, Stenotrachelidae, Ochodaeidae, and Chelonariidae (Arnett et al. 2002).

Flight Interception Trap (FIT) (Fig. 6.3d, e) The FIT was proposed by Chapman and Kinghorn (1955) and adapted by Peck and Davies (1980). It is designed for the capture of active flying insects that fall when encountering an obstacle. This method is very efficient for the capture of beetles (Juillet 1963; Bouget et al. 2008; Lamarre et al. 2012), in particular Histeridae and Scarabaeidae (Andrade et al. 2011; Milhomem et al. 2003), and it is also effective in capturing Stenotrachelidae, Ripiphoridae (Arnett et al. 2002), Cryptophagidae, Endomychidae, Mordellidae, Sphindidae, and Staphylinidae (Klimaszewski and Brunke 2018). A large number of small cryptic flying taxa can be sampled with the FIT (Juillet 1963; Bouget et al. 2008).

The FIT consists of a translucent or camouflaged vertical barrier (glass, acetate sheet, plexiglass, or fine fabric) held upright by stakes in the ground (Fig. 6.3d) or suspended (Fig. 6.3e) (Lamarre et al. 2012). At the bottom of the barrier, a trough or an array of trays, which acts as a collecting apparatus with killing/preserving liquid, is placed so that the beetles that fly toward the panel fall and drown (Bouget et al. 2008; Covell 2009; Almeida et al. 2012). It is indicated that a protection against rain and flooding of the collecting apparatus is coupled to the FIT setup.

There are some models adapted from the origin FIT, encompassing anywhere from single-plane setups (Bouget 2005) to multidirectional "cross-vane" traps with interlocking panels (Hines and Heikkenen 1977). Adaptations can also be made according to the purpose of the study, such as sticky panel, multiple funnel, use of baits, or other innovative combination (Peck and Davies 1980; Bouget et al. 2008; Legros and Beresford 2010; Stone et al. 2010; Maia et al. 2012, 2013). The FIT is particularly useful for canopy sampling (Lamarre et al. 2012).

In addition to the high efficiency in the collection of flying beetles, the FIT is advantageous because it is easily constructed and installed and can be standardized and replicable, and its use is not labor-intensive (Lamarre et al. 2012).

Light Trap (Fig. 6.3f) Light traps can be applied to collect dim-light flying beetles that are phototactic. They are widely used and suitable both for ecological studies (Miyazaki and Dutra 1995; Oliveira et al. 2001; Alcázar-Ruiz et al. 2003) and pest monitoring (Nakano and Leite 2000; Costa et al. 2009). There are several adaptations of light traps, designed to best suit the purpose of the collection and improve efficiency.

In general terms, a light trap is constructed using a funnel with a central luminous source above it and vanes or baffles to intercept approaching flying insects, which fall through the funnel into a collecting jar that usually contains a killing/preserving liquid (Covell 2009; Gibb and Oseto 2020). Funnels are built out of metal, plastic, or heavy paper. Traps are often covered for protection against rainfall (Gibb and Oseto 2020).

The efficacy of a light trap setup depends on a number of factors that include environmental temperature, rainfall and luminosity, the height stratum, moon phase, position and time of installation of the setup, and the light source of choice (Matioli 1986). The main sources of light are ultraviolet light (UV) and mercury vapor light (250 or 500 W) or both combined. The optimal conditions are just before or right after rainfall, windless conditions, and during the waning and new moon phases. The main negative condition is assumed to be a rainy evening (Aguilar Julio 2010).

Families best represented in light trap collections are Scarabaeidae (Nath and Singh 1994; Flechtmann et al. 1995; Kato et al. 2000; Ronqui and Lopes 2006) and Melolonthidae (Dynastinae, Melolonthinae, and Rutelinae) (García-López et al. 2011). Less frequently, Cantharidae (Hernández and Caballero 2016), Lucanidae, Trogidae, Glaresidae, Hybosoridae (Almeida et al. 2012), Curculionidae, Heteroceridae, Chelonariidae, Callirhipidae, Cerophytidae, Throscidae, Elateridae, Phengodidae (males), Ripiphoridae, Ochodaeidae, Chelonariidae, Ptilodactylidae, Throscidae, Synchroidae, Endomychidae, Pythidae, Anthicidae, and Cerophytidae are also recorded (Arnett et al. 2002).

Carvalho-47 Trap (Fig. 6.3g) This trap was proposed by Carvalho (1998) and adapted by Carvalho and Trevisan (2015) to collect wood-damaging beetles (Scolytinae and Platypodinae—Curculionidae, Bostrichidae, and Cerambycidae), but it can also attract Nitidulidae, some Histeridae, and Staphylinidae (Aguilar Julio 2010). For the preparation of the trap, an aperture (or window) is cut in the middle section of a standard 2 L PET bottle to which the fermented fruit soup is added. As bait, a soup of fermented fruit must be prepared. This can be attained by adding sugarcane juice and/or beer to mashed overripe bananas, ca. 72 h prior to insect collection. Beetles that are attracted by the fermented fruit bait will enter the trap through the aperture and then either dip into the soup or become stuck inside the bottle. The trap can be suspended at different heights, using the accessory vegetation and providing a sampling of the differences in the vertical stratification of the fauna (Leivas et al. 2013). It must be checked at least once a day because the decomposition process of captured specimens is accelerated due to the fermented bait. The beetles can be removed with the aid of forceps or by sifting the fermented fruit soup. The Carvalho-47 trap is advantageous because it is easily installed, can be standardized and replicable, is simple to construct, and is not labor-intensive.

One variation of the standard model of this trap entails the use of a thin nylon fabric (like a pantyhose) instead of a bottle (Fig. 6.3h). In this variant, the fermented fruit soup is placed inside the pantyhose, which is then closed at both ends by a knot. The pantyhose containing the bait is then set on a tree trunk or branch. The beetles that are attracted by the odor will land on the fabric, mainly below it, and can thus be easily handpicked (Leivas et al. 2013).

6.2.1.2 Edaphic Beetles

The edaphic fauna is characterized by being directly associated with the soil, where they live their entire life cycle or at least one stage of it (Assad 1997; Aquino 2005). They are categorized into micro-, meso-, or macrofauna with beetles composting the last two divisions (Aquino 2005; Brown et al. 2015). Beetles are commonly collected in soil fauna surveys, and captured specimens belong to species from different trophic categories, such as decomposers, geophages, predators, phytophagous, and scavengers (Nitzu et al. 2008; Brown et al. 2015). Depending on the studied trophic group, different sampling methodologies can be applied, such as active collecting under rocks with tweezers or exhausters (Nitzu 1997), Berlese funnels (Bremner 1990; Sakchoowong et al. 2007; Sabu et al. 2011; Barberena-Arias et al. 2012; Owens and Carlton 2015), Kempson extractor (Adis 1987; Rodrigues 1992), Pitfall traps (Sabu et al. 2011; Hernández et al. 2014; Rosário et al. 2014; Azevedo et al. 2016), and Winkler extractor (Krell et al. 2005; Sakchoowong et al. 2007; Sabu et al. 2011; Owens and Carlton 2015). Edaphic fauna is an appropriate indicator group, and the use of these methods allows the development of a range of studies, such as environmental and soil quality (Brown et al. 2015).

Methods for Collecting Edaphic Beetles

Passive Collecting

Winkler Extractor (Fig. 6.4a) This method, also called Winkler funnel, Winkler/Moczarski eclector, or Winklers, was developed around the same time of the Berlese funnel and has the advantages of not requiring an electrical energy source and being more portable (Holdhaus 1910; Owens and Carlton 2015). It consists of a set of two wooden or metal frames (usually 30 × 50 cm and 33 × 27 cm), in which the upper and lower frames are 42 cm apart and covered more externally with a sturdy white cotton. Inside it has thin mesh bags from 1.5 to 4 mm diameter, depending on the research goal (Holdhaus 1910; Besuchet et al. 1987). Fischer (1996) developed the mini-Winkler extractor, which instead of storing four 2 L samples, as in the standard extractor, holds only one 2 L sample. That structure works as a containment system, and it is efficient for collecting litter organisms (Holdhaus 1910; Almeida et al. 2012). As the entrapped organisms move along the substrate, the bag walls work as a large surface area, used as they leave the substrate, and eventually fall into the collecting jar or plastic bags, in the case of the mini-Winkler, which is filled with a killing solution and is attached to the lower part of the structure (Holdhaus 1910; Fischer 1996; Owens and Carlton 2015). The collecting jar can also be filled with a wet cloth to collect living specimens (Besuchet et al. 1987).

Berlese Funnel (Fig. 6.4b) The Berlese funnel, also known as Berlese-Tullgren funnel, consists of an extractor funnel with a light and heat source placed on top (e.g., a 25 or 40 W incandescent light bulb) to increase the temperature and produce a gradient of soil moisture (Haarløv 1947; Block 1966; Almeida et al. 2012). As a result, the soil-dwelling organisms move away from the light and the heat, migrate to the bottom of the funnel, and fall inside a collecting container filled with a killing/preserving liquid (Haarløv 1947; Block 1966; Barberena-Arias et al. 2012). In this method, the soil or leaf litter is collected and placed inside the extractor funnel that has a mesh in the bottom, which keeps the substrate from falling inside the collecting container (Almeida et al. 2012). The period that the sample remains inside the extractor tunnel varies, although the standard is 15 days (Freire et al. 2015). Some studies applied an extraction time of 48 h (Ferreira and Marques 1998; Ogedegbe and Egwuonwu 2014), 4 days (Souto et al. 2008), 7 days (Barros et al. 2010; São José et al. 2013), and 6 to 9 days (Owens and Carlton 2015). The applicability of this method can be limited because of the need of an electrical energy source. Also, due to the heavy structure of the Berlese funnel, analysis in the field can be challenging (Owens and Carlton 2015).

Kempson Extractor The Kempson extractor is a container with an incandescent 100 W white light bulb and an incandescent 250 W infrared light bulb, which firstly lights up briefly and then progressively increases the time of exposure (Southwood 1978; Nielsen 2003). At the bottom, there are two containers covered with a black nylon mesh in a cold water bath (Southwood 1978; Adis 1987). This technique provides a moisture gradient that pushes soil-dwelling organisms toward the container filled with killing/preserving liquid, usually one part picric acid (preservative) and three parts water and detergent. It is separated from the substrate by a mesh placed on a grid (Southwood 1978; Adis 1987). The time of exposure and extraction is, on average, 8 days, and the Kempson extractor is allegedly more efficient than the Berlese funnel because the heating of the substrate is more gradual (Adis 1987). Thus, it is possible to collect most of the soil-dwelling organisms (Adis 1987; Almeida et al. 2012). Although Southwood (1978) states that this method does not work well for holometabolous insect larvae, Penney (1966) successfully captured Carabidae larvae using it.

Pitfall Trap (Figs. 6.4c, d) It consists of a container (height 12 cm and diameter 15 cm) filled with a 2% solution of water and detergent. It is buried in the soil so that the top is leveled with the ground line and protected by a lid to avoid overflowing from rainwater and litterfall (Newton and Peck 1975; Steyskal et al. 1986; Hernández et al. 2014). Some studies also used 70% ethanol or ethylene glycol inside the container as a killing/preserving liquid (Steyskal et al. 1986; Salomão et al. 2019). Pitfall trap can be used with or without bait, depending on the targeted taxa for sampling (Steyskal et al. 1986; Almeida et al. 2012). It may remain in the field for several days (ideally 5 days) but should be inspected periodically for optimal preservation of captured specimens and bait renewal (in case it is used) (Newton and Peck 1975; Steyskal et al. 1986). This method can work efficiently for the capture of both immature (larvae) and adult beetles (Almeida et al. 2012). Arboreal pitfall is a variation used to capture dung beetles, which forage in canopy attracted by feces from arboreal vertebrates (Davis et al. 1997).

6.2.1.3 Beetles Associated with Animal Nests

Beetles have associations with many terrestrial invertebrates and vertebrates, especially inside their nests and burrows. These nests and burrows provide a very interesting microhabitat, giving opportunity for the occupation of adults and larvae beetles from different trophic groups and interactions, such as predators, saprophagous, or symbionts (Mazur 2005; Kovarik and Caterino 2016; Parker 2016).

Some beetles can be occasional visitors to nest or burrows of vertebrates, while other species are obligate occupants. Usually, the beetles associated with burrows or nests of vertebrates exhibit specialized morphological features, such as elongate legs and antennae (Mazur 2005; Kovarik and Caterino 2016). We can find beetles as occupants of reptile species (Testudinidae) burrows, rodent mammals (Cricetidae, Ctenomyidae, Dipodidae, Heteromyidae, Muridae, Sciuridae, Spalacidae, and Geomyidae), carnivorous mammals (Mustelidae), insectivorous mammals (Talpidae), and lagomorphs (Leporidae and Ochotonidae). Associated with these burrows are the beetles of the families Histeridae (Histerinae, Onthophilinae, and Saprininae), Leiodidae (Coloninae), Ptinidae, Scarabaeidae (Aphodiinae and Scarabaeinae), Staphylinidae (Scydmaeninae), and Tenebrionidae (Kistner 1982; Mazur 2005; Aguilar Julio 2010; Connior 2011; Kovarik and Caterino 2016).

Other beetles can be found in abandoned or live nests of several bird families, such as Ciconiidae, Falconidae, Meropidae, Paridae, Psittacidae, Spheniscidae, Sturnidae, and Tytonidae. Among the beetles associated with these bird nests are Dermestidae, Trogidae, and Histeridae (Histerinae and Saprininae) (Aguilar Julio 2010; Kovarik and Caterino 2016; Nacagava 2019).

Some of the most impressive beetles are those inhabiting ant nests. The biology of these beetles ranges from obligate relationships (like brood predation, kleptoparasitism, phoresis, and symbioses) to facultative associations with the ants (Parker 2016). Other myrmecophile beetles are strongly integrated within their host's society that they are fed by them. These beetles request nutriment from their host by waving their forelegs toward passing ants, and the ants respond by regurgitating liquid that the beetles rapidly ingest (Kovarik and Caterino 2016; Parker 2016). Species from many subfamilies of ants may have beetles as occupants, such as Myrmicinae (Attini, Crematogastrini, Pogonomyrmecini, Solenopsidini), Dolichoderinae (Leptomyrmecini), Formicinae (Camponotini, Formicini, Lasiini, Melophorini, Plagiolepidini), Ponerinae (Ectatommini and Ponerini), and Dorylinae (*Dorylus* Fabricius, *Eciton* Latreilelle, and *Neivamyrmex* Borgmeier) (Mazur 2005; Aguilar Julio 2010; Kovarik and Caterino 2016; Parker 2016).

At least 33 beetle families have been recorded in association with ant colonies (Parker 2016). Among them, we can highlight the beetle families Brentidae, Carabidae (~600 spp. of Paussinae), Cetonidae, Histeridae (~400 spp. of Abraeinae, Chlamydopsinae, Dendrophilinae, Haeteriinae, Histerinae, Ontophilinae, Saprininae, and Tribalinae), Leiodidae (Cholevinae), Scarabaeidae (Aphodiinae and Scarabaeinae), Staphylinidae (mainly species of Aleocharinae, Pselaphinae, and Scydmaeninae), and Tenebrionidae (Pimeliinae) (Kistner 1982; Aguilar Julio 2010; Kovarik and Caterino 2016; Parker 2016). The readers can find a complete list of beetle families associated with ants in Parker (2016).

Several adults and larvae of beetles may have an association with termites. They can occupy termite nests in different conditions: decaying trees infested by termites, living colonies of arboreal and subterranean nests, abandoned nests, as well as detritus piles produced by termites. These beetles can be mycetophagous, saprophagous, or predators of termites in the living colonies or other insects in dead colonies (Aguilar Julio 2010; Costa and Vanin 2010; Kovarik and Caterino 2016). Some species of beetles that live inside nests are luminescent and present a structural modification in the abdomen called physogastry (Costa and Vanin 2010). Termite families reported as beetle hosts are Hodotermitidae, Nasutitermitidae, Rhinotermitidae, and Termitidae (Aguilar Julio 2010; Costa and Vanin 2010; Kovarik and Caterino 2016). The main beetle family occupants of termite nests are Carabidae (Cicindelinae and Harpalinae), Cetoniidae, Elateridae (Elaterinae and Agrypninae), Histeridae (Abraeinae, Chlamydopsinae, Dendrophilinae, Haeteriinae, Histerinae, and Saprininae), Hybosoridae (Ceratocanthinae), Passalidae (Passalinae), Melolonthidae (Dynastinae), Staphylinidae (Aleocharinae, Pselaphinae, and Scaphidiinae), and Tenebrionidae (Kistner 1982; Aguilar Julio 2010; Costa and Vanin 2010; Puker et al. 2012; Ogawa and Maeto 2015; Kovarik and Caterino 2016; Caron et al. 2018).

Other interesting associations of beetles are those with hymenopteran nests. Beetles which are predators, parasites, mycetophagous, and saprophagous are associated with nests of stingless bees (Meliponinae), Africanized honeybees, bumblebees (*Bombus* spp.), and wasps (Kistner 1982; Coletto-Silva et al. 2006; Aguilar Julio 2010; Krüger et al. 2017; Silva-Neto et al. 2019). The main beetle families associated with bee nests are Cetoniidae, Coccinellidae, Cryptophagidae, Histeridae (Histerinae and Saprininae), Nitidulidae (*Aethina tumida* Murray, a worldwide pest), Leiodidae (Leiodinae and Platypsyllinae), and Meloidae (Kistner 1982; Coletto-Silva et al. 2006; Aguilar Julio 2010; Torto et al. 2010; Kovarik and Caterino 2016; Krüger et al. 2017; Silva-Neto et al. 2019). The beetle families associated with wasp nests have been reported in Anobiidae (Ptininae), Cryptophagidae, and Rhipiphoridae (Kistner 1982).

Finally, an unexpected association is between some beetles and spiders. One example is the species *Scapicoelis tibialis* Marseul (Histeridae) reared from an egg sac of a theraphosid spider (*Pamphobeteus* sp.) (Kovarik and Caterino 2016). Other indirect associations with nests, and so we are not going to explore here, are cited to (1) Scarabaeidae with millipedes and land snails and (2) wasps of the families Encryrtidae, Braconidae, Sphecidae, and Cerceridae that due to being beetle hunters (e.g., Buprestidae, Curculionidae, Phalacridae, Nitidulidae, and Chrysomelidae) take them into the nest for food (Aguilar Julio 2010).

Methods for Collecting Beetles from Animal Nests

There is almost no specific bibliography available that describes the sampling of beetles associated with animal nests. Therefore, the beetle appears associated with the vertebrate or invertebrate nest/burrows, and the sampling methods for catching the beetles associated with these microhabitats are from the authors' experiences

(Table 6.1). We recommend two approaches for collecting beetles associated with animal nests: active collecting (collecting directly from the nests) and passive collecting (using traps). The best way to sample beetles associated with nests and burrows is active methods because they bring the richest information about natural history.

Active Collecting

Hand-Collecting The visual inspection is very useful to collect beetles that have different associations with other animals, such as nests of ants, termites, bees, wasps, spiders, birds, and vertebrate burrows. For all these cases, the collector can capture the beetles with forceps, a moistened fine-tipped paintbrush, or an entomological aspirator. However, for collecting beetles associated with abandoned bird nests (e.g., Histeridae, Dermestidae, and Trogidae) (Aguilar Julio 2010), the entomological aspirator is not appropriated because this microhabitat contains other organisms (e.g., fungi and bacteria) that can harm our health when inhaled.

Hand-Collecting by General Visual Inspection (Fig. 6.5a) It represents the first step toward the inventory of beetles associated with animal nests. Different beetles are usually associated with debris or feces near the nest/burrow of birds, ants, termites, and vertebrates (Aguilar Julio 2010; Kovarik and Caterino 2016). For natural stingless bee nests, the collector must inspect the nest entrance, which usually exists in the ground or tree trunks (Aguilar Julio 2010), while for artificial nests, the collector must check the part that contains dead bees and debris (Coletto-Silva et al. 2006; Silva-Neto et al. 2019). The procedures used for stingless bee nests can also be employed to check wasp nests (Kistner 1982). For the Africanized honeybee, the collector must examine the entrance, floor, and honeycomb hives (Krüger et al. 2017). If the intention is to inspect bee and wasp nests, clothing and other safety equipment are necessary to avoid injuries.

Hand-Collecting Using a White Tray or White Cloth This method can make it easier to see the occupants. First, the collector may use a shovel to collect the debris in the vertebrate's burrow or peripherally arranged in a nest of ants (e.g., Attini nests), termites (e.g., *Hodotermes* nests), and bumblebees (*Bombus* spp.). Second, perform a visual examination for beetles using a white tray or white cloth to facilitate the inspection. Also, it is possible to do the same technique with the fungal garden chambers, ant subterranean refuse chambers (leaf debris stored beneath the ground in excavated cavities), and Psittacidae' nest bed after removing from the tree (Kistner 1982; Aguilar Julio 2010; Kovarik and Caterino 2016; Nacagava 2019).

Hand-Collecting by Emigration Trail Visual Inspection (Fig. 6.5b) This technique is useful for ants of the subfamily Dorylinae. When an army ant colony is emigrating (e.g., *Eciton* spp.), the occupants (e.g., Histeridae, Carabidae, and Staphylinidae) tend to follow the ants. This method consists of locating the emigration trail, sitting nearby, and observing the occupants who are running along with the ants or trapped on the ants' bodies (Aguilar Julio 2010; Kovarik and Caterino

2016; Parker 2016). Some groups of beetles (e.g., Histeridae) can appear minutes or hours after the emigration trail finished. Thus, it is recommended that the collector wait a little longer at the site.

Flotation (Fig. 6.5c) This method is widely used in entomology (Almeida et al. 2012) and is useful with some soil ants' colonies (e.g., Solenopsidini). The soil containing the colony should be excavated and placed into a bucket which is gradually filled with water. It is expected that the inquilines (e.g., Histeridae) and hosts will emerge to the water surface. Thus, all material that emerged must be sieved (a traditional sieve like those to strain milk) and placed on a white tray or white cloth to facilitate the beetle's inspection. A second option is completely to remove the nest, place in an airtight bucket, and freeze for 24 h in the laboratory (Almeida et al. 2012). After 24 h of freezing, the collector can perform the flotation with the insects already dead.

Fogging This is a very invasive technique, but it can give satisfactory results for catching beetles inside nests. It consists of killing the anthill using an aerosol insecticide and then looking for the beetles on a white tray or white cloth to aid the inspection (Degallier and Gomy 1983). This inspection can also be conducted by taking the entire dead colony to the laboratory.

Nest Excavation (Fig. 6.5d) This method is more applicable for live or inactive subterranean termites' nests. The collector must excavate the nest, gradually break the nest aggregates, and inspect on a white tray or white cloth. Different beetles (e.g., Histeridae and Staphylinidae) can be found in a gallery network that has an opening on the outside or inside inner chambers (Aguilar Julio 2010; Costa and Vanin 2010; Puker et al. 2012; Kovarik and Caterino 2016; Caron et al. 2018). The same breaking and inspection process can be applied to arboreal termite nests and decaying trees infested by termites.

Sifting Debris It is a versatile method that the collector can utilize in conditions similar to the Winkler extractor or Berlese funnel, but this method has an active effort differently from the others. Thus, the collector must use a shovel to take the soil, debris, or feces peripherally arranged in the vertebrate burrow or invertebrate nest and then sift with sieves for granulometric analysis. The sieves for larger granulometry retain the debris, while the beetles are directed downward and retained in the lower sieves for smaller granulometry. The same procedure can be followed for Psittacidae nest bed, which after being removed from the tree should be placed in the sieve (Nacagava 2019). In the lower sieves, it is possible to put a piece of cloth moistened with ethyl acetate to avoid escaping of groups that run or fly quickly during the sieve checking (e.g., Staphylinidae). The readers can find more details of some sifting debris applications in Aguilar Julio (2010) and Connior (2011).

Passive Collecting

Adhesive Trap This method can be efficient at sampling beetles (e.g., Histeridae, Leiodidae, and Nitidulidae) associated with artificial bee nests. Small corrugated cardboard wrapped into a spiral or sticky board, with aggregation/attractant pheromones (e.g., pollen and propolis), must be placed in the apiary or inside individual hives. However, these methods can also attract apiary pests, so the trap should be exposed for a short time (Aguilar Julio 2010).

Flight Intercept Trap (FIT) (Fig. 6.3d) This method was described in detail in the section "Methods for Collecting Flying Beetles" and is commonly useful to collect different myrmecophile and thermophilic beetles in flight activity (e.g., Histeridae and Staphylinidae) (Leivas et al. 2013). It is recommended to install it close to nests of ants and termites or vertebrate burrows to enhance the usage of this trap. Placing the FIT during some days in front of a broken termite nest can enhance the catch of Hybosoridae (Ceratocanthinae) (Aguilar Julio 2010).

Light Trap (Fig. 6.3f) This method is described in detail in the section "Methods for Collecting Flying Beetles." For the use of this trap, some specificities regarding the activity time of the groups must be considered. Degallier and Gomy (1983) reported that turning on lights 15 min before sunset seems to be efficient for collecting beetles associated with ants in Africa (e.g., Histeridae).

A modification can be performed to catch beetles associated with bees. The collector can install a light at the ground level and above a white tray containing water with detergent (Aguilar Julio 2010). The beetles associated with the nests will be attracted by the light and fell on the tray with water. Although the affinity for this method with any family has not been reported, we believe that potentially all families associated with bee nests can be sampled. However, some groups not associated with the nests may be attracted, as well as many of the bees.

Pitfall Trap (Fig. 6.4c, d) This method was described in detail in the section "Methods for Collecting Edaphic Beetles" and is commonly useful to collect different myrmecophile and thermophilic beetles (e.g., Histeridae and Staphylinidae) that are walking or running on the ground, mainly near ants and termites (Aguilar Julio 2010; Leivas et al. 2013). It is recommended to use the pitfall baited with feces from the host vertebrate and put the pitfall near the entrances of the burrow to enhance the usage of this trap. It is expected to attract the beetles that are inside the burrows of these vertebrates (Aguilar Julio 2010).

Winkler (Moczarski) Extractor and Berlese Funnel (Fig. 6.4a, b) These methods were described in detail in the section "Methods for Collecting Edaphic Beetles," and they are useful to collect different beetles usually present in debris or feces of nests and burrows or the soil close to these places (e.g., Histeridae, Carabidae, and Staphylinidae). *Termite nests.* After digging the termite nests, the aggregates and the debris can be examined with the Winkler extractor and the Berlese funnel (Aguilar Julio 2010). *Ant nests.* The leaf debris piles deposited on the surface of leafcutter ant nests or the subterranean refuse chambers can be explored to collect beetles using these methods. *Bee nests.* The trash from the bee nest (containing dead bees and debris) or the straw that is put into the hives to protect the bees from cold (more commonly used in regions with a temperate climate, Germany) can be investigated through Winkler extractor (Aguilar Julio 2010). *Vertebrate burrows.* The beetles associated with burrows and nests of vertebrates can be caught by putting the soil and debris from a burrow in a Winkler extractor or Berlese funnel (Aguilar Julio 2010).

6.2.1.4 Beetles Associated with Fungi and Myxomycetes

The interaction between beetles and fungi and myxomycetes can be driven by the search for resources, such as food and habitat for the development of the beetle life cycle (Sörensson 1997; Lipkow and Betz 2005). Mycophagy is one of the most widespread feeding habits among Coleoptera. This feeding habit is ancestral among beetles, and approximately half of Coleoptera families are primarily mycophagous or saprophagous (Lawrence 1989). Beetles feed especially on mycelium, plasmodium, and fruit bodies (Persiani et al. 2010).

Among Coleoptera, Cucujiformia and Staphylinoidea are intrinsically related to fungi, using them as food resource or oviposition substrate (Lawrence 1989). Furthermore, ambrosia beetles (Scolytinae, Curculionidae) are completely dependent upon fungi, which they use for the cultivation of their nitrogenous-rich food (Batra 1963). Ambrosia beetles have particular fungi-related behavior, tunneling into the wood surface and forming a substrate for the cultivation of fungi that serve as their food (Beaver 1989; Lévieux and Cassier 1994). Although dung beetles (Scarabaeidae) feed mostly on excrement, there are many species, markedly from the genus *Onthophagus* Latreille, which feed on fresh and decaying fungi (Halffter and Halffter 2009). In this group, it is still unclear if mycophagy is mostly related to specialized or generalist feeding habits (Larsen et al. 2006). Similarly, Scaphidiinae (Staphylinidae), Ptillidae, and Leiodidae have many species specialized in feeding on myxomycetes (Lawrence and Newton 1980; Newton and Stephenson 1990). It is common to observe beetles on the spore-producing structures of the myxomycetes (i.e., sporocarps), mainly on their mature fruit bodies. Also, species of Haeteriinae (Histeridae) and Sphinginae (Sphindidae) are found in myxomycetes. Probably these beetles visit the myxomycetes in search of prey since the species of these families are carnivorous (Lemos et al. 2010).

Methods for Collecting Beetles on Fungi and Myxomycetes

Active Collecting

Hand-Collecting Beetles are hardly found on fungi and myxomycetes, despite their little mobility. The specimens' small size makes it difficult to see with the naked eye (Wheeler and Miller 2005). It is necessary to carefully inspect the fungi and myxomycetes to collect these beetles without removing them. When found, then they must be extracted using forceps (Blackwell 1984; Hanski 1989; Betz et al. 2003). The aspirator is not appropriate due to the risk of spore inhalation (Aguilar Julio 2010).

Passive Collecting

Berlese Funnel (Fig. 6.4b) This method was described in detail in the section "Methods for Collecting Edaphic Beetles." It is very effective in separating beetles from leaf litter, fungi, and other organic matter. The collector places the substrate inside the funnel with a lamp attached. The heat from the lamp induces beetles escaping and thus falling into the vial (Aguilar Julio 2010).

6.2.1.5 Ectoparasite Beetles

Ectoparasite organisms inhabit the fur of their host, and through observations made in the host, it is possible to obtain information on biological and ecological aspects of the parasite (Hopla et al. 1994). Depending on the microhabitat specificity of ectoparasite beetles in the host body (e.g., wing, head, or holes), their ecological aspects may be better understood. For example, the dung beetle of the genus *Bradypodidium* Vaz-de-Mello (Scarabaeidae) apparently has an obligatory relationship with sloths (Vaz-de-Mello 2008). Such dung beetles live attached to the fur near sloth anus, waiting for the mammal to defecate, and thus feed on the mammal excrement (Ratcliffe 1980).

The parasitic habit is unusual in beetles, and few species are ectoparasites on invertebrates and vertebrates, mainly on mammals (Galbreath et al. 2019), and feed on living tissue and organic matter (Peck 1982). In general, ectoparasites are not specific to a host species but regularly exhibit some degree of preference toward a host (Tripet and Richer 1997; Soler et al. 1999; Autino et al. 2011). Host-specific ectoparasites have the advantage of suffering from low levels of competition when compared to host generalist ectoparasites. However, host-specific ectoparasites are sensitive to population fluctuations and the distribution of its host species (Whitaker et al. 2009).

Staphylinoidea comprises one of the main groups of mammal ectoparasites (Timm and Ashe 1987). Among them, Leiodidae is a more specialized ectoparasite family of mammals (rabbit, rodents, and shrew). This group contrasts from the others due to its morphological features that comprise an ectoparasitic syndrome, such as highly reduced eyes and wings, flattened body, being short, and thick legs (e.g., Marshall 1981). Other beetle families also present records as ectoparasites (e.g., Carabidae, Bothrideridae, Passandridae, Rhipiceridae, Elateridae, Cucujidae, Meloidae, and Rhipiphoridae) (Lawrence and Newton 1982; Bouchard et al. 2017). The most common hosts of beetle species are rodents, although marsupials, sloths, and bats have also been reported (Timm and Ashe 1987).

Methods for Collecting Ectoparasite Beetles

Active Collecting

Hand-Collecting The collector can observe hair or fur (e.g., Kunz 1976; Marshall 1981), in various parts of the host body, and the specimens are then actively collected from the host. The search for ectoparasites is done on dead or live animals. In the case of a dead host, ideally, it should be recently dead. When this is not possible, then the animal's body should be put in a freezer for later analysis (Peck 2007; Galbreath et al. 2019). It is also possible to search for ectoparasite beetles on specimens prepared as stuffed or open dried skins. In such cases, ectoparasites must be carefully extracted, since they may be fragile and trapped in the fur of the host. Among the vertebrates, bats are the most likely to be inspected while they are alive. The host must be immobilized, and beetles must be collected with soft or hard forceps, depending on the parasite's mobility. In all cases, the ectoparasites are preserved in 70% ethanol (Barbier and Bernard 2017).

6.2.1.6 Endogean Beetles

This is a small group of species that live inside leaf or stems, usually on galls (see Chap. 16) or mines (see Chap. 17). These beetles include phytophagous (e.g., Curculionidae) and predator (e.g., Mordellidae) species (Tooker and Hanks 2004). Mines are formations on the leaf or stem caused by beetle movement and development in plant tissue, and this behavior may damage and even kill the host plant (Korotyaev et al. 2005). Gall systems can also support several inquiline species, including beetles. Galls are plant formations induced by bacteria, viruses, or some invertebrates, including beetles, that can provide an ideal habitat for the development of these organisms, which are referred to as inquilines (Sanver and Hawkins 2000).

Endogean beetles are crucial to agriculture and forestry, due to the economic damage that they can promote in agriculture. Almost all endophytic species belong to the superfamilies Curculionoidea (Curculionidae) and Chrysomeloidea (Cerambycidae and Chrysomelidae). However, Buprestoidea (Buprestidae), Nitidulidae, and Mordellidae also comprise endophytic beetles (Korotyaev et al. 2005).

Methods for Collecting Endogean Beetles

Active Collecting

Hand-Collecting Endogean larvae can be collected directly from inside the galls or mines. First, the collector needs to locate the site where the larvae feed and then dissect the plant tissue using a scalpel. Second, the beetles are collected using a fine-hair brush, forceps, or an aspirator. Another strategy to find the targeted insects is by packing the galls in plastic pots sealed with a breathable fine mesh fabric. Inside the pot, one can also place a ball of moistened cotton, to maintain the humidity. In time, imagoes will emerge and come out of the galls.

For more detailed methods for these two guilds, see Chaps. 16 and 17.

6.2.1.7 The Use of Beetles of Medicolegal Interest in Forensic Investigations

Forensic entomology is a valuable tool for legal investigations. A good understanding of distribution, biology, ecology, and behavior of insects might facilitate the collection of individuals and the interpretation of results for criminal investigations (Byrd et al. 2010). Beetles are an important group present in the colonization process of corpses (Archer 2013; Merritt and De Jong 2016), capable of accelerating the skeletonization process (Schroeder et al. 2002; Voigt 1965), and are particularly important in the last stages of decomposition (Kulshrestha and Satpathy 2001). The principal families for forensic entomology are Silphidae (carrion beetles), Histeridae (clown beetles), Staphylinidae (rove beetles), Dermestidae (dermestid beetles), Nitidulidae (sap beetles), Cleridae (checkered beetles), and Scarabaeidae (scarab beetles) (Goff and Catts 1990; Schoenly et al. 2007).

The evaluation of life stages of beetles may provide an estimate of time since death (Goff and Flynn 1991; Kulshrestha and Satpathy 2001) as well as may indicate postmortem movements (Ururahy-Rodrigues et al. 2008), use of drugs (Miller et al. 1994; Bourel et al. 2001), and unveil information for identification of suspects by DNA methods (Brown et al. 1997; Benecke 1998). Thus, methods for collecting beetles for criminal investigations shall follow standardization and guideline protocols (Amendt et al. 2007; Lord and Rodriguez 1989).

Methods for Collecting Beetles of Medicolegal Interest in Forensic Investigations

Active Collecting

Hand-Collecting This is the easiest and efficient sampling technique for collecting forensic beetles, but the results vary considerably depending on the collector experience (Schoenly et al. 2007). This active method is low cost and is directed to catch crawling or walking individuals (i.e., larvae and adults) using forceps with different levels of spring tension (Amendt et al. 2007; Lord and Rodriguez 1989). Spoons or shovels are particularly necessary in cases of advanced decay for a careful examination of the soil beneath the corpse (Lord and Rodriguez 1989).

On the other hand, smaller specimens (under 5 mm) or eggs can be collected with a small paintbrush dipped in a preserving fluid (Amendt et al. 2007; Lord and Rodriguez 1989). The researcher should also collect the beetle residues such as feces, pupal skins, or empty puparium, which document previous activity and insect presence (Amendt et al. 2007). During sampling, it is essential to protect the hands with surgical gloves (Lord and Rodriguez 1989).

Aerial Sweep Net This method was described in detail in the section "Methods for Collecting Flying Beetles." An aerial sweep net is an active method used for capturing flying adults, and the process consists of performing zigzag motions of 180° twists (Lord and Rodriguez 1989; Byrd et al. 2010). The effectiveness of collecting flying adults may vary with the ability of the collector, and the use of aerial sweep nets in forensic sampling has shown low efficiency for beetles (Schoenly et al. 2007).

Passive Collecting

Adhesive Trap Sticky cards are glue-based traps used to catch and immobilize insects at crime scene (Haskell and Williams 1990). However, the use of this passive method alone is risky because of its low sampling efficiency for beetles, and the removal process requires more time and can cause damage to parts of the body (Schoenly et al. 2007).

6.2.2 *Immature Beetles*

Many field studies have focused on the beetle adult stages, and immature stages are little explored, and this scenario can be associated with limitations to identify the larvae collected. There are some possible explanations on limitations to identifying field-collected beetle larvae. One, there are few reliable identification keys to all instars, and identification based on their morphological external features is a time-consuming task which requires a level of expertise (Habu and Sadanaga 1961; Ortloff et al. 2014). Two, captive breeding until larvae become adults often requires intensive labor (Erwin et al. 1979). Three, genetic identification of larvae based on COI gene (mitochondrial cytochrome oxidase subunit 1) or other genetic markers is more rapid (DiZinno et al. 2002; Hayashi and Sota 2010) but has cost limitations, and larval instars cannot be defined (Jinbo et al. 2011). However, new tools involving interaction between morphological and molecular characters have been reliable to identify immature stages of beetles collected in the field (Sasakawa 2013).

Most of the studies on larval stages deal with species of economic value (e.g., Melolonthidae, Beaudoin-Ollivier et al. 2000; Curculionidae, Hoff et al. 2002; and Chrysomelidae, Ruesink and Haynes 1973) or forensic importance (i.e., Staphylinidae, Silphidae, Cleridae, Trogidae, Dermestidae, and Histeridae) (Bala and Singh 2015; Ortloff et al. 2014). However, larval sampling can be particularly helpful for surveying species whose adult males or females are not or are rarely attracted to conventional trap systems (Albuquerque et al. 2016; Bedford 1975), understanding ecological functions of species (Chandra et al. 2008; Klimeš and Saska 2010), and accessing valuable phylogenetic information (Beutel and Leschen 2005; Lawrence et al. 2011).

6.2.2.1 Methods for Collecting Immature Beetles

Active Collecting

Hand-Collecting (Fig. 6.4e, f) This is the most straightforward sampling technique for collecting beetle larvae. The researcher in charge of sampling effort and success is entirely dependent on the knowledge regarding the life history of the target beetles (Costa et al. 1998). This active method is low-cost and depends on the manual ability to catch larvae inside their feeding habitats using forceps, brush, or jars (Upton and Mantle 2010). The sampling effort may vary among different collectors and habitat types, meaning that standardization and replicability are hard tasks to achieve. Furthermore, beetles have a wide diversity of larval feeding habits (Albertoni et al. 2014; Donlan et al. 2004; Triplehorn and Johnson 2011), and temperature may influence the mobility rate of insects (Hagstrum and Subramanyam 2010) making it difficult to find the beetles.

Examples of Larval Feeding Habits in Coleoptera Examples include (a) deadwood in terrestrial environments (e.g., Passalidae, Lucanidae, Buprestidae, Elateridae, Dyropidae, Mordellidae, Prostomidae, Synchroidae, Oedemeridae, and Stenotrachelidae); (b) living wood in terrestrial environments (e.g., Buprestidae and Cerambycidae); (c) dung (e.g., Geotrupidae and Scarabaeidae); (d) plants including leaves, seeds, sprouts, and stalks (e.g., Buprestidae, Curculionidae, Nitidulidae, Tenebrionidae, Brentidae, Languriidae, Plecomidae, and Melolonthidae); (e) fungi (e.g., Phalacridae, Synchroidae, Anthribidae, Tenebrionidae, Nitidulidae, and Ciidae); (f) invertebrate animals including millipedes (e.g., Scarabaeidae), aphids (e.g., Coccinellidae), slug (e.g., Lampyridae), grasshopper, and bees (e.g., Meloidae), and wasps (e.g., Ripiphoridae); (g) vertebrate animals including eggs of turtles (e.g., Elateridae); (h) deadwood submerged in aquatic environments (e.g., Elmidae, Lutrochidae, and Ptilodactylidae); (i) seaweed (e.g., Notoridae, Hydraenidae, and Lutrochidae); (j) moss (e.g., Byrrhidae); and (k) bromeliad tissues (e.g., Scarabaeidae).

Aerial Sweep Net This method was described in detail in the section "Methods for Collecting Flying Beetles." An aerial sweep net is an active method used for capturing terrestrial larvae of beetles. Sweep nets with hoop about 38 cm in diameter are used for sweeping larvae on the vegetation (Cothran and Summers 1972; Hoff et al. 2002). The process consists of performing zigzag motions of 180° twists with the opening of the net placed into the foliage at approximately half of its diameter (Hoff et al. 2002). The effectiveness for collecting larvae using this method may vary with the ability of the collector, wind, sun angle, temperature, humidity, plant size, distance between the sweep net rim and the ground, and speed and length of the hit (DeLong 1932; Ruesink and Haynes 1973).

The use of sweep net is prevalent in studies on the management of agricultural pests such as Curculionidae (alfalfa weevil, Hoff et al. 2002) and Chrysomelidae larvae (cereal leaf beetle, Ruesink and Haynes 1973). Pest control studies usually require standardized protocols and are generally based on the delimitation of linear transects in which each collector step corresponds to one sweep, and the collector sweeps through the vegetation until 10 or 20 times (Hoff et al. 2002). Also, recommendations are to start the sampling before the peak of occurrence of the more damaging late instars (Berberet and Hutchison 1994).

Shake-Bucket This method involves systematically and randomly collecting parts of plant ("stem") from various locations in the field (Hoff et al. 2002). The stems are removed by the palm-tip method (Higgins et al. 1991). The palm-tip consists of kindly breaking off or cutting the stem, placing the tip into the bottom of a light-colored plastic bucket, and shaking vigorously against the bucket walls (Hoff et al. 2002). This method has been used to estimate larval density and to define management strategies for Curculionidae larvae (Barney and Legg 1989; Hoff et al. 2002; Pellissier et al. 2017). Shake-bucket stem collecting method is sensitive to detect small larvae and an easy-to-learn technique for the inexperienced collectors (Hoff et al. 2002).

Soil Samples The Tropical Soil Biology and Fertility (TSBF) method or monolith extraction is a standardized method for studying soil organisms (Anderson and Ingram 1993). The area for the removal of soil monolith (25 x 25 cm and 30 cm depth) is made using a sampler made of galvanized iron plates, followed by the excavation and removal of soil layer (Anderson and Ingram 1993). The soil samples are transported to the laboratory in plastic bags and visually searched manually for beetles in the lab. This method allows for catching larvae living underground that are otherwise difficult to find using other methods. Moreover, TSBF allows the sampling of a constant volume of soil, allowing further comparison between different types of soils, habitats, and communities of soil fauna (Araújo et al. 2010; Feller et al. 2012; Marchão et al. 2009). This method has been frequently applied to evaluate both diversity patterns and ecological functions of larvae in different environments, such as is used with Melolonthidae, Cantharidae, Carabidae, and Cicindelidae (Amazonas et al. 2018; Negrete-Yankelevich et al. 2007; Romero-López et al. 2015; Wodika and Baer 2015).

Aquatic Net (Fig. 6.6a) Aquatic nets are used to capture aquatic larvae of beetles with a flat, rectangular, triangular, or D-frame bottom design (Hilsenhoff and Tracy 1985; Spangler 1981). The nets are used to collect in shallow vegetation areas by pushing or sweeping the net through the vegetation and bottom debris until the net is about one-third full of debris (Hilsenhoff 1991; Segura et al. 2012). This method is efficient for sampling Hydrophilidae (e.g., *Hydrobius*, *Tropisternus*, and *Hydrochra*) and Dytiscidae (e.g., *Matus*, *Laccophilus*, *Ranthus*, and *Agabus*), and the capture rate may vary accordingly to the larvae swimming capacity (Hilsenhoff 1991).

Surber Sampler (Fig. 6.6d) This method was described in detail in the section "Methods for Collecting Aquatic and Semiaquatic Beetles."

Passive Collecting

Bottle Trap This method is used for capturing large and mobile adults and larvae of water beetles (e.g., Dytiscidae, Haliplidae, and Hydrophilidae) which are otherwise difficult to collect with net because of their habitats, such as dense vegetation (Hilsenhoff and Tracy 1985; Klečka and Boukal 2011; Turić et al. 2017). The collector can assemble the trap by using two 1.5 L plastic bottles, forming a funnel with the opening end of one bottle, and removing the bottom of another bottle. Thus, insert the funnel, staple it, then fill the trap with water, and submerge horizontally in the desired depth (Hilsenhoff and Tracy 1985; Hilsenhoff 1987). Since the collector places the bottle trap horizontally, one limitation is the unsuitability to capture species that move vertically in the water column (Hilsenhoff 1987). A discrete flag should be used to mark the site of each trap and to avoid attracting the general public. The time required depends on the number of beetles captured and water temperature, while the effectiveness of the trap depends on horizontal swimming habits, nocturnal activity, and acoustic communication of the species studied (Hilsenhoff 1987). There is a variation involving the addition of light inside the bottle trap (Hungerford et al. 1955), but underwater vegetation can restrict light penetration (Hilsenhoff and Tracy 1985). It is more effective to use bottle traps associated with a D-frame net (Hilsenhoff 1991).

6.2.3 *Aquatic and Semiaquatic Beetles*

Aquatic beetles are found practically in all sorts of aquatic habitats and microhabitats, such as ditches, groundwater, lakes, phytotelmata, puddles, rivers, seepages, and springs (Yule and Sen 2004; Jäch and Balke 2008). Although they do not inhabit oceans, some beetles were recorded in hypersaline rock pools of the supralittoral zone (Sharma et al. 2019). It is necessary to understand and carefully study the ontogeny, to recognize whether the beetle is aquatic or not (Jäch and Balke 2008).

According to Jäch (1998), there are six ecological categories to define beetles, which are at some level associated with aquatic habitats: (1) True Water Beetles, adult partly submerged; (2) False Water Beetles, larva submerged but adult stage terrestrial; (3) Phytophilous Water Beetles, feeding and living on water plants (herbivore specialists) in any developmental stage; (4) Parasitic Water Beetles, with aquatic mammal hosts; (5) Facultative Water Beetles, occasionally or regularly actively submerged or actively living on the water surface, for instance, seeking refuge or feeding; and (6) Shore Beetles, beetles that live near riparian areas nearby the water's margin during all developmental stages, and not entering water voluntarily.

Thus, it is essential to know their habitats and behavior to choose the appropriate sampling equipment to maximize efficiency. To avoid sampling bias, the sampling effort (e.g., operator, sampling units, and time) and the quantity of sweeps for handnetting (e.g., D-frame net) should be standardized. Equipment to sample water beetles typically can be used in both standing and running water, but many sampling devices have a limitation because they must be handheld and depend on water depth for use (Hilsenhoff and Tracy 1985; Williams and Feltmate 1992).

6.2.3.1 Methods for Collecting Aquatic and Semiaquatic Beetles

Active Collecting

Hand-Collecting (Fig. 6.6b) Another method is the hand-collecting of substrates (e.g., leaves and woody debris) on marginal habitats or lentic ecosystems, such as puddles, and collecting the specimens using forceps. The use of sieves can facilitate the collection of small beetles on the debris. However, as the hand-netting method, it is essential to make uniform the number of samples and the operator. This method is more suitable to collect intact specimens for taxonomy and genetic studies (Oliveira and Pes 2019). Since hand-collecting requires a great sampling effort, it is more recommended in small and shallow aquatic habitats or microhabitats (e.g., puddles and phytotelmata). It is also recommended for hygropetric microhabitats since other methods are difficult to execute in the thin water film found on the rock and other surfaces, and usually involves turning over the rocks to find hidden specimens (Turner 2007). Due to the need to search for marginal habitats to capture semiaquatic individuals, hand-collecting is the easiest method for this purpose (Turner 2007).

Mark and Recapture There are many approaches to mark and recapture insects to understand mating behavior or differ sex in a sample. In aquatic beetles, several ways were tested to respond to these questions, and the most effective was proposed by Murvosh and Miller (1970) that consist of "tape dots" in tiny dots of colored, marked on elytra. It helps identify large beetles (e.g., *Hydrophilus triangularis* Say, 1823, that has about 40 mm). Similar methods, such as fixing numbers in labels, were also selected to understand the activity patterns or physiological cues that stimulate flight in ponds (Davy-Bowker 2002; Csabai et al. 2012). In this method, the possibility to leave the beetles alive for other studies is a great advantage.

6.2.3.2 Passive Collecting

D-Frame Net (Fig. 6.6a) D-frame net is the most used method to collect aquatic insects, including water beetles. Usually, this net is employed as a qualitative sampling tool, but it might be considered as a semiquantitative method in surveys if carefully applied (Williams and Feltmate 1992; Turić et al. 2017). It is commonly composed of a handle (about 1.2 m) attached to a D-shaped frame with a 60 cm catching net of 0.25 mm mesh (Ghani et al. 2016). This type of trap is more fitting for sampling less mobile and small beetles (Turić et al. 2017). In lentic habitats (e.g., lakes or ponds), the collector can sweep the net back and forth on substrate surface layers or across vegetation at the bottom over the same area during a specific time to capture the water beetles (Hilsenhoff and Tracy 1985). In running waters (e.g., streams), beetles may be sampled by holding the D-net in the opposite water flow direction (Williams and Feltmate 1992; Oliveira and Pes 2019). The application in wide rivers has a drawback because the animals freely floating in the water column can avoid the small net opening (about 40 cm) (Ghani et al. 2016). Another downside is the hardness involved in sampling highly vegetated areas due to the high amount of debris that may clog the net or areas such as swamps because they have less water, decreasing the sampling efficiency (Hilsenhoff and Tracy 1985).

Surber Sampler (Fig. 6.6d) The Surber sampler is a quantitative sampling tool used for benthic macroinvertebrates such as water beetles. The Surber usually has a catching area of 900 cm^2 and a net of 0.25 mm mesh (Silveira et al. 2004). However, the mesh size depends on whether the collector's main goal is to capture the immature or adult stage. The sampler is placed, against the water flow, on the substrate with the opening facing upstream to collect water beetles, and the collector disturbs the substrate in front of the opening frame (Williams and Feltmate 1992). Different from the D-frame net, the Surber has a frame in front of the net delimiting the disturbed substrate area (Oliveira and Pes 2019). In the laboratory, the material should be rinsed in water under sieves with different mesh sizes, according to the collector goal. The use of Surber is most recommended for small lotic environments such as streams and river springs (Silveira et al. 2004). The collector may have difficulties with the Surber, like any other hand-netting method, which is the possibility of net clogging resulting in material escaping. Since a large amount of substrate gets inside the catching net, another downside of the Surber is the long sample processing time (Ghani et al. 2016).

Light Trap (Fig. 6.6c) In aquatic environments, the light trap has a different mode of operation and attractive powers (Carlson 1972). It is commonly used to provide a continuous time series with seemingly little effort. The collector can effectively capture other groups of insects in lentic habits, like stoneflies, mayflies, and caddisflies (Kovats et al. 1996; Radwell and Camp 2009; Klečka and Boukal 2011), and allows selectivity of the target group. Despite the small feasibility, some beetles are lured to it like some semiaquatic adults (e.g., individuals of the subfamily Sphaeridiinae). The low representativeness in some studies could be owing to most species of water beetles avoiding the trap. Thus, it is essential to consider the light source used in the trap according to its wavelength to be suitable to catch the water beetles and the lunar cycle because darker nights have a better result (Oliveira and Pes 2019). Usually, light traps are used with white or black UV lights (Benetti and Hamada 2003; Segura et al. 2012). Nabli et al. (1999) claimed that although a black UV light is more efficient to catch beetles, it could also attract predator beetles making the identification more challenging. In general, the trap consists of a polyethylene bucket 35 cm height, a piece cut out of the bottom, and a 20 cm inverted funnel of white polyethylene fastened inside. The lights stuck in the cover are powered by a small 12 volts motorcycle battery and have a power consumption rate of 0.9 amp. Inside the trap is a device like an aspirator with a bottle with ethyl acetate to kill the insects (Carlson 1971). Collectors can adjust to be more suitable for their goals. Some modifications involve the placement of a white sheet near the water; the use of lamps, flashlights, or lanterns for beetles' attraction (Albuquerque et al. 2016); or the use of Pennsylvania trap adaptations (Frost 1957).

Bottle Trap This method was described in detail in the section "Methods for Collecting Larvae Beetles."

6.2.4 Subterranean Beetles

6.2.4.1 Cave Beetles

Since the methods applied for terrestrial surface habitat are not necessarily suitable for caves (Weinstein and Slaney 1995), sampling invertebrates in these environments is challenging (Zagmajster et al. 2010; Wynne et al. 2019). Major drawbacks are the difficulties in assessing the subterranean environment, the low local species richness, and their limited dispersal abilities (Zagmajster et al. 2010). Biospeleologists have historically applied distinct devices and techniques for better sampling of cave invertebrates' diversity information to deal with such peculiarities. For cave-dwelling arthropods, different methods have been proposed which can be active or passive collecting. The *passive collecting* comprises pitfall traps, Winkler extractor, and Berlese-Tullgren funnel extractor. The *active collecting* includes hand-collecting, timed searches, quadrat sampling, and baiting (Hunt and Millar 2001; Sabu et al. 2011; Pellegrini and Ferreira 2012a; Oliveira 2014; Bichuette et al. 2015; Wynne et al. 2018; Bichuette et al. 2019). It is recommended that the combination of multiple methods and the sampling of unique microhabitats (e.g., bat guano deposits and water bodies) seem to be efficient (Pellegrini and Ferreira 2012a; Wynne et al. 2018; Borges-Filho et al. 2019). However, choosing one better-suited technique might be necessary due to time and cost limitations (Oliveira et al. 2019). In that case, the efficiency of each method must be considered, and this task can be facilitated when having a target taxon such as beetles.

Among the terrestrial fauna adapted to caves (i.e., troglobiotic), the beetles comprise the richest group, with more than 2000 described species (Decu and Juberthie 1998), reaching the mark of about 42% of the subterranean biodiversity in Dinaric karst (Sket et al. 2004; Zagmajster et al. 2008, 2010). Due to their importance for taxonomic and conservation purposes (e.g., Giuseffi et al. 1978; Faille et al. 2014; Pellegrini and Ferreira 2014), the group is used as models in many ecological and evolutionary speleological studies (e.g., Zagmajster et al. 2008, 2010; Sánchez-Fernández et al. 2016). The preferred sampling technique for cave beetles depends on many factors, such as the cave substrate type, the collector experience in collecting subterranean fauna, and cave lithology (Oliveira 2014; Borges-Filho et al. 2019).

Choosing the best method for cave beetle sampling is a crucial issue for conservation purposes. Their restricted distribution in the subterranean makes them highly vulnerable to human activities. Accessing reliable diversity data is necessary for addressing ecological issues and providing information to better guide programs and strategies for the conservation of the cave community (Zagmajster et al. 2010; Jaffé et al. 2018; Wynne et al. 2018).

Methods for Collecting Cave Beetles

Active Collecting

Hand-Collecting (Fig. 6.7a–c) The technique of hand-collecting or active searching is also applied to other environments and was described in detail in the section "Methods for Collecting Beetles from Animal Nests." In the cave environment, there are variations in the preferred microhabitat for beetles searching when compared with the nests (see preferred microhabitats in the topic below). This method is an alternative for minimizing population impacts arising from the traps. The collector performs the hand collection with the aid of forceps, brushes, and aspirators, and the method can be applied in several ways.

General Cave Visual Inspection (Fig. 6.7a) As proposed by Snyder et al. (2006), this technique consists of visually searching for individuals in preferred microhabitat (e.g., leaf litter deposits, under rocks, and trunks) throughout the cave environment accessible to man. This no trapping method reduces the chances of over-collecting invertebrates (Paquin and Provost 2010), the reason why historically this is one of the most widely used technique in Brazilian surveys (e.g., Trajano and Gnaspini-Netto 1990; Souza-Silva et al. 2011; Pellegrini and Ferreira 2012b; Fernandes et al. 2016; Zepon and Bichuette 2017).

Timed Searches The timed searches are an alternative to standardize the sampling method, especially for ecological studies. The cave can be divided into sectors with similar sizes, spending the same time per sector or standardized time searching per square meter (Oliveira 2014; Wynne et al. 2018). The beetles' hand-collecting is performed.

Quadrants Sampling (Figs. 6.7d, e) This alternative method is also preferred to standardize the active and visually searching techniques for ecological studies, and it has been taking place more recently (Sharratt et al. 2000; Pellegrini and Ferreira 2012b; Tobin et al. 2013; Bichuette et al. 2015; Zepon and Bichuette 2017). For the quadrat sampling technique, standardized quadrat in area are demarcated in the cave under which an extensive visual searching is performed (Sharratt et al. 2000). This technique can reveal significantly more beetle species than the active searching by the general cave visual inspection (Bichuette et al. 2015).

Baiting The visual searching in a cave with huge chambers associated with the presence of very large boulders that cannot be turned manually is not recommended. Such features allow beetles to easily escape from collecting and hiding under rocks (Faille et al. 2014). In this case, the use of baits without trapping can be effective, especially for collecting predator carabids. Moss, rotten wood, smelly meat, and cheese can be used as baits for beetles (Faille et al. 2014). The environment created to attract beetles must offer shelter, and the bait may be under some type of cover (e.g., under stones or surrounded by cobbles) (Hunt and Millar 2001). Decomposer species are attracted by the baits and also its predators (ground beetles), which are visually searched and sampled through a sucker tor with wet brushes. This technique allows the identification of known species and their release at the same location, avoiding the oversampling and population damages. It is worth mentioning that the use of baits in caves should be done with reservations, since it can attract surface fauna, configuring themselves as a sampling bias (Castro-Souza et al. 2019).

Passive Collecting

Pitfall Trap (Fig. 6.7c, d) This technique is useful for other environments and was described in detail in the section "Methods for Collecting Edaphic Beetles." The use of baited pitfall traps for cave beetles is common in temperate regions (Borges et al. 2012; Reboleira and Ortuño 2014; Wynne et al. 2018). However, some studies point out that this technique might significantly negatively affect population sizes (Weinstein and Slaney 1995; Sharratt et al. 2000), a major concern for the species often represented by small population sizes. Due to the potential of this technique, over-collecting invertebrates until extinction, Hunt and Millar (2001) were very emphatic in mentioning that "Never leave a baited pitfall trap set in a cave indefinitely."

Winkler Extractor and Berlese Funnel (Fig. 6.7a, b) The use of extractors is less often in caves (Bichuette et al. 2015). *Winkler* and *Berlese-Tullgren funnelextractors* are commonly used to sample insects and other invertebrates that inhabit soil and litter (Krell et al. 2005, Sabu et al. 2011), and more detailed methods are better described in the section "Methods for Collecting Edaphic Beetles." In caves, the captured fauna mainly comprises edaphic invertebrates (Nitzu et al. 2006; Oliveira 2014) and is an interesting method for sampling smaller and/or faster beetles (e.g., some Staphylinidae, Ptiliidae, and Bembidiini species), for which visually searching and hand-collecting can be difficult. Furthermore, in caves, Berlese funnels can be used for soil and litter samples to capture edaphic fauna (Nitzu et al. 2006). However, it is usually applied for invertebrates associated with bat guano deposits, especially to capture mites (Pellegrini and Ferreira 2013).

6.2.4.2 Aquatic Cave Beetle

Despite the high proportion of subsurface freshwater on Earth, about 97% (Gibert et al. 1994), few insects have successfully invaded subterranean aquatic habitats (Howarth 2009). Consequently, there are only five families representing cave aquatic beetles in the world, Dytiscidae, Noteridae, Dryopidae, Elmidae, and Hydrophilidae (Spangler 1996). The most diverse and widespread in caves are the diving beetles (Dytiscidae), and the subterranean dytiscid (i.e., stygobiont) fauna includes hundreds of recently discovered species (Jean et al. 2012). The growing number of findings reflects the lack of studies in subterranean aquatic environments, which have remained in this situation for a long time due to technical difficulties faced in sampling this habitat (Gibert et al. 1994). All the devices proposed for collecting aquatic beetles in detail in the section "Methods for Collecting Aquatic Beetles" can be also applied in the cave aquatic environment.

Methods for Collecting Aquatic Cave Beetles

Active Collecting

Hand-Collecting This method comprises the walking up of the stream and under small ponds with the visual search for aquatic beetles. The beetles are sampled through a sucker that can be a thin glass pipette with rubber bulb to collect small and fast beetles (Hunt and Millar 2001).

Passive Collecting

Bottle Trap This method was described in detail in the section "Methods for Collecting Larvae Beetles." In the subterranean environment, it is especially useful for groundwaters with no human access. Baiting the aquatic trap is a good option for attracting beetles due to the scarcity of food. The possible baits include chunk of stewing meat and smelly cheese, and the trap can be installed in the riverbed for about 5 days (Hunt and Millar 2001).

Aquatic Leaf Litter Packs Packs made with a variable mesh size (0.04–9 mm^2) containing litter inside can live in the riverbed for several weeks to be colonized (Hunt and Millar 2001; Souza-Silva et al. 2013a, 2013b). This method is efficient for sampling Elmidae species (Souza-Silva et al. 2013b).

D-Frame Net and Surber Sampler (Fig. 6.6a, d) Both methods were described in detail in the section "Methods for Collecting Aquatic and Semiaquatic beetles." The Surber sampler is most commonly used in ecological studies of small subterranean streams, and it commonly samples Dytiscidae, Elmidae, Gyrinidae, Hydrophilidae, and Scirtidae species from both adult and larva stages (Gunn et al. 2000; Pellegrini et al. 2018; Martins and Ferreira 2020).

6.2.4.3 Mesovoid Shallow Substratum "Milieu Souterrain Superficiel" (MSS)

The term "Milieu Souterrain Superficiel" (MSS) was defined in the 1980s by Christian Juberthie and is also synonymous with "Mesovoid Shallow Substratum," "Superficial Underground Compartment," "Superficial Subterranean Habitat," and

"Upper Hypogean Zone" (Mammola et al. 2016). The MSS represents the subterranean domain formed by small canaliculi filled by air formed by cracks and a network of fissures in rock fragments beneath the soil surface (Ortuño et al. 2013). The MSS allows the movement of fauna between different environments (Ortuño et al. 2013) and is frequently occupied by a high coleopteran diversity (Rendoš et al. 2012) in which both cave-adapted (i.e., troglobionts) and epigean/edaphic fauna are found, playing a role of an ecotone (Ortuño et al. 2013; Mammola et al. 2016). The suitability for troglobiont fauna and the absence of light, low energy inputs, and relative microclimate stability are ecological features shared by the MSS with caves (Pipan and Culver 2019). Thereunto, sampling the MSS can be an important tool for accessing the cave-dwelling fauna distribution patterns.

Methods for Collecting MSS Beetles

Active Collecting

Dig and Search This method consists of digging from the soil up to the MSS or in turning large stones embedded in the soil. When reaching the MSS, the fauna is hand-collected using the same devices of the cave collecting, with the aid of forceps, brushes, and aspirators (Karaman 2009; Sugaya and Yamasako 2008).

Flotation (Fig. 6.5c) and Sifting Those are variations from the dig and search method; however, instead of visually searching the beetles, the flotation or sifting method is used (Karaman 2009; Mammola et al. 2016). In the flotation, a sample of the rocky substrate can be placed into bucked and gradually filled with water, and the associated invertebrate fauna will emerge and the beetles are collected on the water surface. Similarly, sifting soil and detritus is also an alternative method (Karaman 2009). For more details in using the flotation and sifting, see the section "Methods for Collecting from Animal Nests."

Passive Collecting

Pitfall Trap (Fig. 6.4d) The first step is the vertical digging up to the MSS with the allocation of the pitfall trap, and the opened space must be refilled with stones and soil (Mammola et al. 2016). However, instead of using the ordinary plastic container, glass or rigid plastic jars are preferable for avoiding the pitfall trap deformation. It is recommended to put a signal up to the surface in order to relocate the trap. This technique is useful for other environments and was described in detail in the section "Methods for Collecting Edaphic Beetles."

Subterranean Sampling Devices (SSDs) The SSD consist of a multiperforated PVC cylinder (the fauna enters the trap through the perforations), measuring about 11 cm in diameter and 0.75–1 m long, and with a pitfall trap mounted inside, the technique was first proposed by Gers (1992) and developed by Owen (1995)—see López and Oromí (2010), Ortuño et al. (2013, 2019), and Mammola et al. (2016) for more details. The pitfall trap contains a liquid preservative and it can be baited or not (e.g., smelling cheese) (Ortuño et al. 2019). The SSD is buried vertically in the ground, positioned so that the perforations in the PVC cylinder correspond to the MSS layers. For the external environment isolation, a silicone cap closes the tube on the surface. This device is specially recommended for studies that demand periodic fauna checking. The tube with pitfall traps and the cap allow the device to be revisited with the collection of fauna without mechanically digging the soil each time (López and Oromí 2010; Pipan et al. 2011; Mammola et al. 2016).

6.2.5 Collecting Beetles for Molecular Study

Molecular studies of beetles have contributed to the understanding of relationships between species (e.g., Zhang et al. 2018), the discovery of cryptic species (e.g., Duran et al. 2019), and the association of life stages (e.g., Jeon and Ahn 2007). Furthermore, these studies have also provided interesting data about interactions between host beetle species and their associated microbial ecosystem (e.g., Tagliavia et al. 2014) as well as helped identify parasites (e.g., Zeng et al. 2007) and analyze beetle diets from gut contents (e.g., Seric-Jelaska et al. 2014).

The collection of Coleoptera specimens focused on molecular studies requires a few specifications. As advocated by Nagy (2010) for zoological studies, concerning larger organisms, a tissue sample (e.g., a single leg) is usually collected (Pentinsaari et al. 2019), which insignificantly affects the identification and investigation of the specimen. For small organisms, from which no tissue sample can be removed, the well-identified sampling should be separated into two parts: one for genetic analysis and another for taxonomic studies. Nondestructive methods have been suggested for using the entire specimen (see Gilbert et al. 2007).

Killing agents directly affect the success of extraction and amplification/sequencing of genetic material. In general, methods such as freezing and ethanol have little effect on DNA integrity (Reiss et al. 1995; Bisanti et al. 2009). However, prolonged exposure to ethyl acetate can cause problems in laboratory procedures (Nakahama et al. 2019). In traps, the preservative substances such as propylene glycol and ethanol are useful for DNA conservation, with a recent emphasis on better DNA preservation by propylene glycol rather than ethanol (Nakamura et al. 2020). In contrast, according to Fukatsu (1999), acetone and ethanol are more vulnerable to water contamination in studies of RNA rather than DNA, requiring higher care and attention.

In Coleoptera, commonly used methods for collection storage include ethanol, propylene glycol, acetone, ultralow-temperature freezer at least −20 °C (for further details, see Gibb and Oseto 2020), and other specific products such as RNAlater, which helps preserve the RNA material (e.g., Lord et al. 2016). For the total preservation of large organism specimens, separating the abdomen from the rest of the body is recommended for better penetration of the preservation fluid into the internal parts (e.g., Maddison and Cooper 2014). For Lepidoptera, small holes are pocked into the thorax and abdomen with a micro pin, which allows the preservative liquid to infiltrate (Quicke et al. 1999). However, there are no specific evaluations of this method for Coleoptera.

Low-temperature (at least −20 °C) storage is recommended for long-term storage, even for specimens kept in liquids such as 70–100% ethanol (Quicke et al. 1999). For long-term storage, without freezing, specimens can be stored in propylene glycol, 100% ethanol, and acetone, but 70% ethanol and Carnoy solution are not recommended (Bisanti et al. 2009; Ferro and Park 2013). Methods of dry-mounting and pinning enable mtDNA extraction and amplification capacity for only a few years, with refrigerated methods being recommended for long-term conservation (Gibb and Oseto 2020). Finally, as evidenced by international projects such as the International Barcoding of Life (IBOL), the packaging, labeling, and depositing of the vouchers must be done carefully when the samples are destined for molecular studies. The traceability of each sample is essential when using and depositing sequences in genetic databases (e.g., GenBank, Bold system).

Acknowledgments LI is grateful to Conselho Nacional de Desenvolvimento Científico e Tecnológico-CNPq for the grants PQ2. TGP is grateful to the Fundação de Amparo à Pesquisa do Estado de Minas Gerais (FAPEMIG) and to VALE S.A. for the fellowship financial support. RPS was supported by Programa Nacional de Pós Doutorado/Capes (PNPD/CAPES, Brazil). JCSC thanks the Coordenação de Aperfeiçoamento de Pessoal de Nível Superior-Brasil (CAPES) for the financial support (award #88881.129014/2016-01). The authors are grateful to Filipe Aléssio, Guilherme M. Viana, Jean C. Santos, João C. Regueira, Lucas R. C. Lima, Marconi Souza-Silva, Portal de Zoologia de Pernambuco, Rodrigo L. Ferreira, Sheila P. C. Fernandes, and Tamara G. Carvalho, for providing the photographs. We are grateful to Luana Iannuzzi for diagramming of the plates.

References

Achterberg KV (2009) Can Townes type Malaise trap be improved? Some recent developments. Entomol Ber 69:129–135

Adis J (1987) Extraction of arthropods from neotropical soils with a modified Kempson apparatus. J Trop Ecol 3:131–138. https://doi.org/10.1017/S0266467400001863

Aguilar Julio C (2010) Methods for catching beetles. In: Viedma JB (ed) Naturalia scientific collection. Montevideo, Asuncion

Albertoni FF, Frank-Thorsten K, Steiner J et al (2014) Life history and description of larva and pupa of *Platyphileurus felscheanus* Ohaus, 1910, a scarabaeid feeding on bromeliad tissues in Brazil, to be excluded from Phileurini (Coleoptera, Scarabaeidae, Dynastinae). ZooKeys 389:49–76. https://doi.org/10.3897/zookeys.389.6888

Albuquerque LSC, Grossi PC, Iannuzzi L (2016) Flight patterns and sex ratio of beetles of the subfamily Dynastinae (Coleoptera, Melolonthidae). Rev Bras Entomol 60:248–254. https://doi.org/10.1016/j.rbe.2016.03.002

Alcázar-Ruiz JÁ, Morón-Rios A, Morón MA (2003) Fauna de Coleoptera Melolonthidae de Villa Las Rosas, Chiapas, México. Acta Zool Mex 88:59–86

Alford DV (ed) (2003) Biocontrol of oilseed rape pests. Blackwell Science, Amsterdam

Almeida LM, Ribeiro-Costa CS, Marinoni L (1998) Manual de Coleta, Conservação, Montagem e Identificação de Insetos. Holos, Ribeirão Preto

Almeida LM, Ribeiro-Costa C, Marinoni L (2012) Coleta, montagem, preservação e métodos para estudo. In: Rafael JA, Melo GAR, Carvalho CJB et al (eds) Insetos do Brasil diversidade e taxonomia. Holos, Ribeirão Preto, pp 175–190
Amazonas NT, Viani RAG, Rego MGA et al (2018) Soil macrofauna density and diversity across a chronosequence of tropical forest restoration in Southeastern Brazil. Braz J Biol 78:449–456. https://doi.org/10.1590/1519-6984.169014
Amendt J, Campobasso CP, Gaudry E et al (2007) Best practice in forensic entomology—standards and guidelines. Int J Leg Med 121:90–104. https://doi.org/10.1007/s00414-006-0086-x
Anderson JM, Ingram JSI (1993) Tropical soil biology and fertility: a handbook of methods, 2nd edn. CAB International, London
Andrade RBD, Barlow J, Louzada J et al (2011) Quantifying responses of dung beetles to fire disturbance in tropical forests: the importance of trapping method and seasonality. PLoS One 6:e26208. https://doi.org/10.1371/journal.pone.0026208
Aquino AM (2005) Fauna do Solo e sua Inserção na Regulação Funcional do Agroecossistema. In: Aquino AM, Assis RL (eds) Processos biológicos no sistema solo-planta: ferramentas para uma agricultura sustentável. Embrapa Agrobiologia, Brasília, pp 47–75
Aquino PSR, Jesus FG, Rocha EC et al (2018) Predation rates of a beetle (*Canthon virens*) that kills female leaf-cutting ants (*Atta* spp.). Int J Agric Biol 20:1247–1250. https://doi.org/10.17957/IJAB/15.0613
Araújo VFP, Bandeira AG, Vasconcellos A (2010) Abundance and stratification of soil macroarthropods in a Caatinga Forest in Northeast Brazil. Braz J Biol 70:737–746. https://doi.org/10.1590/S1519-69842010000400006
Archer M (2013) Comparative analysis of insect succession data from Victoria (Australia) using summary statistics versus preceding mean ambient temperature models. J Forensic Sci 59:404–412. https://doi.org/10.1111/1556-4029.12345
Arnett RH Jr, Thomas MC, Skelley PE et al (eds) (2002) American beetles—Polyphaga: Scarabaeoidea through Curculionoidea. CRC, Boca Raton
As S (1984) To Fly or Not to Fly? Colonization of Baltic Islands by Winged and Wingless Carabid Beetles. J Biogeogr 11:413–426
Assad MLL (1997) Fauna do solo. In: Vargas MAT, Hungria M (eds) Biologia dos solos do Cerrado. Embrapa Cerrados, Planaltina, pp 363–443
Auffray T, Frérot B, Poveda R et al (2017) Diel patterns of activity for insect pollinators of two oil palm species (Arecales: Arecaceae). J Insect Sci 17:1–6. https://doi.org/10.1093/jisesa/iex018
Autino AG, Claps GL, Barquez RM et al (2011) Ectoparasitic insects (Diptera: Streblidae and Siphonaptera: Ischnopsyllidae) of bats from Iquitos and surrounding areas (Loreto, Peru). Mem Inst Oswaldo Cruz 106:917–925. https://doi.org/10.1590/S0074-02762011000800004
Azevedo FR, Azevedo R, Santos CAM et al (2016) Análise faunística e sazonalidade de insetos edáficos em ecossistemas da Área de Proteção Ambiental do Araripe em duas estações do ano, Barbalha-CE. Revista Agro@mbiente On-line 10:263–272
Bala M, Singh N (2015) Beetles and forensic entomology: a comprehensive review. J Ent Res 39:293–302. https://doi.org/10.5958/0974-4576.2015.00036.5
Barberena-Arias MF, González G, Cuervas E (2012) Quantifying variation of soil arthropods using different sampling protocols: is diversity affected? In: Sudarshana P (ed) Tropical forests. InTech, Rijeka, pp 51–70. https://doi.org/10.5772/30386
Barbier E, Bernard E (2017) From the Atlantic Forest to the borders of Amazonia: species richness, distribution, and host association of ectoparasitic flies (Diptera: Nycteribiidae and Streblidae) in northeastern Brazil. Parasitol Res 116:3043–3055. https://doi.org/10.1007/s00436-017-5615-7
Barney RJ, Legg DE (1989) Conversion factors for use with the shake-bucket technique for sampling alfalfa weevil (Coleoptera: Curculionidae) larvae. J Econ Entomol 82:825–828. https://doi.org/10.1093/jee/82.3.825
Barros YJ, Melo VF, Sautter KD et al (2010) Indicadores de qualidade de solos de área de mineração e metalurgia de chumbo. II—Mesofauna e plantas. R Bras Ci Solo 34:1413–1426. https://doi.org/10.1590/S0100-06832010000400037

Batra LR (1963) Ecology of ambrosia fungi and their dissemination. Trans Kans Acad Sci 66:213–236

Beaudoin-Ollivier L, Prior RNB, Laup S (2000) Simplified field key to identify larvae of some rhinoceros beetles and associated scarabs (Coleoptera: Scarabaeoidea) in Papua New Guinea coconut developments. Ann Entomol Soc Am 93:90–95. https://doi.org/10.1603/0013-8746(2000)093[0090:SFKTIL]2.0.CO;2

Beaver RD (1989) Insect-fungus relationships in the bark and ambrosia beetles. Paper presented at the insect-fungus interactions. 14th symposium of the Royal Entomological Society of London in collaboration with the British mycological society. In: Wilding N, Collins NM, Hammond PM (eds) Insect-fungus interactions. Academic Press, London, pp 121–143

Bedford GO (1975) Trap catches of the coconut rhinoceros beetle *Oryctes rhinoceros* (L.) (Coleoptera Scarabaeidae, Dynastinae) in New Britain. Bull Entomol Res 65:443–451. https://doi.org/10.1017/S0007485300006118

Bellas TE (1989) Insects as a cause of inhalational allergies: a bibliography 1900e1987. CSIRO Division of Entomology, Canberra

Benecke M (1998) Random amplified polymorphic DNA (RAPD) typing of necrophageous insects (Diptera, Coleoptera) in criminal forensic studies: validation and use in practice. Forensic Sci Int 98:157–168. https://doi.org/10.1016/S0379-0738(98)00150-9

Benetti CJ, Hamada N (2003) Fauna de coleópteros aquáticos (Insect: Coleoptera) na Amazônia central, Brasil. Acta Amazon 33:701–710. https://doi.org/10.1590/S0044-59672003000400015

Benton MJ (2010) The origins of modern biodiversity on land. Philos Trans R Soc Lond B 365:3667–3679. https://doi.org/10.1098/rstb.2010.0269

Berberet RC, Hutchison WD (1994) Sampling methods for insect management in alfalfa. In: Pedigo LP, Buntin GD (eds) Handbook of sampling methods for arthropods in agriculture. CRC, Boca Raton, pp 357–381

Bernhard D, Schmidt C, Korte A (2006) From terrestrial to aquatic habitats and back again-molecular insights into the evolution and phylogeny of Hydrophiloidea (Coleoptera) using multigene analyses. Zool Scr 35:597–606. https://doi.org/10.1111/j.1463-6409.2006.00251.x

Besuchet C, Burckhardt DH, Löbl I (1987) The "Winkler/Moczarski" eclector as an efficient extractor for fungus and litter Coleoptera. Coleopt Bull 41:392–394

Betz O, Thayer MK, Newton AF (2003) Comparative morphology and evolutionary pathways of the mouthparts in spore-feeding Staphylinoidea (Coleoptera). Acta Zool 84:179–238. https://doi.org/10.1046/j.1463-6395.2003.00147.x

Beutel RG, Leschen RAB (2005) Phylogenetic analysis of Staphyliniformia (Coleoptera) based on characters of larvae and adults. Syst Entomol 30:510–548. https://doi.org/10.1111/j.1365-3113.2005.00293.x

Beutel RG, Pohl H, Evgeny HP et al (2019) The phylogeny of Coleopterida (Hexapoda)—morphological characters and molecular phylogenies. Syst Entomol 44:75–102. https://doi.org/10.1111/syen.12316

Bichuette ME, Simões LB, von Schimonsky DM et al (2015) Effectiveness of quadrat sampling on terrestrial cave fauna survey—a case study in a Neotropical cave. Acta Sci Biol Sci 37:345–351. https://doi.org/10.4025/actascibiolsci.v37i3.28374

Bichuette ME, Simões LB, Zepon T et al (2019) Richness and taxonomic distinctness of cave invertebrates from the northeastern state of Goiás, Central Brazil: a vulnerable and singular area. Subterr Biol 29:1–33. https://doi.org/10.3897/subtbiol.29.30418

Bisanti M, Ganassi S, Mandrioli M (2009) Comparative analysis of various fixative solutions on insect preservation for molecular studies. Entomol Exp Appl 130:290–296. https://doi.org/10.1111/j.1570-7458.2008.00821.x

Blackwell M (1984) Myxomycetes and their arthropod associates. In: Wheeler Q, Blackwell M (eds) Fungus—insect relationships—perspectives in ecology and evolution. Columbia University Press, New York, pp 67–90

Blanche R, Cunningham SA (2005) Rain forest provides pollinating beetles for atemoya crops. J Econ Entomol 98:1193–1201. https://doi.org/10.1603/0022-0493-98.4.1193

Blaxter M (2003) Counting angels with DNA. Nature 421:122–124. https://doi.org/10.1038/421122a

Block W (1966) Some characteristics of the Macfadyen high gradient extractor for soil micro-arthropods. Oikos 17:1–9. https://doi.org/10.2307/3564777

Borges PAV, Cardoso P, Amorim IR et al (2012) Volcanic caves: priorities for conserving the Azorean endemic troglobiont species. Int J Speleo 41:101–112. https://doi.org/10.5038/1827-806X.41.1.11

Borges-Filho EL, Audino LD, Bernardi LFO et al (2019) Comparação da eficiência de três métodos de amostragem de invertebrados em cavidades naturais subterrâneas neotropicais. In: Zampaulo RA (ed) 35° Congresso Brasileiro de Espeleologia, Bonito, pp 710–719

Bouchard P, Bousquet Y (2020) Additions and corrections to "Family-group names in Coleoptera (Insecta)". ZooKeys 922:65–139. https://doi.org/10.3897/zookeys.922.46367

Bouchard P, Bousquet Y, Davies AE et al (2011) Family-group names in Coleoptera (Insecta). ZooKeys 88:1–972. https://doi.org/10.3897/zookeys.88.807

Bouchard P, Smith ABT, Douglas HB et al (2017) Biodiversity of Coleoptera. In: Foottit RG, Adler PH (eds) Insect biodiversity: science and society, 2nd edn. Wiley-Blackwell, West Sussex, pp 337–417. https://doi.org/10.1002/9781118945568.ch11

Bouget C (2005) Short-term effect of windstorm disturbance on saproxylic beetles in broadleaved temperate forests. Part I: do windthrow changes induce a gap effect? Forest Ecol Manag 216:1–14. https://doi.org/10.1016/j.foreco.2005.05.036

Bouget C, Brustel H, Brin A et al (2008) Sampling saproxylic beetles with window flight traps: methodological insights. Rev Écol (Terre Vie) 10:21–32

Bourel B, Tournel G, Hédouin V et al (2001) Determination of drug levels in two species of necrophagous Coleoptera reared on substrates containing morphine. J Forensic Sci 46:600–603. https://doi.org/10.1520/JFS15010J

Braithwaite RW (1991) Fauna and habitat surveys as ecological pathfinders. In: Margules CR, Austin MP (eds) Nature conservation: cost effective biological surveys and data analysis, 1st edn. CSIRO, Melbourne, pp 23–35

Bremner G (1990) A Berlese funnel for the rapid extraction of grassland surface macro-arthropods. New Zeal Entomol 13:76–80. https://doi.org/10.1080/00779962.1990.9722596

Brown RJ, Malcolm CA, Mason PL (1997) Genetic differentiation between and within strains of the saw-toothed beetle, *Oryzaephilus surinamensis* (Coleoptera: Silvanidae) at RAPD loci. Insect Mol Biol 6:285–289. https://doi.org/10.1046/j.1365-2583.1997.00184.x

Brown GG, Niva CC, Zagatto MRG et al (2015) Biodiversidade da fauna do solo e sua contribuição para os serviços ambientais. In: Parron LM, Garcia JR, Oliveira EB et al (eds) Serviços Ambientais em Sistemas Agrícolas e Florestais do Bioma Mata Atlântica. Embrapa Florestas, Brasília, pp 122–154

Byrd JH, Lord WD, Wallace JR (2010) Collection of entomological evidence during legal investigations. In: Byrd JH, Castner JL (eds) Forensic entomology: the utility of arthropods in legal investigations, 2nd edn. CRC, Boca Raton, pp 127–176

Carlson D (1971) A method for sampling larval and emerging insects using an aquatic black light trap. Can Entomol 103:1365–1369. https://doi.org/10.4039/Ent1031365-10

Carlson D (1972) Comparative value of black light and cool white lamps in attracting insects to aquatic traps. J Kansas Entomol 45:194–199

Caron E, Bortoluzzi S, Rosa CS (2018) Two new species of obligatory termitophilous rove beetles from Brazil (Coleoptera: Staphylinidae: Termitomorpha Wasmann). Zootaxa 4413:566–578. https://doi.org/10.11646/zootaxa.4413.3.10

Carvalho AG (1998) Armadilha, modelo Carvalho-47. FLORAM 5:225–227

Carvalho AG, Trevisan H (2015) Novo modelo de armadilha para captura de Scolytinae e Platypodinae (Insecta, Coleoptera). FLORAM 22:575–578. https://doi.org/10.1590/2179-8087.105114

Castro-Souza RA, Pellegrini TG, Souza-Silva M et al (2019) Drivers of ant composition, richness, and trophic guilds in Neotropical iron ore cavities. Int J Speleol 48:279–293. https://doi.org/10.5038/1827-806X.48.3.2270

Chandra G, Mandal SK, Ghosh AK et al (2008) Biocontrol of larval mosquitoes by *Acilius sulcatus* (Coleoptera: Dytiscidae). BMC Infect Dis 8:138. https://doi.org/10.1186/1471-2334-8-138
Chapman JA, Kinghorn JM (1955) Window flight traps for insects. Can Entomol 87:46–47
Cherman M, Morón M (2014) Validación de la familia Melolonthidae Leach, 1819 (Coleoptera: Scarabaeoidea). Acta Zool Mex 30:201–220. https://doi.org/10.21829/azm.2014.301139
Coletto-Silva A, Freire DCB (2006) *Hololepta (Leionota) reichii* Marseul (Coleoptera, Histeridae) un nuevo enimigo natural para la meliponicultura en la Amazonia Central, Brasil. Rev Bras Zool 23:588–591. https://doi.org/10.1590/S0101-81752006000200036
Connior MB (2011) Beetles (Coleoptera: Histeridae and Scarabaeidae) from previously unsampled populations of pocket gopher burrows in Louisiana. Insecta Mundi 0153:1–5
Costa C, Vanin SA, Casari-Chen SA (1998) Larvas de Coleoptera do Brasil. Museu de Zoologia, São Paulo
Costa RB, Fernandes PM, Oliveira FS et al (2009) Captura de adultos de *Liogenys fuscus* (Coleoptera: Melolonthidae) com armadilha luminosa em área sob sistema de plantio direto. Biosci J 25:1–8
Costa MS, Silva RJ, Paulino-Neto HF et al (2017) Beetle pollination and flowering rhythm of *Annona coriacea* Mart. (Annonaceae) in Brazilian cerrado: behavioral features of its principal pollinators. PLoS One 12(2):e0171092. https://doi.org/10.1371/journal.pone.0171092
Costa C, Vanin S (2010) Coleoptera Larval Fauna Associated with Termite Nests (Isoptera) with Emphasis on the "Bioluminescent Termite Nests" from Central Brazil. Psyche: A Journal of Entomology:1–13. https://doi.org/10.1155/2010/723947
Cothran WR, Summers CG (1972) Sampling for the Egyptian alfalfa weevil: a comment on the sweep-net method. J Econ Entomol 65:689–691. https://doi.org/10.1093/jee/65.3.689
Covell CV Jr (2009) Collection and preservation. In: Resh VH, Carde RT (eds) Encyclopedia of insects, 2nd edn. Academic Press, Burlington, pp 201–206
Cresswell MJ (1995) Malaise trap: collection attachment modification and collection fluid. Weta 18:10–11
Crowson RA (1981) The biology of the Coleoptera. Academic Press, London
Csabai Z, Kálmán Z, Szivák I et al (2012) Diel flight behaviour and dispersal patterns of aquatic Coleoptera and Heteroptera species with special emphasis on the importance of seasons. Naturwissenschaften 99:751–765. https://doi.org/10.1007/s00114-012-0957-6
Darling DC, Packer L (1988) Effectiveness in Malaise trap in collecting Hymenoptera. The influence of trap design, mesh size and location. Can Entomol 120:787–796. https://doi.org/10.4039/Ent120787-8
Davis AJ, Huijbregts J, Kirk-Spriggs AH et al (1997) The ecology and behaviour of arboreal dung beetles in Borneo. In: Stork NE., Adis J and Didham R (eds) Canopy Arthropods, Chapman & Hall, London pp. 417–432.
Davy-Bowker J (2002) A mark and recapture study of water beetles (Coleoptera: Dytiscidae) in a group of semi-permanent and temporary ponds. Aquat Ecol 36:435–446. https://doi.org/10.1023/A:1016550127986
Degallier N, Gomy Y (1983) Caractères généraux et techniques de récolte des Coléoptères Histeridae. L'Entomologiste 39:9–17
Decu V, Juberthie C (1998) Coleopteres (Generalites et Syntheses). In: Juberthie C, Decu V (eds) Encyclopaedia biospeologica, vol 3. Société de Biospéologie, Bucarest, pp 1147–1162
DeLong DM (1932) Some problems encountered in the estimation of insect populations by the sweeping method. Ann Entomol Soc Am 25:13–17. https://doi.org/10.1093/aesa/25.1.13
Dijkstra K-DB, Monaghan MT, Pauls SU (2014) Freshwater biodiversity and aquatic insect diversification. Annu Rev Entomol 59:143–163. https://doi.org/10.1146/annurev-ento-011613-161958
DiZinno JA, Lord WD, Collins-Morton MB et al (2002) Mitochondrial DNA sequencing of beetle larvae (Nitidulidae:*Omosita*) recovered from human bone. J Forensic Sci 47:1–3
Donlan EM, Townsend JH, Golden EA (2004) Predation of *Caretta caretta* (Testudines: Cheloniidae) eggs by larvae of *Lanelater sallei* (Coleoptera: Elateridae) on Key Biscayne, Florida. Caribb J Sci 40:415–420

Dugdale J, Hutcheson J (1997) Invertebrate values of kanuka (*Kunzea ericoides*) stands, Gisborne region. In: Science for conservation, vol 55. Department of Conservation, Wellington

Duran DP, Herrmann DP, Roman SJ (2019) Cryptic diversity in the North American *Dromochorus* tiger beetles (Coleoptera: Carabidae: Cicindelinae): a congruence-based method for species discovery. Zool J Linn Soc 186:250–285. https://doi.org/10.1093/zoolinnean/zly035

Eitzinger B, Traugott M (2011) Which prey sustains cold-adapted invertebrate generalist predators in arable land? Examining prey choices by molecular gut-content analysis. J Appl Ecol 48:591–599. https://doi.org/10.1111/j.1365-2664.2010.01947.x

Erwin TL, Ball GE, Whitehead DR et al (1979) Carabid beetles: their evolution, natural history, and classification. Springer, Dordrecht

Faille A, Bourdeau C, Deharveng L (2014) Weak impact of tourism activities on biodiversity in a subterranean hotspot of endemism and its implications for the conservation of cave fauna. Insect Conserv Diver 8:205–215. https://doi.org/10.1111/icad.12097

Feller G, Blancharta E, Bernouxa M et al (2012) Soil fertility concepts over the past two centuries: the importance attributed to soil organic matter in developed and developing countries. Arch Agron Soil Sci 58:3–21. https://doi.org/10.1080/03650340.2012.693598

Fernandes CS, Batalha MA, Bichuette ME (2016) Does the cave environment reduce functional diversity? PLoS One 11:e0151958. https://doi.org/10.1371/journal.pone.0151958

Ferreira RL, Marques MMGSM (1998) A Fauna de artrópodes de serrapilheira de áreas de monocultura com *Eucalyptus* sp. e mata secundária heterogênea. An Soc Entomol Brasil 27:395–403. https://doi.org/10.1590/S0301-80591998000300007

Ferro ML, Park JS (2013) Effect of propylene glycol concentration on mid-term DNA preservation of Coleoptera. Coleopt Bull 67:581–586. https://doi.org/10.1649/0010-065X-67.4.581

Fischer BL (1996) Ant diversity patterns along an elevational gradient in the Réserve Naturelle Intégrale d'Andohahela, Madagascar. Fieldiana Zool 85:93–108

Flechtmann CAH, Rodrigues SR, Couto HTZ (1995) Controle biológico da mosca-dos-chifres (*Haematobia irritans irritans*) em Selvíria, Mato Grosso do Sul-4: Comparação entre métodos de coleta de besouros coprófagos (Scarabaeidae). Rev Bras Entomol 39:259–276

Fonseca MM, Lima E, Lemos F et al (2017) Non-crop plant to attract and conserve an aphid predator (Coleoptera: Coccinellidae) in tomato. Biol Control 115:129–134. https://doi.org/10.1016/j.biocontrol.2017.10.005

Freire LR, Araújo ES, Berbara RLL (2015) Tempo de captura de organismos da mesofauna do solo e seus reflexos na interpretação de índices da comunidade edáfica. R Bras Ci Solo 39:1282–1291. https://doi.org/10.1590/01000683rbcs20150117

Frost S (1957) The Pennsylvania insect light trap. J Econ Entomol 50:287–292. https://doi.org/10.1093/jee/50.3.287

Fukatsu T (1999) Acetone preservation: a practical technique for molecular analysis. Mol Ecol 8:1935–1945. https://doi.org/10.1046/j.1365-294x.1999.00795.x

Galbreath KE, Hoberg EP, Cook JA (2019) Building an integrated infrastructure for exploring biodiversity: field collections and archives of mammals and parasites. J Mammal 100:382–393. https://doi.org/10.1093/jmammal/gyz048

García-López A, Micó E, Zumbado MA et al (2011) Sampling scarab beetles in tropical forests: the effect of light source and night sampling periods. J Insect Sci 11:95. https://doi.org/10.1673/031.011.9501

Ghani WMHWA, Rawi CSM, Hamid SA et al (2016) Efficiency of different sampling tools for aquatic macroinvertebrate collections in Malaysian streams. Trop Life Sci Res 27:115–133

Gibb TJ, Oseto C (2020) Insect collection and identification: techniques for the field and laboratory, 2nd edn. Elsevier, London

Gibert J, Danielopol D, Stanford JA (1994) Groundwater ecology. Academic Press, San Diego

Gilbert MTP, Moore W, Melchior L et al (2007) DNA extraction from dry museum beetles without conferring external morphological damage. PLoS One 2:e272. https://doi.org/10.1371/journal.pone.0000272

Giuseffi S, Kane TC, Duggleby WF (1978) Genetic variability in the Kentucky cave beetle *Neaphaenops tellkampfii* (Coleoptera: Carabidae). Evolution 32:679–681. https://doi.org/10.1111/j.1558-5646.1978.tb04614.x

Goff ML, Catts EP (1990) Arthropod basics—structure and biology. In: Catts EP, Haskell NH (eds) Entomology and death—a procedural guide. Joyce's Print Shop, Clemson, pp 46–47

Goff ML, Flynn MM (1991) Determination of postmortem interval by arthropod succession: a case study from the Hawaiian islands. J Forensic Sci 36:607–614. https://doi.org/10.1520/JFS13067J

Gould SJ (2002) The structure of environmental theory. Belknap Press of Harvard University Press, Cambridge and London

Grimaldi D, Engel MS (2005) Evolution of the insects. Cambridge University Press, Cambridge

Gullan PJ, Cranston PS (2005) The insects. An outline of entomology, 3rd edn. Blackwell, Oxford

Gunn J, Hardwick P, Wood PJ (2000) The invertebrate community of the peak–speedwell cave system, Derbyshire, England—pressures and considerations for conservation management. Aquat Conserv 10:353–369. https://doi.org/10.1002/1099-0755(200009/10)10:5%3C353::AID-AQC413%3E3.0.CO;2-S

Haarløv N (1947) A new modification of the Tullgren apparatus. J Anim Ecol 16:115–121

Habu A, Sadanaga K (1961) Illustrations for identification of larvae of the Carabidae found in cultivated fields and paddy-fields. Bull Natl Inst Agric Sci Ser C 13:207–248

Hagstrum DW, Subramanyam B (2010) Immature insects: ecological roles of mobility. Am Entomol 56:230–241. https://doi.org/10.1093/ae/56.4.230

Halffter G, Halffter V (2009) Why and where coprophagous beetles (Coleoptera: Scarabaeinae) eat seeds, fruits or vegetable detritus. Boletín de la S E A 45:1–22

Hammond PM (1979) Wing-folding mechanisms of beetles, with special reference to investigations of adephagan phylogeny (Coleoptera). In: Erwin TL, Ball GE, Whitehead DR et al (eds) Carabid beetles. Springer, Dordrecht, pp 113–180

Hanski I (1989) Fungivory: fungi, insects and ecology. In: Wilding N, Collins NM, Hammond PM et al (eds) Insect–fungus interactions. Academic Press, London, pp 25–68

Hanski I, Cambefort Y (1991) Dung beetle ecology. Princeton University Press, Princeton

Haskell NH, Williams RE (1990) Collection of entomological evidence at the death scene. In: Catts EP, Haskell NH (eds) Entomology and death: a procedural guide. Joyce's Print Shop, Clemson, pp 82–97

Hayashi M, Sota T (2010) Identification of elmid larvae (Coleoptera: Elmidae) from Sanin District of Honshu, Japan, based on mitochondrial DNA sequences. Entomol Sci 13:417–424. https://doi.org/10.1111/j.1479-8298.2010.00404.x

Hernández CXP, Caballero SZ (2016) Temporal variation in the diversity of Cantharidae (Coleoptera), in seven assemblages in tropical dry forest in Mexico. Trop Conserv Sci 9:439–464. https://doi.org/10.1177/194008291600900124

Hernández MIM, Barreto PSCS, Costa VH et al (2014) Response of a dung beetle assemblage along a reforestation gradient in Restinga forest. J Insect Conserv 18:539–546. https://doi.org/10.1007/s10841-014-9645-5

Higgins RA, Rice ME, Blodgett SL et al (1991) Alfalfa stem-removal methods and their efficiency in predicting actual numbers of alfalfa weevil larvae (Coleoptera: Curculionidae). J Econ Entomol 84:650–655. https://doi.org/10.1093/jee/84.2.650

Hilsenhoff WL (1987) Effectiveness of bottle traps for collecting Dytiscidae (Coleoptera). Coleopt Bull 41:377–380

Hilsenhoff WL (1991) Comparison of bottle traps with a D-frame net for collecting adults and larvae of Dytiscidae and Hydrophilidae (Coleoptera). Coleopt Bull 45:143–146

Hilsenhoff WL, Tracy BH (1985) Techniques for collecting water beetles from lentic habitats. Proc Acad Nat Sci Philadelphia 137:8–11

Hines JW, Heikkenen HJ (1977) Beetles attracted to severed Virginia pine (*Pinus virginiana* Mill.). Environ Entomol 6:123–127. https://doi.org/10.1093/ee/6.1.123

Hodek I, Honek A, van Emden HF (eds) (2012) Ecology and behaviour of the ladybird beetles (Coccinellidae). Wiley, Chichester

Hoff KM, Brewer MJ, Blodgett SL (2002) Alfalfa weevil (Coleoptera: Curculionidae) larval sampling: comparison of shake-bucket and sweep-net methods and effect of training. J Econ Entomol 95:748–753. https://doi.org/10.1603/0022-0493-95.4.748

Holdhaus VDK (1910) Die Siebetechnik zum Aufsammeln der Terricolfauna (nebst Bemerkungen über die Oekologie der im Erdboden lebenden Tierwelt). Zeitschr Wiss Insektenbiol 6:44–57

Hopla CE, Durden LA, Keirans JE (1994) Ectoparasites and classification. Rev Sci Tech 13:985–1017. https://doi.org/10.20506/rst.13.4.815

Howarth FG (2009) Cave insects. In: Resh VH, Cardé RT (eds) Encyclopedia of insects, 2nd edn. Academic Press, Burlington, pp 139–143. https://doi.org/10.1016/B978-0-12-374144-8.00047-3

Hungerford HB, Spangler PJ, Walker NA (1955) Subaquatic light traps for insects and other animal organisms. Trans Kans Acad Sci 58:387–407. https://doi.org/10.2307/3626106

Hunt M, Millar I (2001) Cave invertebrate collecting guide. Department of Conservation technical Series, Wellington

Hunt T, Bergsten J, Levkanicova Z et al (2007) A comprehensive phylogeny of beetles reveals the evolutionary origins of a superradiation. Science 318:1913–1916. https://doi.org/10.1126/science.1146954

Hyvärinen E, Kouki J, Martikainen P (2006) A comparison of three trapping methods used to survey forest-dwelling Coleoptera. Eur J Entomol 103:397–407. https://doi.org/10.14411/eje.2006.054

Iannuzzi L, Maia ACD, Nobre CEB et al (2003) Padrões locais de diversidade de Coleoptera (Insecta) em vegetação de caatinga. In: Leal IR, Tabarelli M, da Silva JMC (eds) Ecologia e conservação da caatinga. Editora Universitária da Universidade Federal de Pernambuco, Recife, pp 367–389

Ikeda H, Kagaya T, Kubota K et al (2008) Evolutionary relationships among food habit, loss of flight, and reproductive traits: life-history evolution in the Silphinae (Coleoptera: Silphidae). Evolution 62:2065–2079. https://doi.org/10.1111/j.1558-5646.2008.00432.x

Ikeda H, Nishikawa M, Sota T (2012) Loss of flight promotes beetle diversification. Nat Commun 3:1–8. https://doi.org/10.1038/ncomms1659

Jäch M (1998) Annotated check list of aquatic and riparian/littoral beetle families of the world (Coleoptera). In: Jäch MA, Ji L (eds) Water beetles of China, vol 2. Zoologisch-Botanische Gesellschaft in Österreich and Wiener Coleopterologenverein, Wien, pp 25–42

Jäch MA, Balke M (2008) Global diversity of water beetles (Coleoptera) in freshwater. Hydrobiology 595:419–442. https://doi.org/10.1007/s10750-007-9117-y

Jackman JA, Nelson CR (1995) Diversity and phenology of tumbling flower beetles (Coleoptera, Mordellidae) captured in a malaise trap. Entomol News 106:97–107

Jaffé R, Prous X, Calux A et al (2018) Conserving relics from ancient underground worlds: assessing the influence of cave and landscape features on obligate iron cave dwellers from the eastern Amazon. PeerJ 6:1–23. https://doi.org/10.7717/peerj.4531

Jean A, Telles ND, Gibson JR et al (2012) Description of a new genus and species of stygobiontic diving beetle, *Psychopomporus felipi* Jean, Telles, and Miller (Coleoptera: Dytiscidae: Hydroporinae), from the Edwards-Trinity aquifer system of Texas, USA. Coleopt Bull 66:105–110. https://doi.org/10.1649/072.066.0202

Jeon MJ, Ahn KJ (2007) Descriptions of late instars of three littoral *Cafius* species (Coleoptera: Staphylinidae) by association of life stage with DNA sequences. Fla Entomol 90:465–474. https://doi.org/10.1653/0015-4040(2007)90[465:DOLIOT]2.0.CO;2

Jinbo U, Kato T, Ito M (2011) Current progress in DNA barcoding and future implications for entomology. Entomol Sci 14:107–124. https://doi.org/10.1111/j.1479-8298.2011.00449.x

Jonason D, Smith HG, Bengtsson J et al (2013) Landscape simplification promotes weed seed predation by carabid beetles (Coleoptera: Carabidae). Landsc Ecol 28:487–494. https://doi.org/10.1007/s10980-013-9848-2

Juillet JA (1963) A comparison of four types of traps used for capturing flying insects. Can J Zool 41:219–223. https://doi.org/10.1139/z63-023

Karaman IM (2009) The taxonomical status and diversity of Balkan sironids (Opiliones, Cyphophthalmi) with descriptions of twelve new species. Zool J Linn Soc 156:260–318. https://doi.org/10.1111/j.1096-3642.2009.00446.x

Kato M, Itioka T, Sakai S et al (2000) Various population fluctuation patterns of light-attracted beetles in a tropical lowland dipterocarp forest in Sarawak. Popul Ecol 42:97–104. https://doi.org/10.1007/s101440050014

Kim KC (1993) Biodiversity, conservation and inventory: why insects matter. Biodivers Conserv 2:191–214. https://doi.org/10.1007/BF00056668

Kistner DH (1982) III. Insects. In: Hermann HR (ed) Social insects, vol 3. Academic Press, New York, pp 42–145

Klečka J, Boukal DS (2011) Lazy ecologist's guide to water beetle diversity: which sampling methods are the best? Ecol Indic 11:500–508. https://doi.org/10.1016/j.ecolind.2010.07.005

Klimaszewski J, Brunke AJ (2018) Canada's adventive rove beetle (Coleoptera, Staphylinidae) Fauna: a long-term case study on the detection, origin, introduction pathways, and dynamic distribution of non-native beetles. In: Betz O, Irmler U, Klimaszewski J (eds) Biology of rove beetles (Staphylinidae). Springer, Cham, pp 65–79. https://doi.org/10.1007/978-3-319-70257-5_5

Klimeš P, Saska P (2010) Larval and adult seed consumption affected by the degree of food specialization in *Amara* (Coleoptera: Carabidae). J Appl Entomol 134:659–666. https://doi.org/10.1111/j.1439-0418.2009.01463.x

Korotyaev BA, Konstantinov AS, Lingafelter SW et al (2005) Gall-inducing coleoptera. In: Raman A, Schaefer CW, Withers TM (eds) Biology, ecology, and evolution of gall-inducing arthropods, vol 1. Science Publishers, Enfield, pp 239–271

Kotze DJ, O'Hara RB (2003) Species decline—but why? Explanations of carabid beetle (Coleoptera, Carabidae) declines in Europe. Oecologia 135:138–148. https://doi.org/10.1007/s00442-002-1174-3

Kovarik PW, Caterino MS (2016) Histeridae. In: Beutel RG, Leschen RAB (eds) Handbook of Zoology, Arthropoda: Insecta, Coleoptera, Beetles Volume 1: Morphology and Systematics (Archostemata, Adephaga, Myxophaga, Polyphaga partim), 2nd edition. Walter de Gruyter, Berlin, pp 281–314

Kovats Z, Ciborowski J, Corkum L (1996) Inland dispersal of adult aquatic insects. Freshw Biol 36:265–276. https://doi.org/10.1046/j.1365-2427.1996.00087.x

Krell FT, Chung AYC, DeBoise E et al (2005) Quantitative extraction of macro-invertebrates from temperate and tropical leaf litter and soil: efficiency and time-dependent taxonomic biases of the Winkler extraction. Pedobiologia 49:175–186. https://doi.org/10.1016/j.pedobi.2004.10.005

Krinsky WL (2019) Beetles (Coleoptera). In: Mullen GR, Durden LA (eds) Medical and veterinary entomology, 3rd edn. Academic Press, London, pp 129–143. https://doi.org/10.1016/B978-0-12-814043-7.00009-1

Krüger E, Kahlow C, Leivas FWTL, Schühli G (2017) Scientific note: the histerid beetle *Omalodes foveola* (Coleoptera: Histeridae) found as a Melittophile, co-inhabiting Africanized honeybee hives in Brazil. Apidologie 48:572–574. https://doi.org/10.1007/s13592-017-0492-8

Kulshrestha P, Satpathy DK (2001) Use of beetles in forensic entomology. Forensic Sci Int 120:15–17. https://doi.org/10.1016/S0379-0738(01)00410-8

Kunz TH (1976) Observations on the Winter ecology of the bat Fly *Trichobius corynorhim* Cockerell (Diptera: Streblidae). J Med Entomol 12:631–636. https://doi.org/10.1093/jmedent/12.6.631

Lamarre GPA, Molto Q, Fine PVA et al (2012) A comparison of two common flight interception traps to survey tropical arthropods. Zookeys 216:43–55. https://doi.org/10.3897/zookeys.216.3332

Lambert DM, Baker A, Huynen L et al (2005) Is a large-scale DNA-based Inventory of ancient life possible? J Hered 96:279–284. https://doi.org/10.1093/jhered/esi035

Larsen TH, Lopera A, Forsyth A (2006) Extreme trophic and habitat specialization by peruvian dung beetles (Coleoptera: Scarabaeidae: Scarabaeinae). Coleopt Bull 60:315–324. https://doi.org/10.1649/0010-065X(2006)60[315:ETAHSB]2.0.CO;2

Larsen TH, Lopera A, Forsyth A et al (2009) From coprophagy to predation: a dung beetle that kills millipedes. Biol Lett 5:152–155. https://doi.org/10.1098/rsbl.2008.0654

Lawrence JF (1989) Mycophagy in the Coleoptera: feeding strategies and morphological adaptations. In: Wilding N, Collins NM, Hammond PM et al (eds) Insect fungus interactions. Academic Press, London, pp 1–23

Lawrence JF, Newton AF (1980) Coleoptera associated with the fruiting bodies of slime molds (Myxomycetes). Coleopt Bull 34:129–143

Lawrence JF, Newton AF (1982) Evolution and Classification of Beetles. Ann Rev Ecol Syst 13:261–90. https://doi.org/10.1146/annurev.es.13.110182.001401

Lawrence JF, Newton AF (1995) Families and subfamilies of Coleoptera (with selected genera, notes, references and data on family-group names). In: Pakaluk J, Ślipiński SA (eds) Biology, phylogeny, and classification of Coleoptera: papers celebrating the 80th birthday of Roy A. Crowson. Muzeum i Instytut Zoologii PAN, Warszawa, pp 779–1006

Lawrence JF, Ślipiński A, Seago AE et al (2011) Phylogeny of the Coleoptera based on morphological characters of adults and larvae. Ann Zool (Wars) 61:1–217. https://doi.org/10.3161/000345411X576725

Lawrence JF, Ślipiński A, Beutel RG et al (2013) A possible larva of *Lepicerus inaequalis* Motschulsky (Coleoptera: Myxophaga: Lepiceridae) from Panama. Zootaxa 3701:393–400. https://doi.org/10.11646/zootaxa.3701.3.8

Legros DL, Beresford DV (2010) Aerial foraging and sexual dimorphism in burying beetles (Silphidae: Coleoptera) in a Central Ontario forest. J Entomol Soc Ont 141:3–10

Leivas FWT, Grossi PC, Almeida LM (2013) Histerídeos (Staphyliniformia: Coleoptera: Histeridae) dos Campos Gerais, Paraná, Brasil. Biota Neotrop 13:196–204. https://doi.org/10.1590/S1676-06032013000200019

Lemos DBN, Agra LANN, Iannuzzi L et al (2010) Co-existence of myxomycetes and beetles in an Atlantic rainforest remnant of Pernambuco, Brazil, with emphasis on staphylinids (Coleoptera: Staphylinidae). J Nat Hist 44:1365–1376. https://doi.org/10.1080/00222931003632724

Lévieux J, Cassier P (1994) Vection de champignons pathogènes des résineux par les xylophages forestiers européens. Approche synthétique. Ann Biol 33:19–37

Linzmeier AM, Ribeiro-Costa CS (2013) Padrão sazonal de Chrysomelidae (Coleoptera) no estado do Paraná, sul do Brasil. Biota Neotrop 13:153–162. https://doi.org/10.1590/S1676-06032013000100018

Lipkow E, Betz O (2005) Staphylinidae and fungi. Faun-Ökol Mitt 8:383–411

López H, Oromí P (2010) A type of trap for sampling the mesovoid shallow substratum (MSS) fauna. Speleobiol Notes 2:7–11

Lord WD, Rodriguez WC (1989) Forensic entomology: the use of insects in the investigation of homicide and untimely death. PRO 22:41–48

Lord NP, Plimpton RL, Sharkey CR et al (2016) A cure for the blues: opsin duplication and subfunctionalization for short-wavelength sensitivity in jewel beetles (Coleoptera: Buprestidae). BMC Evol Biol 16:107. https://doi.org/10.1186/s12862-016-0674-4

Losey JE, Vaughan M (2006) The economic value of ecological services provided by insects. BioScience 56:311–323. https://doi.org/10.1641/0006-3568(2006)56[311:TEVOES]2.0.CO;2

Lövei GL, Sunderland KD (1996) Ecology and behavior of ground beetles (Coleoptera: Carabidae). Annu Rev Entomol 41:231–256. https://doi.org/10.1146/annurev.en.41.010196.001311

Maddison DR, Moore W, Baker MD et al (2009) Monophyly of terrestrial adephagan beetles as indicated by three nuclear genes (Coleoptera: Carabidae and Trachypachidae). Zool Scr 38:43–62. https://doi.org/10.1111/j.1463-6409.2008.00359.x

Maddison DR, Cooper KW (2014) Species delimitation in the ground beetle subgenus *Liocosmius* (Coleoptera: Carabidae: Bembidion), including standard and next-generation sequencing of museum specimens. Zool J Linn Soc 172:741–770. https://doi.org/10.1111/zoj.12188

Maia ACD, Dötterl S, Kaiser R et al (2012) The key role of 4-methyl-5-vinylthiazole in the attraction of scarab beetle pollinators: a unique olfactory floral signal shared by Annonaceae and Araceae. J Chem Ecol 38:1072–1080. https://doi.org/10.1007/s10886-012-0173-z

Maia ACD, Gibernau M, Dotterl S et al (2013) The floral scent of *Taccarum ulei* (Araceae): attraction of scarab beetle pollinators to an unusual aliphatic acyloin. Phytochemistry 93:71–78. https://doi.org/10.1016/j.phytochem.2013.03.005

Malaise RA (1937) A new insect-trap. Entomol Tidskr 38:148–160

Mammola S, Giachino PM, Piano E et al (2016) Ecology and sampling techniques of an understudied subterranean habitat: the Milieu Souterrain Superficiel (MSS). Sci Nat 103:88. https://doi.org/10.1007/s00114-016-1413-9

Marchão RL, Lavelle P, Celini L et al (2009) Soil macrofauna under integrated crop-livestock systems in a Brazilian Cerrado Ferralsol. Pesq Agropec Bras 44:1011–1020. https://doi.org/10.1590/S0100-204X2009000800033

Marinoni RC (2001) Os grupos tróficos em Coleoptera. Rev Bras Zool 18:205–224. https://doi.org/10.1590/S0101-81752001000100024

Marinoni RC, Dutra RRC (1997) Famílias de Coleoptera capturadas com armadilha Malaise em oito localidades do Estado do Paraná, Brasil. Diversidades alfa e beta. Rev Bras Zool 14:751–770. https://doi.org/10.1590/S0101-81751997000300021

Marshall AG (1981) The ecology of ectoparasitic insects. Academic Press, London

Martins VM, Ferreira RL (2020) Limiting similarity in subterranean ecosystems: a case of niche differentiation in Elmidae (Coleoptera) from epigean and hypogean environments. Hydrobiologia 847:593–604. https://doi.org/10.1007/s10750-019-04123-x

Matioli JC (1986) Armadilhas luminosas: uma tentativa no controle de pragas. Informe Agropecuário 12:36–38

Mazur S (2005) Ecology and habitats of Histeridae. In: Proceedings on taxonomy and faunistics of beetles (Coleoptera) dedicated to the 100th birthday of the Latvian entomologist Mihails Stiprais (1905–1990), p 29–36

Merritt RW, De Jong GD (2016) Arthropod communities in terrestrial environments. In: Benbow EM, Tomberlin JK, Tarone AM (eds) Carrion ecology, evolution, and their applications. CRC, Boca Raton, pp 65–91

Milhomem MS, Vaz-de-Mello FZ, Diniz IR (2003) Técnicas de coleta de besouros copronecrófagos no Cerrado. Pesqui Agropecu Bras 38:1249–1256. https://doi.org/10.1590/S0100-204X2003001100001

Miller ML, Lord WD, Goff ML (1994) Isolation of amitriptyline and nortriptyline from fly puparia (Phoridae) and beetle exuviae (Dermestidae) associated with mummified human remains. J Forensic Sci 39:1305–1313. https://doi.org/10.1520/JFS13717J

Miyazaki RD, Dutra RRC (1995) Famílias de Coleópteros capturadas com armadilhas luminosas em oito localidades do Paraná, Brasil. Rev Bras Zool 12:321–332. https://doi.org/10.1590/S0101-81751995000200010

Monné ML, Monné MA, Mermudes JRM (2009) Inventory of the Cerambycinae species (Insecta, Coleoptera, Cerambycidae) of the Parque Nacional do Itatiaia, RJ, Brazil. Biota Neotrop 9:283–312. https://doi.org/10.1590/S1676-06032009000300027

Mora C, Tittensor DP, Adl S, Simpson AGB, Worm B (2011) How many species are there on earth and in the ocean? PLoS Biol 9:e1001127. https://doi.org/10.1371/journal.pbio.1001127

Murvosh CM, Miller B (1970) A method of tagging large aquatic beetles. Entomol News 81:253–255

Nabli H, Bailey WC, Necibi S (1999) Beneficial insect attraction to light traps with different wavelengths. Biol Control 16:185–188. https://doi.org/10.1006/bcon.1999.0748

Nacagava VAF (2019) Artropofauna em ninhos de arara-azul no Pantanal do Miranda. Dissertation, Universidade Federal do Mato Grosso do Sul

Nagy ZT (2010) A hands-on overview of tissue preservation methods for molecular genetic analyses. Org Divers Evol 10:91–105. https://doi.org/10.1007/s13127-010-0012-4

Nakahama N, Isagi Y, Ito M (2019) Methods for retaining well-preserved DNA with dried specimens of insects. Eur J Entomol 116:486–491. https://doi.org/10.14411/eje.2019.050

Nakamura S, Tamura S, Taki H et al (2020) Propylene glycol: a promising preservative for insects, comparable to ethanol, from trapping to DNA analysis. Entomol Exp Appl 168:158–165. https://doi.org/10.1111/eea.12876

Nakano O, Leite CA (2000) Armadilhas para insetos: pragas agrícolas e domésticas. FEALQ, Piracicaba

Nath P, Singh J (1994) Impact of weather factors on light-trap catches of scarabaeid beetles (Col, Scarabaeidae). Giendg 7:137–141

Negrete-Yankelevich S, Fragoso C, Newton AC et al (2007) Successional changes in soil, litter and macroinvertebrate parameters following selective logging in a Mexican Cloud Forest. Appl Soil Ecol 35:340–355. https://doi.org/10.1016/j.apsoil.2006.07.006

Newton AF (1998) Phylogenetic problems, current classification and generic catalog of world Leiodidae (including Cholevidae). In: Giachino PM, Peck SB (eds) Phylogeny and evolution of subterranean and Endogean Cholevidae (= Leiodidae). Proceedings of XX international congress of entomology, Florence, vol 1996. Museo Regionale di Scienze Naturali, Torino, pp 41–178

Newton A, Peck SB (1975) Baited pitfall traps for beetles. Coleopt Bull 29:45–46

Newton AF, Stephenson SL (1990) A beetle/slime mold assemblage from northern India (Coleoptera: Myxomycetes). Orient Insects 24:197–218. https://doi.org/10.1080/00305316.1990.11835535

Nichols E, Spector S, Louzada J et al (2008) The Scarabaeinae Research Network. Ecological functions and ecosystem services provided by Scarabaeinae dung beetles. Biol Conserv 141:1461–1474. https://doi.org/10.1016/j.biocon.2008.04.011

Nielsen V (2003) Métodos Para Recolectar Insectos. Rev Agr Trop 33:59–68

Nielsen ES, Mound LA (1999) Global diversity of insects: the problems of estimating numbers. In: Raven PH, Williams T (eds) Nature and human society: the quest for a sustainable world. National Academy Press, Washington, DC, pp 213–222

Nitzu E (1997) Edaphicolous, endogeous and subterranean Coleoptera from the movile Karstic area (southern Dobrogea, Romania). Trav Inst Spéol 46:73–98

Nitzu E, Nae A, Popa I (2006) Eco-faunistic study on the invertebrate fauna (Araneae, Collembola and Coleoptera) from the Vârghis Gorge Natural Reserve (Eastern Carpathians, Romania), with special note on the micro-refugial role of the subterranean habitats. Trav Inst Spéol 45:31–50

Nitzu E, Nae A, Popa I (2008) The fauna of soil beetles (edaphic Coleoptera) as a sensitive indicator of evolution and conservation of ecosystems. A study on the altitudinal gradient in the Rodnei Mountains Biosphere Reserve (the Carpathians). In: Makarov SE, Dimitrijević RN (eds) Advances in arachnology and developmental biology: papers dedicated to prof. Dr. Božidar Ćurčić. University of Belgrade, Belgrade, pp 405–416

Ogawa R, Maeto K (2015) The Termitophilous Scaphidiinae (Coleoptera: Staphylinidae) from Sulawesi, Indonesia. Coleopts Bull 69:301–304. https://doi.org/10.1649/0010-065X-69.2.301

Ogedegbe ABO, Egwuonwu IC (2014) Biodiversity of soil arthropods in Nigerian Institute for Oil Palm Research (NIFOR), Nigeria. J Appl Sci Environ Manag 18:377–386

Oliveira MPA (2014) Os métodos de coleta utilizados em cavernas são eficientes para a amostragem da fauna subterrânea? Dissertation, Universidade Federal de Lavras

Oliveira VC, Pes AMO (2019) Inventário da fauna de insetos aquáticos: coleta, preservação e criação. In: Hamada N, Nessimian JL, Querino RB (eds) Insetos aquáticos na Amazônia: Taxonomia, biologia e ecologia. Editora INPA, Manaus, pp 151–167

Oliveira HG, Zanuncio TV, Zanuncio JC et al (2001) Coleópteros associados à eucaliptocultura na região de Nova Era, Minas Gerais, Brasil. Flora 8:52–60

Oliveira MPA, Bastos-Pereira R, Torres SHS et al (2019) Choosing sampling methods for Chilopoda, Diplopoda and Isopoda (Oniscidea): a case study for ferruginous landscapes in Brazilian Amazonia. Appl Soil Ecol 143:181–191. https://doi.org/10.1016/j.apsoil.2019.07.012

Ortloff A, Zanetti N, Centeno N et al (2014) Ultramorphological characteristics of mature larvae of *Nitidula carnaria* (Schaller 1783) (Coleoptera: Nitidulidae), a beetle species of forensic importance. Forensic Sci Int 239:1–9. https://doi.org/10.1016/j.forsciint.2014.03.010

Ortuño VM, Gilgado JD, Jiménez-Valverde A et al (2013) The "alluvial mesovoid shallow substratum", a new subterranean habitat. PLoS One 8:e76311. https://doi.org/10.1371/journal.pone.0076311

Ortuño VM, Ledesma E, Jiménez-Valverde A et al (2019) Studies of the mesovoid shallow substratum can change the accepted autecology of species: the case of ground beetles (Coleoptera, Carabidae) in the Sierra de Guadarrama National Park (Spain). Anim Biodiv Conserv 42:213–226. https://doi.org/10.32800/abc.2019.42.0213

Owens BE, Carlton CC (2015) "Berlese vs. Winkler": comparison of two forest litter Coleoptera extraction methods and the ECOLI (extraction of Coleoptera in litter) protocol. Coleopt Bull 69:645–661. https://doi.org/10.1649/0010-065X-69.4.645

Paquin P, Provost P (2010) Modification of the mouth aspiration for collection of cave arthropods. Speleobiol Notes 2:1–3

Pardal PPO, Silva CTC, Monteiro WM (2016) Dermatitis after contact with *Pheropsophus* sp (Coleoptera, Carabidae, Brachininae) in the Pará State, Brazilian Amazon. Rev Soc Bras Med Trop 49:799–801. https://doi.org/10.1590/0037-8682-0218-2016

Parker J (2016) Myrmecophily in beetles (Coleoptera): Evolutionary patterns and biological mechanisms. Myrmecol News 22:65–108

Peck S (1982) A review of the ectoparasitic *Leptinus* beetles of North America (Coleoptera: Leptinidae). Can J Zool 60:1517–1527. https://doi.org/10.1139/z82-201

Peck SB (2007) Distribution and biology of the ectoparasitic beetles Leptinillus validus (Horn) and L. aplodontiae Ferris of North America (Coleoptera: Leiodidae: Platypsyllinae). Insecta Mundi 3:1–7. https://digitalcommons.unl.edu/insectamundi/55

Peck SB, Cook J (1992) Use of "car-nets" to sample flying micro-Coleoptera. Can Entomol 124:745–749. https://doi.org/10.4039/Ent124745-5

Peck SB, Davies AE (1980) Collecting small beetles with large-area "window" traps. Coleopt Bull 34:237–239

Pellegrini TG, Ferreira RL (2012a) Metodologias diferenciadas aumentam a eficiência de inventários faunísticos em cavernas? Arquivos do Museu de História Natural e Jardim Botânico da UFMG 21:111–121

Pellegrini TG, Ferreira LR (2012b) Management in a neotropical show cave: planning for invertebrates conservation. Int J Speleol 41:359–366. https://doi.org/10.5038/1827-806X.41.2.19

Pellegrini TG, Ferreira RL (2013) Structure and interactions in a cave guano–soil continuum community. Eur J Soil Biol 57:19–26. https://doi.org/10.1016/j.ejsobi.2013.03.003

Pellegrini TG, Ferreira RL (2014) Ultrastructural analysis and polymorphisms in *Coarazuphium* caatinga (Coleoptera: Carabidae, Zuphiini), a new Brazilian troglobitic beetle. Zootaxa 3765:526–540. https://doi.org/10.11646/zootaxa.3765.6.2

Pellegrini TG, Pompeu PS, Ferreira RL (2018) Cave benthic invertebrates in South-Eastern Brazil: are there 'key' factors structuring such communities? Mar Freshw Res 69:1762–1770. https://doi.org/10.1071/MF18025

Pellissier ME, Nelson Z, Jabbour R (2017) Ecology and management of the alfalfa weevil (Coleoptera: Curculionidae) in Western United States Alfalfa. J Integr Pest Manag 8:1–7. https://doi.org/10.1093/jipm/pmw018

Penney MM (1966) Studies on certain aspects of the ecology of *Nebria brevicollis* (F.) (Coleoptera, Carabidae). J Anim Ecol 35:505–512. https://doi.org/10.2307/2488

Pentinsaari M, Anderson R, Borowiec L et al (2019) DNA barcodes reveal 63 overlooked species of Canadian beetles (Insecta, Coleoptera). ZooKeys 894:53–150. https://doi.org/10.3897/zookeys.894.37862

Persiani AM, Audisio P, Lunghini D et al (2010) Linking taxonomical and functional biodiversity of saproxylic fungi and beetles in broad-leaved forests in southern Italy with varying management histories. Plant Biosyst 144:250–261. https://doi.org/10.1080/11263500903561114

Pipan T, Culver DC (2019) Shallow subterranean habitats. In: White WB, Culver DC, Pipan T (eds) Encyclopedia of caves, 3rd edn. Academic Press, London, pp 896–908

Pipan T, López H, Oromí P et al (2011) Temperature variation and the presence of troglobionts in terrestrial shallow subterranean habitats. J Nat Hist 45:253–273. https://doi.org/10.1080/00222933.2010.523797

Puker A, Lopes-Andrade C, Rosa C, Grossi P (2012) New records of termite hosts for two species of *Hoplopyga*, with notes on the life cycle of *Hoplopyga brasiliensis* (Coleoptera: Scarabaeidae: Cetoniinae). Ann Entomol Soc Am 105:872 –878. https://doi.org/10.1603/AN12068

Quicke DLJ, Lopez-Vaamonde C, Belshaw R (1999) Preservation of hymenopteran specimens for subsequent molecular and morphological study. Zool Scr 28:261–267. https://doi.org/10.1046/j.1463-6409.1999.00004.x

Radwell AJ, Camp NB (2009) Comparing chemiluminescent and LED light for trapping water mites and aquatic insects. Southeast Nat 8:733–738. https://doi.org/10.1656/058.008.0414

Rafael JA (2002) A amostragem protocolo e técnicas de captura de Diptera. In: Costa C, Vanin SA, Lobo JM et al (eds) Proyecto de Red Iberoamericana de Biogeografía y Entomología Sistemática PrIBES, vol 2, pp 301–304

Ratcliffe BC (1980) New species of Coprini (Coleoptera: Scarabaeidae: Scarabaeinae) taken from the pelage of three toed sloths (*Bradypus tridactylus* L.) (Edentata: Bradypodidae) in Central Amazonia with a brief commentary on scarab-sloth relationships. Coleopt Bull 34:337–350

Reboleira ASPS, Ortuño VM (2014) A new species of *Duvalius* from world's deepest cave (Coleoptera: Carabidae). Zootaxa 3784:267–274. https://doi.org/10.11646/zootaxa.3784.3.6

Reiss RA, Schwert DP, Ashworth AC (1995) Field preservation of coleoptera for molecular genetic analyses. Environ Entomol 24:716–719. https://doi.org/10.1093/ee/24.3.716

Rendoš M, Mock A, Jászay T (2012) Spatial and temporal dynamics of invertebrates dwelling karstic mesovoid shallow substratum of Sivec National Nature Reserve (Slovakia), with emphasis on Coleoptera. Biologia 67:1143–1151. https://doi.org/10.2478/s11756-012-0113-y

Rodrigues JMG (1992) Abundância e distribuição vertical de coleópteros do solo em Capoeira de Terra Firme na região de Manaus—AM, Brasil. Acta Amazon 22:323–333. https://doi.org/10.1590/1809-43921992223333

Roff DA (1990) The evolution of flightlessness in insects. Ecol Monogr 60:389–421. https://doi.org/10.2307/1943013

Romero-López AA, Rodríguez-Palacios E, Alárcon-Gutiérrez E et al (2015) Effects of white grubs on soil water infiltration. Neotrop Entomol 44:134–139. https://doi.org/10.1007/s13744-015-0273-x

Ronqui DC, Lopes J (2006) Composição e diversidade de Scarabaeoidea (Coleoptera) atraídos por armadilha de luz em área rural no norte do Paraná. Brasil Iheringia Sér Zool 96:103–108. https://doi.org/10.1590/S0073-47212006000100018

Rosário VSV, Batista TFV, Provenzano R et al (2014) Edaphic insect fauna associated with reforestation with *Schizolobium parahyba* Barneby in Amazonia. Rev Cienc Agrar 57:373–381. https://doi.org/10.4322/rca.1659

Ruesink WG, Haynes DL (1973) Sweepnet sampling for the cereal leaf beetle, *Oulema melanopus*. Environ Entomol 2:161–172. https://doi.org/10.1093/ee/2.2.161

Russo L, Stehouwer R, Heberling JM et al (2011) The composite insect trap: an innovative combination trap for biologically diverse sampling. PLoS One 6:e21079. https://doi.org/10.1371/journal.pone.0021079

Sabu TK, Shiju RT, Vinod KV et al (2011) A comparison of the pitfall trap, Winkler extractor and Berlese funnel for sampling ground-dwelling arthropods in tropical montane cloud forests. J Insect Sci 11:1–19. https://doi.org/10.1673/031.011.0128

Sakchoowong W, Nomura S, Ogata K et al (2007) Comparison of extraction efficiency between Winkler and Tullgren extractors for tropical leaf litter macroarthropods. Thai J Agric Sci 40:97–105

Salomão RP, Brito LC, Iannuzzi L et al (2019) Effects of environmental parameters on beetle assemblage in a fragmented tropical rainforest of South America. J Insect Conserv 23:111–121. https://doi.org/10.1007/s10841-018-00120-y

Sánchez-Fernández D, Rizzo V, Cieslak A et al (2016) Thermal niche estimators and the capability of poor dispersal species to cope with climate change. Sci Rep 6:1–8. https://doi.org/10.1038/srep23381

Sanver D, Hawkins BA (2000) Galls as habitats: the inquiline communities of insect galls. Basic Appl Ecol 1:3–11

São José JB, Rieff GG, Sá ELS (2013) Mesofauna edáfica e atividade microbiana em diferentes sistemas de manejo do solo na cultura do tabaco. Curr Agric Sci Technol 19:56–66

Sasakawa K (2013) A novel technique for identifying the instar of field-collected insect larvae. PLoS One 8:e57836. https://doi.org/10.1371/journal.pone.0057836

Scheller HV (1984) The role of ground beetles (Carabidae) as predators on early populations of cereal aphids in spring barley. Z Angew Entomol 97:451–463. https://doi.org/10.1111/j.1439-0418.1984.tb03775.x

Schoenly KG, Haskell NH, Hall RD et al (2007) Comparative performance and complementarity of four sampling methods and arthropod preference tests from human and porcine remains at the Forensic Anthropology Center in Knoxville, Tennessee. J Med Entomol 44:881–894. https://doi.org/10.1093/jmedent/44.5.881

Scholtz CH, Davis ALV, Kryger U (2009) Evolutionary biology and conservation of dung beetles. Pensoft, Sofia

Schowalter TD (2012) Ecology and management of bark beetles (Coleoptera: Curculionidae: Scolytinae) in southern pine forests. J Integr Pest Manag 3:1–7. https://doi.org/10.1603/IPM11025

Schroeder H, Klotzbach H, Oesterhelweg L et al (2002) Larder beetles (Coleoptera, Dermestidae) as an accelerating factor for decomposition of a human corpse. Forensic Sci Int 127:231–236. https://doi.org/10.1016/S0379-0738(02)00131-7

Segura M, Fonseca-Gessner A, Spies M et al (2012) Water beetles in mountainous regions in southeastern Brazil. Braz J Biol 72:311–321. https://doi.org/10.1590/S1519-69842012000200012

Seric-Jelaska L, Franjevic D, Jelaska SD et al (2014) Prey detection in carabid beetles (Coleoptera: Carabidae) in woodland ecosystems by PCR analysis of gut contents. Eur J Entomol 111:631–638. https://doi.org/10.14411/eje.2014.079

Sharma S, Sharma G, Pir FA (2019) Diversity and habitat selection of aquatic beetles (Coleoptera). IOSR J Pharm Biol Sci 14:31–37. https://doi.org/10.9790/3008-1401013137

Sharratt NJ, Picker MD, Samways MJ (2000) The invertebrate fauna of the sandstone of the caves of the Cape Peninsula (South Africa): patterns of endemism and conservation priorities. Biodivers Conserv 9:107–143. https://doi.org/10.1023/A:1008968518058

Sheikh AH, Thomas M, Bhandari R et al (2016) Malaise trap and insect sampling: mini review. Biol Bull 2:35–40

Silva FAB, Vaz-de-Mello FZ, Barclayc MVL (2018) An updated key to the millipede-hunting subgenus *Aganhyboma* Kolbe, 1893 of the genus *Deltochilum* Eschscholtz, 1822 (Coleoptera: Scarabaeidae: Scarabaeinae), with description of a new species from Bolivia and Peru. Insect Syst Evol 49:231–240. https://doi.org/10.1163/1876312X-00002173

Silveira M, Queiroz J, Boeira R (2004) Protocolo de coleta e preparação de amostras de macroinvertebrados bentônicos em riachos. Embrapa Meio Ambiente—Comunicado Técnico, Jaguariúna

Silveira FAO, Santos JC, Mello L et al (2006) Predation on *Atta laevigata* (Smith 1858) (Formicidae Attini) by *Canthon virens* (Mannerheim 1829) (Coleoptera Scarabaeidae). Trop Zoo 19:1–7

Sket B, Paragamian K, Trontelj P (2004) A census of the obligate subterranean fauna of the Balkan peninsula. Balkan biodiversity. In: Griffiths HI, Krystufek B, Reed JM (eds) Pattern and process in Europe's biodiversity hotspot. Kluwer Academic, Dordrecht, pp 309–322

Skvarla MJ, Dowling APG (2017) A comparison of trapping techniques (Coleoptera: Carabidae, Buprestidae, Cerambycidae, and Curculionoidea excluding Scolytinae). J Insect Sci 17:1–28. https://doi.org/10.1093/jisesa/iew098

Ślipiński SA, Leschen RAB, Lawrence JF (2011) Order Coleoptera Linnaeus, 1758. In: Zhang Z-Q (ed) Animal biodiversity: an outline of higher-level classification and survey of taxonomic richness. Zootaxa 3148:203–208. https://doi.org/10.11646/zootaxa.3148.1.39

Silva-Neto CM, Brito PVA, Freitas PVDX (2019) Predador de abelhas sem ferrão no Cerrado brasileiro: primeiro registro de *Hololepta (Leionota)* reichii Marseul (Coleoptera, Histeridae) em colônia de *Melipona quadrifasciata*. Tecnia 4:194–200

Snyder BA, Draney ML, Sierwald P (2006) Development of an optimal sampling protocol for millipedes (Diplopoda). J Insect Conserv 10:277–288. https://doi.org/10.1007/s10841-006-6699-z

Soler JJ, Martínez JG, Soler M et al (1999) Host sexual selection and cuckoo parasitism: an analysis of nest size in sympatric and allopatric magpie *Pica pica* populations parasitized by the great spotted cuckoo Clamator glandarius. Proc R Soc Lond B 266:1765–1771. https://doi.org/10.1098/rspb.1999.0844

Sörensson M (1997) Morphological and taxonomical novelties in the world's smallest beetles, and the first Old World record of Nanosellini (Coleoptera: Ptiliidae). Syst Entomol 22:257–283. https://doi.org/10.1046/j.1365-3113.1997.d01-40.x

Southwood TRE (1978) Ecological methods: with particular reference to the study of insect populations, London. Chapman and Hall, London

Souto PC, Souto JS, Miranda JRP et al (2008) Comunidade microbiana e mesofauna edáficas em solo sob caatinga no semi-árido da Paraíba. Rev Bras Ciênc Solo 32:151–160. https://doi.org/10.1590/S0100-06832008000100015

Souza-Silva M, Martins RP, Ferreira RL (2011) Cave lithology determining the structure of the invertebrate communities in the Brazilian Atlantic Rainforest. Biodivers Conserv 20:1713–1729. https://doi.org/10.1007/s10531-011-0057-5

Souza-Silva M, Júnior AS, Ferreira RL (2013a) Food resource availability in a quartzite cave in the Brazilian montane Atlantic Forest. J Cave Karst Stud 75:177–188. https://doi.org/10.4311/2010JCKS0158

Souza-Silva M, Rezende RKS, Ferreira RL (2013b) Detritus processing in lentic cave habitats in the neotropics. Subterr Biol 11:3–14. https://doi.org/10.3897/subtbiol.11.5107

Spangler PJ (1981) A durable, lightweight net and a manual aspirator for collecting aquatic organisms. Pan-Pac Entomol 57:245–250

Spangler PJ (1996) Four new stygobiontic beetles (Coleoptera: Dytiscidae; Noteridae; Elmidae). Insecta Mundi 10:241–259

Stejskal V, Hubert J, Aulicky R et al (2015) Overview of present and past and pest-associated risks in stored food and feed products: European perspective. J Stored Prod Res 64:122–132. https://doi.org/10.1016/j.jspr.2014.12.006

Steyskal GC, Murphy WL, Hoover EM (eds) (1986) Insects and mites: techniques for collection and preservation. U.S. Department of Agriculture, Washington

Stone C, Goodyer G, Sims K et al (2010) Beetle assemblages captured using static panel traps within New South Wales pine plantations. Aust J Entomol 49:304–316. https://doi.org/10.1111/j.1440-6055.2010.00769.x

Stork NE (2018) How many species of insects and other terrestrial arthropods are there on earth? Annu Rev Entomol 63:31–45. https://doi.org/10.1146/annurev-ento-020117-043348

Stork NE, McBroom J, Gely C et al (2015) New approaches narrow global species estimates for beetles, insects, and terrestrial arthropods. Proc Natl Acad Sci U S A 112:7519–7523. https://doi.org/10.1073/pnas.1502408112

Sugaya K, Yamasako J (2008) A new species of the genus *Ishikawatrechus* (Coleoptera, Trechinae) from Japan. Zootaxa 3768:189–195. https://doi.org/10.11646/zootaxa.3768.2.6

Tagliavia M, Messina E, Manachini B et al (2014) The gut microbiota of larvae of *Rhynchophorus ferrugineus* Oliver (Coleoptera: Curculionidae). BMC Microbiol 14:136. https://doi.org/10.1186/1471-2180-14-136

Tandon R, Manohara TN, Nijalingappa BHM et al (2001) Pollination and Pollen-pistil Interaction in Oil Palm, Elaeis guineensis. Ann Bot 87:831–838. https://doi.org/10.1006/anbo.2001.1421

Timm RM, Ashe JS (1987) Host and elevational specificity of parasitic beetles (*Amblyopinus* Solsky) (Coleoptera: Staphylinidae) in Panama. Proc Biol Soc Wash 100:13–20
Tobin BW, Hutchins BT, Schwartz BF (2013) Spatial and temporal changes in invertebrate assemblage structure from the entrance to deep-cave zone of a temperate marble cave. Int J Speleo 42:203–214. https://doi.org/10.5038/1827-806X.42.3.4
Tripet F, Richner H (1997) The coevolutionary potential of a 'generalist' parasite, the hen flea *Ceratophyllus gallinae*. Parasitology 115:419–427. https://doi.org/10.1017/S0031182097001467
Torto B, Fombong AT, Mutyambai DM, Muli E, Arbogast RT, Teal PEA (2010) *Aethina tumida* (Coleoptera: Nitidulidae) and *Oplostomus haroldi* (Coleoptera: Scarabaeidae): occurrence in Kenya, distribution within honey bee colonies, and responses to host odors. Ann Entomol Soc Am 103:389–396. https://doi.org/10.1603/AN09136
Tooker JF, Hanks LM (2004) Impact of prescribed burning on endophytic insect communities of prairie perennials (Asteraceae: *Silphium* spp.). Biodivers Conserv 13:1875–1888. https://doi.org/10.1023/B:BIOC.0000035872.24201.30
Townes HK (1962) Design of a Malaise trap. Proc Entomol Soc Wash 64:253–262
Trajano E, Gnaspini-Netto P (1990) Composição da fauna cavernícola brasileira, com uma análise preliminar da distribuição dos táxons. Rev Bras Zool 7:383–407. https://doi.org/10.1590/S0101-81751990000300017
Triplehorn CA, Johnson NF (2011) Estudo dos insetos. Cengage Learning, São Paulo
Turić N, Temunović M, Vignjević G et al (2017) A comparison of methods for sampling aquatic insects (Heteroptera and Coleoptera) of different body sizes, in different habitats using different baits. Eur J Entomol 114:123–132. https://doi.org/10.14411/eje.2017.017
Turner C (2007) Collecting water beetles: an introduction. In: Staals R, de Moor IJ (eds) Guides to the freshwater invertebrates of southern Africa. Aquatic Coleoptera, vol 10. Water Research Commission of South Africa, Pretoria, pp 193–204
Upton MS, Mantle BL (2010) Methods for collecting, preserving and studying insects and other terrestrial arthropods. Paragon Printers Australasia, Canberra
Ururahy-Rodrigues A, Rafael JA, Wanderley RF et al (2008) *Coprophanaeus lancifer* (Linnaeus, 1767) (Coleoptera, Scarabaeidae) activity moves a man-size pig carcass: relevant data for Forensic Taphonomy. Forensic Sci Int 182:19–22. https://doi.org/10.1016/j.forsciint.2008.09.009
Vane-Wright RI, Humphries CJ, Willams PH (1991) What to protect? Systematics and the agony of choice. Biol Conserv 55:235–254. https://doi.org/10.1016/0006-3207(91)90030-D
Vaz-de-Mello FZ (2008) Synopsis of the new subtribe Scatimina (Coleoptera: Scarabaeidae: Scarabaeinae: Ateuchini), with descriptions of twelve new genera and review of Genieridium, new genus. Zootaxa 1955:1–75. https://doi.org/10.11646/zootaxa.1955.1.1
Viviani VR, Rocha MY, Hagen O (2010) Bioluminescent beetles (Coleoptera: Elateroidea: Lampyridae, Phengodidae, Elateridae) in the municipalities of Campinas, Sorocaba-Votorantim and Rio Claro-Limeira (SP, Brazil): biodiversity and influence of urban sprawl. Biota Neotrop 10:103–116. https://doi.org/10.1590/S1676-06032010000200013
Voigt J (1965) Specific post-mortem changes produced by larder beetles. J Forensic Med 12:76–80
Wang Y, Liu X, Garzón-Orduña IJ et al (2017) Mitochondrial phylogenomics illuminates the evolutionary history of Neuropterida. Cladistics 33:617–636. https://doi.org/10.1111/cla.12186
Weinstein P, Slaney D (1995) Invertebrate faunal survey of Rope Ladder Cave, Northern Queensland: a comparative study of sampling methods. J Aust Entomol Soc 34:233–236. https://doi.org/10.1111/j.1440-6055.1995.tb01329.x
Wheeler QD, Miller KB (2005) Slime-mold beetles of the genus *Agathidium* Panzerin North and Central America, part I. Coleoptera: Leiodidae. Bull Am Mus Nat Hist 290:1–95
Whitaker JO Jr, Ritzi CM, Dick CW (2009) Collecting and preserving bat ectoparasites for ecological study. In: Kunz TH, Parsons S (eds) Ecological and behavioral methods for the study of bats, 2nd edn. Johns Hopkins University Press, Washington, pp 806–827

White RE (1983) A field guide to the beetles of North America: text and illustrations. Houghton Mifflin, Boston

Williams DD, Feltmate BW (1992) Aquatic insects. C.A.B. International, Wallingford

Wilson WO (2000) On the future of conservation biology. Conserv Biol 14:1–3

Wodika BR, Baer SG (2015) If we build it, will they colonize? A test of the field of dreams paradigm with soil macroinvertebrate communities. Appl Soil Ecol 91:80–89. https://doi.org/10.1016/j.apsoil.2015.02.005

Wynne JJ, Sommer S, Howarth FG et al (2018) Capturing arthropod diversity in complex cave systems. Divers Distrib 24:1478–1491. https://doi.org/10.1111/ddi.12772

Wynne JJ, Howarth FG, Sommer S et al (2019) Fifty years of cave arthropod sampling: techniques and best practices. Int J Speleol 48:33–38. https://doi.org/10.5038/1827-806X.48.1.2231

Yule CM, Sen YH (2004) Freshwater invertebrates of the Malaysian region. Academy of Sciences Malaysia, Kuala Lumpur

Zagmajster M, Culver DC, Sket B (2008) Species richness patterns of obligate subterranean beetles (Insecta: Coleoptera) in a global biodiversity hotspot–effect of scale and sampling intensity. Divers Distrib 14:95–105. https://doi.org/10.1111/j.1472-4642.2007.00423.x

Zagmajster M, Culver DC, Christman MC et al (2010) Evaluating the sampling bias in pattern of subterranean species richness: combining approaches. Biodivers Conserv 19:3035–3048. https://doi.org/10.1007/s10531-010-9873-2

Zeng Y, Giblin-Davis R, Weimin Y et al (2007) *Bradynema listronoti* n. sp. (Nematoda: Allantonematidae), a parasite of the carrot weevil Listronotus oregonensis (Coleoptera: Curculionidae) in Quebec, Canada. Nematology 9:608–622. https://doi.org/10.1163/156854107782024749

Zepon T, Bichuette ME (2017) Influence of substrate on the richness and composition of Neotropical cave fauna. An Acad Bras Ciênc 89:1615–1628. https://doi.org/10.1590/0001-3765201720160452

Zhang SQ, Che LH, Li Y et al (2018) Evolutionary history of Coleoptera revealed by extensive sampling of genes and species. Nat Commun 9:1–11. https://doi.org/10.1038/s41467-017-02644-4

Chapter 7
Sampling Methods for Adult Flies (Diptera)

Brian V. Brown

7.1 The Study of Diptera and Its Importance

Diptera have been studied for more than 200 years, but what we know is still less than what is still to be discovered (Jaschhof and Jaschhof 2014; Borkent et al. 2018; Brown et al. 2018; Srivathsan et al. 2019). With over 160,000 described species, and estimates ranging to one million or more existing species, it is little wonder that flies are infrequently used to assess biodiversity (Basset et al. 2012; Borkent and Brown 2015). For many general inventories, the challenge of dealing with a largely unknown Diptera fauna forces researchers to fall back on better-known groups of organisms, such as birds, butterflies, and plants. Inclusion of flies in biodiversity studies, however, allows an extremely fine-grained view of species diversity over space and time.

Flies are predators, parasitoids, parasites, herbivores, scavengers, and fungivores, filling nearly every available niche on the planet, with one notable exception: the marine environment (Brown et al. 2009; Marshall 2012; Marshall and Kirk-Spriggs 2017). There, flies are found mostly on the periphery, on beaches, in mangrove swamps, and on rocky shorelines, with few exceptions (Cheng 1976). In terrestrial and freshwater habitats, however, Diptera abound (Adler and Courtney 2019).

The practical importance of Diptera to humans can scarcely be overstated, with mosquito-borne diseases such as malaria causing thousands of deaths per year, while another fly, *Drosophila melanogaster*, is a model organism whose easy lab culture has allowed our understanding of many genetic processes. Aside from their immediate practical and commercial values, flies are also important pollinators,

B. V. Brown (✉)
Entomology Section, Natural History Museum of Los Angeles County,
Los Angeles, CA, USA
e-mail: bbrown@nhm.org

J. C. Santos, G. W. Fernandes (eds.), *Measuring Arthropod Biodiversity*,
https://doi.org/10.1007/978-3-030-53226-0_7

decomposers, and controllers of other insect populations, delivering ecosystem services often taken for granted (Larson et al. 2001; Raguso 2020).

7.2 Sampling Goals

A clear understanding of the sampling goals should be the first consideration. Sampling can be done to provide ecological information, to search for specific taxonomic groups, or to attempt to inventory a site. Only the latter is discussed in detail in this chapter. Ecological sampling has a large associated literature (Henderson and Southwood 2016) that will not be surveyed here. Instead, I concentrate on sampling for inventory, based especially on my work in tropical rain forests, although most of these techniques work elsewhere. Sampling for specific taxonomic groups involves a subset of inventory techniques, requiring specialized knowledge and innovation.

An inventory is expensive, both in resource and time commitments. Obviously, the more complete the inventory required, the greater the costs. Rarely is a full inventory needed or possible, although enthusiastic researchers often set out to attain this goal. The overwhelming diversity of families such as Cecidomyiidae, Phoridae, Sciaridae, and Sphaeroceridae (Janzen and Hallwachs *in review*), make the selection of smaller subsets of taxa a more feasible goal.

A preliminary inventory can be the assessment of the families present at a site. This can be accomplished by a few researchers over a short period (about 1 week), with a high probability of encountering most of the families present. Targeted collecting will be necessary to capture some of the more specialized families (such as bat and bird parasites) and to associate specimens with life histories, but a combination of hand collecting, Malaise traps, light traps, Shannon traps, and pan traps will collect most of the families present.

To obtain more detailed inventories, such as at the genus or species level, I suggest extending the family-level inventory with longer-term collecting with Malaise traps. Operation of one or more such traps over an entire year, in addition to the collecting for the family-level trapping, would be optimal.

In rare cases, an entire faunal inventory can be attempted, such as the Costa Rican ZADBI project (Borkent and Brown 2015). Such an effort requires dozens of collaborators and funding of about $1 million and may still undersample a diverse site (Borkent et al. 2018; Brown et al. 2018).

7.3 Killing and Preservation Substances

In order to produce acceptable or even excellent specimens for research purposes, flies must be properly preserved. There are many opinions as to the best way to prepare specimens, depending on the type of fly, the history for preparing specimens in a group, and the need for future use of the specimen. As a general rule, small specimens are collected into alcohol and larger specimens kept dry.

Collecting smaller specimens into alcohol preserves body structure and allows for later treatment with chemical drying, slide mounting, or sequencing. Dry pinning of small flies, often by preparing double mounts (micro pins that are themselves pinned into small pieces of foam), is popular with some researchers of small acalyptrate flies, at least partly because they can be prepared immediately in the field. The flies prepared in this manner will shrivel and distort, but this is acceptable for these researchers, who are accustomed to working with such specimens. I do not recommend double mounting of specimens, but it is such an ingrained part of the culture of some workers, that I doubt its use will disappear.

In order to produce better, undistorted specimens, small flies collected into alcohol can be critical-point-dried (Gordh and Hall 1979) or chemically dried using hexamethyldisilazane (HMDS, Brown 1993) or ethyl acetate (Vockeroth 1966). Another possible chemical drying agent is amyl acetate, which is less toxic than HMDS and does not damage DNA (Austin and Dillon 1997), although I have not investigated the quality of specimens treated by this chemical. The difference between air-dried and treated specimens can be astonishing and often represents the difference between a specimen that will be used and one that will be rejected. Long-term storage in alcohol is not recommended, as drying and slide mounting do not work as well on older specimens, and DNA will degrade over time (whereas I suspect HMDS might help preserve DNA). If storage in alcohol is required for some period of time, however, the material should be kept at as low a temperature as possible in a refrigerator or, preferably, a freezer.

The taxonomy of some small flies (e.g., Psychodidae, Chironomidae, and Ceratopogonidae) requires that they be dissected and slide mounted. This is a meticulous and slow process that requires skill and training. Researchers have preferences as to the way dissections are conducted and how specimens are arranged on slides, but one constant is that they should be preserved in a permanent mounting medium. In the past this has not always been done, with often lamentable results, as large collections of important specimens mounted with easy-to-use, nonpermanent mounting media have eventually become almost worthless through drying and cracking. I cannot recommend more highly the importance of using Canada balsam as a permanent mounting medium. In the past, this was a difficult material with which to work, requiring a series of dehydrations of increasing alcohol percentages, followed by soaking in xylene, to fully dehydrate specimens, making them ready to mount. Now, however we can transfer specimens from 70 to 90% ethanol into the easily available clove oil as a single-step, relatively nontoxic dehydrating treatment prior to mounting. Once the specimens have cleared in clove oil (a process that can be hastened by heating in a microwave oven for 10 s bursts), they can be immediately mounted in Canada balsam.

Larger flies can be kept dry, since they do not tend to shrivel and distort so extensively. Indeed, the color patterns and setae of such specimens are better kept completely dry, such that the difficult-to-obtain and relatively dangerous chemical potassium cyanide is the best killing agent, followed by direct pinning. If cyanide is not available, however, a small amount of ethyl or amyl acetate in a killing jar is acceptable as a killing agent.

Large specimens that are collected into alcohol, through trapping or other wet collecting methods (below), can be preserved in a reasonable state by pinning them while still wet (in alcohol), blotting off the alcohol, and putting the pinned specimen in a container of amyl or ethyl acetate for 4–8 h. The pinned specimens can then be removed from the ethyl acetate using forceps on the pin, blotted dry, and puffed up by blowing air on it. Note that for several hours the pin head will be not be secure, and flies should be allowed to dry for several hours in a fume hood (to avoid the fumes) before handling the pin. Also, note that ethyl acetate (but not amyl acetate) will possibly damage DNA and render specimens unusable for DNA sequencing.

Another possible mounting method for large flies is direct pinning while still fresh, followed by putting the pinning container in a freezer for a few weeks to months. This freeze-drying technique is slow, but avoids the use of chemicals.

Any specimens for which sequencing is a possibility should be kept dry or in high percentage alcohol (95%) and stored as cold as possible.

7.4 Basic Collecting

Flies are ubiquitous, and any general collecting methods used to sample insects will collect some (Marshall et al. 1994; Danks 1996; Häuser and Riede 2015). Some methods produce more flies per unit effort, however, and specialized techniques are necessary to collect meaningful numbers of harder-to-find species (Disney et al. 1982; Chandler 2010; Borkent and Brown 2015).

The most basic method of collecting flies is by searching for them visually and catching the discovered specimens with a net or aspirator (pooter). Insect nets for Diptera should be lightweight and can have extremely fine weave bags for retaining minute forms (the smallest flies in the world are about 0.4 mm long, Brown 2018) or more open mesh to help reduce wind drag when attempting to catch only larger species.

Many flies are extremely fast fliers and can only be collected by clamping the net on the ground in a vigorous manner. Specimens are commonly lost through slow or indecisive swings of the net. Once the net is clamped to the ground, the bag should be lifted with one hand while the other hand keeps the net firmly on the ground. Flies usually move up into the net, where they can be isolated and transferred to a collecting container (vial, bottle, bag).

Sweeping is swinging an insect net above or through vegetation, over soil, mud, streams, or even in the air without any particular target. Flies are more fragile than many other groups of insects, such that the net needs to be checked frequently to prevent damage to specimens with repeated pounding against vegetation or the net bag. A wet net bag is especially damaging to dipteran specimens, which will form a worthless pulpy mass under these conditions. Different types of vegetation have to be swept with different degrees of vigor, including dense bunch grasses that must be almost pummeled with the net to yield the flies hiding in the bases.

Special forms of sweeping include mechanized suction traps (Favret et al. 2019), which have been used for aphids. The dipteran fauna of these samples is yet to be studied.

Another variation on sweeping is the use of car nets: large net bags mounted on a car or other vehicle, which is then driven through suitable habitat (Sommerman and Simmet 1965; Loy et al. 1968; Barnard 1979). As with other methods, the catch must be protected from excessive battering during the collection process.

When there is no room to sweep, a net can be placed over specimens, which will generally fly up into the net bag carefully placed over them. Sometimes it is necessary to disturb the flies, which might be occupied with feeding or other activities.

Often, a net is unnecessary or a hindrance to collecting. In these situations, flies can be aspirated directly either into an intermediate container or into a rigid tube and then blown into a collecting vial.

7.5 Mass Collection Devices (Traps)

Although sweeping and hand collecting can catch many flies, using traps is necessary to encounter many smaller and difficult to find species.

7.5.1 Malaise Traps: The Ultimate Diptera Sampling Devices

There is a large literature associated with Malaise traps (Steyskal 1981), beginning with the original proposal of the device (Malaise 1937). Basically, a Malaise trap is a small to large, open-sided tent with a collecting bottle near the top. Inside the bottle is a killing agent, usually ethanol, but users who study flies that they prefer to keep dry (such as calyptrates) often use cyanide.

The Malaise trap takes advantage of the tendency for most insects to be positively phototropic (to be attracted to the light) and negatively geotropic (to move upward) when they encounter an obstruction. Sometimes, pan traps (plastic trays with water, soap (to break the surface tension), and a preservative (salt, glycol)) are placed under the central panel of a Malaise trap to collect insects that drop when they hit a barrier. A trap that consists of just a central panel and pan traps is called as FIT (flight interception trap, Masner and Goulet 1981). Both FITs and Malaise pans have been shown to collect different taxa than the Malaise trap collecting bottle (Disney et al. 1982).

There have been many types of Malaise trap designs proposed, from a folded sheet with an attached soda bottle to sophisticated, self-supporting models (van Achterberg 2009). The two most commonly used variations are the Townes lightweight design (Fig. 7.1; Townes 1972) and the larger, two-headed Gressitt model

Fig. 7.1 Townes-style lightweight Malaise trap

(Fig. 7.2; Gressitt and Gressitt 1962). The Townes model is often preferred for smaller species and the Gressitt for catching larger, stronger fliers. Mesh size for the traps can be an issue, with the smallest flies in the world being less than one-half millimeter in length, but coarser mesh traps reputedly are reportedly better for larger insects (Darling and Packer 1988).

Placing a large, dark object in the Malaise trap can increase the number of biting flies attracted (Roberts 1972). A balloon or ball spray-painted black and hung in the trap attracts many more tabanids, for example.

One potential disadvantage of Malaise traps, noted by Disney et al. (1982), is that they collect large numbers of insects and samples take a long time to sort. Whether sorting large samples is tiresome or exhilarating depends on the catch and the sorter's point of view, but large samples can be time-consuming to separate. The process can be improved by gently screen sorting the catch to produce subsamples of more uniform specimen size. A mesh of 1 cm can produce a "coarse" fraction, one of 5 mm a "medium" fraction, and 1 mm a "fine" fraction. Fine fractions often contain rare or easily overlooked species that are hidden by larger insects in mixed samples. To avoid damaging specimens, I suggest gently swirling successive fractions in a tub of alcohol. A net made out of extremely fine mesh material is needed to strain out all insects from the alcohol for examination under a stereoscope.

Fig. 7.2 Gressitt-style Malaise trap

Preventing larger insects from entering the trap by placing a mesh screen over the Malaise trap bottle entrance is a viable option for those working exclusively on smaller flies. The advantages of such screening is that overall sample size and time needed to sort the sample are greatly reduced, damage to smaller specimens due to the thrashing of larger insects is minimized, and losses of smaller specimens that tend to get irretrievably caught in the bodies of larger insects, especially ones with dense setae, are reduced. One proposal for this treatment is given by Martens et al. (2019) and Martens and Johnson (2019), who found that the number and diversity of minute wasps recovered by screened traps was greatly enhanced compared to unscreened samples.

Recent collecting in the Brazilian Amazon has convinced me that the Gressitt-style trap has some advantages over the smaller Townes-style traps. Even for phorid and sphaerocerid flies, which are notorious for their ground-level activity, a single Gressitt-style trap outcollected six Townes-style traps in terms of number of specimens collected (unpublished observation). It is not yet determined whether the six Townes-style traps collected a greater diversity of species, as might be imagined through their sampling of more different microhabitats. Regardless, the cost of one Gressitt-style trap is equal to that of about three Townes-style traps and might be a more cost-effective option per specimen in a relatively small collecting area.

7.5.2 Light Traps

In a survey of insects at a tropical cloud forest, Borkent et al. (2018) found that light trapping collected the greatest complementary diversity of Diptera to Malaise traps, although collecting effort between the two methods is difficult to compare. Different lights provide different catches; extra-bright mercury vapor lights that moth and beetle collectors use will attract flies as well, but it might be true that many flies do not approach such a bright light. Using an ultraviolet light over a pan trap (see below) with a screen over top to exclude larger insects captures many species of flies too tiny to be noticed in the flurry of mercury vapor collecting (Brown 2012). Light trapping near water often results in huge catches of chironomids, but light trapping is worthwhile even in extremely dry habitats like deserts, where diurnal activity is often limited. Some parasitic flies, such as those associated with nocturnal or crepuscular army ant colonies, are collected by light traps, as are the nocturnal parasitoids of the tachinid genus *Ormia* and the families Pyrgotidae and Ctenostylidae.

The CDC (Centers for Disease Control and Prevention, Sudia and Chamberlain 1962) light trap uses dim lights, and when used in dark forests it is effective for collecting flies less attracted to the brighter lights (Jaschhof and Jaschhof 2014; Borkent et al. 2018). A small fan provides the suction to draw in the flies attracted to the lights.

7.5.3 Pan Traps

Many dipterists are avid users of pan traps (also called water traps). The pan trap is a colored dish partly filled with water, to which some detergent (such as a few drops of dishwashing liquid) is added to reduce surface tension. Without this detergent, specimens often float or bounce off the surface, sometimes escaping. Additionally, the water can be saturated with salt (sodium chloride) to prevent swelling and to slow decomposition of specimens. For traps that will be left for many days, a trapping mixture including ethylene glycol (often used in car window washing fluid) has been used because it does not evaporate. This endangers wildlife and pets, however, which will drink that fluid and poison themselves. Use of picric acid to make the solution extremely bitter is advised by some researchers, but in general less toxic alternatives, such as swimming pool antifreeze, are suggested to prevent evaporation of liquids in longer-term trapping (Thomas 2008).

Any water-based traps should be frequently emptied, preferably at least daily, to prevent softening, swelling, bloating, or bursting of specimens. Additionally, decomposition of DNA is extremely rapid in water, so pan-trapped specimens might not be suitable for any molecular work. On the other hand, swollen specimens often have diagnostic characters more easily visible, with the swelling exposing overlapping tergites or sometimes everting hidden genitalia.

Table 7.1 Summary of pan trap color preferences

Taxon	Color preferred	Reference
Syrphidae	Blue	Campbell and Hanula (2007)
Chironomidae	Orange, brown	Disney et al. (1982)
Most families	White, yellow	Disney et al. (1982) and Vrdoljak and Samways (2012)
Agromyzidae	Yellow	Disney et al. (1982)
Phoridae	White	Disney et al. (1982)
Empidoidea	White, yellow	Pollet and Grootaert (1994)
Pollinating Diptera	Yellow	Schleuter and Stewart (2011)

Pan traps possibly mimic light pools that occur in forests and elsewhere or just tend to intercept ground inhabiting insects. If the former, then using many small pan traps should be more productive than an equal area sampled by larger pan traps, since each trap would represent a single light pool. This idea was tested by Brown and Feener (1995), who found smaller traps were significantly more productive per unit area.

The color of the pans used for pan trapping can produce significantly different results and has been studied frequently, especially for bees (Hymenoptera), but few results are available for Diptera. Colors most commonly used are yellow, white, and blue, with studies showing various types of preference (Table 7.1).

7.5.4 *Pitfall Traps*

Pitfall traps are small containers with steep sides that are dug into the ground, such that their rim is level with the surrounding soil. A killing and preserving fluid, often water with a small amount of detergent to break the surface tension, is usually used to capture the specimens (see more on killing fluids under pan traps). More commonly used for beetles, ants, and spiders, pitfall traps are sometimes used in conjunction with baits to collect flies.

7.5.5 *Emergence Traps*

These are traps that cover a given area (1 square meter is the usual size) and collect all specimens that emerge from whatever is underneath. They can be placed on the ground to compare the emergence from specific soil types (e.g., pasture versus forest) or even over water (Kimerle and Anderson 1967; Cadmus et al. 2016). The advantage of emergence traps is the specificity of the substrate from which they collect.

The number of specimens collected by emergence traps can be startling. In Europe, Froese (1992) and Prescher (1992) recorded several thousand Diptera per m^2 in fields of various crops, with comparable numbers found in less disturbed habitats (Szadziewski 1977). More recent researches on terrestrial emergence trapping to study general diversity, especially in tropical regions, appear to be lacking.

7.5.6 Sticky Traps

Sticky traps are usually yellow-colored cards with a thin layer of adhesive material on its surface. Insects are attracted by the color of the card or just randomly land on it and are trapped. Specimens collected by sticky traps can be removed, if at all, only by solvents that dissolve the adhesive, are often damaged or dessicated, are difficult to identify to species, and generally make poor museum specimens. Sticky traps are extremely useful tools for quickly sampling a habitat or environment when concerns about specimen quality are not a high priority. After exposing the traps for a few hours or days, they can quickly be scanned under a microscope for general characterization of the captured fauna. Replicated sampling can easily be done and scored for some or all of the fauna. They are most commonly used in agricultural studies, often in concert with baits or lures.

Larger-scale sticky trapping for dolichopodid flies is done with plastic sheets wrapped around tree trunks. The sticky material is applied to the plastic sheets, which are later soaked in solvent to release specimens (Bickel and Tasker 2004).

7.5.7 Bait Traps

The variety of baited traps that can be used to collect Diptera is only limited by the imagination of researchers. Usually, one of the above sampling methods (pitfall traps, Malaise traps, pan traps) is combined with some kind of bait that is attractive to the target insects. Dung-baited pitfall traps, for instance, are a mainstay of sphaerocerid fly collectors. A small amount of dung (human dung is often the most consistently available) is suspended in a small cloth bag over the fluid in a pitfall trap. Other baits that could be used this way include carrion (not overlooking insect carrion as a bait), decaying fungi, CO_2 (for biting flies), and vegetable material (potato peelings, other kitchen waste). Such baits can also be placed in Malaise traps or over pan traps. Scattering baits, especially dung, on the leaves of undergrowth plants provides the opportunity to observe and collect flies of families that are frequently found on feces of monkeys and birds, especially sarcophagids and micropezids.

Bait traps of the type used for butterflies (Lucci Freitas et al. 2014) are used to collect carrion flies such as calliphorids and sarcophagids. The bait is placed on a small platform, over which is suspended a cylinder of cloth mesh, closed at the top.

Fig. 7.3 Shannon trap

These traps can be hung off the ground or even high in the forest canopy to sample different faunas. The flies collected in the traps are not killed, so a collector can choose which individuals to preserve (if any).

A more extreme trap of this type is the Shannon trap, as implemented by some researchers in Brazil and demonstrated to me by Dalton Amorim (Fig. 7.3). The trap is a room-sized (3 m square, with walls about 2 m high) cube of black, fine-meshed material, sewn into four walls and a roof. This cube is suspended over the ground, such that there is a small gap of about 6 in. between the bottom of the walls and the surface of the ground. In the center of the trap, a mound of vegetable and animal material, such as fruits, chicken meat and viscera, a fish, oatmeal, and human urine, is placed in a shallow pit. Over several days the decomposing bait becomes extraordinarily fragrant and attracts thousands of flies and other insects that congregate in the top of the trap. A regular insect net can be used to gather the specimens, although care is needed to avoid being stung by the multitudes of vespid wasps also attracted.

Ants are often attracted to a bait or resource exposed to attract flies. They frequently disturb or even steal the baits, especially in tropical climates. The use of a pan trap in which the bait is kept separate from the ants by a liquid moat works well and has the additional virtue of collecting specimens.

Spraying a mixture of honey, water, and optionally other sweet substances (such as cola drinks) on the leaves of vegetation at a convenient spot for collecting can attract large numbers of flies, bees, and bee-parasitizing flies (Brown 2001). As much honey as possible should be used without causing the sprayer to clog, typically a mixture of about 10 parts water to one part honey. The amount of honey used should be reduced if honey bees start to arrive in large numbers, as there is a threshold below which they will not recruit in large numbers, but which is still attractive to flies. Frequently attracted flies include tachinids, syrphids, and muscids, but almost any type of fly can be attracted to honey spray, probably because it mimics a common food source, honeydew (Downes and Dahlem 1987).

7.5.8 McPhail Traps

The McPhail trap apparently dates back to a design proposed in the late 1800s (Steyskal 1977). It is a glass (now more often plastic) inverted bell jar with a collecting fluid in a moat around a raised central opening, below which is some kind of bait. Today, they are most often used with pheromone baits to monitor pest species of Tephritidae, but they could be used for other purposes.

7.5.9 Frog Calls

Recordings of frog vocalizations are effective in attracting corethrellid flies, which feed upon the blood of frogs. Interestingly, the call of eastern North American frogs is effective in many places in the world, although more recent studies have shown differentiation of the species attracted based on different frog calls (Grafe et al. 2008; Ambrozio-Assis et al. 2018) and that even synthesized versions are effective.

7.5.10 Smoke

A small but interesting fauna of Diptera is attracted to wood smoke and can easily be collected by sweeping a fine-meshed net above campfires. Often, the flies will be above head level, sometimes using a person as a marker for a swarm. The main smoke flies are *Microsania* species (Platypezidae), although *Hypocerides* (Phoridae), *Hormopeza* (Empididae), and *Anabarhynchus* (Therevidae) are also attracted (Klocke et al. 2011).

7.5.11 Hilltops

Hilltops are well-known aggregation sites for certain butterflies that use them as mating sites, but some flies are also strongly attracted to hilltops (Skevington 2008). In particular, sarcophagids, tachinids, and empidids can be extremely numerous, and some oestrids can scarcely be found otherwise. Often there are swarms of flies aggregating on or above the leaves of the highest plants of the hilltop, a situation that can be magnified by spraying a mixture of honey and water (above) on the plants.

7.5.12 Army Ants and Other Social Insects

The resources provided by social insects, including their own bodies, are attractive to some flies, especially those of the family Phoridae.

Many insects are associated with New-World army ants (Rettenmeyer et al. 2011), including many flies. Large parasitoid flies of the families Tachinidae and Stylogasteridae that attack the insects flushed out by army ant raids are particularly noticeable, and the sound of their buzzing can often be used to locate the raiding front. There are additional smaller parasitoid flies of the family Phoridae that attack either ants flushed out by the army ants or the army ants themselves (Brown and Feener 1998). The garbage dump of the flies is patronized by many wingless or brachypterous phorids and sphaerocerids that are sometimes also found running along the trails of the army ants, especially when they are changing their nest sites (Akre and Rettenmeyer 1968; Rettenmeyer and Akre 1968). The Old-World army ant fauna has not been studied in as much detail, and the fly fauna is much less well known.

Many other ants are parasitized by phorid flies, which can be collected by cutting open (when they are in dead wood) or digging up the ant colonies. The alarm pheromones of the disturbed ants are probably the main attractant to the flies, which frequently arrive in large numbers. Some phorids are also collected in the fungus gardens of the New-World leaf-cutter ants, although some specimens can be collected as they wander in and out of the nest entrances. Similarly, nest entrances of some *Crematogaster* ants are the breeding sites of some milichiid flies (personal observation).

Termites are attractive to many flies when they are crushed, providing feeding opportunities for muscids and phorids. The fungus galleries of Old-World termites are difficult to study, requiring excavation of the often enormous nests, but are home to a range of bizarre phorids as well (Disney and Kistner 1995, 1997).

Crushing other potential hosts, including stingless bees, millipedes, and snails has also been productive for collecting specialized phorids (Brown 1997; Hash et al. 2017; Brown and Vendetti 2020).

7.5.13 Mass Sampling and Molecular Sorting

As documented above, it is relatively easy to collect large, diverse samples of Diptera. Sorting and identifying these samples in a timely manner has been problematic, however, and most museums have freezers full of unsorted or partly sorted mass samples. Often these samples have been sorted for a few groups of interest to the original collector, with the rest left to be identified at some unspecified time in the future. Sadly, many such samples dry up or are discarded when space is needed.

Recent methodological advances allow the flies collected in mass-sampling campaigns to be used in biodiversity studies to an unprecedented extent. By allowing sorting using DNA barcodes (Wang et al. 2018), rather than looking at the specimens, vast samples of the type generated by Malaise traps can be processed much more quickly. The barcodes can be saved in databases, even if they are not immediately identified to named species, and the data from the mass sampling will be available. The largest such effort is the Barcode of Life project (Ratnasingham and Hebert 2007), but there are even more accelerated inventory efforts. In metabarcoding projects (Gibson et al. 2014; Morinière et al. 2016), entire samples are sequenced, rather than examining the specimens within them; in the process, the specimens are usually destroyed. Some attempts have been made to sequence specimens from the alcohol preservative from trap samples (Marquina et al. 2019). Another approach is to extract DNA from individual specimens from samples, add DNA tags, and then pool the DNA for efficient next-generation sequencing (Meier et al. 2016; Kutty et al. 2018; Srivathsan et al. 2018, 2019; Wang et al. 2018). The specimens are retained, sorted according to the results of sequencing, with some fraction of the identifications checked visually (preferably by experienced taxonomists familiar with the group), and a representative of each species photographed.

It is argued by some that most biodiversity work in the near future needs to be molecular, as the traditional taxonomic process is far too slow to deliver results necessary to finish the job of describing the world's species and to inform conservation efforts. If so, then Diptera taxonomists need to be prepared to describe species based largely on sequence divergence through BINs (Biodiversity Index Numbers—Ratnasingham and Hebert 2013) or other DNA segregates rather than morphology (or with a lesser morphological input). The future of biodiversity research on Diptera may largely consist of DNAbarcoding of Malaise trap samples and its interpretation.

Acknowledgments For discussions and camaraderie in the field, I thank Dalton Amorim, Art Borkent, Giar-Ann Kung, Steve Marshall, Wendy Porras, José Rafael, and Mike Sharkey. For issues pertaining to DNA barcoding, I thank Emily Hartop and Rudolph Meier.

References

Adler PH, Courtney GW (2019) Ecological and societal service of aquatic Diptera. Insects 10:70. https://doi.org/10.3390/insects10030070

Akre RD, Rettenmeyer CW (1968) Trail-following by guests of army ants (Hymenoptera: Formicidae: Ecitonini). J Kansas Entomol Soc 41:165–174

Ambrozio-Assis A, Cortês Lopes B, Amaral AP et al (2018) Preferences for anuran calls in hematophagous corethrellids (Diptera: Corethrellidae) from southern Brazil. Aust Entomol 58:622–628. https://doi.org/10.1111/aen.12376

Austin A, Dillon N (1997) Extraction and PCR of DNA from parasitoid wasps that have been chemically dried. Aust J Entomol 36:241–244. https://doi.org/10.1111/j.1440-6055.1997.tb01461.x

Barnard DR (1979) A vehicle-mounted insect trap. Can Entomol 111:851–854

Basset Y, Cizek L, Cuénaud PD et al (2012) Arthropod diversity in a tropical forest. Science 338:1481–1484. https://doi.org/10.1126/science.1226727

Bickel DJ, Tasker EM (2004) Tree trunk invertebrates in Australian forests: conserving unknown species and complex processes. In: Lunney D (ed) Conservation of Australia's Forest Fauna, 2nd edn. Royal Zoological Society of New South Wales, Mosman, pp 888–898. https://doi.org/10.7882/FS.2004.888

Borkent A, Brown BV (2015) How to inventory tropical flies (Diptera)-one of the megadiverse orders of insects. Zootaxa 3949:301–322. https://doi.org/10.11646/zootaxa.3949.3.1

Borkent ART, Brown BV, Adler PH et al (2018) Remarkable fly (Diptera) diversity in a patch of costa Rican cloud forest: why inventory is a vital science. Zootaxa 4402:53–90. https://doi.org/10.11646/zootaxa.4402.1.3

Brown BV (1993) A further chemical alternative to critical-point-drying for preparing small (or large) flies. Fly Times 11:10

Brown BV (1997) Systematics and fossil evidence of host-parasitoid relationships of *Calamiscus* Borgmeier (Diptera: Phoridae). J Nat Hist 31:1253–1259. https://doi.org/10.1080/00222939700770681

Brown BV (2001) Beginning to study the bee-killing flies, the parasitoid genus *Melaloncha* Brues. Phorid Newsl 9:1–5

Brown BV (2012) A simple light trap for collecting small insects. Entomol News 122:188–191. https://doi.org/10.3157/021.122.0213

Brown BV (2018) A second contender for "world's smallest fly" (Diptera: Phoridae). Biodivers Data J 6:e22396. https://doi.org/10.3897/BDJ.6.e22396

Brown BV, Feener DH (1995) Efficiency of two mass sampling methods for sampling phorid flies (Diptera: Phoridae) in a tropical biodiversity survey. Contrib Sci 459:1–10

Brown BV, Feener DH (1998) Parasitic phorid flies (Diptera: Phoridae) associated with army ants (hymenoptera: Formicidae: Ecitoninae, Dorylinae) and their conservation biology. Biotropica 30:482–487. https://doi.org/10.1111/j.1744-7429.1998.tb00084.x

Brown BV, Vendetti JE (2020) *Megaselia steptoeae* (Diptera: Phoridae): specialists on smashed snails. Biodivers Data J 8:e50943. https://doi.org/10.3897/BDJ.8.e50943

Brown BV, Marxhall SA, Wood DM (2009) Natural history. In: Brown BV, Borkent A, Cumming JM et al (eds) Manual of central American Diptera. NRC Research Press, Ottawa, pp 51–63

Brown BV, Borkent A, Adler PH et al (2018) Comprehensive inventory of true flies (Diptera) at a tropical site. Commun Biol 1:8. https://doi.org/10.1038/s42003-018-0022-x

Cadmus P, Pomeranz J, Kraus J (2016) Low-cost floating emergence net and bottle trap: comparison of two designs. J Freshwater Ecol 31:653–658. https://doi.org/10.1080/02705060.2016.1217944

Campbell JW, Hanula JL (2007) Efficiency of Malaise traps and colored pan traps for collecting flower visiting insects from three forested ecosystems. J Insect Conserv 11:399–408. https://doi.org/10.1007/s10841-006-9055-4

Chandler PJ (ed) (2010) Dipterist's handbook, 2nd edn. The Amateur Entomologist's Society, Manningtree

Cheng L (1976) Marine insects. North Holland, Amsterdam

Danks HV (1996) How to assess insect biodiversity without wasting your time. Biological survey of Canada (terrestrial arthropods), Document series 5, Ottawa

Darling C, Packer L (1988) Effectiveness of Malaise traps in collecting hymenoptera: the influence of trap design, mesh size, and location. Can Entomol 120:787–796. https://doi.org/10.4039/Ent120787-8

Disney RHL, Kistner DH (1995) Revision of the Afrotropical Termitoxeniinae (Diptera: Phoridae). Sociobiology 26:117–225

Disney RHL, Kistner DH (1997) Revision of the oriental Termitoxeniinae (Diptera: Phoridae). Sociobiology 29:3–118

Disney RHL, Erzinclioglu YZ, Henshaw DJC et al (1982) Collecting methods and the adequacy of attempted fauna surveys, with reference to the Diptera. Field Stud 5:607–621

Downes WL, Dahlem GA (1987) Keys to the evolution of Diptera: the role of Homoptera. Environ Entomol 16:847–854. https://doi.org/10.1093/ee/16.4.847

Favret C, Lessard V, Trépanier A (2019) Voegtlin-style suction traps measure insect diversity and community heterogeneity. Insect Conserv Diver 12:373–381. https://doi.org/10.1111/icad.12368

Froese A (1992) Vegleichende Untersuchungen zur Biologie und Ökologie der Dipteran auf integriert und konventionell bewirtschafteten Feldern. Eberhard-Karls-Universität, Tubigen, Germany

Gibson J, Shokralla S, Porter TM et al (2014) Simultaneous assessment of the macrobiome and microbiome in a bulk sample of tropical arthropods through DNA metasystematics. Proc Natl Acad Sci U S A 111(22):8007–8012. https://doi.org/10.1073/pnas.1406468111

Gordh G, Hall JC (1979) A critical point drier used as a method of mounting insects from alcohol. Entomol News 90:57–59

Grafe TU, Saat HBM, Hagen N et al (2008) Acoustic localisation of frog hosts by blood-sucking flies *Corethrella* Coquillet (Diptera: Corethrellidae) in Borneo. Aust J Entomol 47:350–354. https://doi.org/10.1111/j.1440-6055.2008.00667.x

Gressitt JL, Gressitt MK (1962) An improved Malaise trap. Pac Insects 4:87–90

Hash JM, Millar JG, Heraty JM et al (2017) Millipede defensive compounds are a double-edged sword: natural history of the millipede-parasitic genus *Myriophora* Brown (Diptera: Phoridae). J Chem Ecol 43(2):198–206. https://doi.org/10.1007/s10886-016-0815-7

Häuser CL, Riede K (2015) Field methods for inventorying insects. In: Watson MF, Lyal CHC, Pendry CA (eds) Descriptive taxonomy: the Foundation of Biodiversity Research. The systematics association special volume, vol 84. Cambridge University Press, Cambridge, pp 190–213

Henderson P, Southwood T (2016) Ecological methods, 4th edn. Wiley, Chichester

Jaschhof M, Jaschhof C (2014) *Zadbimyia*, a new genus of asynaptine Porricondylinae (Diptera: Cecidomyiidae) with twenty-two new species from the cloud forest of Costa Rica. Zootaxa 3866:1–29. https://doi.org/10.11646/zootaxa.3866.1.1

Kimerle RA, Anderson NH (1967) Evaluation of aquatic insect emergence traps. J Econ Entomol 60:1225–1259. https://doi.org/10.1093/jee/60.5.1255

Klocke D, Schmitz A, Schmitz H (2011) Fire-adaptation in *Hypocerides nearcticus* Borgmeier and *Anabarhynchus hyalipennis hyalipennis* Marquart and new notes about the Australian "smoke Fly" *Microsania australis* Collart (Diptera: Phoridae, Therevidae and Platypezidae). Open Entomol J 5:10–14. https://doi.org/10.2174/1874407901105010010

Kutty SN, Wang W, Ang Y et al (2018) Next-generation identification tools for nee soon freshwater swamp forest, Singapore. Gard Bull Singapore 70:155–173. https://doi.org/10.26492/gbs70(suppl.1).2018-08

Larson BMH, Kevan PG, Inouye DW (2001) Flies and flowers: taxonomic diversity of anthophiles and pollinators. Can Entomol 133:439–465. https://doi.org/10.4039/Ent133439-4

Loy VA, Barnhart CS, Therrien AA (1968) A collapsible, portable vehicle-mounted insect trap. Mosq News 28:84–87

Lucci Freitas AV, Agra Iserhard C, Pereira Santos J et al (2014) Studies with butterfly bait traps: an overview. Rev Colomb Entomol 40:203–212
Malaise R (1937) A new insect-trap. Entomologisk Tidskrift 58:148–160
Marquina D, Esparza-Salas R, Roslin T et al (2019) Establishing insect community composition using metabarcoding of soil samples, and preservative ethanol and homogenate from Malaise trap catches: surprising inconsistencies between methods. BioRxiv. https://doi.org/10.1101/597302
Marshall SA (2012) Flies: the natural history and diversity of Diptera. Firefly Books, New York
Marshall SA, Kirk-Spriggs AH (2017) Natural history of Diptera. In: Kirk-Spriggs AH, Sinclair BJ (eds) Manual of afrotropical Diptera, vol 1. SANBI Graphics & Editing, Suricata 4, Pretoria, pp 135–152
Marshall SA, Anderson RS, Roughley RE et al (1994) Terrestrial arthropod biodiversity: planning a study and recommended sampling techniques. A brief prepared by the biological survey of Canada (terrestrial arthropods). Entomol Soc Can Bull Suppl 26:1–33
Martens AP, Johnson PJ (2019) Avoiding the riff-raff: modified Malaise sampling for the collection of aphid parasitoids. In: Entomological collections network annual meeting entomological collections network, St. Louis, pp 9–10
Martens AP, Johnson PJ, Boe A et al (2019) The diversity of aphid parasitoids (Hymenoptera: Braconidae: Aphidiinae) in soybean fields of eastern South Dakota. Proc S D Acad Sci 98:77–89
Masner L, Goulet H (1981) A new model of flight-interception trap for some hymenopterous insects. Entomol News 92:199–202
Meier R, Wong W, Srivathsan A et al (2016) $1 DNA barcodes for reconstructing complex phenomes and finding rare species in specimen-rich samples. Cladistics 32:100–110. https://doi.org/10.1111/cla.12115
Morinière J, Araujo BC, Lam AW et al (2016) Species identification in Malaise trap samples by DNA barcoding based on NGS technologies and a scoring matrix. PLoS One 11:e0155497. https://doi.org/10.1371/journal.pone.0155497
Pollet M, Grootaert P (1994) Optimizing the water trap technique to collect Empidoidea (Diptera). Stud Dipterol 1:33–48
Prescher S (1992) Ökologie und Biologie der Diptera, insbesondere der Brachycera, eines klärschlamm gedüngten Ackerbodens. Technischen Universität Carolo-Wilhelmina zu Braunschweig, Braunschweig
Raguso RA (2020) Don't forget the flies: dipteran diversity and its consequences for floral ecology and evolution. Appl Entomol Zool 55:1–7. https://doi.org/10.1007/s13355-020-00668-9
Ratnasingham S, Hebert P (2007) BOLD: the barcode of life data system (www.barcodinglife.org). Mol Ecol Notes 7:355–364. https://doi.org/10.1111/j.1471-8286.2007.01678.x
Ratnasingham S, Hebert PD (2013) A DNA-based registry for all animal species: the barcode index number (BIN) system. PLoS One 8(7):e66213. https://doi.org/10.1371/journal.pone.0066213
Rettenmeyer CW, Akre RD (1968) Ectosymbiosis between phorid flies and army ants. Ann Entomol Soc Am 61:1317–1326
Rettenmeyer CW, Rettenmeyer ME, Joseph J et al (2011) The largest animal association centered on one species: the army ant *Eciton burchellii* and its more than 300 associates. Insect Soc 58:281–292. https://doi.org/10.1007/s00040-010-0128-8
Roberts RH (1972) The effectiveness of several types of Malaise traps for Tabanidae and Culicidae. Mosq News 32:542–547
Schleuter MA, Stewart NG (2011) Which trap type and trap color work best in collecting different groups of bees (family: Apidae) and pollinating flies (order: Diptera)? In: Paper presented at the annual meeting of the Entomological Society of America, Reno, NV
Skevington JH (2008) Hilltopping. In: Capinera JL (ed) Encyclopedia of entomology, vol 4, 2nd edn. Springer, Dordrecht, pp 1799–1807
Sommerman KM, Simmet RP (1965) Car-top insect trap with terminal cage in auto. Mosq News 24:172–182

Srivathsan A, Baloglu B, Wang W (2018) A MinION™-based pipeline for fast and cost-effective DNA barcoding. Mol Ecol Resour 18:1035–1049. https://doi.org/10.1111/1755-0998
Srivathsan A, Hartop E, Puniamoorthy J et al (2019) Rapid, large-scale species discovery in hyperdiverse taxa using 1D MinION sequencing. BMC Biol 17:96. https://doi.org/10.1186/s12915-019-0706-9
Steyskal GC (1977) History and use of the McPhail trap. Fla Entomol 60:11–16
Steyskal GC (1981) A bibliography of the Malaise trap. P Entomol Soc Wash 83:225–229
Sudia WD, Chamberlain RW (1962) Battery-operated light trap, an improved model. Mosq News 22:126–129
Szadziewski R (1977) Geofilne muchówki (Diptera) rezerwatu "Las Piwnicki" kolo Torunia. Acta Universitatis Nicolai Copernici Biologia 19:107–115
Thomas DB (2008) Non-toxic antifreeze for insect traps. Entomol News 119:361–365. https://doi.org/10.3157/0013-872X-119.4.361
Townes H (1972) A light-weight Malaise trap. Entomol News 83:239–247
van Achterberg K (2009) Can Townes type Malaise traps be improved? Some recent developments. Entomologische Berichten 69(4):129–135
Vockeroth JR (1966) A method of mounting insects from alcohol. Can Entomol 98:69–70
Vrdoljak SM, Samways MJ (2012) Optimising coloured pan traps to survey flower visiting insects. J Insect Conserv 16:345–354. https://doi.org/10.1007/s10841-011-9420-9
Wang WY, Srivathsan A, Foo M et al (2018) Sorting specimen-rich invertebrate samples with cost-effective NGS barcodes: validating a reverse workflow for specimen processing. Mol Ecol Resour 18:490–501. https://doi.org/10.1111/1755-0998.12751

Chapter 8
Sampling Methods of True Fruit Flies (Tephritidae)

Vicente Hernández-Ortiz, Mónica Hernández-López, and José F. Dzul-Cauich

8.1 Introduction

Diptera is the third most diversified order of insects being particularly abundant in temperate and tropical landscapes. Even though they are involved in a variety of ecological functions, including decomposition, predation, or pollination, the biology and diversity of these insects are poorly understood (Hughes et al. 2000). The superfamily Tephritoidea consists of nine families of flies: Lonchaeidae, Richardiidae, Piophilidae, Ulidiidae, Platystomatidae, Ctenostylidae, Tephritidae, Pyrgotidae, and Tachiniscidae (sensu Wiegmann et al. 2011). While it comprises a variety of groups with different habits including saprophagy and parasitism, its highest diversity is represented by phytophagous species. Members within the family Tephritidae are commonly known as "true fruit flies" represented by 4716 species worldwide (Pape et al. 2011), of which around a quarter occurs in the Americas. Except for certain groups in the subfamily Phytalmiinae (Old World and Tropical Asia) with saprophytic larvae, all other members possess phytophagous maggots feeding on different plant tissues including fruits, seeds, flower buds, leaves, stems, or even roots (Christenson and Foote 1960; Freidberg 1984; Hardy 1986; Headrick and Goeden 1998).

8.2 Recognition and Natural History

Adult flies are ranging from 2 to 35 mm in length. The body is usually colored with wing patterns consisting of stripes or spots forming conspicuous designs, mostly associated with elaborate mating courtships of adults. Females possess a very

V. Hernández-Ortiz (✉) · M. Hernández-López · J. F. Dzul-Cauich
Red de Interacciones Multitróficas, Instituto de Ecología A.C., Xalapa, Veracruz, Mexico
e-mail: vicente.hernandez@inecol.mx

J. C. Santos, G. W. Fernandes (eds.), *Measuring Arthropod Biodiversity*,
https://doi.org/10.1007/978-3-030-53226-0_8

specialized structure called aculeus used to insert the eggs into plant tissues of many different plants. Both the wing features and aculeus forms, along with other morphological traits, are used for the taxonomic identification of the species (Norrbom 2010; Hernández-Ortiz et al. 2020).

The larvae of frugivorous species often feed on the pulp of fruits until complete full development; however, some species may also feed upon seeds. Therefore, some specific taxa are of major economic significance for fruit growing systems around the world, since they constitute dangerous pests resulting in quarantine restrictions for many countries (Qin et al. 2015). Meanwhile, non-frugivorous larvae are found within flower buds, inflorescences, achenes, and stems; some others are gall inducers forming specialized structures, from which they obtain food and protection. Most of the species belonging to the subfamily Tephritinae feed mainly on plants of the family Asteraceae, but also on other families as Verbenaceae and Lamiaceae. Due to this close relationship and diversification within Asteraceae, they are the most specialized taxa of the Tephritidae (Headrick and Goeden 1998).

After larvae achieve full development and maturity, they leave the fruit to be buried underground at variable depths, according to soil features. Conversely, other species growing on flowers, seeds, or stems usually are kept inside during pupation until adult emergence. The puparium of fruit flies is heavily sclerotized, turning creamy white, reddish-brown, or even black. Given this remarkable range of specialized feeding habits, the monitoring and searching of fruit flies demand a combination of techniques for their screening to assess the diversity of fruit flies in different biomes or environments (Fig. 8.1a, b).

8.3 Indirect Collecting Methods

The indirect sampling techniques are defined as such, since there is no certainty of finding target groups because it is not exclusively aimed to find them. Un-baited traps are distinguished because do not require any lure, since they work by the interception of flying insects, though some of them can also pick up other invertebrates from the ground and leaf litter. Given these features, the right option of the traps to be used and their location becomes extremely important, as their efficiency is mainly dependent on this.

8.3.1 Aerial or Sweeping Net

Netting is undoubtedly one of the best sampling methods because potential sites can be selected where specific target insect groups occur. However, certain taxa can be harder to sample due to their particular habits and behavior. Several types of entomological nets are available for capturing adult flies in herbaceous vegetation and lower parts of trees. Those consist of a metallic ring with a conical mesh net with an

Fig. 8.1 Fruit fly of the genus *Hexachaeta* feeding on a drop of water (**a**) (photo courtesy Maurilio López-UV); female of the genus *Rhynencina* exploring a flowerhead for oviposition (**b**)

aluminum handle of variable length. However, collapsible models with telescopic handles exist, facilitating collection at higher levels (Fig. 8.2a). The netting is advantageous since it can be made on the potential substrates where the flies occur, or over herbaceous vegetation where likely host plants are found, especially during early morning hours which would be the periods of highest activity. A disadvantage would be a significant time and effort expended, while efficiency will also depend ultimately upon the skills and expertise of the collector.

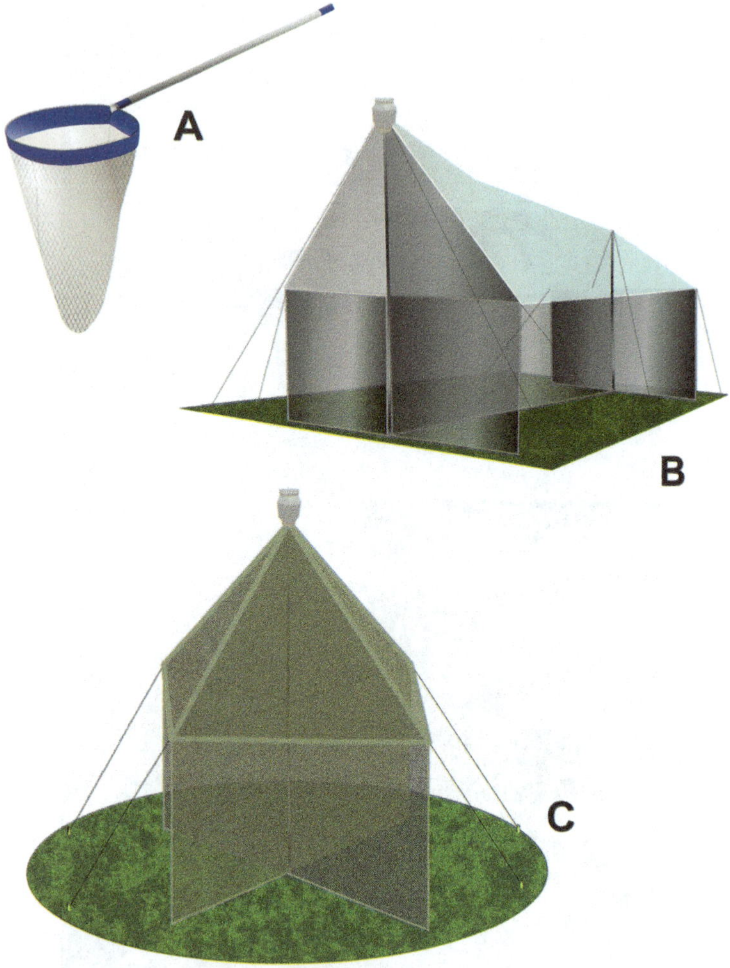

Fig. 8.2 Typical aerial net for sweeping in herbaceous vegetation and lower branches of trees (**a**); classic Malaise trap of three walls (**b**); symmetrical Malaise trap constructed with four walls (**c**)

8.3.2 Malaise Trap

The Swedish René Edmond Malaise formerly designed this device for trapping Hymenoptera (Malaise 1937). Historically, this type trap had extensively used as a sampling device for biodiversity studies in nearly every region of the world. However, further studies proved that Malaise traps are most efficient for capturing several flying insect groups, especially Diptera, Hymenoptera, and Lepidoptera (Moeed and Meads 1987; Brown 2005; Boulter et al. 2011).

Although distinct types of Malaise traps are available, most of the configurations consist of three basic components: (a) trap body, made up of three or four walls; (b) top shell, characterized by an uppermost central or lateral peak; and (c) collecting receptacle, situated in the upper part. The first two are made of mesh screen, and the last one is of plastic material fitted an invagination below or lateral for the entrance of insects (Martin 1977) (Fig. 8.2b, c).

Since the original design up to date, many variants were developed with changes in size, number of entries and position, and number of collecting containers (e.g., Townes, Gressitt, land and air Malaise) (Achterberg 2009; Sheikh et al. 2016). Those variations provide options for several applications, even targeting levels in the canopy of tropical forests.

Trap operating Malaise trap exploits the positive phototropism of flies in the understory herbaceous strata, since they are prone to flying upwards seeking the light and caught in the collecting jar. The container uses ethanol 75% as preserving or killing agent; however, it can be replaced by a mixture of propylene glycol/water (1:1) to prevent desiccation, adding a few drops of liquid soap. It is particularly useful in hot environments with continuous trap operation for long-time exposures (Dilling et al. 2007; Savage et al. 2011).

This trap can be used in a wide variety of environments, depending on the goals pursued by the research. It can become a handy monitoring device for understory species of tropical and temperate forests, mainly where an extensive herbaceous stratum occurs. However, it is also an excellent sampling method in open areas dominated by herbaceous vegetation where non-frugivorous tephritids are common. Overall, the choice of trap location strongly impacts the captures, since the traps placed at the forest edge are known to catch significantly more individuals of Diptera than those set inside the forest at very shady areas (Brown 2009).

This trap does not use any kind of attractant, so minimal effort is needed for sampling. It can be used for long continuous periods, subject to weather conditions and the long life of manufacturing materials. This design is not specific for fruit flies, however maximizes the possibility of capturing insects regardless of flight direction, typically catching large amounts of insects of various sizes. It is highly efficient in monitoring population diversity and abundance of major families of Diptera, but also those rare or whose numbers are low and difficult to collect using other sampling methods (Brown et al. 2018; Karlsson et al. 2020).

8.3.3 Baited Traps

The baited traps can be useful for the species monitoring and inventory, as well as for assessing either relative abundance or seasonal phenology of adults in a specific region or biome. Diversity studies in wild environments can be made using this approach since most of the fruit flies are strongly attracted to food baits and sex lures. Nevertheless, climatic features of the biome under study should be taken into consideration, as they may influence the response to different lure types, including the species captured, sex, and age of adults, among others (Díaz-Fleischer et al.

2014). The traps are baited with different lures, based on the nutritional demands of adults, their sexual behavior, or response to chemical and visual stimuli, which should compete successfully with a broad spectrum of environmental stimuli (Epsky et al. 2014).

8.3.3.1 Sex-Based Attractants

Both sex pheromones and male lures play a crucial role in fruit fly mating systems of many economically important species (Sivinski and Calkins 1986). Even though sex-based attractants were primarily designed to capture males, they are also attractive to females (Drew and Hooper 1981; Montoya et al. 2006). For instance, the methyl eugenol and cuelure are for the attractiveness of *Bactrocera* species; the trimedlure is for *Ceratitis* species; and the ammonium acetate is for the *Anastrepha* and *Rhagoletis* species (Beroza et al. 1961; FAO/IAEA 2018). Given such levels of sensitivity, this kind of lures would not be a suitable option for extensive surveys of the tephritid diversity inventories.

8.3.3.2 Food-Based Attractants

Early studies were conducted by using nature-based lures made of fermented sugar solutions, fruit juices, vinegar, and hydrolyzed proteins (McPhail 1937). Later on, further developments of water-based liquid proteins or synthetic protein baits have been made, mostly from vegetable products, animal proteins, bacteria, or even droppings and urine from mammals (Drew et al. 1983; Piñero et al. 2003; IAEA 2007). Currently, the volatile compounds released by those substances are the basis for baits developed for some fruit fly pest species (Epsky et al. 2014; Biasazin et al. 2018). Synthetic lures are somewhat specific to certain fruit fly pests (Heath et al. 1995; Katsoyannos et al. 1999; Raghu 2007; Thomas et al. 2008); meanwhile food lures are generic by nature; hence protein-based liquids attract an extensive range of tephritid flies, including other families of Diptera (Thomas 2003; FAO/IAEA 2018). The efficiency of this type of attractants is highly dependent upon the time required to start the fermentation process, which releases ammoniacal or other volatile compounds attractive for adults (Buttery et al. 1983; Heath et al. 1997; Jang et al. 2007).

8.3.3.3 McPhail-Type Traps

It was formerly conceived late of the nineteenth century in Europe, for trapping flying scavenger insects (Dahl 1896). Some 50 years later, M. McPhail conducted studies in Mexico and popularized the use of this device in the Americas for monitoring fruit flies (Steyskal 1977). Mainly it consists of a combined designed for the retention of specimens with an attractant. Such traps are currently known as "McPhail traps," being valuable instruments to carry out biodiversity inventories for fruit flies, but also many other groups of Diptera (Thomas 2003). The manufacturing process was firstly a bottle-shaped glass container with an invaginated hole at the bottom to permit the access of flies, employing using liquid mixtures of sweetened or fermented plant agents (Fig. 8.3a). Later designs incorporated additional features, including vessels to be used with dry matters, different access features, use of colors, or other shapes and designs, even handmade traps manufactured with plastic beverage bottles (Fig. 8.3b).

The most used design consists of a two-piece plastic trap that can be separated making it easy for bait placement and inspection (Thomas et al. 2001). The upper section is cylinder-shaped and translucent hanging up, with a built-in device for dry attractants; the bottom part is either yellow-, green-, or red-colored, consisting of a fluid container holding about 250 mL, provided with a conical invagination below, allowing permanent access of flies. Most used attractants for tephritids include two types: (a) sexual attractants (parapheromones and pheromones) and (b) food attractants (natural or synthetics) (Fig. 8.3c).

McPhail-trap operating The fruit flies are sensitive for a variety of chemical and physical stimuli, especially those related to behavior, physiology, and the quality of the lures used (Tan et al. 2014; Díaz-Fleischer et al. 2014). Nutritional needs of the females rely on sugars and water, but primarily on amino acids to reach sexual maturation and egg development (Drew and Yuval 1999). Based upon this rationale, an extensive menu of natural and synthetic compounds has been developed based on feeding, mating, and host-plant seeking (Epsky et al. 1995, 2014). However, fly densities on the traps might be influenced by the environmental conditions, as liquid attractants in dry environments appear highly efficient in the tephritid captures, as they provide a source of protein and sugar, but also a source of water.

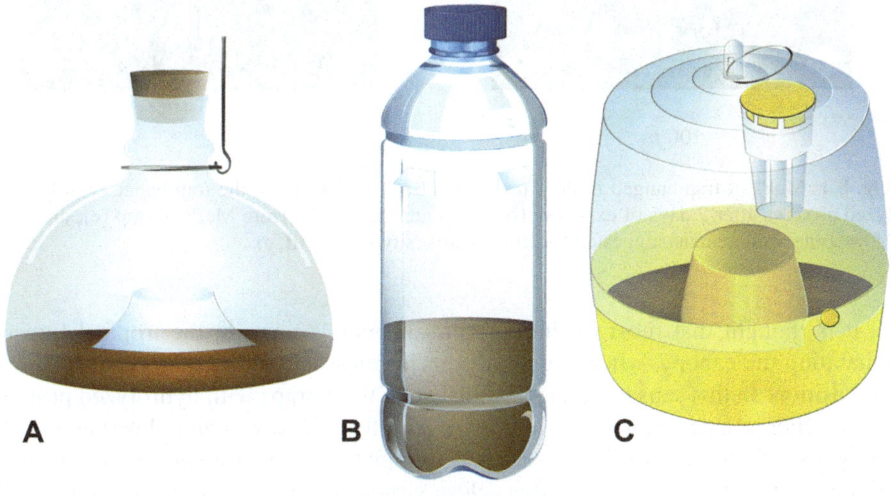

Fig. 8.3 Original McPhail trap glass-constructed (**a**); handmade trap with recycled materials (**b**); new design of McPhail trap of two-piece plastic-constructed (**c**)

Fig. 8.4 McPhail trap hanged in the canopy of a tree (**a**); service of the trap baited with hydrolyzed proteins after 7 days of exposure (**b**); specimens recovered from McPhail trap (**c**); drawing of trapping arrangement suggested for fruit fly diversity studies (**d**, **e**)

It is thought that most of the tephritid diversity occurs in the tropical forests inhabiting the canopy, where frugivore species mate and locate their host fruits for oviposition. In this sense, we suggest the use of these traps with hydrolyzed protein baits, either as a device to the inventory of fruit fly diversity or an indirect approach to assess their density and seasonal populations in tropical environments. Furthermore, it also targets other closely related taxa, such as Lonchaeidae, Ulidiidae, and Richardiidae, among others (Fig. 8.4a).

The bait for each trap should be prepared with a mixture of water (235 mL), concentrate of hydrolyzed proteins (15 mL) of commercial use, and sodium borate (borax 5 g) recommended as a preserver for the samples. The aqueous medium operates both as a bait and catching mechanism. Such proteins become more attractive after several days of exposure in the trap, with peak levels on the fourth to tenth day. Thus, the amount of protein in the mixture should be standardized (Fig. 8.4b, c). Traps should be placed in the foliage of trees, avoiding the obstruction of the bottom entrance, and in tropical forests which are preferably suspended at approximately 5 m height over the ground, although they may be a right choice for canopy inventories of fruit flies, as they could be placed higher up.

In terms of the spatial setting thereof, they feature a limited operating ratio, established in one trap per hectare for economically significant species (FAO/IAEA 2018). Nevertheless, we suggested the use of 3–5 traps per hectare for intensive inventories of fruit fly diversity, providing synergies among action ratios, enhancing their attractiveness (Fig. 8.4d, e).

The efficiency of different food attractants is mainly dependent upon the time needed to start their fermentation process, because protein hydrolysis releases amino acids and additional volatiles, primarily ammonia compounds highly attractive to the flies (Epsky et al. 1999; Mazor et al. 2002). A restriction of these traps is that baits should be replaced periodically. Traps with hydrolyzed proteins or yeasts must be renewed every 7 days and retrieve specimens kept within, while servicing with synthetic attractants may last approximately 6–10 weeks. Besides, if dry lure is used, trap should be supplemented with a liquid medium in the body of the trap (~200 mL) acting as a retaining medium. This can be a mixture of propylene glycol/water (1:1) to prevent evaporation, few drops of liquid soap, and sodium borate as a preserver of the sample (5 g).

8.4 Direct Collecting Methods

8.4.1 Adults

Fruit fly adults are captured outdoors employing trapping systems. Nonetheless, there are also focal points or site-specific habitats of opportunity where flies can be found. Adult flies are observed feeding on exudates of flowers, nectaries, or even fluids from vertebrate droppings (especially from birds), but also on fallen ripening fruits in the rotting process. Therefore, adults of many families of Diptera, especially different genera of Tephritidae, are assiduous visitors of the weed flowers, making these sites very suitable places to gather adults directly, where they obtain food resources, but also to perform egg laying in their host plants.

8.4.2 Collecting Fruits and Flowers

Because members of this family possess obligatory phytophagous habits, the sampling of potential host fruits is a valuable source for research purposes. Although fruit sampling demands much effort, it provides remarkable and poorly known data on the tritrophic interactions among the flies, hosts, and their associated parasitoids. Breeding is an approach deserving of further attention, whether in the field or the laboratory, though highly useful for understanding insights on their natural history and for identifying interspecific trophic relationships.

Most of the frugivorous larvae develop on fleshy fruits, so those having an advanced maturity stage with soft areas, dark stains, rot, orifices, or injuries that might have originated from female oviposition or larval feeding activities are targeted for inspection. Several authors have reported diverse host plant families that are frequently exploited by tephritids, which would be suitable for search larval infestation (e.g., Allwood et al. 1999; Norrbom 2015; Hernández-Ortiz et al. 2019). The sample size mainly depends upon fruit availability and the state of ripeness, since unripe fruits or with early development may result in non-infested or having eggs or larvae in early instars; thus, lab rearing will take longer time and care. Instead, the mature fruits usually contain fully developed larval instars.

Fruits may be collected directly from the tree at intermediate or advanced ripeness, ideally with signs of oviposition. Fruits fallen on the ground can be also suitable samples to search for larvae; therefore, we suggest avoiding fruits opened or at advanced rotting conditions, since larvae usually leave fruits to pupate. Furthermore, the fruits need to be stored and shipped in mesh bags, keeping them in a cool place until processing at the laboratory. The origin of sampled plants should be recorded through the acquisition of voucher botanical samples.

Flower collecting is carried out on the prospect host plant, preferably from different individuals taking mature flowers following the anthesis, where larvae have kept a longer feeding period; thus, there are better probabilities of finding late larval instars to enable adult emergence. Picked flowers are placed in mesh bags (near 40 × 30 cm), to provide aeration avoiding higher temperatures, which may lead to the death of larvae or leaving the flowers. If fieldwork takes more than 1 day, we suggested bringing a thermic container to keep the samples fresh (Fig. 8.5a, b).

8.4.3 Recovering Larvae from Fruits and Flowers

At the laboratory, both the fruits and flowerheads gathered should be removed carefully from the bags, looking for mature larvae that had left their feeding substrate. According to the goals of the study, samples from each host are separated into individual breeding chambers or into small groups within a single chamber allowing proper handling of the samples. A single chamber consists of a plastic container

Fig. 8.5 Larva of *Dictyotrypeta* feeding in achenes on development (**a**); mature larva of *Tetreuaresta* feeding within the host inflorescence (**b**)

(250 mL approx.) with top aeration. The size of this may differ with the size of the fruits or flowers and the number of samples put on it.

Breeding chambers are supplied by a bed of sterilized river sand used as pupation substrate; the samples are inspected daily, dissecting any fruit or blossom in the last ripening period for removal of larvae and pupae. Following separation of larvae, the feeding substrate should be discarded, keeping only pupae on the sand, and regularly moistened with a water/sodium benzoate solution 1%, to prevent fungal growth until adult emergence. Upon hatching, the adults are kept alive for 2–3 days feeding flies with a water/honey solution (3:1) for full pigmentation, then can be stored in

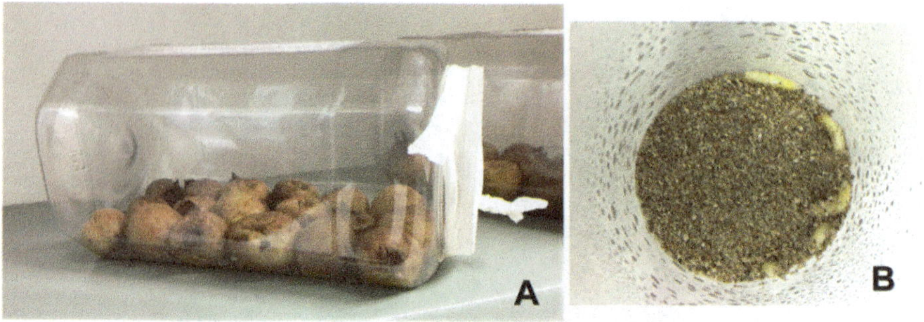

Fig. 8.6 Breeding chamber used for fruit sampled (**a**); mature larvae extracted from the fruits for pupation (**b**)

ethanol 75%, or killed in a lethal chamber using ethyl acetate vapors for further dry storage on pins (Fig. 8.6a, b).

8.4.4 Collecting and Rearing Gall Inducers

Some insect orders are gall inducers acting on meristematic tissues of their host plants to provide their offspring with food and shelter (Fernandes et al. 2014). The true fruit flies comprise various feeding habits, but especially members of the subfamily Tephritinae are gall inducers on many species of Asteraceae; however, they also occur on other plant families such as Lamiaceae, Acanthaceae, Goodeniaceae, and Verbenaceae (Freidberg 1997; Headrick and Goeden 1998; Hernández-Ortiz et al. 2020).

The galls are spheroid-shaped structures produced primarily on the stems and flowerheads in which one or more larval cells may be present, depending on the fly species (Fig. 8.7a–d). Several studies in the Americas described the role of these flies by using different host plants, their parasitic relationships, or the impact on host architecture and seed viability (Lalonde and Shorthouse 1984; Abrahamson et al. 1989; Emlen 1992; Fernandes and Price 1994; Silva et al. 1996).

Therefore, the focus of search should be on weeds, which grow in secondary forests or disturbed areas, where they are very common. This kind of studies becomes highly time-consuming due to the intensive field sampling and location of the host plants. However, knowledge of such specialized feeding habits remains scarce particularly in tropical environments, even though a great diversity of gall-inducing tephritids occur in more than 48 plant genera (Freidberg 1997), some of which (e.g., *Eutreta xanthochaeta*, *Procecidochares utilis*, and *Urophora solstitialis*) have been used as biocontrol agents for some weeds (Turner 1996).

Stem galls are collected by a cross-cutting, leaving near 4 cm on each side to prevent early desiccation of the structure. The mature galls are usually the largest, whose larvae are more likely to reach the adult stage. The galls of the flowerheads

Fig. 8.7 Central stem gall with a single cell produced by *Eutreta* sp. (**a**, **b**); lateral stem gall produced by *Procecidochares* sp. showing two cells (**c**, **d**)

commonly appear as unusual protrusions basally, so the whole structure is collected making a basal cross-cutting from 4 cm. Transportation to the laboratory should be done in mesh bags to preserve samples fresh avoiding the larval mortality (Fig. 8.8a–c).

For laboratory breeding there are two options. The first one is by cutting the gall for the extraction of larva, then setting it in a single breeding chamber, and putting down a base of absorbent paper to facilitate pupation. However, if the larva has not yet completed the feeding process, it is likely that pupation and adult stages will not be accomplished by this method. The second option consists of keeping the gall in a single rearing chamber on a floral foam pad, saturated with a sodium benzoate/water solution 1%, ensuring continuous moisturization and full development of the larva. The breeding chamber is stored under environmental conditions, ideally at 22–26 °C and 50–70% of relative humidity. It should be daily monitored until the emergence of flies or parasitoids. Afterwards, flies should be nourished with a

Fig. 8.8 Breeding chamber hatching a stem gall (**a**); breeding chamber with a flower (**b**); pupae extracted for adult emergence (**c**)

honey/water solution (3:1) for 2–3 days to fill staining and sclerotization. Gall dissection is recommended to retrieve puparia.

8.5 Storage and Preserving Samples

8.5.1 Preservation in Liquids

The adult flies either recovered from the sample traps, or reared from their host plants, both can be preserved in ethanol 75%. If specimens are required for barcoding studies, they should be stored safely in ethanol 100% and chilled to prevent fluid loss and damage of the genetic material. For the preservation of the immature stages, the larvae of different instars should be fixed avoiding degradation, by putting the alive individuals into a small vessel with hot water (nearly boiling temperature) and adding some salt grains (sodium chloride). The sample should be kept at that temperature for 3–5 min. Afterwards, let it cool down and rinse with distilled water to remove any salt excess and deposit specimens in ethanol 75%. This technique prevents any change of color of the specimens when fixed directly in ethanol.

8.5.2 Dry Preservation of Flies

Most popular liquid killing agents include ethyl acetate, carbon tetrachloride, ether, chloroform, and ammonia water solution. However, ethyl acetate is recommended by many entomologists as the most satisfactory liquid killing agent. The fumes of ethyl acetate are less toxic to humans than the fumes of all other substances (Gibb and Oseto 2020). Adult specimens obtained from breeding hosts, which need to be dry preserved for ultimate storage at a scientific collection, must be killed in a dry lethal chamber. The adults are placed in a lethal glass chamber with absorbent paper inside, moistened with ethyl acetate as insecticide agent for approximately 20 min.

Usually, the specimens stored in alcohol for a long time tend to dry out and shrink the body and sometimes be collapsed, often losing their natural colors and appearance. Therefore, dry mounting of specimens formerly stored in ethanol needs firstly a treatment consisting of setting specimens into pure ethyl acetate for 24 h; afterwards, removes the excess substance from the body, drying on absorbent paper for no more than 1 min; mount immediately using entomological pins, taking into account that the specimen becomes fully hardened once the fluid has evaporated from the body. The smaller specimens should be preserved by double mounted, pinning fly in a stainless-steel minute pin (12 mm long), itself supported by a standard entomological pin. Instead, the fly can be mounted by gluing with a drop of Euparal (soluble in ethanol) onto a small cardboard triangle.

Acknowledgments We thank Maurilio López (UV-Mexico) for sharing with us the picture of Fig. 8.1a. We appreciate feedback and suggestions from the editors on the early draft of the manuscript. This chapter is a contribution to the long-term project of VHO "Biodiversity and Systematics of Mexican Diptera" (BIODIMEX) (Ref: 20030-10128).

References

Abrahamson WG, McCrea KD, Anderson SS (1989) Host preference and recognition by the goldenrod ball gallmaker *Eurosta solidaginis* (Diptera: Tephritidae). Am Midl Nat 121:322. https://doi.org/10.2307/2426036

Achterberg KV (2009) Can Townes type Malaise trap be improved? Some recent developments. Entomolog Ber 69:129–135

Allwood A, Chinajariyawong A, Kritsaneepaiboon S et al (1999) Host plant records for fruit flies (Diptera: Tephritidae) in Southeast Asia. Raffles Bull Zool 47(Suppl 7):1–92

Beroza M, Green N, Gertler SI et al (1961) Insect attractants, new attractants for the Mediterranean fruit fly. J Agric Food Chem 9:361–365. https://doi.org/10.1021/jf60117a007

Biasazin T, Chernet H, Herrera S et al (2018) Detection of volatile constituents from food lures by tephritid fruit flies. Insects 9(3):119. https://doi.org/10.3390/insects9030119

Boulter SL, Lambkin CL, Starick NT (2011) Assessing the abundance of seven major arthropod groups along an altitudinal gradient and across seasons in subtropical rainforest. Mem Queensl Mus 5:303–313

Brown BV (2005) Malaise trap catches and the crisis in neotropical dipterology. Am Entomol 51:180–183. https://doi.org/10.1093/ae/51.3.180

Brown BV (2009) Introduction. In: Brown BV, Borkent A, Cumming JM et al (eds) Manual of central American diptera: volume I. NRC Research Press, Ottawa, p 714

Brown BV, Borkent A, Adler PH et al (2018) Comprehensive inventory of true flies (Diptera) at a tropical site. Commun Biol 1:8. https://doi.org/10.1038/s42003-018-0022-x

Buttery RG, Ling LC, Teranishi R et al (1983) Insect attractants: volatiles of hydrolyzed protein insect baits. J Agric Food Chem 31:689–692. https://doi.org/10.1021/jf00118a003

Christenson LD, Foote RH (1960) Biology of fruit flies. Annu Rev Entomol 5:171–192. https://doi.org/10.1146/annurev.en.05.010160.001131

Dahl F (1896) Vergleichende untersuchungen über die lebensweise wirbelloser aasfresser. Sitzungsberichte der Königlich Preussischen Akademie der Wissenschaften zu Berlin 1(32):17–30

Díaz-Fleischer F, Piñero JC, Shelly TE (2014) Interactions between tephritid fruit fly physiological state and stimuli from baits and traps: looking for the pied piper of hamelin to lure pestiferous fruit flies. In: Shelly TE, Epsky N, Jang EB et al (eds) Trapping and the detection, control, and regulation of tephritid fruit flies: lures, area-wide programs, and trade implications. Springer, Dordrecht, pp 145–172. https://doi.org/10.1007/978-94-017-9193-9_5

Dilling C, Lambdin P, Grant J et al (2007) Insect guild structure associated with eastern hemlock in the southern Appalachians. Environ Entomol 36:1408–1414. https://doi.org/10.1603/0046-225x(2007)36[1408:Igsawe]2.0.Co;2

Drew RAI, Hooper GHS (1981) The responses of fruit fly species (Diptera: Tephritidae) in Australia to various attract ants. Aus J Entomol 20:201–205. https://doi.org/10.1111/j.1440-6055.1981.tb01032.x

Drew RA, Yuval B (1999) The evolution of fruit fly feeding behavior. In: Aluja M, Norrbom A (eds) Fruit flies (Tephritidae): phylogeny and evolution of behavior. CRC Press, Boca Raton, pp 749–768

Drew RAI, Courtice AC, Teakle DS (1983) Bacteria as a natural source of food for adult fruit flies (Diptera: Tephritidae). Oecologia 60:279–284. https://doi.org/10.1007/bf00376839

Emlen DJ (1992) Observations of a high altitude population of the sage stem-galling fly *Eutreta diana* (Diptera: Tephritidae) and its associated parasitoids. J Kans Entomol Soc 65:203–207

Epsky ND, Heath RR, Guzman A et al (1995) Visual cue and chemical cue interactions in a dry trap with food–based synthetic attractant for *Ceratitis capitata* and *Anastrepha ludens* (Diptera: Tephritidae). Environ Entomol 24:1387–1395. https://doi.org/10.1093/ee/24.6.1387

Epsky ND, Hendrichs J, Katsoyannos BI et al (1999) Field evaluation of female-targeted trapping systems for *Ceratitis capitata* (Diptera: Tephritidae) in seven countries. J Econ Entomol 92:156–164. https://doi.org/10.1093/jee/92.1.156

Epsky ND, Kendra PE, Schnell EQ (2014) History and development of food-based attractants. In: Shelly TE, Epsky N, Jang EB et al (eds) Trapping and the detection, control, and regulation of tephritid fruit flies: lures, area-wide programs, and trade implications. Springer, Dordrecht, pp 75–118. https://doi.org/10.1007/978-94-017-9193-9_3

FAO/IAEA (2018) In: Enkerlin WR, Reyes-Flores J (eds) Trapping guidelines for area-wide fruit fly programmes, 2nd edn. Rome, Italy, p 65

Fernandes GW, Price PW (1994) Life history, courtship, and mating behavior of the gall-forming *Aciurina trixa* (Diptera: Tephritidae) on *Chrysothamnus nauseosus hololeucus* (Asteraceae). Proc Entomol Soc Wash 96:301–307

Fernandes GW, Coelho MS, Santos JC (2014) Neotropical insect galls: status of knowledge and perspectives. In: Fernandes GW, Santos JC (eds) Neotropical insect galls. Springer, Dordrecht, pp 1–14. https://doi.org/10.1007/978-94-017-8783-3_1

Freidberg A (1984) Gall Tephritidae (Diptera). In: Ananthakrishnan TN (ed) Biology of gall insects. Oxford and IBH Publishing Co, London, pp 129–167

Freidberg A (1997) Tephritd galls and gall Tephritidae revisited, with specieal emphasis on Myopitine galls. In: Csoka G, Mattson W, Stone G et al (eds) The biology of gall-inducing arthropods. USDA: United States Department of Agriculture Matrafüred, Matrafured, pp 36–43

Gibb TJ, Oseto C (2020) Insect collection and identification, techniques for the field and laboratory, 2nd edn. Academic Press, New York. https://doi.org/10.1016/C2018-0-00136-X

Hardy DE (1986) Fruit flies of the subtribe Acanthonevrina of Indonesia, New Guinea, and the Bismarck and Solomon Islands (Diptera: Tephritidae: Trypetinae: Acanthonevrini). Pac Insects Monograph 42:1–191

Headrick DH, Goeden RD (1998) The biology of nonfrugivorous Tephritid fruit flies. Annu Rev Entomol 43:217–241. https://doi.org/10.1146/annurev.ento.43.1.217

Heath RR, Epsky ND, Guzman A et al (1995) Development of a dry plastic insect trap with food-based synthetic attractant for the Mediterranean and Mexican fruit flies (Diptera: Tephritidae). J Econ Entomol 88:1307–1315. https://doi.org/10.1093/jee/88.5.1307

Heath RR, Epsky ND, Dueben BD et al (1997) Adding methyl-substituted ammonia derivatives to a food-based synthetic attractant on capture of the Mediterranean and Mexican fruit flies (Diptera: Tephritidae). J Econ Entomol 90:1584–1589. https://doi.org/10.1093/jee/90.6.1584

Hernández-Ortiz V, Barradas-Juanz N, Díaz-Castelazo C (2019) A review of the natural host plants of the *Anastrepha fraterculus* complex in the Americas. In: Pérez-Staples D, Díaz-Fleischer F, Montoya P et al (eds) Area-wide management of fruit fly pests, 1st edn. CRC Press, Boca Raton, pp 89–122. https://doi.org/10.1201/9780429355738

Hernández-Ortiz V, Hernández-López M, Steck GJ (2020) Morfología y taxonomía de Tephritidae: Especies de importancia económica y cuarentenaria en América. In: Montoya P, Toledo J, Hernández E (eds) Moscas de la fruta: fundamentos y procedimientos para su manejo. SyG Editores, Mexico, pp 71–116

Hughes JB, Daily GC, Ehrlich PR (2000) Conservation of insect diversity: a habitat approach. Conserv Biol 14:1788–1797. https://doi.org/10.1111/j.1523-1739.2000.99187.x

IAEA (2007) Development of improved attractants and their integration into fruit fly SIT management programmes. In: IAEA-TECDOC-1574 (ed) Proceedings of the research coordination meeting, Vienna, Austria. IAEA, p 73

Jang EB, Holler TC, Moses AL et al (2007) Evaluation of a single-matrix food attractant tephritid fruit fly bait dispenser for use in federal trap detection programs. Proc Hawaii Entomol Soc 39:1–8

Karlsson D, Hartop E, Forshage M et al (2020) The Swedish Malaise trap project: a 15 year retrospective on a countrywide insect inventory. Biodivers Data J 8:1–35. https://doi.org/10.3897/BDJ.8.e47255

Katsoyannos BI, Heath RR, Papadopoulos NT et al (1999) Field evaluation of Mediterranean fruit fly (Diptera: Tephritidae) female selective attractants for use in monitoring programs. J Econ Entomol 92:583–589. https://doi.org/10.1093/jee/92.3.583

Lalonde RG, Shorthouse JD (1984) Developmental morphology of the gall of *Urophora cardui* (Diptera, Tephritidae) in the stems of Canada thistle (*Cirsium arvense*). Can J Bot 62:1372–1384. https://doi.org/10.1139/b84-187

Malaise RE (1937) A new insect trap. Entomol Tidskr 58:148–160

Martin JEH (1977) The insects and arachnids of Canada. Part 1: collecting, preparing, and preserving insects, mites, and spiders. Publ 1643. Canada Department of Agriculture, Ontario

Mazor M, Peysakhis A, Reuven G (2002) Release rate of ammonia—a key component in the attraction of female Mediterranean fruit fly to protein-based food lures. IOBC/WPRS Bull 25:323–328

McPhail M (1937) Relation of time of day, temperature and evaporation to attractiveness of fermenting sugar solution to Mexican fruitfly. J Econ Entomol 30:793–799. https://doi.org/10.1093/jee/30.5.793

Moeed A, Meads MJ (1987) Seasonality of arthropods caught in a Malaise trap in mixed lowland forest of the Orongorongo Valley, New Zealand. N Z J Zool 14:197–208

Montoya P, Toledo J, Flores-Breceda S (2006) Traps and attractants for detection and monitoring of fruit fly Populations (Diptera: Tephritidae). In: Barrera JF, Montoya P (eds) Simposio sobre Trampas y atrayentes en detección, monitoreo y control de plagas de importancia económica. Sociedad Mexicana de Entomología y El Colegio de la Frontera Sur, Manzanillo, pp 17–25

Norrbom AL (2010) Tephritidae (fruit flies, Moscas de frutas). In: Brown BV, Borkent A, Cumming JM et al (eds) Manual of central American diptera, vol 2. NRC Research Press, Ottawa, pp 909–954

Norrbom AL (2015) Compendium of fruit fly host information (CoFFHI). Edn 5.0. USDA-APHIS-CPHST. https://coffhi.cphst.org/. Accessed Apr 2020

Pape T, Blagoderov V, Mostovski MBJZ (2011) Order Diptera Linnaeus, 1758. In: Zhang Z-Q (ed) Animal biodiversity: an outline of higher-level classification and survey of taxonomic richness. Zootaxa 3148:222–229

Piñero J, Aluja M, Vázquez A et al (2003) Human urine and chicken feces as fruit fly (Diptera: Tephritidae) attractants for resource-poor fruit growers. J Econ Entomol 96:334–340. https://doi.org/10.1603/0022-0493-96.2.334

Qin Y, Paini DR, Wang C et al (2015) Global establishment risk of economically important fruit fly species (Tephritidae). PLoS One 10:1–8. https://doi.org/10.1371/journal.pone.0116424

Raghu S (2007) Functional significance of phytochemical lures to dacine fruit flies (Diptera: Tephritidae): an ecological and evolutionary synthesis. Bull Entomol Res 94:385–399. https://doi.org/10.1079/ber2004313

Savage J, Wheeler TA, Moores AMA et al (2011) Effects of habitat size, vegetation cover, and surrounding land use on Diptera diversity in temperate Nearctic bogs. Wetlands 31:125–134. https://doi.org/10.1007/s13157-010-0133-8

Sheikh AH, Thomas M, Bhandari R et al (2016) Malaise trap and insect sampling: mini review. Bio Bull 2:35–40

Silva IM, Andrade GI, Fernandes GW et al (1996) Parasitic relationships between a gall-forming insect *Tomoplagia rudolphi* (Diptera: Tephritidae) and its host plant (*Vernonia polyanthes*, Asteraceae). Ann Bot 78:45–48. https://doi.org/10.1006/anbo.1996.0093

Sivinski JM, Calkins C (1986) Pheromones and parapheromones in the control of tephritids. Fla Entomol 69:157–168. https://doi.org/10.2307/3494757

Steyskal GC (1977) History and use of the McPhail trap. Fla Entomol 60:11. https://doi.org/10.2307/3494476

Tan KH, Nishida R, Jang EB et al (2014) Pheromones, male lures, and trapping of tephritid fruit flies. In: Shelly TE, Epsky N, Jang EB et al (eds) Trapping and the detection, control, and regulation of tephritid fruit flies: lures, area-wide programs, and trade implications. Springer, Dordrecht, pp 15–74. https://doi.org/10.1007/978-94-017-9193-9_2

Thomas DB (2003) Nontarget insects captured in fruit fly (Diptera: Tephritidae) surveillance traps. J Econ Entomol 96:1732–1737. https://doi.org/10.1093/jee/96.6.1732

Thomas DB, Holler TC, Heath RR et al (2001) Trap-lure combinations for surveillance of *Anastrepha* fruit flies (Diptera: Tephritidae). Fla Entomol 84:344–351

Thomas DB, Epsky ND, Serra CA et al (2008) Ammonia formulations and capture of *Anastrepha* fruit flies (Diptera: Tephritidae). J Entomol Sci 43:76–85. https://doi.org/10.18474/0749-8004-43.1.76

Turner CE (1996) Tephritidae in the biological control of weeds. In: McPheron BA, Steck GJ (eds) Fruit fly pests. St. Lucie Press, Delray Beach, pp 157–164

Wiegmann BM, Trautwein MD, Winkler IS et al (2011) Episodic radiations in the fly tree of life. Proc Natl Acad Sci 108:5690–5695. https://doi.org/10.1073/pnas.1012675108

Chapter 9
Sampling Methods for Dragonflies and Damselflies

Rodrigo Roucourt Cezário, Paloma Pena Firme, Gabrielle C. Pestana, Diogo S. Vilela, Leandro Juen, Adolfo Cordero-Rivera, and Rhainer Guillermo

9.1 Introduction

The Odonata are amphibiotic insects (i.e., part of their life cycle occurs in the aquatic environment) of striking importance for the conservation of water resources. These organisms have cosmopolitan distribution, except for Antarctica, with greater abundance in tropical regions. There are 5912 described species in the Global Species Database of Odonata (van Tol 2019), and estimates indicate the existence of 6650 species (Trueman 2007). Odonates are carnivorous predators at all life stages and play a major role on aquatic trophic cascades (Corbet 1999).

The order has three suborders, namely, Anisoptera, with 2984 species; Anisozygoptera, with two to three species; and Zygoptera, with 2928 species (van Tol 2019). The suborder Anisoptera, interpreted as the sister group of Epiophlebiidae, is a very specious taxon, with its representatives commonly known as dragonflies. Anisopterans usually perch with wings spread, except for *Zenithoptera* (Anisoptera, Libellulidae) and *Cordulephya* Selys, 1870 (Anisoptera, Corduliidae). It includes 11 taxonomic families, many of which need review. In Epiophlebiidae, two species belonging to *Epiophlebia* are recognized (*E. laidlawi* Tillyard 1921, *E. superstes* Selys 1889) which have been recently reexamined and validated (Büsse 2016). The status of other species described from China (e.g., Li et al. 2012) needs further

R. R. Cezário · D. S. Vilela
LESTES Lab, Federal University of São Carlos, São Carlos, SP, Brazil

University of São Paulo—USP, Ribeirão Preto, SP, Brazil

P. Pena Firme · G. C. Pestana · R. Guillermo (✉)
LESTES Lab, Federal University of São Carlos, São Carlos, SP, Brazil

L. Juen
Laboratory of Ecology and Conservation, Federal University of Pará, Belém, PA, Brazil

A. Cordero-Rivera
ECO-EVO Lab, University of Vigo, Pontevedra, Spain

J. C. Santos, G. W. Fernandes (eds.), *Measuring Arthropod Biodiversity*,
https://doi.org/10.1007/978-3-030-53226-0_9

research (Büsse 2016). This relic taxon (Nel 1993) has a restricted distribution, inhabiting cold streams of coniferous and deciduous forests (Li et al. 2012). *Epiophlebia* nymphs have anal pyramid with gills inside. Unlike anisopterans, Epiophlebiidae species have functional ovipositors like in Zygoptera.

Zygoptera is an equally speciose group, with their representatives known as damselflies. Zygopterans have separate compound eyes and usually fold the wings when at rest, with exceptions in at least eight families with species that land with wings unfolded (Paulson 2004). In the World Odonata List, 24 taxonomic families are recognized as belonging to Zygoptera (Schorr and Paulson 2019).

Odonata is a widely used taxon in studies involving biomonitoring (Miguel et al. 2017a). These organisms have been used in assessments of water quality, the level of anthropization, and degradation of environments and studies involving climate change (Miguel et al. 2017b). Odonates have become a model for environmental assessments, as they have unique ecological characteristics, such as wide distribution in the aquatic ecosystem and the presence of eurytopic (species which are more resilient to environmental changes) and stenotopic species (species able to adapt only to a small range of environmental disturbance) (Corbet 1999). It is essential to study this group of animals because of their status for conservation science and policy making. Odonates were the first insect group ever assessed in terms of global risk of extinction, and mainly because of its cosmopolitan occurrence and their response to changes on habitat structure, the order can work as a proxy of global biodiversity (Clausnitzer et al. 2009).

Furthermore, odonates play a key role in the trophic chains of riparian forests, acting as top predators in areas next to water resources (Knight et al. 2005). Concern for this group has been around since 2009, when the IUCN also began to address threatened species of dragonflies and damselflies, viewed as key conservation species and as bioindicators of pollution and degradation (Clausnitzer et al. 2009). Growing urbanization over natural environments impacts the entire habitat of these animals, which show different sensitivities at each stage of their life cycle (Monteiro-Junior et al. 2013; Villalobos-Jiménez et al. 2016).

Given the importance of these insects, this chapter aims to introduce methods for sampling Odonata to facilitate understanding and raise the interest of new scientists to this exciting and enigmatic group that has much to tell about its ecology. It is important to make the correct choice of sampling method so that there is a suitable experimental design for each type of research with Odonata.

9.2 Study Methods and Sampling Techniques

At the moment of collection, all the excitement of being outside searching for specimens may drive us away from some important details. Collecting is not only about taking your net or whatever equipment you need and randomly collect/store the animals. You must be respectful with them, and some cares can be easily taken before you head to the field. Storing the specimens is one of the most imperative tasks one must learn before collecting. Knowing how to store specimens prevents one from damaging the individuals, making identification difficult (sometimes impossible). Therefore, before leaving in a collecting trip, be sure to have the following items in hands (besides the nets and traps): little paper bags, tweezers, a hard box, and a bag (a money belt or a vest with many pockets is also quite good).

Paper bags are cheap, if you choose to buy them, and easy to assemble if you choose to make them. It is recommended to use paper instead of plastic bags to store the specimens, because the latter may damage them due to friction and may accelerate the degradation of dead specimens and fungi growth. You can easily find envelopes on the Internet (philately and stationery stores), and they usually came in square shapes (Fig. 9.1), besides being available in different sizes (e.g., usually 6 × 9 cm for Zygoptera and 7 × 11 cm for Anisoptera), which come in handy when you want to sample both suborders. Another choice is the classic triangular envelopes. They are better to assemble using vegetal paper with a grammage of 90/95 g^2, which confers more stability on the folding (see Sect. 9.6).

Active or passive sampling techniques can be used to capture odonates. Active methods involve the search for insects by the researcher. It is easy to collect adult damselflies, which fly slowly, using a hand net (40 cm in diameter, 65 cm deep, 90 cm long). For perching or flying dragonflies, the best strategy is to approach them from behind, then move the net fast to catch the insect, and finally rotate the handle to close the net. This method can damage large aeshnids if they touch the net support: they easily lost the head by the impact. Therefore, refrain from trying to capture a flying dragonfly, unless it is necessary for the research. Passive traps are more dependent on insect habits, are generally less selective, and can be left in the field for relatively longer periods, when compared to active methods. Entomological nets are the most used method to capture adults in flight (Fig. 9.2). In Japan, a traditional method to trap flying Anisoptera (called "buri" or "toriko") is the use of small weights united by a thread, which are thrown out to the sky in front of a swarm of dragonflies. The dragonflies get trapped when trying to capture the weights, confused by their similarity to a flying prey (Hatto 1994). Malaise traps can also be

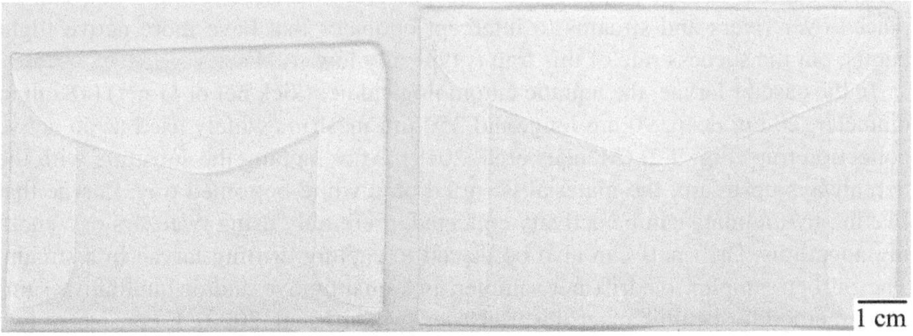

Fig. 9.1 Paper bags of 6 × 9 (left) and 7 × 11 cm (right), used for specimen storage in the field. After placing the specimen within the bag, be assured to store them in a hard case, to avoid smashing the specimens

Fig. 9.2 An example of sampling using the aquatic entomological net (kick net or D-net)

used to capture adults. Malaise traps are passive, nonselective traps that can be placed over rivers and streams to intercept odonates that have more active flight habits, but the success rate of this trap is typically low.

In the case of larvae, the aquatic entomological net (kick net or D-net) (18 cm in diameter, 20 cm deep, 90 cm long, and 250-μm mesh) is widely used as an active collection trap (Fig. 9.2) (Mendes et al. 2019). After tapping the substrate with the net, always upstream, the material is sorted on a white-bottomed tray. Larvae that live in phytotelmata can be actively collected, preferably using tweezers or a sucking apparatus. Drift nets can also be placed to capture drifting larvae in a stream. The Surber sampler, or drift-net sampler, is a quantitative and/or qualitative sampling method for benthic macroinvertebrates (Malard et al. 2002). It consists of two perpendicular squares of the same area and a filter net in one of the squares. The portion without the filter should be placed parallel to the substrate, and the content under that area should be swept and filtered. There are samplers that cover different areas, which can be better used according to the objectives of sampling. Surber samplers are used in lotic environments and less commonly in lentic environments. However, some species are inhabitants of lentic microhabitats in rivers and cannot be sampled with a Surber (e.g., Macromiidae). Alternatively, small hand sieves (diameter around 15–30 cm) can also be used to sample in shallow ponds or small places that the other larger equipment are not capable to reach. The Surber sampler can also be used to study the spatial distribution of odonate larvae (Hawking and New 1999), but only larval sampling is not a good indicator of population dynamics; hence, exuviae should also be sampled (Raebel et al. 2010). These methods have their limitations, as odonate larvae that live in macrophytes or have nektonic habits (i.e., larvae that swim freely in the water column) will hardly be collected. It is therefore crucial that all the types of substrates are sampled, particularly tree roots, where many species live hidden as larvae.

Some species exhibit evening twilight (eocrepuscular) and twilight activity (crepuscular) (Corbet 1999), especially aeshnids. Thus, the use of light traps can be an effective method for attracting and capturing these organisms (Umar et al. 2012). Light traps consist of a collecting chamber, a funnel-shaped lid, a light source, and a shower protect top (Ramamurthy et al. 2010). The light source used can be a mercury light, black light, or UV, depending on the object of study, although Ramamurthy et al. (2010) showed a higher efficiency by mercury light trap in catching odonates. The light trap can also be used to capture larvae, in order to characterize local diversity (Schilling et al. 2009). This aquatic light trap consists of a rectangular box containing a gill-breathing airspace, a sloping wall (25°) toward the center of the trap, a movable gate wall, and a side containing two panels separated by a screen fiberglass with L-shaped brackets for fitting the light chamber. The bottom panel contains a gate for removal of the contents of the collection. The light chamber contains a 12-V 8″ fluorescent tube connected to a car battery (Aiken 1979).

To collect odonates in their last stage of aquatic life, an emergence trap can be used, which is placed at the edges of the rivers and streams, close to riparian forests. Collection can only be done by placing the trap and leaving it on the bank in contact with water, as it can be placed where exuviae were previously seen (Camargo et al. 2015). In order to be able to account for the emergence of larvae, one can use a trap consisting of a large open-bottom model with a plastic base and a structured ceiling lined with a mesh (Voigts 1973). Variations of Voigt's trap are also used in other studies, depending on the species to be desired and the purpose of the study (McLaughlin and Harris 1990). These are collections of the representative type (Corbet 1964), which require first prior knowledge of the species, understanding their way of life, and emergence patterns for an honest collection. It is important that the trap remains for a recorded time so that the sample is standardized (Corbet 1964).

Exceptionally, giant damselflies (Zygoptera, Pseudostigmatidae), such as the helicopter damselfly *Megaloprepus caerulatus*, and some aeshnids (Anisoptera, Aeshnidae) (e.g., *Gynacantha*) (Fincke 1984) lay their eggs in water-filled tree holes, also known as phytotelmata. In this microhabitat, determinants such as larval body size, developmental time, and the extension of the breeding period make these odonate larvae one of the most abundant predators and an important model in ecological studies (Fincke 1999). For sampling purposes, egg-laying traps made from plastic bottles make a good method to assess abundance and/or to sample these phytotelma dwellers (Sahlén and Hedström 2005). In this case, the plastic bottles must be placed in the tree hole, preferentially at about 1 or 2 m height, and should be inspected regularly (Hedström and Sahlén 2007).

The collected individuals can be kept alive and later taken to the laboratory until emergence for morphological studies and association of the exuviae with adults or may be placed in 80% alcohol for identification and/or use in environmental monitoring analyzes. After the emergence of the adult, the "skin" of the larva is left on the substrate (vegetation, rocks, the bare soil, or even spider webs; Cordero-Rivera et al. 1999). This is known as the exuvia and is a direct evidence that the species has completed the life cycle in the habitat. Therefore, sampling exuviae is truly relevant for ecological purposes. Recently, there has been much debate on how biased samples of communities are, when based only on adults or only on exuviae (Raebel et al. 2010; Bried et al. 2012). The fact is that in some habitats exuviae are easily found, especially for large species, but in other cases, are extremely difficult to detect, like in the rainforest streams. Da Silva (2007) compared the species list of Portuguese rivers using adults, larvae, and exuviae, and he found more species when working with exuviae than with adults or larvae and suggested to combine at least adults and exuviae in future studies.

9.3 Rearing Larvae

Another way to sample taxa is by rearing larvae. This is particularly a priority in understudied regions. In the lab, it is possible to rear larvae for the purpose of obtaining the adults. This is not an easy task for some species as they need a very specific environment to survive and grow (Piersanti et al. 2015). To rear larvae, the first thing you need to do is collect them in their natural environment. The focus of the collection is on large larvae that exhibit wing pads and therefore are close to emergence. After collection, larvae are kept in plastic vials with small strips of wet filter paper. It is important to keep these vials under shade or inside coolers to prevent overheating (e.g., Bright 1999).

An important observation to increase larval survival during transport is not to fully fill vials with water. Instead, fill no more than 1/4 of the vial, or mix the larvae with damp vegetation with little free water, and the larvae should make it safely to the lab (ACR, DSV, and RG personal observations). This will ensure that the larva has oxygen until it gets to its new home. It is better to place the vial in a Styrofoam container and put it in the shade if you remain at the collection site for long periods (more than 30 min).

After transport, it is important to renew the filter paper, as the previous one is likely to contain feces and water from the collection site. This new paper should be moistened with water from a spring source (or mineral water) and stored at room temperature. Do not use tap water as it has too much chlorine, fluoride, and other chemicals. However, it is possible to obtain chlorine-free water if tap water is left for at least 1 day in an aquarium and used afterward. Some pipes are made of iron or copper, which can kill the animals as well. Add some mineral water to the vial until it covers the whole animal and about 2 cm of water column is left. Change this water daily or at most every 2 days. For species of lotic environments, it is better to place the animal in a small aquarium filled with water and aerate with an air pump. In either case, there should be a support for the animal to hold. We usually use wood sticks, but it is important to remember to change them every 2 days as they may rot or release substances in the water that will harm the animal. An alternative is to use plastic spoons or glass sticks or even plastic plants (for aquaria) which reduce the chance of contamination (Piersanti et al. 2015; Tavares et al. 2017).

Every time you change the water in the container, remember to change the filter paper and sanitize or change the glass. In the case of aquaria, sanitize them as organic matter and light can favor the growth of algae and cyanobacteria that can kill the larvae. Leave the animals in a cool, light-accessible environment so that the photoperiod favors larval growth and adult emergence (Lutz 1968; Corbet 1999). For food, we usually raise brine shrimp or collect mosquito larvae to provide *ad libitum*. It is noteworthy that last instar larvae will stop feeding once the metamorphosis is close. Where adult emergence is sought, last instar larvae should be the focus and avoid collecting larvae in other instars. The last and second to last instar larvae are easily recognized by the developed wing shoots (Pritchard and Pelchat 1977). For the emergence of adults, it is interesting to generate some stress on larva so that it decides to leave the unfavorable environment and emerge (Stoks et al. 2008). For example, low water levels, food shortages (or no food at all), the presence of conspecifics or predators, or even rearing the larva in a bright environment can accelerate adult emergence.

It is also possible to collect and rear eggs. For this, a detailed methodology is provided by Gutiérrez et al. (2015). Eggs can be found by observing tandem pairs or females at the time of oviposition usually in aquatic plants. These plants can be placed in vials and kept under laboratory conditions (25 ± 2 °C, 70 ± 5% relative humidity and 12 h photoperiod). After the prolarvae emerge, it is advisable to separate them into individual vials. The same result can be obtained by collecting females and allowing them to oviposit on wet filter paper (Sanmartín-Villar and Cordero-Rivera 2016).

9.4 Mark and Recapture

Populations have different ecological properties, such as abundance, distribution, density, demographic coverage, and dispersion (Neal 2004). These properties can be estimated using different collection methods, to highlight processes that are occurring in the environment. However, in the case of winged insects that fly long distances, some of which have a relatively short life span, assessing these properties becomes difficult. In the case of odonates, surveys involving marking and recapture methods to study population dynamics are the most common (Cordero-Rivera and Stoks 2008).

The mark and recapture method is useful in individual- or population-level studies (Fig. 9.3). The odonates can be captured with entomological nets, marked often by writing a unique number with permanent pencil, commonly in the non-pigmented part of the wings, and then returned to the population. Subsequently a second survey can be performed which will allow us to estimate the population size (Cordero-Rivera and Stoks 2008). Marking is also a necessary protocol to assess individual microhabitat, behavior, and other traits that are part of biodiversity. For consistency sake, we usually measure body size and wing length and write it down in a table along with date and collection site (Table 9.1). Odonata is also a target group for migration studies. In this case, radio transmitters are good tools to assess the movement of migratory individuals (Wikelski et al. 2006), which can be attached to the adult thorax by using different types of glues.

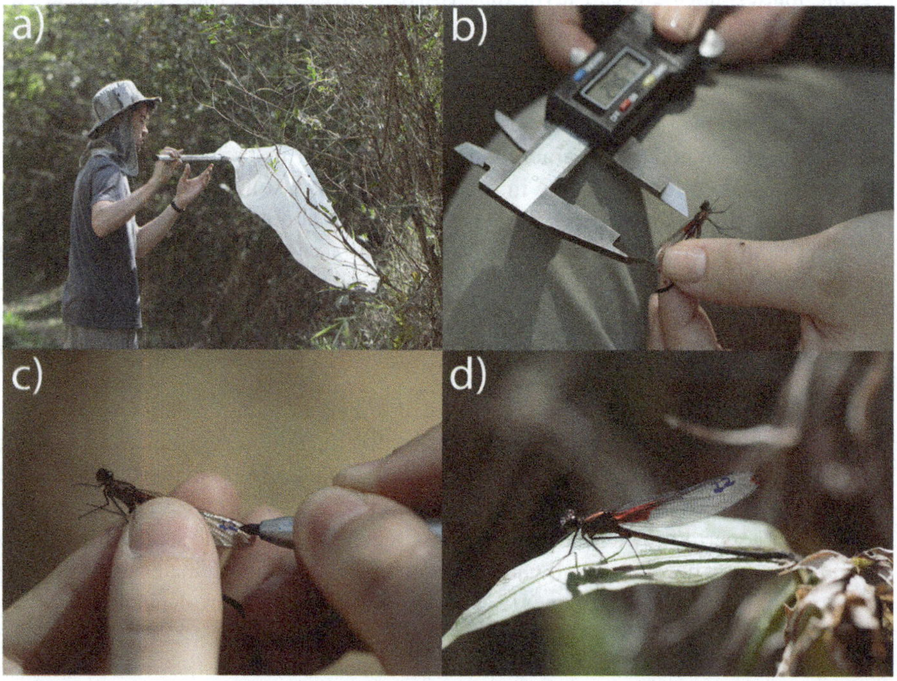

Fig. 9.3 Marking and recapture method. (**a**) Step 1: capturing with an entomological net. (**b**) Step 2: measure the body and wings. (**c**) Step 3: write a unique number with permanent pencil, preferably in the non-pigmented part of the wings. (**d**) Step 4: release the animal

Table 9.1 An example of the traits collected from the sampled specimens in the field

Individual	Sex	Body size (mm)	Wing length (mm)	Date	Site	Observation
07	Male	37.11	24.87	03/11/19	A	
12	Female	39.43	27.81	04/11/19	B	

9.5 Specific Traits

Biodiversity assessments of odonates must consider particular traits of some species. For instance, some are shade seekers and hide in places where most people would not consider collecting while sampling an odonate assemblage. In other cases, some have crepuscular habits and will not come to the water until dusk. Some species occupy the canopy, while others inhabit or seek shelter away from the water inside the forest, like members of the genus *Philogenia* (Vilela et al. 2019). When collecting larvae, this notion is even more crucial to pay close attention. Larvae exhibit site selection and habitat occupancy behaviors. Some are claspers and hold onto plants, leaves, and other substrates, others are burrowers and bury themselves into mud or sand, and some other species are sprawlers and crawl or swim on the surface (Corbet 1999). The most remarkable case lies on the larvae that live in phytotelmata and tree holes (Fincke 1984). In all cases they occupy different microhabitats and different niches. That is why collecting dragonflies and damselflies, adults and larvae, involves meticulous ahead planning and the selection of specific microhabitats to sample the taxa of interest. It is wise not to sample the same substrates or patches, but to variate sampling sites, techniques, and methods to cover assemblages as more effectively as possible.

9.6 The Odonata Scanning Protocol (OSP)

To conduct sampling in a way that data are comparable or to test hypotheses, methods must be standardized in space and time. For odonates, one of the most used protocols is the observation and scanning in fixed areas, or the OSP, in which the researcher sets a 100 m patch of a water body and walks on its margins observing and capturing all possible dragonflies and damselflies (Juen and De Marco 2011). To make sampling easier, this area can be divided in 20 segments of 5 m each, making it possible to detect, identify, and, if necessary, collect most individuals (Fig. 9.4). In each stream or river, to guarantee sample independence, it is necessary to establish a single patch.

Sampling is conducted by one collector for an hour (or two collectors for 30 min each); thus, the mean time a researcher must collect inside each 5 m segment is 3 min. Researchers often use entomological nets to capture specimens (40 cm in diameter, 65 cm deep, 90 cm long) (Fig. 9.3). Due to the differences and ecophysiological requirements within the order Odonata, collections should preferably be performed between 9:00 and 14:00 on sunny days (Oliveira-Junior and Juen 2019a, b). If the weather becomes cloudy or rainy, the collection should be interrupted and resumed when conditions improve from the segment where it was interrupted. If the study does not aim to assess the effect of seasonality, we recommend that collections be performed during the dry season for four main reasons: (i) it is due to the ecophysiological requirements of the Odonata (see May 1976, 1991; Corbet 1999);

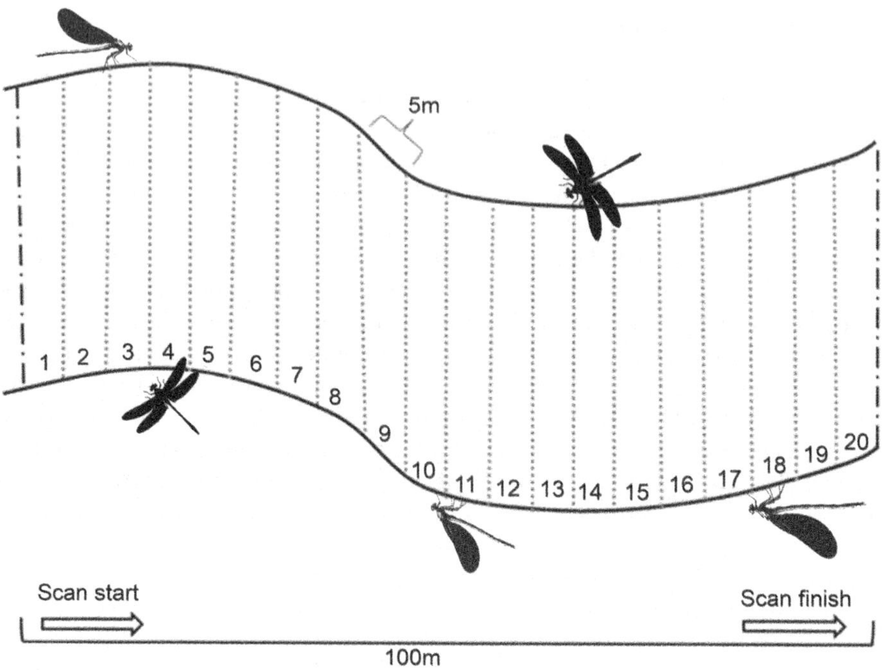

Fig. 9.4 Representation of the OSP in streams

(ii) sampling in a single seasonal period may reduce sample "noise" in the analyses and results (see Heino 2014); (iii) several studies demonstrate that greater diversity of Odonata can be accessed during the dry season (see Baptista et al. 2001; Fulan and Henry 2007); and (iv) the shallow depth of the water during this period makes these insects concentrate in smaller areas, making it easier to find and capture them (see Oliveira-Junior and Juen 2019a).

This protocol (OSP) has been used very successful in several studies (De Marco et al. 2015; Oliveira-Junior et al. 2015; Calvão et al. 2016; Juen et al. 2016; Brasil et al. 2017, 2018; Carvalho et al. 2018; Oliveira-Junior et al. 2017; Brasil et al. 2019; Oliveira-Junior and Juen 2019a, b). Furthermore, because of the division into segments, it can be executed in a modeling manner that can be easily adapted to other protocols using a larger sampling, such as the Environmental Protection Agency (EPA; Peck et al. 2006), which uses a varied footage according to the width of the stream. In the Amazon, when we use the EPA's environmental sampling protocol, we standardize the size of the stream patch at 150 m, because the streams are very narrow. Therefore, the 150 m stretches are subdivided into ten sections of 15 m each, following the instructions of Peck et al. (2006). Each section is subdivided into three segments of 5 m each, and the Odonata collection is performed in the first two segments. The third segment of each section is used to measure limnological variables (e.g., dissolved oxygen) (Fig. 9.5).

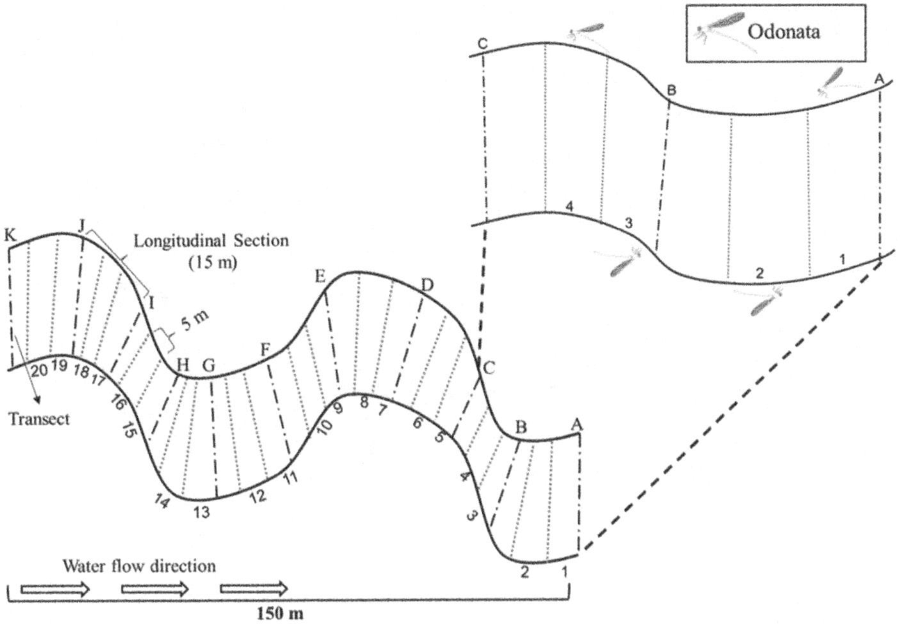

Fig. 9.5 Representation of the OSP in streams, adapted from Peck et al. (2006), named *Field Operations Manual for Wadeable Streams*—FOMWS

The collected material is stored in an entomological envelope of parchment paper, where the name of the species or morphospecies and the segment number and name of the collected stream are noted (with preferential use of waterproof ink pen or pencil) (Fig. 9.6). It is useful to write down also the time of the day and to store together mating pairs. All storage process follows the protocol described by Lencioni (2005, 2006). To facilitate the recording of specimens that were only sighted but not collected, we use Table 9.2, where each segment is a row and each species or morphospecies collected/observed is a column. Often it is not possible to identify or know the name of the species in the field, so only the name of the morphospecies can be entered to be confirmed or identified in the laboratory. There is already plenty of literature material for identification, for example (Lencioni 2005, 2006; Garrison et al. 2006, 2010; Von Ellenrieder 2013; Pessacq 2014).

At the end of the study, it is possible to perform analysis by assessing the variation of species richness or abundance within the stream (in which case we would consider each segment as a pseudoreplicate) or by summing the 20 segments to consider the stream as a sample, to perform analysis or hypothesis testing involving comparisons between streams. In statistical analysis, we can follow two lines: we can work qualitatively or quantitatively. In the qualitative form, we aim at comparing environments classified into distinct categories, such as species richness (number of species) found in preserved streams compared with species richness found in altered streams. In the quantitative form, it sought to evaluate gradients of

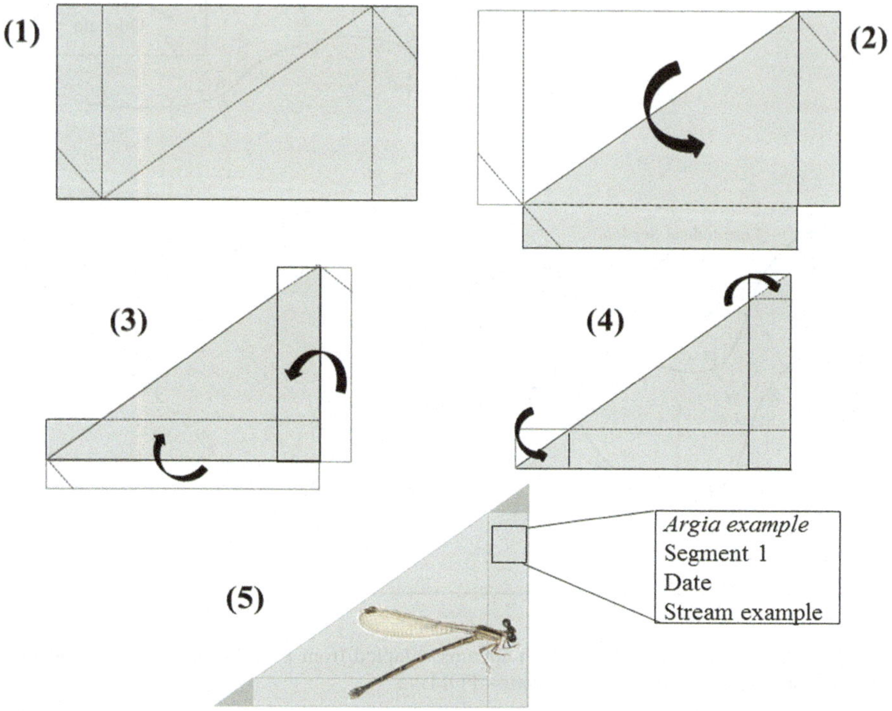

Fig. 9.6 Steps for assembling the entomological paper bag, used for adult Odonata storage (steps 1–5), highlighting important information to be written on it. (Drawings by José Max B. Oliveira Júnior)

environmental change, such as testing the effect of riparian forest width variation on species richness. Most qualitative tests work using mean comparisons considering the variance of data deviations or variances. Therefore, in order to conduct a qualitative study, we need to have at least three samples from each category, and using the example cited above, we would need at least three preserved streams and three altered streams so that we could be able to calculate the mean and the deviation of each treatment. This minimum sample number (three) would allow us to perform the tests; however, the greater the variability in our samples, the greater the number of streams (samples) we need to estimate population parameters with higher precision. Therefore, it is recommended that you work with a larger number of samples in each category whenever possible. As the objective of the study is to evaluate quantitatively, as mentioned above the example of the effect of riparian forest width variation on Odonata species richness, a larger number of samples are recommended, and trying to contemplate the entire variation gradient, the larger the gradient, the greater the number of samples needed. For shorter environmental gradients, generally 10–15 streams (samples) would already be sufficient to measure the effects on the Odonata community. For a review of the main methods of univariate

Table 9.2 Data sheet for collected specimens in the field, in each of the 20 segments

	Species	Environmental data sheet					
		Site:					
1		Time:					
2		Temperature:					
3		Humidity:					
4		Locality:					
5		GPS:					
6		Date:					
7		Abiotic data					
8		Segments	1	5	10	15	20
9		Conductivity					
10		Turbidity					
11		Dissolved O_2					
12		pH					
13		Water flow					
14		Depth					
15		Width					
16		Observations:					
17							
18							
19							
20							

analysis, it is recommended to read Zar's statistical book (1999). Spatially closer biological communities are expected to be more similar than more distant communities, because of the greater ease of movement and dispersal of species in the landscape over shorter spatial distances. This effect is called spatial autocorrelation (Legendre 1993). Therefore, in order to verify how much the similarity of the communities is influenced or explained simply by the spatial distribution of the sampling points, as well as to try to avoid the spatial biases, it is important that the collections have a certain geographical distance from each other. It is complicated to establish a minimum distance; however, we recommend sampling different water bodies to ensure data independence. Hence, geographic information is collected from each collection point, such as latitude and longitude.

We strongly recommend that, in addition to biological data, in this same area where the transect for Odonata collection was established, environmental data collection is also performed. In addition to assessing whether there is a difference in species distribution, it is also possible to test which environmental variables are affecting the variation in community distribution. Studies have shown that vegetation modifications (Samways and Steytler 1996, Carvalho et al. 2013, 2018; Oliveira-Junior et al. 2017), conservation status of the water body (Stewart and Samways 1998; Oliveira-Junior et al. 2015), physical and chemical habitat variables (Jacob et al. 1984; Oliveira-Junior et al. 2015), and anthropogenic changes (Oertli

2008; Carvalho et al. 2013; Monteiro-Júnior et al. 2013; Oliveira-Junior et al. 2015; Silva et al. 2015;) have a major effect on Odonata communities. Therefore, for studies to have greater discussion power, it is interesting that at least some of these metrics be evaluated.

Some examples of these analytical protocols can be easily found using odonates as model organisms. Monteiro-Júnior et al. (2016) tested whether there was a difference in species richness and abundance of odonates between streams inside and outside conservation areas. To test their hypotheses, the authors used a Student's t test. In the study by Calvão et al. (2016), the authors were interested in assessing whether the type of timber harvesting affected species richness and Odonata abundance. To this end, the authors conducted studies on streams that were draining three different landscape conditions: streams that drained conventional logging areas where environmental legislation was not obeyed, creeks that drained low-impact logging area that complied with environmental legislation, and lastly, creeks that drained control areas where there was no impact.

Since the objective was to evaluate the effect of a categorical variable of three levels of variation (conventional cutoff streams, managed-area streams, and control-area streams) on a quantitative variable (species richness and/or abundance), the authors used one-way analysis of variance (one-way ANOVA). In the study by Oliveira-Junior et al. (2017), the authors evaluated the effect of environmental integrity on species richness. In this study, environmental integrity was measured by the Nessimian et al.'s (2008) Physical Integrity Index, which is a quantitative variable whose values range from 0 to 1, and the higher the value, the more integral the system. As the authors were interested in evaluating the effect of a quantitative variable (integrity index) on a quantitative variable (species richness), they used linear regression analysis. Other interesting analyses that can serve as models for other studies can be found in Juen and De Marco (2011), De Marco et al. (2015), Rodrigues et al. (2016), Brasil et al. (2017, 2018), Calvão et al. (2018), Alves-Martins et al. (2018), Brasil et al. (2019), Carvalho et al. (2018), and Mendes et al. (2019). All these studies used the observation and scanning method with fixed areas, as described above.

Acknowledgments We thank the National Council for Scientific and Technological Development (CNPq) for productivity grants to LJ (process 304710/2019-9) and RGF (process 307836/2019-3) and grants to RRC, GCP, and DSV. ACR was funded by grants from the Spanish Ministry of Science, which included FEDER funds (CGL2014-53140-P and PGC2018-096656-B-I00). We thank the editors and two anonymous reviewers for their valuable comments.

References

Aiken R (1979) A size selective underwater light trap. Hydrobiologia 65:65–68. https://doi.org/10.1007/BF00032720

Alves-Martins F, Calatayud J, Medina NG, De Marco P, Juen L, Hortal J et al (2018) Drivers of regional and local diversity of Amazonian stream Odonata. Insect Conserv Diver 12:251–261. https://doi.org/10.1111/icad.12327

Baptista DF, Dorvillé LFM, Buss DF et al (2001) Spatial and temporal organization of aquatic insects assemblages in the longitudinal gradient of a tropical river. Rev Bras Biol 61:295–394. https://doi.org/10.1590/S0034-71082001000200012

Brasil LS, Vieira TB, de Oliveira-Junior JMB et al (2017) Elements of metacommunity structure in Amazonian Zygoptera among streams under different spatial scales and environmental conditions. Ecol Evol 7:3190–3200. https://doi.org/10.1002/ece3.2849

Brasil LS, Oliveira-Júnior JM, Calvão LB, Carvalho FG et al (2018) Spatial, biogeographic and environmental predictors of diversity in Amazonian Zygoptera. Insect Conserv Diver 11:174–184. https://doi.org/10.1111/icad.12262

Brasil LS, Silverio DV, Cabette HSR et al (2019) Net primary productivity and seasonality of temperature and precipitation are predictors of the species richness of the Damselflies in the Amazon. Basic Appl Ecol 35:45–53. https://doi.org/10.1016/j.baae.2019.01.001

Bried JT, D'Amico F, Samways MJ (2012) A critique of the dragonfly delusion hypothesis: why sampling exuviae does not avoid bias. Insect Conserv Divers 5:398–402. https://doi.org/10.1111/j.1752-4598.2011.00171.x

Bright E (1999) Sampling protocol for Odonata larvae. Michigan Odonata survey technical note

Büsse S (2016) Morphological re-examination of *Epiophlebia laidlawi* (Insecta: Odonata) including remarks on taxonomy. Int J Odonat 19:221–238. https://doi.org/10.1080/13887890.2016.1257442

Calvão LB, Juen L, Oliveira-Júnior JMB et al (2018) Land use modifies Odonata diversity in streams of the Brazilian Cerrado. J Insect Conserv 22:657–685. https://doi.org/10.1007/s10841-018-0093-5

Calvão LB, Nogueira DS, de Assis Montag LF et al (2016) Are Odonata communities impacted by conventional or reduced impact logging? Forest Ecol Manag 382:143–150. https://doi.org/10.1016/j.foreco.2016.10.013

Camargo AJA, Oliveira CM, Frizzas MR et al (2015) Coleções entomológicas: legislação brasileira, coleta, curadoria e taxonomia para as principais ordens, Brasília, Embrapa

Carvalho FG, Pinto NS, Oliveira-Junior JMB et al (2013) Effects of marginal vegetation removal on Odonata communities. Acta Limnol Bras 25:10–18. https://doi.org/10.1590/S2179-975X2013005000013

Carvalho FG, Oliveira Roque F, Barbosa L et al (2018) Oil palm plantation is not a suitable environment for most forest specialist species of Odonata in Amazonia. Anim Conserv 21:526–533. https://doi.org/10.1111/acv.12427

Clausnitzer V, Kalkman VJ, Ram M et al (2009) Odonata enter the biodiversity crisis debate: the first global assessment of an insect group. Biol Conserv 142:1864–1869. https://doi.org/10.1016/j.biocon.2009.03.028

Corbet PS (1964) Temporal patterns of emergence in aquatic insects. Can Entomol 96:264–279. https://doi.org/10.4039/Ent96264-1

Corbet PS (1999) Dragonflies: behavior and ecology of Odonata. Comstock Publishing Associates, Ithaca

Cordero-Rivera A, Stoks R (2008) Mark-recapture studies and demography. In: Córdoba-Aguilar (ed) Dragonflies and damselflies: model organisms for ecological and evolutionary research. Oxford Press, Oxford, pp 7–20. https://doi.org/10.1093/acprof:oso/9780199230693.003.0002

Cordero-Rivera A, Utzeri C, Santolamazza-Carbone S (1999) Emergence and adult behaviour of Macromia splendens (Pictet) in Gsalicia, northwestern Spain (Anisoptera: Corduliidae). Odonatologica 28:333–342

Da Silva G (2007) Conservación e manexo de odonatos ameazados: ecoloxía, xenética poboacional e dispersión. PhD thesis, University of Vigo

De Marco P, Batista JD, Cabette HSR (2015) Community assembly of adult odonates in tropical streams: an ecophysiological hypothesis. PLoS One 10(4):e0123023. https://doi.org/10.1371/journal.pone.0123023

Fincke OM (1984) Giant damselflies in a tropical forest: reproductive biology of *Megaloprepus coerulatus* with notes on *Mecistogaster* (Zygoptera: Pseudostigmatidae). Adv Odonatol 2:13–27

Fincke OM (1999) Organization of predator assemblages in Neotropical tree holes: effects of abiotic factors and priority. Ecol Entomol 24:13–23. https://doi.org/10.1046/j.1365-2311.1999.00166.x
Fulan JÁ, Henry R (2007) Temporal distribution of immature Odonata (Insecta) on Eichhornia azurea (Kunth) stands in the Camargo Lake, Paranapanema River, Sao Paulo. Rev Bras Entomol 51:224–227. https://doi.org/10.1590/S0085-56262007000200013
Garrison RW, Von Ellenrieder N, Louton JA (2006) Dragonfly genera of the New Word: an illustrated and annotated key to the Anisoptera. The Johns Hopkins University Press, Baltimore, p 368
Garrison RW, Von Ellenrieder N, Louton JA (2010) Damselfly genera of the New World. Baltimore, an illustrated and annotated key to the Zygoptera. The Johns Hopkins University Press, Baltimore
Gutiérrez Y, Freitas HL, Oliveira EE (2015) *Acanthagrion viridescens* (Odonata: Coenagrionidae): description of the final larval stadium and biological notes. Zootaxa 4057.1:125–134. https://doi.org/10.11646/zootaxa.4057.1.8
Hatto Y (1994) "Buri" or "toriko", a traditional Japanese method of catching dragonflies. Odonatologica 23:283–289
Hawking JH, New TR (1999) The distribution patterns of dragonflies (Insecta: Odonata) along the Kiewa River, Australia, and their relevance in conservation assessment. Hydrobiologia 392:249–260. https://doi.org/10.1023/A:1003510514937
Hedström I, Sahlén G (2007) The dry season governs the reproduction of three pseudostigmatid zygopterans in Costa Rica (Odonata: Pseudostigmatidae). Int J Odonatol 10:53–63. https://doi.org/10.1080/13887890.2007.9748288
Heino J (2014) Taxonomic surrogacy; numerical resolution and responses of stream macroinvertebrate communities to ecological gradients: are the inferences transferable among regions? Ecol Indic 36:186–194. https://doi.org/10.1016/j.ecolind.2013.07.022
Jacob U, Walther H, Klenke R (1984) Aquatic insect larvae as indicators of limiting minimal contents of dissolved oxygen-part II. Aquat Insects 6:185–190. https://doi.org/10.1080/01650428409361183
Juen L, De Marco P (2011) Odonate biodiversity in terra-firme forest streamlets in Central Amazonia: on the relative effects of neutral and niche drivers at small geographical extents. Insect Conserv Divers 4:265–274. https://doi.org/10.1111/j.1752-4598.2010.00130.x
Juen L, Cunha EJC, Carvalho FG et al (2016) Effects of oil palm plantations on the habitat structure and biota of streams in eastern amazon. River Res Appl 22:1085–1095. https://doi.org/10.1002/rra.3050
Knight TM, McCoy MW, Chase JM, McCoy KA, Holt RD et al (2005) Trophic cascades across ecosystems. Nature 437(7060):880. https://doi.org/10.1038/nature03962
Legendre P (1993) Spatial autocorrelation: trouble or new paradigm? Ecology 74:1659–1673. https://doi.org/10.2307/1939924
Lencioni FAA (2005) The Damselflies of Brazil: an illustrated guide—the non Coenagrionidae families. All Print Editora, São Paulo, 332pp
Lencioni FAA (2006) The Damselflies of Brazil: an illustrated guide—Coenagrionidae. All Print Editora, São Paulo, 419pp
Li JK, Nel A, Zhang XP et al (2012) A third species of the relict family Epiophlebiidae discovered in China (Odonata: Epiproctophora). Syst Entomol 37:408–412. https://doi.org/10.1111/j.1365-3113.2011.00610.x
Lutz PE (1968) Effects of temperature and photoperiod on larval development in *Lestes eurinus* (Odonata: Lestidae). Ecology 49:637–644. https://doi.org/10.2307/1935529
Malard F, Dole-Olivier MJ, Mathieu J, et al (2002). Sampling manual for the assessment of regional groundwater biodiversity. Protocols for the Assessment and Conservation of Aquatic Life in the Subsurface (PASCALIS). http://www.pascalis-project.com/results/samplingmanual.html
May ML (1976) Thermoregulation in adaptation to temperature in dragonflies (Odonata: Anisoptera). Ecol Monogr 46:1–32. https://doi.org/10.2307/1942392
May ML (1991) Thermal adaptations of dragonflies, revisited. Adv Odonatol 5:71–88

McLaughlin DB, Harris HJ (1990) Aquatic insect emergence in two Great Lakes marshes. Wetl Ecol Manag 1(2):111–121. https://doi.org/10.1007/BF00177285

Mendes TM, Benone NL, Juen L (2019) To what extent can oil palm plantations in the Amazon support assemblages of Odonata larvae? Insect Conserv Divers 12(5):448–558. https://doi.org/10.1111/icad.12357

Miguel TB, Calvão LB, Vital MVC et al (2017a) A scientometric study of the order Odonata with special attention to Brazil. Int J Odonatol 20:27–42. https://doi.org/10.1080/13887890.2017.1286267

Miguel TB, Oliveira-Junior JMB, Ligeiro R et al (2017b) Odonata (Insecta) as a tool for the biomonitoring of environmental quality. Ecol Indic 81:555–566. https://doi.org/10.1016/j.ecolind.2017.06.010

Monteiro-Júnior CS, Couceiro SEM, Hamada N et al (2013) Effect of vegetation removal for road building on richness and composition of Odonata communities in Amazonia, Brazil. Int J Odonatol 16:135–144. https://doi.org/10.1080/13887890.2013.764798

Monteiro-Júnior CS, Esposito MC, Juen L (2016) Are the adult odonate species found in a protected area different from those present in the surrounding zone? A case study from eastern Amazonia. J Insect Conserv 20:643–652. https://doi.org/10.1007/s10841-016-9895-5

Neal D (2004) Introduction to population biology. Cambridge University Press, Cambridge, 393pp

Nel A (1993) Les "Anisozygoptera" fossiles: phylogenie et classification (Odonata). Martinia 3:1–311

Nessimian JL, Venticinque E, Zuanon J et al (2008) Land use, habitat integrity, and aquatic insect assemblages in Central Amazonian streams. Hydrobiologia 614:117–131. https://doi.org/10.1007/s10750-008-9441-x

Oertli B (2008) The use of dragonflies in the assessment and monitoring of aquatic habitats. In: Córdoba-Aguilar A (ed) Dragonflies and Damselflies: model organisms for ecological and evolutionary research. Oxford University Press, Oxford, pp 79–95. https://doi.org/10.1093/acprof:oso/9780199230693.003.0007

Oliveira-Junior JMB, Juen L (2019a) The Zygoptera/Anisoptera ratio (Insecta: Odonata): a new tool for habitat alterations assessment in Amazonian streams. Neotrop Entomol 48:552–560. https://doi.org/10.1007/s13744-019-00672-x

Oliveira-Junior JMB, Juen L (2019b) Structuring of dragonfly communities (Insecta: Odonata) in Eastern Amazon: effects of environmental and spatial factors in preserved and altered streams. Insects 10:1–18. https://doi.org/10.3390/insects10100322

Oliveira-Junior JMB, Shimano Y, Gardner TA et al (2015) Neotropical dragonflies (Insecta: Odonata) as indicators of ecological condition of small streams in the eastern Amazon. Austral Ecol 40:733–744. https://doi.org/10.1111/aec.12242

Oliveira-Junior JMB, De Marco P, Dias-Silva K et al (2017) Effects of human disturbance and riparian conditions on Odonata (Insecta) assemblages in eastern Amazon basin streams. Limnologica 66:31–39. https://doi.org/10.1016/j.limno.2017.04.007

Paulson DR (2004) Why do some zygopterans (Odonata) perch with open wings? Int J Odonatol 7:505–515. https://doi.org/10.1080/13887890.2004.9748235

Peck DV, Herlihy AT, Hill BH et al (2006) Environmental monitoring and assessment program–surface waters western pilot study: field operations manual for Wadeable Streams. EPA/620/R–06/003. U.S. Environmental Protection Agency, Washington, DC

Piersanti S, Rebora M, Salerno G et al (2015) A method for rearing a large number of damselflies (*Ischnura elegans*, Coenagrionidae) in the laboratory. Int J Odonatol 18:125–136. https://doi.org/10.1080/13887890.2015.1015179

Pritchard G, Pelchat B (1977) Larval growth and development of *Argia vivida* (Odonata: Coenagrionidae) in warm sulphur pools at Banff, Alberta. Can Entomol 109:1563–1570. https://doi.org/10.4039/Ent1091563-12

Raebel EM, Mercls T, Riordan P et al (2010) The dragonfly delusion: why it is essential to sample exuviae to avoid biased surveys. J Insect Conserv 14:523–533. https://doi.org/10.1007/s10841-010-9281-7

Ramamurthy VV, Akhtar MS, Patankar NV et al (2010) Efficiency of different light sources in light traps in monitoring insect diversity. Munis Entomol Zool 5:109–114

Rodrigues ME, Roque FO, Quintero JMO et al (2016) Nonlinear responses in damselfly community along a gradient of habitat loss in savanna landscape. Biol Conserv 194:113–120. https://doi.org/10.1016/j.biocon.2015.12.001

Sahlén G, Hedström I (2005) The larva of *Mecistogaster linearis*, with notes on its abundance in lowland rainforest of Costa Rica (Odonata: Pseudostigmatidae). Int J Odonatol 8:59–66. https://doi.org/10.1080/13887890.2005.9748243

Samways MJ, Steytler NS (1996) Dragonfly (Odonata) distribution patterns in urban and forest landscapes, and recommendations for riparian management. Biol Conserv 78:279–288. https://doi.org/10.1016/S0006-3207(96)00032-8

Sanmartín-Villar I, Cordero-Rivera A (2016) The inheritance of female colour polymorphism in *Ischnura genei* (Zygoptera: Coenagrionidae), with observations on melanism under laboratory conditions. PeerJ 4:e2380. https://doi.org/10.7717/peerj.2380

Schilling EG, Loftin CS, Huryn AD (2009) Macroinvertebrates as indicators of fish absence in naturally fishless lakes. Freshwater Biol 54:181–202. https://doi.org/10.1111/j.1365-2427.2008.02096.x

Schorr M, Paulson D (2019) World Odonata List, Slater Museum of Natural History. University of Puget Sound, Tacoma. https://bit.ly/2GnIQW4. Accessed 23 Apr 2020

Silva DP, De Marco P, Resende DC (2015) Adult odonate abundance and community assemblage measures as indicators of stream ecological integrity: a case study. Ecol Indic 10:744–752. https://doi.org/10.1016/j.ecolind.2009.12.004

Stewart DAB, Samways MJ (1998) Conserving dragonfly (Odonata) assemblages relative to river dynamics in an African savanna game reserve. Conserv Biol 12:683–692. https://doi.org/10.1046/j.1523-1739.1998.96465.x

Stoks R, Johansson F, De Block M (2008) Life-history plasticity under time stress in damselfly larvae. Dragonflies and damselflies. Model organisms for ecological and evolutionary research. Oxford University Press, Oxford, pp 39–51. https://doi.org/10.1093/acprof:oso/9780199230693.003.0004

Tavares RIS, Mandelli AM, Mazão GR et al (2017) The relsationship between habitat complexity and emergence time in damselflies. Limnologica 65:1–3. https://doi.org/10.1016/j.limno.2017.04.009

Trueman JWH (2007) A brief history of the classification and nomenclature of Odonata. Zootaxa 1668:381–394. https://doi.org/10.11646/zootaxa.1668.1.20

Umar DM, Marinov MG, Schorr M et al (2012) Odonata attracted by light: a new topic for mythbusters. Int Dragonf Fund 43:1–52

Van Tol J (2019) Odonata: global species database of Odonata (version Dec 2011). In: Roskov Y, Ower G, Orrell T, Nicolson D, Bailly N, Kirk PM, Bourgoin T, DeWalt RE, Decock W, van Nieukerken E, Zarucchi J, Penev L (eds) Species 2000 & ITIS catalogue of life, 2019 annual checklist. www.catalogueoflife.org/annual-checklist/2019. Accessed 4 Oct 2019

Vilela DS, Guillermo-Ferreira RH, Encalada AC et al (2019) *Philogenia gaiae* sp. nov. (Zygoptera: Philogeniidae) and description of the female of *P. macuma* Dunkle, 1986, two species from the Ecuadorean lowland rainforest. Zootaxa 4683(3):412–420. https://doi.org/10.11646/zootaxa.4683.3.5

Villalobos-Jiménez G, Dunn A, Hassall C (2016) Dragonflies and damselflies (Odonata) in urban ecosystems: a review. Eur J Entomol 113:217–232. https://doi.org/10.14411/eje.2016.027

Voigts DK (1973) An odonate emergence trap for use in marshes. Proc Iowa Acad Sci 80:67–68

Wikelski M, Moskowitz D, Adelman JS, Cochran J, Wilcove DS, May ML et al (2006) Simple rules guide dragonfly migration. Biol Lett 2:325–329. https://doi.org/10.1098/rsbl.2006.0487

Zar JH (1999) Biostatistical analysis. Prentice-Hall, Englewood Cliffs

Chapter 10
Sampling Methods for Termites (Insecta: Blattaria: Isoptera)

Reginaldo Constantino

10.1 Introduction

Termites are eusocial, highly derived cockroaches. Although termites have been classified as order Isoptera, their relationship with cockroaches and mantises (the Dictyoptera) has been known for a long time. The hypothesis of a sister-group relationship between termites and *Cryptocercus*, a genus of wood-feeding roaches, was first presented by Cleveland et al. (1934), was endorsed by Wilson (1971), and has been corroborated by all recent phylogenies (Inward et al. 2007; Legendre et al. 2015).

All termite species live in colonies with distinct morphological castes: imagoes (winged adults and reproductives), soldiers, and workers (Roisin 2000). A typical colony includes a single reproductive pair (queen and king) and a large number of workers, soldiers, and immatures.

Most termites feed on dead plant matter, and termite assemblages play an important role in the decomposition process in tropical terrestrial ecosystems (Lavelle et al. 1997). They are the main decomposers of cellulose in many forests and savannas, accelerating the mineralization of dead plant matter. They also have a physical effect on the environment, building nests and tunnels, and moving soil particles vertically (Holt and Lepage 2000). As members of the soil fauna, their activity has a positive effect on soil properties and fertility. Termites are an important food item for many vertebrates (Redford 1987). They are also important urban and agricultural pests, damaging cellulosic materials such as wood and paper (Su and Scheffrahn 2000) and several crops (Rouland-Lefèvre 2011).

There are 2968 known termite species, belonging to 300 genera and 9 families (Krishna et al. 2013; Constantino 2019). The estimated total number is considerably higher, probably more than 5000 species and 700 genera (Constantino 2018).

R. Constantino (✉)
Departamento de Zoologia, Universidade de Brasília, Brasilia, Brazil
e-mail: constant@unb.br

J. C. Santos, G. W. Fernandes (eds.), *Measuring Arthropod Biodiversity*,
https://doi.org/10.1007/978-3-030-53226-0_10

Termite abundance and diversity are higher in tropical regions, declining with latitude, elevation, and aridity (Jones and Eggleton 2011). Species richness in local assemblages is highly variable, reaching a maximum around 130 species in some African equatorial forests (Jones and Eggleton 2000).

10.2 Termite Nesting Habits

Anyone sampling termites needs to acquire a basic knowledge of termite habits and polymorphism. Termite colonies, and their nests, vary greatly in size and complexity. Nests of some *Macrotermes* species may reach 30 m in diameter and contain millions of individuals, while nests of some drywood termites may contain fewer than 100 individuals living in tunnels excavated in wood. Nests may be epigeic, subterranean, or arboreal.

Termite nests may be divided into four different types (Fig. 10.1):

1. Single-piece nests, which are just tunnels excavated in wood. This type of nest occurs in all species of Kalotermitidae, Stolotermitidae, Archotermopsidae, and a few Rhinotermitidae. They may occur in dead branches of standing trees, tree stumps, or logs on the ground. These termites do not build any external structure and do not forage outside their nest. The only way to collect them is to open the wood and remove them from the tunnels using tweezers or aspirators.
2. Intermediate nests, also excavated within wood, but these termites do forage out of their nest and build external structures. This type of nest occurs in Rhinotermitidae and Mastotermitidae.
3. Separate nests, which are independent of wood and may be arboreal (Fig. 10.2a), epigeic (Fig. 10.2b), or subterranean (Fig. 10.2c). This type of nest occurs in Termitidae, Hodotermitidae, and a few Rhinotermitidae.

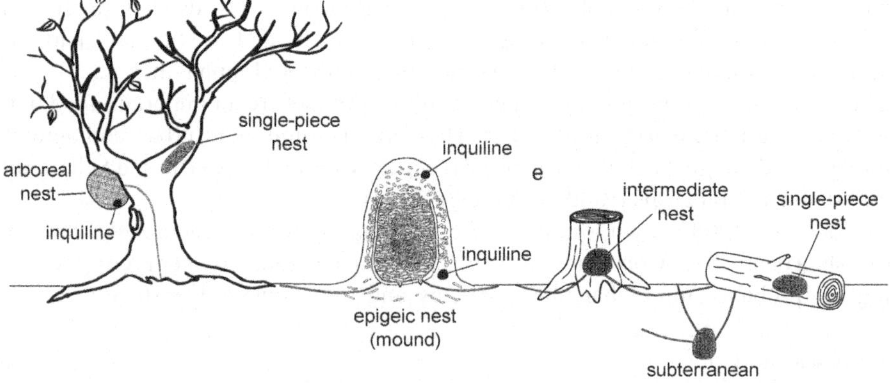

Fig. 10.1 Different types of termite nests: single-piece, intermediate, epigeic, subterranean, arboreal, and inquiline

Fig. 10.2 Termite nests. (**a**) Arboreal nest of *Nasutitermes macrocephalus*. (**b**) Epigeic nest of *Cornitermes cumulans*. (**c**) Foraging workers of *Syntermes* sp., leaving the subterranean nest through a small hole on the surface. (**d**) *Serritermes serrifer*, inquiline inside the nest of *Cornitermes cumulans*

4. Inquilines, which build their nests inside termitaria built by other species (Fig. 10.2d). They are very common in epigeic mounds and may also occur in arboreal nests. Some large and old termite mounds may contain dozens of inquiline colonies belonging to 15 or more species. Some of these inquilines, such as *Inquilinitermes* spp. and *Serritermes serrifer*, are very specialized and occur only in nests built by one or a few species and do not forage outside the host nest. Cohabitation in termite nests adds considerable difficulty to sampling protocols.

Many samples are not collected from nests but rather from foraging groups in the soil, litter, wood, or tunnels. In many cases it is impossible to determine where the nest is located and which foraging groups belong to the same colony. Foraging individuals may be found at more than 70 m from their nest (Su et al. 1993).

10.3 Castes and Polymorphism

Polymorphism is highly variable among termites, and it is important to become familiar with the various morphs present before collecting. Most colonies have a single reproductive pair, which may be difficult to find inside the nest. The queen can be recognized by the enlarged abdomen. Alates may be present or not, depending on the age of the colony and the season. Imagoes are, in most cases, not required for identification but should be collected if possible.

The soldier caste is usually the most important for taxonomic identification. Samples without soldiers are very difficult or impossible to identify based on morphology. Some species have very few soldiers per colony, which may result in soldierless samples. Soldier polymorphism varies considerably among termites, as listed below:

1. Monomorphic soldier. Most species have a single soldier morph.
2. Dimorphic soldier. Several species have two types of soldiers. Sometimes the difference is only in size, but in some species, e.g., in Rhinotermitinae, the soldiers are so different that may seem to belong to different species (Fig. 10.3a, b). Usually the smaller morph is more abundant, but the large one may be necessary for identification.
3. Trimorphic soldier. A few species of Macrotermitinae and Nasutitermitinae have three distinct soldiers, e.g., the neotropical genera *Diversitermes* (Fig. 10.3c, e) and *Velocitermes*. Sometimes the identification is only possible with the examination of all morphs, or at least two of them.
4. Soldierless termites. Soldiers are completely absent in the *Anoplotermes* group (Apicotermitinae) and in a few species of Termitinae. Soldierless termites are abundant and diverse in the neotropical and Ethiopian regions. The workers in these species are usually monomorphic.

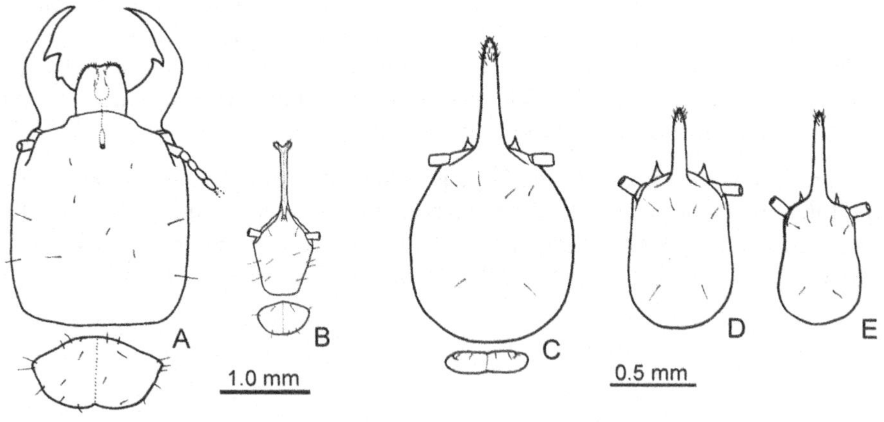

Fig. 10.3 Examples of termites with polymorphic soldiers. *Rhinotermes marginalis*: (**a**) major soldier and (**b**) minor soldier. *Diversitermes diversimiles*: (**c**) major soldier, (**d**) intermediate soldier, and (**e**) minor soldier

10.4 Feeding Groups

Functional diversity may be more important than taxonomic diversity because it is related to ecosystem services. The termite assemblage is delimited by classification, which reflects phylogeny, not ecology. Termites are usually divided into four to five functional groups based on their feeding habits. They feed on dead plant matter along a humification gradient, ranging from wood feeders to soil feeders. Donovan et al. (2001) divided all termites into four feeding groups based on the humification gradient only:

Wood feeders. Termites feeding on sound and moderately decayed wood. This group includes the so-called lower termites (all families except Termitidae), and some genera of Termitidae such as *Nasutitermes* and *Microcerotermes*. Corresponds to group I and part of group II of Donovan et al. (2001).

Litter feeders. Termites foraging for leaf and grass litter and small woody items. They include species that forage on the surface and store litter inside their nests such as *Syntermes* spp. and *Velocitermes* spp. They correspond to part of group II of Donovan et al. (2001).

Intermediate feeders. Termites feeding on decayed wood and litter, which do not ingest significant amounts of mineral soil. They correspond to group III of Donovan et al. (2001) and to soil/wood interface feeders of Eggleton et al. (1995).

Soil feeders. Termites feeding on soil organic matter which ingest large amounts of mineral soil. They correspond to group IV of Donovan et al. (2001).

Specialized feeders. Some termites have very specialized diets. For instance, *Hospitalitermes* and *Constrictotermes* feed on micro-epiphytes (e.g., algae, lichens, fungi) on tree bark.

10.5 Termite Collecting

A "sample" is a set of specimens from the same place (e.g., nest, wood piece, or soil sample). Ideally, each sample should contain a good number of specimens from a single colony, including all castes present (Fig. 10.4). However, samples taken from the soil, litter, large mounds, and large pieces of wood often contain more than one species. These mixed samples will need careful sorting later in the laboratory, which usually takes considerable time and effort.

Because termites are soft-bodied insects, they should always be killed and preserved in alcohol. If the samples will be used only for morphological study, the termites may be killed and kept in 80–85% ethanol. Specimens to be used for DNA extraction should be killed and kept in strong alcohol (95% or higher) and stored in a freezer as soon as possible. DNA degrades rapidly in specimens preserved in diluted alcohol at room temperature. However, specimens preserved in strong alcohol become dehydrated and brittle and tend to lose hairs, antennae, and legs. The best strategy is to kill the termites in 95% alcohol and later divide the sample into two subsamples, one kept in 95–99% ethanol and another in 80–85% ethanol.

Fig. 10.4 Sample of *Microcerotermes indistinctus*. (**a**) Nest. (**b**) Vial with sample (80% ethanol). (**c**) Castes present in the same sample, from left to right: soldier, workers, and imago

It is important to clean the samples and replace the alcohol a few days after collecting. The presence of soil and debris may physically damage the termites and alter pigmentation. The usual procedure is to place the sample in a petri dish and use forceps to transfer the termites to another petri dish or a clean vial. Flotation may be used to speed up the cleaning process.

All samples should be clearly and carefully labeled. Careless and disorganized labeling often results in data loss and waste of time and effort. Recommendations for labels are:

1. Add a unique, easily readable alphanumeric identifier to each sample. For instance, unique field labels may be prepared with a combination of [initials from the collector's name] [year]-[sequential number], e.g. "RC2019-001." This identifier will be permanently kept with the sample and can be used to associate it with field notes. These labels can be printed in advance and used in the field (Fig. 10.5a). A unique identifier is also necessary for databases.
2. Add a standard specimen label with locality (country, state or province, locality, geographic coordinates), date, and name of the collector for each sample (Fig. 10.5b).
3. All labels should be small and easily readable. Avoid large handwritten labels with a lot of data.
4. Place labels inside the vial. Do not stick the labels on the outside surface of the vial.

The following information should be recorded for each sample, if possible: (1) habitat and vegetation type; (2) elevation; (3) soil type; (4) nest type with picture; or (5) if specimens were not collected from nest, microhabitat (e.g., soil, wood, litter, foraging tunnels on trees). Take notes in a field notebook using a waterproof and alcohol-proof pen. All this information needs to be properly organized and permanently associated with the specimen samples.

Each sample should be stored in a separate vial. Good vials are transparent, are resistant, and have watertight caps. A 5-mL vial is adequate for most samples. A few larger vials (10–20 mL) may be necessary for large samples. To save time during

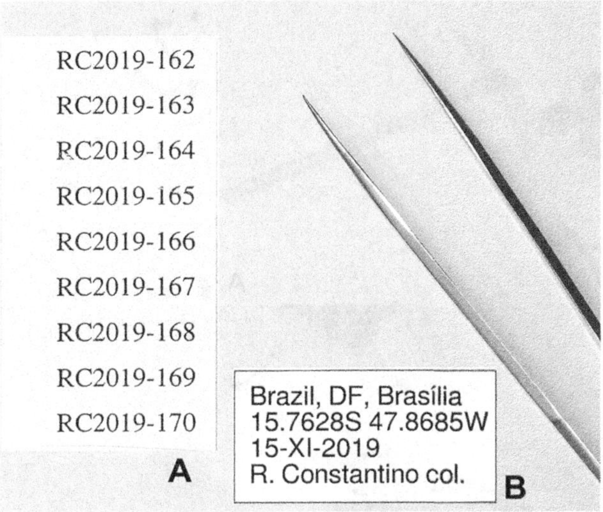

Fig. 10.5 Labels. (**a**) Set of laser-printed field labels, which can be cut and added to vials as needed. (**b**) Specimen label with locality, date, and collector's name

fieldwork, it is best to carry an adequate number of vials previously filled with alcohol.

Several different tools may be necessary to collect termites:

1. Featherweight forceps (Fig. 10.6a, b): useful for manipulating soft-bodied specimens without damaging them.
2. Hatchet: to collect termites from large wood pieces.
3. Aspirator (Fig. 10.6a, c, d): useful to extract termites from wood and some nests.
4. Large pickaxe: to excavate large mounds.
5. Gardening tools such as a small hoe (Fig. 10.6a) and small shovel.
6. Plastic trays (Fig. 10.6a): useful for sorting termites from nest pieces, soil, and litter.

10.6 Sampling Methods

10.6.1 Belt Transects

This has been the most popular method in recent years. The most used protocol was developed by Jones and Eggleton (2000) using 100 m × 2-m transects divided into 20 segments of 2 m × 5 m (Fig. 10.7a). The sampling effort is standardized, with two collectors working for 30 min in each segment. Therefore, the collecting effort will be 1 man-hour per section and 20 man-hours per transect. In each section, the collectors search for termites in 12 samples of surface soil, in nests, tunnels, wood,

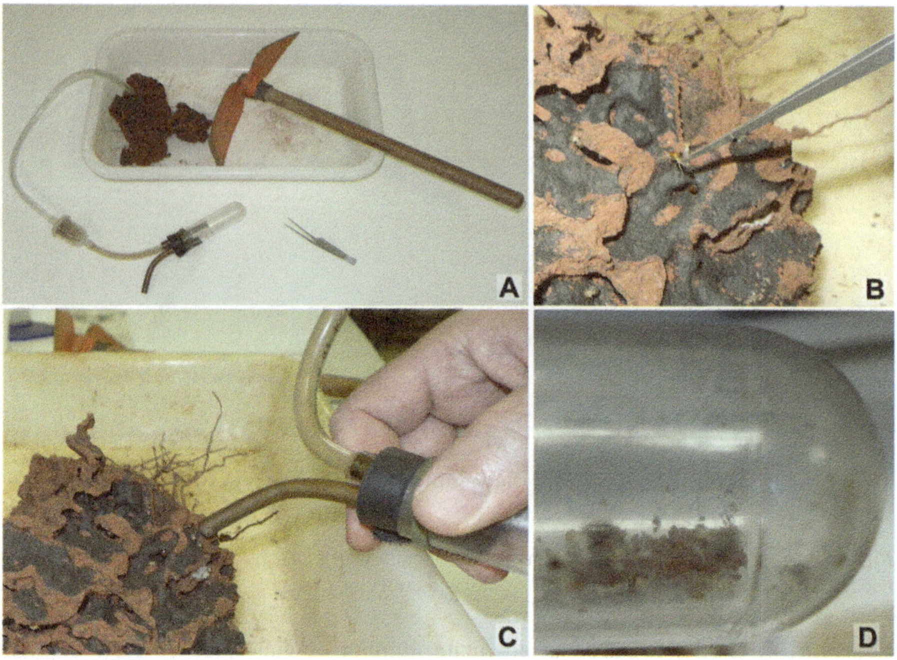

Fig. 10.6 Termite collecting. (**a**) Plastic tray with a nest piece, hoe, forceps, and aspirator. (**b**) Using a featherweight entomological forceps. (**c**) Using an aspirator. (**d**) Bottom of aspirator tube with termites

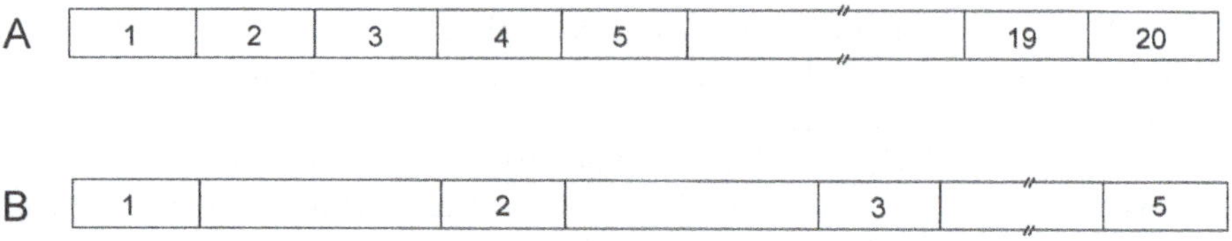

Fig. 10.7 Belt transects. (**a**) A 100-m transect used in the protocol described by Jones and Eggleton (2000). (**b**) Alternative transect with a 10-m space between sampling segments (Cancello et al. 2014)

and litter up to 2 m above ground. The 30-min time limit includes all collecting activities, data recording, and specimen labeling. At least three transects should be sampled in each site or treatment, with a minimum distance of 200 m from each other.

For data analysis, each species should be recorded only once per section, which is called an "encounter" (Jones et al. 2003). The rationale is that most samples are collected from foraging groups and samples of the same species in the same sections are most likely from the same colony.

Several modified versions of this protocol have been used by different authors. For instance, Cancello et al. (2014) used 65-m transects divided into five sections of

5 × 2 m, with a 10-m space between sections (Fig. 10.7b). The spacing between sections prevents repeated sampling of the same colonies in contiguous sections. The collecting procedures for each section were the same as in the original protocol.

Advantages This method is rapid, flexible, and efficient and will provide semi-quantitative data suitable for statistical analyses.

Disadvantages This method is focused on soil termites and is not adequate for Kalotermitidae, which will be undersampled. In forest habitats, many kalotermitids are found in dead branches above 2 m. Many arboreal nests are also above 2 m. It is not possible to estimate termite abundance using this method.

10.6.2 Large Plots

Various protocols using square or rectangular plots of different sizes have been used to sample termite assemblages. Large plots may be used to estimate the density of mounds. For instance, Sands (1965) used 20 × 20-m quadrats to sample *Trinervitermes* mounds, which are relatively small, and Wood et al. (1982) used 40 × 40-m plots divided into four sections of 20 × 20 m. Eggleton et al. (1999) used 50 × 50-m plots divided into 10 × 10-m sections to estimate termite biomass in mounds, wood, litter, and soil.

Advantages This method may be used to estimate biomass and density of mounds.

Disadvantages This method may be more laborious and time-consuming than transects. It is difficult to lay out and sample in dense vegetation. It is not recommended as a general sampling method for termite assemblages, except for quantitative sampling.

10.6.3 Soil Monoliths

Soil monoliths are often used to sample soil macrofauna (Bignell et al. 2008), including termites. The standard TSBF monolith is 25 × 25 cm wide and 30 cm deep, divided into 3 horizontal layers of 10 cm each (Fig. 10.8). It is necessary to dig a trench on two sides of the monolith, collect the layers, and transfer them to trays for hand sorting, which is laborious and requires the work of a team of about six people. At least five soil monoliths should be sampled in each site.

Advantages Abundance and biomass can be estimated for soil fauna.

Disadvantages Because termites are social insects, most individuals are aggregated inside their nests. The monolith will sample foraging termites only. Typical samples from monoliths contain individuals of several different species, mostly workers. Variation is usually high, with many samples with a few specimens or none

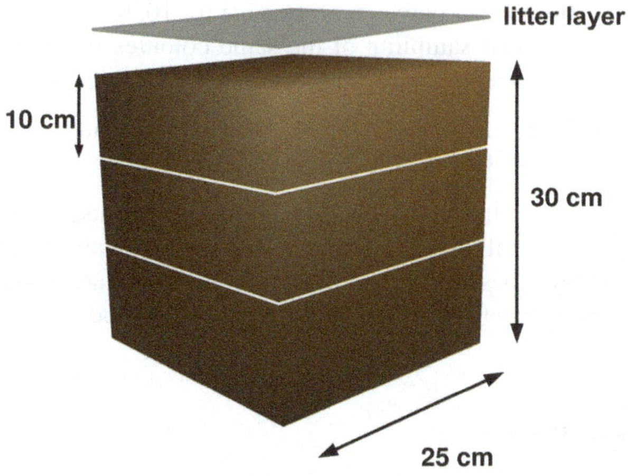

Fig. 10.8 A TSBF soil monolith used to sample soil macrofauna. The monolith is divided into four layers: litter, 1–10, 10–20, and 20–30 cm

and some with a large number. Taxonomic identification is difficult because many samples lack soldiers. This method is not recommended if the purpose is to sample termites only.

10.6.4 Baiting

Baiting is most often used to monitor subterranean termite population or to study foraging behavior (e.g., La Fage et al. 1973). Baits may be made of wood, paper, or carton (Fig. 10.9a) and are usually distributed in the field on grids. They must be inspected at regular intervals, and termites present should be collected and preserved in alcohol for identification. Abensberg-Traun (1993) recorded a higher termite diversity in naturally occurring timber than in baits (wooden stakes and toilet paper rolls) in an Australian woodland. Davies et al. (2013) compared transect sampling with toilet paper baiting in African savannas and concluded that baiting was adequate for drier habitats only.

Advantages Baits are good to sample subterranean wood feeders and may be used as complement to other sampling methods. They are easy to use and are not affected by the experience of the collectors.

Disadvantages Because baits need to be left in the field for a long time, this method is inadequate for short-term surveys. All types of baits are selective and will usually sample only a fraction of the local assemblage. They typically attract more wood feeders such as *Heterotermes* spp. and some litter foragers. Soil feeders are rarely collected in baits and only when they are left in the field for a long time.

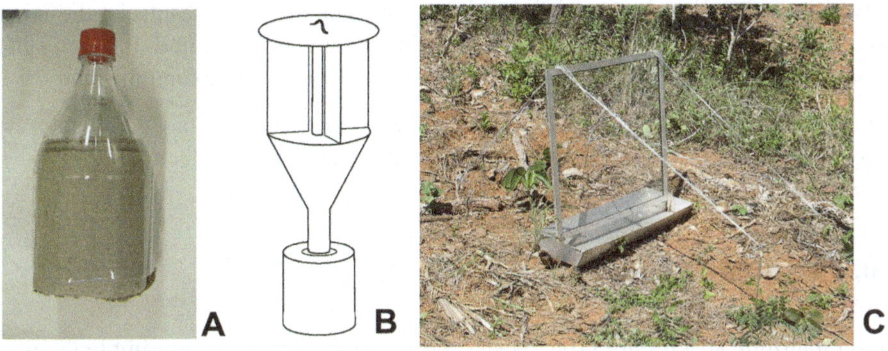

Fig. 10.9 Baits and traps. (**a**) Cellulosic bait (corrugated carton), for subterranean termites. (**b**) Light trap, which can be used to sample alates. (**c**) Flight intercept trap, also for alates

10.6.5 Alate Traps

Alate traps are a good complement to conventional sampling because they collect species that are undersampled by other methods, especially drywood termites and subterranean species. For instance, Rebello and Martius (1994) collected a large number of alates of Kalotermitidae using light and flight interception traps in Amazonian forests. These termites are rarely recorded by other sampling methods. Bourguignon et al. (2009) found similar results in a rainforest in Panama, also using light and flight interception traps (Fig. 10.9b, c).

Advantages If the goal is to document the whole termite assemblage, this is a good complementary method. Trapping does not require special training, and the results are not affected by the collector's experience.

Disadvantages Termite flights are seasonal and usually occur during a few months of the year. Traps need to be left in the field for a long time, at least 1 year, and checked every few days. These traps will also capture many other groups of insects, and it is necessary to sort the material, which requires considerable work. Also, it is much more difficult to identify alates than soldiers. Therefore, the taxonomic work will take more time and effort.

10.6.6 Qualitative Surveys

In this case, the collectors will search for termites in all microhabitats possible. This method can be standardized based on the number of people collecting and the time spent on each site (Scheffrahn et al. 2018).

Advantages This method is best for taxonomic sampling and results in the most complete species list for each locality compared to any other method. It also results in samples of higher quality, which will be more adequate for taxonomic study.

Disadvantages Usually it does not produce data adequate for statistical analyses. The results will be strongly affected by the experience of the collectors.

10.7 Sampling in Urban Areas

Several termite species are urban pests, and it is often necessary to sample them in and around buildings. However, because cities and buildings are too variable, it is difficult to define a simple sampling protocol. Sampling is usually qualitatively looking for the pest species and their nests, which may be inside or outside the building. Common urban termite pests include drywood termites (Kalotermitidae), subterranean termites (Rhinotermitidae), and some nasute termites (Termitidae: Nasutitermitinae). Collecting is usually difficult because termites are often hidden inside the structure of buildings.

Drywood termites are hidden inside wood, but may be detected by their fecal pellets which they discard through small holes, and form small piles of granulated frass. To collect them it is necessary to use a hatchet or similar tool and an aspirator. Baiting does not work for drywood termites.

Subterranean and nasute termites may be detected by their covered tunnels, but sometimes these tunnels are hidden inside the building structure. Cellulosic baits can be used to detect and collect these termites. The nests of nasute termites such as *Nasutitermes corniger* are usually arboreal and may be found in nearby trees, but may also be under the roof.

10.8 Identification

Termite identification requires training and takes a lot of time and effort. Accurate identification is essential for data integrity. Poor taxonomy may affect the quality of the dataset and lead to erroneous conclusions. Before starting the identification, all samples should be clean and properly labeled and contain specimens of a single species. Mixed samples should be separated into as many vials as necessary before starting the identification.

All samples collected should be kept for some time after the identification because they may need to be reexamined for confirmation. Representative vouchers should be deposited in a permanent collection or museum.

The following equipment and materials are recommended for the taxonomic work: (1) a good stereomicroscope capable of magnifications of 50× or higher, with both incident and transmitted light and an ocular micrometer; (2) a compound microscope; (3) petri dishes; (4) fine sand or glass beads; (5) tweezers; (6) wash bottle filled with 80% ethanol; (7) microscope slides and slide covers; (8) mounting medium (Hoyer's or similar); and (9) dissecting tools.

A petri dish half-filled with fine white sand or glass beads is convenient to help position the specimens for examination and to take measurements. Hairs are easier to see in a petri dish without sand, with dark-field illumination. The specimens should always be kept immersed in alcohol.

The usual steps for identification are the following: (1) genus-level identification using keys; (2) sorting of samples into morphospecies; and (3) species identification. After some training, one may be able to identify the genera without using keys.

For people with little or no experience in termite taxonomy, the sorting of morphospecies may be challenging. It is necessary to learn external morphology and to recognize the different castes and the taxonomic characters useful for identification. Specimens should be carefully compared under the stereomicroscope.

Species identification may be conducted using one of the following methods: (1) use of identification keys, usually present in taxonomic revisions and monographs; (2) comparison with taxonomic descriptions; (3) comparison with reference collections; and (4) sending specimens to an expert taxonomist. DNA barcoding is also a possible alternative, but the costs involved may be high and the specimens need to be adequately preserved for DNA extraction. Barcoding is limited to the available database, which is far from complete.

The list of the relevant taxonomic literature for each genus and species can be found in Krishna et al. (2013) and the Termite Database (Constantino 2019). Digitized copies of older taxonomic works are available from the Biodiversity Heritage Library (https://www.biodiversitylibrary.org).

The identification of soldierless termites usually requires the examination of the enteric valve armature, which needs to be dissected and mounted on microscope slides (Sands 1972). The gut segment with the valve should be cut and removed, slit longitudinally, and cleaned (Fig. 10.10). It is important to remove the muscle layer and mount the cuticle only, especially if the valve is weakly sclerotized. Water-based mounting media such as Hoyer's are easier to use and provide good results. After mounting, the slide should be left drying for about a week at 50 °C. The examination of details of unsclerotized valves requires the use of phase contrast.

Acknowledgments This work was supported by a grant from Conselho Nacional de Desenvolvimento Científico e Tecnológico (CNPq no. 311016/2016-2).

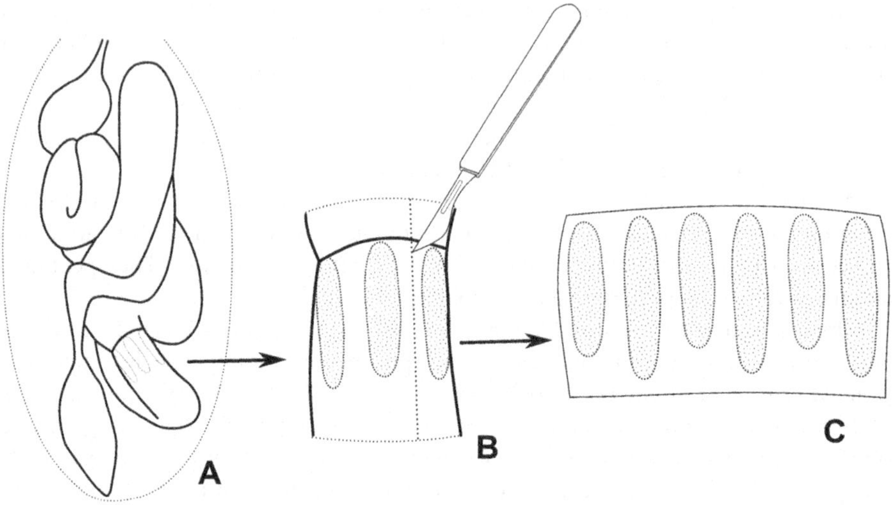

Fig. 10.10 Schematic drawings illustrating the dissection of the enteric valve. (**a**) Worker gut of a soldierless termite (*Anoplotermes* sp.) in situ (right view). (**b**) Gut segment with the enteric valve. (**c**) Mounted enteric valve, showing a typical armature with six plates

References

Abensberg-Traun M (1993) A comparison of two methods for sampling assemblages of subterranean, wood-eating termites (Isoptera). Aust J Ecol 18:317–324

Bignell D, Constantino R, Csuzdi C et al (2008) Macrofauna. In: Moreira FMS, Huising EJ, Bignell DE (eds) A handbook of tropical soil biology. Earthscan, London, pp 43–83

Bourguignon T, Leponce M, Roisin Y (2009) Insights into the termite assemblage of a neotropical rainforest from the spatio-temporal distribution of flying alates. Insect Conserv Divers 2:153–162

Cancello EM, Silva RR, Vasconcellos A et al (2014) Latitudinal variation in termite species richness and abundance along the Brazilian Atlantic forest hotspot. Biotropica 46:441–450. https://doi.org/10.1111/btp.12120

Cleveland LR, Hall SR, Sanders EP, Collier J (1934) The wood-feeding roach *Cryptocercus*, its protozoa, and the symbiosis between protozoa and roach. Mem Am Acad Arts Sci 17:184–342

Constantino R (2018) Estimating global termite species richness using extrapolation. Sociobiology 65:10–14. https://doi.org/10.13102/sociobiology.v65i1.1845

Constantino R (2019) Termite database. http://termitologia.unb.br. Accessed 30 Dec 2019

Davies AB, Eggleton P, van Rensburg BJ, Parr CL (2013) Assessing the relative efficiency of termite sampling methods along a rainfall gradient in African savannas. Biotropica 45:474–479. https://doi.org/10.1111/btp.12030

Donovan SE, Eggleton P, Bignell DE (2001) Gut content analysis and a new feeding group classification of termites. Ecol Entomol 26:356–366

Eggleton P, Bignell DE, Sands WA et al (1995) The species richness of termites (Isoptera) under differing levels of forest disturbance in the Mbalmayo Forest Reserve, southern Cameroon. J Trop Ecol 11:85–98

Eggleton P, Homathevi R, Jones DT et al (1999) Termite assemblages, forest disturbance and greenhouse gas fluxes in Sabah, East Malaysia. Philos Trans R Soc Lond B Biol Sci 354:1791–1802

Holt JA, Lepage M (2000) Termites and soil properties. In: Abe T, Bignell DE, Higashi M (eds) Termites: evolution, sociality, symbioses, ecology. Kluwer Academic Publishers, Dordrecht, pp 389–407

Inward D, Beccaloni G, Eggleton P (2007) Death of an order: a comprehensive molecular phylogenetic study confirms that termites are eusocial cockroaches. Biol Lett 3:331–335. https://doi.org/10.1098/rsbl.2007.0102

Jones DT, Eggleton P (2000) Sampling termite assemblages in tropical forests: testing a rapid biodiversity assessment protocol. J Appl Ecol 37:191–203

Jones DT, Eggleton P (2011) Global biogeography of termites: a compilation of sources. In: Bignell DE, Roisin Y, Lo N (eds) Biology of termites: a modern synthesis. Springer Netherlands, Dordrecht, pp 477–498

Jones DT, Susilo FX, Bignell DE et al (2003) Termite assemblage collapse along a land-use intensification gradient in lowland central Sumatra, Indonesia. J Appl Ecol 40:380–391

Krishna K, Grimaldi DA, Krishna V, Engel MS (2013) Treatise on the Isoptera of the world. Bull Am Mus Nat Hist 377:1–2704. https://doi.org/10.1206/377.1

La Fage JP, Haverty MI, Nutting WL (1973) Desert subterranean termites: a method for studying foraging behavior. Environ Entomol 2:954–956. https://doi.org/10.1093/ee/2.5.954

Lavelle P, Bignell D, Lepage M et al (1997) Soil function in a changing world: the role of invertebrate ecosystem engineers. Eur J Soil Biol 33:159–193

Legendre F, Nel A, Svenson GJ et al (2015) Phylogeny of Dictyoptera: dating the origin of cockroaches, praying mantises and termites with molecular data and controlled fossil evidence. PLoS One 10:e0130127. https://doi.org/10.1371/journal.pone.0130127

Rebello AMC, Martius C (1994) Dispersal flights of termites in Amazonian forests (Isoptera). Sociobiology 24:127–146

Redford KH (1987) Ants and termites as food. In: Genoways HH (ed) Current mammalogy. Springer US, Boston, pp 349–399

Roisin Y (2000) Diversity and evolution of caste patterns. In: Abe T, Higashi M, Bignell DE (eds) Termites: evolution, sociality, symbioses, ecology. Kluwer Academic Publishers, Dordrecht, pp 95–119, xxii + 466 p

Rouland-Lefèvre C (2011) Termites as pests of agriculture. In: Bignell DE, Roisin Y, Lo N (eds) Biology of termites: a modern synthesis. Springer, Dordrecht, pp 499–517

Sands WA (1965) Termite distribution in man-modified habitats in West Africa, with special reference to species segregation in the genus *Trinervitermes* (Isoptera, Termitidae, Nasutitermitinae). J Anim Ecol 34:557–571

Sands WA (1972) The soldierless termites of Africa (Isoptera: Termitidae). Bull Br Mus Nat Hist Entomol Suppl 18:1–244

Scheffrahn RH, Chase JA, Mangold JR, Hochmair HH (2018) Relative occurrence of the family Kalotermitidae (Isoptera) under different termite sampling methods. Sociobiology 65:88. https://doi.org/10.13102/sociobiology.v65i1.2097

Su NY, Scheffrahn RH (2000) Termites as pests of buildings. In: Abe T, Higashi M, Bignell DE (eds) Termites: evolution, sociality, symbioses, ecology. Kluwer Academic Publishers, Dordrecht, pp 437–453, xxii + 466

Su N-Y, Ban PM, Scheffrahn RH (1993) Foraging populations and territories of the eastern subterranean termite (Isoptera: Rhinotermitidae) in Southeastern Florida. Environ Entomol 22:1113–1117

Wilson EO (1971) The insect societies. Harvard University Press, Cambridge

Wood TG, Johnson RA, Bacchus S et al (1982) Abundance and distribution of termites in a riparian forest in the southern Guinea savanna vegetation zone of Nigeria. Biotropica 14:25–39

Chapter 11
Measuring Orthoptera Diversity

Carlos Frankl Sperber, Edison Zefa, Elliott Centeno de Oliveira, Lucas Denadai de Campos, Márcio Perez Bolfarini, Marcos Fianco, Marcos Gonçalves Lhano, Natállia Vicente, Neucir Szinwelski, Pedro Guilherme Barrios de Souza Dias, Riuler Corrêa Acosta, and Victor Mateus Prasniewski

C. F. Sperber (✉) · N. Vicente
Laboratório de Orthopterologia, Departamento de Biologia Geral, Universidade Federal de Viçosa, Viçosa, MG, Brazil
e-mail: sperber@ufv.br

E. Zefa
Departamento de Zoologia e Genética, Universidade Federal de Pelotas, Capão do Leão, RS, Brazil

E. C. de Oliveira
Programa de Pós-Graduação em Entomologia, Universidade de São Paulo, Ribeirão Preto, SP, Brazil

L. D. de Campos
Programa de Pós-Graduação em Zoologia, Instituto de Biociências, Universidade de São Paulo, São Paulo, SP, Brazil

M. P. Bolfarini
Departamento de Ecologia e Biologia Evolutiva, Universidade Federal de São Carlos, São Carlos, SP, Brazil

M. Fianco
Programa de Pós-Graduação em Entomologia, Universidade Federal do Paraná, Curitiba, PR, Brazil

M. G. Lhano
Centro de Ciências da Natureza, Universidade Federal de São Carlos, São Carlos, SP, Brazil

N. Szinwelski
Centro de Ciências Biológicas e da Saúde, Universidade Estadual do Oeste do Paraná, Cascavel, PR, Brazil

P. G. B. de Souza Dias
Departamento de Entomologia do Museu Nacional, Universidade Federal do Rio de Janeiro, Museu Nacional, Rio de Janeiro, RJ, Brazil

R. C. Acosta
Programa de Pós-Graduação em Biologia Animal, Universidade Federal do Rio Grande do Sul, Porto Alegre, RS, Brazil

V. M. Prasniewski
Programa de Pós-Graduação em Ecologia e Conservação da Biodiversidade, Universidade Federal de Mato Grosso, Cuiabá, MT, Brazil

J. C. Santos, G. W. Fernandes (eds.), *Measuring Arthropod Biodiversity*,
https://doi.org/10.1007/978-3-030-53226-0_11

11.1 Introduction

Orthoptera are insects with enlarged hind legs for jumping, well developed compound eyes, a large prothorax with a shield-like pronotum curving downward laterally, and mandibulate mouthparts for biting and chewing. This order occurs worldwide, in grasslands and forest areas, from the Arctic to tropical and desert regions. They are an ancient insect order, dating from the Paleozoic age, 300 million years ago (Grimaldi et al. 2005), and have 28,500 described species (Bidau 2014; Cigliano et al. 2020). The order has two suborders: (1) Ensifera, which comprehend true crickets (Gryllidae, Fig. 11.1a), mole crickets (Gryllotalpidae, Fig. 11.1b), camel crickets (Raphidophoridae), bush crickets or katydids (Tettigoniidae, Fig. 11.1c), grigs (Prophalangopsidae), wetas (Anastostomatidae, Fig. 11.1d), and Cooloola monsters (Coolooidae) and are characterized by long antennae (usually longer than the body), swordlike or needlelike ovipositors, and auditory organs, the tympani, on the front tibiae, and (2) Caelifera, which comprehend grasshoppers (Acrididae, Fig. 11.1e, f), ground-hoppers (Tetrigidae, Fig. 11.1g), and pygmy mole crickets (Tridactylidea), characterized by a shorter antennae, ovipositor with only two valvular pairs, and abdominal tympana.

Except for the subfamily Oecanthinae, all Orthoptera have a hypognathous head, with mouthparts directed downward and the head vertical in relation to the body axis (Johnson and Triplehorn 2005). The jaws of Orthoptera are large and very distinct between the groups, reflecting their varied eating habits. The presence of wings is prevalent. The first pair of wings comprehends the tegmina, which, in addition to protecting the second pair of wings, is related to sound production (stridulation) in Ensifera and some Caelifera. The second pair of wings is membranous, used for flight. The presence of developed anterior and posterior wings is not a rule in Orthoptera, since there are brachypterous, micropterous, and apterous species (Bidau 2014). Orthoptera presents paurometabolous development, i.e., from each egg hatches an immature, called nymph, that looks like an adult, but smaller, without wings and with fewer antennal articles (Sperber et al. 2012). The first and the second pair of legs are usually cursor-like, but in predatory species there are modifications with strong and curved spines (Sperber et al. 2012). In mole crickets and some species of grasshoppers (Tridactylidae), the first pair of legs is fossorial, allowing the excavation of burrows and galleries (Johnson and Triplehorn 2005; Bidau 2014).

In this chapter, we will present the diversity of habitats and behaviors of Orthoptera and the manner in which this diversity affects their sampling, presenting

Fig. 11.1 (continued) P.G.B.S. Dias. (**e**) Grasshopper *Chromacris speciosa* (Thunberg, 1824) (Orthoptera: Caelifera: Romaleidae) from Belterra, Pará State, Brazil. Photo: L.D. Campos and P.G.B.S. Dias. (**f**) Unidentified grasshopper (Orthoptera: Caelifera: Acrididae) from Itatiaia, Rio de Janeiro State, Brazil; (**g**) Ground-hopper (Orthoptera: Caelifera: Tetrigidae) from Belterra, Pará State, Brazil. Photo: L.D. Campos and P.G.B.S. Dias. (**h**) Pterochrozinae (Orthoptera: Ensifera: Tettigoniidae) from Belterra, Pará State, Brazil. Photo: L.D. Campos and P.G.B.S. Dias

Fig. 11.1 (**a**) Cricket *Miogryllus* sp. (Orthoptera: Ensifera: Gryllidae), from Belterra, Pará State, Brazil. Photo: L.D. Campos and P.G.B.S. Dias. (**b**) Mole cricket (Orthoptera: Ensifera: Gryllotalpidae) from Belterra, Pará State, Brazil. Photo: L.D. Campos and P.G.B.S. Dias. (**c**) Katydid *Neoconocephalus* sp. (Orthoptera: Ensifera: Tetigoniidae) from Boracéia, São Paulo State, Brazil. Photo: L.D. Campos and P.G.B.S. Dias. (**d**) Weta *Apotetamenus* sp. (Orthoptera: Ensifera: Anostostomatidae) from Veredas, Minas Gerais State, Brazil. Photo: L.D. Campos and

specific methods for Orthoptera sampling according to habitat, taxon, and study aims. We will dedicate a special section to studies on orthopteran behavior and breeding, bioacoustics, chromosomes, and DNA, as well as recommendations for specimen preservation for scientific collections. We will also discuss sampling design and cautions for data analysis in ecological studies on orthopteran populations and communities.

11.2 Habitats and Behaviors of Orthopterans

In general, Ensifera (crickets, katydids, and alike) are active at night, while Caelifera (grasshoppers and alike) are active during the day. Crickets are commonly seen walking and jumping on soil and litter but can also occur on tree trunks and shrub foliage, under fallen branches and trunks, in natural crevices, and on rocks. Several crickets use surfaces such as tree trunks or shrub foliage to stridulate, mostly males calling females for reproduction. Crickets are also an important component of natural cave faunas. Most species of katydids present arboreal habits, occupying from brushes of the understory to canopy; however, there are katydids that live under fallen wood trunks, litter, and soil or associated with macrophytes of continental aquatic ecosystems (Gwynne 2001; Rentz 2010). Most katydids are active at night (Rentz 2010), but there are species that are active during the day, as some species of *Conocephalus* (Tettigoniidae: Conocephalinae) (Chamorro-Rengifo et al. 2018). Grasshoppers are more often found in grasslands and at forest edges and ecotone habitats. There are semiaquatic orthopterans, such as semiaquatic grasshoppers, that feed on aquatic macrophytes.

The feeding habit of Orthoptera is highly diversified and comprises species that go from exclusively predatory to exclusively herbivorous. Omnivory, however, is the most common feeding habit in orthopterans (Gangwere 1961), except Caelifera, which are predominantly herbivorous. Popularly, the omnivorous eating habit was grounded in the laboratory, since insects housed in lab are fed with dog food, cat food, vegetables, fruits, and other food sources (Huber et al. 1989; Bidau 2014). Analysis of the proventricular and fecal content concluded that there are few oligophagous (that eat few species) or monophagous (that eat only a single species) orthopterans; even exclusively herbivorous species generally consume a wide variety of plant families (Gangwere 1961).

Grasshoppers are herbivores, with some species feeding on at least 14 plant species, from 11 different families (Sperber 1996). Herbivore grasshoppers may damage various crops, such as soybeans, corn, pastures, and grass. In North America and Africa, there are records of large grasshopper swarms, which consume about 60% of crops (Nerney 1960; Gallo et al. 2002; Simon 2020). Mole crickets are predominantly herbivores and may cause great economic impact on golf and football fields (Bello 2020); they dig tunnels and feed on roots of grass, killing the aboveground plant tissue (Hertl et al. 2001); they may also present predation habits, especially on sessile or little moving prey, such as eggs, pupae, aphids, mealybugs, and larvae of insects and earthworms (Huber et al. 1989). In katydids, feeding habit is diverse, with subfamilies that feed on pollen or nectar, called "winged flower lovers," as katydids of the subfamily Zaprochilinae. There are katydids that feed on flowers and leaves (Phaneropterinae, Mecopodinae, and Pseudophyllinae) and on arthropods (Saginae, Lipotactinae, Tympanophorinae, and Listroscelidinae), but only in the subfamily Listroscelidinae all species are exclusively predatory (Gwynne 2001). Most species of katydids are omnivorous, with a tendency to being herbivorous; this trend is corroborated by the morphology of their mouthpieces, with adaptations that allow grinding of leafs (Gwynne 2001). In forest environments, crickets are abundant, especially on the litter. They coexist in this high productivity environment without evident segregation of trophic niches (Jesus 2015). Individuals of the same genus present a greater overlap of food resources than individuals of different genera, and as the number of coexisting species increases, there seems to be a narrowing of the trophic niche; in addition, there is great plasticity in the diet, since the same species can explore different resources in different areas (Jesus 2015).

Camouflage is a widespread characteristic in katydids, their appearance resembling the substrate where they rest (Gwynne 2001). Several Phaneropterinae and Conocephalinae are green. The wings of Phaneropterinae may resemble dicotyledon leaves (Magnoliopsida), whereas Conocephalinae can be polymorphous in color (as in some *Neoconocephalus*, with greenish and brownish individuals in the same population) with wings resembling monocotyledon leaves (Liliopsida, see Fianco 2019). Phaneropterines are generally associated with dicotyledons of forest environments, while conocephalines are generally associated with meadows and grasslands. Pseudophyllinae and Pterochrozinae of the Neotropics are generally brownish or with greenish stains, resembling tree trunks and associated biota (as mosses and lichen) or resembling dry leaves (Fig. 11.1h), and their habitat is the substratum they resemble.

11.3 Sampling Methodologies According to Habitat and Behavior

11.3.1 Sampling Ground Dwelling Orthopterans

Orthoptera that move on the ground or litter can be sampled with pitfall traps (Fig. 11.2a–d). These traps may capture these organisms passively when there is no bait associated to the trap (Fig. 11.2b, d) or with baits, such as cotton soaked with sugary solution (Fig. 11.2c). The pitfall traps might be installed with covering, to reduce entrance of water during heavy rains, thus avoiding dilution of the killing solution, but at the same time covering might reduce capture of orthopterans.

There are several cricket species that spend most of their time moving through or above forest litter, as is the case of flightless crickets as the Nemobiinae and Phalangopsidae. The efficiency of the pitfall trap depends on its ability to capture

Fig. 11.2 (**a**) Installing pitfall traps. Photo: L.D. Campos and P.G.B.S. Dias; (**b**) unbaited pitfall trap without covering. Photo: N. Szinwelski; (**c**) baited pitfall trap with covering. Photo: M.G. Lhano; (**d**) unbaited pitfall trap with covering. Photo: M.G. Lhano. The covering reduces entrance of water during heavy rains, avoiding dilution of the killing solution, but at the same time covering might reduce the capture of orthopterans

organisms that fall into the trap, and once the organism has fallen into the trap, the trap preventing the organism from escaping. Sperber et al. (2003) showed that using a solution containing alcohol increases the capture efficiency of orthopterans in pitfall traps, because the alcohol accelerates the cricket sinking and killing (Szinwelski et al. 2013), thus avoiding its escape. In sites where ethanol fuel is available, it might be used as a killing solution, preserving cricket's DNA (Szinwelski et al. 2012a, b). Sugarcane juice, which can be substituted by sugar or sugarcane syrup diluted in water, sprinkled on the vegetation or used in soaked cotton attached to the pitfall trap (Fig. 11.2c) attracts orthopterans and can be used as bait, to increase pitfall capture efficiency. Sugarcane syrup functions as an increase in local resource availability, promoting cricket aggregation and increasing local diversity (Szinwelski et al. 2015).

Sperber et al. (2007) showed that litter vibration through walking alters the number of captured orthopterans, probably because the substrate vibration provokes them to jump. The number of captures in pitfalls presents a nonlinear response to disturbance frequency, increasing in low to middle disturbance frequencies but saturating and eventually decreasing with high disturbance frequencies (Sperber et al. 2007). For studies that aim to estimate naturally occurring densities and diversities, we recommend avoiding the use of baits and minimizing the effects of litter vibration. The number of orthopteran captures per trap is normally low (Ribas et al. 2005), so placing at least ten sets of pitfall traps parallel to each other at 15 m intervals, with each set consisting of a line of five traps 1 m apart (Szinwelski et al. 2012a, b) for each site, is recommend. More recently, we were using a larger sampling effort, of 150 sets of 5 traps each (Vargas 2013, 2018) per site, being each set at least 30 m apart. The original idea was to consider each set of five traps as a replicate, based on the assumption that 30 m distance would be sufficient to achieve independence. However, we realized that the drivers of orthopteran abundance, diversity, and occurrence might predominate at still larger spatial scales, requiring replicates to be further away. Thus, when studies focus on large-scale drivers, replicates have to be at least 10 km away to be considered real replicates, if landscape or biome drivers are being evaluated (Sperber et al. 2013). This does not apply to geographically restricted areas, such as mountain tops.

11.3.2 Sampling Shrub Dwelling Orthopterans

For the quantitative sampling of shrub dwelling orthopterans, a commonly used methodology is sweeping with an entomological net, hitting the vegetation, and capturing the insects that are located on leaves and twigs directly with the net. This method has its shortcomings, however. Captured insects might be damaged and relevant specimens might escape. Thus, such methodology is of little use for inventories. Sweeping with the entomological net might be useful for homogenizing sampling effort, when the number of beatings per time and distance walked by the collector is standardized (Janzen and Schoener 1968). In this case, the collector has also to be standardized, to enable estimates of abundance or species richness comparisons among sites, habitats, or experimental treatments. For inventories, the most common sampling methodology is active searching, in which the researcher localizes the specimen and then makes a directed effort to collect it. This assumes that the researcher is focused on the taxon of interest.

11.3.3 Sampling Cave Orthopterans

Cave crickets, along with other invertebrates (beetles, spiders, and harvestman) and vertebrates (fish and bats), are frequent in cave fauna (Gnaspini and Trajano 1994). Cave crickets are mainly represented by two families, with different geographical distributions: Rhaphidophoridae, mainly tropical (Asian and Australian), although there are few species of rainforest in Europe and North America, and Phalangopsidae, which occur along almost all Neotropical, African, and some Asian and Australian regions (Rampini et al. 1983).

Studies of cave fauna are of extreme importance, because in addition to the intrinsic fragility presented by the subterranean environment, these environments have a key role in environmental licensing. Below are the most used collection methods for collecting cricket in underground environments:

Direct Intuitive Searching (DIS) Direct intuitive searching is targeting a microhabitat or environmental zone to address a research question (Souza-Dias et al. 2014; Bolfarini and Bichuette 2015). These microhabitats include flood detritus, penetrating tree roots hanging from ceilings or walls, guano deposits, edges of drip pools and ponds, muddy banks, and animal or insect carcasses. These microhabitats are likely to support a high diversity or present specific functional groups (e.g., guanophiles) (Wynne et al. 2019). This methodology involves searching for the greatest possible diversity of environments found inside the cavities, giving priority to sites where the specimens can be captured manually, with the aid of tweezers and brush. In this way, it is possible to quantify the number of individuals collected and thus estimate the impacts that the removal of their habitats may have.

Quadrat Method Quadrat sampling involves the visual inspection of delimited sample areas, optimizing collection of subterranean fauna. This method allows estimating sampling effort and enabling comparisons among subterranean systems (Krebs 1999; Bichuette et al. 2015). Sampling by the quadrat method has been used for a long time in ecological studies, mainly in plant ecology (Weaver 1918; Gleason 1920). In quadrat sampling, it is necessary to inform the sampling time and the number of captured individuals. Sampling locations can be qualitative, targeted to specific microhabitats (e.g., guano, carcasses).

Leaf Litter Traps This collecting technique was developed for cave ecology studies (Humphreys 1991; Weinstein 1994) and involves placing a permeable sheet on the floor to collect the leaf litter produced along a period of time and transporting it to the laboratory, where the litter can be weighted and screened for organisms. Leaf litter traps are constructed with rectangular plastic sheet containers by firstly screwing a wooden support into each corner of the container lid and next suspending a mosquito net between these four supports. A leaf litter trap of standard volume can be thus placed into the container base and transported into the cave as a sealed unit. These traps can be exposed for up to 7 days to accumulate material; a shortcomming of this method is that it often captures very young individuals who are difficult to identify, and might require its rearing to adulthood.

Pitfall Traps and Vulcan Traps Pitfall traps are also usually employed in caves (with or without bait, Fig. 11.2), but this method should be used with caution since it can impact the whole terrestrial fauna, oversampling some taxonomic groups, such as collembolans, orthopterans, and cockroaches (Sharratt et al. 2000). The traps are made of low-rimmed dishes or buried bottles (on unconsolidated terrestrial substrates), installed in places with and without guano on the substrate of the cave, with or without the aid of bait, in places where there are no natural pools, or these occur only on a seasonal basis. The traps are distributed in several locations in the caves. For consolidated substrates, the solution for fall traps are "vulcan traps," which are pitfalls with access ramps for fauna, so they do not need to be buried.

The traps must not exceed 48 h of exposure, because with time the traps accumulate dead animals and attract saprophages.

11.3.4 Sampling Semiaquatic Orthopterans

Worldwide, about 80 Orthoptera species are considered semiaquatic and another 110 are water-dependent, from which about 50 species occur in the Neotropics, mainly represented by Acridomorpha, Tetrigoidea, Gryllidae, and Tettigoniidae. This semiaquatic habit seems to have evolved independently in the history of Orthoptera as a result of adaptive radiations, especially in Latin America, where there are important large river systems with considerable variations in water levels that generate immense surfaces of freshwater habitats (Amédégnato and Devriese 2008; Cover and Bogan 2015; Nunes-Gutjahr and Braga 2018). Semiaquatic orthopterans generally present a strict relationship with aquatic host plants (especially for host-specific oviposition, feeding habits, and nymph development). Despite its ability to swim and stay long periods under water, semiaquatic grasshoppers maintain the ability to leave water and perform similar ecological functions to their terrestrial relatives.

Semiaquatic grasshoppers are mainly associated to riparian habitats, living in marshes, floodplains, swamps, bogs, inundated grasslands, margins of rivers, lakes, flatwoods, ditch banks, inundation zones, and shores of running water, where they can complete their life cycles in macrophytes and herbaceous or graminaceous plants (Amédégnato and Devriese 2008; Nunes-Gutjahr and Braga 2018).

Like terrestrial Caelifera, semiaquatic grasshoppers are mostly diurnal, and some species are active both during day and at night. Semiaquatic orthopterans have physiological, ethological, and morphological adaptations that allow them to live permanently in the aquatic habitats on floating or rooted plants. Aquatic orthopterans might be able to swim, using paddle-shaped hind tibiae, and they may use a plastron respiration (a special adaptation to store oxygen, Thorpe 1950) when submerged, as well as morphological and chromatic mimicry, modified tibial spurs bearing hydrophobic bristles, modified ovipositor valves for egg insertion into or on the surface of aquatic plants (endophytic or epiphytic oviposition) (Braker 1989), and possess diving behavior in order to feed on submerged plants and to escape predators (Nunes-Gutjahr and Braga 2018).

Generally, semiaquatic Orthoptera can swim, walk, or skate over the water film surface. When attacked or in danger, they fly away or swim rapidly, even against currents greater than 1 m per second, hiding among roots of aquatic plants or deep-diving as long as 10 m and staying motionless under the water for more than 30 min until the danger has gone. Within Tetrigoidea, individuals of Scelimini are fully aquatic and can swim effectively (Rentz and Su 2003), feeding on aquatic weeds, aquatic algae, lichens, mosses, and debris (Kočárek et al. 2008). Besides, although Tridactylidae and Ripipterygidae are excellent jumpers, they are commonly found in galleries in moist soil, close to rivers, lakes, and beaches.

Given the relative scarcity of sampling of this semiaquatic fauna, it is very likely that this fauna is underestimated. The difficulty in collecting these grasshoppers is mainly due to access to wet habitats (often due to the vegetation associated with those habitats), the escape behavior of these insects (as they can be in both terrestrial and aquatic environments, changing habitats according to the situation), and not being suited for being captured in traps.

Semiaquatic orthopterans may be sampled through active collecting with sweep netting, transect counts, and hand capture. Passive methods for sampling are rarely used and don't provide reliable data for population or diversity estimates. On the other hand, sweep netting performance varies according to the surveyors or environmental conditions (O'Neill et al. 2003).

For a successful sampling, the essential is to answer these good and concise questions: what do I want to sample? What is my hypothesis? Where do I want to sample? With those question answered, it is a natural process to select the proper method to assess the semiaquatic fauna. For example, sweep netting is a good methodology in general, but for some Leptysminae as *Cornops aquaticum* (Bruner 1906), which swim swiftly, it is not the best choice. For semiaquatic Caelifera in general, the most common technique for sampling is sweep netting, which provides quick, low-cost, and accurate estimates of the grasshopper community composition. The entomological sweeping net is made of a sturdy structure, with a strong handle, and a strong net bag that does not snag on sharp vegetation (Samways et al. 2010). Standardized methods use an entomological net measuring 70 cm in diameter and a collecting bag of 100 cm in length, with easy handling and robust enough for the sampled vegetation. Typical methods include sweeping some predetermined number of times (Onsager and Henry 1977), along a predetermined time, and made by a predetermined number of collectors. According to O'Neill et al. (2003), the specific technique used in sweep sampling can vary among samplers and even among different samples taken by the same person. For them, including the net speed could affect the ability to capture grasshoppers since they have rapid escape responses. Differences among species composition among sets of samples could result from species-specific difference in escape behavior. O'Neill et al. (2003) recommend that the sampling method is described more accurately in the published studies, including the procedures taken for standardization.

For sweep netting standardization, the researcher may establish the distance from the vegetation, sweeping speed, intensity, height, angle of sampling, and standards the number of sweep beatings per meter, when the vegetation is homogeneous. In addition, transects can be standardized for ecological studies. A transect can consist in a number of sweeps taken on a number of consecutive strides, sweeping generally the net once with each step, keeping a consistent sweeping speed. It's highly recommended having a large sampling effort and reduced variability among samples, which might be achieved by standardizing the number and identity of the collectors, as well as the time of the day that the sampling is undertaken. Simultaneous collectors should sample simultaneously along parallel transects and keep a regular distance from each other (the optimal distance depends on the habitat where the sampling is made) (Onsager 1977; Onsager and Henry 1977).

Grasshoppers are generally difficult to sample due to their jumping and flying behavior (Browde et al. 1992). Adults jump only at higher temperatures; thus, sampling should be done in the early morning, when temperatures are not so high but sufficient for grasshoppers to begin their activity, or in the late evening, when temperatures are not too low and orthopterans haven't started to look for shelter to spend the night. In many wetlands, a boat should be used to access the vegetation, to help sampling and displacement in the study area.

To sample semiaquatic grasshoppers which live on macrophytes, Vieira and Adis (1992) developed an aluminum cage to increase sampling effectiveness. It is a 1 m^3 cage, with two openings (at the top and at the bottom), where grasshoppers can be trapped and removed manually afterwards. After a cautious approach with the boat, the previously assembled cage must be thrown (with the bottom open) by two collectors, which will capture macrophytes and grasshoppers inside the cage. After capture, the cage must be closed quickly and individuals can be collected through the opening at the top. This methodology was used in studies made in the Amazon and Pantanal and was proven to be efficient (Lhano 2002). The sampling effort for this methodology can include standardized techniques as quantity the number of samples, sampling area, etc.

Other focused methodologies can be used for specific groups, like for Tridactylidae and Ripipterygidae, which are insects easily attracted by light. Some grasshoppers, as the Leptysminae *Guetaresia lankesteri* Rehn, 1929, have arboreal habits in tropical wet forests and can be collected using canopy fogging.

After sampling, the recommended method to kill specimens is to place them in a freezer for at least 8 h and avoid using chemical jars. Besides, it is highly recommended to keep one hind leg in a labeled tube containing 95–99% ethanol for future genetic studies.

11.4 Sampling Methodologies According to Taxon

11.4.1 Sampling Katydids

To collect katydids, the two most often sampling methods used are active searching and light traps (Rentz 2010). Active searching is done both during the day (4 h before sunset) and during the night (4 h after sunset), walking through trails inside or at the border of the vegetation, both searching visually and hearing attentively (Montealegre-Z 1997). Through visual searching, we detect individuals on the vegetation, while hearing enables us to first locate the region where the singing individual is active, following the produced acoustic sounds. During this localization, the space of observation is reduced, and this restriction of searching space enables visual detection of the singing katydid. Thus, hearing-oriented search leads generally to capturing males, since the males are the ones that produce more conspicuous sounds. During active searching we use a robust entomological net that must be more resistant than the ones used to capture butterflies. This net is composed of a rim of ca. 40 cm of diameter attached to a cable of about 1.5 m; a tapered bag is fixed on the rim (the fabric must be resistant) with a tulle bottom and with a length of a little less than the length of the collector's arm, so that the collector can reach those individuals that are there deposited, after collecting them with the net. When the individual is located, it must be captured with the help of the entomological net, passing it rapidly and in an abrupt movement from under the vegetation upward, where the organism is located, as far as a common escape strategy of these organisms is to fall down to the floor. The net movement has to be swift, because several katydids fly away when perceiving the vibration of the vegetation. This method of collection is the most recommended for ecological purposes, since it's cheap and can be applied in areas which are inaccessible for other types of collection. We recommend collecting both during day and night, to better access the complete katydid fauna of the environment. For standardization, we suggest establishing a transect to be traveled in a predetermined time, and to standardize both time and distance traveled in each sampling unit, if possible replicating the sampling units, so as to have homogeneous sampling effort among sampling units.

A passive sampling method is using light traps, composed of a light source positioned in front of a white cloth that must be perpendicular to the soil (Fig. 11.3). This kind of trap must be mounted in a clearing or forest gap, surrounded by vegetation, so that the light can dissipate into a considerable distance and in this way attract a higher number of katydids (Rentz 2010). The light must be turned on just before sunset and stays on for about 6 h. The best dates to use this kind of sampling are during the darkest nights, with new moon, because katydids will be more active then. It's recommended to use sodium or mercury vapor lamps and/or black light. To standardize this sampling method we recommend homogenizing the time of sampling, carrying out at least 2 days of collection effort (sampling units) on each site.

Another sampling method for katydids are bat detectors, which are less frequently used because of their high cost. This equipment works complementing searching through hearing, as it transforms the acoustic signals that humans are not able to hear (above 20 kHz) to signals situated in a frequency band that humans can hear (Rentz 2010). The detector is directed toward the vegetation, capturing the high-frequency sounds and transforming them, thus enabling auditory orientation of the researcher toward the acoustic source, where the collector proceeds the manual capture of the stridulating individuals. To standardize this sampling method, we recommend doing it at the same time of the day, with the same sampling effort among sampling units.

A further device for sampling katydids are Malaise traps, rarely used for orthopterans; they have the advantage to enable capturing rare species of katydids (see Fianco et al. 2019). At ground level, it is apparently not effective, but the suspended mounting tends to collect rare species, especially species with sustained flight or flightless species that inhabit the understorey or canopy that make contact with the trap.

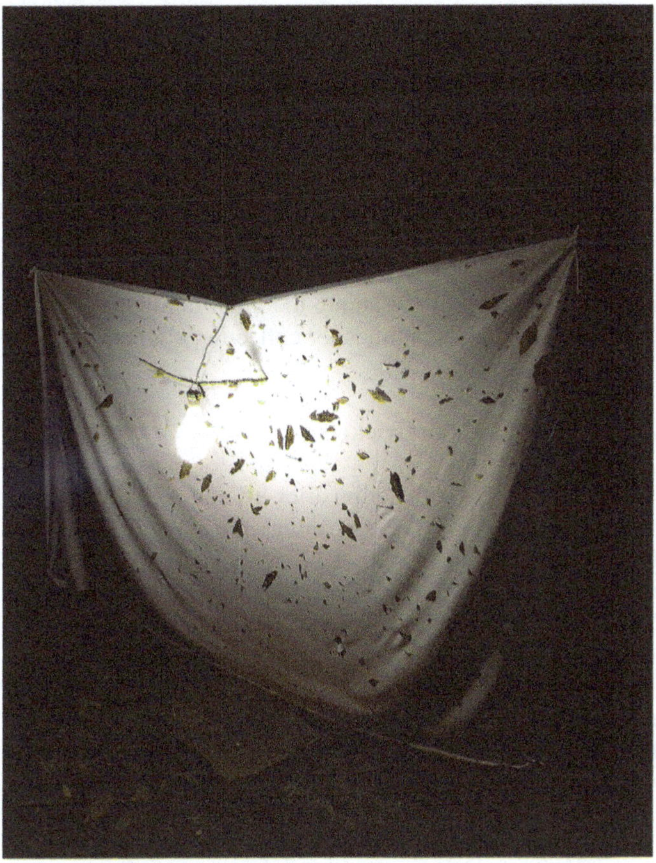

Fig. 11.3 Active light collecting at white sheet. Photo: L.D. Campos and P.G.B.S. Dias

11.4.2 Sampling Crickets

For practical reasons, we will refer to crickets *sensu lato*, which includes the monophyletic superfamilies Gryllotalpoidea (Gryllotalpidae and Myrmecophilidae) and Grylloidea. Crickets *sensu stricto* (or true crickets) are grouped into one superfamily, Grylloidea, one of the main lineages of Orthoptera, with more than 5,600 described species (Chintauan-Marquier et al. 2015; Cigliano et al. 2020). Grylloidea is subdivided into four families: Gryllidae, Mogoplistidae, Phalangopsidae, and Trigonidiidae. The relationships among cricket families and subfamilies are discussed in Chintauan-Marquier et al. (2015).

Crickets occupied almost all available habitats in terrestrial ecosystems, from grasslands to dense forests, from underground to forest canopies. Several lineages are adapted to inhabit different habitats in their environments: ground and leaf litter (Gryllidae, Phalangopsidae, Gryllotalpidae, Trigonidiidae), cavities under and aboveground level (holes, caves, crevices) (Gryllotalpidae, Gryllidae, Phalangopsidae, Myrmecophilidae), living and dead tree trunks (Gryllidae, Mogoplistidae, Phalangopsidae), shrubs (Gryllidae, Mogoplistidae, Trigonidiidae), branches (Gryllidae, Trigonidiidae), and canopy (Gryllidae, Phalangopsidae, Trigonidiidae).

Crickets *sensu lato* are generally omnivorous, although there are some predatory and herbivorous species in Gryllotalpidae. Most crickets are nocturnal, spending daytime hidden in their shelters. The wide variety of habitats and feeding habits makes sampling crickets a challenge, with no specific trap for them. Thus, one must use different sampling methods to sample the main families of crickets in a given area. The main methods and techniques for collecting crickets are presented below.

11.4.2.1 Active Searching for Crickets

The most effective method for collecting crickets, and the only one that can be specific to a clade, family, or species, is active searching (Fig. 11.4). This kind of sampling can focus on a large area, with a broad microhabitat range, such as litter, cavities, bushes, tree trunks, and foliage, or with a more narrow and restricted focus, such as specific sites and microhabitats, like cavities or caves. It is also possible to focus on adults instead of juveniles (nymphs), since juveniles are not useful for taxonomic purposes and are of difficult identification. Yet, if the objective is sampling tissue or raising species for behavioral or other laboratory studies, nymphs are welcome. The best period for sampling crickets is nighttime, where almost all crickets are active. However, daytime searches are important to find the shelters of nocturnal crickets and to collect diurnal species, such as trigs and nemobiines (Trigonidiinae), crickets of the genera *Lerneca* (Phalangopsidae, Luzarinae), and many eneopterines (Gryllidae, Eneopterinae).

There are some entomological equipment and techniques that can be used in active sampling of crickets:

Entomological Net This is the most common equipment for active sampling of insects and highly effective to collect grasshoppers (Caelifera). Considering the nocturnal habits of crickets, low density of populations, and their habitats (small cavities, tree trunks, leaf litter), this equipment is not recommended. Small nets used in fishkeeping are a good solution for collecting species inside cavities, trunks, or any other situation in which you are close to the cricket.

Plastic Tubes and Pots A highly effective way of catching a cricket is using plastic tubes and pots (transparent are preferable) and searching actively for these organisms, on the ground, cavities, bushes, etc.

Entomological Umbrella (Fig. 11.5) This is recommended for collecting crickets from shrubs and branches, as Mogoplistidae, Trigonidiidae (Trigonidiinae), and Gryllidae (Oecanthinae, Tafaliscinae, Eneopterinae, Podoscirtinae). This equipment

Fig. 11.4 Active searching for crickets with plastic tube and jar. Photo: L.D. Campos and P.G.B.S. Dias

Fig. 11.5 An alternative design for entomological umbrella. Photo: L.D. Campos and P.G.B.S. Dias

can be used in the traditional way, placing it above a branch, beating the branch with a stick to drop the specimens into the umbrella, and then capturing them.

Light Sampling (Fig. 11.3) Active sampling at strong light sources, that attract nocturnal insects, is effective for crickets, such as Gryllotalpidae, Trigonidiidae (Trigonidiinae), and Gryllidae (Eneopterinae, Oecanthinae, Podoscirtinae, Tafaliscinae, Gryllinae), besides other insect orders. The most commonly used technique is placing a light source near a white sheet.

Baits Using baits, laid along forest trails, is effective to attract some groups of crickets such as Gryllidae, Phalangopsidae, and Nemobiinae (Trigonidiidae). The baits can be oat or biscuit bran, small pieces of fruits, or even a molasses solution sprayed on the vegetation or on the ground. After some time (30 min or more), some individuals can be seen feeding on the baits. Using oat as bait on the ground has an additional advantage for night sampling as it helps focusing the collectors' eye.

11.4.2.2 Passive Sampling Crickets

For Grylloidea and Stenopelmatoidea, it is recommended to use bird, banana, or watermelon baits, placed on the soil, at night, for direct manual collection or with the aid of wide mouth flasks. During the day, these organisms may be found under rocks, fallen trunks, or tree bark, where they remain hidden until night. The most frequently used passive sampling method for crickets are pitfall traps. In this case, special care should be taken to keep the region maintained without human access, as far as vibration caused in the substrate alters the frequency of capture (Sperber et al. 2007).

Further devices and techniques that can be used for the passive sampling of crickets are:

Malaise Very frequently used in entomological surveys for sampling flying insects such as Diptera and several families of Hymenoptera and Coleoptera, Malaise traps are little effective for crickets. The main reason is that crickets are not good flyers. Thus, few cricket specimens are caught in Malaise surveys. Nevertheless, Malaise can collect specimens of Gryllidae (Eneopterinae, Oecanthinae, Podoscirtinae, Tafaliscinae) and Trigonidiidae (Trigonidiinae).

Moericke (Fig. 11.6) Yellow pan traps (Moericke traps) are not much used in crickets' surveys. However, in recent expeditions, this method has been efficient for capturing ground crickets such as Nemobiinae (Trigonidiidae) and Luzarinae (Phalangopsidae) and small flying crickets such as Trigonidiinae (Trigonidiidae).

Pitfall Traps (Fig. 11.2) Pitfall traps are the most common method of sampling ground crickets. It is widely used in crickets' surveys and taxonomical and ecological studies. Those traps are cheap and easy to construct and operate. Usually, each pitfall trap is filled with a liquid, which must kill and preserve the specimens. The chosen liquid also depends on the time that the traps will remain in the field and the resources available. It can be water + detergent (detergent is important for reducing waters surface tension) for short periods (~24 h), water + detergent + salt (salt is an alternative way of preserving DNA for short periods), and ethanol in different concentrations (for longer periods). The best killing solution is ethanol or ethanol fuel because it kills the specimens quickly, avoiding their escaping the trap, and preserves DNA (Szinwelski et al. 2012a, b, 2013). One can also use attractive liquids such as cane molasses diluted in water as a way to increase sampling efficiency.

11.4.3 Sampling Grasshoppers

For terrestrial grasshoppers, sweep net sampling is the most common sampling technique used to determine species composition present in a specific area. The captured grasshoppers are manually placed in labeled plastic bags containing local vegetation and transported in thermoplastic boxes in order to avoid overheating and damage to the material. The sweeping net's fabric must be resistant to the type of vegetation present in the area where the study will be performed, and spare nets must always be carried to the fieldwork.

To avoid varying sampling effort and biasing ecological analyses, standardized methods of sweep sampling must include predetermined number of times of

Fig. 11.6 Moericke traps. Photo: L.D. Campos and P.G.B.S. Dias

sweeping, duration of the sampling, number of collectors, distance among the collectors, and who will perform the sampling—it must be the same person for all sampling along the study. Typical methods used to standardize are two consecutive periods of 60 min with 10-min break (for density population studies, a minimum of 50 individuals is frequently used independently of time), two collectors, and at least 10 m distance from each other during the samplings. It is vital to put the collected grasshoppers of each sweeping period (or whatever the sampling unit established) in separate and adequattely labeled plastic bags, so as to be able to estimate density and diversity per sampling unit. For taxonomic and inventory studies, it is not necessary to control those variables, and just walking around the area and using the entomological net to capture the specimen after visual contact are efficient enough. However, absence of sampling unit control eliminates information on within area variation, and thus prevents comparisons among areas and any extrapolation of the results.

Different kinds of entomological nets (sweep net, aerial net, combination of aerial and sweep net, aquatic net, etc.) can be used according to habitat and taxa which will be sampled. For example, Carvalho (2010) designed a very efficient net to use in grasslands, 1.5 m long and a 1.3 m high, hold by two persons, and with the net almost touching the ground, walking a previously predetermined distance (5, 10 m, etc.).

We reccomend that, as soon as the fieldwork is completed, the sampled specimens are killed in a freezer at −20 °C. Before killing, we recommend keeping one hind leg in a labeled tube containing 90% alcohol for future genetic studies. After killed, each specimen must be prepared for conservation in a dry environment. For taxonomic studies, it is recommended to carry out an evisceration of the specimen, removing the internal organs with a longitudinal cut in the region of the cervical membrane. After evisceration, specimens must be carefully filled with antiseptic talc (neutral and unscented) and borax (1:1 proportion). Lastly, the grasshoppers must be pinned and mounted and then taken to the drying oven for 12 h at 40 °C or at room temperature. Using high temperature might damage the grasshoppers coloration. An alternative, quicker, method is to maintain the grasshopper specimens in the freezer for 8–12 days, mounting the specimens when they are still soft and flexible and maintaining them to dry at room temperature for 1 month. The advantage of this is avoiding the procedure of taking the guts out, which is an arduous work that takes a whole evening for each specimen.

11.5 Sampling Methodologies According to Study Aims

11.5.1 Sampling for Breeding and Behavioral Studies

Behavioral studies depend directly on massive breeding of these insects to carry out the experiments. After collecting the individuals, it is recommended that the orthopterans be taken to the laboratory for maintenance at a temperature of around 23 °C ± 1, in a terrarium containing moist substrate, hiding places, a source of moisture and food ad libitum. We use sifted earth or sand, egg cartons, damp cotton to maintain humidity around 60%, and fish food to feed the insects. The moist substrate or cotton serves as a substrate for oviposition. Air humidifier might be necessary to maintain humidity.

For the experiments, it is ideal that the individuals be isolated before the last molt, thus guaranteeing the individuals' virginity. This separation should be done when males and females are already showing visible wing buds.

11.5.2 Sampling for Bioacoustic Studies

Bioacoustic diversity might be studied using a set of sound recorders (Riede 1994; Chesmore 2004). Orthopterans have been a source of wonderment for their songs since Paleolithic times (Grimaldi et al. 2005). Orthopteran organs for the production of acoustic signals and hearing appeared independently in the two lineages that make up the group, Caelifera and Ensifera. Grasshoppers stridulate by scraping a row of denticles present in the inner part of the posterior leg femur on the lateral region of the anterior wings or rubbing the anterior wing on denticles present on the posterior wing. Crickets and katydids produce sound by scratching the pallet, on the side of one of the anterior wings, against a row of denticles on the opposite wing.

The repertoire of orthopteran songs is diverse, but most species produce sounds of calling, courtship, and aggressiveness, the first being the most studied for its use in species taxonomy and for issues related to sexual selection. The acoustic signals emitted by orthopterans are classified as pure tones due to their emission in a species-specific frequency band. In crickets, the frequency values vary between 3 and 9 kHz, while in katydids, much higher frequencies are found, even reaching ultrasound ones. The values shown in the frequencies depend on the rigidity of the exoskeleton, body size, and the flexibility of the emitting organ, the tegmina.

The recording of the acoustic signals can be done in the field or in the laboratory, preferably with high-quality digital recorders and directional microphones. With the advancement of technology, simple equipment, even cell phones, enable sound recordings with enough quality for scientific analyses.

Here we suggest a protocol that should be followed before, during, and after sound recording, so as to meet minimum scientific requirements:

1. Locate the insect in the field and approach it without causing disturbance to the environment—care must be taken to avoid touching the vegetation.
2. When visualizing the insect, position the microphone and start recording, avoiding sudden movements, so that the stridulation is not interrupted.
3. After about 3 min of sound recording (the recording time may vary according to the situation), press the "pause" button on the recorder.
4. Photograph and collect the individual for later recognition of the species.
5. Immediately after collecting the individual, measure air temperature in the location as close as possible to the stridulation site, since temperature influences several parameters of the orthopteran sound.

After following the steps described above, some information should be verbalized following the recording of the sound produced by the insect; thus, release the "pause" key and pronounce the following information:

1. Individual code—it will be necessary to create a code for each registered individual; the same code must be kept with the collected specimen.
2. Location—country, state, city, etc.: if possible, insert geographic coordinates.
3. Description of the exact location of the stridulation.
4. Date.
5. Recording time.
6. Temperature.
7. Brand of the equipment used for recording.
8. Other information may be inserted as it deems pertinent, such as the position or behavior of the insect during stridulation; the type of the stridulation site, such as a leaf, trunk, branch, or crevice; if the insect uses the curvature of the leaf to amplify the sound; etc.

Software for analyzing and editing sounds can be found on the Internet, such as Avisoft, CoolEdit, Adobe Audition, Raven, Audacity, etc. Some of these are freeware or open source. With the advancement of technology and the emergence of portable digital recorders coupled with the creation of indexes used to reflect acoustic activities from audio files, passive fauna monitoring studies have become increasingly accessible and reliable. Passive acoustic monitoring is a technique used to monitor animals indirectly, using the individual acoustic communication to record data about the community, being minimally invasive to fauna and flora.

Among the devices currently used, two passive acoustic recorders, SongMeter (Wildlife Acoustics, Inc.) and AudioMoth, stand out, both enabling the recording of acoustic signals from the environment, obtaining good acoustic information quality. Both devices can be configured with recording protocols allowing total control of the sampling time, which can be adjusted according to the researcher's aim. The frequency bands covered by these recorders capture signals from 28 Hz to 48 kHz, enough to register crickets and grasshoppers. Depending on the species of katydid, special equipment that captures ultrasound is necessary, usually equipment developed for the bioacoustic studies of bats.

Orthopterans emit two types of sounds, trills (continuous sequence of pulses) and chirps (groups of pulses) (Fig. 11.7). Pulse is a chain of sound cycles produced during inward movement of the forewings. Here are some parameters for the description of the sound of a species:

(a) **Trill** (1) dominant frequency, highest intensity spectral component of the song; (2) pulse rate, number of pulses per second; (3) pulse period, elapsed time from the pulse's first sound wave, up to the beginning of the subsequent pulse; (4) pulse duration, elapsed time from the first to the last sound wave of a pulse; (5) pulse interval, elapsed time from the pulse's last sound wave, up to the beginning of the subsequent pulse; and (6) number of sound waves per pulse.
(b) **Chirps** (1) dominant frequency; (2) chirp rate, number of chirps per minute; (3) chirp period, elapsed time from the chirp's first pulse to the beginning of the subsequent chirp; (4) chirp duration, elapsed time from the first to the last pulse of a chirp; (5) chirp's interval, elapsed time from the chirp's last sound wave, up to the beginning of the subsequent chirp; (6) number of pulses per chirp; (7) pulse period, elapsed time from the pulse's first sound wave, up to the beginning of the subsequent pulse; (8) pulse duration, elapsed time from the first to the last sound wave of a pulse; (9) pulse interval, elapsed time from the pulse's last sound wave, up to the beginning of the subsequent pulse; and (10) number of sound waves per pulse.

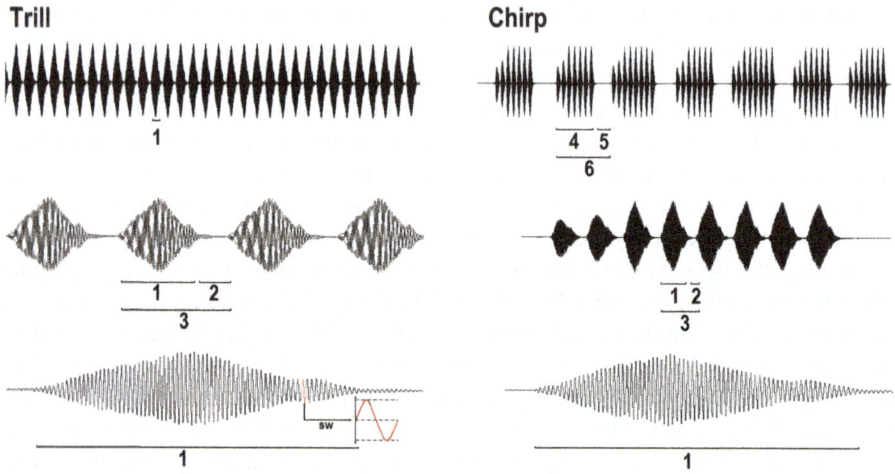

Fig. 11.7 Calling song of the cricket *Miogryllus itaquiensis* Orsini et al. 2017 (Orthoptera: Ensifera: Gryllidae), composed of trill and chirp sequences. 1, pulse; 2, pulse interval; 3, pulse period; 4, chirp duration (chirp with seven pulses); 5, chirp interval; 6, chirp period; SW, sound waves (pulse with around 72 sound waves, each one produced by a tooth stroke)

11.5.3 Sampling for Chromosomal Studies

Studies of chromosomes in Orthoptera have been widespread in recent centuries, since their karyotypes have a relatively small number of chromosomes, in general with a large size, facilitating their observation with optic microscope. For this reason, many important discoveries about mitotic and meiotic processes were made through the study of the chromosomes of these insects. The methodology for obtaining chromosomes from orthopterans (and also from other organisms) demands that individuals be collected and kept alive until the dissection process is carried out to obtain and fix the tissues that will be used for cytogenetic techniques.

The most practical way to keep individuals alive for a few hours until they are taken to the laboratory is to pack them in plastic bags with leaves and dry branches so that they remain sheltered in the bags so as to not collapse. It is not recommended to put green leaves in the plastic bags, as the leaves' perspiration may accumulate water droplets inside the plastic bag, which can be harmful to individuals. Plastic or glass tubes, about 13 cm long and 2.5 cm in diameter, are also good options for transporting individuals; the tube must be closed with a dry cotton plug, compacted firmly in the tube's opening.

Katydids can be collected on grasses and shrubs, following the same methodology described for grasshoppers. Night collections can be carried out passively, with light traps, or actively, searching the bushes and lower branches of the trees by means of manual picking, or with the aid of an entomological umbrella. Katydids can be put into plastic bags, as described for grasshoppers; however, large individuals should be packed individually in glass tubes, as they usually have well-developed jaws, and it is not uncommon for them to damage other individuals if kept together.

The method of collecting and storing crickets is similar to the procedure described for katydids. However, unlike katydids, crickets are very abundant on soil litter, and in this environment, collections should be made with pitfall traps, which can be made with a 1–2 L pet bottle, with the tip cut and inverted forming a funnel. Considering the objective of getting the individuals alive, it is recommended to install pitfall traps in the late afternoon and remove them in the early morning of the following day, since coleopterans, ants, centipedes (Scolopendridae), spiders, and scorpions may fall into the trap preying or damaging the crickets. It is recommended to use oat flakes to attract them, as the oat flakes are dry, preventing the crickets from becoming trapped and dying, as occurs when using moist baits such as cane molasses, banana, or other fruits.

When orthopterans are collected for chromosomal analysis, preference is given to adult males, as the group's taxonomic features are mainly based on male terminalia. In addition, males have well-developed testicles, composed of follicles where intense mitotic and meiotic activity occurs. Thus, even with conventional chromosomal analysis, it is possible to determine the diploid number, the morphology of the bivalent chromosomes in mitotic metaphase (spermatogonial metaphase), as well as the sex determination mechanism, since most species have X0♂-XX ♀ sex system, the X chromosome being univalent and easily observed in prophase and metaphase I. On the other hand, collecting living adult females is important, as they are usually fertilized and can be raised in the laboratory until they lay eggs. The offspring can supply a large amount of material for obtaining chromosomes, with the advantage that some individuals can be kept until adulthood so that males are used for species determination. In addition, nymphs offer good material for obtaining mitotic chromosomes, which are generally more distended.

The technique for obtaining chromosomes from grasshoppers, katydids, and crickets is relatively simple. For mitotic chromosomes, a 0.05% solution of colchicine must be injected into the individual (male, female, or nymph). The individual must remain alive and after 3–5 h be dissected to remove the cecum and midgut. The tissues must be submitted to the hypotonic solution of KCl 0.075 M (filtered or distilled water can be used instead) for 5–10 min and transferred to an Eppendorf tube with the Carnoy I fixative solution (3 ethylic alcohol:1 glacial acetic acid). The gonads of male and female nymphs also provide good mitotic chromosomes. To obtain meiotic chromosomes, testicles from adult or preadult males are used. In this case, individuals do not need to be injected with colchicine, and the procedures for hypotonization and fixation are as described above. The Eppendorf tubes with the fixed material should be kept in the refrigerator, preferably at 4 °C. From these procedures, the tissues can be used for any type of cytogenetic technique, whether with conventional or differential staining.

11.5.4 Sampling for DNA Studies

DNA sampling within Orthoptera follows basic protocols from molecular methods. First, DNA quality is an important factor for future analyses and will depend on collection and appropriate preservation of samples. Sampling methods like pitfall traps must use preservation liquid like fuel ethanol (Szinwelski et al. 2012a, b) or >96% ethanol as a killing solution. Using detergent spoils all DNA. For active searching, specimens or samples must be immediately stored in preservation liquid. For conservation during long periods, DNA samples must be conserved in >96% ethanol, frozen, and stored at −20 °C. Research dealing with stored DNA must avoid freeze/thaw cycles of samples, since this will affect the stability of the genomic DNA, leading to degradation (Shao et al. 2012). Generally, we use entire hind femora of smaller specimens or fragments of femora of larger specimens for DNA samples of Orthoptera. If the specimen is too small, we must use the entire individual to obtain a sufficient amount of DNA for further analysis. After sample preparation, DNA must be extracted from the nucleus of the cells. This can be done by either ready-to-use DNA extraction kits or conventional protocols; the steps are the same.

DNA Extraction First, cells must be separated from each other by a physical means or vortexing and put into a solution containing salt. A detergent is then added to break down the lipids in the cell membrane and nuclei for DNA release. DNA is then separated from proteins and other cellular debris using a protease and cleaning reagents like phenol chloroform. Finally, we use ice-cold alcohol (either ethanol or isopropanol) that, together with salt, precipitates the DNA. After cleaning, the precipitate is resuspended in a slightly alkaline and ready-to-use buffer.

DNA Amplification and Sequencing After extraction, fragments of DNA are amplified and sequenced. Polymerase chain reaction (PCR) is a DNA amplification technique that uses interleaved cycles of specific temperature for strand denaturation and primer binding and extension by Taq polymerase. These fragments can now be sequenced by either Sanger or next-generation sequencing (NGS) for molecular studies of taxonomy, diversity, and evolution in Orthoptera.

11.5.5 Preservation of Specimens for Scientific Collections

Preservation methods depend on the group to which they belong. Caelifera and Tettigoniidae are preserved dry, remembering that the faster the drying, the better the color preservation of the integument. Katydids must the dissected, in order to remove internal organs; for this, it's necessary to make a cut using scissors in the pleura of the abdomen (of ca. 5 mm); the internal organs must be removed with tweezers (if possible the male genitalia must be kept in vials with glycerin). After being removed, it's necessary to stuff the internal part of the katydid's body with cotton and then fill the body with borax or cotton. After this, we mount the individuals: the forelegs should be positioned forward and the rest back; male cerci must be mounted opened, for a better visualization. At least one individual of each morphospecies must be mounted with the left tegmen opened. After being mounted, the individuals must be dried in an oven, at 40 °C, for 48 h. After being dried, they must be conserved in boxes with mothball or camphor, in places with relatively low temperatures (less than 20 °C). Grylloidea and Stenopelmatoidea are preferably preserved in humid way, inside tubes with 70% alcohol or in ethanol fuel: however, hydration is needed to manipulate these organisms. When collected in pitfall traps, it is recommended to use a more concentrated alcoholic solution, due to the soaking and consequent softening of the specimens, when it rains in the field. For molecular purposes, we recommend that part of individual tissues (e.g., foreleg) be transferred to a tube with >96% alcohol and be stored in a fridge. For grasshoppers, killing and preserving in freezer (−20 °C) is the most efficient method to preserve the coloration. If it is necessary to kill in the field, use insect killing jars, fueled with ethyl acetate or cyanide, avoiding direct contact of the killing agent with the specimen. For drying, avoid using laboratory oven: let them dry at room temperature, in a dry room. After killing, remove the grasshopper's guts through a small incision along the specimen's neck, under the pronotum, using a small forceps. Introduce antiseptic talc (neutral and unscented) and borax (1:1 proportion), and, if necessary, stuff the specimen with cotton. An alternative, quicker method is to maintain the grasshopper specimens in the freezer for 8–12 days, mounting the specimens when they are still soft and flexible and maintaining them to dry at room temperature for 1 month. The advantage of this last technique is that it avoids taking the guts out, which is an arduous work that takes a whole evening for each specimen.

11.6 Sampling Design

If you have a strict hypothetical-deductive approach, you have to control sampling design so as to guarantee a homogenous sampling effort among sampling units, as well as to guarantee independence of sampling replicates, and you have to relate your data to the corresponding effort and environmental variables relevant to your tested hypotheses. If, on the other hand, you aim at an inventory, your focus will be on the broadness of the sampled fauna, aiming completeness and you may disregard control of sampling effort or of environmental variables. Of course, you may have a mixture of both approaches. The behavior of your focus organisms affects sampling efficiency and bias. Thus, you have to plan carefully your sampling according to microhabitat and behavior of your focus organisms, so as not to overextrapolate your conclusions.

Most commonly sampling aiming at a hypothetical-deductive approach involves passive capture of organisms, such as pitfall traps for litter-dwelling crickets. Such sampling has shortcomings, since their efficiency depends on the organisms' mobility, and as such it has low efficiency for broad-scale inventories. The addition of attractive baits may enhance capture efficiency, but it generates bias in the spectrum of captured organisms, which may invalidate the test of ecological hypotheses. For inventories, it is common to combine several simultaneous sampling techniques, including active day and night sampling.

11.7 Cautions for Data Analysis

When the studied orthopterans do not occur in swarms, they can be analyzed as unitary organisms; therefore, when analyzed as response variables (Y), being it the number or density of individuals or the number of species, the appropriate distribution of the data is the Poisson distribution, eventually corrected for overdispersion or substituted by negative binomial distribution (Crawley 2013; Zuur et al. 2009), instead of the classical normal distribution. Thus, it is necessary to adjust generalized linear models (GLMs), which can be done using free software such as R (R Core Team 2019). When the data contain many zero values, which may occur, e.g., with cricket individuals captured in pitfall traps, it might be necessary to use zero-inflated Poisson (ZIP) models (Zuur et al. 2012), or, alternatively, work with presence-absence data, using binomial distribution. Frequently, however, data with many zeros indicate insufficient sampling effort, maybe associated with incorrect spatial scale. Species occurrence analyses should privilege large spatial scales, with appropriate geographical data analyses.

To deal with nested data, great care must be taken. For example, you may have a single information on the explanatory variables of your model, e.g., if you are interested in the effects of forest remnant regeneration time on cricket species richness (Szinwelski et al. 2012a, b), you might pool several sampling units within each forest remnant, so as to avoid pseudoreplication (sensu Hurlbert 1984), or you might use mixed effects models to include the nested structure of your data into the model, which is essential if you have different spatial scales for local and regional explanatory variables. Most frequently mixed effects models consider random intercepts (Zuur et al. 2012), which means that you admit random variation in the intercept of the effect of your explanatory variable, e.g., the effect of the canopy cover on cricket species richness, where some remnants have higher richness than others, given a fixed canopy cover, but the slope of crickets' species richness driven by the canopy cover value is always the same (Farias-Martins et al. 2017). More complex mixed effects assume random intercept and random slope (Gelman and Hill 2006), which means that you admit random variation both in the intercept and in the slope of the effect of your explanatory variable, e.g., the effect of the canopy cover on cricket species richness. In this more complex, and probably overcomplex model, the effect of your explanatory variable on your response variable, e.g., of the canopy cover unto cricket species richness, varies randomly, among forest remnants, both in the intercept and in the slope, meaning that in some forest remnants, cricket species richness increases more steeply with canopy cover than in other remnants.

An important issue to consider in field studies is spatial autocorrelation, because this process leads to greater similarity between the closest sites due to factors that are spatially correlated, such as environmental conditions (soil type, temperature, slope of our altitude) (Fleishman and Mac Nally 2006) and biological attributes (dispersion, home-range size, and biotic interactions) (Wilkinson and Edds 2001). To minimize this problem, it is possible to use statistical analysis dealing with spatial autocorrelation, and for univariate-dependent variables (e.g., species richness, density, abundance), generalized least squares (GLS) models are used to identify and control spatial autocorrelation (available in "nmle" R package, Pinheiro et al. 2020), while for multivariate-dependent variables (e.g., species, phylogenetic and functional composition), restricted ordinations (partial redundancy analysis, partial RDA; partial canonical correlation analysis, partial CCA) are used (available in "vegan" R package; Oksanen et al. 2019). However, modeling the spatial autocorrelation can be complex, so we suggest opting for sample designs that lead to greater sample independence, so as to avoid pseudoreplication. When possible, we suggest *a priori* sampling to identify the spatial connectivity of the variables in the study regions in order to elaborate a sample design with the smallest possible spatial autocorrelation.

To evaluate biodiversity metrics (e.g., richness, species diversity, abundance, density) at broad geographic scales, it is necessary to know the availability of these metrics in the studied space (e.g., continent, biome, biogeographical province) and the gradient of the predictors of biodiversity (e.g., climate, average annual temperature; landscape, forest cover; topography, slope and elevation). Thus, samples should be proportionally arranged on the geographical and environmental variation of the predictors, in order to sample the most representative sites both in geographical and environmental space. After this step, the spatial modeling of biodiversity metrics can be performed in two different ways: (1) direct modeling of biodiversity metrics, based on the correlation between diversity metrics, estimated at known sites, and environmental variables (macroecological modeling (MEM)) (see Gotelli et al. 2009), and (2) species distribution modeling in isolation, with subsequent overlap, to predict species richness index (stacked species distribution modeling (SSDM)) (see Guisan and Rahbek 2011). MEM usually uses regression models that depend on the metric used: count data, e.g., richness and abundance use (Generalized Linear Models (GLM) or Generalized Additive Models (GAM) with Poisson distribution or machine learning classification models; random forest (RF) and support vector machine (SVM)), and continuous data, e.g., diversity and density (GLM/GAM with Gaussian distribution, SVM, RF). SSDMs can be based on different statistical algorithms depending on the characteristic of the biodiversity metric used: presence data (Bioclim, Gower, Domain, Mahalanobis) and presence and absence data (GLM/GAM with binomial distribution, RF, SVM). Note that to make the spatial prediction of metrics, it is necessary to have raster layer maps (maps with matrix structure, composed of pixels which have values of spatial variable ending, e.g., temperature, altitude) with the values of the predictor variables for the entire region of interest, and the resolution of this raster depends on the size of the study region and the processes and factors to be studied, so that the spatial pattern of biodiversity metrics is demonstrated on a map of the study region.

Acknowledgment Utmost thanks goes to Kátia Matiotti da Costa (PUC-RS) for precious comments. Financial support: Conselho Nacional de Desenvolvimento Científico e Tecnológico (CNPq) financed CFS' productivity research grant (Processes No 310032/2015-6 and 40774 5/2017-3). This study was partially supported by research grants by CNPq, Coordenação de Aperfeiçoamento de Pessoal de Nível Superior (CAPES), Fundação de Apoio à Pesquisa do Estado de Minas Gerais (FAPEMIG), Fundação de Amparo à Pesquisa do Estado de São Paulo (FAPESP), and finantial support by SISBIOTA-Brasil (CNPq/FAPEMIG—5653360/2010-0). Authors would like to thank the Programa de Pós-Graduação em Ecologia of the Universidade Federal de Viçosa/EDITAL FORTIS/FAPEMIG 01/2013—for funding the English revision.

References

Amédégnato C, Devriese H (2008) Global diversity of true and pygmy grasshoppers (Acridomorpha, Orthoptera) in freshwater. Hydrobiologia 595:535–543. https://doi.org/10.1007/s10750-007-9132-z

Bello LD (2020) To fight locusts, historic rivals India and Pakistan team up. https://undark.org/2020/04/20/india-pakistan-locusts/?utm_source=Nature+Briefing&utm_campaign=527dbb4637-briefing-dy-20200428&utm_medium=email&utm_term=0_c9dfd39373-527dbb4637-43886253. Accessed 5 May 2020

Bichuette ME, Simões L, Von Schimonsky D et al (2015) Effectiveness of quadrat sampling on terrestrial cave fauna survey—a case study in a Neotropical cave. Acta Scient Biol Sc 37:345–351. https://doi.org/10.4025/actascibiolsci.v37i3.28374

Bidau CJ (2014) Patterns in Orthoptera biodiversity. I. Adaptations in ecological and evolutionary contexts. J Insect Biodiv 2:1–39. https://doi.org/10.12976/jib/2014.2.20

Bolfarini MP, Bichuette ME (2015) *Endecous peruassuensis* n. sp. (Orthoptera: Grylloidea: Phalangopsidae) from caves of Eastern Brazil: evidence of isolation in the subterranean realm and discussion about troglom. Zootaxa 4032:297. https://doi.org/10.11646/zootaxa.4032.3.5

Braker E (1989) Evolution and ecology of oviposition on host plants by acridoid grasshoppers. Biol J Linn Soc 38:389–406. https://doi.org/10.1111/j.1095-8312.1989.tb01584.x

Browde JA, Pedigo LP, Degooyer TA et al (1992) Comparison of sampling techniques for grasshoppers (Orthoptera: Acrididae) in soybean. J Econ Entomol 85:2270–2274. https://doi.org/10.1093/jee/85.6.2270

Bruner L (1906) Synoptic list of Paraguayan Acrididae, or locusts, with the description of new forms. Proc US Nation Mus 30(1461):663

Carvalho NL (2010) Análise faunística de gafanhotos (Orthoptera, Acridoidea: Acrididae, Romaleidae e Proscopiidae), no município de São Sepé, RS. M.Sc. thesis, Federal University of Santa Maria, Santa Maria, Brazil

Chamorro-Rengifo J, Silva BC, Olivier RS et al (2018) Meadow katydids (Orthoptera: Tettigoniidae: Conocephalini) from the Central-West Region of Brazil: morphological, bioacoustic and cytogenetic study. Zootaxa 4388:347–372. https://doi.org/10.11646/zootaxa.4388.3.3

Chesmore D (2004) Automated bioacoustic identification of species. An Acad Bras Cienc 76:436–440. https://doi.org/10.1590/S0001-37652004000200037

Chintauan-Marquier IC, Legendre F, Hugel S et al (2015) Laying the foundations of evolutionary and systematic studies in crickets (Insecta, Orthoptera): a multilocus phylogenetic analysis. Cladist 32:54–81. https://doi.org/10.1111/cla.12114

Cigliano MM, Braun H, Eades DC et al (2020). Orthoptera species file. Version 5.0/5.0. http://Orthoptera.SpeciesFile.org. Accessed 2 May 2020

Cover MR, Bogan MT (2015) Minor insect orders. In: Ecology and general biology: Thorp and Covich's freshwater invertebrates, vol I, 4th edn. Academis Press/Elsevier, Amsterdam, pp 1059–1072. https://doi.org/10.1016/B978-0-12-385026-3.00041-3

Crawley MJ (2013) The R book, 2nd edn. Wiley, New York

Farias-Martins F, Sperber CF, Albeny-Simões D et al (2017) Forest litter crickets prefer higher substrate moisture for oviposition: evidence from field and lab experiments. PLos One 12:e0185800. https://doi.org/10.1371/journal.pone.0185800

Fianco M (2019) Tettigoniidae (Orthoptera: Ensifera) do Parque Nacional do Iguaçu, BR: diversidade, bioacústica e descrição de novas espécies. Universidade Federal da Integração Latino-Americana. Dissertação de Mestrado, 265 pp

Fianco M, Preis H, Szinwelski N et al (2019) On brachypterous phaneropterine katydids (Orthoptera: Tettigoniidae: Phaneropterinae) from the Iguaçu National Park, Brazil: three new species, new record and bioacoustics. Zootaxa 4652:240–264. https://doi.org/10.11646/zootaxa.4652.2.2

Fleishman E, Mac Nally R (2006) Patterns of spatial autocorrelation of assemblages of birds, floristics, physiognomy, and primary productivity in the central Great Basin, USA. Divers Distrib 12:236–243. https://doi.org/10.1111/j.1366-9516.2006.00240.x

Gallo D, Nakano O, Neto SS et al (2002) Entomologia agrícola. FEALQ, Piracicaba

Gangwere SK (1961) A monograph on food selection in Orthoptera. Trans Am Entomol Soc 87:67–230

Gelman A, Hill J (2006) Data analysis using regression and multilevel/hierarchical models. Cambridge University Press, Cambridge

Gleason HA (1920) Some applications of the quadrat method. Bull Torrey Bot Club 47:21–33. https://doi.org/10.2307/2480223

Gnaspini P, Trajano E (1994) Brazilian cave invertebrates, with a checklist of troglomorphic taxa. Rev Bras Entomol 38:549–584

Gotelli NJ, Anderson MJ, Arita HT et al (2009) Patterns and causes of species richness: a general simulation model for macroecology. Ecol Lett 12:873–886. https://doi.org/10.1111/j.1461-0248.2009.01353.x

Grimaldi D, Engel MS, Engel MS (2005) Evolution of the insects. Cambridge University Press, Cambridge

Guisan A, Rahbek C (2011) SESAM—a new framework integrating macroecological and species distribution models for predicting spatio-temporal patterns of species assemblages. J Biogeogr 38:1433–1444. https://doi.org/10.1111/j.1365-2699.2011.02550.x

Gwynne DT (2001) Katydids and bush-crickets: reproductive behavior and evolution of the tettigoniidae. Cornell University Press, New York

Hertl PT, Brandenburg RL, Barbercheck ME (2001) Effect of soil moisture on ovipositional behavior in the southern mole cricket (Orthoptera: Gryllotalpidae). Environ Entomol 3:466–473. https://doi.org/10.1603/0046-225X-30.3.466

Huber F, Moore TE, Loher W (1989) Cricket behavior and neurobiology. Cornell University Press, New York

Humphreys WF (1991) Re-establishment of pulse driven populations in a terrestrial troglobite community. J Anim Ecol 60:609–623. https://doi.org/10.2307/5301

Hurlbert SH (1984) Pseudoreplication and the design of ecological field experiments. Ecol Mon 54:187–211. https://doi.org/10.2307/1942661

Janzen DH, Schoener TW (1968) Differences in insect abundance and diversity between wetter and drier sites during a tropical dry season. Ecology 49:96–110

Jesus FM (2015) Nicho trófico de grilos de serrapilheira florestal: uma análise utilizando isótopos estáveis de carbono e nitrogênio. PhD thesis, Universidade Federal de Viçosa

Johnson NF, Triplehorn CA (2005) Borror and Delong's introduction to the study of insects, 7th edn. Cengage, Boston

Kočárek P, Grucmanová S, Filipcová Z et al (2008) Bryophagy in the groundhopper *Tetrix ceperoi* (Orthoptera: Tetrigidae): analysis of alimentary tract contents. Scrip Facult Rer Nat Universit Ostrav 186:348–352

Krebs CJ (1999) Ecological methodology. Addison-Welsey Educational Publishing, New York

Lhano MG (2002) Aspectos biológicos e ecológicos de *Cornops aquaticum* (Bruner, 1906) (Orthoptera: Acrididae) em *Eichhornia azurea* (Swartz) Kurth (Pontederiaceae) no Pantanal de Pocone, Mato Grosso. Dissertation, Federal University of Mato Grosso

Montealegre-Z F (1997) Estudio de la fauna de Tettigoniidae (Orthoptera: Ensifera) del Valle del Cauca. Ph.D. thesis, Universidad del Valle

Nerney NJ (1960) Grasshopper damage on short-grass rangeland of the San Carlos Apache Indian reservation, Arizona. J Econ Entomol 53:640–646. https://doi.org/10.1093/jee/53.4.640

Nunes-Gutjahr AL, Braga CESB (2018) Order Orthoptera. In: Hamada N, Thorp JH, Rogers DC (eds) Thorp and Covich's freshwater invertebrates, 4th edn. Academic Press, New York, pp 143–168

O'Neill KM, Larson DP, Kemp WP (2003) Sweep sampling technique affects estimates of the relative abundance and community composition of grasshoppers (Orthoptera: Acrididae). J Agric Urb Entomol 19:125–131

Oksanen J, Blanchet FG, Friendly M et al (2019) Vegan: community ecology package. R package version 2.5-6. https://CRAN.R-project.org/package=vegan

Onsager J (1977) Comparison of five methods for estimating density of rangeland grasshoppers. J Econ Entomol 70:187–190. https://doi.org/10.1093/jee/70.2.187

Onsager JA, Henry JE (1977) A method for estimating the density of rangeland grasshoppers (Orthoptera, Acrididae) in experimental plots. Acrida 6:231–237

Orsini MP, Costa MKM, Szinwelski N, Martins LP, Corrêa RC, Timm VF, Zefa E (2017) A new species of *Miogryllus* Saussure, 1877 and new record of *Miogryllus piracicabensis* Piza, 1960 (Orthoptera: Gryllidae) from State of Rio Grande do Sul, Brazil, with calling song and chromosome complement. Zootaxa 4291:361–372

Pinheiro J, Bates D, DebRoy S et al (2020) R Core Team. nlme: linear and nonlinear mixed effects models. R package version 3.1-147. https://CRAN.R-project.org/package=nlme

R Core Team (2019) R: a language and environment for statistical computing. R Foundation for Statistical Computing, Vienna, Austria. https://www.R-project.org

Rampini M, Carchini G, Sbordoni V (1983) Population ecology of the cave cricket Dolichopoda geniculata from an artificial cave at Rome (Italy). Fragm Entomol 17:85–100

Rehn JAG (1929) Studies in Costa Rican Dermaptera and Orthoptera. Trans Amer Entomol Soc 55:32

Rentz DCF (2010) A guide to the Katydids of Australia, 1st edn. CSIRO Publishing, Melbourne

Rentz DCF, Su YN (2003) Orthoptera. In: Resh VH, Cardé RT (eds) Encyclopedia of insects. Academic Press, Amsterdam, pp 827–839

Ribas CR, Sobrinho TG, Schoereder JH et al (2005) How large is large enough for insects? Forest fragmentation effects at three spatial scales. Acta Oecol 27:31–41. https://doi.org/10.1016/j.actao.2004.08.008

Riede K (1994) Bioacoustic diversity and resource partitioning in tropical calling communities. In: Tropical biodiversity and systematics. Proceedings of the international symposium on biodiversity and systematics in tropical ecosystems, Bonn, pp 275–280

Samways MJ, McGeoch M, New TR (2010) Insect conservation: a handbook of approaches and methods, 1st edn. Oxford University Press, New York

Shao W, Khin S, Kopp WC (2012) Characterization of effect of repeated freeze and thaw cycles on stability of genomic DNA using pulsed field gel electrophoresis. Biopreserv Biobank 10:4–11. https://doi.org/10.1089/bio.2011.0016

Sharratt NJ, Picker MD, Samways MJ (2000) The invertebrate fauna of the sandstone caves of the Cape Peninsula (South Africa): patterns of endemism and conservation priorities. Biodivers Conserv 9:107–143. https://doi.org/10.1023/A:1008968518058

Simon M (2020) Africa's huge locust swarms are growing at the worst time. https://www.wired.com/story/africas-huge-locust-swarms-are-growing-at-the-worst-time/. Accessed 5 May 2020

Souza-Dias PGB, Bolfarini MP, Nihei SS et al (2014) *Endecous apterus*: a new species of cave cricket from northeast Brazil, with comments on the use of subterranean habitats by Luzarinae

crickets (Orthoptera: Grylloidea: Phalangopsidae: Luzarinae). Zootaxa 3784:120–130. https://doi.org/10.11646/zootaxa.3784.2.2
Sperber CF (1996) Field diet of the grasshopper *Abracris dilecta* Walker (Orthoptera, Acrididae). Revta Bras Zool 13:127–135. https://doi.org/10.1590/S0101-81751996000100012
Sperber CF, Vieira GH, Mendes MH (2003) Aprimoramento da amostragem de grilos de serapilheira (Orthoptera: Gryllidae) por armadilha. Neotrop Entomol 32:733–735. https://doi.org/10.1590/S1519-566X2003000400030
Sperber CF, Soares LG, Pereira MR (2007) Litter disturbance and trap spatial positioning affects the number of captured individuals and genera of crickets (Orthoptera: Grylloidea). J Orthop Res 16:77–83. https://doi.org/10.1665/1082-6467(2007)16[77:LDATSP]2.0.CO;2
Sperber CF, Mews CM, Lhano MG et al (2012) Orthoptera. In: Rafael JA, Melo GAR, Carvalho CJB (eds) Insetos do Brasil: diversidade e taxonomia. Holos Editora, Ribeirão Preto
Sperber CF, Szinwelski N, Yotoko K et al (2013) Environmental drivers of litter cricket diversity in Central Amazon forest. In: Abstracts of 11th international congress of orthopterology, Kunming, Yunnan, China, 11th to 15th of August 2013
Szinwelski N, Fialho V, Yotoko K et al (2012a) Ethanol fuel improves arthropod capture in pitfall traps and preserves DNA. ZooKeys 196:11–22. https://doi.org/10.3897/zookeys.196.3130
Szinwelski N, Rosa CS, Schoereder JH et al (2012b) Effects of forest regeneration on crickets: evaluating environmental drivers in a 300-year chronosequence. Int J Zool 2012:1–13. https://doi.org/10.1155/2012/793419
Szinwelski N, Yotoko K, Solar R et al (2013) Ethanol fuel improves pitfall traps through rapid sinking and death of captured orthopterans. Environ Entomol 42:752–758. https://doi.org/10.1603/EN13030
Szinwelski N, Rosa C, Solar R et al (2015) Aggregation of cricket activity in response to resource addition increases local diversity. PLoS One 10:e0139669. https://doi.org/10.1371/journal.pone.0139669
Thorpe WH (1950) Plastron respiration in aquatic insects. Biol Rev 25:344–390
Thunberg CP (1824) Grylli: monographia, illustrata. Académie impériale des Sciences, Mem Acad Imp Sci St Peterburg 3:404
Vargas JMS (2013) Variáveis ambientais que influenciam a abundância de grilos (Orthoptera, Grylloidea) na Amazônia Central. Dissertation, Universidade Federal de Viçosa
Vargas JMS (2018) Influência de variáveis ambientais na abundância de grilos (Orthoptera: Grylloidea) de liteira em cinco diferentes biomas brasileiros. Ph.D. thesis, Universidade Federal de Viçosa
Vieira MF, Adis J (1992) Abundância e biomassa de *Paulinia acuminata* (De Geer, 1773) (Orthoptera: Pauliniidae) em um lago de várzea da Amazônia Central. Amazon 12:337–352
Weaver JE (1918) The quadrat method in teaching ecology. The Plant World 21:267–283
Weinstein P (1994) Behavioural ecology of tropical cave cockroaches: preliminary field studies with evolutionary implications. Austr J Entomol 33:367–370
Wilkinson CD, Edds DR (2001) Spatial pattern and environmental correlates of a Midwestern stream fish community: including spatial autocorrelation as a factor in community analyses. Am Midl Nat 146:271–289. https://doi.org/10.1674/0003-0031(2001)146[0271:SPAECO]2.0.CO;2
Wynne JJ, Sommer S, Howarth FG et al (2019) Capturing arthropod diversity in complex cave systems. Diver Dist 24:1478–1491. https://doi.org/10.1111/ddi.12772
Zuur A, Ieno EN, Walker N et al (2009) Mixed effects models and extensions in ecology with R. Springer Science & Business Media, New York. https://doi.org/10.1007/978-0-387-87458-6
Zuur AF, Saveliev AA, Ieno EN (2012) Zero inflated models and generalized linear mixed models with R. Highland statistics

Chapter 12
Hemiptera Sampling Methods

Cristiano F. Schwertner, Renan Carrenho, Felipe F. F. Moreira, and Gerasimos Cassis

12.1 Introduction

Hemiptera (bugs) is the fifth most speciose insect order, comprising about 120,000 described living species, nested in more than 150 families (Larivière et al. 2011; Henry 2017; Bartlett et al. 2018; Hardy 2018; Cassis 2019). Hemipterans are ubiquitous in the world (except the polar regions) and are important functional groups in terrestrial and aquatic ecosystems (Moir and Brennan 2007); they are found in tropical and temperate forests, grasslands, savannas, lakes, ponds, and rivers (Dietrich 2009; Gullan and Martin 2009; Schaefer 2009). Bugs are also important crop pests, and a few species are vectors of human diseases or are nuisance species in the built environment (Brambila and Hodges 2008). Hemipterans are also successful biocontrol agents in forestry and agriculture, with both phytophagous and predaceous beneficials known.

C. F. Schwertner (✉)
Departamento de Ecologia e Biologia Evolutiva, Instituto de Ciências Ambientais, Químicas e Farmacêuticas, Universidade Federal de São Paulo (UNIFESP), Diadema, SP, Brazil

Museu de Zoologia, Universidade de São Paulo, São Paulo, SP, Brazil
e-mail: schwertner@unifesp.br

R. Carrenho
Museu de Zoologia, Universidade de São Paulo, São Paulo, SP, Brazil

F. F. F. Moreira
Instituto Oswaldo Cruz, Laboratório de Biodiversidade Entomológica, Laboratório de Biodiversidade Entomológica, Fundação Oswaldo Cruz, RJ, Brazil

G. Cassis
Evolution & Ecology Research Centre, School of Biological, Earth and Environmental Sciences, University of New South Wales, Sydney, Australia
e-mail: gcassis@unsw.edu.au

J. C. Santos, G. W. Fernandes (eds.), *Measuring Arthropod Biodiversity*,
https://doi.org/10.1007/978-3-030-53226-0_12

Piercing-sucking mouthparts are the best defining character of the Hemiptera (Dietrich 2009; Schaefer 2009); they comprise paired elongate mandibular and maxillary stylets that are interlocked to form food and salivary canals. The stylets are enclosed by the labium, a usually three- or four-segmented tube, which together are referred to as the rostrum and vary in position and length. Unlike other paraneopteran taxa, they lack maxillary and labial palps.

Hemipteran mouthparts are primarily adapted for plant-feeding (phytophagy), on either vascular, mesophyllic, or reproductive tissues (seeds or fruits). In comparison with other hemipteran suborders, Heteroptera display far greater diversity in feeding strategies and food preferences; they also feed on animal tissue (carnivory), vertebrate blood (hematophagy), fungi (mycetophagy), and dead and decaying material (saprophagy). It is orthodoxy that phytophagy is the ancestral feeding strategy in the Hemiptera, with animal- and fungal-feeding-derived conditions that have evolved independently on numerous occasions in the Heteroptera and soma Auchenorrhyncha.

12.1.1 Hemiptera Classification: Species, Habitats, and Ecological Diversity

The order is currently classified into four suborders (Table 12.1), which are well supported as monophyletic groups. These suborders are typically classified into infraorders and superfamilies (Grazia et al. 2012; Bartlett et al. 2018), which facilitate the recognition, study, and discussion of the diversity of hemipteran families and subordinate taxa.

The suborder **Auchenorrhyncha** is comprised of ca. 43,000 described species, in almost 6000 genera (Dietrich 2009; Cavichioli and Takiya 2012; Bartlett et al. 2018), all of which are terrestrial and mostly phytophagous, with a few mycetophagous or saprophagous species. Diagnostic characteristics for the suborder include the aristoid antennae, rostrum arising from the posteroventral region of the head, and the tymbal apparatus (a sound-producing structure). They range from 3 to 100 mm in size. The most common groups are the cicadas (Cicadidae, Fig. 12.1a), the froghoppers and spittlebugs (Cercopoidea), the leafhoppers and treehoppers (Membracoidea, Fig. 12.1b), and the planthoppers (Fulgoroidea, Fig. 12.1c).

Adult and nymphal auchenorrhynchans are similar in morphology and habits and are usually very active, aside from the sessile nymphs of cercopoids that live within a frothy secretion. Both adults and most nymphs of auchenorrhynchans feed on aboveground parts of their hosts. There are also highly specialized groups within this suborder; these include cicada and cercopid nymphs, as well as some fulgoroid families and a few leafhopper genera, which feed on roots. Nymphs of the fulgoroid families Achilidae and Derbidae live under bark or in litter, whereas other fulgoroid nymphs are cavernicolous (cave inhabiting). Nymphs of the family Tettigometridae are inquilines in ant nests. Other notable behaviors include ant mutualism, nymphal

Table 12.1 Suborders and infraorders of Hemiptera and number of species (no. spp.), habitat (H), and feeding strategies (FS) of selected well-known families

Suborder/infraorder	Superfamily	Family	No. spp.[a]	H	FS
Auchenorrhyncha					
Cicadomorpha					
	Cicadoidea	Cicadidae	2,900	T	Ph
	Cercopoidea	Cercopidae	1,480	T	Ph
	Membracoidea	Cicadellidae	21,350	T	Ph
		Membracidae	3,481	T	Ph
Fulgoromorpha					
	Fulgoreoidea	Delphacidae	2,033	T	Ph
		Derbidae	1,682	T	Ph
		Dictyopharidae	687	T	Ph
		Flatidae	1,302	T	Ph
		Fulgoridae	709	T	Ph
Coleorrhyncha					
		Peloridiidae	37	T	Ph
Heteroptera					
Cimicomorpha					
	Cimicoidea	Anthocoridae	445	T	Pr
		Cimicidae	110	T	H
		Plokyophilidae	13	T	Pr
		Polyctenidae	32	T	H
	Miroidea	Miridae	11,101	T	Ph, Pr
		Tingidae	2,124	T	Ph
	Reduvoidea	Reduviidae	6,878	T	Pr
Dipsocoromorpha					
		Dispocoridae	51	T	Pr
		Ceratocombidae	52	T	Pr
		Schizopteridae	229	T	Pr
Enicocephalomorpha					
		Enicocephalidae	405	T	Pr
Leptopodormopha					
	Saldoidea	Saldidae	335	S, T	Pr
Gerromorpha					
	Gerroidea	Gerridae	751	S	Pr
		Veliidae	962	S	Pr
	Hidrometroidea	Hidrometridae	126	S	Pr
Nepomorpha					
	Corixoidea	Corixidae	607	A	Pr
	Naucoroidea	Naucoridae	391	A	Pr
	Nepoidea	Belostomatidae	160	A	Pr
		Nepidae	268	A	Pr

(continued)

Table 12.1 (continued)

Suborder/infraorder	Superfamily	Family	No. spp.[a]	H	FS
	Notonectoidea	Notonectidae	400	A	Pr
	Ochteroidea	Gelastocoridae	111	S	Pr
Pentatomomorpha					
	Aradoidea	Aradidae	1,931	T	M
	Coreoidea	Coreidae	3,178	T	Ph
	Lygaeoidea	Blissidae	436	T	Ph
		Lygaeidae	968	T	Ph
		Rhyparochromidae	1,850	T	Ph
	Pentatomoidea	Acanthosomatidae	317	T	Ph
		Cydnidae	765	T	Ph
		Pentatomidae	5,000	T	Ph, Pr
		Plataspidae	604	T	Ph
		Scutelleridae	531	T	Ph
Sternorrhyncha					
	Aphidoidea	Aphididae	4,800	T	Ph
	Aleyrodoidea	Aleyrodidae	1,556	T	Ph
	Coccoidea	Coccidae	1,286	T	Ph
		Diaspididae	2,643	T	Ph
		Pseudococcidae	2,023	T	Ph
	Psylloidea	Psyllidae	1,205	T	Ph

[a], Martin and Mound (2007); Burckhardt (2009); Larivière et al. (2011); Garcia-Morales et al. (2016); Bartlett et al. (2018); Cassis (2019); Favret (2020); Ouvrard (2020)

A aquatic, H hematophagous, M mycetophagous, Ph phytophagous, Pr predatory, S semiaquatic, T terrestrial

and adult aggregation, and parental care behavior in treehoppers (Membracidae) and fulgoroids. Nymphal development in cicadas can be among the longest in insects, with so-called periodic cicadas taking up to 17 years to complete development (Williams and Simon 1995).

Auchenorrhynchans exhibit tight associations with their host plants (Dietrich 2009). The majority of species feed on phloem. There are also obligate xylem-feeders, as in the cicadas, cercopids, and a few cicadellid species. The majority of Cicadellidae species are mesophyll-feeders. Host range is often narrow and many species are host plant specific or restricted to species in a family or genus; cicadellids are also known to develop on alternate hosts. As in other hyperdiverse insect lineages, there are also highly polyphagous species. Hosts include species belonging to most higher green plant groups, with the majority feeding on flowering plants. Species of Cicadellidae (Membracoidea) and Delphacidae (Fulgoroidea) are dominant herbivores in grasslands (Dietrich 2009; Cavichioli and Takiya 2012). Leafhoppers and planthoppers are among the main insect vectors of plant pathogens, transmitting viruses, bacteria, and mycoplasmas (Dietrich 2009; Bartlett et al.

Fig. 12.1 Examples of hemipteran diversity. (**a–c**) Auchenorrhyncha: (**a**) Cicadomorpha: Cicadoidea: *Guyalna bonariensis* (Cicadidae); (**b**), Cicadomorpha: Membracoidea: *Adippe alliacea* (Membracidae); (**c**),Fulgoromorpha: Fulgoroidea: *Rhynchomitra microrhina* (Dictyopharidae). (**d–m**) Heteroptera: (**d**), Cimicomorpha: Miroidea: *Lygus lineolaris* (Miridae); (**e**), Cimicomorpha: Reduvoidea: *Debilia* sp. (Reduviidae); (**f**), Dipsocoromorpha: *Schizoptera* sp. (Schizopteridae); (**g**), Gerromorpha: Gerroidea: *Paravelia itatiayana* (Veliidae); (**h**), Leptopodomorpha: *Saldula* sp. (Saldidae); (**i**), Nepomorpha: Nepoidea: *Belostoma lutarium* (Belostomatidae); (**j**), Nepomorpha: Naucoroidea: *Picrops usingeri* (Naucoridae); (**k**), Pentatomomorpha: Lygaeoidea: *Ischnodemus bosqi* (Blissidae); (**l**), Pentatomomorpha: Coreoidea: *Acanthocephala* sp. (Coreidae); (**m**), Pentatomomorpha: Pentatomoidea: *Edessa rufomarginata* (Pentatomidae). (**n–p**), Sternorrhyncha: (**n**), Aphidoidea: *Aphis gossypii* (Aphidae), attended by ants; (**o**), Coccoidea: *Ceroplastes* sp. (Coccidae); (**p**), *Planococcus* sp. (Pseudococcidae). Credits. Photos **a**, **b**, **l**, **m**, **o**: C.F. Schwertner. Photos (**c**, **d**): K. Schulz (iNaturalist Website CC BY 4.0). Photo (**e**): V. Castro; Photos (**f**, **h**, **k**): by Renan Carrenho. Photo (**g**): F. Albertoni. Photo (**i**): F. Estefanello. Photo (**j**): by H. Rodrigues. Photo (**n**): J. Rorabaugh (iNaturalis Website CC0 1.0). Photo (**p**): F.F. Sarabia (iNaturalist Website CC BY 4.0)

2018). There are few gall inducers in the suborder, all included in the leafhopper family Cicadellidae (Matsukura et al. 2010; Rakitov and Appel 2012).

The suborder **Coleorrhyncha**, also known as moss bugs, comprise 17 genera and 37 species of minute (2–5 mm) and morphologically autapomorphic bugs (Burckhardt 2009; Burckhardt et al. 2011; Larivière et al. 2011). These species are nested in a single family, the Peloridiidae (Table 12.1). Peloridiids are exclusively phytophagous and terrestrial and are a textbook case of a southern Gondwanan group, recorded from the Australian region (Australia, New Caledonia, and New Zealand) and southern South America (Patagonia in Argentina and Chile). These cryptic hemipterans live in temperate and subantarctic rainforests, particularly those dominated by *Nothofagus* (Nothofagaceae), where they feed on mosses and probably hepatics, lichens, sphagnum bogs, and ferns (China 1962; Evans 1982). They are mostly ground-dwelling and have been collected in leaf litter, wet rock faces, surface roots, tree trunks, branches, and logs. Peloridiids are found in lowland, montane, and subalpine habitats and have low dispersal capability; they are nearly all flightless, with jumping behavior recorded in an Australian species (Burrows et al. 2007).

The suborder **Heteroptera** (true bugs) comprise 45,000+ species, in more than 6000 genera, making it the most speciose lineage of hemimetabolous insects. Most species are terrestrial, but there are also multiple aquatic and semiaquatic lineages, as well as ectoparasitic and inquilinous lineages (Schuh and Slater 1995; Henry 2017; Cassis 2019). Lifestyle diversification is a standout feature of the Heteroptera, differentiating them from the other hemipteran suborders. Diagnostic characteristics of the Heteroptera include the labium arising from the front of the head, the presence of a distinct gula (post-buccal region), the forewing typically modified into hemelytron (tegminal or reduced in some), scent glands openings on the abdominal terga in nymphs, metathoracic scent glands in adults, and enlarged and typically triangular scutellum (often covering most of the abdomen). True bugs have a broad range in body length (0.5–130 mm), including some of the largest living insects (Belostomatidae: *Lethocerus*). Common groups include assassin bugs (Reduviidae, Fig. 12.1e), bed bugs (Cimicidae), chinch bugs and relatives (Lygaeoidea, Fig. 12.1k), flat bugs (Aradidae), giant water bugs (Belostomatidae, Fig. 12.1i), lace bugs (Tingidae), leaf-footed bugs (Coreidae, Fig. 12.1l), minute pirate bugs (Anthocoridae), plant bugs (Miridae, Fig. 12.1d), shore bugs (Saldidae, Fig. 12.1h), stink bugs and relatives (Pentatomoidea, Fig. 12.1m), water striders (Gerridae and Veliidae, Fig. 12.1g), water boatmen (Corixidae) and water scorpions (Nepidae). There are also cryptozoic lineages, which are less encountered, such as the unique-headed bugs (Enicocephalidae) and the minute litter bugs (Dipsocoridae, Schizopteridae, Fig. 12.1f).

Within the Hemiptera, the true bug suborder is the most diverse group ecologically, as evidenced by their food preferences, habitation, and ecomorphological adaptations (Cassis 2019). True bugs live in a remarkably wide range of habitats: on plants, above- and belowground, leaf litter, bird nests, spider and embiid webs, caves, at the surface or submerged in freshwater, on the surface of oceans, on riparian habitats, as well as in the built environment (Schuh and Slater 1995; Weirauch et al. 2019). This ecological diversity is in undoubtedly correlated with morphological variability in the Heteroptera (e.g., Fig. 12.1d–m).

As with many hemimetabolous groups, nymphs resemble adults in structure and coloration and live in similar environments; however, in some cases such as in the stink bug family (Pentatomidae), early instars can exhibit aposematic coloration and are thereby differentiated from conspecific adults. Adults and nymphs also have similar behavior and are generally active; there are exceptions such as bat bugs (Polyctenidae) and some lace bugs that are more sedentary (Schuh and Slater 1995). Brooding behavior and parental care is common in some groups, as well as nymphal and adult aggregation. An unusual form of parental care exists in giant water bugs (Belostomatidae) where females lay their eggs on the back of males, which in turn attend them. Species in several heteropteran families are associated with ants and ant nests. In comparison with other hemipterans, ant mutualism is uncommon, recorded in but a few coreids and pentatomoids (e.g., Guerra et al. 2011).

Other specialized adaptations in the suborder include obligatory association with termites (Termitaphididae, Reduviidae: Tegeinae) and living in webs of spiders and embiopterans (Plokiophilidae, Reduviidae: Emesinae in part and one genus in Miridae and Nabidae, respectively). One of the most specialized behaviors occurs in the infraorder Cimicomorpha, where traumatic insemination has evolved independently in three separate lineages (Schuh and Slater 1995). As in other insects, true bugs modulate their behavior through chemical communication, with the production and release of aggregation and sex pheromones, as well as antipredatory behavior with the release of defensive allomones (Schuh and Slater 1995; Cassis 2019).

Food preferences in Heteroptera are remarkably broad, and unlike their allies, they feed on plants, fungi, and animals. Their feeding habits are often uniform or almost so at higher taxonomic levels (Weirauch et al. 2019). For example, the less speciose and earliest diverging infraorders (Enicocephalmorpha, Dipsocoromorpha, Gerromorpha, Nepomorpha, and Leptopodomorpha) are almost exclusively predaceous (some water boatmen feed on algae). The majority of superfamilies and/or families in the land bug infraorders Cimicomorpha or Pentatomoidea are either uniformly predaceous, phytophagous, or mycetophagous, with notable exceptions such as the hyperdiverse family Miridae (Table 12.1), where all prey types are utilized, with some lineages mixed-feeders (e.g., Dicyphini).

Predatory groups primarily feed on other arthropods (including other hexapods, crustaceans, millipedes, mites, and spiders), as well as snails, and occasionally vertebrates. Giant water bugs are frequently observed preying on fishes, frogs, turtles, and even snakes. Hematophagy has arisen at least three times among predaceous groups (Cimicidae, Polyctenidae, and Reduviidae: Triatominae) and one time within a phytophagous clade (Rhyparochromidae: Cleradini) (Schuh and Slater 1995; Weirauch et al. 2019). Host specificity is atypical among predatory groups and most are generalist feeders. Some predatory bugs specialize on higher groups of arthropods. For example, the assassin bug tribe Holoptilinae is one of the few groups of insects that are ant-feeders. Members of the reduviids are also known to specialize on termites (Tegeinae) or millipedes (Ectrichodinae). This is also the case for hametophagous lineages, such as the bedbug family Cimicidae that usually feed on bats or birds, aside from two bedbug species that specialize on humans. The evolutionary transition to predation within a phytophagous clade is known in the stink bug (Pentatomidae) subfamily Asopinae, many of which are generalist predators of lepidopteran larvae. Secondary predatory behavior also occurs in the Miridae (e.g., Deraeocorinae) and the Lygaeoidea (Geocorinae), including the arachnophagy in the lygaeoid family Berytidae.

Phytophagy is the main feeding strategy in the Heteroptera, accounting for c. two-thirds of species in the suborder (Schuh and Slater 1995; Weirauch et al. 2019). Like their subordinal allies, many heteropterans feed on sap mobilized in plant vessels, particularly in phloem, with only a few xylem-feeders. Many of the smaller bodied phytophagous lineages are mesophyll-feeders, such as most species of the Miroidea (Thaumastocoridae, Tingidae, and Miridae). Many of the pentatomomorphan phytophagous lineages are adapted to feed on tissues richer in nitrogen than sap, preferring developing fruits and seeds, as well as somatic tissues that mobilize nitrogen (Schaefer 2009).

Host plant specialization is well known in the phytophagous lineages of the Cimicomorpha and Pentatomomorpha (Cassis 2019). Most species feed on rosids and asterids, which account for about two-thirds of all living flowering plants, with secondary specialization on grasses and their allies (Poales). Food preferences vary at taxonomic levels, with species in the Miridae and Tingidae known to be host plant specific or oligophagous with preferential host switching to related plants. Grass-feeding lineages are well known in the Miridae and within numerous families of the Lygaeoidea, also in some groups of Pentatomidae (Schuh and Slater 1995; Weirauch et al. 2019). In many groups of the Pentatomoidea, host plant utilization is broad, and there are many generalist feeders. This is best seen in the Pentatomidae, which includes a number of highly pestiferous species that are pod-feeders on a broad range of human food crops, such as the green vegetable bug (McPherson 2018). In comparison with sternorrhynchans, there are few gall inducers, with most notable examples in the lace bug family Tingidae (Schaefer 2005); interestingly, three species of the stilt bug family Berytidae were recorded feeding on a Cecidomyiidae gall (Fernandes et al. 1987). In general, phytophagous heteropterans are not notable vectors of plant pathogens (Mitchell 2004), and their economic importance is a result of direct feeding on seeds or fruits.

Mycetophagy is restricted in the Heteroptera, having evolved independently in the Aradidae and the mirid subfamily Cylapinae, where both primarily feed on hyphae (Schuh and Slater 1995). Saprophagy is reportedly far less common and is known in sea skaters on open oceans as well as occasional instances in land bugs (e.g., Oxycarenidae; Gossner and Damken 2018), where it is likely of a facultative nature.

All species of the suborder **Sternorrhyncha** are terrestrial, plant-dwelling, and phytophagous (Gullan and Martin 2009; Wolff 2012; Hardy 2018). The suborder comprises c. 19,000 species in c. 2,150 genera. Diagnostic characteristics of the suborder include the labium inserted at the base of the head, between the procoxae;

one- or two-segmented tarsi; and the elongate and filiform antennae. Most species are small, with the body length ranging between 1 and 11 mm.

Sternorrhyncha are classified into four superfamilies (Table 12.1): aphids (Aphidoidea, Fig. 12.1n), jumping plant lice or psylloids (Psylloidea), scale insects (Coccoidea, Fig. 12.1o, p), and white flies (Aleyrodoidea). These abundant and ubiquitous insects are recognized by the position of the rostrum, apparently arising between the fore coxae (and hence, the origin of the name of the suborder) and the two segmented tarsi. However, these characters are not unique within Hemiptera, and usually the recognition of each of the four superfamilies is easier. Aphids have a soft oval body; a pair of abdominal cornicles (= siphunculi), projected posterodorsally; females are mostly wingless, at least in alternate generations; and males are winged. Jumping plant lice resemble small cicadas and leafhoppers, typically with multisegmented antennae, typical wing venation, and hind saltatorial legs. Scale insects and relatives are the only sternorrhynchans with pronounced sexual dimorphism; the female body being undifferentiated (without tagmosis); with paedomorphic development; usually with well-developed mouthparts; and body with a waxy protective covering ("scale" that is detached or adherent to the body). In contrast, males of scale insects have standard body organization (head-thorax-abdomen); reduced mouthparts; normal development; and membranous fore wings and hind wings, which are usually reduced (balancers or halteres), akin to small dipteran flies; males sometimes are apterous. Whiteflies are analogous to very small moths, with the body and wings usually covered with a white waxy secretion.

Adult and immature sternorrhynchans are usually strongly gregarious, living in small groups or colonies on their host plants; only some species are solitary (Gullan and Martins 2009; Wolff 2012; Hardy 2018). Aphids and immatures of jumping plant lice live on fresh plant growth, including buds, stems, and leaves; they rarely feed on roots. Scale insects are found in different plant parts, whereas whiteflies usually live and develop on the underside of leaves. Species of each higher clade show distinctive life history traits. For instance, aphids have annual or biannual life cycles, using either one (monoecious) or more than one (heteroecious) host plant, with or without alternation between sexual or asexual generations (holocyclic or anocyclic species, respectively). In these species, winged adults are produced when colony resources are scant. Aphids and some other groups of Sternorrhyncha may increase explosively due to parthenogenetic viviparity and telescoping generations.

Jumping plant lice do not have such complex life cycles, and the nymphs can be free-living, gall-inducing, or live inside a sugary or starchy cover excreted by the nymphs (= lerps). In scale insects and whiteflies, first instars are mobile and later instars are sessile. They undergo a form of complete development, which includes one or two nonfeeding pupa-like stages. Females of scale insects are sessile and may live for months to several years, whereas males live for a few days and are more mobile but do not feed. The pupa-like stage of whiteflies has been used to recognize genera and species; this stage is sometimes confused with scale insects; however, in the latter, they complete development into winged males and females. A widespread and very well-known characteristic among Sternorrhyncha is the excretion of a sugary liquid called honeydew, which often attracts ants (e.g., Fig. 12.1n), that may protect them from natural enemies; this kind of symbiont relationship is usually facultative, but several obligate cases are known in aphids and coccoids (Delabie 2001); in comparison, few species of whiteflies are ant-attended. Gall-inducing species are recorded among aphids and psyllids.

Nearly all sternorrhynchans are phloem-feeders, with a few species feeding on xylem or parenchyma (Gullan and Martins 2009; Hardy 2018). Psylloids include the majority of non-phloem-feeding species, while Diaspididae (Coccoidea), the most speciose family of scale insects, feed exclusively on parenchyma cells. Host specificity is common in all major sternorrhynchan groups, with oligophagous or polyphagous species. There are also highly polyphagous species, particularly in aphids. Psylloids exhibit an interesting difference in host breadth by life stage; nymphs tend to be specialist feeders, whereas adults of the same species may be oligophagous or polyphagous. Certain sternorrhynchan species are also known specialists on plant structures or growth stages. The economic importance of sternorrhynchans results from the loss of plant sap via continuous feeding or transmission of pathogens (Gullan and Martins 2009; Hardy 2018). In some cases, the overload of honeydew allows the spread of mold on host leaves, diminishing photosynthetic capacity. Scale insects are also sources of wax, shellac, and natural dyes, whereas some species have been successfully used as biocontrol agents of invasive plants (Hardy 2018).

12.2 Sampling Methods

Because of the ecological diversity of hemipterans, and the microhabitats and ecosystems they occupy, no single method is suitable for all subordinate taxa (Moir et al. 2005; Moir and Brennan 2007; Diez et al. 2018). It is therefore important to choose sampling methods that target the known or predicted habitat, distribution, or species interaction of a taxon (Table 12.2). The methods used are also predetermined by research parameters, which can vary by the nature of the study, if it has a taxonomic, ecological, or applied purpose (e.g., Barcellos 2006; Cohnstaedt et al. 2012; McCravy 2018). In the case of ecological studies, there are requirements such as survey design, adequacy and frequency of sampling, and the statistical analyses to be performed. For a quarantine investigation, researchers may target an invasive species through the use of visual inspection or pheromone lures, where presence or absence data are of importance. For taxonomic studies, field work focuses on maximizing the collection of species and specimens, including immature and adult life stages, and is not dictated by quantitative requirements other than collection of adequate number of specimens.

The collection of hemipteran species can be subdivided into aquatic and terrestrial habitats (Table 12.2), each requiring specialized equipment. Only the suborder Heteroptera has aquatic-adapted bugs (aquatic and semiaquatic), and it is typical for researchers to specialize on them. Aquatic heteropterists also work on a handful of heteropteran lineages that are interzonal (inshore, riparian species). There are two

Table 12.2 Sampling methods to collect aquatic, ground-dwelling, and plant-dwelling Hemiptera

Sampling method	Habitat			
	Aquatic	Ground	Plant-dwelling (canopy)	Plant-dwelling (understorey)
Aspirator	X	X		X
Beating sheets				X
Berlese funnel		X		
Canopy fogging			X	
Grass tufting		X		
Insect net	X		X	X
Intercept traps	X	X	X	X
Light trap	X	X	X	X
Malaise trap		X		
Manual collecting	X	X		X
Pitfall trap		X		
Pyrethrum knockdown		X		
Sieving litter		X		
Suction devices		X		X
Winkler		X		

Bartlett et al. (2018), Burckhardt (2009), Cassis (2019), Favret (2020), García Morales et al. (2016), Larivière et al. (2011), Martin and Mound (2007), Ouvrard (2020)

additional heteropteran infraorders (Dipsocoromorpha, Enicocephalomorpha) that are usually cryptic and ground-dwelling, associated with leaf litter, soil, fallen logs, and rotting organic matter. The majority of the remaining Heteroptera and the other suborders are mostly phytophagous and are associated directly or indirectly with plants, either on plants or on the ground foraging for fallen seeds or fruits. There are also highly specialized taxa that require considerable effort and time to capture, like cave and nest dwellers, root-feeders, and bugs that lives on bats.

Hemipteran taxonomy is chiefly based on adult morphology, although identification of immatures is possible at supraespecifi levels (Schuh and Slater 1995; Dietrich 2009; Gullan and Martin 2009); one exception is that of whiteflies, in which the taxonomy of species and genera level are based on the morphology of the fourth instar nymph (Gullan and Martin 2009). To maximize the capture of adults, collectors must rely on existing collection event data and knowledge about seasonality and number of generations per year. As for other insect groups, terrestrial hemipterans living in tropical regions are typically multivoltine, and their emergence is usually mediated by rainfall (rainy season); in dry conditions, individuals are usually scant or aestivating in aggregations. Latitude is an important factor for collecting bugs in subtropical and temperate regions, where seasonality (temperature, photoperiod) determines emergence and number of generations. Temperate hemipterans are often uni- or bivoltine, and development is also strongly influenced by biotic factors, such as resource availability and species interactions. Other important conditions for the collection of terrestrial hemipterans in both tropical and temperate regions include host preference and specialization, microhabitat availability, vegetation type, and the effect of altitude.

Most tropical aquatic and semiaquatic bugs are multivoltine, making it possible to collect them throughout the whole year. However, water bug specialists prefer collecting in the dry season, when populations are more stable due to lower flow of streams and rivers or lower volume of lakes and ponds. Under such conditions, individuals tend to be more concentrated, facilitating the collection of a larger number of specimens. Microhabitat conditions are also important indicators of taxon distribution, such as the presence of vegetation or riparian debris. There is also taxon differentiation across freshwater and marine gradients. In marine systems, collecting techniques are usually more specialized, sometimes requiring access to boats, as well as accommodating tides for inshore taxa.

The majority of collection techniques and taxon-appropriate equipment fall into two major categories, direct and indirect. Below follows the description of these sampling methods employed by hemipterists to maximize the capture of specimens.

12.2.1 Direct Methods

Direct methods refer to sampling by hand or through the use of a handheld equipment such as an insect net and a suction device or the use of insecticide applicators, from ground level to canopy (Table 12.2). Direct methods are designed to capture species in situ with collection of live specimens that are typically euthanized in the field. It has the benefit of allowing collectors to observe specimens alive and record natural history data. It targets a species microhabitat as well as the ecosystem where the microhabitats are predicted to exist.

The most widely used technique for catching bugs by the direct method is through the use of insect nets and beating vegetation, which are used in both aquatic and terrestrial habitats (e.g., Coscarón et al. 2009; Dem 2015). Specific equipment commonly used for direct collecting of hemipterans include forceps of different kinds and sizes, pins, sample vials or tubes with screw tops (empty or containing alcohol or other preservatives), small to medium containers (filled with soft paper to avoid damage to specimens), and small envelops for temporary storage of specimens (Cassis et al. 2016; Gibb and Oseto 2006). Other collecting equipment include cutting (e.g., hand pruners) and digging (e.g., trowel) tools for plant material (including stems and buds, galls, seed pods, roots). When searching for cryptic species, sifters and brushes are used to dislodge individuals from microhabitats and then place them in white trays, from where minute specimens can be searched for by eye or with a hand lens, in a contrasting white tray or sheet. Bags of different sizes, preferably cloth bags, are used for the collection and temporary storage of plant material, leaf litter samples, and other organic material, which can then be returned to the laboratory for further inspection (Gibb and Oseto 2006).

Fig. 12.2 Examples of Hemiptera sampling methods and equipment. (**a, b**) sweeping net; (**c**) beating sheet; (**d**) light sheet trapping, assembled close to the water; (**e, f**) malaise trapping (**e**, standard trap; **f**, double-headed trap, assembled to capture aquatic insects). Credits. Photo (**a**): J. Mourão. Photos (**b, d**): J.S. Prando. Photo (**c**): B. Genevcius. Photos (**e, f**): C.C. Gonçalves

Arguably the most indispensable item to efficiently collect small to mid-size hemipterans is the aspirator (or pooter), a tool that is used to selectively capture hemipterans of interest (Schuh and Slater 1995). It is typically used to collect live specimens that are difficult to capture otherwise and are most often used in conjunction with a sweep net (Fig. 12.2b) or beating sheet or directly from moss and hepatics, leaf mold, and rocks. It is also a useful tool for collecting using indirect methods, such as at light sheets and other traps. Fit for purpose equipment are also used for the group of Hemiptera under study. For example, ziplock bags are used for field collection of galls. Hand collecting for nocturnal species requires a flashlight or head lamp. Hereafter we divide direct collection methods into those for plant-dwelling, ground-dwelling, aquatic and semiaquatic species.

12.2.1.1 Plant-Dwelling Species

Hemipterans associated with land plants are typically found feeding on vegetative (leaves, stems, tree trunks and branches, on or under bark) or reproductive tissues. Other plant-dwelling hemipterans include predators of herbivores, mycetophagous species, and gall-inducing species (see Chap. 16). Although most hemipteran phytophagous species are active, there are numerous less mobile lineages such as sessile females of scale insects.

Manual Collecting This involves the detection of specimens by visual inspection of the vegetation (Moir et al. 2005; Coscarón et al. 2009). This can be achieved by searching for evidence of plant damage, such as chlorosis caused by lace bug

(Tingidae) mesophyllic feeding. It can also be achieved through prior knowledge of tissue preferences, such as searching on seed pods for granivores (e.g., Hemiptera: Lygaeoidea in part).

Less mobile or sessile species (e.g., scale insects and immatures of psylloids) are best collected by turning leaves or examination of stems; this is a standard method where abundances are low and beating may be less effective. This is also true for rainforest species which feed on broadleaf trees, where few individuals are encountered. Underside of leaves is the favored habitat for certain hemipteran groups such as lace bugs.

There are many heteropteran species found under bark, where they can be either active, aestivating, or overwintering (e.g., some assassin bugs, bark bugs, plant bugs, stink bugs, and relatives). They are best detected by stripping bark into a white plastic tray or on a drop sheet around the base of a tree where fallen bugs can be dislodged away from the stripped bark.

Manual collecting can be used to sample insects from other cryptic microhabitats found on plants, e.g., wet moss and hepatics in tree branches and trunks. Spraying alcohol from a squeeze bottle can be helpful to collect very active species that live in these microhabitats, which are immobilized and then collected using a soft forceps, a paint brush, or an aspirator. For larger cryptic species, specimens need to be collected directly into a vial using more robust forceps or by hand.

Insect Sweep Nets (Figs. 12.2a, b) These are generally the most productive direct method tools for capturing active plant-dwelling hemipterans on low and soft vegetation, particularly forbs and monocots (grasses, sedges) (e.g., Stewart 2002; Moir et al. 2005; Barcellos 2006; Chanthy et al. 2013; Diez et al. 2018). Sweeping can also be used to catch bugs in flight, using a back-and-forth motion. Different types of net can be used, depending on the environment, characteristics of the vegetation, and taxa to be sampled. This method is highly effective in capturing auchenorrhyncans and active terrestrial true bugs.

Beating Sheets (Fig. 12.2c) Beating sheets coupled with a beating stick are the most useful equipment for capturing bugs from leaves and phyllodes typical of woody perennial shrubs and trees (e.g., Moir et al. 2005; Chanthy et al. 2013; Diez et al. 2018). The beating sheet is placed below a tree or shrub, which is struck multiple times with the beating stick. As a result, bugs fall onto the beating sheet and then collected. Beating sheets can be of different types; the most common is made of a square of white thick cloth material, held by two lateral or crossed wooden bars. Another type is the beating net, an oval bag of short depth with a short handle. In both types, bugs are best collected using an aspirator or directly into a vial if the bug width is greater than the diameter of the inlet tube of the aspirator.

Suction Devices This type of equipment is used for collecting ground-dwelling hemipterans but also those near to the ground in association with vegetation (e.g., Stewart 2002; Diez et al. 2018). It is commonly used in sampling grasslands to search and count hemipterans of economic importance in agroecosystems.

Pyrethrum Knockdown This is an insecticide-based method that focuses on cryptozoic heteropteran species, from microhabitats such as fallen logs and rock ledges. It can also be used to collect bugs associated with bark and other plant microhabitats. This method works similarly to bark peeling, placing a sheet under the selected site where the insecticide will be sprayed; the bugs fall onto the drop sheet and can then be collected. This method should be used with caution, since it can cause unwanted bycatch and the unnecessary killing of nontarget taxa.

Canopy Fogging Knockdown methods were originally designed for collecting insects in tree canopies, not accessible at ground level (see Chap. 18). This involves the use of industrial scale foggers or misters hoistered into the canopy, with large conical collectors at ground level. This technique is designed for general insect collecting and has resulted in capturing functional groups and taxa rarely or not previously seen, including hemipterans (e.g., Coscarón et al. 2009). Despite its efficiency when the environmental conditions are favorable, this method has not been widely used because of its costs and challenges with obtaining approval by regulatory authorities. Natural pyrethrum is the best insecticides to be applied as it quickly degrades within a few hours after application, leaving no lasting residues.

12.2.1.2 Ground-Dwelling Species

Hemipteran species live at different layers of the ground profile, from the superface and shallow layers of leaf litter to more deep layers of compacted soil; these include microhabitats such as over and below fallen logs and rocks, rotten bark and wood, leaf mold, rock surfaces and ledges, ground mosses and hepatics, turfs, grass tussocks, and vegetation debris (Schuh and Slater 1995; Dietrich 2009; Gullan and Martin 2009). Functional bug groups that inhabit such microhabitats include predators, as well as mycetophagous, saprophagous, and phytophagous species. They can feed on post-dispersal seeds, roots, and other plant parts and structures near ground. Predatory and hematophagous species, living near and in nests, caves, or human households, are usually collected using the sampling methods described in this section.

Manual Collecting This method can be used to collect bugs from cryptic microhabitats; for example, under rocks, ground plants, and litter (e.g., Coscarón et al. 2009). In the case of economically important species such as burrower bugs (Cydnidae), manual collection can be used as there is prior knowledge of the host plant (e.g., Avila et al. 2016). However, in general, it is difficult to target the collection of lower levels of the ground profile and where taxa such as cydnids and cicada nymphs are known to inhabit but difficult to find. Cydnids are more readily collected at light and cicada nymphs once they have emerged.

Sieving Litter This method is used for collecting ground-dwelling insects. Litter is gathered by hand and placed into a sifter, which comprises a nylon bag and mesh, and sieved materials are deposited in a white plastic tray from which specimens are collected or placed in an extraction funnel. The final step is accomplished using an aspirator or a forceps (also see Berlese/Tullgren funnel below). Different groups of Hemiptera have been collected using this method (e.g., Kozár and Miller 1999; Kment et al. 2016).

Grass Tufting This method is used to collect cryptozoic hemipterans that live in leaf sheaths. This method is useful in capturing minute litter bugs, some groups of lace bugs, leaf-footed bugs, and lygaeoids (G Cassis, unpublished).

Pyrethrum Knockdown This is similar to the method applied to collect plant-dwelling species; it is used to collect cryptozoic species that live in ground-related microhabitats.

12.2.1.3 Aquatic and Semiaquatic Species

True bugs (Nepomorpha, Gerromorpha) in aquatic environments include predatory species, occupying different microhabitats in either slow- (lentic) or fast (lotic)-flowing waters (Moreira et al. 2018; Cassis 2019). Many aquatic taxa specialize by water flow, although some occur in both lentic and lotic waters. Microhabitat conditions are also important indicators of taxon distribution, such as the presence of vegetation (creeping water bugs, marsh treaders) or riparian debris (water bugs of the genus *Lethocerus*). Aquatic and semiaquatic true bugs can also include some species that live on phytotelma, caves, coastal marine environments (e.g., coral), as well as the open ocean (Andersen and Cheng 2004).

Collecting methods for aquatic bugs are various and habitat dependent but usually include the use of simple nets and active searching (see Chap. 20). A major exception is for oceanic research, since research vessels and neuston nets are used. Several species of shore bugs and some groups of semiaquatic bugs (e.g., Macroveliidae: *Chepuvelia*; Mesoveliidae: *Cryptovelia*; Veliidae: *Altavelia*), despite their common name, are actually terrestrial, being found away from water bodies, and can be collected by the methods traditionally used for other land hemipterans.

Manual Collecting This method can be used to actively sample the madicolous, hygropetric, or marginal fauna, including leptopodomorphans, ochterids, gelastocorids, and many gerromorphans. These true bugs live on surfaces where using nets or sieves is not possible and must be searched for and sampled individually with forceps, brushes, or aspirators.

Aquatic Insect Nets More robust nets and sieves with larger mesh size are useful typically for larger specimens, such as belostomatids, nepids, and gerrids. Smaller nets and sieves are used for minute bugs, such as helotrephids, pleids, and some veliids, and are particularly useful for reaching into small cracks and crevices. Important features of aquatic nets are size of the mesh and handle extension to enable longer reach. When sampling on the surface or on the margins of water bodies, specimens can be collected almost free of debris and easily caught from the nets. However, for those that live submerged, either on the water column or on different substrates, content from the nets should be transferred to a plastic tray with a bit of water. This type of collecting requires patience to ensure that the sampling of the diversity and abundance of these bugs is fully realized. For example, nepids tend to become immobile in the tray and can be easily confused with branches or sticks, whereas minute bugs such as pleids, helotrephids, potamocorids, and many corixids are more easily viewed when swimming; the latter can be difficult to detect as they typically hide among sand, gravel, or vegetable matter.

12.2.2 Indirect Methods

Indirect methods focus on the collection of specimens by trapping or extraction methods that either intercept, attract, or involve the removal of habitat material from which specimens are extracted into a collecting container by light or gravity. Attracting bugs to light or other media (e.g., yellow pan traps or sticky traps) is a conventional and effective method for specimen collection and is useful for smaller heteropterans belonging to the Auchenorrhyncha and Sternorrhyncha. Where collectors are targeting a particular taxon (e.g., stink bugs in agroecosystems), pheromone lures can be used; such lures are however primarily developed for species of economic significance. Intercept traps are designed to capture insects in motion, either walking, running, or flying, with a variety of commercial traps available, which can also be handmade at a more cost-effective rate. These include traps above ground level (e.g., malaise traps, flight intercept traps) or at ground level (e.g., pitfall traps).

Below, we describe the indirect methods most used for collecting and sampling hemipteran bugs on plants and at ground level:

12.2.2.1 Plant-Dwelling Species

Branch Traps These are used for sampling trees at different heights, from the canopy to ground level, focusing on bugs associated with tree branches. The trap consists of a collar tightly fitted to the branch, a plastic funnel placed beneath the collar (about 5 cm), and a container with a preservation fluid (Koponen et al. 1997; Koponen 2004). This sampling method was successfully used by Gossner (2006) to complement the collection of Heteroptera in canopies.

12.2.2.2 Ground-Dwelling Species

Pitfall Traps These are used to collect active ground-dwelling true bugs and some groups of auchenorrhynchans (e.g., Stewart 2002; Coscarón et al. 2009; Michelotto et al. 2019; Salomão et al. 2019). The trap consists of small plastic buckets or cups that are filled with water and set in the ground with the rim flush to the surface. To avoid flooding during rainy days and minimize drying out when the weather is warm, it is often topped with a roof such as a pot plant saucer or a disposable plastic plate. The traps can be either partially filled with a preservative such as propylene glycol (less subject to evaporation) or added with liquid soap that lowers the water's surface tension. Daily removal of the captured insects and changing its content are ideal for traps without preservatives.

Berlese/Tullgren Funnel These devices are designed to separate or extract specimens from substrate such as leaf litter, shore detritus, or dung (Gibb and Oseto 2006). Similar to a sifter, with a cone at the bottom of the cylinder, it is connected to a container (usually fill with preservation fluid). The top part of the funnel is fitted with a light bulb, which, when turned on, drives bugs into the bottom collecting container. The extraction time typically lasts for 24–48 h. These methods have been efficient to collect different groups of hemipterans (e.g., Stewart 2002; Coscarón et al. 2009).

Winkler Funnels These work without a heat source and extraction of the specimens is accomplished by gravity, assisted by occasional turning of the substrate material by hand. These funnels are ideal for collecting moss bugs, minute litter bugs, and unique-headed bugs in the field (Emsley 1969; Wygodzinsky and Schmidt 1991; Burckhardt 2009; Larivière et al. 2011).

12.2.2.3 Indirect Methods Used for Collecting Active Hemipterans

Light Sheet Trapping (Fig. 12.2d) This method has been successfully applied to capture a wide range of families of Hemiptera (e.g., Stewart 2002; Lucas et al. 2016; Michelotto et al. 2019), including aquatic true bugs. The method consists of a light bulb (mercury vapor, black light) hung in front of a white cloth (usually 2 m × 2 m) or wall; attracted specimens are collected by hand or with an aspirator. There are ideal conditions for using light traps that include a combination of high humidity, high temperatures, and still and moonless nights. Aquatic and semiaquatic bugs have been also collected by using light traps set up close to water; commonly collected taxa include genera of Belostomatidae, Corixidae, Gerridae, Hebridae, Hydrometridae, Veliidae, Nepidae, Notonectidae, and Saldidae.

Passive Light Traps These are designed for collecting specimens over a long period and left unattended. Several designs of these traps are available, including simple devices such as light bucket traps (e.g., Cassis et al. 2016). There are also more elaborate passive light traps, which can be self-assembled or are available commercially. Disadvantages of these passive traps include the collecting of nontarget insects, which results in time-consuming processing and damage to specimens; specimens are often coated with lepidopteran scales.

Pheromone Lures These are used to attract hemipterans, such as stink bugs and mealybugs. In the main, this method is still poorly explored for the vast majority of bug taxa. It is, however, a commonly used method for detection and monitoring of species of economic importance in agroecosystems (e.g., Penca et al. 2020). It is efficient where collecting is targeted for particular taxa and in non-heterogeneous environments. The development of pheromone lures requires significant research and investment and is not commonly used to collect hemipterans in natural settings.

Baited Traps Such traps are used to lure insects with food additives. Although these methods have been poorly reported for collecting Hemiptera, there are examples in the literature for the detection and monitoring of important groups such as kissing bugs (Reduviidae: Triatominae) (Noireau et al. 2002; Arias et al. 2012).

Sticky Traps These are commercially available and are known to attract pest insects, such as aphids and other sternorrhynchans and auchenorrhynchans (e.g., Stewart 2002; Moir et al. 2005; Trebicki et al. 2010). Other hemipterans occasionally are captured in sticky traps in natural or modified environments. The traps may also include some type of bait to improve attractiveness and selectivity.

Yellow Pan Traps These are small bowl-shaped plastic containers that are yellow in color and three-fourths filled with water and a drop or two of detergent. These are useful for collecting Fulgoroidea and Membracoidea (Trebicki et al. 2010). They are typically left out for a few hours, and specimens are strained and then transferred to a vial with alcohol.

Flight Interception Traps This type of traps is designed to intercept insects in flight, regardless of the habitat they live in (Gibb and Oseto 2006). The most popular of these are malaise traps (Fig. 12.2e, f), which are bottomless tentlike traps made of mesh material, designed to capture insects that stereotypically move upward once intercepted and fall into top-positioned containers with some preservative fluid. This is an efficient method to capture active auchenorrhynchans (Stewart 2002; Cavichioli and Takiya 2012) but is of little use in capturing true bugs. Other flight interception traps are useful in catching flying hemipterans. Cross-vane traps have been successfully used to sample Heteroptera in different environments and conditions (e.g. Gossner 2006, 2008, 2009; Vialatte et al. 2010; Stubbins et al. 2016).

12.3 Preparation, Curation of Specimens, and Identification

Once collected in the field, hemipterans can be euthanized through the application of a killing agent or by immersion in alcohol for wet preservation. If specimens are brought alive to a facility, freezing is also an option for killing specimens. Vials, tubes, or other containers used to store and transport specimens should be properly labeled with the collection event data, backstopped with a field note book or electronic device. Label data for each collection event usually include the following minimum data: detailed locality data including georeferenced spatial data, altitude (meters above sea level), date, collector(s) with the family name (essential) and initials for given names (optional), collecting technique, habitat, host plant data preferably with a binomial or at least genus-group name as well as a family-group name in brackets, and a field-based voucher code if used.

For systematic hemipterists that collect plant-associated taxa, it is essential to also collect vouchers of host plants, which can be identified by botanists and vouchered in registered collections. As such, it is very important to collect material in a manner that meets the curatorial requirements of an herbarium.

Adults are usually dry-mounted on pins or point mounts, for both terrestrial and most aquatic bugs. For more delicate species and immatures, wet preservation in alcohol at 70–80% concentration is used for most groups; aphids need to be preserved in alcohol concentration of 90–95%. Specimens collected for DNA sampling, as usual for other insects, should be preserved in alcohol at 95–100% concentration or other DNA preservative. Alcohol preservation for plant bug (Miridae) specimens is highly discouraged, unless for DNA sampling. For scale insects and other sessile bugs collected together with host plant parts, the specimens may be left to dry in the plant press; the vouchers of the hosts are stored with the specimens in glass vials, card, or wood boxes. Aphids, scale insects, and other coccoid families can also be mounted and preserved (and studied) in permanent slide mounts.

In nearly all groups of Hemiptera, it is often necessary to examine the genitalia to identify species. Material in alcohol is more easily examined because the specimen is soft; however, in pinned specimens, it is necessary to soften the insect abdomen before dissection. The pinned specimen can be relaxed using Barber's fluid or by immersion in 70% ethyl alcohol for 6–10 h (larger specimens requiring up to 24 h). A faster option is to heat water to a temperature of approximately 80 °C and then immerse the specimen for a few minutes. To examine taxonomic features, it is generally not necessary to clear the genitalia. However, if observation of internal structures is necessary, a cold solution of 10% KOH can be used (referred to as maceration). The maceration time is dependent on the degree of sclerotization. Robust specimens can be left for up to 10 h, whereas small soft species will require far less time (as little as 5 min for soft-bodied taxa such as most mirids). One can heat the KOH solution to speed up the process of maceration, but one needs to take care as this can often result in overclearing of the specimen.

After maceration, the whole abdomen or the structure dissected is then washed in clean water containing a small amount of glacial acetic acid, and then pure water, and immersed in glycerine; washing series in alcohol at different concentrations before the immersion in glycerine is also applied for some groups. After examination, the cleared structure is stored in a glass or plastic microvial pinned beneath the rest of the specimen.

12.4 Concluding Remarks

Hemiptera is the most speciose group of hemimetabolous insects and an important part of terrestrial and aquatic ecological communities, and also includes some of the most important crop pests (e.g., Brown Marmorated Stink Bug) and vectors of human diseases (e.g., Chagas disease; see Chap. 21) (Bartlett et al. 2018; Hardy 2018; Cassis 2019). It is a major component of global species diversity, with species often ubiquitous and with high abundances (Moir and Brennan 2007). They are found in most levels of habitat strata and are important functional groups in food webs. The important economic and medical groups cause billions of dollars of loss on an annual basis. They display a wide array of feeding strategies and are sensitive to habitat disturbance.

We have summarized the array of sampling methods used to capture this remarkable order of insects. It is a fundamental requirement to adequately design and continually refine these methods so that bugs can be efficiently collected. Because of the diversity of habits, behavior, and habitats of hemipterans, it is important to plan any sampling study in advance, defining the aim of the study, the target groups of interest, which habitats will be sampled, as well as the landscape and environment settings.

In this chapter, we summarize sampling methods employed by hemipterists to maximize the capture of specimens, in different habitats and conditions (Table 12.2). We describe the individual equipment commonly used, following the description of the main sampling methods by habitat types (plant-dwelling, ground-dwelling, and aquatic habitats) and other methods used regardless of the habitat that they are found in.

Our contribution provides a modern overview of the collection methods for the entire hemipteran order, relevant to environmental and economic issues (e.g., Hallmann et al. 2017; van Klink et al. 2020). Hemiptera are still a group that requires substantial basic taxonomic research, and in all likelihood less than half of the species that exist in nature are described globally. This taxonomic impediment hampers our broad understanding of the evolution, bionomics, economics, and management of this fascinating group of hemimetabolous insects. Our intent is for this contribution to support and encourage other studies of the Hemiptera, under any condition.

Acknowledgments The authors would like to thank the editors for the opportunity to participate in this project and for all their patience throughout the process to finalize the manuscript. This work was partially funded by FAPESP (Proc. 2015/50322-0) and CNPq (Proc. 421413/2017-4). CFS thanks the CNPq research Fellowship program (313214/2017-4), and RC thanks the scholarship from Coordenação de Aperfeiçoamento de Pessoal de Nível Superior—BRASIL (CAPES)—Finance Code 001. FFFM benefited from grants provided by the State of Rio de Janeiro Research Foundation, FAPERJ (#210.508/2016 and #203.207/2017), and the National Council for Scientific and Technological Development, CNPq (#301942/2019-6). Tatiana Ruschel and Wanessa da Silva Costa are gratefully acknowledged for their help in the identification of Fig. 12.1a, f, respectively. We also thank the following colleagues who kindly provided photos to illustrate this chapter: Clayton C. Gonçalves, Fabiano Stefanello, Fernando Albertoni, Higor Rodrigues, Jádila S Prando, Juliana Mourão, and Valentina Castro.

References

Andersen NM, Cheng L (2004) The marine insect *Halobates* (Heteroptera: Gerridae): biology, adaptations, distribution, and phylogeny. Oceanogr Mar Biol Annu Rev 42:119–180. https://doi.org/10.1201/9780203507810.ch5

Arias AR, Abad-Franch F, Acosta N et al (2012) Post-control surveillance of *Triatoma infestans* and *Triatoma sordida* with chemically-baited sticky traps. PLoS One 6:1–10. https://doi.org/10.1371/journal.pntd.0001822

Avila CJ, Xavier LMS, Santos V (2016) Fluctuation and vertical distribution of a population of brown root stink bug *Scaptocoris castanea* (Hemiptera: Cydnidae) in the soil profile in Mato Grosso do Sul State, Brazil. Entomotrop 31:40–47

Barcellos A (2006) Hemípteros terrestres. In: Becker FG, Ramos RA, Moura LA (eds) Região da Lagoa do Casamento e dos Butiazais de Tapes, Planície Costeira do Rio Grande do Sul. Ministério do Meio Ambiente, Brasil, pp 198–209

Bartlett CR, Deitz LL, Dmitriev DA, Sanborn AF, Soulier-Perkins A, Wallace MS (2018) The diversity of the true hoppers (Hemiptera: Auchenorrhyncha). In: Foottit RG, Adler PH (eds) Insect biodiversity: science and society, vol 2. Wiley, Oxford, pp 501–590. https://doi.org/10.1002/9781118945582.ch19

Brambila J, Hodges G (2008) Bugs (Hemiptera). In: Capinera JL (ed) Encyclopedia of entomology, 2nd edn. Springer Science & Business Media, Dordrecht, pp 591–611

Burckhardt D (2009) Taxonomy and phylogeny of the Gondwanan moss bugs or Peloridiidae (Hemiptera, Coleorrhyncha). Dtsch Entomol Z 56:173–235. https://doi.org/10.1002/mmnd.200900019

Burckhardt D, Bochud E, Damgaard J et al (2011) A review of the moss bug genus *Xenophyes* (Hemiptera: Coleorrhyncha: Peloridiidae) from New Zealand: systematics and biogeography. Zootaxa 2923:1–26. https://doi.org/10.11646/zootaxa.2923.1.1

Burrows M, Hartung V, Hoch H (2007) Jumping behaviour in a Gondwanan relict insect (Hemiptera: Coleorrhyncha: Peloridiidae). J Exp Biol 210:3311–3318. https://doi.org/10.1242/jeb.007914

Cassis G (2019) True bugs (Insecta: Hemiptera: Heteroptera): evolution, classification, biodiversity and biology. Reference module in life sciences. Elsevier, Amsterdam, pp 1–10. https://doi.org/10.1016/B978-0-12-809633-8.20710-3

Cassis G, Reid C, Symonds C et al (2016) Entomology methods and diagnostics manual. biodiversity capacity building in Papua New Guinea and sustainable development of its primary industries, University of New South Wales, Sydney. https://www.discoverlife.org/png/. Accessed 1 May 2020

Cavichioli RR, Takiya DM (2012) Hemiptera. In: Rafael JA, Melo GAR, Carvalho CJB et al (eds) Insetos do Brasil, diversidade e taxonomia. Holos Editora, Ribeirão Preto, pp 359–368

Chanthy P, Martin RJ, Gunning RV et al (2013) Arthropod survey on soybean crops in Cambodia: a comparison of the sweep netting and beat sheeting collection methods for estimating arthropod diversity and species richness. Aust J Entomol 52:299–308. https://doi.org/10.1111/aen.12035

China WE (1962) South American Peloridiidae (Hemiptera-Homoptera:Coleorrhyncha). Trans R Entomol Soc Lond 114:131–161. https://doi.org/10.1111/j.1365-2311.1962.tb01077.x

Cohnstaedt LW, Rochon K, Duehl AJ et al (2012) Arthropod surveillance programs: basic components, strategies, and analysis. Ann Entomol Soc Am 105:135–149. https://doi.org/10.1603/AN11127

Coscarón MDC, Melo MC, Coddington J et al (2009) Estimating biodiversity: a case study on true bugs in Argentinian wetlands. Biodivers Conserv 18:1491–1507. https://doi.org/10.1007/s10531-008-9515-0

Delabie JH (2001) Trophobiosis between Formicidae and Hemiptera (Sternorrhyncha and Auchenorrhyncha): an overview. Neotrop Entomol 30:501–516. https://doi.org/10.1590/S1519-566X2001000400001

Dem FF (2015) Community parameters of Auchenorrhyncha (Hemiptera: Insect) along altitudinal gradients in subtropical and tropical rainforests in Australia. Dissertation, Griffith University

Dietrich CH (2009) Auchenorrhyncha (Cicadas, Spittlebugs, Leafhoppers, Treehoppers, and Planthoppers). In: Resh VH, Cardé RT (eds) Encyclopedia of insects, 2nd edn. Academic Press, Amsterdam, pp 56–64. https://doi.org/10.1016/B978-0-12-374144-8.00015-1

Diez F, Cornelis M, Pall JL et al (2018) Community assembly and inventory efficiency at a regional scale: the case of terrestrial Heteroptera (Insecta) in northern Patagonia, Argentina. Syst Biodivers 16:565–575. https://doi.org/10.1080/14772000.2018.1465485

Emsley MG (1969) The Schizopteridae (Hemiptera: Heteroptera) with the description of new species from Trinidad. Mem Am Entomol Soc 25:1–154

Evans JW (1982) A review of present knowledge of the family Peloridiidae and new genera and new species from New Zealand and New Caledonia (Hemiptera: Insecta). Rec Aust Mus 34:381–406

Favret C (2020) Aphid species file, version 5.0. http://Aphid.SpeciesFile.org. Accessed 1 May 2020

Fernandes GW, Martins RP, Tameirão-Neto E (1987) Food web relationships involving *Anadiplosis* sp. galls (Diptera: Cecidomyiidae) on *Machaerium aculeatum* (Leguminosae). Rev Brasil Bot 10:117–123

García Morales M, Denno BD, Miller DR et al (2016) ScaleNet: a literature-based model of scale insect biology and systematics.http://scalenet.info. Accessed 1 May 2020. https://doi.org/10.1093/database/bav118

Gibb TJ, Oseto C (2006) Insect collection and identification: techniques for the field and laboratory. Academic Press, London

Gossner M (2006) Phenological activity patterns of imaginal Heteroptera in the canopy of different tree species in Bavaria, Germany. Denisia 19:1055–1094

Gossner M (2008) Heteroptera (Insecta: Hemiptera) communities in tree crowns of beech, oak and spruce in managed forests: diversity, seasonality, guild structure, and tree specificity. In: Floren A, Schmidl J (eds) Canopy arthropod research in Central Europe: basic and applied studies from the high frontier. Bioform Entomology, Nurnberg, pp 119–143

Gossner M (2009) Light intensity affects spatial distribution of Heteroptera in deciduous forests. Eur J Entomol 106:241–252. https://doi.org/10.14411/eje.2009.032

Gossner MM, Damken C (2018) Diversity and ecology of saproxylic Hemiptera. In: Ulyshen MD (ed) Saproxylic insects. Springer, Cham, pp 263–317. https://doi.org/10.1007/978-3-319-75937-1_9

Grazia J, Cavichiolli R, Wolff V et al (2012) Hemiptera. In: Rafael JA, Melo GAR, Carvalho CJB et al (eds) Insetos do Brasil, diversidade e taxonomia. Holos Editora, Ribeirão Preto, pp 348–405

Guerra TJ, Camarota F, Castro FS, Schwertner C, Grazia J (2011) Trophobiosis between ants and *Eurystethus microlobatus* Ruckes (Hemiptera: Heteroptera: Pentatomidae) a cryptic, gregarious and subsocial stinkbug. J Nat Hist 45:1101–1117. https://doi.org/10.1080/00222933.2011.552800

Gullan PJ, Martin JH (2009) Sternorrhyncha (jumping plant–lice, whiteflies, aphids, and scale insects). In: Resh VH, Cardé RT (eds) Encyclopedia of insects, 2nd edn. Academic Press, Amsterdam, pp 957–967. https://doi.org/10.1016/B978-0-12-374144-8.00253-8

Hallmann CA, Sorg M, Jongejans E et al (2017) More than 75 percent decline over 27 years in total flying insect biomass in protected areas. PLoS One 12:1–21. https://doi.org/10.1371/journal.pone.0185809

Hardy NB (2018) The biodiversity of Sternorrhyncha: scale insects, aphids, psyllids, and whiteflies. In: Foottit RG, Adler PH (eds) Insect biodiversity: science and society, vol 2. Wiley, Oxford, pp 591–625. https://doi.org/10.1002/9781118945582.ch20

Henry TJ (2017) Biodiversity of Heteroptera. In: Foottit RG, Adler PH (eds) Insect biodiversity: science and society, vol 2. Wiley, Oxford, pp 279–335. https://doi.org/10.1002/9781118945568.ch10

Kment P, Hemala V, Baňař P (2016) *Rhyparoclava pyrrhocoroides*, a new genus and species of autapomorphic Rhyparochromidae with clavate antennae from Madagascar (Hemiptera: Heteroptera). Acta Entomol Mus Nat Pragae 56:517–545

Koponen S (2004) Arthropods from high oak branches—comparison of two trap types, with a special reference to spiders. Latv entomol 41:71–75

Koponen S, Rinne V, Clayhills T (1997) Arthropods on oak branches in SW Finland, collected by a new trap type. Entomol Fenn 8:177–183. https://doi.org/10.33338/ef.83940

Kozár F, Miller DR (1999) Observations on collecting scale insects (Hemiptera: Coccoidea). Entomologica 33:243–250. https://doi.org/10.15162/0425-1016/842

Larivière MC, Burckhardt D, Larochelle A (2011) Peloridiidae (Insecta: Hemiptera: Coleorrhyncha). Fauna N Z 67:1–78. https://doi.org/10.7931/J2/FNZ.67

Lucas M, Forero D, Basset Y (2016) Diversity and recent population trends of assassin bugs (Hemiptera: Reduviidae) on Barro Colorado Island, Panama. Insect Conserv Diver 9:546–558. https://doi.org/10.1111/icad.12191

Martin JH, Mound LA (2007) An annotated check list of the world's whiteflies (Insecta: Hemiptera: Aleyrodidae). Zootaxa 1492:1–84

Matsukura K, Matsumura M, Tokuda M (2010) Both nymphs and adults of the maize orange leafhopper induce galls on their host plant. Commun Integr Biol 3:388–389

McCravy KW (2018) A review of sampling and monitoring methods for beneficial arthropods in agroecosystems. Insects 9:1–27. https://doi.org/10.3390/insects9040170

McPherson J (2018) Invasive stink bugs and related species (Pentatomoidea): biology, higher systematics, semiochemistry, and management. CRC Press, Boca Raton

Michelotto MD, Rincao RO, Bolonhezi D et al (2019) Flutuação populacional do percevejo-preto, *Cyrtomenus mirabilis* (Perty, 1836) (Hemiptera: Cydnidae) com o uso de armadilhas em amendoim. Braz J Dev 5:22529–22570. https://doi.org/10.34117/bjdv5n10-371

Mitchell PL (2004) Heteroptera as vectors of plant pathogens. Neotrop Entomol 33:519–545. https://doi.org/10.1590/S1519-566X2004000500001

Moir ML, Brennan KEC (2007) Using bugs (Hemiptera) as ecological and environmental indicators in forest ecosystems. In: Muñoz SI (ed) Forest ecology research progress. Nova Science Publishers, New York, pp 79–115

Moir ML, Brennan KE, Majer JD et al (2005) Toward an optimal sampling protocol for Hemiptera on understorey plants. J Insect Conserv 9:3–20. https://doi.org/10.1007/s10841-004-2351-y

Moreira FFF, Rodrigues HDD, Sites RW et al (2018) Order Hemiptera. In: Hamada N, Thorp JH, Rogers DC (eds) Keys to Neotropical Hexapoda: Thorp and Covich's freshwater invertebrates, vol 3. Academic Press, London, pp 175–216. https://doi.org/10.1016/B978-0-12-804223-6.00007-X

Noireau F, Abad-Franch F, Valente SA et al (2002) Trapping Triatominae in silvatic habitats. Mem Inst Oswaldo Cruz 97:61–63. https://doi.org/10.1590/S0074-02762002000100009

Ouvrard D (2020) Psyl'list the world Psylloidea database. https://www.hemiptera-databases.org/psyllist/. Accessed 1 May 2020. https://doi.org/10.5519/0029634

Penca C, Hodges AC, Leppla NC et al (2020) Trap-based economic injury levels and thresholds for *Euschistus servus* (Hemiptera: Pentatomidae) in Florida peach orchards. J Econ Entomol 113(3):1347–1355. https://doi.org/10.1093/jee/toaa044

Rakitov R, Appel E (2012) Life history of the camelthorn gall leafhopper, *Scenergates viridis* (Vilbaste) (Hemiptera, Cicadellidae). Psyche 930975:1–19. https://doi.org/10.1155/2012/930975

Salomão RP, Santacruz JB, Favila ME (2019) Diversity of edaphic Heteroptera (Hemiptera) over a heterogeneous neotropical landscape. J Insect Conserv 23:909–920. https://doi.org/10.1007/s10841-019-00172-8

Schaefer CW (2005) Gall-inducing heteropterans (Hemiptera). In: Raman A, Schaefer CW, Withers TW (eds) Biology, ecology, and evolution of gall-inducing arthropods. CRC Press, New Hampshire, pp 231–238

Schaefer CW (2009) Prosorrhyncha (Heteroptera and Coleorrhyncha). In: Resh VH, Cardé RT (eds) Encyclopedia of insects, 2nd edn. Academic Press, Amsterdam, pp 839–855. https://doi.org/10.1016/B978-0-12-374144-8.00220-4

Schuh RT, Slater JA (1995) True Bugs of the World (Hemiptera: Heteroptera). Cornell University Press, Ithaca

Stewart AJ (2002) Techniques for sampling Auchenorrhyncha in grasslands. Denisia 4:491–512

Stubbins FL, Greene JK, Toews MD et al (2016) Assessment of a cross-vane trap as a tool for sampling the invasive *Megacopta cribraria* (Hemiptera: Plataspidae) in soybean with associated evaluations of female reproductive status. Environ Entomol 45:1262–1270. https://doi.org/10.1093/ee/nvw113

Trebicki P, Harding RM, Rodoni B et al (2010) Diversity of Cicadellidae in agricultural production areas in the Ovens Valley, northeast Victoria, Australia. Aust J Entomol 49:213–220. https://doi.org/10.1111/j.1440-6055.2010.00758.x

van Klink R, Bowler DE, Gongalsky KB et al (2020) Meta-analysis reveals declines in terrestrial but increases in freshwater insect abundances. Science 368:417–420. https://doi.org/10.1126/science.aax9931

Vialatte A, Bailey RI, Vasseur C et al (2010) Phylogenetic isolation of host trees affects assembly of local Heteroptera communities. P Roy Soc B-Biol Sci 277:2227–2236. https://doi.org/10.1098/rspb.2010.0365

Weirauch C, Schuh R, Cassis G et al (2019) Revisiting habitat and lifestyle transitions in Heteroptera (Insecta: Hemiptera): insights from a combined morphological and molecular phylogeny. Cladistics 35:67–105. https://doi.org/10.1111/cla.12233

Williams KS, Simon C (1995) The ecology, behavior, and evolution of periodical cicadas. Ann sRev Entomol 40:269–295

Wolff VRS (2012) Subordem Sternorrhyncha Amyot & Serville, 1843. In: Rafael JA, Melo GAR, Carvalho CJB et al (eds) Insetos do Brasil, diversidade e taxonomia. Holos Editora, Ribeirão Preto, pp 349–359

Wygodzinsky PW, Schmidt K (1991) Revision of the New World Enicocephalomorpha (Heteroptera). Bull Am Nat Hist 200:1–265

Chapter 13
Collecting and Sampling Methods for Thrips

Estevão Alves Silva, Elison Fabricio B. Lima, Rita Marullo, and Arturo Goldaracena Lafuente

13.1 What Are Thrips?

Thrips are tiny (generally <2 mm in length) and furtive insects which occur all over the world, except on the poles. These insects are opportunistic and explore intermittently occurring environments, such as blooming flowers and expanding leaves. Most thrips are *r*-selected and usually possess few natural enemies (Funderburk et al. 2000). Thrips exhibit varied lifestyles (Morse and Hoddle 2006)—as predators, facultative predators, leafhopper ectoparasites, phytophagous, gallers, and fungus feeders (Mound and Marullo 1996; Mound 2005). In some cases, a single species can occupy distinct niches, such as the *Pseudophilothrips* that inhabits Neotropical Malpighiaceae. Both adults and larvae feed on all aerial plant parts and are found on leaf buds, leaves, flower buds, flowers, and fruits (Alves-Silva and Del-Claro 2016).

E. A. Silva (✉)
Instituto Federal de Educação, Ciência e Tecnologia Goiano, Campus Urutaí, Urutaí, GO, Brazil

Universidade do Estado de Mato Grosso, Nova Xavantina, MT, Brazil

E. F. B. Lima
Universidade Federal do Piauí, Campus Amílcar Ferreira Sobral—CAFS, Floriano, PI, Brazil
e-mail: efblima@ufpi.edu.br

R. Marullo
Dipartimento di Agraria, Università degli Studi Mediterranea di Reggio Calabria, Reggio Calabria, Italy
e-mail: rmarullo@unirc.it

A. G. Lafuente
Earth and Life Institute—ELIB, Université catholique de Louvain, Louvain La Neuve, Belgium
e-mail: arturo.goldaracena@uclouvain.be

J. C. Santos, G. W. Fernandes (eds.), *Measuring Arthropod Biodiversity*,
https://doi.org/10.1007/978-3-030-53226-0_13

Although thrips exhibit such diverse habits, the Thysanoptera are best known for the economic damage they cause in several crops. In phytophagous species, the thrips puncture the epidermis of plant tissues to feed on sap and cell contents; as a result, plant tissues turn brownish and blackish, and these necrosis marks are evidence of feeding by thrips. Thrips can be taken to long distances, either by wind or by the introduction of infested plant (Mound 1997). Only about 2% of the described thrips species are recorded as pests (Moritz et al. 2004), and studies in general focus on these pest species. Thus, our knowledge of sampling, monitoring, identification, and life stories are widely known for these species, in comparison to non-pest species.

13.2 Objective of the Chapter

This chapter aims to cover the main sampling methods for thrips and the need for both detailed field studies and meticulous monitoring. We reviewed widely used methods for sampling thrips, and we provide the main characteristics of each. Our research covers pest species only, since most sampling techniques have been developed to monitoring this group, especially species in the genera *Frankliniella* and *Thrips*. A detailed understanding of the advantages (or the lack thereof) of each sampling technique not only helps to prevent potential risks to crops but also to control thrips populations after outbreaks.

13.3 Why Sample Pest Species?

Sampling of pest thrips is the first and most important step to prevent, anticipate, and monitor potential damaging species. Samplings also allow identification of the main characteristics that must be taken into consideration when performing actions to control thrips, such as: *(1) functional groups*—fruit, flower, and leaf feeders; *(2) immigrant or exotic species,* permitting a rapid evaluation of invasive species that might become pests, so actions can be anticipated to prevent a burst of populations (Morse and Hoddle 2006); and (iii) *host exploitation,* niche overlap among species, temporal and spatial distribution in plants (Lee and Wen 1982; Tappan 1986).

13.4 Economic Importance of Thrips

Despite early conviction that thrips were small and were not a concern in some crops (e.g., soybean—Irwin and Yeargan 1980), current research shows that even at exceptionally low densities thrips can cause considerable yield losses. Regardless of the use of biological control agents (expensive and environmentally risky),

Fig. 13.1 Thrips infesting *Enterolobium* sp. (Fabaceae) in a greenhouse in Brazil. In a short period of time (less than a month), thrips devastated the plants, leaving several scars on the leaf surface. Herbivory by thrips eventually resulted in defoliation. The last photo shows the larvae of thrips (circles). Because of their small size, thrips can be seen in plants only when it is too late, as in this *Enterolobium* individual. Photos by E.A.S.

pesticides, and integrated pest management (IPM) programs (Helyer and Brobyn 1992), thrips have been a source of damage to crops causing severe losses in yield (Welter et al. 1990; Walsh et al. 2012). Their agricultural importance involves direct damage to plant tissues during oviposition and feeding and/or transmission of phytopathogenic agents, especially viruses (Lewis 1997; Pereira et al. 2017). The role and ability of thrips to vector a wide diversity of viruses (currently labelled as orthotospoviruses) is often the most serious damage from thrips, especially those in the family Thripidae (Jones 2005).

Thrips feeding on plants can damage all aboveground parts, such as fruits, leaves, and shoots. The insects can also cause serious injury in young plants, which seem to be more susceptible to thrips attack (Fig. 13.1) (Dreistadt 2001).

13.5 Literature Survey

We conducted an extensive survey of the literature to investigate the methods that have been used so far to collect thrips, which crops were targeted for thrips sampling, and the thrips species that were commonly sampled. The search was made using Google Scholar (as it combines several databases and papers from many publishers), with the words *thrips sampling* and *thrips monitoring* and included papers published until December 2017. The reference list at the end of papers was also used as a source of information to enrich our survey. Each paper was carefully read, and the pertinent information was retrieved and systematically organized. This resulted in 64 published papers in which authors had studied, compared, and/or developed sampling techniques for thrips.

The samplings included thrips in fields (containing plants of many species at once) and in over 39 specific plants. The most important plants, in terms of associated thrips, are shown in Fig. 13.2, which also indicates the richness of thrips,

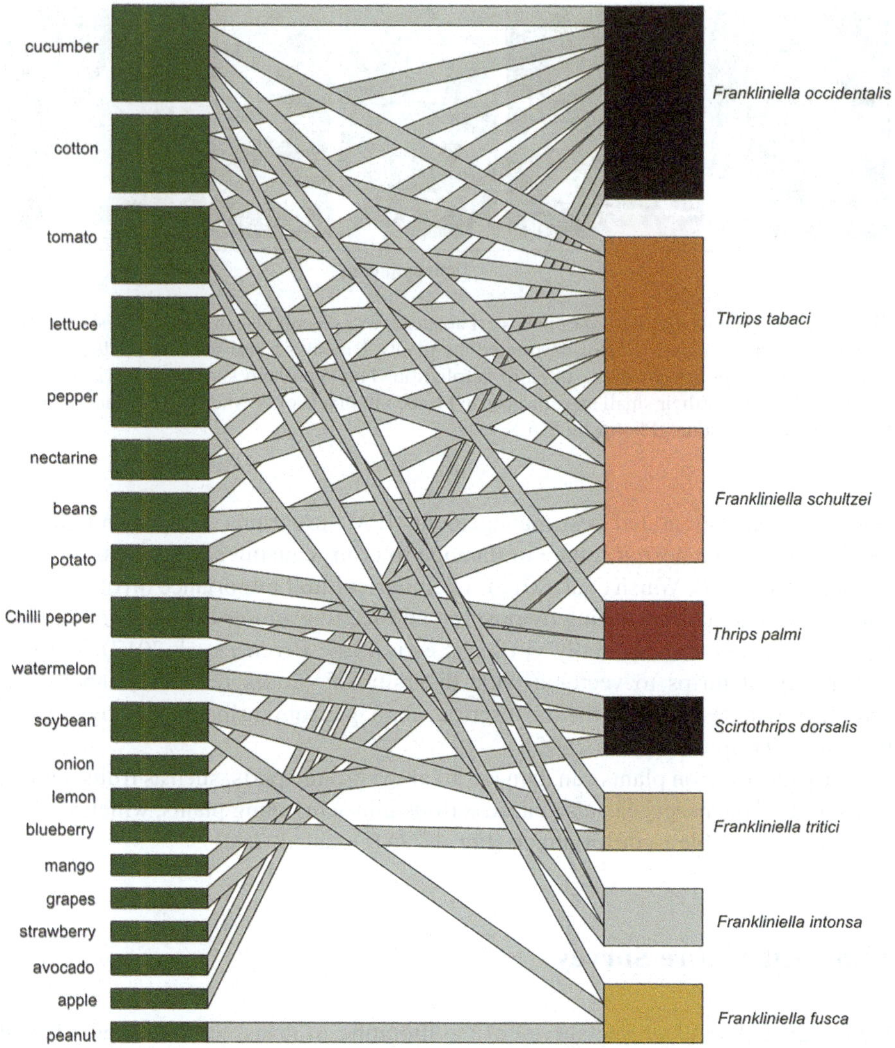

Fig. 13.2 Bipartite representation of thrips and their host plants. *Frankliniella occidentalis*, *F. schultzei*, and *Thrips tabaci* were found in many crops and can be considered highly polyphagous. The likelihood of their attacking several crops is thus high. The data comprises part of a large dataset that was extracted from 64 studies that focused on thrips sampling and monitoring. The figure was made using *R* statistical software (bipartite package), and we show only thrips species that were sampled in three or more different crops

especially in the genera *Frankliniella* and *Thrips*, associated with plant species. Species in these two genera were consistently sampled and monitored (Fig. 13.3) (according to the data obtained from the literature), and the richness of *Frankliniella* outnumbered the other genera. Together, both *Frankliniella occidentalis* and *Thrips tabaci* were found to be pervasive in the investigations to sample and monitor thrips (Fig. 13.4).

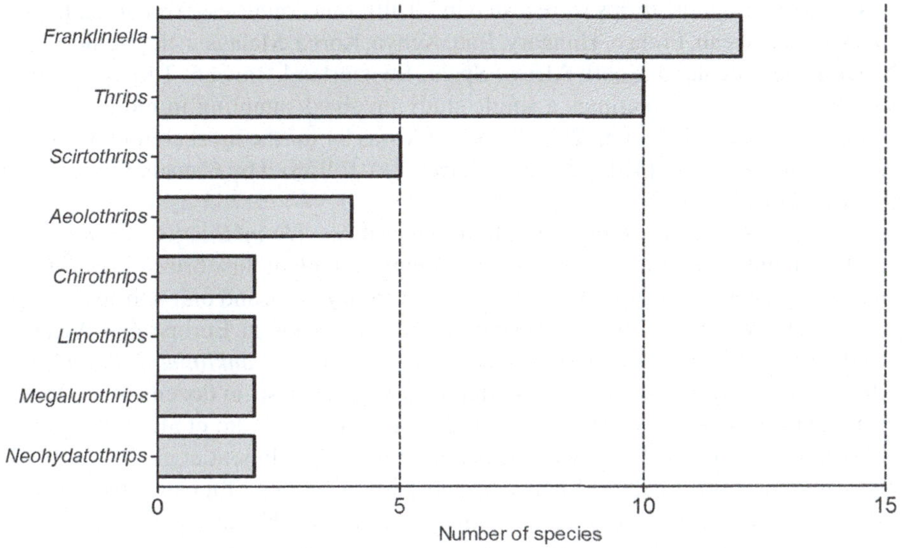

Fig. 13.3 Pervasiveness of *Frankliniella* and *Thrips* as the genera with more species in samplings of thrips

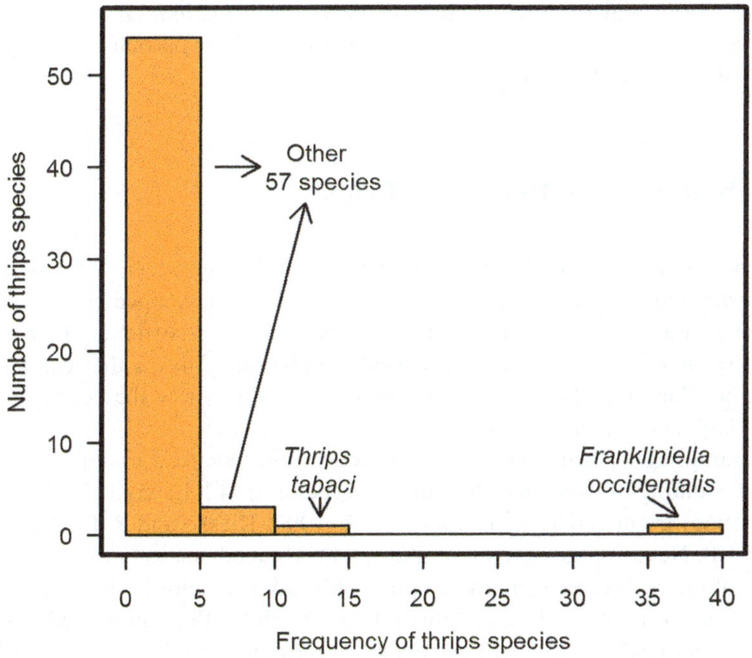

Fig. 13.4 Frequency distribution of the thrips species in samplings of thrips in crops

Methods to sample thrips were tested in 20 different countries (Argentina, Brazil, Canada, Caribbean, France, Hungary, Iraq, Kenya, Korea, Malaysia, the Netherlands, Nigeria, New Zealand, South Africa, Spain, Switzerland, Taiwan, Turkey, the UK, and the USA), and sometimes a single study involved sampling in more than one country (e.g., Abdullah et al. 2015). The USA was by far the most common country where thrips were sampled (42% of occurrences), followed by Canada (7%), Brazil (6%), and France (6%).

Several pest species, such as western flower thrips (*Frankliniella occidentalis*), banded greenhouse thrips (*Hercinothrips femoralis*), gladiolus thrips (*Taeniothrips simplex*), greenhouse thrips (*Heliothrips haemorrhoidalis*), and onion thrips (*Thrips tabaci*), are relatively common species in nursery crops in Europe and America (Hoddle et al. 2008). Major pest species in the genus *Frankliniella* and *Thrips*, which were fairly common in our literature survey, are pests in dozens of economic important crops (e.g., rice [Cavalleri et al. 2010], tomato [Cho et al. 1995], potato [Cho et al. 2000], peach, strawberry, persimmon, grape [Pinent et al. 2007, 2008]) and can attack practically all aboveground parts of plants (Tappan 1986). *Thrips tabaci* is a major pest of onion and garlic; *T. palmi* attacks various vegetables and fruit trees; *Frankliniella occidentalis* has great importance in horticultural and ornamental plants; *F. schultzei* is highly polyphagous (Lewis 1997). Viruses transmitted by thrips have been recorded in over 1000 plant species (Riley et al. 2011).

Our data shows that thrips attack economically important crops that are grown all over the world, and the most important pests are *Frankliniella* and *Thrips*. The development of control measures has thus focused on these particular thrips species, as they attack a variety of crops.

13.6 Sampling Methods for Thrips

Thrips populations are mainly found in the soil (including litter), vegetation, or air. Several methods for sampling have accordingly been developed, then compared, and consistently used for sampling thrips in the field and to estimate their richness, density, economic injury level (the lowest population of thrips that can yet cause damage to plants), and economic threshold (a density below the economic injury level) (Nault and Shelton 2010).

The sampling of thrips has been reviewed and has evolved to optimize the capture and counting of these insects in the field (Pelley 1942; Lewis 1959; Powelll and Landis 1965; Irwin and Yeargan 1980; Lewis 1997; Rieske and Raffa 2003; Reisig et al. 2010). Nevertheless, even nowadays, with great advances in the knowledge of thrips biology and ecology, many scientists still face problems in dealing with accurate methods to monitor thrips populations in the field (Pearsall and Myers 2000). In applied entomology, absolute or relative surveys of pest thrips are described for several crops, although the action levels (when available) for implementing control tactics may differ in distinct areas of the globe.

Sampling and knowledge of the species (i.e., which and where) are a first step to establishing the most appropriate strategy for controlling thrips and preventing further economic cropping losses. Trustworthy sampling and monitoring protocols, established through detailed observations (and even experimentation), and previous experience in working with thrips are vital for management decisions (Nault and Shelton 2010). For instance, the presence of some species of *Frankliniella* (e.g., *occidentalis, schultzei*) above a determined threshold can be seen as an early warning sign, because of the potential of some thrips to transmit orthotospoviruses (Cho et al. 1995; Riley et al. 2011).

The success of any sampling strategy depends upon knowledge of the distribution of thrips in time and space (Cho et al. 1995). Since each species presents its own behavior in plant occupation (e.g., leaves, flowers, fruits), feeding preferences (pollen, leaf sap, cannibalism) (Mound and Marullo 1996), and biology (Reisig et al. 2010), this imposes a difficulty in establishing a single sampling procedure for all thrips which can be found even in a single plant species (Irwin and Yeargan 1980). In crops where more than one species of pest thrips occurs concurrently (Fig. 13.2), it might be necessary to perform several samplings and compare different methods to make a good estimation of thrips richness (Cho et al. 1995). Economic thresholds must be established for pest species, to anticipate whether thrips densities are in a level that incurs significant yield losses (MacIntyre-Allen et al. 2005; Nault and Shelton 2010). This is a key step for IPMs and helps farmers to take control measures (Abdullah et al. 2015; Pereira et al. 2017).

Ideally, the sampling plans for thrips should follow important phases, including the economic injury levels associated with the pest, the control of costs, and the validation of the sampling. A complete step-by-step procedure is given by Pereira et al. (2017). These authors determined the control costs by surveying the pesticides and equipment used for spraying watermelons (~US$ 8000 ha^{-1}), inspected plants in the field to establish a relationship between thrips densities and yield losses (t ha^{-1}; R^2 = 0.97), investigated the crop production value (US$ from 62.5 to 218.75 t^{-1}), analyzed the economic loss caused by thrips (36%) and the economic injury level (0.02–0.09 insects per sample), estimated the costs of controlling thrips (2% of the production value), and provided information about the decision to control thrips based on plant phenological stage.

13.7 Sampling Plans for Thrips: Parameters

A total of 20 sampling methods for thrips were recorded in our literature survey, and the main characteristics of each are described below. The samplings can be classified according to distinct parameters (Table 13.1). Two types of sampling are (con) currently used to monitor thrips: the relative, when only a proportion of thrips infesting the plants are inferred, and the absolute, when all thrips per plant are

Table 13.1 A brief description of the parameters used to sample thrips

Parameters of sampling	Description
Relative or absolute	The relative infers only a proportion of thrips infesting plants and crops, while the absolute permits the estimation of all thrips
Destructive or nondestructive	The destructive sampling requires the removal of plant parts, while the no-destructive permits the evaluation of thrips density without the need to extract plant structures (e.g., leaves, flowers, stems)
Disturbing or non-disturbing	Whenever a sampling method requires touching and shaking the plant and/or removal of plant parts, it is disturbing, because some thrips may fly away or drop from plant during this process
Selective or nonselective	Whether the sampling permits to capture a target thrips species or if it captures several species at once
Direct or delayed	Some techniques permit the counting of thrips in the field (in situ), whereas in others it is necessary to take the sample to the laboratory to count these insects
Limited or unlimited	Depending on the method, the samples can be saturated and no thrips can be further collected. Some other techniques permit unlimited sampling of thrips
Complete or incomplete	Whenever the sampling accounts for adults and immatures of thrips, it is complete. However most sampling targets adults only
Repetition or non-repetition	Sometimes a sampling method can be repeated in the same plant structure or crop to monitor thrips in time and space, but in other occasions, this cannot be done, for instance when parts of the plant, or even the whole, are extracted

counted. An early classification of samplings is given by Irwin and Yeargan (1980), who classified samplings as either delayed (i.e., removal of plant parts in the field and processing in the laboratory) or direct counting (counting thrips directly in the field) of thrips infesting soybean.

Methods can be destructive (e.g., delayed counting) or nondestructive (e.g., direct counting) (Pearsall and Myers 2000; Muvea et al. 2014). In the former, plant structures are collected or cut off, whereas in the latter this is not necessary. Sampling plans can also be selective or nonselective (Lewis 1959). For instance, the use of sticky traps and odors is a selective method, as the response of thrips to colors and odors is species-specific. The beating method is nonspecific as it captures all thrips species foraging on a plant structure.

Many of the methods used to sample thrips were originally created for other insects (e.g. pitfalls, sweep nets, beating, light traps), and guidelines have evolved to account for most thrips exceptionalities (phenotypic variation, life stages, within-plant distribution) (Chu et al. 2006; Fedor et al. 2007; Reisig et al. 2010). The characteristics of each method are summarized below and are a compilation of the sampling strategies used and developed throughout the years to provide the best estimate of thrips in several distinct scenarios.

13.8 Absolute Sampling

Absolute counting involves the extraction of the whole plant from the soil and the counting of all thrips (Joost and Riley 2004; Liu and Chu 2004). It is made with low-sized crops or seedlings (e.g., lettuce, onion) (Palumbo 2003). Bagged plants are filled with ethanol and shaken to detach thrips from the plant parts. The ethanol is then filtered and thrips are counted. Freezing plants to immobilize thrips, instead of the use of bags with alcohol, is also described (Al-karboli and Al-Anbaki 2014). Plants can also be washed off in the laboratory to remove the thrips (Liu and Chu 2004). The absolute method can sometimes be used as a reference for comparisons with other sampling techniques (Joost and Riley 2004; Aliakbarpour and Rawi 2010). Instead of collecting the whole plant, researchers can also collect a given plant structure, take it to the laboratory, and then estimate thrips density in that plant structure (Aliakbarpour and Rawi 2010).

13.9 Relative Samplings

Binomial sampling This technique is not based on the complete enumeration of thrips in plants or plant structures, but rather on the presence or absence of these insects. This method is preferred over the others when thrips are difficult to count (Ugine et al. 2011). Action is undertaken when the proportion of plants with thrips exceeds a given threshold (e.g., 50%) (Cho et al. 1995; Laudonia et al. 2000). Further samples of the incidence of thrips can be made to record their density (Navas et al. 1994).

Direct counts Plants are carefully inspected and thrips abundance is estimated directly on the aboveground structures; adults, and sometimes larvae, are recorded (Palumbo 2003). This is acknowledged as the best method to assess the density of *Frankliniella schultzei* in watermelon (Pereira et al. 2017).

Beating Plant structures are shaken over a sheet of white paper/plastic (or a counting board, Powelll and Landis 1965), which may be moist to avoid thrips from flying. In a variation of the method, plant structures are beat vigorously against a screened pan for a predetermined time, and the dislodged thrips are retained via a sticky trap located below the pan (Palumbo 2003). The plant part can also be shaken inside a plastic cup (Reisig et al. 2010; Joost and Riley 2004). Beating permits the collection/counting of adults and larvae (Powelll and Landis 1965). Thrips can be collected with a fine brush and placed in vials for further identification. This method is used in plants where thrips are abundant and rich, and it permits the identification of all thrips species infesting plants, as well the temporal monitoring of their populations. This procedure was used to monitor thrips in sweet pepper in Hungary, and the results revealed the occurrence of 21 thrips species infesting greenhouse crops. It was also possible with this method to unravel different temporal variations of thrips species and their rise and collapse in the course of years (Aliakbarpour and Rawi 2010; Orosz et al. 2017).

Sticky traps/cards/boards Sheets (e.g., 10 cm × 25 cm, Abdullah et al. 2015; 13 × 10 cm, Demirel and Yildirim 2008; 16 × 19.5, Coli et al. 1992) with a sticky substance and different colors (yellow, red, blue, black—Liu and Chu 2004), depending on the thrips species to be sampled, are placed near or above plants, and flying thrips are intercepted by these traps. Sheets are gridded with 1 cm^2 squares, which are used as a proxy to estimate thrips numbers. This is also used to investigate the flight activity of thrips (Aliakbarpour and Rawi 2010). Counting can begin after a predetermined period (Palumbo 2003). This method takes advantage of the high mobility of thrips species, especially the pest species. Traps are often replaced to record the abundance of thrips along the cropping period; thrips are removed from the sticky cards with kerosene and collected with a fine brush (Muvea et al. 2014). This is one of the most common monitoring methods to sample economically important thrips (Bergant et al. 2005). It is commercially available (Fig. 13.5).

Plant part extraction Plant structures (leaves, inflorescences, branches) are cut off and placed in vials that can be empty or contain a preserving solution. When empty vials are used, this method permits thrips to be counted in vivo in the laboratory. Thrips usually detach from the plant structures and wander on the vial, from where they can be collected with a fine brush. This method is the most appropriate for observations of thrips behavior in the laboratory, as it conserves the original plant structure and the associated thrips. If, instead, the plant structure is put in a vial containing a preserving solution (Aliakbarpour and Rawi 2010), it can be poured into a Petri dish or filtered using a fine cloth. In the case of flowers, dissection can be necessary to extract thrips that become stuck inside the corolla (Cho et al. 1995; Moreira et al. 2017; Orosz et al. 2017). A variation of the method includes positioning the plant structure over a table and brushing it to remove thrips (e.g., potato plants—Powelll and Landis 1965). A pole pruner or a tree climber can be used to collect samples on tall plants (Werner et al. 2004).

Washing off the plant part In the field, a plant structure is submerged into a jar containing ethanol and shaken; the ethanol is then filtered using a fine mesh and the thrips are counted (Aliakbarpour and Rawi 2010).

Odor baits and olfactory response Chemical attractants are spread in sticky cards with the function of luring thrips (Muvea et al. 2014). They are commercially available (Abdullah et al. 2015). A variation consists of volatile compounds mixed with water and placed in plastic containers (0.5 L) suspended on wooden stakes (Rieske and Raffa 2003).

LEDs in sticky traps Evaluated by Chen et al. (2004) as a monitoring method for western flower thrips. Light-emitting diodes (LEDs) are attached to sticky cards.

Caged sticky cards Cards are covered with a nylon mesh screen to prevent the capture of large insects, which might saturate the sticky cards and prevent the capture of thrips (Chen et al. 2004).

Fig. 13.5 (**a**) Yellow and (**b**) blue sticky traps used to monitor thrips inside a greenhouse (**c**). In some cases, odor baits can be added to the sticky traps, either to maximize thrips capture or to trap specific pest species. Photos by R.M.

Immobilization with CO_2 Plant parts containing thrips are enclosed in a plastic bag, and carbon dioxide is released inside the bag. After a period of 30 s, CO_2 incapacitates the thrips, which fall to the bottom of the plastic bag and are then taken to the laboratory (Aliakbarpour and Rawi 2010).

Light traps As presented by Fedor et al. (2007), the light trap makes use of the positive phototaxy and negative geotropism of bark-dwelling insects. The trap consists of an inverted funnel that is fixed to a PVC tube containing a conserving solution. This PVC tube acts as a collecting vessel. Above, and involving the PVC tube, another tube is placed to prevent the escape of insects. The trap is wired to the bark of trees.

Tanglefooted tile This method is described by Irwin and Yeargan (1980) to collect flying thrips, but it seems to be more appropriate to sampling aphids (Irwin 1980). The method comprises a horizontally oriented ceramic tile covered in a thin layer of tanglefoot (the same wax used in ant-exclusion experiments) placed at canopy level of soybeans. Thrips attached to the wax are removed with acetone.

Aspirator or suction traps Sucks up thrips from plants (Joost and Riley 2004).

Electric vacuum device An electric suction tube is used to collect thrips (Joost and Riley 2004). D-vacuum insect nets are a variation (Rhodes and Morse 1989).

Sweep nets A funneled net of approximately 38 cm diameter radius is swung from side to side in a 180° arc (Reisig et al. 2010).

Ground traps, pupation papers, or carton trap Sticky traps are placed under plants to collect pupating larvae that drop off plants to pupate in the soil (Tanigoshi and Moreno 1981; Rhodes and Morse 1989). A variation consists of folded papers that are distributed among horizontal twigs, above the ground, or inside the plant canopy; it captures immature thrips (Grout et al. 1986; Rhodes and Morse 1989).

Shotgun Used, and not recommended, for sampling thrips by Werner et al. (2004). It was designed to sample branches from tall trees.

Flotation Used to sample thrips from soil and litter; after filtering the soil using different meshes, heptane is used to float thrips (Parker et al. 1992).

13.10 Comments on Some Methods

All these methods have advantages and disadvantages, and the choice of which to use may depend on the objectives of the sampling (e.g., to be sure of the presence of thrips, to identify the biodiversity of thrips, to identify damage to plants), costs, and time involved. The relationship between sampling method and sampling parameters can aid in the choice of the most appropriate method to sample thrips (Table 13.2). Some methods might be impractical (use of shotgun), costly (immobilization with CO_2), or time demanding (absolute counting), and researchers must thus find a balance between sampling, parameters, and costs.

A comparison of three relative methods (direct counts, beating tray, and sticky traps) revealed that absolute counting was by far the best for estimating thrips populations in lettuce; the other methods either failed to make correct estimates of thrips or provided low numbers of thrips larvae. In some occasions, the beat pan yielded comparable results to absolute sampling and could be used as an alternative to such highly invasive methods (Palumbo 2003). Of two sampling methods used in mango orchards, the beating tray and immobilization using CO_2, the latter provided the best estimates of thrips abundance, compared with absolute counting (Aliakbarpour and Rawi 2010). No difference was found in beans between sticky traps, collecting, and tapping leaves and flowers (Muvea et al. 2014); in cotton, the washing technique yielded more thrips in comparison to visual estimation (Parajulee et al. 2006), and in roses the yellow sticky traps were accurate enough to be used in greenhouses (Pizzol et al. 2010).

Absolute sampling demands more effort but is advantageous because the researcher or farmer can be confident that the species sampled are indeed in the study plant, even though some species may be only visitors. Absolute sampling also provides information about immature individuals and brachypterous or apterous specimens that are rarely caught on sticky traps (Reisig et al. 2010). This method may require the removal of the whole plant from the soil/crop, so it cannot be used

Table 13.2 Main characteristics of the methods used to sample thrips

Sampling techniques	Parameters of sampling						
	Destructive	Disturbing	Specific	Direct	Limited	Complete	Repetition
Binomial sampling	No	No	Yes	Yes	No	Yes	Yes
Direct counts	No	No	Yes	Yes	No	Yes	Yes
Beating	No	Yes	No	Yes	No	Yes	Yes
Sticky traps	No	No	No	Yes	Yes	No	Yes
Plant part extraction	Yes	Yes	No	No	No	Yes	No
Washing off the plant part	No	Yes	No	No	No	Yes	Yes
Odor baits	No	No	No	Yes	Yes	No	Yes
LEDs in sticky traps	No	No	No	Yes	Yes	No	Yes
Caged sticky cards	No	No	No	Yes	Yes	No	Yes
Immobilization with CO_2	No	Yes	No	Yes	No	Yes	Yes
Light traps	No	Yes	No	No	No	Yes	Yes
Tanglefooted tile	No	No	No	Yes	Yes	No	Yes
Aspirator or suction traps	No	Yes	No	No	No	Yes	Yes
Electric vacuum device	No	Yes	No	No	No	Yes	Yes
Sweep nets	No	Yes	No	No	No	Yes	Yes
Pupation papers	No	No	No	Yes	Yes	Yes	Yes
Shotgun	Yes	Yes	No	No	No	Yes	No
Flotation	No	No	No	No	No	Yes	No

A brief description of each parameter can be found in Table 13.1

for large plants. Relative methods infer only a proportion of thrips infesting the plants, and thus they should be used with this drawback in mind.

Sticky traps are by far the most widely used method to sample thrips. The colors used to attract thrips species may vary between different crops, which might be related to the contrast between the crop and trap colors. Although the range of colors varies for each species, growers in agricultural areas prefer to use yellow sticky traps due to their cost-effectiveness, as other pests can also be attracted by this color, including whiteflies and aphids (Lewis 1997). These traps can give a good information about the species composition of a given area but have some disadvantages. Although commonly used, they are affected by wind speeds, and thrips from other areas, carried by the wind, could be collected. Sticky traps also have the disadvantage of damaging thrips when they strongly attach to the trap material, thus jeopardizing their identification. Sometimes the species collected on cards are not actually associated with the plant from which thrips are being sampled. Cho et al. (1995) tested different colors of sticky traps to sample thrips from tomato fields and noted that *Frankliniella tritici* had been collected in high populations from colored traps, although very few specimens were collected directly from flowers.

The beating method may remove not only thrips but also other insects from the plant. When made on flowers, beating can cause damage to corollas, loss of pollen, and even flower abscission. This method causes disturbance to the plant and to the associated arthropods. Once disturbed, some thrips might simply fly away, and thus this method might underestimate thrips abundance.

Direct counts are widely used in ecological studies, and this seems to be the less invasive method, as it requires neither thrips collection nor plant extraction, so that the system remains relatively undisturbed. This method should be undertaken by an experienced researcher who is able to reliably tell thrips from other insects. Another issue is that this method is not appropriate for sampling thrips in plants that are rich in these insects, as the identification of thrips species in these cases might be difficult. Moreover, tiny thrips larvae cannot be reliably estimated in this method, nor thrips that are highly spatially distributed in the plant, inconspicuous individuals, or those that hide in small cracks. It can therefore eventually underestimate the number of thrips, jeopardizing the economic injury level and economic threshold estimation. This method can also be time-consuming, should large plants and several structures be inspected (Fig. 13.6).

13.11 Relationship Between Sampling Technique and Time

The choice among different sampling techniques must take into account several factors, such as personnel, budget available, time, species of thrips, and characteristics of the target crop. For example, in a comparison of the beating tray and flower collection (in vineyards), the latter was the best sampling method, yielding three times more thrips than the former method; however, the costs (14-fold higher) and time (1.4-fold higher) associated with it were considerably elevated. The authors then estimated the most economically effective sampling and concluded that ten inflorescences per hectare were ideal to provide reliable indicators of thrips abundance in vineyards (Moreira et al. 2017). In another study, four sampling methods were evaluated (absolute counting, direct counting, water cups, and sticky cards), and the authors found a great difference with regard to the time required for processing (preparation to counting all thrips) the samples. Absolute counting required longer periods of processing, yielding a 15-fold difference in comparison to direct counting, but was the best sampling technique (Liu and Chu 2004).

Two formulas are usually used in studies that compare sampling techniques and costs of sampling, namely, *relative variation*(RV) and *relative net precision* (RNP). The former evaluates the precision of a sampling method; values <25% are most appropriate for general estimates. The latter compares the best sampling efficiency. The higher the *RNP*, the better the sampling efficiency (more details in Pearsall and Myers 2000; Joost and Riley 2004). The formulas are given below:

Fig. 13.6 Direct counts of the thrips *Heterothrips peixotoa*. (**a**) Thrips within the stamens of *Banisteriopsis stellaris* are seen under a stereomicroscope with 10× magnification. (**b**) Several thrips foraging on the petals, the chamber, and the stamens of *B. stellaris*, thus making it difficult to estimate their density. (**c**) Thrips hiding in between the leaves of *B. malifolia*. A non-experienced researcher might miss these insects or may even not distinguish them from other insects. (**d**) A single individual thrips foraging on a flower bud—in these cases the number of thrips can be recorded with precision. (**e**) High density of thrips in a petal of *Cochlospermum regium*. Thrips' small size and sensitivity to disturbance calls for caution when estimating their numbers in this situation. Photos by E.A.S.

$$RV = \left(\frac{SEM}{mean}\right).100 \quad RNP = \left(\frac{1}{RV.Cs}\right).100$$

where RV = relative variation; it is calculated by SEM/mean and is estimated with the number of thrips in each sampling unit. Cs = costs in hours of labor of collecting and managing a given number of samples.

In staked tomatoes, no difference in time was found for sampling effort regardless of whether two, five, or ten flowers were collected, but the time to process the samples, including the counting and identification of thrips, was significantly different between samples. The sampling of ten flowers per unit was almost double that of two flowers, but the relationship between cost and sampling efficiency (*relative net rprecision*) was higher for the 10-unit flower survey (Cho et al. 1995).

13.12 Monitoring Thrips in the Crop

There are no treatment guidelines based on thrips density for most crops. Plants highly susceptible to viruses should be kept virtually free of thrips. Deciding when to treat crops is difficult and subjective because often there is no simple relationship between thrips density and plant damage, and thrips density in traps does not always correlate with thrips density in crops. However, growers who regularly inspect plants and consistently employ well-maintained traps can establish thresholds over the long term by judging the historic monitoring records. Whenever there is evidence of thrips infesting and damaging plants, both adult and larvae can be sampled using any of the methods described here. A rapid assessment could be made by beating or shaking vegetation onto a sheet of paper, and a long estimation could require the use of sticky traps (Broughton and Harrison 2012). The addition of lures baited with aggregation pheromones or kairomones to sticky traps can be used as an addition tool to capture specific thrips species (Broughton et al. 2015). Plants suspected of being infected by orthotospoviruses can be reliably identified from symptomatic parts sent to a laboratory that tests for plant pathogens. The thrips species can be identified only by an expert taxonomist, because individuals are small and polymorphic and some characters are hard to be seen by naïve taxonomists (Mound 2005).

13.13 Difficulties in Sampling and Identification

To the best of our knowledge it has so far been impossible to devise a standard procedure to sample thrips in all plants. Since thrips are highly abundant in plants, with a great capacity for moving from plant structures, both active and passive flying, and with an overlap of different species in the same individual plant, it is necessary to judge which sampling methodology is the most appropriate in each type of study (Werner et al. 2004). For instance, Nault and Shelton (2010) decided to visually count only thrips larvae in onions, as they claimed that adults moved between plants and plots. Since larvae are more abundant than adults in many cases, are voracious feeders, and transmit orthotospoviruses (Navas et al. 1994; Pearsall 2000; Riley et al. 2011; Moreira et al. 2017), this strategy can be worthwhile, depending on the objective.

Sampling plans are not easy to establish. The density of thrips is affected by weather, time, plant phenology and age, the presence of alternative hosts, natural enemies, prey, and competitors (Pickett et al. 1988; Tamo et al. 1993; Al-karboli and Al-Anbaki 2014; Orosz et al. 2017). Depending on the plant species and time of sampling, larvae can outnumber adults (Werner et al. 2004; Moreira et al. 2017), thus imposing restraints on the type of sampling to be employed, since larvae are wingless and, in most cases, cannot be identified to species.

For most methods, sampling the plant structure on which thrips are most abundant or expected provides maximum efficiency. Thrips are minute and furtive insects, which makes them very difficult to sample, and some may be unnoticed at first (Mehle and Trdan 2012). Behaviors such as hiding in small cracks, inside flowers, and under leaves are common in several species (Kirk 1984; Del-Claro et al. 1997; Mound and Terry 2001), thus making sampling difficult (Frey et al. 1994; Ugine et al. 2011) or even impractical in relation to time and budget restrictions. For instance, thrips can be very abundant in flowers where they are very active, coming in and out of the corolla frequently or wandering on the petals; these thrips have a tendency to escape, hide, or drop from plants in response to disturbance (Powelll and Landis 1965; Alves-Silva and Del-Claro 2016), and thus the mere presence of the researcher can have a big effect and affects a precise estimation of thrips abundance (Werner et al. 2004; Ugine et al. 2011). A sampling plan must undoubtedly take these considerations into account.

Thrips are not randomly distributed within plants, thus imposing a restriction on the methods that can potentially be used or requiring a preliminary study with observation of the host exploitation by thrips (Navas et al. 1994; Navarro-Campos et al. 2012). A failure to recognize this can result in elevated sampling error, underestimation of the actual density of thrips, and an inability to anticipate which changes in population may cause economic injury (Pearsall and Myers 2000). According to Irwin and Yeargan (1980), as well as thrips characteristics such as the distribution within plants, it is necessary to accurately include the time (period of day, the periodicity of the sampling, phenological events of plants), the sample unit (leaves, flowers, the whole plant), and the number of sample units in a sampling program.

For taxonomic studies involving collections of thrips from plants or wood, active sampling methods are usually the best option. Only in this way will the researcher gather accurate details on the origin of the specimens to be provided as biological data in species descriptions or other publications. A widely used method involves beating leaves, flowers, or dead twigs using a wooden stick and collecting specimens with a soft brush from a plastic tray put under the area.

The identification of male and female thrips can be made only with specimens mounted on slides for microscopic visualization, and, depending on the method used, a large quantity of thrips must be required (several methods are reviewed in Mehle and Trdan 2012). For larvae, only the second instar onward is usually used for thrips species identification (Orosz et al. 2017). This might be a problem for plants that harbor many larvae (Moreira et al. 2017), since identification is imprecise; thus, a plant might sustain many species of thrips, but they cannot be reliably identified from larvae. For a few thrips, identification from first instar larvae is possible (e.g., *Pezothrips kellyanus* (Bagnall)), but even so, raising larvae until the adult stage provides the best accuracy in the identification of thrips (Navarro-Campos et al. 2012). The reliable identification of thrips is necessary whenever a researcher aims to determine actions to control potential pests, because some species, even within the same genus (e.g., *Frankliniella*), are capable of transmitting a virus, whereas other species are not (Navas et al. 1994).

13.14 Comments on Native Species of Brazil

The Brazilian fauna of Thysanoptera comprises over 600 species, representing about 10% of the approximately 6200 described valid worldwide (Lima 2019; Cavalleri et al. 2019). This richness makes Brazil as the country with the largest known thrips fauna in the Neotropics (Alves-Silva and Del-Claro 2010; Mound 2014), although the total richness represents less than one third of the estimated (Monteiro and Mound 2012) and comes largely from of studies conducted in southern and southeastern Brazil. In the Atlantic Forest biome, mostly present in the south and southeast and occupying about 13% of the national territory, there are 433 recorded species. On the other hand, the biomes such as the Amazon and the Cerrado (Brazilian savanna), mostly present in the northern and central zones and occupying around 50% and 25% of the national territory, respectively, account for approximately 100 species recorded each (Lima 2019).

Even with this discrepant scenario, there is a great diversity of habits among Brazilian species. Although the most known thrips are approximately 45 pest species (mostly in the genera *Frankliniella* and *Thrips*), of which seven are reported as *Orthotospovirus* vectors, the majority of Brazil's thrips is composed of fungivorous species that live especially on dry branches or leaves (Monteiro 2002; Lima 2019). In addition to these, there are four species of *Aulacothrips*, the only genus of ectoparasite thrips, that live in association with leafhoppers (Cavalleri and Kaminski 2014); several species of *Scolothrips*, *Karnyothrips*, *Franklinothrips*, and *Leptothrips*, generalist predators of small arthropods (Mound and Marullo 1996; Monteiro 2002; Lima 2019); predators in the genus *Mirothrips*, which feed on social wasp eggs (Cavalleri et al. 2013); and pollinators of endangered species such as *Frankliniella gardenia*, who dwells on flowers of *Ocotea porosa* (Lauraceae) (Danieli-Silva and Varassin 2013).

13.15 Conclusion

Thrips, notably the pest species, are highly polyphagous and can attack several crop species. Our survey shows that dozens of tools have been developed to monitor thrips, but actions to control their populations depend on reliable taxonomy, since even within sympatric thrips pest species, differences in niches, behavior, and resistance to pesticides must be considered in order to achieve the goal of controlling their populations. Most pest thrips that occur in crops, and are consequently sampled, belong to the genera *Frankliniella* and *Thrips*, and in fact these are the most studied genera all over the world when it comes to thrips. We note that most studies have been conducted in the USA, and thus the results of the methods to sample thrips are geographically biased. Thrips exhibit a series of interactions with the environment and other insects (prey/food, competitors, predators), and samplings in different parts of the world might provide very different results in terms of species sampled and appropriate methods with which to estimate the density of thrips.

Acknowledgments We are deeply grateful for the invitation to write this chapter and contribute with the study of thrips. Our most sincere acknowledgments go to editors (J. C. Santos and G. W. Fernandes) and to professor Laurence Mound, for his kindness and attention to the study of thrips worldwide. EAS also thanks the CNPq and PELD (Programa Ecológico de Longa Duração) for funding and to professor K. Del-Claro for introducing him to the study of thrips ecology when he wanted to study spiders.

References

Abdullah ZS, Greenfield BPJ, Ficken KJ et al (2015) A new attractant for monitoring western flower thrips, *Frankliniella occidentalis* in protected crops. Springerplus 4:89. https://doi.org/10.1186/s40064-015-0864-3

Aliakbarpour H, Rawi CSM (2010) Diurnal activity of four species of thrips (Thysanoptera: Thripidae) and efficiencies of three nondestructive sampling techniques for thrips in mango inflorescences. J Econ Entomol 103:631–640. https://doi.org/10.1603/EC09167

Al-Karboli HH, Al-Anbaki HA (2014) Efficacy of tow sampling methods for monitoring, control and estimating seasonal abundance of onion thrips, *Thrips tabaci* Lindeman (Thripidae: Thysanoptera) on onion in Iraq. Int J Agric Technol 10:243–251

Alves-Silva E, Del-Claro K (2010) Thrips in the Neotropics: what do we know so far? Trends Entomol 6:77–88

Alves-Silva E, Del-Claro K (2016) On the inability of ants to protect their plant partners and the effect of herbivores on different stages of plant reproduction. Austral Ecol 41:263–272. https://doi.org/10.1111/aec.12307

Bergant K, Trdan S, Žnidarčič D et al (2005) Impact of climate change on developmental dynamics of *Thrips tabaci* (Thysanoptera: Thripidae): can it be quantified? Environ Entomol 34:755–766. https://doi.org/10.1603/0046-225X-34.4.755

Broughton S, Harrison J (2012) Evaluation of monitoring methods for thrips and the effect of trap colour and semiochemicals on sticky trap capture of thrips (Thysanoptera) and beneficial insects (Syrphidae, Hemerobiidae) in deciduous fruit trees in Western Australia. Crop Prot 42:156–163. https://doi.org/10.1016/j.cropro.2012.05.004

Broughton S, Cousins DA, Rahman T (2015) Evaluation of semiochemicals for their potential application in mass trapping of *Frankliniella occidentalis* (Pergande) in roses. Crop Prot 67:130–135. https://doi.org/10.1016/j.cropro.2014.10.011

Cavalleri A, Kaminski LA (2014) Two new ectoparasitic species of *Aulacothrips* Hood, 1952 (Thysanoptera: Heterothripidae) associated with ant-tended treehoppers (Hemiptera). Syst Parasitol 89:271–278. https://doi.org/10.1007/s11230-014-9526-z

Cavalleri A, Mendonça MS Jr, Rodrigues ENL (2010) Thrips species (Thysanoptera, Terebrantia) inhabiting irrigated rice and surrounding habitats in Cachoeirinha, state of Rio Grande do Sul, Brazil. Rev Bras Entomol 54:501–504. https://doi.org/10.1590/S0085-56262010000300023

Cavalleri A, de Souza AR, Prezoto F, Mound LA (2013) Egg predation within the nests of social wasps: a new genus and species of Phlaeothripidae, and evolutionary consequences of Thysanoptera invasive behaviour. Biol J Linn Soc 109:332–341. https://doi.org/10.1111/bij.12057

Cavalleri A, Masumoto M, Minaei K, et al (2019) ThripsWiki - providing information on the World's thrips. https://thrips.info/wiki/Main_Page. Accessed 5 Sep 2019

Chen T-Y, Chu C-C, Fitzgerald G et al (2004) Trap evaluations for thrips (Thysanoptera: Thripidae) and hoverflies (Diptera: Syrphidae). Environ Entomol 33:1416–1420. https://doi.org/10.1603/0046-225X-33.5.1416

Cho K, Eckel CS, Walgenbach JF, Kennedy GG (1995) Spatial distribution and sampling procedures for *Frankiniella* spp. (Thysanoptera: Thripidae) in staked tomato. J Econ Entomol 88:1658–1665. https://doi.org/10.1093/jee/88.6.1658

Cho K, Kang SH, Lee GS (2000) Spatial distribution and sampling plans for *Thrips palmi* (Thysanoptera: Thripidae) infesting fall potato in Korea. J Econ Entomol 93:503–510. https://doi.org/10.1603/0022-0493-93.2.503

Chu C-C, Ciomperlik MA, Chang N-T et al (2006) Developing and evaluating traps for monitoring *Scirtothrips dorsalis* (Thysanoptera: Thripidae). Florida Entomol 89:47–55. https://doi.org/10.1653/0015-4040(2006)89[47:daetfm]2.0.co;2

Coli WM, Hollingsworth CS, Maier CT (1992) Traps for monitoring pear thrips (Thysanoptera: Thripidae) in maple stands and apple orchards. J Econ Entomol 85:2258–2262. https://doi.org/10.1093/jee/85.6.2258

Danieli-Silva A, Varassin IG (2013) Breeding system and thrips (Thysanoptera) pollination in the endangered tree *Ocotea porosa* (Lauraceae): implications for conservation. Plant Species Biol 28:31–40. https://doi.org/10.1111/j.1442-1984.2011.00354.x

Del-Claro K, Marullo R, Mound LA (1997) A new Brazilian species of *Heterothrips* (Insecta: Thysanoptera) co-existing with ants in the flowers of *Peixotoa tomentosa* (Malpighiaceae). J Nat Hist 31:1307–1312. https://doi.org/10.1080/00222939700770731

Demirel N, Yildirim AE (2008) Attraction of various sticky color traps to *Thrips tabaci* Lindeman (Thysanoptera: Thripidae) and *Empoasca decipies* Paoli (Homoptera: Cicadellidade) in cotton. J Entomol 1:1–6. https://doi.org/10.3923/je.2008.389.394

Dreistadt SH (2001) Integrated pest management for floriculture and nurseries. State Integrated Pest Management. Project University of California Division of Agriculture and Natural Resources publication 3402

Fedor P, Dubovský M, Majzlan O (2007) Tree photoeclector in sampling bark-dwelling thrips (Thysanoptera). Thysanopteron Pismo Entomol 3:14–16

Frey JE, Cortada RV, Helbling H (1994) The potential of flower odours for use in population monitoring of western flower thrips *Frankliniella occidentalis* Perg. (Thysanoptera: Thripidae). Biocontrol Sci Technol 4:177–186. https://doi.org/10.1080/09583159409355325

Funderburk J, Stavisky J, Olson S (2000) Predation of *Frankliniella occidentalis* (Thysanoptera: Thripidae) in field peppers by *Orius insidiosus* (Hemiptera: Anthocoridae). Environ Entomol 29:376–382. https://doi.org/10.1603/0046-225X(2000)029

Grout TG, Morse JG, O'Connell NV et al (1986) Citrus thrips (Thysanoptera: Thripidae) phenology and sampling in the San Joaquin Valley. J Econ Entomol 79:1516–1523. https://doi.org/10.1093/jee/79.6.1516

Helyer NL, Brobyn PJ (1992) Chemical control of western flower thrips (*Frankliniella occidentalis* Pergande). Ann Appl Biol 121:219–231. https://doi.org/10.1111/j.1744-7348.1992.tb03434.x
Hoddle MS, Mound LA, Paris DL (2008) Thrips of California. CBIT Publishing, Queensland
Irwin ME (1980) Sampling aphids in soybean fields. In: Kogan M, Herzog DC (eds) Sampling methods in soybean entomology. Springer, New York, pp 239–259
Irwin ME, Yeargan KV (1980) Sampling phytophagous thrips on soybean. In: Kogan M, Herzog DC (eds) Sampling methods in soybean entomology. Springer, New York, pp 283–304
Jones DR (2005) Plant viruses transmitted by thrips. Eur J Plant Pathol 113:119–157. https://doi.org/10.1007/s10658-005-2334-1
Joost PH, Riley DG (2004) Sampling techniques for thrips (Thysanoptera: Thripidae) in preflowering tomato. J Econ Entomol 97:1450–1454. https://doi.org/10.1603/0022-0493-97.4.1450
Kirk WDJ (1984) Ecological studies on *Thrips imaginis* Bagnall (Thysanoptera) in flowers of *Echium plantagineum* L. in Australia. Aust J Ecol 9:9–18. https://doi.org/10.1111/j.1442-9993.1984.tb01613.x
Laudonia S, Pedata PA, Viggiani G (2000) Development of a sequential sampling program for *Frankliniella occidentalis* Pergande (Thysanoptera: Thripidae) on strawberry in plastic tunnels in Southern Italy. Bull IOBC/WPRS 23:201–204
Lee HS, Wen HC (1982) Seasonal occurrence of and injury caused by thrips and their control on mangoes. Plant Prot Bull 24:179–187. https://doi.org/10.1111/1750-3841.12171
Lewis T (1959) A comparison of water traps, cylindrical sticky traps and suction traps for sampling thysanopteran populations at different levels. Entomol Exp Appl 2:204–215. https://doi.org/10.1111/j.1570-7458.1959.tb00434.x
Lewis T (1997) Pest thrips in perspective. In: Lewis T (ed) Thrips as crop pests. CAB International, Wallingford, pp 1–13
Lima EFB (2019) Thysanoptera in Catálogo Taxonômico da Fauna do Brasil. PNUD. <http://fauna.jbrj.gov.br/fauna/faunadobrasil/316>. Accessed 25 Nov 2019
Liu TX, Chu CC (2004) Comparison of absolute estimates of *Thrips tabaci* (Thysanoptera: Thripidae) with field visual counting and sticky traps in onion field in south Texas. Southwest Entomol 29:83–89
MacIntyre-Allen JK, Scott-Dupree CD, Tolman JH, Harris CR (2005) Evaluation of sampling methodology for determining the population dynamics of onion thrips (Thysanoptera: Thripidae) in Ontario onion fields. J Econ Entomol 98:2272–2281. https://doi.org/10.1603/0022-0493-98.6.2272
Mehle N, Trdan S (2012) Traditional and modern methods for the identification of thrips (Thysanoptera) species. J Pestic Sci 85:179–190. https://doi.org/10.1007/s10340-012-0423-4
Monteiro RC (2002) The Thysanoptera fauna of Brazil. In: Marullo R, Mound LA (eds) Thrips and Tospoviruses: proceedings of the 7th international symposium on Thysanoptera. Australian National Insect Collection, Canberra, pp 325–340
Monteiro RC, Mound LA (2012) Thysanoptera. In: Rafael JA, Melo GAR, Carvalho CJB et al (eds) Insetos do Brasil: Diversidade e Taxonomia. Holos Editora, Ribeirão Preto, pp 407–422
Moreira AN, Oliveira JV, Oliveira JEM et al (2017) Sampling technique for thrips in vineyards. Rev Bras Frutic 39:1–13. https://doi.org/10.1590/0100-29452017582
Moritz GB, Mound LA, Morris DC, Goldarazena A (2004) Thrips ID: pest thrips of the world. An interactive identification and information system. Canberra, ACIAR. CD-ROM
Morse JG, Hoddle MS (2006) Invasion biology of thrips. Annu Rev Entomol 51:67–89. https://doi.org/10.1146/annurev.ento.51.110104.151044
Mound LA (1997) Thrips as crop pests. In: Lewis T (ed) Biological diversity. CAB International, Wallingford, pp 197–215
Mound LA (2005) Thysanoptera: diversity and interactions. Annu Rev Entomol 50:247–269. https://doi.org/10.1146/annurev.ento.49.061802.123318
Mound LA (2014) Austral Thysanoptera: 100 years of progress. Austral Entomol 53:18–25. https://doi.org/10.1111/aen.12054

Mound LA, Marullo R (1996) The thrips of Central and South America: an introduction (Insecta: Thysanoptera). Associated Publishers, Gainesville

Mound LA, Terry I (2001) Thrips pollination of the central Australian cycad, *Macrozamia macdonnellii* (Cycadales). Int J Plant Sci 162:147–154. https://doi.org/10.1086/317899

Muvea AM, Waiganjo MM, Kutima HL et al (2014) Attraction of pest thrips (Thysanoptera: Thripidae) infesting French beans to coloured sticky traps with Lurem-TR and its utility for monitoring thrips populations. Int J Trop Insect Sci 34:197–206. https://doi.org/10.1017/S174275841400040X

Nault BA, Shelton AM (2010) Impact of insecticide efficacy on developing action thresholds for pest management: a case study of onion thrips (Thysanoptera: Thripidae) on onion. J Econ Entomol 103:1315–1326. https://doi.org/10.1603/EC10096

Navarro-Campos C, Aguilar A, Garcia-Marí F (2012) Aggregation pattern, sampling plan, and intervention threshold for *Pezothrips kellyanus* in citrus groves. Entomol Exp Appl 142:130–139. https://doi.org/10.1111/j.1570-7458.2011.01204.x

Navas VES, Funderburk JE, Mack TP et al (1994) Aggregation indices and sample size curves for binomial sampling of flower-inhabiting *Frankliniella* species (Thysanoptera: Thripidae) on tomato. J Econ Entomol 87:1622–1626. https://doi.org/10.1093/jee/87.6.1622

Orosz S, Éliás D, Balog E, Tóth F (2017) Investigation of thysanoptera populations in Hungarian greenhouses. Acta Univ Sapientiae Agric Environ 9:140–158. https://doi.org/10.1515/ausae-2017-0013

Palumbo JC (2003) Comparison of sampling methods for estimating western flower thrips abundance on lettuce. College of Agriculture and Life Sciences, University of Arizona, Tucson

Parajulee MN, Shrestha RB, Leser JF (2006) Sampling methods, dispersion patterns, and fixed precision sequential sampling plans for western flower thrips (Thysanoptera: Thripidae) and cotton fleahoppers (Hemiptera: Miridae) in cotton. J Econ Entomol 99:568–577. https://doi.org/10.1603/0022-0493-99.2.568

Parker BL, Grehan JR, Skinner M (1992) Method for extracting pear thrips (Thysanoptera: Thripidae) from forest soil. J Econ Entomol 85:865–869. https://doi.org/10.1093/jee/85.3.865

Pearsall IA (2000) Damage to nectarines by the western flower thrips (Thysanoptera: Thripidae) in the interior of British Columbia, Canada. J Econ Entomol 93:1207–1215. https://doi.org/10.1603/0022-0493-93.4.1207

Pearsall IA, Myers JH (2000) Evaluation of sampling methodology for determining the phenology, relative density, and dispersion of western flower thrips (Thysanoptera: Thripidae) in nectarine orchards. J Econ Entomol 93:494–502. https://doi.org/10.1603/0022-0493-93.2.494

Pelley RH (1942) A method of sampling thrips populations. Bull Entomol Res 33:147–148. https://doi.org/10.1017/S0007485300026420

Pereira PS, Sarmento RA, Galdino TVS et al (2017) Economic injury levels and sequential sampling plans for *Frankliniella schultzei* in watermelon crops. Pest Manag Sci 73:1438–1445. https://doi.org/10.1002/ps.4475

Pickett CH, Wilson LT, Gonzalez D (1988) Population dynamics and within-plant distribution of the western flower thrips (Thysanoptera: Thripidae), an early-season predator of spider mites infesting cotton. Environ Entomol 17:551–559. https://doi.org/10.1093/ee/17.3.551

Pinent SMJ, Pinent CED, Botton M, Redaelli LR (2007) Thrips species (Thysanoptera) on strawberry, persimmon and grape in the Gaucho Highlands, Rio Grande do Sul State, southern Brazil. J Insect Sci 7:1–44

Pinent SMJ, Mascaro F, Botton M, Redaelli LR (2008) Thrips (Thysanoptera: Thripidae, Phlaeothripidae) damaging peach in Paranapanema, São Paulo State, Brazil. Neotrop Entomol 37:486–488. https://doi.org/10.1590/S1519-566X2008000400019

Pizzol J, Nammour D, Hervouet P et al (2010) Comparison of two methods of monitoring thrips populations in a greenhouse rose crop. J Pestic Sci 83:191–196. https://doi.org/10.1007/s10340-010-0286-5

Powelll DM, Landis BJ (1965) A comparison of two sampling methods for estimating population trends of thrips and mites on potatoes. J Econ Entomol 58:1141–1144. https://doi.org/10.1093/jee/58.6.1141

Reisig DD, Godfrey LD, Marcum DB (2010) Grass thrips (*Anaphothrips obscurus*) (Thysanoptera: Thripidae) population dynamics and sampling method comparison in timothy. Environ Entomol 39:1617–1625. https://doi.org/10.1603/EN09329

Rhodes AA, Morse JG (1989) *Scirtothrips citri* sampling and damage prediction on California navel oranges. Agric Ecosyst Environ 26:117–129. https://doi.org/10.1016/0167-8809(89)90022-4

Rieske LK, Raffa KF (2003) Evaluation of visual and olfactory cues for sampling three thrips species (Thysanoptera: Thripidae) in deciduous forests of the northern United States. J Econ Entomol 96:777–782. https://doi.org/10.1603/0022-0493-96.3.777

Riley DG, Joseph SV, Srinivasan R, Diffie S (2011) Thrips vectors of tospoviruses. J Integr Pest Manag 2:1–10. https://doi.org/10.1603/IPM10020

Tamo M, Baumgärtnera J, Delucchia V, Herrena HR (1993) Assessment of key factors responsible for the pest status of the bean flower thrips *Megalurothrips sjostedti* (Thysanoptera: Thripidae) in West Africa. Bull Entomol Res 83:251–258. https://doi.org/10.1017/S000748530003474X

Tanigoshi LK, Moreno DS (1981) Traps for monitoring populations of the citrus thrips, *Scirtothrips citri* (Thysanoptera: Thripidae). Can Entomol 113:9–12. https://doi.org/10.4039/Ent1139-1

Tappan WB (1986) Relationship of sampling time to tobacco thrips (Thysanoptera: Thripidae) numbers in peanut foliage buds and flowers. J Econ Entomol 79:1359–1363. https://doi.org/10.1093/jee/79.5.1359

Ugine TA, Sanderson JP, Wraight SP et al (2011) Binomial sampling of western flower thrips infesting flowering greenhouse crops using incidence-mean models. Environ Entomol 40:381–390. https://doi.org/10.1603/EN10085

Walsh B, Maltby JE, Nolan B, Kay I (2012) Seasonal abundance of thrips (Thysanoptera) in capscium and chilli crops in South-East Queensland, Australia. Plant Prot Q 27:19–22

Welter SC, Rosenheim JA, Johnson MW et al (1990) Effects of *Thrips palmi* and western flower thrips (Thysanoptera: Thripidae) on the yield, growth, and carbon allocation pattern in cucumbers. J Econ Entomol 83:2092–2101. https://doi.org/10.1093/jee/83.5.2092

Werner SM, Nordheim EV, Raffa KF (2004) Comparison of methods for sampling Thysanoptera on basswood (*Tilia americana* L.) trees in mixed northern hardwood deciduous forests. For Ecol Manage 201:327–334. https://doi.org/10.1016/j.foreco.2004.07.014

Part II

Chapter 14
Techniques for Collection and Sampling of Pseudoscorpions (Arthropoda: Arachnida)

Edwin Bedoya-Roqueme and Everton Tizo-Pedroso

14.1 Introduction

Pseudoscorpions are small arachnids (Fig. 14.1), varying in length between 0.5 and 10.0 mm (Weygoldt 1969; Del-Claro and Tizo-Pedroso 2009). Pseudoscorpions inhabit all the terrestrial ecosystems, with exception of the poles (Weygoldt 1969; Judson 1994; Harvey 2013). These arachnids are relatively more abundant in the tropics and subtropics, while some species exhibit a circumpolar distribution (Weygoldt 1969; Harvey 1986). The fossil origin of the pseudoscorpions date from the middle of the Devonian, and their fossil records are estimated from 390 million years ago (Shear et al. 1989; Harms and Dunlop 2017; Benavides et al. 2019). Currently, this mesodiverse order of arachnids, comprises 3850 species and 454 genera, belongs to 26 extant families (and 1 extinct family, with 49 known species) (Harms and Dunlop 2017; Benavides et al. 2019).

Pseudoscorpion species thrive in the leaf litter, under rocks, under the barks of trees and decaying trunks, and in the forests of tropical and temperate regions (Weygoldt 1969; Adis and Mahnert 1985; Mahnert and Adis 2002; Aguiar et al. 2006). Pseudoscorpions are associated with many microhabitats such as bat guano, nests of ants and bees, and different parts of plants and are known to take shelter in the nests of birds, insects, and mammals (Weygoldt 1969; Zeh and Zeh 1992a, b; Cole et al. 1994; Poinar et al. 1998; Aguiar and Bührnheim 2003; Francke and Villegas-Guzmán 2006; Gonzalez et al. 2007). However, the mechanisms by which

E. Bedoya-Roqueme
Programa de Pós-graduação, Recursos Naturais do Cerrado, RENAC, Universidade Estadual de Goiás, Anápolis, GO, Brazil

E. Tizo-Pedroso (✉)
Centro de Ensino e Aprendizagem em Rede, Laboratório de Ecologia Comportamental de Aracnídeos, Programa de Pós-Graduação em Recursos Naturais do Cerrado, Universidade Estadual de Goiás, Anápolis, GO, Brazil
e-mail: tizopedroso@ueg.br

J. C. Santos, G. W. Fernandes (eds.), *Measuring Arthropod Biodiversity*,
https://doi.org/10.1007/978-3-030-53226-0_14

Fig. 14.1 Females of the pseudoscorpion *Paratemnoides nidificator* (Balzan 1888) (Atemnidae) taking care of nymphs of the social parasite pseudoscorpion *Parachernes melanopygus* Beier, 1959 (based on Tizo-Pedroso and Del-Claro 2014)

pseudoscorpions are able to survive in these environments are little known, and data pertaining to ecological factors are very scarce, despite the fact that significant studies are being carried out on these arachnids in intertidal microhabitats (Weygoldt 1969; Mahnert and Schuster 1981; Gonzalez et al. 2007; Harvey 2007, 2009; Mahnert et al. 2014).

14.2 Order Pseudoscorpiones

From a systematic point of view, the classification of the order Pseudoscorpiones is based on the apomorphic characters of each body division, among which the structure of the chelicerae, shape of the carapace, number of eyes, and the presence or absence of a venom apparatus are considered for classification purposes (Weygoldt 1969; Harvey 1992). The group is currently subdivided into four suborders (†Palaeosphyronida, Heterosphyronida, Atoposphyronida, and Iocheirata), the latter comprising several infraorders and microorders (Harvey 2002; Harvey and Muchmore 2013; Benavides et al. 2019).

The phylogenetic position of pseudoscorpions among other arachnids includes a relationship to Solifugae (=Haplocnemata/Apatellata) (Giribet et al. 2002; Shultz 2007), or to Scorpiones (Garwood et al. 2017). Few phylogenetic hypotheses have been proposed to date, and the scheme of classification presented by Chamberlin (1931) is mostly conserved. Some modifications have been proposed by later authors, including the addition and deletion of families, but the general scheme has essentially remained the same (Harvey 2002; Harvey and Muchmore 2013). Harvey (1992) presented a phylogenetic hypothesis based on the cladistic analysis of 126 characters, and later Murienne et al. (2008) proposed the first molecular phylogenetic analysis for the order. A recent study by Benavides and collaborators (Benavides et al. 2019) discusses the latest updates in pseudoscorpion phylogeny proposing four suborders.

Phylum: Arthropoda.
Subphylum: Chelicerata.
Class: Arachnida.
Order: Pseudoscorpiones.
Suborder: †Palaeosphyronida.
Suborder: Heterosphyronida.
Suborder: Atoposphyronida.
Suborder: Iocheirata.

The classification scheme is proposed by Benavides et al. (2019).

14.2.1 General Morphology of Pseudoscorpiones

The body of pseudoscorpions is divided into two well-differentiated structures (Fig. 14.2), the prosoma and opisthosoma, which are united along their entire width, without intermediate narrowing (Chamberlin 1931; Mahnert and Adis 2002; Zaragoza 2015). Prosoma consists of six symmetrically arranged metamers with a pair of appendices each: the biarticulate chelicerae on the anterior region and on

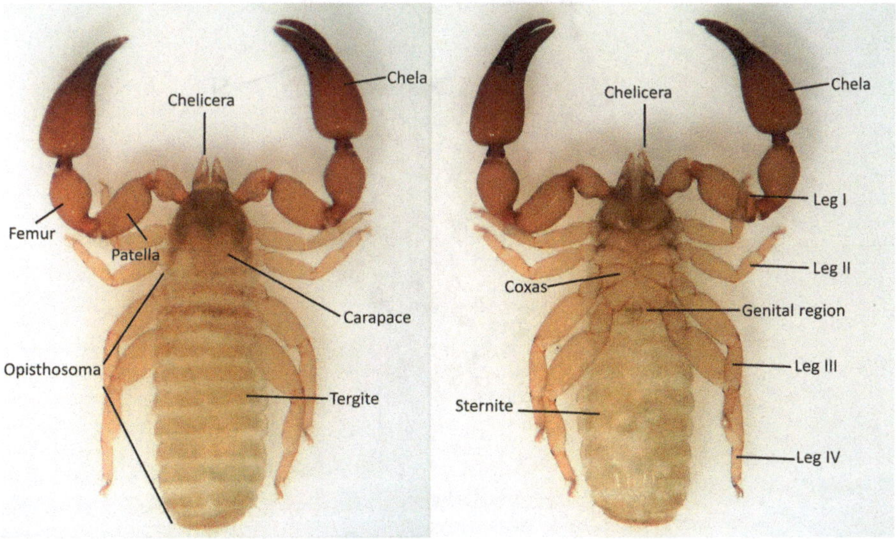

Fig. 14.2 General morphology of a pseudoscorpion showing the dorsal (left) and ventral (right) views. Photos by Edwin Bedoya-Roqueme

both sides of the pedipalps and four pairs of walking legs (Fig. 14.2). The dorsal aspect of the prosoma is composed of a single plate called the carapace. The ventral face is occupied by the coxae. The family Sternophoridae possesses a vestige of a sternum or pseudo-sternum (Harvey 1992).

The carapace has a trapezoidal or subrectangular appearance and extends from the narrowed front portion of the eyes in the cucullus to the widely rounded portion in the anterior half (Fig. 14.2). One or two pairs of simple eyes are present on both sides of the anterior half of the carapace; in some species the eyes are absent or reduced to eyespots, and the cave species present a marked ocular recession in adaptation to the subterranean environment (Chamberlin 1931; Harvey 1992; Mahnert and Adis 2002; Weygoldt 1969; Zaragoza 2015).

The chelicerae consist of biarticulated appendages (fixed fingers and movable fingers) (Fig. 14.3), located in the anterior region of the body. The fixed finger widens toward the base in one hand, and the rallum is located ventrally, in addition to be equipped with a visible external lamina and an internal lamina (Fig. 14.3)

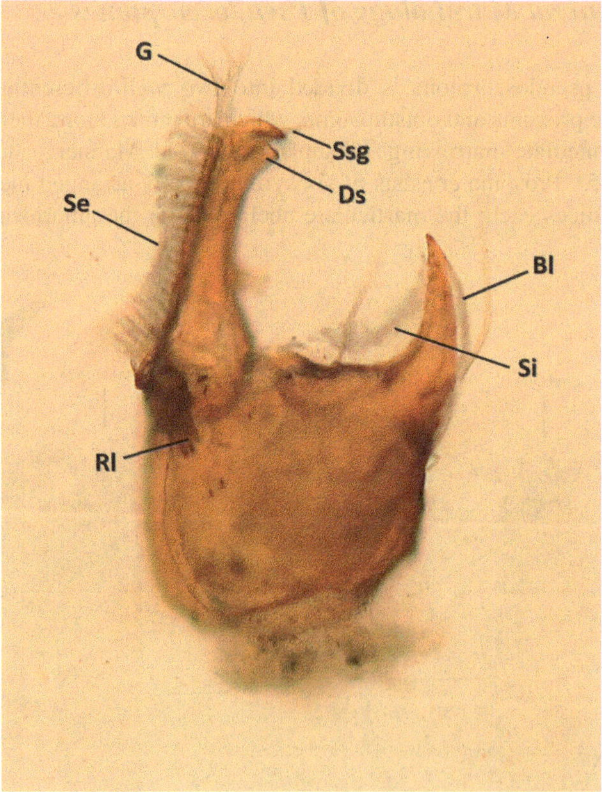

Fig. 14.3 Dorsal aspect of pseudoscorpion chelicerae. *Ds* subapical tooth, *G* galea, *Ssg* subgaleal seta, *Le* external blade, *Se* external serrula, *Si* internal serrula, *Rl* rallum. Photo by Edwin Bedoya-Roqueme

(Chamberlin 1931; Harvey 1992; Mahnert and Adis 2002; Weygoldt 1969; Zaragoza 2015). The movable finger is equipped with a serrula exterior, usually pectinate, which is comparable in some way to that of the fixed finger. There is a single silk gland and numerous teeth in the subdistal position (Fig. 14.3). At the distal end, there is a transparent structure, the galea, where the channels of the silk glands open, and in some families, it is replaced by a simple and opaque sericigen tubercle (Chamberlin 1931; Harvey 1992; Mahnert and Adis 2002; Zaragoza 2015).

The pedipalps are appendices well-developed and culminate in a robust chela and consist of several parts, including the coxa, trochanter, femur, patella, and a chela that is articulated to movable and fixed fingers (Fig. 14.4). The internal aspect of the fingers is armed with a row of teeth, although some species have some extra accessory teeth in the margins of the dental row. The distal ends of the fingers end in strong teeth, either crossed or uncrossed by the venom glands that begin with a small widening called the nodus ramosus, which may be expanded or unexpanded, and flow into the end of the fingers, whose absence or presence in either both or one finger or another is a character of maximum importance in group systematics (Chamberlin 1931; Harvey 1992; Mahnert and Adis 2002; Zaragoza 2015).

The palpal chela is equipped with trichobothria, which are sensorial setae inserted into the fingers and are rarely present on the hand. The number of sensorial setae is usually 12, with 8 on the fixed finger and 4 on the movable finger, although some families can have numerous trichobothria. The relative position of the trichobothria is based on a nomenclatural system established by Chamberlin (1931), which is used to define the families, genera, and species (Chamberlin 1931; Harvey 1992; Mahnert and Adis 2002; Weygoldt 1969; Zaragoza 2015).

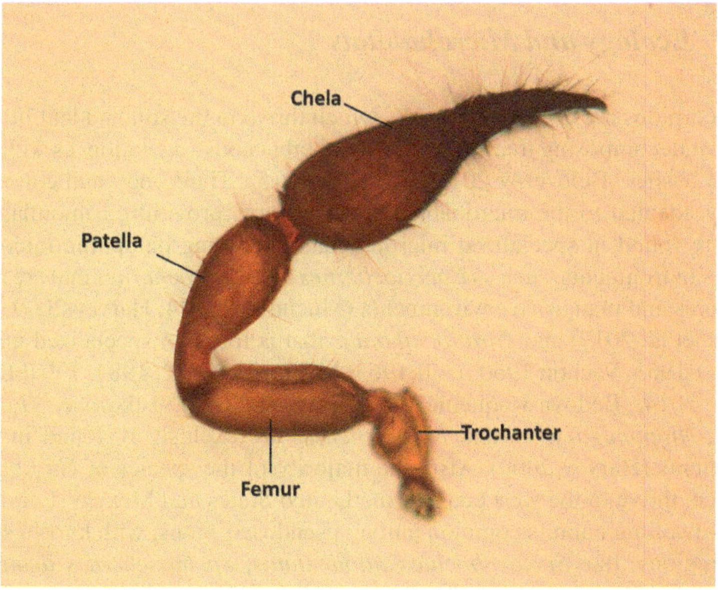

Fig. 14.4 Dorsal aspect of a pseudoscorpion pedipalp. Photo by Edwin Bedoya-Roqueme

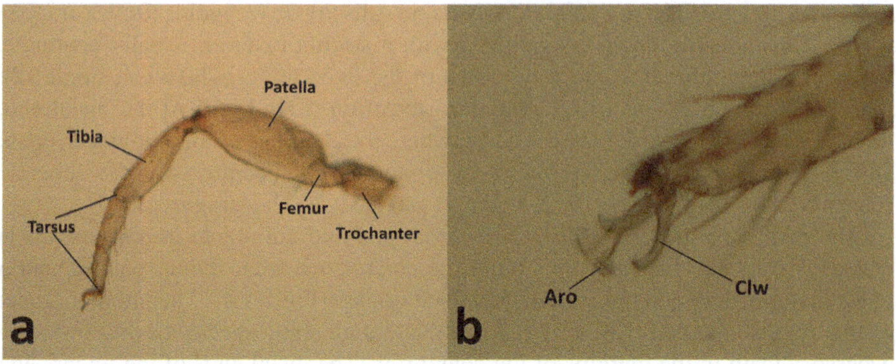

Fig. 14.5 Aspect of a pseudoscorpion leg IV. (**a**) Leg IV. (**b**) Distal end of a pseudoscorpion leg. Aro, arolia; Clw, claws. Photos by Edwin Bedoya-Roqueme

The legs are composed of the following segments: coxa, trochanter, femur, patella, tibia, and tarsus, of which the tarsus can have one or two segments (tarsus I, tarsus II) (Fig. 14.5a). Similarly, on the outer side of the tarsus can be present a tactile seta. The legs possess a pretarsal at the distal ends, which is a tiny segment carrying a pair of smooth or serrated nails, and an arolium (Fig. 14.5b), which is a transparent adhesive organ that facilitates locomotion and which, unlike in other families, exhibits distal expansions (Weygoldt 1969; Chamberlin 1931; Harvey 1992; Mahnert and Adis 2002; Zaragoza 2015).

14.2.2 Ecology and Microhabitats

Pseudoscorpions are cryptic arachnids, which thrive in the soil and leaf litter, under the bark of decomposing trees and trunks, in herbaceous vegetation, as well as in the trees and bushes (Belozerov 2013; Zaragoza 2015). Their body and coloration are generally adapted to the microhabitats they occupy, providing camouflage. Some genera are found in specialized microhabitats, including halophilic microhabitats and cave environments, such as *Epactiochernes* and *Mucrochernes* that are restricted to seashores and mangrove environments (Muchmore 1974; Harvey 2009; Bedoya-Roqueme et al. 2017) and *Paraliochthonius* that is found in specialized microhabitats (Beier 1963; Vachon 1960; Hoff 1963; Muchmore 1972, 1984, 1994; Lee 1979; Mahnert 2014; Bedoya-Roqueme 2015; Judson 2016). Likewise, *Halobisium*, *Parahya*, *Nipponogarypus*, and *Nannochelifer* are exclusively found in seashore environments (Harvey 2009). Also, the majority of the species of *Garypus* genus, Garypidae, thrive in the west coast of the United States and Mexico (Lee 1979).

Cave-dwelling habit is common among pseudoscorpions, with known species in several regions, like *Spelaeobochica allodentatus*, *Spelaeochernes dentatus*, and *Tenebriochernes concavus* in South America and *Protochelifer cavernarum* in

Australia. There are many cavernicolous species in Brazil (Schimonsky and Bichuette 2019). The cosmopolitan synanthropic species, *Chelifer cancroides* (Cheliferidae), thrives in human dwellings, particularly among books (Weygoldt 1969; Belozerov 2013). Some species of pseudoscorpions are found in more specialized habitats, such as bird nests and burrows of mammals. For instance, *Juxtachelifer fructuosus*, *Lustrochernes grossus*, *Serianus dolosus*, and *Tychochernes inflatus* thrive in the burrows of mammals, while *Lechytia chthoniiformis* and *Chelifer cancroides* can be found under the elytra of large beetles (Aguiar and Bührnheim 1992, 2003; Cole et al. 1994; Poinar et al. 1998; Andrade and Gnaspini 2002; Villegas-Guzmán and Pérez 2005a, b; Francke and Villegas-Guzmán 2006; Gonzalez et al. 2007).

Due to their small size and limited modes of dispersal, pseudoscorpions have acquired some dispersal strategies like phoresy and rafting (Lee 1979; Martínez et al. 2018; Poinar et al. 1998; Weygoldt 1969). There are reports on the phoresy of pseudoscorpions, including as vectors 44 insect families and 3 arachnid families. The majority of phoretic vectors are insects (Diptera, Coleoptera, and Lepidoptera) (Santos et al. 2005; Lira et al. 2014; Lira and Tizo-Pedroso 2017). Although phoresy is considered an evolutionary behavioral pattern (Shear et al. 1989; Zeh and Zeh 1992b), there are records on a limited number of families. Pseudoscorpions have been reported to disperse by rafting in flooded areas and employ vertical migration in rainy seasons owing to the high humidity (Lee 1979; Adis and Mahnert 1985; Adis et al. 1988; Aguiar and Bührnheim 1998; Battirola et al. 2017a, b; Bedoya-Roqueme et al. 2016).

14.3 Scientific Production Involving Pseudoscorpions

Although pseudoscorpions represent the third more diversified order of Arachnida, more scientific studies on pseudoscorpions need to be conducted. Recently, more studies are being dedicated to broadening the understanding about pseudoscorpions. We estimated the number of studies on pseudoscorpions published between 1945 and 2018 on the Web of Science (WS) (www.webofknowledge.com), and the results revealed that the number of scientific articles on pseudoscorpions in the WS considerably increased in the last 15 years (Fig. 14.6), and the data are summarized in Table 14.1.

We observed that scientific articles on pseudoscorpions are relatively well distributed all over the continents and countries. However, five countries were responsible for 47.5% of all the studies published during the period of investigation (Fig. 14.7), and these countries are also among the most cited (Fig. 14.8). A major proportion of the published articles deal with taxonomic studies (55%), followed by behavioral (10%), ecological (7%), and molecular (7%) studies. This brief estimate suggested that pseudoscorpions represent a potential taxonomic group for ecological, behavioral, and genetic researches.

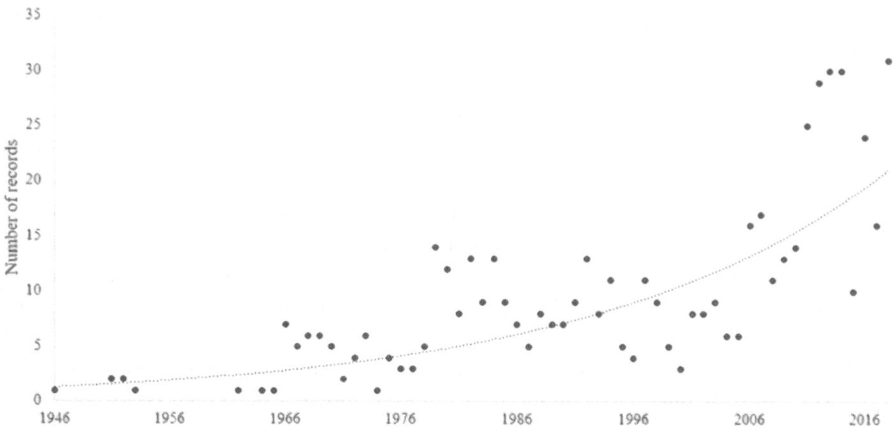

Fig. 14.6 Number of scientific articles on pseudoscorpions published between 1945 and 2018 in the Web of Science

Table 14.1 Summary of the data related to the scientific articles on pseudoscorpions published in the Web of Science between 1945 and 2018

Description	Results
Number of articles	560
Period of investigation	1945–2018
Mean citations per published paper	8
Number of authors	505
Authors of single-authored papers	59
Multiauthored papers	446
Single-authored papers	229
Papers per author	1.11
Number of co-authors per paper	2.4
Collaboration index	1.35

14.4 Field Collection Techniques

14.4.1 Collection and Preservation of Pseudoscorpiones

Pseudoscorpions are found in almost all ecological niches, although their abundance is discretely reduced in comparison with that of the other groups. Specialized tools are generally not required for collecting pseudoscorpion specimens, although this depends on the search method employed and the habitat from which the specimens are being collected, and specimens of certain families are captured on the basis of their degree of adaptation (Chamberlin 1931; Mahnert and Adis 2002; Barrientos 2004).

For the species that are subjected to direct capture, such as those thriving beneath rock crevices, or cave-dwelling and anthropophilic species, among others, only a brush, soft tweezers, a vacuum cleaner, and tubes filled with 70% alcohol are

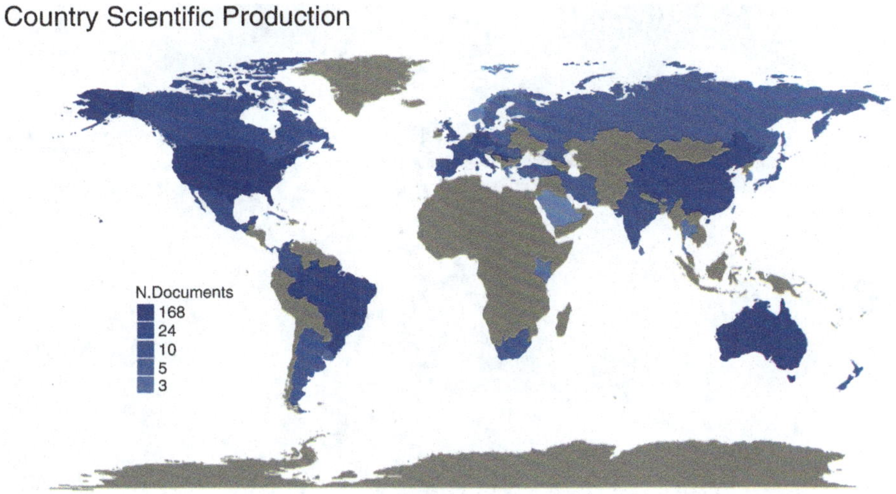

Fig. 14.7 Country-wise distribution of the number of scientific publications on pseudoscorpions published in the Web of Science between 1945 and 2018

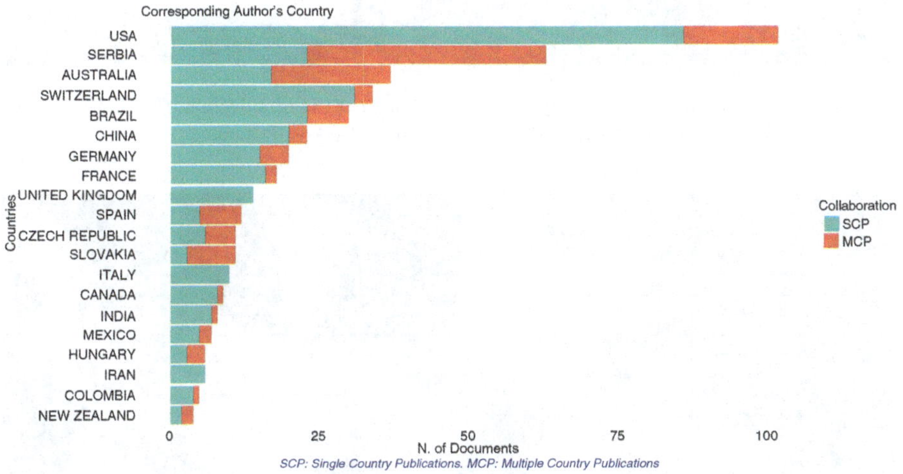

Fig. 14.8 Participation of researchers as first or second authors in the articles published on pseudoscorpions between 1945 and 2018 in the Web of Science

required for retaining and preserving the specimens (Figs. 14.9 and 14.10). The use of Winkler-type sieves is necessary for collecting endogenous species and species inhabiting mossy regions, detritus, and forest leaf litter, followed by selection using Berlese funnels; likewise, a quadrant can be used (Fig. 14.9a, d), depositing all the material in sealed bags to prevent moisture loss (Fig. 14.9b, c) (Parr and Chown 2001; Ivanov and Keiper 2009). The same method is used for the collection of species that are found in the burrows of mammals, bird nests, and stables, after

Fig. 14.9 Litter sampling using the quadrant technique: (**a**) Litter collection. (**b**) Humus collection. (**c**) Sealed bags. (**d**) Quadrant used. Photos by Edwin Bedoya-Roqueme

Fig. 14.10 Sampling of pseudoscorpions thriving in the bark of trees: (**a**) Direct search. (**b**) Collection with tweezers. Photos by Edwin Bedoya-Roqueme

collecting the humus covering the ground (Fig. 14.9a, c). A second visit to the collection site or continued studies in the same area are very advantageous for the use of fall traps at different levels of the substrate (Chamberlin 1931; Weygoldt 1969; Gabbutt 1970; Mahnert and Adis 2002).

The use of entomological sleeves or umbrellas is suitable for collecting the genera that inhabit the upper parts of bushes or tree branches in certain seasons of the year. Depending on the degree of adaptation, markedly halophilic species can be found in the bushes near the coastlines, in dunes, and in the cracks of rocks that are bathed in waves or flooded in cycles of tides, for which manual collection is recommended for direct sampling (Fig. 14.10a, b) (Mahnert and Adis 2002; Battirola et al. 2017a, b; Dambros et al. 2018).

14.4.2 Collection and Housing of Live Pseudoscorpions

Due to their small size and cryptic habits, it is difficult to directly observe and study the behavior of pseudoscorpions in their natural habitats (Fig. 14.11a). For example, several species of Atemnidae and Chernetidae live beneath the barks of living trees, making it difficult to observe the animals without damaging their microhabitat (Fig. 14.11b). Therefore, the collection of living specimens for laboratory studies represents an important task in the understanding of pseudoscorpion biology and behavior, for example, reproduction, parental care, and hunting and feeding behavior (Lira et al. 2014; Tizo-Pedroso and Del-Claro 2005, 2014; Zeh and Zeh 2013).

Field collection methods require a few materials but demand a lot of patience and skill. In general, a thin brush, tweezers, and small transparent plastic (or acrylic) vials are enough for specimen collection. Pseudoscorpions thriving under rocks or tree barks can be collected by touching their pedipalps with the bristles of the brush (Fig. 14.10a, b). Pseudoscorpions tend to grab the bristles, at which point they can be removed and transferred to the vials. A small piece of humid paper is introduced into the vials for maintaining the humidity during transportation, and for providing a substrate for the pseudoscorpions to hold onto and hide, for avoiding stress. Since pseudoscorpions are commonly cannibalistic, the individual specimens must be kept separately in the vials of 5–15 mL.

Some species of pseudoscorpions are naturally more tolerant to laboratory conditions than others. The individuals can be housed in relatively small containers (Fig. 14.12a). Apart from this, it is important to keep a small fragment of the natural substrate in the container, for example, a piece of dried leaf, a piece of bark, or a piece of filter paper could be used instead (Fig. 14.12b). Pseudoscorpions can be housed in transparent 15 mL vials or inside small petri dishes (Fig. 14.12c). If feasible, the container could also be maintained in a seed germination chamber for regulating the luminosity, temperature, and humidity.

Eventually, studies on arthropod inventories may represent an opportunity to find pseudoscorpions. These arachnids can be obtained by pitfall traps, using the Winkler and Berlese sampling methods. However, pseudoscorpions can be found attached to some insects such as beetles and flies (Santos et al. 2005; Lira et al. 2014; Karpinski et al. 2017; Lira and Tizo-Pedroso 2017; Christophoryová et al. 2018; Krajcovicova et al. 2018) or associated with plant galls (Maia and Carvalho-Fernandes 2016; Maia and Silva 2016). This behavior is known by phoresy, a biological interaction

Fig. 14.11 Common habitat structure used by pseudoscorpions under the bark of trees (**a**) *Vatairea macrocarpa* (Leguminosae). (**b**) *Caesalpinia pluviosa* var. *peltophoroides* (Fabaceae); both from Brazilian Cerrado. Photos by Everton Tizo-Pedroso

Fig. 14.12 Pseudoscorpion colonies, housed under laboratory conditions. (**a**, **b**) Colonies housed on glass plates of dimensions 10 cm × 10 cm and 15 cm × 20 cm, respectively, and housed inside transparent plastic boxes. (**c**) Pseudoscorpions housed in a petri dish (10 cm)

Fig. 14.13 Pseudoscorpion phoresy. Multiple phoresy of *Paratemnoides nidificator* by attachment to a Reduviidae (Hemiptera). Photo by Everton Tizo-Pedroso

in which a small animal gets physically transported by a larger one (Szymkowiak et al. 2007; Bartlow et al. 2016). Phoresy is widely known among pseudoscorpions (Fig. 14.13), and the sampling of Coleopteran, Hemipteran, or Dipteran fauna may offer an opportunity to observe this phenomenon (Fig. 14.13) (Beier 1948, Poinar et al. 1998; Reyes-Castillo and Hendrichs 1975). These findings are remarkably interesting with respect to understanding the biology and behavior of pseudoscorpions.

14.5 Techniques for Mounting and Laboratory Study

14.5.1 Dissection Process

Pseudoscorpion size ranges between 0.5 and 10 mm. Due to their small size, pseudoscorpions are dissected under a stereomicroscope. In this process, the specimens should initially be rinsed with distilled water, following which they are deposited in a petri dish containing 70% alcohol. The specimen is submerged and placed in ventral view, and by general convention, it is recommended to dissect all the appendices on the right side, as described hereafter, including leg I, leg IV, a chelicera, and a pedipalp, the chelae of which are separated (Fig. 14.14a, b). The separate parts must subsequently be deposited in a microvial within the same vial that contains the

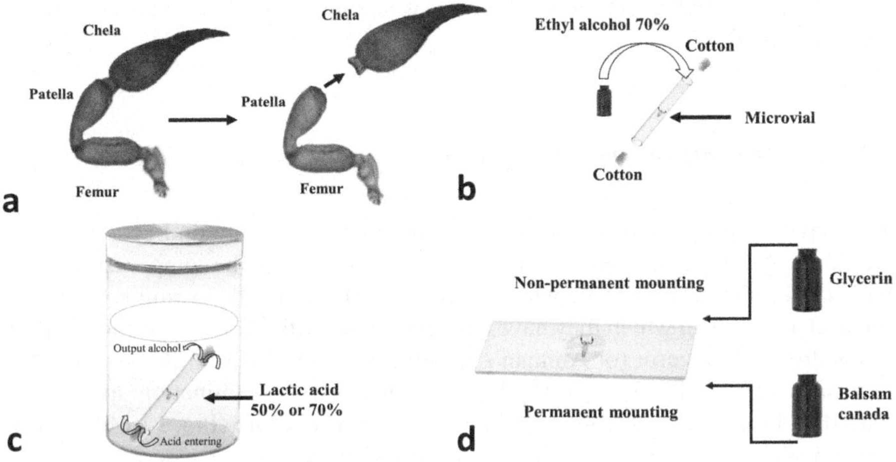

Fig. 14.14 Laboratory techniques for mounting and studying. (**a**) Process of pedipalp and chela dissection. (**b**) Process of specimen preparation. (**c**) Slow diffusion in lactic acid. (**d**) Temporary and permanent mounting processes for observation

study specimen, for future mounting and observation (Chamberlin 1931; Harvey 1992; Judson 1992a; Mahnert and Adis 2002).

14.5.2 Process of Specimen Preparation

Pseudoscorpions are arachnids whose sclerotization and lipid content largely depend on the microhabitat and sex of the individual (Weygoldt 1969; Judson 1992a). Following dissection of the different appendages, 5–10% KOH solution is used as a lightening solution for strongly sclerotized specimens, and the duration of exposure is varied according to the species being studied. Authors such as Harvey (1992) recommend making an incision in the pleural membrane and exposing the specimen to 10% KOH, followed by dehydration in alcohol at different concentrations. Judson (1992b) suggests a similar slow diffusion process using 50–70% lactic acid for minimizing damage to the study material (Fig. 14.14b, c). Zaragoza (2002) also advises the use of 50–70% lactic acid or Amman's lactophenol as lightening solutions.

By general agreement, the use of 5–10% KOH solution is recommended for studying the genitalia and oral parts and for dissolving the internal tissues (Judson 1994). In case of specimens with high fat content (especially female specimens), it is recommended to soak the specimen in 10% acetone solution for a few hours, for dissolving the fat and allowing the observation of the spermatheca, following which the specimen is rinsed with 70% alcohol. Mahnert and Adis (2002) suggested that once the material has been rinsed with the aforementioned products, it should be washed with abundant amounts of acetic water and passed through alcohols of increasing gradation.

14.5.3 Mounting Process and Microscopic Observation

Two types of preparative methods are used for the assembly and observation of pseudoscorpions, depending on the purpose of the study. The temporary method, which is the simplest and most widely used, consists of directly examining the specimens in the vials in which they have been preserved with alcohol. For this purpose, a few drops of glycerin (or Amman's lactophenol) are placed in the bottom of an excavated object holder, into which the animal or the examination specimen is placed in the desired orientation and covered with a coverslip (Fig. 14.14d). By this method, the specimen can be directly observed under the microscope. Once the examination is complete, the specimen is washed in alcohol and reintroduced in the tube. The advantage of this method lies in the fact that the animal can be rotated for observing the details on the lateral sides of the specimen and can be arranged in the desired orientation (Barrientos 2004; Harvey 1992; Judson 1994). The small portion of a fishing nail could be useful for mounting the specimens on microscopic slides with 10 or 12 mm coverslips supported by small sections of fishing lines of 0.25, 0.35, or 0.50 mm diameter (Fig. 14.14d) (Harvey 1992; Harvey et al. 2012).

In the permanent preparation method, following rinsing with the aforementioned solutions, the specimen is washed with abundant quantities of acetic water, passed through alcohols of increasing concentration, and soaked in xylol prior to fixing with balsam of Canada, André, or Swann liquid, according to the preference of the researcher (Fig. 14.14d). Prior to preparation, the specimen is cut along one side of the pleura, so that the dorsal and ventral parts of the opisthosoma can be opened like a book. The specimen is mounted in this orientation, together with the appendages placed in a desired orientation, with a few drops of preservative liquid placed in an excavated portal and covered with a coverslip (Fig. 14.14d). The disadvantages of the permanent method are that the specimens are fixed and thus its orientation cannot be altered and there are chances of progressive lightening and possible deformation of the specimens due to pressure or alteration of the preservative (Harvey 1992; Judson 1994; Barrientos 2004).

14.6 Identification and Key to Families

From a systematic point of view, the classification of the order includes the consideration of phylogenetic characters, variations during postembryonic development, and the apomorphies. By this strategy, the main characters that are to be taken into account include the presence of a venom apparatus in the fingers of the palpal chela, the structure of the chelicerae, the aspect of the carapace, and the number and disposition of the trichobothria in the pedipalps.

The postembryonic developmental stages can be easily identified from the number of trichobothria on the movable finger on the chela of the pedipalp. The protonymph possesses one trichobothrium, the deutonymph has two trichobothria, the tritonymph stage has three trichobothria, and the adult has four trichobothria (Harvey 1992; Mahnert and Adis 2002).

Terminologies and mensuration mostly follow the scheme provided by Chamberlin (1931), except for the naming of the pedipalps and legs, and some minor terminological modifications for the trichobothria (Harvey 1992), chelicerae (Harvey 2007; Judson 2007), and the faces of the appendages (Harvey et al. 2012). The ratios are given (×) as length/width for carapace, chelicerae, and pedipalps and length/depth for the legs. When two articles are compared, the length/length ratio index is considered.

Here, we present an adaptation and extension of the identification keys proposed by Harvey (1992) and Mahnert and Adis (2002) for pseudoscorpion families. To provide an update, we considered the description made by Harms and Harvey (2013) for family Pseudotyrannochthoniidae; and Harvey and Šťáhlavský (2010) for families Garypinidae and Larcidae; and the modifications made by Benavides et al. (2019); as follows:

1. Chelal fingers without venom apparatus; front legs with one tarsal segment; hind legs with two tarsal segments; and trichobothrium*xs* present	2
One or both chelal fingers with venom apparatus; all walking legs with the same number of tarsal segments; trichobothrium *xs* absent (suborder Iocheirata)	6
2. Carapace with anterior lobes; tergite XI and sternite XI fused and surrounding anal cone; all legs with metatarsi and tarsi fused (suborder Atoposphyronida); superfamily Feaelloidea	3
Carapace without anterior lobes; tergite XI and sternite XI not fused and neither surrounding anal cone; and only legs I and I1 with metatarsi and tarsi fused (suborder Heterosphyronida); superfamily Chthonioidea	4
3. Carapacal alae absent and carapace and abdomen with unusual articulation	Family Feaellidae
Carapacal alae present and carapace and abdomen with normal articulation	Family Pseudogarypidae
4. Trichobothria*ib* and *isb* situated distally near the base of the fixed finger of the pedipalpal hand so that the hand appears to be devoid of trichobothria	Family Pseudotyrannochthoniidae.
Trichobothria *ib* and *isb* which are situated on the dorsum of chelal hand	Family Chthoniidae
(a) Carapace of adults with numerous (80–140) setae; galea of nymphs consist of 2–3 stylets	Subfamily Tridenchthoniinae
Carapace of adults with 16–18 large setae and galea of nymphs simple and dome-shaped	b.

(b) Trichobothria*eb* and *esb* on the dorsum of hand and blades of flagellum in a straight line, with penultimate blade curved and recumbent	Subfamily Lechytiinae
Trichobothria *eb* and *esb* at the base of the fingers and blades of rallum arranged in two rows	Subfamily Chthoniinae
6. Movable cheliceral finger with several teeth; *gs* submedial; serrula exterior attached for 1/2 to 2/3 of its length; and chelicerae without velum or lamina exterior (infraorder Hemictenata), superfamily Neobisioidea	7
Movable cheliceral finger with 1–2 subapical teeth; gs subdistal; serrula exterior attached along the entire length; and chelicerae with velum and lamina exterior (infraorder Panctenata)	13
7. Venom ducts very short, not extending past et and nodus ramosus expanded	8
Venom ducts long, extending well past et; and nodus ramosus not expanded	11
8. Venom apparatus only present in the fixed chelal finger	9
Venom apparatus only present in the movable chelal finger	Family Gymnobisiidae
9. Apex of pedipalpal coxa with two setae	10
Apex of pedipalpal coxa with three or more setae	Family Neobisiidae
10. Arolium without distal extensions and trichobothrium t lanceolate	Family Syarinidae
Arolium with distal extensions and trichobothrium t acuminate	Family Parahyidae
11. Chelae with 30–45 trichobothria	Family Ideoroncidae
Chelae with 12 trichobothria	12
12. Femurs I and II with small sub-basal mound and female anterior genital operculum with minute setae	Family Hyidae
Femurs I and II without small sub-basal mound and female genital operculum with normal setae	Family Bochicidae
13. Carapace subtriangular; abdomen subovate; and eyes present situated away from the anterior margin of the carapace	14
Carapace subrectangular; abdomen subrectangular; and eyes absent and/or eyespots present situated near the anterior margin of the carapace	22
14. Paired and enlarged dorsal anterior glands of the male present (superfamily Garypinoidea)	15
Paired and enlarged dorsal anterior glands of the male absent	16
15. Arolia bifurcate; coxa IV approximately the same width as coxa I; and sternite XI without desclerotized region surrounding anal cone	Family Garypinidae
Arolia undivided; coxa IV much wider than coxa I; and sternite XI with desclerotized region surrounding anus	Family Larcidae
16. Two eyes (superfamily Cheiridioidea)	17
Four eyes (superfamily Garypoidea)	18
17. Tergites chevron-shaped and coxa IV much enlarged posteriorly	Family Pseudocheiridiidae

Tergites not chevron-shaped and coxa IV not enlarged posteriorly	Family Cheiridiidae
18. Rallum cheliceral with three blades and fixed chelal finger without pitlike structures on external margin	19
Rallum cheliceral with a blade; fixed chelal finger with pitlike structures on external margin; spiracular helix present; and bipartite median genital sac of males absent	Family Geogarypidae
19. Spiracular helix present and bipartite median genital sac of males absent	20
Spiracular helix absent and bipartite median genital sac of males present	Family Garypidae
20. Venom apparatus present in both chelal fingers; fixed chelal finger with at most 8 trichobothria; pedipalpal femur with 1–2 trichobothria, usually in basal half; and specialized articulation joint absent between coxae II and III	21
Venom apparatus only present in the fixed chelal finger; fixed chelal finger with 11 trichobothria; pedipalpal femur without trichobothria in basal half; and specialized articulation joint present between coxae II and III	Family Menthidae
21. Venom ducts short on chelal fingers and rarely extend past trichobothria *et*	Family Olpiidae
Venom ducts long on chelal fingers, which are usually extremely long and extend past trichobothria *et*	Family Hesperolpiidae
22. Pseudosternum present and spermathecae absent (superfamily Sternophoroidea)	Family Sternophoridae
Pseudosternum absent and spermathecae present (superfamily Cheliferoidea)	23
23. Junction of femur/patellae I and II perpendicular and male sternites generally with discrete patches of sensory setae	Family Withiidae
Junction of femur/patellae I and II oblique and male sternites without discrete patches of sensory setae	24
24. Venom apparatus present in both chelal fingers and males generally with coxal sacs and ram's horn organs	Family Cheliferidae
Venom apparatus generally absent from at least one chelal finger and males without coxal sacs or ram's horn organs	25
25. Venom apparatus generally present in the movable chelal finger only; tarsi with proximal raised slit sensillum; and chelal fingers normally with at least one accessory tooth	Family Chernetidae
Venom apparatus present in the fixed chelal finger only; tarsi without proximal raised slit sensillum; and chelal fingers without accessory teeth	Family Atemnidae

Acknowledgments E Bedoya-Roqueme thanks to Coordenação de Aperfeiçoamento de Pessoal de Nível Superior (CAPES) for the scholarship. E Tizo-Pedroso thanks to Conselho Nacional de Desenvolvimento Científico e Tecnológico (CNPq) (408977/2016-7) and Fundação de Amparo à Pesquisa do Estado de Goiás (FAPEG) (201610267001020) for the funding grants.

References

Adis J, Mahnert V (1985) On the natural history and ecology of Pseudoscorpiones (Arachnida) from an Amazonian blackwater inundation forest. Amazoniana 9:297–314

Adis J, Mahnert V, de Morais JW, Rodrigues JMG (1988) Adaptation of an Amazonian pseudoscorpion (Arachnida) from dryland forests to inundation forests. Ecology 69:287–291. https://doi.org/10.2307/1943185

Aguiar NO, Bührnheim PF (1992) Pseudoscorpiones (Arachnida) in phoretic association with Passalidae (Insecta, Coleoptera) in the Amazon State, Brazil. Amazoniana 12:187–205

Aguiar NO, Bührnheim PF (1998) Phoretic pseudoscorpions associated with flying insects in Brazilian Amazônia. J Arachnol 26:452–459

Aguiar NO, Bührnheim PF (2003) Pseudoescorpiões (Arachnida) da vegetação de sub-bosque da floresta primária tropical de terra firme (Coari, Amazonas, Brasil). Acta Amazon 33:515–526. https://doi.org/10.1590/S0044-59672003000300016

Aguiar NO, Gualberto TL, Franklin E (2006) A medium-spatial scale distribution pattern of Pseudoscorpionida (Arachnida) in a gradient of topography (altitude and inclination), soil factors, litter in central Amazonia forest reserve, Brazil. Braz J Biol 66:791–802. https://doi.org/10.1590/S1519-69842006000500004

Andrade R, Gnaspini P (2002) Feeding in *Maxchernes iporangae* (Pseudoscorpiones, Chernetidae) in captivity. J Arachnol 30:613–617. https://doi.org/10.1636/0161-8202(2002)030[0613:FIMIPC]2.0.CO;2

Barrientos JA (2004) Curso práctico de entomología. Universidad Autonoma De Barcelona, Barcelona

Bartlow AW, Villa SM, Thompson MW et al (2016) Walk or ride? Phoretic behaviour of amblyceran and ischnoceran lice. I J Parasitol 46:221–227. https://doi.org/10.1016/j.ijpara.2016.01.003

Battirola LD, Brizzola GS, Meurer E et al (2017a) Soil and canopy Pseudoscorpiones (Arthropoda, Arachnida) in a monodominant forest of *Attalea phalerata* Mart. (Arecaceae) in the Brazilian Pantanal. Stud Neotrop Fauna E 52:87–94. https://doi.org/10.1080/01650521.2017.1282210

Battirola LD, Rosado-Neto GH, Batistella DA et al (2017b) Vertical and time distribution of Pseudoscorpiones (Arthropoda: Arachnida) in a floodplain forest in the Brazilian Pantanal. Rev Biol Trop 65:445–459

Bedoya-Roqueme E (2015) Una nueva especie de *Paraliochthonius* (Pseudoscorpiones: Chthoniidae: Tyrannochthoniini) de Colombia. Rev Iber Aracnol 26:49–54

Bedoya-Roqueme E, Salleg-Pérez G, Quirós-Rodríguez JA (2016) Sobre la ecología de pseudoescorpiones (Arachnida: Pseudoscorpiones) en bosques de manglar del sur del golfo de Morrosquillo, Caribe colombiano. Rev Iber Aracnol 28:65–74

Bedoya-Roqueme E, Pérez-Agudelo M, Zaragoza JA et al (2017) Nuevos reportes de Pseudoescorpiones (Arachnida) de bosques de manglar en Córdoba, Caribe Colombiano. Rev Iber Aracnol 30:25–36

Beier M (1948) Phoresie and phagophilie bei pseudoscorpionen. Oesterr Zool Zeitschr 1:441–497

Beier M (1963) Ordnung Pseudoscorpionidea (Afterskorpione). In: d'Aguilar J (ed) Bestimmungsbücher zur Bodenfauna Europas. Akademie–Verlag, Berlin, pp vi, 1–vi313

Belozerov VN (2013) Seasonal aspects of the life cycles of pseudoscorpions (Arachnida, Pseudoscorpiones). Entomol Rev 93:634–652. https://doi.org/10.1134/s0013873813050126

Benavides LR, Cosgrove JG, Harvey MS et al (2019) Phylogenomic interrogation resolves the backbone of the Pseudoscorpiones tree of life. Mol Phylogenet Evol 139:1–14. https://doi.org/10.1016/j.ympev.2019.05.023

Chamberlin JC (1931) The arachnid order Chelonethida. Stanford University, Oxford University Press, California

Christophoryová J, VidliČka Ľ, KrajČoviČová K (2018) New cases of phoresy of *Lamprochernes nodosus* (Pseudoscorpiones: Chernetidae) on Diptera observed in Slovakia. Biharean Biol 12:114–115

Cole DC, Elgar MA, Harvey MS (1994) Associations between Australian pseudoscorpions and ants. Psyche 101:221–227. https://doi.org/10.1155/1994/23982

Dambros J, Vindica VF, Delabie JHC et al (2018) Canopy ant assemblage (Hymenoptera: Formicidae) in two vegetation formations in the northern Brazilian pantanal. Sociobiology 65:358–369. https://doi.org/10.13102/sociobiology.v65i3.1932

Del-Claro K, Tizo-Pedroso E (2009) Ecological and evolutionary pathways of social behavior in pseudoscorpions (Arachnida: Pseudoscorpiones). Acta Ethol 12:13–22. https://doi.org/10.1007/s10211-009-0052-y

Francke OF, Villegas-Guzmán GA (2006) Symbiotic relationships between pseudoscorpions (Arachnida) and packrats (Rodentia). J Arachnol 34:289–298. https://doi.org/10.1636/04-36.1

Gabbutt PD (1970) Sampling problems and the validity of life history analyses of pseudoscorpions. J Nat Hist 4:1–15. https://doi.org/10.1080/00222937000770011

Garwood RJ, Dunlop JA, Knecht BJ et al (2017) The phylogeny of fossil whip spiders. BMC Evol Biol 17:105. https://doi.org/10.1186/s12862-017-0931-1

Giribet G, Edgecombe GD, Wheeler WC et al (2002) Phylogeny and systematic position of Opiliones: a combined analysis of chelicerate relationships using morphological and molecular data. Cladistics 18:5–70. https://doi.org/10.1006/clad.2001.0185

Gonzalez VH, Mantilla B, Mahnert V (2007) A new host record for *Dasychernes inquilinus* (Arachnida, Pseudoscorpiones, Chernetidae), with an overview of pseudoscorpion-bee relationships. J Arachnol 35:470–474. https://doi.org/10.1636/H06-62.1

Harms D, Dunlop JA (2017) The fossil history of pseudoscorpions (Arachnida: Pseudoscorpiones). Foss Rec 20:215–238. https://doi.org/10.5194/fr-20-215-2017

Harms D, Harvey MS (2013) Review of the cave-dwelling species of *Pseudotyrannochthonius* Beier (Arachnida: Pseudoscorpiones: Pseudotyrannochthoniidae) from mainland Australia, with description of two troglobitic species. Aust J Entomol 52:129–143. https://doi.org/10.1111/aen.12009

Harvey MS (1986) The systematics and biology of pseudoscorpions. In: Austin AD, Heather NW (eds) Australian arachnology. Australian Entomological Society, Brisbane, pp 75–85

Harvey MS (1992) The phylogeny and classification of the Pseudoscorpionida (Chelicerata: Arachnida). Invertebr Taxon 6:1373–1435. https://doi.org/10.1071/IT9921373

Harvey MS (2002) The neglected cousins: what do we know about the smaller arachnid orders? J Arachnol 30:357–372. https://doi.org/10.1636/0161-8202(2002)030[0357:TNCWDW]2.0.CO;2

Harvey MS (2007) The smaller arachnid orders: diversity, descriptions and distributions from Linnaeus to the present (1758 to 2007). Zootaxa 1668:363–380. https://doi.org/10.11646/zootaxa.1668.1.19

Harvey MS (2009) The first Australasian species of the genus pseudoscorpion halophilic *Paraliochthonius* (Pseudoscorpiones: Chthoniidae). Rec West Aust Mus 25:329–344. https://doi.org/10.18195/issn.0312-3162.25(3).2009.329-344

Harvey MS (2013) Order Pseudoscorpiones. In: Zhang ZQ (ed) Animal biodiversity: an outline of higher-level classification and survey of taxonomic richness. Zootaxa 3703:34–35. https://doi.org/10.11646/zootaxa.3703.1.8

Harvey MS, Muchmore WB (2013) The systematics of the pseudoscorpion family Ideoroncidae (Pseudoscorpiones: Neobisioidea) in the New World. J Arachnol 41:229–290. https://doi.org/10.1636/k13-42.1

Harvey MS, Šťáhlavský F (2010) A review of the pseudoscorpion genus *Oreolpium* (Pseudoscorpiones: Garypinidae), with remarks on the composition of the Garypinidae and on pseudoscorpions with bipolar distributions. J Arachnol 38:294–308. https://doi.org/10.1636/A09-53.1

Harvey MS, Ratnaweera PB, Udagama PV et al (2012) A new species of the pseudoscorpion genus *Megachernes* (Pseudoscorpiones: Chernetidae) associated with a threatened Sri Lankan rainforest rodent, with a review of host associations of *Megachernes*. J Nat Hist 46:2519–2535. https://doi.org/10.1080/00222933.2012.707251

Hoff CC (1963) The pseudoscorpions of Jamaica. Part 2. The genera *Pseudochthonius*, *Paraliochthonius*, *Lechytia*, and *Tridenchthonius*. Bull Inst Jam Sci Ser 10:1–35

Ivanov K, Keiper J (2009) Effectiveness and biases of Winkler litter extraction and pitfall trapping for collecting ground-dwelling ants in Northern temperate forests. Environ Entomol 38:1724–1736. https://doi.org/10.1603/022.038.0626

Judson M (1992a) A simple, slow-diffusion method for clearing small arthropods. Newsl Br Arachnol Soc 64:6–7

Judson MLI (1992b) The pseudoscorpion subgenus *Chthonius* (*Sigmodactylus*) Hadzi, 1930: type fixation and consequent synonymy with *Pseudochthonius* Balzan, 1982 (Chelonethi, Chthoniidae). Bull Br Arachnol Soc 9:64

Judson MLI (1994) Les Pseudoscorpions. Pénélope 12:6–37

Judson MLI (2016) Pseudoscorpions (Arachnida, Chelonethi) in Mexican amber, with a list of extant species associated with mangrove and Hymenaea trees in Chiapas. Bol Soc Geol Mex 68:57–79. https://doi.org/10.5281/zenodo.50595

Judson MLI (2007) A new and endangered species of the pseudoscorpion genus *Lagynochthonius* from a cave in Vietnam, with notes on chelal morphology and the composition of the Tyrannochthoniini (Arachnida, Chelonethi, Chthoniidae). Zootaxa: 53–68. https://doi.org/10.11646/zootaxa.1627.1.4

Karpinski L, Rutkowski T, Szczepanski WT (2017) First record of phoresy of *Dendrochernes cyrneus* (L. Koch, 1873) (Pseudoscorpiones, Chernetidae) on *Cerambyx cerdo* Linnaeus, 1758 (Coleoptera, Cerambycidae) and their potential value as bioindicators. Anim Biodiv Conserv 40:187–192. https://doi.org/10.32800/abc.2017.40.0187

Krajcovicova K, Matyukhin AV, Christophoryova J (2018) First comprehensive research on pseudoscorpions (Arachnida: Pseudoscorpiones) collected from bird nests in Russia. Turk J Zool 42:480–487. https://doi.org/10.3906/zoo-1801-47

Lee VF (1979) The marine pseudoscorpions of Baja California, Mexico (Arachnida: Pseudoscorpionida). Calif Acad Sci 131:1–38

Lira AFA, Tizo-Pedroso E (2017) Report of *Sphenochernes camponoti* (Beier, 1970) (Pseudoscorpiones, Chernetidae) in phoresy on Fanniidae (Diptera). Acta Sci 39:449–454. https://doi.org/10.4025/actascibiolsci.v39i4.36373

Lira AFA, Tizo-Pedroso E, Albuquerque CMR (2014) Phoresy by *Americhernes* Aff. *incertus* (Pseudoscorpiones: Chernetidae) on a tropical fly *Fannia canicularis* (Diptera: Fanniidae) in a fragment of the Atlantic forest, Brazil. Entomol News 124:24–28. https://doi.org/10.3157/021.124.0103

Mahnert V (2014) Pseudoscorpions (Arachnida: Pseudoscorpiones) from the Galapagos Islands (Ecuador). Rev Suisse Zool 121:135–210

Mahnert V, Adis J (2002) Pseudoscopiones. In: Adis J (ed) Amazonian Arachnida and Myriapoda: identification keys, to all classes, orders, families, some genera, and lists of known terrestrial species. Pensoft Publishers, Sofia, pp 367–380

Mahnert V, Schuster R (1981) *Pachyolpium atlanticum* n. sp., ein Pseudoskorpion aus der Gezeitenzone der Bermudas: Morphologie und Ökologie (Pseudoscorpiones: Olpiidae). Rev Suisse Zool 88:265–273

Mahnert V, Sharaf M, Aldawood AS (2014) Further records of pseudoscorpions (Arachnida, Pseudoscorpiones) from Saudi Arabia. Zootaxa 3764:387–393. https://doi.org/10.11646/zootaxa.3764.3.8

Maia VC, Carvalho-Fernandes SP (2016) Insect galls of a protected remnant of the Atlantic Forest tableland from Rio de Janeiro State (Brazil). Rev Bras Entomol 60:40–56. https://doi.org/10.1016/j.rbe.2015.09.001

Maia VC, Silva LO (2016) Insect galls of Restinga de Marambaia (Barra de Guaratiba, Rio de Janeiro, RJ). Braz J Biol 76:787–795. https://doi.org/10.1590/1519-6984.05314

Martínez RJ, Quirós DI, Emmen D et al (2018) Primer reporte de foresia de pseudoscorpiones (Arachnida: Pseudoscorpiones: Cheiridiidae) por *Panstrongylus geniculatus* (Latreille, 1811) (Insecta: Hemiptera: Reduviidae). Rev Iber Aracnol 32:127–130

Muchmore WB (1972) The pseudoscorpion genus *Paraliochthonius* (Arachnida, Pseudoscorpionida, Chthoniidae). Entomol News 83:248–256

Muchmore WB (1974) Pseudoscorpions from Florida. 3. *Epactiochernes*, a new genus based upon *Chelanops tumidus* Banks (Chernetidae). Fla Entomol 57:397–407. https://doi.org/10.2307/3493504

Muchmore WB (1984) Further data on *Mucrochernes hirsutus* (Banks) (Pseudoscorpionida, Chernetidae). Pan-Pac Entomol 60:20–22

Muchmore WB (1994) On four species of pseudoscorpions from California described by E. Simon in 1878 (Pseudoscorpionida: Neobisiidae, Chernetidae, Cheliferidae). J Arachnol 22:60–69

Murienne J, Harvey MS, Giribet G (2008) First molecular phylogeny of the major clades of Pseudoscorpiones (Arthropoda: Chelicerata). Mol Phylogenet Evol 49:170–184. https://doi.org/10.1016/j.ympev.2008.06.002

Parr CL, Chown SL (2001) Inventory and bioindicator sampling: TESTING pitfall and Winkler methods with ants in a South African savanna. J Insect Conserv 5:27–36. https://doi.org/10.1023/A:1011311418962

Poinar GOJ, Curcic BPM, Cokendolpher JC (1998) Arthropod phoresy involving pseudoscorpions in the past and present. Acta Arachnol 47:79–96. https://doi.org/10.2476/asjaa.47.79

Reyes-Castillo P, Hendrichs J (1975) Pseudoescorpiones asociados con pasálidos. Acta Politéc Mex 16:129–133

Santos JC, Tizo-Pedroso E, Fernandes GW (2005) A case of phoresy of *Semeiochernes armiger* Balzan, 1892 (Pseudoscorpiones: Chernetidae) on the giant tropical fly *Pantophthalmus tabaninus* Thunberg, 1819 (Diptera: Pantophthalmidae) in an Amazonian rainforest, Pará. Lundiana 6:11–12

Schimonsky DMV, Bichuette ME (2019) Distribution of cave-dwelling pseudoscorpions (Arachnida) in Brazil. J Arachnol 47:110–123. https://doi.org/10.1636/0161-8202-47.1.110

Shear WA, Schawaller W, Bonamo PM (1989) Record of palaeozoic pseudoscorpions. Nature 341:527–529. https://doi.org/10.1038/341527a0

Shultz JW (2007) A phylogenetic analysis of the arachnid orders based on morphological characters. Zool J Linn Soc 150:221–265. https://doi.org/10.1111/j.1096-3642.2007.00284.x

Szymkowiak P, Górski G, Bajerlein D (2007) Passive dispersal in arachnids. Biol Lett 44:75–101

Tizo-Pedroso E, Del-Claro K (2005) Matriphagy in the neotropical pseudoscorpion *Paratemnoides nidificator* (Balzan 1888) (Atemnidae). J Arachnol 33:873–877. https://doi.org/10.1636/s03-61.1

Tizo-Pedroso E, Del-Claro K (2014) Social parasitism: emergence of the cuckoo strategy between pseudoscorpions. Behav Ecol 25:335–343. https://doi.org/10.1093/beheco/art114

Vachon M (1960) Sur une nouvelle espèce halophile de Pseudoscorpions de l'Archipel de Madère: *Paraliochthonius hoestlandti* (Fam. des Chthoniidae). Bull Mus Natl Hist Nat 32:331–337

Villegas-Guzmán GA, Pérez TM (2005a) Hallazgo Pseudoscorpiones (Arachnida: Pseudoscorpiones) foréticos de *Felis catus* Linnaeus, 1758, en la ciudad de México. Folia Entomol Mex 44:85–87

Villegas-Guzmán GA, Pérez TM (2005b) Pseudoscorpiones (Arachnida: Pseudoscorpionida) asociados a nidos de ratas del Género *Neotoma* (Mammalia: Rodentia) del altiplano Mexicano. Acta Zool Mex 21:63–77

Weygoldt P (1969) The biology of pseudoscorpions. Harvard University Press, Cambridge

Zaragoza JA (2002) Dos nuevos *Roncocreagris* cavernícolas de Galicia (Arachnida, Pseudoscorpiones, Neobisiidae). Rev Iber Aracnol 5:91–98

Zaragoza JA (2015) Orden Pseudoscorpiones. Revista I D E A 20:1–10

Zeh DW, Zeh JA (1992a) Dispersal generated sexual selection in a beetle-riding pseudoscorpion. Behav Ecol Sociobiol 30:135–142. https://doi.org/10.1007/BF00173949

Zeh DW, Zeh JA (1992b) Failed predation or transportation—causes and consequences of phoretic behavior in the pseudoscorpion *Dinocheirus arizonensis* (Pseudoscorpionida, Chernetidae). J Insect Behav 5:37–49. https://doi.org/10.1007/BF01049156

Zeh JA, Zeh DW (2013) On the threshold of dispersal: hitchhiking on a giant fly favours exaggerated male traits in a male-dimorphic pseudoscorpion. Biol J Linn Soc 108:509–520. https://doi.org/10.1111/Bij.12011

Chapter 15
Standardized Sampling Methods and Protocols for Harvestman and Spider Assemblages

Ana Lúcia Tourinho and Nancy Lo-Man-Hung

15.1 Why Use Arachnids in Biodiversity Assessments?

Arachnids are conspicuous and very charismatic invertebrates and are very present in daily human life as an important component of our culture and history. Just to cite some examples, arachnids are represented as major zoomorphic deities in many different cultures and religions across time, such as the goddess weaver named Arachne (spider in Greek) in ancient mythology. A number of scorpion goddesses existed in the Egyptian pantheon as well: Hedetet, Ta-Bitjet, Serket, and the Aztec Malinalxochitl and the Mesopotamian Ishara, to cite a few. An impressive design of a large spider is among the 70 zoomorphic figures in the group of geoglyphs present in the desert of Nazca, Palpa, and Jumana, in southern Peru.

Arachnids are also contemporaneous superheroes of our times (Spider-Man, Venom, Scorpion) (Da-Silva et al. 2014); they are scary, wise, and mysterious characters in awarded movies and books [Charlotte, Charlotte's Web (White 1957); Shelob, Lord of the Rings (Tolkien 1968); Aragog and Amblypygi, Harry Potter (Rowling 1997–2007)], and in his oil canvas—Daddy Longlegs of the Evening-Hope!—the famous artist Salvador Dalí painted a harvestman in reference to a French peasant legend which says that the sighting of a harvestman in the evening hour is a sign of good luck (Pinto-da-Rocha et al. 2007).

Indeed, invertebrates rule the world, they are no less than 95% of all animal species described (Wilson 1987), and their importance for providing stability to the

A. L. Tourinho (✉)
NEBAM—Núcleo de Estudos da Biodiversidade da Amazônia Matogrossense, Instituto de Ciências Naturais, Humanas e Sociais, Universidade Federal do Mato Grosso, Sinop, MT, Brazil

N. Lo-Man-Hung
Instituto de Biociências, Universidade de São Paulo, São Paulo, SP, Brazil

J. C. Santos, G. W. Fernandes (eds.), *Measuring Arthropod Biodiversity*,
https://doi.org/10.1007/978-3-030-53226-0_15

ecosystem and maintaining ecological balance must be highlighted. Yet unfortunately, conservation studies often fail to include reliable invertebrate data; they are still underrepresented in environmental licensing studies and largely understudied in most ecosystems around the Earth. This is commonly attributed to taxonomic bias (there are no specialists for all groups), small body sizes (they are difficult to collect and identify), narrow distribution ranges or difficult sites to reach diversity, and many other factors (Hopkins and Freckleton 2002; Drew 2011; Pearson et al. 2011). But this bias may be overcome by massively adopting standard protocols (Cardoso et al. 2011a, b).

Arthropods are extremely important for humankind because they are fundamental components of all terrestrial and freshwater food webs and they are key contributors to ecosystem services influencing the cycling of carbon and other nutrients (Yang and Gratton 2014; Eisenhauer et al. 2019), affecting cross-ecosystem interactions and subsidies (e.g., see Romero and Srivastava (2010) for terrestrial predator effect on aquatic invertebrates and nutrient dynamics within water-filled bromeliads and Schuldt et al. (2017) for ecosystem functioning dependence on numerous species interactions above and below ground including spiders) and playing an important foundational role as a keystone for vegetation structure and diversity (e.g., see Whitney (2004) for the mutually beneficial relationship between spiders and an Australian shrub and Vleminckx et al. (2019) for how trophic interactions, which include spiders, shape community assembly in tropical forests).

Given the nature of their interactions and relationships with other living organisms and abiotic systems, several invertebrates groups, such as spiders and harvestmen, are highly and quickly affected by environmental changes, being impacted quickly to forest structure, fragmentation, and microclimate changes (Koivula et al. 2019; Tourinho et al. 2019). It is disturbing that declines in invertebrate biomass, diversity, and communities are already being documented for various insects (Forster et al. 2019; Siebold et al. 2019), e.g., bees (Brown and Paxton 2009), beetles (Gardner et al. 2008a), butterflies (Barlow et al. 2007), and several other groups of arthropods including arachnids (Siebold et al. 2019).

Humans are excessively using environmental resources and degrading wildlife; the situation is even more dire given that many species of arachnids, including some species of spiders and most species of harvestmen, are highly endemic to restricted areas (Pinto-da-Rocha et al. 2007; Lo-Man-Hung et al. 2008; Resende et al. 2012) (e.g., Figs. 15.1, 15.2, and 15.3). Habitat loss and other negative environmental impacts are sadly high in the iconic Amazon rainforest and in other regions of the world suffering high anthropogenic pressure which are still largely undersampled [e.g., Cerrado and Caatinga in Brazil (Oliveira et al. 2016, 2019), Borneo (Bowne et al. 2019), Temperate Forest (Fischer et al. 2013); Boreal Forest (Ficken et al. 2019)].

Unfortunately, species of arthropods may become extinct before experts know they ever existed according to recently published results (Pimm and Raven 2000; Zalasiewicz et al. 2010; Lees and Pimm 2015; WWF 2018). This dramatic and silent loss of invertebrates calls for urgent attention especially to tropical forests

Fig. 15.1 *Acanthoscurria geniculata* (CL Koch, 1841), also known as "giant whiteknee," "Brazilian whiteknee tarantula," and "whitebanded tarantula," endemic to the Amazon basin of northern Brazil (Photo: Pimenta F.)

Fig. 15.2 *Megaphobema teceae* (Pérez-Miles et al. 2006), endemic to the northeastern region of South America. It is a relatively large spider (5–7 cm of total length, excluding chelicerae and spinnerets). The first individuals (Pérez-Miles et al. 2006) of this species were collected in pitfall traps for herpetological inventory (30 cm diameter by 42 cm height), because the pitfall traps used to collect spiders (usually 9 cm diameter by 14 cm height) can undersample large specimens, such as theraphosid spiders (Photo: Pimenta F.)

which still have large fractions of biodiversity but unfortunately, at the same time, have very high rates of deforestation (Hansen et al. 2013) due to extremely unsustainable, ecologically destructive, and shortsighted political decisions and policies (Carvalho et al. 2019). New international biodiversity assessments, monitoring

Fig. 15.3 Some endemic harvestmen. (**a**) *Stygnomma* sp. from Manaus, Amazonas—Brazil. It is a minute harvestman mostly collected using litter sifting and manual sorting. (**b**) *Poecilaemula aff. lavarrei* Goodnight and Goodnight, 1942, from Manaus, Amazonas—Brazil. (**c**) *Stygnobates barbiellinii* Mello-Leitão, 1927. (**d**) *Saramacia lucasae* (Jim and Soares 1991) from Manaus, Amazonas. (**e**) *Phareicranaus manauara* (Pinto-da-Rocha 1994), from Manaus, Amazonas. F. *Cobania picea* (Bertkau 1880) from Itatiaia, Rio de Janeiro—Brazil. *P. aff. Lavarrei*, *S. barbiellinii*, *P. manauara*, *Cobania picea*, and *S. lucasae* are larger in size and most easily detected and collected during active nocturnal searches. (Photos: **a**, **e**, Carvalho L.S.; **b**, **d**, Dias S.C.; **c**, **f**, Pinto-da-Rocha R.)

initiatives, and action plans must be developed and implemented quickly before it is too late (Eisenhauer et al. 2019).

Harvestmen, spiders, ants, beetles, and many other arthropod groups are effective bioindicators (Folgarait 1998; Bragagnolo et al. 2007; Greenslade 2007;

Fig. 15.4 Lycosidae. The distribution of arachnid fauna can be influenced by vegetation structure that provides microhabitats as suitable refuges and substrata for web attachment (Wise 1993; Indicatti et al. 2005). The "wolf spiders" from family Lycosidae are well known to be favored by simplified habitats such as grasslands (Jocqué and Alderweireldt 2005) as well as in forests with shallow litter cover (Uetz 1999; Lo-Man-Hung et al. 2008) (Photo: Carvalho L.S.)

Colmenares et al. 2016; Tourinho et al. 2019) (e.g., Figs. 15.3 and 15.4). Spiders, as an example, have potential use for biological control of arthropod pests through predation (Maloney et al. 2003; Nyffeler and Sunderland 2003; Harwood et al. 2004; Hooks et al. 2006), as agrobiont species (the dominating species in various geographical areas) (Samu and Szinetár 2002; Pekár and Haddad 2005; Thorbek and Topping 2005; De Meester and Bonte 2010), and through assessment of guild structure (Gibson et al. 1992; Uetz 1999; Lo-Man-Hung et al. 2011; Cardoso et al. 2011b). Harvestmen and spiders are also important in competition and interactions (Polis and McCornick 1986; Marshall and Rypstra 1999; Wise et al. 1999; Colmenares et al. 2016; Tourinho et al. 2019), diversity, and distribution (Samu et al. 1999; Pinto-da-Rocha et al. 2005; Seyfulina 2010; Samu et al. 2016).

Arachnids represent about 2% of the described species worldwide (Adis 2002) and 12% of soil fauna (Decaëns et al. 2006); both spiders and harvestmen provide valuable information on environmental quality at both small and large scales on the basis of rapid surveys (Maleque et al. 2006; Bragagnolo et al. 2007; Melo et al. 2015; Colmenares et al. 2016; Tourinho et al. 2019).

Thus periodic reviews and adaptations of the criteria used for species inclusion in the IUCN Red List are urgent, also the inclusion of invertebrates in environmental licensing studies prior to excecution of development projects (Maleque et al. 2006; Cardoso et al. 2011a, b; Melo et al. 2015). Only by doing so, with reasonable, adequate information, that it will be possible to assess all taxa, from the smallest invertebrate to the largest vertebrate, in an unbiased and most objective way possible, ensuring the biodiversity necessary for a sustainable future (Martín et al. 2010).

15.2 What Should Be Considered Before Selecting Methods?

15.2.1 What Are the Questions Posed?

An accurate species list of a given area and also an estimate of their abundances and assemblages composition (Fisher 1999; Cardoso 2009; Tourinho et al. 2018) are highly desirable to understand the distribution of species across different sites and scales and their relationship with the environment. Indeed, a successful biodiversity inventory may be divided into two main categories: a strict inventory and a community characterization (Longino and Colwell 1997). A strict inventory focus is to obtain a reliable list of species in a given area, where the species identification is of more interest than their relative abundances. On the other hand, the community characterization focuses on estimating the distributions of species abundances, community species richness, and their complementarity (Colwell and Coddington 1994) with other communities.

The applications of such categories also vary given the goals of different studies, and the strict inventory focus is most useful when some unique trait of organisms have some potential value to human welfare (e.g., agricultural researchers, pharmaceutical companies, natural history studies, etc.), while community characterizations are more often used to set conservation priorities and monitor areas (Longino and Colwell 1997).

Structured inventories integrating important components of both strict inventories and community characterization (Oliver and Beattie 1996; Longino and Colwell 1997) improve the quality of results and the inventory efficiency and harbor more information in the most efficient way possible with less cost (Tourinho et al. 2018).

Having the study purpose in mind, researchers will then choose whether to use an optimized or a standardized protocol. An optimized protocol spreads the number of samples among methods to obtain a maximum number of species, while a standardized protocol is used to be suitable to all sites of the same habitat and to provide comparative data across sites (Malumbres-Olarte et al. 2016).

Standardized protocols may be rich in repertory and possibilities; combinations of several sampling methods are generally used in studies of diversity and in monitoring arachnids (Bonaldo et al. 2009; Tourinho et al. 2018). The idea behind this is that using a wide variety of methods combined should then increase accuracy and inventory efficiency, providing more reliable estimations of species richness, and allow significant comparisons between areas (Cardoso et al. 2009).

The practice of using the very same method to sample different orders of arachnids, however, may result in several biases, different metrics, and high asymmetry in terms of effort efficiency, congruence, and financial cost for harvestmen and spiders (Tourinho et al. 2018). This happens because for some widely used methods, accuracy likely varies naturally with habitat complexity (Privet et al. 2020), and so does efficiency within taxa and in different regions (see Azevedo et al. 2016; Tourinho et al. 2018). This was demonstrated in a study comparing the number of species and composition of spiders and harvestmen simultaneously sampled using three of the most commonly used sampling methods (beating tray, active nocturnal search, and Winkler apparatus) in an Amazonian upland forest (Tourinho et al. 2018). Some methods when used to describe the diversity of different orders of arachnids having different natural histories can be either redundant or even ineffective in their cost-benefit performance.

Additionally the very same methods of sampling the same target organisms may achieve different results and performances depending on the study design (Tourinho et al. 2014; Porto et al. 2016; Bueno et al. 2019) (e.g., Fig. 15.2), the region and environment they are applied to as has been documented for spiders in the Amazon (Tourinho et al. 2018) and in the Atlantic Forest (Azevedo et al. 2013), and even in the same region, results and performances may vary within the same area due to habitat complexity variation (Privet et al. 2020). Then it is always desirable to have in advance detailed knowledge of the natural history of species, geographic region, habitat to be sampled, and the comparisons of the methods to be used; all these factors together can lead to the best use of complementary methods (Cardoso 2009; Cardoso et al. 2009; Azevedo et al. 2013; Tourinho et al. 2018).

Therefore, defining the questions posed and the main scope of the study should be the first step when planning for any project; as mentioned above, the same protocol is not capable of producing the same results for a large spectrum of different studies (Longino and Colwell 1997, Porto et al. 2016), and the same protocol is never going to be able to cover successfully systematics, taxonomy, biogeography, macroecology, community ecology, functional ecology, monitoring, and behavior. Unfortunately, there is no such enchanted protocol to attend to all questions in all types of areas of study.

15.2.2 Landscape and Other Environmental Factors

As taxonomy and taxonomists have become a rare and endangered profession over the decades (Hopkins and Freckleton 2002; Boxshall 2020), the need for professional taxonomists, parataxonomists, and even amateur researchers capable of producing lists of species is always desirable (Hopkins and Freckleton 2002; Drew 2011; Pearson et al. 2011).

The consequences of a shrinking community of taxonomists are likely to impact future conservation efforts in many places (Hopkins and Freckleton 2002). In tropical forests, as an example, many taxa are only identified and collected by experienced taxonomists, e.g., harvestmen individuals belonging to the Phalangiodea of the suborder Eupnoi and the minute members of Escadabiidae, Samoidae, Stygnommatidae (Fig. 15.3a), and Zalmoxidae of the suborder Laniatores can only be accurately included in a proper family and/or subfamily by experts after detailed examination of the genital characters of the male under a microscope.

Experts are way more efficient as well in collecting some specific species of mygalomorph and araneomorph spiders. Species, such as the trapdoors spiders, have cryptozoic and fossorial habits, so finding and recognizing some of their shelters is difficult especially when the trapdoor is closed, because of the silk, plant, and soil material used to camouflage it. Only experienced professionals are able to easily identify it.

Our tropical forests are populated with astonishingly high numbers of species unknown to science, and we are still very ignorant of the distributions of most known species and the quantitative and causal relationships between forest biodiversity and ecosystem services. Thus, ecological and monitoring studies are extremely urgent to advance environmental mitigation planning and policies in the age of the sixth mass extinction and climate crisis (Barnosky et al. 2011; Eisenhauer et al. 2019).

However, as pointed to above, the best combination of methods, effort, and the sampling design to tackle questions on inventory, conservation priorities, and monitoring is completely different depending on regional geography, climate, and the environment structure of the sampling site and on the target taxa (Cardoso 2009; Cardoso et al. 2009; Azevedo et al. 2013; Tourinho et al. 2018). For instance, active nocturnal searches are the most efficient methods for both spider and harvestmen surveys in the tropics (Azevedo et al. 2013; Tourinho et al. 2018), while the use of passive sampling methods, such as pitfall traps, is by far the most common sampling technique used for spider surveys in temperate ecosystems (Černecká et al. 2017; Mihál and Černecká 2017; McCravy 2018), mostly because it is a method that can sample high abundance and diversity depending on specimen activity levels (McCravy 2018).

The differences in diversity of tropical forests compared with temperate forests have intrigued scientists for decades. Several gradients of diversity describe the pattern of higher species diversity in tropical areas compared to temperate areas, such as the latitudinal gradient and other tropical-temperate gradients in arthropod diversity (Pianka 1966; Willig et al. 2003). Recent results of a standardized tropical-temperate forest comparison of species richness and evenness for spiders show that high diversity of these predators is associated not only with higher diversity of prey in tropical ecosystems but also increased diet breadth (Privet and Petillon 2018).

Methods clearly reflect the natural history of the focal group. As generalist predators, harvestmen and spiders are often more dependent on the physical structure of their habitat than on the availability of prey (Srivastava et al. 2008). Spiders have developed a wide range of hunting strategies through their evolutionary trajectory and also explore a variety of habitats and vegetation strata (Privet et al. 2020). All arachnids, in general, are dependent on humid habitats, they are susceptible to desiccation, and consequently majority have nocturnal activity. Environmental variation thus has a major influence in their community distributions at both small and regional scales (Lo-Man-Hung et al. 2011; Colmenares et al. 2016; Tourinho et al. 2019).

Regular disturbance and/or high levels of land management can decrease harvestmen and spider species richness and density and negatively influence turnover of species (Lo-Man-Hung et al. 2011; Černecká et al. 2017; Mihál and Černecká 2017; Tourinho et al. 2019). Indeed, several other studies have shown that spider and harvestmen diversity and density is impacted by reduction in habitat complexity (a reduction of plant species richness) (Jeanneret et al. 2003; Perner and Malt 2003;

Lo-Man-Hung et al. 2008; Haddad et al. 2009; Lafage et al. 2019; Tourinho et al. 2019; Privet et al. 2020), having more impact on their patterns of species composition and abundance than more commonly predictable biogeographic variables such as geographic distance (Tourinho et al. 2019).

15.2.3 Costs and Benefits

Financial cost is a major limiting factor for any research project, and a balance between the costs, effort, and time consumed is highly required and important in modern biodiversity studies (Gardner et al. 2008b; Souza et al. 2012; Porto et al. 2016; McCravy 2018; Tourinho et al. 2018). Due to complex logistics, long distances, and the difficulty accessing some areas, the costs of systematic biodiversity research are high in tropical forests (Gardner et al. 2008b; Tourinho et al. 2018), and these factors have strongly influenced the collection bias on biodiversity knowledge (Oliveira et al. 2016).

On the other hand, even with scientists protesting against science budget cuts, the global economic crisis has severely impacted the available scientific budget restraining funding to many types of research in the last years (Porto et al. 2016; Carvalho et al. 2019; Mega 2019; Malakoff and Mervis 2020). Rather than using all protocols and sampling all animals possible, adjusting and optimizing our survey protocols is a good option. Protocols can be optimized to focus on target indicator groups and species that may be more efficient to characterize and answer questions on the patterns of diversity (Caro and O'Doherty 1999). For example, soil traps are very efficient in sampling epigeic arachnids (Lo-Man-Hung et al. 2008; McCravy 2018), and instead of analyzing the results taking into account the species determination, guilds can be established (Dias et al. 2010; Tourinho et al. 2018) and optimized protocols can quickly respond to changes in environmental conditions (as exemplified for wandering spiders by Jocqué et al. (2005) and Rego et al. (2005, 2007) and for spider guilds by Cardoso et al. (2011b) and Lo-Man-Hung et al. (2011) and for harvestmen by Tourinho et al. (2014) and Porto et al. (2016)). The selection of target taxa is important to accurately and successfully answer the research question, indicating the most efficient set of methods and its best cost-effectiveness performance (Tourinho et al. 2018).

Time spent in the field is also a limiting factor for research activities; it is estimated that a collector needs 60–90 min to remove spiders from a 0.5 m^2 plot with a mean vegetation height of 15 cm (Duffey 1974). Reducing the overall time and effort for surveys while still retaining as much information about species richness and communities as possible can be challenging in many ways.

The ideal is to find a protocol that is cheap, allows rapid surveys, collects as many species as possible, and is also optimized to meet requirements for both ecological studies and monitoring species occurrence (Porto et al. 2016). Using reduced versions of traditional methods may reduce up to 87% of financial costs depending on the concept of reduction applied (see Porto et al. 2016). Porto et al. (2016) evaluated beating tray methods and three versions of active nocturnal search (traditional, modified, and cryptic active search) to sample harvestmen across a tropical forest area in the Amazon basin. The authors found out that from all methods tested by them, active cryptic nocturnal search could register all taxa with the maximum effort (Porto et al. 2016). They also concluded that using the reduced version of the active nocturnal search could reduce 87% of financial costs (Porto et al. 2016). When the three sampling methods (Winkler, beating tray, and active nocturnal) were compared in a study done in the Amazonian upland forest (Tourinho et al. 2018), it was estimated that Winkler is the most time-consuming field method (55% of time spent in the field) while beating tray and active nocturnal each took 22% of time spent in the field. Similar results were estimated for the financial costs of these methods, Winkler requiring 46% of financial costs, followed by beating tray (29%) and active nocturnal search (25%) (Tourinho et al. 2018). Gardner et al. (2008a) studied 13 taxa across 5 sites in northeastern Brazilian Amazonia with a total budget of US$145,550.00, including field and laboratory equipment and salary costs of project personnel involved in sampling and identification. About 66% of it was spent on salaries, while salary cost for invertebrates was lower when compared with vertebrates [54% (proportion of total invertebrates cost) vs. 71%]. Vertebrates also required more investment in-field equipment than invertebrates (25% vs. 12% of total respective costs), but invertebrates required greater costs to identification and laboratory costs (34% vs. 4% of total respective costs) (Gardner et al. 2008a).

So far, understand that any sampling method will have inherent advantages and limitations and it is essential for researchers to be aware of these so that results can be properly interpreted. Also, time, logistic constraints, and financial costs may limit the application of sampling and analysis. It is highly encouraged to have very clear research goals and the use of an objective evidence-based approach to select which taxa to study to avoid bias and improve the cost-effectiveness of biodiversity research.

15.3 Overview of Most Used Protocols

15.3.1 Beating Tray

Beating tray (Fig. 15.5a) was first devised by the English entomologist George Carter Bignell (1906) when he designed a unique folding screen for collecting insects falling from the beaten branch (MacGregor 2018). Nowadays, the device usually consists of a cotton white cloth stretched out across holding sticks of 1 m × 1 m. The device is held under a tree or shrub and the foliage is shaken with a 1-m-long stick. Invertebrates on the leaves or stems of the branch fall from it and land on the white cloth. They can then be examined and hand-sorted collected using the hands or a vial, i.e., hand-sorting is the process by which animals are visually separated into different groups and collected as needed. Each sample is composed of the total number of invertebrates collected in a determined time, space, or number of scrubs/trees accessed (see sampling design for effort standardization).

Fig. 15.5 The most common methods used to sample spiders and harvestmen. (**a**) Beating tray is placed under the vegetation and foliage is shaken with a stick. (**b**) Active nocturnal searching is performed by collector hand searching using a headlamp. (**c**) Collecting larger spiders can be done with the help of forceps. (**d**) Pitfall traps can be dug with the help of a posthole digger. (**e**) Plastic cup buried into the ground; there is preservative solution inside cup and the trap is covered by a small plastic plate suspended by sticks (Photos: **a**, Candiani D.F.; **b**, Tourinho A.L.; **c**, Pimenta F.; **d**, Bastos N.; **e**, Lo-Man-Hung N.)

Beating tray focuses on the arachnids present in the understory vegetation habitats, and it was the second most expensive method tested in a study in Amazonian upland forests comparing three sampling methods (Tourinho et al. 2014, 2018; Porto et al. 2016). Besides the Amazon, beating tray was also tested in the Atlantic Forest (Azevedo et al. 2013) and in a tropical montane forest (Malumbres-Olarte et al. 2016). For all sites, it is among the two most productive methods to sample spiders. On the other hand, it was not as efficient for collecting harvestmen (Porto et al. 2016; Tourinho et al. 2018). Beating tray also can be highly complementary with the nocturnal search for spiders and largely redundant with Winkler apparatus (Azevedo et al. 2013; Tourinho et al. 2014, 2018; Porto et al. 2016).

15.3.2 Active Searching

Active searching sampling accesses arachnids present in all kinds of habitats in the forest except the canopy, from understory vegetation to leaf litter and cryptic habitats. It may be performed during the day and/or at night. For arachnids, the most common period is during the night because the nocturnal active search is the most effective and the most economical method for collecting both spiders and harvestmen in the New World and should be used solely if budget is very limited and one single method must be chosen (Azevedo et al. 2013; Tourinho et al. 2014, 2018; Porto et al. 2016; Privet et al. 2020).

It is performed by collector hand searching using a headlamp (Fig. 15.5b), vials, tweezers, and transects (Fig. 15.5c). Collector searches (looking up and looking down) are on bushes, under rocks, on logs, on wrapped retreats tucked into leaves, in leaves, and in all possible habitats including artificial habitats as bricks, walls, fences, window, etc. Usually, a sample vial is composed of the total number of animals collected in a determined time and space.

There are many variations of nocturnal search methods proposed and tested. It may be performed to target arachnids living in environments occurring at the height of the collector's knee to as high as she/he can reach to exclusively from ground to knee level, called looking up and looking down methods, respectively, and they are performed separately for 1 h each (Coddington et al. 1991).

For harvestmen, the active cryptic nocturnal search focuses essentially on environments that likely contain cryptic species (such as fallen logs, beneath tree bark, small cavities and holes, suspended leaf litter, soil leaf litter, cavities, and roots) and is the most efficient and cheapest for collecting harvestmen (Porto et al. 2016). Such habitats are those favored by harvestmen, providing ideal conditions of humidity and temperature for their development (Curtis and Machado 2007). Usually, active hand sampling during the day or night time does not affect estimates for spiders' diversity (Vedel et al. 2015); however, it affects harvestmen as they can hardly be seen or found during the day.

Given most arachnids have nocturnal habits, collecting during the night increases the chances of an encounter, and also at night the headlamp can shine into the eyes of some arachnids such as spiders and reflect the tapetum lucidum. Unfortunately, this does not happen for harvestmen.

15.3.3 *Pitfall Traps*

The pitfall trap is used to access ground habitats. Usually focused on median and small arachnids, the size of animals collected using this method will depend on the size of the traps. It consists of a container (usually 500 mL of a plastic cup) buried into the ground with its rim at the surface level (Fig. 15.5d). Inside the trap, there is a solution to kill and preserve the trapped animals (usually formaldehyde, or ethanol; some drops of detergent may be added to break surface tension).

The opening of the trap is usually covered by a lid (e.g., a small plastic plate suspended by two to three sticks) to avoid rain through or debris inside the trap (Fig. 15.5e). There are several factors on traps that could significantly affect the impact on the arachnid catches, for example, the color of the plastic plate (Buchholz and Möller 2018), collection period time, trap placement, design, number of traps, liquid preservative, etc. (Uetz and Unzicker 1976).

Pitfall trapping is highly influenced by habitat structure, rendering this sampling method incomplete for complex tropical environments (Privet et al. 2020); thus, its efficiency varies in many different studies (Work et al. 2002; Azevedo et al. 2013; Tourinho et al. 2014, 2018; Malumbres-Olarte et al. 2016; Privet et al. 2020). Pitfall traps are a popular method for spider surveys in temperate regions being used even instead of nocturnal searches because it may sample high abundances (McCravy 2018); also it can be effectively conducted in a research requiring longer durations with cheap maintenance and low labor requirements (Bali et al. 2019).

Pitfall is not an efficient method for harvestmen; compared with sifting and litter manual sorting, it consumes more resources and time in the field and collects low numbers of individuals and species (Tourinho et al. 2014, 2018; Pinto-da-Rocha and Bonaldo 2006). For spiders, however, pitfall traps sample of a distinct assemblage and is a complementary method that may be combined with nocturnal search and beating tray successfully (Azevedo et al. 2013; Tourinho et al. 2018; Privet et al. 2020).

15.3.4 Sifting and Litter Manual Sorting

Litter sifting is performed using a sifter with mesh (size of 5 mm or less); the bag sieve of the Winkler apparatus (Fig. 15.6a, b) may be also used (see below a description of this method). Sifted litter and soil are transported to the laboratory in bags, or they are hand-sorted in the field by using tweezers and small brushes (Fig. 15.6c–g), and the arachnids are then stored in alcohol (70–100%) or other fluid preservation (see below detailed descriptions of specimen preparation and conservation).

The samples can be delimited by depth of the sifted soil layers (e.g., sifted sample of original size of 25 cm long, 25 cm wide, and 5 cm dug) or according to the volume of sifted material (e.g., two liters of sifted material form favorable microsite as decaying stub) or defining quadrats as units or by time hand-sorted in each sifted sample and even by combining some of those.

Manual sorting may involve some steps such as sifting the same litter sample using sifters with different sizes of mesh to segregate the larger and faster specimens first and gradually the smaller specimens. Thanatotic specimens are very common in litter samples (a protective response where the animal remains immobile for a lengthy period and consequently is difficult to detect in the litter), and for those, a variation of sifting and manual sorting called hot dancing method was created (Novak et al. 2016). It basically consists of applying heat and illumination to trigger invertebrates to start moving, and it can be done with handheld, heat-emitting electric light (a halogen light) directed from a height of 10–15 cm on the sample during hand-sorting (Novak et al. 2016).

Sifting and sorting litter manually are among the two most effective methods for harvestmen surveys and best combined with nocturnal search (Pinto-da-Rocha and Bonaldo 2006; Tourinho et al. 2014).

Fig. 15.6 Sifting leaf and litter manual sorting. (**a**) Litter sifting can be performed using the Winkler sieve. (**b**) Or, using a sifter. (**c**) Sifting above a laboratory tray. (**d**) Sifted litter. (**e**) Hand-sorting using small brushes. (**f**) Hand-sorting can be done in the field with a laboratory tray. (**g**) Or hand-sorting can be done above a cotton white cloth (Photos: **a**, **g**, Pérez-González A.; **b–f**, Lo-Man-Hung N.)

15.3.5 Winkler Apparatus

Winkler extractor is an efficient method of sampling small leaf litter arachnids. One square meter of leaf litter is sifted by vigorously shaking a bag sieve (a bag above a coarse mesh size 5 mm screen) for about 1 min for each portion (Fig. 15.7a). A quadrat sample, or several, is taken every 5–10 min. After sifting, samples are transferred into a bag for transportation (Fig. 15.7b). Each sifted leaf litter is hanged in thin mesh sacks (Fig. 15.7c), then put into the Winkler bag, and closed and suspended at the top (Fig. 15.7d), usually in a row, in a sheltered place for 2–3 days; usually, no further hand-sorting is performed after the suspended days. The invertebrates will fall from the litter mesh bag into a vial containing alcohol or another preservative substance.

Winkler trap is the most expensive method when compared with active nocturnal search, beating tray, and sifting leaf litter; this method consumes nearly half of the financial resources (46%) and time in the field (55%) (Tourinho et al. 2018). It usually exclusively samples litter macroarthropod species while at the same time being highly redundant with the beating tray method according to a study that compared the number of species and composition of harvestmen collected with both methods in an Amazonian area (Tourinho et al. 2018).

Winkler virtually removes all the microhabitat available at the sampling location and thus should offer more realistic information on abundances and biomass of arachnids per unit area. This however should be interpreted with caution because this method is selecting species in advance either way, when sampled leaf litter is passed through a sieve before it is hanged to dry out and later when it sits hanged for extraction, since it has a second sieve inside the Winkler traps. So it may be covering biomass and abundances indeed but exclusively of those small-sized species with cryptic behavior (Tourinho et al. 2018).

Overall considering the explanations above, we may say that for spiders in the Amazon, combining Winkler and beating tray may give more comprehensive information about the abundance and species richness, while Winkler plus nocturnal search will give a good estimation of whole species assemblage while reducing costs (Tourinho et al. 2014, 2018).

15.3.6 Other Methods

15.3.6.1 Sweep Nets

This method was designed to sweep the vegetation and collect arthropods or other invertebrates. It is similar to a butterfly net, an aerial net that is deliberately built lightly (Fig. 15.7e). Unlike aerial nets, the sweep nets are very robust, with short, thick handles and heavy, inflexible hoops. The bag is generally constructed from more sturdy material than aerial nets, for example, sweep nets are often constructed with canvas cloth or nylon bag. The collector has to sweep the substrate with some

Fig. 15.7 Winkler apparatus and sweep nets. Winkler (**a–d**). (**a**) Winkler, 1 square meter leaf litter is sifted by a bag sieve. (**b**) Bag for transportation, litter, and mesh sacs. (**c**) Litter is placed in mesh sack into Winkler bag. (**d**) The Winkler bag is closed for usually 2 days. Sweep nets (**e, f**). (**e**) Sweeping the vegetation. (**f**) Hand-sorting invertebrates collected in sweeping nets (Photos: **a**, **d**, Candiani D.F.; **b**, **c**, Dias S.C.; **e, f**, Pérez-González A.)

force capable of collecting the invertebrates in the net. The samples consist of invertebrates collected in a specific area, time, or number of sweeps. The specimens can be collected by hand-sorting in the field (Fig. 15.7f), put on a bag for later hand-sorting, or collected from the net using an aspirator. After hand-sorting, the arachnids are stored in a preserving solution.

15.3.6.2 Pipe Traps

This is a method developed to catch specimens into shallow subterranean habitats. It is a trap composed of plastic pipes, about 75 cm long and 11 cm in diameter, perforated along its surface by many oblique small holes (5–7 mm in diameter), and a vial containing preservation fluid lowered inside the trap, on a nylon cord [see López and Oromí (2010) Figs. 15.1, 15.2, and 15.3 for pipe trap design]. These traps are usually installed vertically into holes in the subterranean habitats (López and Oromí 2010).

15.3.6.3 Berlese-Tullgren Funnels

Berlese has undergone a few modifications over time, but the principle is still the same: a funnel is filled with soil or leaf litter is somehow heated to drive the invertebrates down toward an alcohol-filled vial placed at the base of the apparatus. Berlese-Tullgren funnels were first designed by Berlese (1905); his trap consisted of a metal funnel with a water jacket that could be heated, using a Bunsen burner. Trägårdh (1910) made the trap more portable using spirit lamps to heat the water. Then, Krausse (1916) restricted the heated water jacket to the top part of the apparatus, and finally, Tullgren (1917) decided to use an electric lamp to heat the surface of the soil or litter sample (Fig. 15.8a–d), allowing a temperature gradient and unidirectional escape for invertebrates into vial collection. Berlese is faster than Winkler in extracting the animals but will require a field station with infrastructure as well as available electric energy or batteries.

15.3.6.4 Leaf Blower-Vac

Collecting arachnids may be different depending on the environment and habit in focus. Some places are difficult to collect arachnids, for example, cracks or slots in walls or flooded crop fields; thus, some researchers suggested a modification of a leaf blower-vac for field sampling of arthropods (Zou et al. 2016). This method consists of machines modified to reverse function, i.e., to suck rather than blow. It is a nondestructive method; the specimens can be collected alive and hand-sorted from the sampling net to be stored in a preservative substance. Some limitations of this method are people operate the machine (at least two people) and loud noise can disturb the site and the specimens (Zou et al. 2016).

Fig. 15.8 Berlese-Tullgren funnels. (**a**) Berlese-Tullgren modern apparatus, vial collection is placed to fall from bag. (**b**) A lamp is placed on top of the apparatus. (**c**) There is a sieve in the middle of the bag. (**d**) Litter, leaves are placed inside Berlese-Tullgren; with lamp turned on, the bag is closed for a few hours (Photos: **a**, **c**, **d**, Pérez-González A.; **b**, Lo-Man-Hung N.)

15.3.6.5 Fogging

One of the first records of canopy invertebrate quantification occurred in 1929 on an expedition to Guyana with O.W. Richards' pioneering efforts; he hoisted light traps on the canopy (Basset et al. 2003). Beginning in the 1970s, canopy invertebrate collection techniques have progressed and diversified over the years (see Roberts 1973; Erwin and Scott 1980; Sutton and Hudson 1980; Kitching et al. 1993). Currently one of the most widely used techniques for accessing canopy invertebrate fauna is the canopy fogging method originally developed for pest control; the first arthropod survey using canopy fogging was performed by Roberts (1973) in Costa Rica and Gagné (1979) in Hawaii; since then, this technique has been employed and refined over the years.

At each site sampled, white sheets of synthetic cloth (2 × 4 m each) are spread in the understory to support the arachnid fauna that fell from the canopy (Fig. 15.9a, b). The fogging is done with a non-residual pyrethroid (Cautini) plus diesel oil and knockdown substance (at a concentration of 10%) with a 20–30-min on-floor operated thermal fogger (Fig. 15.9c, d). Each sheet can be numbered and mapped, enabling the location and analysis of spatial distribution, or all the sheets can be considered a sample; it will depend on the questions asked.

Fig. 15.9 Fogging. (**a**) White sheets of cloth are spread in the understory. (**b**) Top view of sheets. (**c**) Thermal fogger being operated. (**d**) Fogging will rise to the canopy (Photos: **a**, **b**, **d**, Lo-Man-Hung N.; **c**, Dias S.C.)

There are still few studies comparing canopy arachnid fauna. Otto and Floren (2007) and Mupepele et al. (2014) studied the spider fauna of tree canopies in Białowieża Forest. Höfer et al. (2008) used the fogging technique to collect spiders from tree canopies in the Central Amazon. Thorn et al. (2016) investigate the canopy of conifer-dominated forests of the Northern Hemisphere; however, Thorn et al. (2016) did not use the fogging method to collect the epigeal spiders and harvestmen but windthrows located under an open and closed canopy, and none of the authors provide pieces of information about costs or effectiveness of the method for arachnids.

15.3.6.6 Amazonas-Trap

The Amazonas-trap is a recently described method developed to exhaustively sample plant-inhabiting arthropod communities in tropical forest understory, and its main goal is to describe the plant-arthropod interactions in these environments (Lopes et al. 2019). The method was tested in Amazon rainforest sites at Ducke Reserve and is designed to sample free-standing plants of 1–3 m high and at least 0.5 m from neighboring plants.

The trap consists of a light fabric made of white 3-m-long and 3-m-wide polyester having a velcro strip in its two lateral sides extending along its entire length. The bottom part of the fabric is sewn to form a hem of 5 cm wide through which a weldable PVC pipe of 2 cm diameter and 1.5 m long is inserted. The PVC pipe has a three-way T-shaped connector (2 cm diameter) attached with a bolt and nut in one of its ends. The other end is then installed manually being inserted into the connector and forming a circle surrounding the plant. The connector also has an aluminum tube (2 cm diameter, 2.5 m long) attached to the connector, and a 5 m rope with a loop knot at the tip is inserted into a hem sewn 15 cm below the top of the collecting cloth.

It is placed on the ground around the lowest branch of the chosen plant with its two end sections placed together to form a circle around the plant. The velcro is then tightly closed throughout the entire length, top to bottom to avoid the escaping of animals. The end of the rope will then be passed through the loop knot, while the aluminum tube is joined into the connector.

After installation, the entire structure of the trap is pulled up quickly and energetically, enveloping the plant completely while the rope is pulled to close the trap and at the same time trapping all the animals inside. The collector may insert his hand inside through a small opening in the velcro, hold the trunk of the plant, and shake it causing the fall of all inhabiting arthropods into the lower part of the trap or resting on the fabric from where they may be collected easily with an aspirator in the trap.

Amazonas-trap had a better performance than beating tray and active manual collecting with which it was compared (Lopes et al. 2019); it collected more spiders than the other two methods, and although it was more time-consuming, still it was more successful for all the parameters tested in the study. This method combined with the variation of nocturnal searches used by Colmenares et al. (2016) are of a great potential to investigate the relationships of spiders and harvestmen with the vegetation.

15.3.6.7 Aspirator (or "Pooter")

This is a method used for sucking up arachnids (and other invertebrates) using aspirator devices called aspirators or "pooters," named after its inventor William Poos (1929). It consists of a small vial, the lid of which is penetrated by two flexible plastic or rubber tubings (Upton and Mantle 2010; Toth 2000). On the inner end of one tube, fine mesh is attached, and this tube leads to the user's mouth (usually connected by a long, flexible piece of tubing). The end of the second tube projects into the collecting chamber, and its far end can then be placed over an insect or other small organism; the user sucks on the first tube, and the specimen is drawn into the collecting chamber or fixed bottle through the other. Unpleasant odors, fine spores from fungi, and pathogens from, for example, bird nests will not be excluded by the filters on the suction tubes, so it needs a lot of caution when used (Upton and Mantle 2010).

15.4 Sampling Design

As discussed above, both optimized and standardized protocols offer valuable information and powerful results on biodiversity in studies, but they depend on the questions to be answered and hypotheses being tested. An optimized protocol will focus on exhaustively collecting the maximum number of species possible, while for a standardized protocol, the purpose is to provide comparative data (Malumbres-Olarte et al. 2016). The problem formulation, questions asked, and hypotheses to be tested are fundamental for planning discussions, because it will provide a foundation upon which best protocol and methods are to be used.

Designing a field research study requires considering and controlling survey effort, and to accomplish that, both effort and samples must be comparable. The collecting effort must be standardized, equivalent, and constant between samples and sampled sites based on the concepts of redundancy and complementarity of sampling methods, especially for ecological studies (Tourinho et al. 2018). Additionally, the sampling protocol chosen should account for species captured per sampling method as well as per unit area or time, or both, to infer the effectiveness (Verberk 2011). When designing a fieldwork study, we use a broader concept of survey effort control as follows:

15.4.1 Number of Data Collection

Species richness estimation is an important measure in conservation and biological inventory assessment in multiple contexts. However, in tropical arthropod surveys, the frequency of singletons (= species represented by a single individual) is usually anomalously high (average, 32%) (Coddington et al. 2009).

Undersampling is suggested as the cause of this systematic negative bias in species richness patterns (Coddington et al. 2009). A way to overcome this bias and produce more realistic species richness estimates is by increasing drastically the sampling effort, e.g., using several replicates in the same site or plot (Coddington et al. 2009) as generally used in many of the inventories undertaken in the tropics (Brescovit et al. 2004; Indicatti et al. 2005; Bonaldo et al. 2009; Bonaldo and Dias 2010; Ricetti and Bonaldo 2008).

Unfortunately, as already commented above, the budgets for research are becoming smaller and scarcer each year. Forty years ago, the Conservation International (CI) organization created the Rapid Assessment Program (RAP) for rapid surveys including teams of experts on short expeditions to evaluate biodiversity and ecosystems as a way to keep the inventory quality while reducing costs.

As the urge for the use of RAP surveys rather than extensive sampling quickly grew, many RAP programs have emerged including many different protocols since then. The RAP is sometimes a choice rather than an extensive sampling in some regions (Magnusson et al. 2005, 2013; McCullough et al. 2007; Hero et al. 2010;

Pezzini et al. 2012). It can provide a preliminary characterization of the fauna; however, it often uses qualitative or semiquantitative method (McCullough et al. 2007), an approach most used when fauna is unknown or when some information is needed urgently to support effective decision-making. As far as resources, time, and objectives allow, researchers should consider more quantitative approaches.

How much data is needed to achieve successful results will then depend on the type of question tackled. Large datasets generally require the use of several collecting methods and replicates to estimate the species richness and describe assemblage composition (Coddington et al. 2009, Tourinho et al. 2018).

Those large datasets are in general desirable for long-term biodiversity monitoring, to understand biogeographic and ecological patterns and processes as they require an extensive spatial and temporal sampling (Coddington et al. 2009; Malumbres-Olarte et al. 2016). Large sampling efforts in inventories for taxonomic purposes are traditionally devoted to collect the largest number of species with reduced singleton frequency in a single visit (Coddington et al. 2009). However, for studies focused on both species richness and composition, and even taxonomic studies, this may not be always necessary; using fewer methods may be more efficient and less expensive than several redundant ones depending on the target taxa and/or on the question (Tourinho et al. 2014, 2018; Porto et al. 2016).

Our study in an Amazonian forest suggests that optimizing survey methods for harvestmen and spider assemblages improved the results (Tourinho et al. 2018). Singleton frequency used to be higher in megadiverse groups of arachnids (e.g., spiders) than in groups with moderate diversity (harvestmen). Indeed, harvestmen are not likely to be easily subject to a singleton phenomenon when protocols are adequately adjusted (Tourinho et al. 2014, 2018; Porto et al. 2016).

On the other hand, a reduced effort will produce reduced datasets, and its impact on the taxonomic and numerical resolution in a given study must be evaluated together with assemblage congruence among different taxonomic groups, especially if there are multiple questions to be answered (detailed explanations on effort reduction below in c. Target Organisms).

15.4.2 Area and Time

The species-area relationship (SAR) is one of the oldest known and best-documented patterns in ecology. The first model was presented one century ago (Arrhenius 1921, Gleason 1922). It generally describes the pattern of increase in species richness with increasing area of observation. Many aspects and mechanisms may influence SAR, e.g., patterns of nested or scattered observations across space, differences in sampling sizes and shapes, and other intrinsic aspects of sampling design influence the shape and parameters of the species-area relationship through their different sensitivities to some species biological, environmental, and biogeographic processes (Tessel et al. 2015).

The SARs are then susceptible to sampling effect/design and are often quantified differently, depending on the goals of a study. This is of major importance when planning a research survey (see Bueno et al. 2019; Tourinho et al. 2019).

Both time spent collecting and area covered by each sampling method in the field are among the most important metrics to infer effectiveness and for effort control in an inventory. Biodiversity composition and richness change across space, time, or both.

The analysis of species accumulation curves is a successful method to evaluate the effectiveness of surveys employing several sampling methods. In general, species accumulation curves are a graph of the cumulative number of species collected within a defined area or time as a function of an effort measure (Soberón and Llorente 1993; Colwell and Coddington 1994; Chao and Chiu 2016).

The species accumulation curves are produced randomly and repeatedly resampling a pool of individuals or samples and plotting the number of species represented by increasing numbers of individuals (Colwell and Coddington 1994). Because the actual number of harvestmen and spiders (or other taxa) in any particular area is unknown, the accumulation curve can be used as an estimation on how complete the sampling effort is.

Also, by comparing the curves of different sets of methods, it is possible to estimate how efficient was the capturing of species and individuals by examining if the curves approach an asymptote or not based on values determined by species richness (Colwell and Coddington 1994; Longino and Colwell 1997; Fisher 1999).

Defining the standardized metrics of area and time of a study is very important because they are also effort parameters in species accumulation curves. This can be done by planning the use of standardized plots to control area and time when designing the study.

In Neotropics, most in South America, the studies held in the past used 30-m-long transects and several replicates in the same area and survey site as a strategy to increase effort (Porto et al. 2016). Methods such as beating tray, active nocturnal search, sifting, and sweep net were standardized by time (1 h) as well, and then each transect was surveyed for 1 h by one collector.

This design was also applied in the inventories held in Brazil by the program Research Program on Biodiversity Characterization, Conservation, Restoration, and Sustainable Use BIOTA-FAPESP, launched in 1999 by the São Paulo Research Foundation/FAPESP to discover, map, and analyze the origins, diversity, and distribution of the flora and fauna in the State of São Paulo (e.g., Candiani et al. 2005; Indicatti et al. 2005; Pinto-da-Rocha et al. 2005; Bragagnolo et al. 2007), although the program expanded to sample arachnids in other Brazilian States which harbored the Atlantic Forest sites (Staley 2001), and this design was replicated in other biomes as well (Rego et al. 2005, 2009, 2007; Tourinho et al. 2019).

However, covering target areas adequately can be very costly in some regions (i.e., to delimit standardized plots or transects), even small areas. For example, the Amazon basin forests cover about 7 million km^2, and 700 sampling sites would be needed to cover a single sample site in an area of 10,000 km^2 (Magnusson et al. 2005).

In order to adequately accommodate the most important designing metrics and financial cost limitations, researchers from the National Institute of Amazonian Research (INPA), in Manaus, Brazil, developed the long-term ecological research (LTER) system RAPELD design; the name of the method comes from a combination of RAP surveys and the Brazilian acronym for LTER (Magnusson et al. 2005).

RAPELD system is a modular grid of 5 km × 5 km or 5 km × 1 km of both trails and plots used to sample large areas cost-effectively (Magnusson et al. 2005). It is a robust sampling design that can be replicated in any ecosystem; it also integrates multidisciplinary surveys for all taxa and allows comparisons with samples taken over larger areas. The system is appropriate for long-term ecological studies but also allows rapid surveys to evaluate biotic complementarity and land-use planning (Magnusson et al. 2005, 2014).

This specific design is used by the Program for Planned Biodiversity Ecosystem Research (PPBio) which is among the largest programs for biodiversity supported by federal funding agencies in Brazil. It has been implemented in Brazil with 70 grids and modules of RAPELD monitoring sites in PPBio Western and Meridional Amazonia, Pantanal, Atlantic Forest, and PPBio Australasia with 52 grids in Australia and Nepal (Hero et al. 2010; Pezzini et al. 2012).

For arachnids, to save time in placing subplots, the long RAPELD plots of 250 m per 2 m width that follow isoclines are then surveyed by one collector for 1 h long (Tourinho et al. 2014, 2018; Porto et al. 2016; Colmenares et al. 2016).

Long, thin plots tend to contain more species than square or circular plots because they tend to occur in higher densities in some habitats than others, and long plots with increasing distance have more chance of intercepting different habitats than short plots (Magnusson et al. 2013). This long design proved to be more efficient for arachnids in tropical forests than the short transects (Porto et al. 2016).

Another optimized and standardized protocol using arachnids as models for studies regarding megadiverse communities in tropical forests was designed using large datasets of semiquantitative field data from different continents (Malumbres-Olarte et al. 2016): the COBRA-TF (Conservation Oriented Biodiversity Rapid Assessment for Tropical Forest). It uses a combination of six methods [three variations of active searching: aerial hand collecting (from the knee level to as high as one can reach), ground hand collecting, cryptic searching; vegetation beating; sweep netting; and pitfall trapping].

Time control for COBRA-TF (except pitfall traps) comprised 1 h/sample and an aspirator could be used. The results suggested the use of a large proportion of aerial night and beating day sampling which is consistent with the other protocols tested above, but with the addition of small proportions of both day and night ground searching. Regarding the sampling area control, plots of 0.25 ha (50 × 50 m) were established, and sampling was undertaken during the dry season or at least to collect shortly after the rainy season when costs spent in expeditions were lower (Malumbres-Olarte et al. 2016).

Regarding methods tested, nocturnal search and beating tray are qualitative methods, and both can be standardized by area and time and both provide comprehensive coverage of species that are present at a site. However, nocturnal search and beating tray are not the best methods for quantitative sampling of tiny species that inhabit leaflitter, their abundances or biomass per unit of area (Tourinho et al. 2018).

Other methods such as Winkler and Berlese are quantitative methods that virtually remove all the microhabitat available (litter) (Krell et al. 2005), and as commented above, they may provide more realistic information on species abundances and biomass per unit of area. On the other hand, they cannot be standardized by time as the active searching methods can be.

Standardizing traps (e.g., pitfall traps) by area will require collecting the same quantity of litter samples and installing the pitfalls in the studied plots as well. The size of litter quadrats may vary, usually from 1 to 2 m^2 depending on the study design, both pitfalls placing and litter quadrats may be either randomly chosen within the plot (Tourinho et al. 2018), collected along with the 250 m plot with a 5 m distance in between them (Tourinho et al. 2018) or each trap may distance 10–15 m from other traps to avoid trap-to-trap interference (Digweed et al. 1995; Work et al. 2002).

The diameters of pitfall traps were tested by Work et al. (2002) to determine optimal size for studying litter-dwelling arthropod biodiversity. They tested five different diameters (4.5, 6.5, 11, 15, and 20 cm) and suggested that the capture rates are not directly proportional to trap size (Work et al. 2002). Usually larger circumferences can collect more individuals and more species. However for monitoring purposes, using small pitfall traps tends to be the most efficient sampling technique to assign dominant epigaeic arthropods, and including some large pitfall could collect rare species references (Work et al. 2002).

15.4.3 The Budget of Field and Lab Research

The costs and benefits of surveying different taxa are also a layer to be considered as a survey effort dimension when designing an experiment or a research collecting trip; it will reflect in the overall time spent in each plot collecting and in the laboratory processing an extra amount of material.

Clearer goals and choices combined with objective evidence-based selection of the study taxa improve the cost-effectiveness of biodiversity research. Detailed and exhaustive biodiversity surveys are time-consuming, very expensive, and as a consequence produce large datasets usually resulting from highly redundant methods (Tourinho et al. 2014, 2018).

In our recent study, we followed Souza et al. (2012), Gardner et al. (2008b), and Porto et al. (2016) to evaluate the project costs and consider financial costs and time spent collecting in the field and sorting and identifying material in the laboratory (Tourinho et al. 2018). The summed time of all sampling techniques together is set up as maximum effort (100%), and we then calculated the fractions of maximum effort for the number of methods employed for each combination of sampling techniques (Tourinho et al. 2018).

Overall the results from testing the most common methods have shown that Winkler traps consume nearly half of the financial costs and time for the study dedicated to the survey in the field. Nocturnal search is the cheapest and fastest method; only 25% of financial resources were consumed and 31% of the total time was dedicated to the study. Beating consumed only 29% of the financial resources; however, it was as fast as nocturnal sampling in the field but consumed 42.87% of the time in the lab sorting and identifying the material surveyed.

An economic and efficient alternative protocol is composed of active nocturnal search and beating tray for spiders, which saves almost 50% of the costs and is efficient in describing both species richness and composition (Tourinho et al. 2018). For harvestmen, one method, the active cryptic nocturnal search, recovered all taxa sampled with the maximum effort (sum of all methods); also the ecological patterns found by the more intensive methods saved 87% of the costs when compared to the maximum effort (Porto et al. 2016).

15.4.4 Target Organisms

Target taxa analysis increases the taxonomic coverage and the spatial information streamlining the inventories. Usually, the chosen taxa represent the pattern of species distributions and also how they correlate with patterns of distributional and environmental heterogeneity of species in unrelated taxonomic groups (Kremen 1994). Their use is especially appropriate to avoid issues like lack of time, funds, or appropriate sampling collecting.

Harvestmen and spiders are target taxa that have often been sampled together, yet recent publications have shown how a reasonable inventory for harvestmen is accomplished with much less effort than is needed for spiders (Tourinho et al. 2018).

A broad concept of effort reduction and sampling design optimization may be applied with success for harvestmen in tropical forests, especially in community ecology studies, boosting their collected diversity by reducing the number of data collected, area sampled, time in the field and in the laboratory, and the entire research budget (Tourinho et al. 2014; Porto et al. 2016). The smaller datasets resulting from methods more directed to the distribution and habitats used by the harvestmen fulfill the taxonomical, ecological, and financial needs of the survey with higher success and lower costs in the Amazon rainforest (Porto et al. 2016).

Additionally, different taxonomic groups such as harvestmen and spiders should not be surveyed simultaneously in tropical forests. Both groups should be considered as independent target taxa, having appropriate protocols dedicated to their survey and independent collectors, such as cryptic nocturnal searches (Tourinho et al. 2018).

It is nearly impossible to assess a complete spider species inventory in the tropics (Coddington et al. 1991). Spider inventories are large, and most protocols developed for its surveys use a combination of multiple methods, generally Winkler and/or pitfalls and active nocturnal search together with a beating tray.

Unfortunately, those combinations are inadequate for harvestmen in terms of efficiency, congruence, and financial cost, especially for monitoring or ecological impact studies (Tourinho et al. 2018). Yet, the most used 4 spider sampling protocol is not the most efficient strategy for studies of assemblage associations with environmental variables and ecological impact either, because it is exhaustive, including multiple methods and several replicates are used and yet the accumulation curves are also still not asymptotic (Coddington et al. 1991; Bonaldo et al. 2009; Cabra-García et al. 2012; Azevedo et al. 2013; Tourinho et al. 2018).

15.5 Processing Collected Material

15.5.1 Sorting, Specimen Preparation, and Conservation in the Field and Laboratory

Hand-sorting the sampled material in the field is a great option if researchers have the required material and infrastructure. For that, a stereomicroscope (usually with 20×, 40×, 60×, or 100× total magnification) is needed, and improvised working stations may also be installed in the field sites where no previous infrastructure exists as well. To save time and keep material organized, the labels may be prepared and printed beforehand; labels may contain information such as country, state, station, site, plots, methods employed, date, collector, and even collection numbers.

Arachnids may be counted and previously identified in the field, and later in the laboratory, they are conferred and studied using other techniques and equipment needed to advance further with specimen identification into the inferior taxonomic levels. The arachnids are stored in ethanol or other fluid preservation required for specific studies (e.g., RNAlater). Some researchers still store samples in 70–80% ethanol; however, especially for those in scientific collections focusing on taxonomy and systematic studies, most samples are usually stored in 95–100% ethanol.

Most taxonomic, systematic, and even some ecological studies demand molecular analysis of the specimens, as for DNA Sanger sequencing or RNA samples for next-generation sequencing methods. Those require an appropriate preservation of the tissue samples and/or DNA sample, and ideally, this preservation has to be done during fieldwork.

The choice of optimal preservation technique will depend on the study purpose, the type of analysis it will require, the duration of storage, the requirements for sampling transportation (if freezing samples is mandatory, the samples must be stored with dry ice and transported in isolated styrofoam boxes), and the targeted quality of the sample (e.g., mtDNA, nuclear DNA, RNA- or protein-based analyses), and it will also depend on the requirements of a nondestructive or little-invasive preservation technique (Nagy 2010).

The most important in tissue preservation is avoiding enzymatic activity (i.e., avoid damage and degradation of nucleic acid) and oxidative degradation (Prendini et al. 2002). Nagy (2010) summarizes all main tissue preservation techniques, appropriate methods, and relevant aspects such as cost supply. The most common tissue preservation techniques for arachnids according to Nagy are as follows:

(a) Ethanol: if possible, a non-denatured ethanol should be used; the optimal concentration is about 95–99%, but there are also alternatives as EDTA added to 95% ethanol or 70% ethanol diluted with 1×TE buffer or 70–80% ethanol with 1–3% glycerine; if absolute ethanol is used, be aware of traces of benzene which could affect DNA preservation. Dry or frozen samples can be transferred to ethanol. The removal of ethanol prior to DNA extraction will be necessary.
(b) Freezing: low temperature inhibits enzymatic activity and DNA degradation; it may be done by freezing at −20 °C, also using additional techniques (ethanol or buffers); ultralow temperature at −80 °C, under fieldwork conditions, can be reached with dry ice (frozen CO_2), a well-isolated box, and small amounts of samples.
(c) Preservation in liquid nitrogen: it is a method applicable to DNA, RNA, and protein samples, but continuous maintenance is expensive.
(d) RNAlater can be used for RNA and DNA preservation, DNA/RNA microarray analyses, and gene expression studies. For short periods, tissues can be preserved in RNAlater at room temperature and to store, samples must be frozen. It is an expensive fluid and it is mostly used when sample quality is important. It will need an extraction kit from the same manufacturer of RNAlater to process samples preserved in RNAlater.

15.5.2 Species Identification

Identification is made using the identification keys (see, e.g., Adis 2002), taxonomic scientific publications, and samples deposited in scientific collections. Furthermore, some arachnids can be identified by specialists by sending the samples to their research centers or when these specialists are visiting the collections where the samples are deposited. A number of morphospecies are used when the specific names are unknown or more commonly when they are of undescribed taxa. Undescribed taxa frequencies are very high in tropical regions; in some locations, they are 70–80% of the taxa sampled in the study, and not rarely genera and even families are new to science and in need of description.

The process of morphotyping can be tricky and needs years of experience and expertise, because most diagnosed characters are based on structures of reproductive morphology. For most of the analysis, only the adult is used (i.e., individuals susceptible to correct morphological identification).

Acknowledgments We are grateful to Professor Dr. Jean Carlos Santos for inviting the first author to participate in this project contributing to the knowledge of survey methods and protocols for arachnids. We thank enormously our colleagues and friends Nayane Bastos, David Candiani Leonardo Carvalho, Sidclay Dias, Abel Pérez-González, Flavio Pimenta, and Ricardo Pinto-da-

Rocha, for sending and allowing us to use many of their photos that are now part of the plates in this chapter. We thank Thomas Gundersen for detailed and careful English proofreading and the reviewer for giving many great suggestions and comments that improved this chapter. During this work, ALT had a Coordination of Improvement of Higher Education Personnel-CAPES grant, and NLMH had fellowships from the Brazilian National Council for Scientific and Technological Development-CNPq (#142192/2017-1), São Paulo Research Foundation-FAPESP (#2017/19616-2 and #2019/12282-7)

References

Adis J (2002) Amazonian Arachnida and Myriapoda. Pensoft Publishers, Sofia

Arrhenius O (1921) Species and area. J Ecol 9:95–99

Azevedo GHF, Faleiro BT, Magalhães ILF et al (2013) Effectiveness of sampling methods and further sampling for accessing spider diversity: a case study in a Brazilian Atlantic rainforest fragment. Insect Conserv Diver 7(4):381–391. https://doi.org/10.1111/icad.12061

Azevedo GHF, Ott R, Griswold CE et al (2016) A taxonomic revision of the ground spiders of the genus Apopyllus (Araneae: Gnaphosidae). Zootaxa 4178(3):301–327. https://doi.org/10.11646/zootaxa.4178.3.1

Bali L, Andrési D, Tuba K et al (2019) Comparing pitfall trapping and suction sampling data collection for ground-dwelling spiders in artificial forest gaps. Arachnol Mitt 58(1):23–28. https://doi.org/10.30963/aramit5808

Barlow J, Overal WL, Araujo IS et al (2007) The value of primary, secondary and plantation forests for fruit-feeding butterflies in the Brazilian Amazon. J Appl Ecol 44:1001–1012. https://doi.org/10.1111/j.1365-2664.2007.01347.x

Barnosky AD, Matzke N, Tomiya S et al (2011) Has the Earth's sixth mass extinction already arrived? Nature 471:51–57. https://doi.org/10.1038/nature09678

Basset Y, Novotny V, Miller SE et al (2003) Arthropods of tropical forests: spatio-temporal dynamics and resource used in the canopy. Cambridge University Press, Cambridge, p 490

Bertkau P (1880) Verzeichnis der von Prof. Ed. von Beneden auf seiner im Auftrage der Belgischen regierung unternommenen wissenschaftlichen Reise nach Brasilien und La Plata i. J. 1872-1875 gesammelten Arachniden. Mém cour Acad 43:1–120

Berlese A (1905) Apparecchio per raccogliere persto en in gran numero piccoli artopodi. Redia 2:85–89

Bignell GC (1906) Insects. In: Page W (ed) OrthopteraThe Victoria History of the County of Devon. Archibald Constable & Company Limited, London, pp 163–164

Bonaldo AB, Dias SD (2010) A structured inventory of spiders (Arachnida, Araneae) in natural and artificial forest gaps at Porto Urucu, Western Brazilian Amazonia. Acta Amazon 40(2):352–372. https://doi.org/10.1590/S0044-59672010000200014

Bonaldo AB, Carvalho LS, Pinto-da-Rocha R et al (2009) Inventário e história natural dos aracnídeos da Floresta Nacional de Caxiuanã. In: Lisboa PLB (ed) Caxiuanã: Desafios Para a Conservação De Uma Floresta Na Amazônia. Museu Paraense Emílio Goeldi, Belém, pp 577–621

Bowne N, Braoun C, Mcllwain J et al (2019) Borneo coral reefs subject to high sediment loads show evidence of resilience to various environmental stressors. PeerJ 7:e7382. https://doi.org/10.7717/peerj.7382

Boxshall GA (2020) Self-help for taxonomists: three things we must do for taxonomy to survive. Megataxa 1(1):39–42. https://doi.org/10.11646/megataxa.1.1.7

Bragagnolo C, Nogueira AA, Pinto-da-Rocha R et al (2007) Harvestmen in an Atlantic forest fragmented landscape: evaluating assemblage response to habitat quality and quantity. Biol Conserv 139(3–4):389–400. https://doi.org/10.1016/j.biocon.2007.07.008

Brescovit AD, Bertani R, Pinto-da-Rocha R et al (2004) Aracnídeos da Estação Ecológica da Jureía-Itatins: Inventário Preliminar e História Natural. In: Marques OAV, Duleba W (eds) Estação Ecológica da Jureía-Itatins: ambiente físico, flora e fauna. Holos Editora, Ribeirão Preto, pp 198–221

Brown MJF, Paxton RJ (2009) The conservation of bees: a global perspective. Apidologie 40(3):410–416. https://doi.org/10.1051/apido/2009019

Buchholz S, Möller M (2018) Assessing spider diversity in grasslands—does pitfall trap color matter? J Arachnol 46(2):376–379. https://doi.org/10.1636/JoA-S-16-062.1

Bueno AS, Masseli GS, Kaefer IL et al (2019) Sampling design may obscure species–area relationships in landscape-scale field studies. Ecography 42:1–12. https://doi.org/10.1111/ecog.04568

Cabra-García J, Bermúdez-Rivas C, Osorio AM et al (2012) Cross-taxon congruence of α and ß diversity among five leaf litter arthropod groups in Colombia. Biodivers Conserv 21:1493–1508. https://doi.org/10.1007/s10531-012-0259-5

Candiani DF, Indicatti RP, Brescovit AD (2005) Composição e diversidade da araneofauna (Araneae) de serapilheira em três florestas urbanas na cidade de São Paulo, São Paulo, Brasil. Biota Neotrop 5(1a):111–123. https://doi.org/10.1590/S1676-06032005000200010

Cardoso P (2009) Standardization and optimization of arthropod inventories—the case of Iberian spiders. Biodivers Conserv 18:3949–3962. https://doi.org/10.1007/s10531-009-9690-7

Cardoso P, Crespo LC, Carvalho R et al (2009) Ad-hoc vs. standardized and optimized arthropod diversity sampling. Diversity 1(1):36–51. https://doi.org/10.3390/d1010036

Cardoso P, Borges PAV, Triantis KA et al (2011a) Adapting the IUCN Red List criteria for invertebrates. Biol Conserv 144:2432–2440. https://doi.org/10.1016/j.biocon.2011.06.020

Cardoso P, Pekár S, Jocqué R et al (2011b) Global patterns of guild composition and functional diversity of spiders. PLos One 6(6):e21710. https://doi.org/10.1371/journal.pone.0021710

Caro TM, O'Doherty G (1999) On the use of surrogate species in conservation biology. Conserv Biol 13:805–814. https://doi.org/10.1046/j.1523-1739.1999.98338.x

Carvalho WD, Mustin K, Hilário RR et al (2019) Deforestation control in the Brazilian Amazon: a conservation struggle being lost as agreements and regulations are subverted and bypassed. Perspect Ecol Conser 17(3):122–130. https://doi.org/10.1016/j.pecon.2019.06.002

Černecká Ľ, Mihál I, Jarčuška B (2017) Response of ground-dwelling harvestman assemblages (Arachnida: Opiliones) to European beech forest canopy cover. Eur J Entomol 114:334–342. https://doi.org/10.14411/eje.2017.042

Chao A, Chiu CH (2016) Species richness: estimation and comparison. Wiley StatsRef 1–26. https://doi.org/10.1002/9781118445112.stat03432.pub2

Coddington JA, Griswold CE, Dávila DS et al (1991) Designing and testing sampling protocols to estimate biodiversity in tropical ecosystems. In: Dudley EC (ed) The unity of evolutionary biology: proceedings of the 4th ESCE, pp 44–60

Coddington JA, Agnarsson I, Miller JA et al (2009) Undersampling bias: the null hypothesis for singleton species in tropical arthropod surveys. J Anim Ecol 78(3):573–584. https://doi.org/10.1111/j.1365-2656.2009.01525.x

Colmenares PA, Baccaro FB, Tourinho AL (2016) Assessing the relationship of vegetation structure and harvestmen assemblage in an Amazonian Upland Forest. bioRxiv. https://doi.org/10.1101/078220

Colwell RK, Coddington JA (1994) Estimating terrestrial biodiversity through extrapolation. Philos Trans R Soc 345:101–118. https://doi.org/10.1098/rstb.1994.0091

Curtis DJ, Machado G (2007) Ecology. In: Pinto-da-Rocha R, Machado G, Giribet G (eds) Harvestmen: the biology of Opiliones. Harvard University Press, Massachusetts, pp. 280–308

Da-Silva ER, Coelho LBN, Müller de Campos TR et al (2014) Marvel and DC characters inspired by Arachnids. J Comic Scholarship 4(1):Art.1. https://doi.org/10.5334/cg.aw

De Meester N, Bonte D (2010) Information use and density-dependent emigration in an agrobiont spider. Behav Ecol 21:992–998. https://doi.org/10.1093/beheco/arq088

Decaëns T, Jiménez JJ, Gioia C et al (2006) The values of soil animals for conservation biology. Eur J Soil Biol 42(S1):S23–S38. https://doi.org/10.1016/j.ejsobi.2006.07.001

Dias SC, Carvalho LS, Bonaldo AB et al (2010) Refining the establishment of guilds in Neotropical spiders (Arachnida: Araneae). J Nat Hist 44:219–239. https://doi.org/10.1080/00222930903383503
Digweed SC, Currie CR, Carcamo HA et al (1995) Digging out the "digging-in effect" of pitfall traps: influences of depletion and disturbance on catches of ground beetles. Pedobiology 39:561–576
Drew LW (2011) Are we losing the science of Taxonomy? As need grows, numbers and training are failing to keep up. BioScience 61(12):942–946. https://doi.org/10.1525/bio.2011.61.12.4
Duffey E (1974) Comparative sampling methods for grassland spiders. Bull Br Arachnol Soc 3:34–37
Eisenhauer N, Bonn A, Guerra A (2019) Recognizing the quiet extinction of invertebrates. Nat Commun 10:50. https://doi.org/10.1038/s41467-018-07916-1
Erwin TL, Scott JC (1980) Seasonal and size patterns, trophic structure and richness of Coleoptera in the tropical arboreal ecosystem: the fauna of the tree Luehea seemannii Triana and Planch in the Canal Zone of Panamá. Coleopt Bull 34:305–322
Ficken CD, Cobbaert D, Rooney RC (2019) Low extent but high impact of human land use on wetland flora across the boreal oil sands region. Sci Total Environ 693:133647. https://doi.org/10.1016/j.scitotenv.2019.133647
Fischer A, Marshall P, Camp A (2013) Disturbances in deciduous temperate forest ecosystems of the northern hemisphere: their effects on both recent and future forest development. Biodivers Conserv 22:1863–1893. https://doi.org/10.1007/s10531-013-0525-1
Fisher BL (1999) Improving inventory efficiency: a case study of leaf-litter ant diversity in Madagascar. Ecol Appl 9:714–731. https://doi.org/10.2307/2641157
Folgarait PJ (1998) Ant biodiversity and its relationship to ecosystem functioning a review. Biodivers Conserv 7:1221–1244. https://doi.org/10.1023/A:1008891901953
Forster ML, Pelton EM, Black SH (2019) Declines in insect abundance and diversity: we know enough to act now. Conserv Sci Prac 1:80. https://doi.org/10.1111/csp2.80
Gagné WC (1979) Canopy-associated arthropods in Acacia koa and Metrosideros tree communities along an altitudinal transect on Hawaii Island. Pac Insects 21:56–82
Gardner TA, Barlow J, Araujo IS et al (2008a) The cost-effectiveness of biodiversity surveys in tropical forests. Ecology 11:139–150. https://doi.org/10.1111/j.1461-0248.2007.01133.x
Gardner TA, Hernández MIM, Barlow J et al (2008b) Understanding the biodiversity consequences of habitat change: the value of secondary and plantation forests for neotropical dung beetles. J Appl Ecol 45:883–893. https://doi.org/10.1111/j.1365-2664.2008.01454.x
Gibson CWD, Hambler C, Brown VK (1992) Changes in spider (Araneae) assemblages in relation to succession and grazing management. J Appl Ecol 29:132–142. https://doi.org/10.2307/2404356
Gleason HA (1922) On the relation between species and area. Ecology 3:158–162. https://doi.org/10.2307/1929150
Goodnight JC, Goodnight ML (1942) New and little known Phalangida from Mexico. Am Mus Novit 1163:1–16
Greenslade P (2007) The potential of Collembola to act as indicators of landscape stress in Australia. Aust J Exp Agric 47:424–434. https://doi.org/10.1071/EA05264
Haddad NM, Crutsinger GM, Gross K et al (2009) Plant species loss decreases arthropod diversity and shifts trophic structure. Ecology 12:1029–1039. https://doi.org/10.1111/j.1461-0248.2009.01356.x
Hansen MC, Potapov PV, Moore R et al (2013) High-resolution global maps of 21st-century forest cover change. Science 342(6160):850–853. https://doi.org/10.1126/science.1244693
Harwood JD, Sunderland KD, Symordson WOC (2004) Prey selection by linyphiid spiders: molecular tracking of the effects of alternative prey on rates of aphid consumption in the field. Mol Ecol 13:3549–3560. https://doi.org/10.1111/j.1365-294X.2004.02331.x
Hero JM, Castley JG, Malone M et al (2010) Long-term ecological research in Australia: innovative approaches for future benefits. Aust Zool 35(2):90–102. https://doi.org/10.7882/AZ.2010.010

Höfer H, Brescovit AD, Adis J et al (2008) The spider fauna of neotropical tree canopies in Central Amazonia: first results. Stud Neotrop FAuna Environ 29(1):23–32. https://doi.org/10.1080/01650529409360913

Hooks CRR, Pandey RR, Johnson MW (2006) Effects of spider presence on Artogeia rapae and host plant biomass. Agric Ecosyst Environ 112:73–77. https://doi.org/10.1016/j.agee.2005.07.004

Hopkins G, Freckleton RP (2002) Declines in the numbers of amateur and professional taxonomists: implications for conservation. Anim Conserv 5(3):245–249. https://doi.org/10.1017/S1367943002002299

Indicatti RP, Candiani DF, Brescovit AD et al (2005) Diversidade de aranhas (Arachnida, Araneae) de solo na Bacia do Reservatório do Guarapiranga, São Paulo, São Paulo, Brasil. Biota Neotrop 5:151–162. https://doi.org/10.1590/S1676-06032005000200013

Jeanneret P, Schüpbach B, Luka H (2003) Quantifying the impact of landscape and habitat features on biodiversity in cultivated landscapes. Agric Ecosyst Environ 98:311–320. https://doi.org/10.1016/S0167-8809(03)00091-4

Jim RLS, Soares HEM (1991) Novo gênero e espécie de Prostygninae do Brasil (Opiliones, Gonyleptidae). Revta bras. Ent 35(4):799–802

Jocqué R, Alderweireldt M (2005) Lycosidae: the grassland spiders. Acta Zool Bulg 1:125–130

Jocqué R, Samu F, Bird T (2005) Density of spiders (Araneae: Ctenidae) in Ivory Coast rainforests. J Zoo 266:105–110. https://doi.org/10.1017/S0952836905006746

Kitching RL, Bergelson JM, Lowman MD et al (1993) The biodiversity of arthropods from Australian rainforest canopies: general introduction, methods, sites and ordinal results. Austr J Ecol 18:181–191

Koch CL (1841) Die Arachniden: Getreu nach der Natur abgebildet und beschrieben CH Zeh'sche Buchhandlung Nürnberg, Achter Band, pp. 41–131, pl. 265–288 (f. 621–694); Neunter Band, pp. 1–56, pl. 289–306 (f. 695–726). https://doi.org/10.5962/bhl.title.43744

Koivula M, Virta T, Kuitunen M et al (2019) Effects of undergrowth removal and edge proximity on ground beetles and vascular plants in urban boreal forests. J Urban Ecol 5(1):1–13. https://doi.org/10.1093/jue/juz007

Krausse A (1916) Ein neuer automatischer *Ausleseapparta besonder* für terrikole Insekten un Milben. Z Angew Entomol 3:303–304

Krell FT, Chung AYC, DeBoise E et al (2005) Quantitative extraction of macro-invertebrates from temperate and tropical leaf litter and soil: efficiency and time-dependent taxonomic biases of the Winkler extraction. Pedobiologia 49:175–186. https://doi.org/10.1016/j.pedobi.2004.10.005

Kremen C (1994) Biological inventory using target taxa: a case study of the butterflies of Madagascar. Ecol Appl 4(3):407–422. https://doi.org/10.2307/1941946

Lafage D, Djoudi EA, Perrin G et al (2019) Responses of ground-dwelling spider assemblages to changes in vegetation from wet oligotrophic habitats of Western France. Arthropod-Plant Inte 13:653–662. https://doi.org/10.1007/s11829-019-09685-0

Lees AC, Pimm SL (2015) Species, extinct before we know them? Curr Biol 25(5):R177. https://doi.org/10.1016/j.cub.2014.12.017

Lo-Man-Hung NF, Gardner TA, Ribeiro-Junior MA et al (2008) The value of primary, secondary, and plantation forests for Neotropical epigeic arachnids. J Arachnol 36:394–401. https://doi.org/10.1636/CT07-136.1

Lo-Man-Hung NF, Marichal R, Candiani DF et al (2011) Impact of different land management on soil spiders (Arachnida, Araneae) in two Amazonian areas in Brazil and Colombia. J Arachnol 39:296–302. https://doi.org/10.1636/CP10-89.1

Longino JT, Colwell RK (1997) Biodiversity assessment using structured inventory: capturing the ant fauna of a tropical rainforest. Ecol Appl 7:1263–1277. https://doi.org/10.1890/1051-0761(1997)007[1263:BAUSIC]2.0.CO;2

Lopes MC, Lamarre GPA, Baraloto C et al (2019) The Amazonas-trap: a new method for sampling plant-inhabiting arthropod communities in tropical forest understory. Entomol Exp Appl 167(6):534–543. https://doi.org/10.1111/eea.12797

López H, Oromí P (2010) A pitfall for sampling the mesovoid shallow substratum (MSS) fauna. Speleobiology Notes 2:7–11

MacGregor A (ed) (2018) Naturalists in the field: collecting, recording and preserving the natural world from the fifteenth to the twenty-first. Brill, Century, 999pp

Magnusson WE, Lima AP, Luizão F et al (2005) RAPELD: a modification of the gentry method for biodiversity surveys in long-term ecological research sites. Biota Neotrop 5(2):1–6. https://doi.org/10.1590/S1676-06032005000300002

Magnusson WE, Braga-Neto R, Pezzini F et al (eds) (2013) Biodiversidade E Monitoramento Ambiental Integrado—Biodiversity and integrated environmental monitoring, 1st edn. Áttema Editorial, Santo André, p 356

Magnusson WE, Lawson B, Baccaro F et al (2014) Multi-taxa surveys: integrating ecosystem processes and user demands. In: Verdade LM, Lyra-Jorge MC, Piña CI (eds) Applied ecology and human dimensions on biological conservation. Springer-Verlag, Heidelberg, pp 177–187. https://doi.org/10.1007/978-3-642-54751-5_12

Malakoff D, Mervis J (2020) Trump's 2021 budget drowns science agencies in red ink, again. https://www.sciencemag.org/news/2020/02/trump-s-2021-budget-drowns-science-agencies-red-ink-again. Accessed 16 Mar 2020

Maleque MA, Ishii HT, Maeto K (2006) The use of arthropods as indicators of ecosystem integrity in forest management. J For 10(3):113–117. https://doi.org/10.1303/aez.2009.1

Maloney D, Drummond FA, Alford R (2003) Spider predation in agroecosystems: can spiders effectively control pest populations? MAFES 190:1–32

Malumbres-Olarte J, Scharff N, Pape T et al (2016) Gauging megadiversity with optimized and standardized sampling protocols: a case for tropical forest spiders. Ecol Evol 7:494–506. https://doi.org/10.1002/ece3.2626

Marshall S, Rypstra AL (1999) Spider competition in structurally simple ecosystems. J Arachnol 27:343–350

Martín JL, Cardoso P, Arechavaleta M et al (2010) Using taxonomically unbiased criteria to prioritize resource allocation for oceanic island species conservation. Biodivers Conserv 19:1659–1682. https://doi.org/10.1007/s10531-010-9795-z

McCravy KW (2018) A review of sampling and monitoring methods for beneficial arthropods in agroecosystems. Insects 9(4):170. https://doi.org/10.3390/insects9040170

McCullough J, Alonso LE, Naskrecki P et al (eds) (2007) A rapid biological assessment of the Atewa Range Forest Reserve, Eastern Ghana. RAP Bull Biol Assess 47:1–191. https://doi.org/10.1896/978-1-934151-09-9

Mega E (2019) Financial crisis looms at Brazilian science agency. Science 365(6455):731. https://doi.org/10.1126/science.365.6455.731

Mello-Leitão CF (1927) Generos novos de Gonyleptideos (Nota previa). Bol Mus Nac Rio de Janeiro 3(2):13–22

Melo TS, Peres MCL, Andrade ARS, Oliveira MLT, Tinoco MS (2015) Artrópodes Terrestres no Licenciamento Ambiental: Um Modelo Para Utilização Por Empresas De Consultoria. Lat Am J Bus Man 6(3):126–144

Mihál I, Černecká Ľ (2017) Structure of harvestmen (Arachnida, Opiliones) communities in different, anthropically disturbed beech ecosystems (Western Carpathians, Slovakia). Vestn Zool 51(3):259–270. https://doi.org/10.1515/vzoo-2017-0032

Mupepele AC, Müller T, Dittrich M et al (2014) Are temperate canopy spiders tree-species specific? PLos One 9(2):e86571. https://doi.org/10.1371/journal.pone.0086571

Nagy GT (2010) A hands-on overview of tissue preservation methods for molecular genetic analyses. Org Divers Evol 10:91–105. https://doi.org/10.1007/s13127-010-0012-4

Novak T, Novak LS, Kozel P (2016) Hot-Dancing Method for Extracting Thanatotic Arachnids from a Substrate. Entomol News 126(2):121–127. https://doi.org/10.3157/021.126.0207

Nyffeler M, Sunderland KD (2003) Composition, abundance and pest control potential of spider communities in agroecosystems: a comparison of European and US studies. Agric Ecosyst Environ 95:579–612. https://doi.org/10.1016/S0167-8809(02)00181-0

Oliveira U, Paglia AP, Brescovit AD et al (2016) The strong influence of collection bias on biodiversity knowledge shortfalls of Brazilian terrestrial biodiversity. Divers Distrib 22:1232–1244. https://doi.org/10.1111/ddi.12489

Oliveira U, Soares-Filho BS, Santos AJ et al (2019) Modelling highly biodiverse areas in Brazil. Sci Rep 9(1):6355. https://doi.org/10.1038/s41598-019-42881-9

Oliver I, Beattie AJ (1996) Invertebrate morphospecies as surrogates for species: a case study. Conserv Biol 10(1):99–109. https://doi.org/10.1046/j.1523-1739.1996.10010099.x

Otto S, Floren A (2007) The spider fauna (Araneae) of tree canopies in the Białowieża Forest. Fragm Faun 50(1):57–70. https://doi.org/10.3161/00159301FF2007.50.1.057

Pearson DL, Hamilton AL, Erwin TL (2011) Recovery plan for the endangered taxonomy profession. BioScience 61(1):58–63. https://doi.org/10.1525/bio.2011.61.1.11

Pekár S, Haddad CR (2005) Can agrobiont spiders (Araneae) avoid a surface with pesticide residues? Pest Manag Sci 61:1179–1185. https://doi.org/10.1002/ps.1110

Pérez-Miles F, Miglio LT, Bonaldo AB (2006) *Megaphobema teceae* n. sp. (Araneae, Theraphosidae), a new theraphosine spider from Brazilian Amazonia. Zootaxa 1115:61–68. https://doi.org/10.5281/zenodo.171589

Perner J, Malt S (2003) Assessment of changing agricultural land use: response of vegetation, ground-dwelling spiders and beetles to the conversion of arable land into grassland. Agric Ecosyst Environ 98:169–181. https://doi.org/10.1016/S0167-8809(03)00079-3

Pezzini FF, Melo PHA, Oliveira DMS et al (2012) The Brazilian Program for Biodiversity Research (PPBio) Information System. B&E 4:265–274. https://doi.org/10.7809/b-e.00083.265

Pianka ER (1966) Latitudinal gradients in species diversity: a review of concepts. Am Nat 100:33–46

Pimm SL, Raven P (2000) Extinction by numbers. Nature 403:843–845. https://doi.org/10.1038/35002708

Pinto-da-Rocha R (1994) A new species of Santinezia from central Amazonia, with systematic comments on the genus (Arachnida, Opiliones, Cranaidae). Zootaxa 13:29–32. https://doi.org/10.11646/zootaxa.3135.1.1

Pinto-da-Rocha R, Bonaldo AB (2006) A structured inventory of harvestmen (Arachnida: Opiliones) at Juruti River plateau, State of Pará, Brazil. Rev Iber Aracnol 13:155–162

Pinto-da-Rocha R, da Silva MB, Bragagnolo C (2005) Faunistic similarity and historic biogeography of the harvestmen of southern and southeastern Atlantic Rain Forest of Brazil. J Arachnol 33(2):290–299. https://doi.org/10.1636/04-114.1

Pinto-da-Rocha R, Machado G, Giribet G (eds) (2007) Harvestmen: the biology of Opiliones. Harvard University Press, Cambridge. 597 pp

Polis GA, McCornick SJ (1986) Scorpions, spiders and solpugids: predation and competition among distantly related taxa. Oecologia 71:111–116. https://doi.org/10.1007/BF00377328

Porto W, Pequeno P, Tourinho AL (2016) When less means more: reduction of both effort and survey methods boosts efficiency and species diversity of harvestmen in a tropical forest. Ecol Indic 69:771–779. https://doi.org/10.1016/j.ecolind.2016.05.014

Poos FW (1929) Leaf hopper injury to legumes. J Econ Entomol 22:146–153

Prendini L, Hanner R, DeSalle R (2002) Obtaining, storing and archiving specimens and tissue samples for use in molecular studies. In: DeSalle R, Giribet G, Wheeler W (eds) Techniques in molecular systematics and evolution. Birkhäuser, Basel, pp 176–248

Privet K, Petillon J (2018) Differences in tropical vs. temperate diversity in arthropod predators provide insights into causes of latitudinal gradients of species diversity. BioRxiv. https://doi.org/10.1101/283499

Privet K, Vedel V, Fortunel C et al (2020) Relative efficiency of pitfall trapping vs. nocturnal hand collecting in assessing soil-dwelling spider diversity along a structural gradient of Neotropical habitats. Diversity 12(2):81. https://doi.org/10.3390/d12020081

Rego FNAA, Venticinque EM, Brescovit AD (2005) Densidades de aranhas errantes (Ctenidae e Sparassidae: Araneae) em uma floresta fragmentada. Biota Neotrop 5(1a):1–8. https://doi.org/10.1590/S1676-06032005000200004

Rego FNAA, Venticinque EM, Brescovit AD (2007) Effects of forest fragmentation on four Ctenus spider populations (Araneae: Ctenidae) in central Amazonia, Brazil. Stud Neotrop Fauna E 42:137–144. https://doi.org/10.1080/01650520600935082

Rego FNAA, Venticinque EM, Brescovit AD et al (2009) A contribution to the knowledge of the spider fauna (Arachnida: Araneae) of the floodplain forests of the main Amazon River channel. Rev Iber Aracnol 17:85–89

Resende LPAR, Pinto-da-Rocha R, Bragagnolo C (2012) The harvestmen fauna (Arachnida: Opiliones) of the Parque Estadual Carlos Botelho, and the Floresta Nacional de Ipanema, São Paulo, Brazil. Biota Neotrop 12(4):146–155. https://doi.org/10.1590/S1676-06032012000400016

Ricetti J, Bonaldo AB (2008) Spiders diversity and richness estimates in four vegetations types of Serra do Cachimbo, Pará, Brazil. Iheringia 98(1):88–99. https://doi.org/10.1590/S0073-47212008000100013

Roberts HR (1973) Arboreal Orthoptera in the rain forest of Costa Rica collected with insecticide: a report on the grasshoppers (Acrididae) including new species. Proc Acad Nat Sci Philadelphia 125:46–66

Romero GQ, Srivastava DS (2010) Food-web composition affects cross-ecosystem interactions and subsidies. J Anim Ecol 79:1122–1131. https://doi.org/10.1111/j.1365-2656.2010.01716.x

Rowling JK (1997-2007) Harry Potter. Novel series. Bloomsburry Publishing and Pottermore, London

Samu F, Szinetár C (2002) On the nature of agrobiont spiders. J Arachnol 30(2):389–402. https://doi.org/10.1636/0161-8202(2002)030[0389:OTNOAS]2.0.CO;2

Samu F, Sunderland KD, Szinetár C (1999) Scale-dependent dispersal and distribution patterns of spiders in agricultural systems: a review. J Arachnol 27:325–332

Samu S, Elbanna SM, Orabi GM et al (2016) Assessing the potential role of spiders as bioindicators in Ashtoum El Gamil Natural Protected area, Port Said, Egypt. Indian J Arachnol 5(1–2):100–112

Schuldt A, Bruelheide H, Buscot F et al (2017) Belowground top-down and aboveground bottom-up effects structure multitrophic community relationships in a biodiverse forest. Sci Rep 7(4222):1–10. https://doi.org/10.1038/s41598-017-04619-3

Seyfulina RR (2010) The spider assemblage (Arachnida, Aranei) in agroecosystems of Kuban Plain: species composition, spatial distribution, and seasonal dynamics. Entomol Rev 90:494–510. https://doi.org/10.1134/S001387381004010X

Siebold S, Gossner MM, Simons NK et al (2019) Arthropod decline in grasslands and forests is associated with landscape-level drivers. Nature 574:671–674. https://doi.org/10.1038/s41586-019-1684-3

Soberón JM, Llorente J (1993) The use of species accumulation functions for the prediction of species richness. Conserv Biol 7:480–488. https://doi.org/10.1046/j.1523-1739.1993.07030480.x

Souza JLP, Baccaro FB, Landeiro VL et al (2012) Tradeoffs between complementarity and redundancy in the use of different sampling techniques for ground-dwelling ant assemblages. Appl Soil Ecol 56:63–73. https://doi.org/10.1016/j.apsoil.2012.01.004

Srivastava DS, Trzcinski MK, Richardson BA et al (2008) Why are predators more sensitive to habitat size than their prey? Insights from Bromeliad Insect Food Webs. Am Nat 172(6):761–771. https://doi.org/10.1086/592868

Staley JT (2001) BIOTA/FAPESP, Brazil's model program for assessing global biodiversity. Biota Neotrop 1(1–2):1–4. https://doi.org/10.1590/S1676-06032001000100002

Sutton SL, Hudson PJ (1980) The vertical distribution of small flying insects in the lowland rain forest of Zaire. Zool J Linn Soc 68:111–123. https://doi.org/10.1111/j.1096-3642.1980.tb01921.x

Tessel SM, Palmquist KL, Peet R (2015) Species-area relationships. In: Gibson D (ed) Oxford bibliographies in ecology. Oxford University Press, New York

Thorbek P, Topping CJ (2005) The influence of landscape diversity and heterogeneity on spatial dynamics of agrobiont linyphiid spiders: an individual-based model. BioControl 50:1–33. https://doi.org/10.1007/s10526-004-1114-8

Thorn S, Bubler H, Fritze MA et al (2016) Canopy closure determines arthropod assemblages in microhabitats created by windstorms and salvage logging. For Ecol Manage 381:188–195. https://doi.org/10.1016/j.foreco.2016.09.029

Tolkien JRR (1968) The lord of the rings. Series. Allen & Unwin, London, 1178pp
Toth F (2000) Aspirator gun—a new device for sampling spiders and insects. Ekologia 19(3):279–280
Tourinho AL, Lança LS, Baccaro FB et al (2014) Complementarity among sampling methods for harvestman assemblages. Pedobiologia 57:37–45. https://doi.org/10.1016/j.pedobi.2013.09.007
Tourinho AL, Dias SC, Lo-Man-Hung NF et al (2018) Optimizing survey methods for spiders and harvestmen assemblages in an Amazonian upland forest. Pedobiologia 67:35–44. https://doi.org/10.1016/j.pedobi.2017.11.002
Tourinho AL, Benchimol M, Porto W et al (2019) Marked compositional changes in harvestmen assemblages in Amazonian forest islands induced by a mega dam. Insect Conserv Diver 1–13. https://doi.org/10.1111/icad.12398
Trägårdh I (1910) Om Berlese's apparat för snann och effecktiv insamling af små leddjur. Entomologisk Tiddskrift 31:35–37
Tullgren A (1917) En enkel apprat för automatiskt vittjande av sällgods. Entomologisk Tidskrift 38:797–100
Uetz GW (1999) The influence of variation in litter on spider communities. Oecologia 40:29–542. https://doi.org/10.1007/BF00388808
Upton MS, Mantle BL (2010) Methods for collecting, preserving and studying insects and other terrestrial arthropods. Paragon Printers Australasia, Canberra
Uetz GW, Unzicker JD (1976) Pitfall trapping in ecological studies of wandering spiders. J Arachnol 3:101–111
Vedel V, Cerdan A, Martinez Q et al (2015) Day-time vs. night-time sampling does not affect estimates of spider diversity across a land use gradient in the Neotropics. J Arachnol 23(3):413–416. https://doi.org/10.1636/0161-8202-43.3.413
Verberk W (2011) Explaining general patterns in species abundance and distributions. Nat Educ Knowledge 3(10):38
Vleminckx J, Schimann H, Decaëns T et al (2019) Coordinated community structure among trees, fungi and invertebrate groups in Amazonian rainforests. Sci Rep 9:11337. https://doi.org/10.1038/s41598-019-47595-6
White EB (1957) Charlotte's web. Harper & Brothers, New York, 192pp
Whitney KD (2004) Experimental evidence that both parties benefit in a facultative plant-spider mutualism. Ecology 85(6):1642–1650. https://doi.org/10.1890/03-0282
Willig MR, Kaufman DM, Stevens RD (2003) Latitudinal gradients of biodiversity: pattern, process, scale, and synthesis. Annu Rev Ecol Evol Syst 34:273–309
Wilson EO (1987) The little things that run the world (The importance and conservation of invertebrates). Conserv Biol 1(4):344–346
Wise DH (1993) Spiders in ecological webs. Cambridge University Press, Cambridge. 328 pp
Wise DH, Snyder WE, Tuntibunpakul P et al (1999) Spiders in decomposition food webs of agroecosystems: theory and evidence. J Arachnol 27:363–370
Work TT, Buddle CM, Korinus LM et al (2002) Pitfall trap size and capture of three taxa of litter-dwelling arthropods: implications for biodiversity studies. Environ Entomol 31(3):438–448. https://doi.org/10.1603/0046-225X-31.3.438
World Wide Fund for Nature (WWF) (2018) Living planet report 2018. https://www.worldwildlife.org/pages/living-planet-report-2018. Accessed 24 Feb 2020
Yang LH, Gratton C (2014) Insect as drivers of ecosystem processes. Insect Sci 2:26–32. https://doi.org/10.1016/j.cois.2014.06.004
Zalasiewicz J, Williams M, Steffen W et al (2010) The new world of the anthropocene. Environ Sci Technol 44(7):2228–2231. https://doi.org/10.1021/es903118j
Zou Y, Telgen MD, Chen J et al (2016) Modification and application of a Leaf Blower-vac for field sampling of arthropods. J Vis Exp 114:54655. https://doi.org/10.3791/54655

Part III

Chapter 16
Sampling Galls and Galling Arthropods

Walter Santos de Araújo, Maria Virgínia Urso-Guimarães, Milton de Souza Mendonça Jr, and Jean Carlos Santos

16.1 Galls and Galling Arthropods

Galls are plant structures induced by biological agent, as viruses, bacteria, nematodes, or arthropods (Shorthouse and Rohfritsch 1992; Raman et al. 2005; Carneiro et al. 2009). Gall-inducing arthropods are considered the most specialized herbivores (Shorthouse et al. 2005), given that insects and mites can induce galls in plant tissues (Araújo et al. 2019b). Most gall inducers possess interactions species-specific with their host plants due to genotypical association between the involved organisms (Mani 1964; Abrahamson and McCrea 1986; Abrahamson and Weis 1987). This highly specialized interaction results in an abnormal increase by hypertrophy or hyperplasia of plant cells (Fernandes et al. 2012), tissues, or organs as an answer to a specific stimulation of an inductor (Stone and Schönrogge 2003; Raman 2011). Each species of inducer produces galls anatomically and physiologically different from those induced by other related species (Stone and Schönrogge 2003).

W. S. de Araújo (✉)
Departamento de Biologia Geral, Centro de Ciências Biológicas e da Saúde, Universidade Estadual de Montes Claros, Montes Claros, Minas Gerais, Brazil

M. V. Urso-Guimarães
Departamento de Biologia, Laboratório de Sistemática de Diptera, Universidade Federal de São Carlos, Sorocaba, São Paulo, Brazil

M. d. S. Mendonça Jr
Departamento de Ecologia, Instituto de Biociências, Universidade Federal do Rio Grande do Sul, Porto Alegre, Rio Grande do Sul, Brazil

Department of Ecology and Evolutionary Biology, UCLA, Los Angeles, CA, USA

J. C. Santos
Departamento de Ecologia, Universidade Federal de Sergipe, São Cristóvão, Sergipe, Brazil

J. C. Santos, G. W. Fernandes (eds.), *Measuring Arthropod Biodiversity*,
https://doi.org/10.1007/978-3-030-53226-0_16

Evidence of insect damage preserved in fossilized plant parts from the Carboniferous (~358–298 mya) indicates ancient associations among insects and plants. The diversity of types of marks in leaves or stems made by a variety of mouthparts of insects indicated the association with tree and seed ferns from Late Carboniferous coal deposits, as weevils (Coleoptera) and thrips (Thysanoptera) with cycads (Gullan and Cranston 2014). Nevertheless, the major diversification of insects came later, in the Cretaceous period (~145–66 mya). Labandeira et al. (2007) recognized nine major categories of arthropod damage to or physical interaction with plants: external foliage feeding, piercing and sucking, leaf mining, seed predation, boring, palynivory, nectarivory, ovipositioning, and gall formation. Krassilov and Shuklina (2008) made taxonomic assignments of gall inducers from fossils of the Upper Cretaceous (93–89 mya), in which they found the leading role of Diptera (cecidomyiids, 45%), Hemiptera (aphids, 22%; physillids, 11%; coccoids, 5.5%), and Hymenoptera (tenthredinids, 11%; cynipids, 5.5%). The leadership has not changed then, but in addition, some species of mostly tropical Thysanoptera (thrips), Coleoptera (mostly weevils), and Lepidoptera (small moths) induce galls. This great phylogenetic and taxonomic diversity of galling arthropods ends up reflecting an enormous functional and morphological diversity of galls in nature (Fig. 16.1).

Presently around 13,000 species of gall inducers arthropods are known worldwide (Raman et al. 2005). However, Espírito-Santo and Fernandes (2007) estimated the value to nearly 130,000 species of gall maker insects, and Hebert et al. (2016) extrapolated to one million the number of cecidomyiid species of the world based on DNA barcoding sequences found in the Canadian fauna. Due to its great diversity, studying the interactions between gall inducers, host plants, and associated fauna is fundamental to understand structural patterns of diversity in terrestrial environments. In this sense, several studies have discussed the importance of studying arthropod galls and formulated hypotheses to explain the diversity of gall inducer species (review in Miller III and Raman 2019). From the applied point of view, the study of galling arthropods is important because some species of gallers can be considered agricultural pests, such as gall midges (Harris 1966; Maia 2018) and galling mites (Lillo et al. 2018). In addition, some studies also point to the applicability of galling insects as habitat quality bioindicators in environmental monitoring and licensing studies (Araújo and Espírito-Santo Filho 2012; Santana and Isaias 2014; Toma et al. 2014). The recent increment in studies on galls and gallers, especially in the Neotropical region (Grandez-Rios et al. 2016) and in Brazil (Araújo 2018), there is still a lack of work compiling the methodological aspects regarding the field sampling and laboratorial processing of arthropod galls.

Fig. 16.1 Examples of morphological diversity in arthropod galls. Note the variety of shapes, sizes, colors, and positions in the plant organ and indumentum (hairs). (Photos: JC Santos)

16.2 Sampling Methods and Patterns in Gall Diversity: The Case of Xeric vs. Mesic Environments

One of the most traditional hypotheses in the literature about insect galls is the hygrothermal stress hypothesis proposed by Fernandes and Price (1988), which predicts a higher diversity of galling species in stressed environments. Testing this hypothesis several studies have compared mesic (non-sclerophyllic) and xeric (sclerophyllic) vegetation (Fig. 16.2), and many have pointed to a higher galling

Fig. 16.2 Sampling of arthropod galls in different environment types. (**a**) A typical xeric environment of Neotropical savanna in Serra do Cabral, Brazil, characterized by open and short vegetation. (**b**) The sampling of galls in xeric environments is generally easy, due to the small size of the trees. (**c**) A humid tropical forest in the Atlantic Forest, Brazil. (**d**) In xeric environments, gall sampling is easy in the forest understory, but in the upper strata such as the canopy, more adequate equipment and methods are needed to access the galls. (Photos: WS Araújo (**a**, **b**) and JC Santos (**c**, **d**))

insect richness in xeric habitats, both along latitudinal and altitudinal gradients (Price et al. 1998; Wright and Samways 1998; Gonçalves-Alvim and Fernandes 2001; Cuevas-Reyes et al. 2004; Julião et al. 2014). On the other hand, a smaller number of studies have found results that contradict the expectation of greater gall diversity in xeric environments when compared to mesic environments (e.g., Blanche 2000; Veldtman and McGeoch 2003). Xeric and mesic vegetations tend to be distributed in the landscape in a gradient of increasing structural complexity of the vegetation (Gonçalves-Alvim and Fernandes 2001), which requires an adequate space-based sampling design (see Sect. 16.3.2). Generally, mesic vegetation has greater structural complexity than xeric habitats (see Fig. 16.2), including a higher canopy stratum, which complicates sampling in these environments (see methods for the sampling of arthropod galls on vertical stratification—Sect. 16.3.4). As most studies do not explicitly present sample sufficiency analysis for different environments, it is not possible to rule out that sampling problems or subsampling carried out in the mesic environments have resulted in this pattern observed in the literature. This questioning about one of the most pivotal patterns for the galling species diversity indicates the importance of correct determination of the sampling methods for gall collection, especially when the goal is to study environmental gradients.

16.3 Field Sampling Methods

Sampling galls for biodiversity research has been considered a relatively easy task, since galls are sessile and can be both visually differentiated from normal plant tissue (sometimes even conspicuous) and rich in species in any given environment (Fernandes et al. 2014). Also, because most gall inducers are specific on their host plant species and produce recognizable and repeatable structures (Carneiro et al. 2009), morphotyping galls instead of identifying galling insects has become common practice, simplifying post-fieldwork lab efforts (Fig. 16.3). Gall morphotyping is accepted in the literature due to the lack of taxonomic information for many galling species, with most species not yet described, especially in the Neotropical region (Espírito-Santo and Fernandes 2007; Araújo 2018; Araújo et al. 2019b). Given that the number of taxonomists specializing in galler groups is scarce, this is an initiative to enable galling studies to go forward, until taxonomic work is further developed to catch up (Carneiro et al. 2009). The positive side of these assertions is that once an adequate sampling design has been chosen, it is likely that the majority of galls on a given area/region might be sampled sufficiently. Thus, galls appear to be useful systems for evaluating diversity, as productive as they have been recognized for other ecology subjects such as ecological interactions and evolution (Abrahamson and Weis 1997).

Fig. 16.3 Sampling galls in the field. (**a**) Sampling galls with pruning shears. (**b**) Gall collection with 10-m-long telescoping aluminum pole. (**c**) Storage of galls in plastic bags for transportation. (**d**) Processing of plants and galls in the laboratory. (Photos: JC Santos (**a**, **c**), WS Araújo (b), BM Oliveira (**d**))

16.3.1 Sampling and Morphotyping: A Call to Maximize Information Provided

Gall morphotyping usually resorts to observing a few gall attributes combined to host plant identification and (usually) superficial galler identification. Troubles and solutions to morphotyping have been explored in detail elsewhere (Isaias et al. 2013), but since this is an important issue for sampling, we would like to add a few indications. First, the more gall characteristics (or attributes) are chosen to describe a morphotype, the better the ability to distinguish between similar galls (see Fig. 16.1). Of course, a trade-off must be reached, but the effort in using more attributes is certainly small. When choosing attributes, if these can coincide with those considered functional gall attributes or traits, morphotyping usefulness could go beyond telling galls apart, opening up the possibility for the analysis of functional diversity. Finally, the ease with which photography can be used nowadays means multiple high-quality pictures of gall morphotypes can be taken and would provide a safeguard against future difficulties in distinguishing among morphotypes (see Sect. 16.4.4 for further details). This can be combined with large spaces for additional information usually provided by scientific journals today, so that more information available increases the usefulness and representativeness of morphotyping. Regional-scale work, employing detailed gall morphotyping, could certainly reveal as yet unknown diversity patterns.

Because most galling insect inducers outside the northern temperate zone cannot be named and placed taxonomically (Espírito-Santo and Fernandes 2007), inducer information is normally vague and restricted to the family level. However, any information on the inducer (or anyway on the gall occupant) is useful in distinguishing galls and providing more ecological and evolutionary detail to biodiversity assessments. As above, we suggest future work on gall diversity using morphotyping could try and include more detail on gall occupants by providing simple information such as larval numbers, size, color, and obvious morphological characteristics, along with illustration by pictures. A good example of morphological detail is the Cecidomyiidae (Diptera, the most common and diverse galler group) prothoracic spatula, whose size, color, and number of teeth can be both informative for morphotyping and also a taxonomic character (Gagné and Jaschhof 2017). We understand this suggestion and the one in the previous paragraph increase lab efforts in a way that might run counter to what we would like to suggest to simplify morphotyping, but researchers will need to rely on common sense to solve this trade-off.

Finally, host plant information cannot be overstated as a useful form of not only establishing a morphotype but also on providing concrete taxonomic data when identifying gall inducers is usually so difficult. Plant identification will usually require consultation with botanical specialists and can result in museum material (vouchers) to guarantee fidelity and authenticity. We even suggest that branches of the plant with galls are herborized and mounted on exciccates for depositing in the collections. Further, although the majority of gall inducers is indeed species-specific, a few species remain that can induce galls on different plants, especially phylogenetically close ones; thus morphotyping must not rely on host plant too heavily as it is not an absolutely safe way to distinguish between galling species.

Although gall usefulness for ecology and specially for biodiversity monitoring is undeniable, there is a negative side: the generality of the assertions about sampling galls and gall inducers (sessile, conspicuous, and species-rich) is, as is the rule with complex biological systems, limited. The question to address is how limited they are and how much they could impact sampling. First, galls are built on plant tissues, and although that makes them sessile, they are not necessarily permanent. Gall tissues could decay and either render galls unidentifiable or even lead them to drop from the plant (Uechi et al. 2004). Alternatively, galls may be shed from the plant because of plant reactions such as leaf abscission (Mendonça and Romanowski 2002; Fernandes et al. 2008). No gall would last forever, but the central issue is that some galls may consistently last more than others, potentially biasing field sampling. The first possibility is that shoot galls could be much persistent than leaf galls, exactly because plants may commonly shed leaves but not usually shoots, either as an anti-herbivore reaction or as a normal response to environmental seasonality. The second possibility is that pedunculated galls (or extraorgan galls, those attached to the host plant organ by a narrow structure instead of being induced within the organ) could also be at a greater risk of being dropped than intraorgan galls. This could come about by a series of effects, such as plant tissue reactions, action by other organisms (fungal decay, predator or parasitoid deliberate attack), or even sheer casualty, such as larger animals "scraping" galls by just brushing against the plant. To the best of our knowledge, the literature is moot about this, and no proper assessments of these possibilities for biodiversity sampling have been developed.

The second perhaps not so general assertion about galls is on the conspicuity. Although bright or evident colors on galls have even been described as an antipredator strategy (Inbar et al. 2010), most galls present colors exactly like or very similar to the normal organ tissues. The same can be said of gall shapes, although there is a much larger variety of these, which can even make morphotyping a harder task (Isaias et al. 2013). Again, frequently galls have fusiform shapes when induced on shoots, and lenticular when on leaves (Mendonça et al. 2014), rendering them not so easily detected in the field. Changes in shape and color during gall development are another problem when sampling unknown galls, and although they are usually predictable and repeatable in structure, galls may not only vary from one organism to the other but also because of other influences (as parasitoids) and in time (Isaias et al. 2014; but see Carneiro et al. 2009).

16.3.2 Sampling Field Methods Based on Space

Sampling gall diversity in the field has involved different designs and schemes (Fig. 16.4). A complete survey would be beyond the scope of this section, but we wish to present a few examples of how people have organized sampling and how ideally this should be accomplished. A perhaps simple but useful way of

Fig. 16.4 Sampling arthropod galls in the field. (**a**) Assembling a sampling plot. (**b**) Finding plants with galls within a plot. (**c**) Inspecting galled leaves. (Photos: JC Santos)

understanding sampling schemes is to differentiate between designs based on space (for which time is not overly important) and those based on time (for which space is not overly important as the basis and interpretation of sampling). As galls are sessile on host plants, individual plants have also been used as sample size measures or even sample units. There are positive and negative aspects to any of these designs, and certainly logistic research limitations are paramount in deciding which to employ; however we would suggest pure space-based sampling produces more comparable results than work based on sampling time and/or individual plants.

Samples of gall diversity based purely on space dimensions have been employed rather uncommonly in the literature (e.g., Blanche 2000), and they are usually further restricted to a certain number of plant organs per individual plant (e.g., Julião et al. 2014) or to a given plant form and size (e.g., trees above 3 cm in diameter; Araújo et al. 2014a). This imposed limitation is unsurprising given that for vegetation-dense environments such as forests, even a small space can contain an immense plant surface to sample galls from. Large sample units may help better characterize plant and gall diversity for diverse environments, but they would be too time-consuming to sample with the usual perspective of visually searching for any and all galls, even inconspicuous ones, on any and all plants.

Space sampling actually involves first a choice among how many spatial dimensions to consider. One-dimensional samples (line transects) can be rather simple to implement in the field, but they are not used for gall sampling explicitly, although many time-based samples report using trails or line transects (Dalbem and Mendonça 2006). Trails through any form of vegetation anyway sit in an uncomfortable position in this case, being a relatively safe and fast way of assessing an environment but at the same time posing a danger in blindingly including unknown edge effects on data, caused by the amount of change in the original environment due to the opening represented by the trail. Two-dimensional samples have been more common. With varied lengths, rectangular setups of up to tens of meters (e.g., 20 × 50 m, Blanche 2000; 5 × 50 m, Cuevas-Reyes et al. 2004) have been used most usually, some of them approaching a "line" situation (5 × 200 m, Oyama et al. 2003) and others resorting to squares (10 × 10 m, Araújo et al. 2013; Araújo et al. 2014a; Araújo 2017). The choice between squares and long rectangles depends on a series of questions, as always, and although squares are usually logistically easier to set up during fieldwork, the longer the rectangle used, the more environmental heterogeneity can be captured, and potentially the higher gall diversity could be.

16.3.3 Sampling Field Methods Based on Vertical Stratification

Another question regarding space-based samples for gall diversity (which we choose to treat separately because of peculiarities) is about the third dimension—height, which is increasingly important as one goes from short (grasslands) to tall (forest) vegetation types being evaluated. Some research indicates specifically that only (human) accessible plants can be examined for galls (e.g., up to 2 m tall plants; Oyama et al. 2003), which is something also reported for time-based samples (e.g., Mendonça and Stiling 2018). Although this can standardize sampling and reveals the true volume used in samples, it is also potentially problematic since the taller the forest, the less representative samples might be. In forest environments, especially in tropical forests, the diversity of galling arthropods shows vertical stratification (Ribeiro et al. 2014). This might be due to the high levels of hygrothermal stress and leaf sclerophylly in the forest canopy (Ribeiro and Basset 2007; Ribeiro et al. 2014). In this sense, some studies have indicated that galling species are more diverse in the canopy than in the understory (Paniagua et al. 2009; Julião et al. 2014); most commonly, though, studies in forest environments only sample forest undergrowth.

Sampling the canopy can be difficult and expensive (see Chap. 18—Canopy Insect Sampling). Blanche (2000) reports that for sampling galls in the canopy it was necessary to climb the tree or stand on the cabin roof of a four-wheel drive vehicle." Several methods have been recorded in the literature for canopy gall sampling, such as sampling within a canopy crane perimeter (Paniagua et al. 2009), use of pin cylinder for sampling within a volumetric space of 1 m in diameter (Ribeiro and Basset 2007), use of climbing techniques, and help of telescoping aluminum pole of 10 m or more (see Fig. 16.3b) (Julião et al. 2008, 2014). Some of these efforts are laudable and efficient to reveal gall diversity and should continue to be used, but there is an obvious limitation when comparing haphazardly sampling a finite number of canopy tree branches to what meticulous examination of plants near the ground can accomplish.

Perhaps the most thorough and representative method to sample galls from the ground level to the canopy is the pin-cylindrical transect sampling unit proposed by Ribeiro and Basset (2007). In this case, access to the canopy is provided vertically for a sampling person, who collects all galls present within 1 m of the central vertical climbing rope, thus effectively generating a cylinder; the authors start the cylinder from 3 m above the ground and continue up to the upper canopy. This method is perhaps complementary with that discussed in the above paragraph: while including all biological material within a known space (as is possible on the ground, but difficult in the canopy) and thus allowing one to calculate densities with much higher confidence, the cylinder is restricted in terms of the environmental heterogeneity revealed in a given sample unit, being usually restricted to sampling a few species of trees (Ribeiro et al. 2014); also, the length of the cylinder is dependent upon canopy height, with a fixed starting point (3 m above ground) but no fixed final point. The specific choice of the technique used for canopy sampling depends on the availability of equipment and trained people, but we emphasize that in forest environments, canopies should be sampled, whenever possible, so that the diversity of galls is not underrepresented.

16.3.4 Sampling Field Methods Based on Time

Time-based samples for galling insect diversity have been employed in different contexts and vegetation types and for research interests. People have employed sample durations from 45 (e.g., Carvalho-Fernandes et al. 2016) to 135 min (Yukawa et al. 2001; actually a work with no time standardization), but usually these last between 60 (Price et al. 1998) and 90 min (e.g., Mendonça et al. 2010). The main allure of this method is its simplicity: a chronometer only may be enough to allow sampling. As usual, simpler may mean less standardization and more unknowns, leading to more difficult comparisons, either among different environments or, even worse, among different protocols and research teams. For different environments, such as grasslands, scrublands, and forests, different plant densities and heights mean that searching for galls and walking upright might result in very distinct outcomes.

One of the main hurdles for time-based samples for gall diversity may be adequate and detailed reporting of methods. Protocols including stopping to collect galls, observe them, open galls to reveal contents and confirm insects inside, store and label them, then picture galls, jolt field notes, and collect and press host plants—all would lead to a very high handling time for each gall or kind of gall found. This is especially true when compared to a simpler option like just picturing galls as they are found—which would lead also to less important information being gathered. Richer environments would probably end up underrepresented, and gall diversity distributions could be flattened out. Apart from protocol simplicity, sampling team size is also of concern. Not only how many people get involved in sampling a given area/trail at a given time is of paramount importance, but also sampler experience and division of labor could affect recording gall richness. The number of people sampling should be always informed, and also the amount of field training each one has experienced, along with roles played by the samplers, if they differ (division of labor in this case could mean one person searching for galls, another taking pictures, and yet others storing galls and plants, etc.). Even though all these things could be informed in as much detail as possible, standardization could still be hard to ascertain. Because people inherently differ in their abilities and division of labor may not mean each person doing what they do best, comparing complex sampling setups would be daunting—how would sampling for 30 min with three persons and complete division of labor compare with a single-person sampling for 90 min?

16.3.5 Architectural Sampling Field Methods

The "architectural" sampling method for gall diversity has been first suggested by Fernandes and Price (1988) and ever since employed and adapted in many different places and situations. It compounds a space-based method (sampling occurs within a 10-m-wide transect) with a plant-based approach—1000 individual herbaceous plants, 100 shrubs, and 45 trees must be sampled at each point. The authors demonstrated that asymptotic species richness was obtained with lower numbers of plants than those above for either tropical or temperate samples (Price et al. 1998). This compound method has been taken up and used across the world (Veldtman and McGeoch 2003, South Africa) but also modified in terms of plant forms and plant numbers sampled (Wright and Samways 1998). It also appears to have inspired other plant-based methods, such as that of Blanche and Westoby (1995, Australia) which more simply differentiates between short (< 50 cm) and tall (>50 cm) plants instead of using plant forms (and uses a 20 × 50 m "site" inside which sampling occurs).

Although a combination of space and plant structure as criteria for sampling galls is fruitful, it derives from the original authors' interest in exploring the effect plant structure and architecture has on the evolution of galling insects (Fernandes and Price 1988). Because the plant forms occur in varied proportions across different terrestrial environments, sampling the fixed amount of plants would not always cover the same area or take a similar time. Veldtman and McGeoch (2003) already pointed out that in South Africa both shrubs and trees were too scattered and keeping to a 10-m-wide transect was not practical. It seems that the numbers decided on by Fernandes and Price (1988) may be adequate for scleromorphic vegetation (Arizona, USA, and Minas Gerais, Brazil) but may not have been ideal for the non-scleromorphic vegetation examined by Veldtman and McGeoch (2003). As the latter authors also point out, a separate question is on exactly how to avoid small shrubs and trees being included as herbs when too young to be properly identified.

The argument behind using an architectural sample is that by explicitly taking plant numbers as a measure of sampling, one does away with plant density effects and better represents the resource basis for gallers (Veldtman and McGeoch 2003). Although it includes a space-based rule to sampling (10-m-wide transect), the transect length depends only on plant density, and actually there could be three lengths to the transect, one for each plant form, with consequences for how sampling is representing the vegetation being characterized and environmental heterogeneity. Thus, we would argue this method should not be considered space-based. This is not a weakness per se, but space may be the less difficult aspect to take into consideration in diversity sampling when establishing comparisons across regions and research teams, before time and plant structure.

16.3.6 Sampling Gall Diversity with Field Methods

If we should strive to achieve standardization in sampling galls, it seems to be easy to derive that space is easier to deal with, having no strong subjectivity in its measurement of sampling effort (unlike time) and representing a universal unit for sampling (unlike plant forms), and to derive spatial densities (Table 16.1). We would then suggest using long rectangles as a way of depicting possible environmental heterogeneity (e.g., increasing the chances of including more plant species and thus

Table 16.1 Comparison of the main field sampling methods

Field method	Advantages	Limitations/biases	References
Space-based sampling	Comparable across vegetation types	Large plant volume to sample from even small sizes	Blanche (2000)
One dimension (1-D)—transects	Easy to implement	Edge effects when using trails	Oyama et al. (2003)
Two dimensions (2-D)—plots	Relatively easy to implement	Low environment heterogeneity captured when using squares	Araújo (2017)
Three dimensions (3-D)—volume	Full consideration of space axes	Difficult logistically	Ribeiro et al. (2014)
Time-based sampling	Very easy to implement	Highly dependent on sampler experience, field team size	Dalbem and Mendonça (2006)
Architectural-based sampling—plant forms: herbs, shrubs, and trees	Easy to implement	Plant form proportions dependent on vegetation; different areas sampled for each plant form	Fernandes and Price (1988)

more hosts in the sample). If we were to take all the arguments above into consideration for a final decision on which is the best method to sample gall diversity in the field, however, there still appears to be scope for putting logistic and project-specific questions in the first place before attempting an absolute generalization.

16.4 Collecting, Preparing, and Rearing Procedures

16.4.1 Collecting Galls and Processing Botanical Material

Galls can be induced in any plant organ, as stems, leaves, flowers, fruits, and even roots. Usually, whole branches of the plants with galls are collected and taken, individualized, and labeled (with the following information: place and date of collection, collector, and identification of the host plant) in plastic bags to be analyzed in a laboratory (Fig. 16.5a). It is very important to collect additional material of the host plant with branches to examine leaves, leaf phyllotaxy, galls, and reproductive organs (flowers and fruits) to be herborized in exsiccates. These will be vouchers to posterior identification of the plant species by specialists. Exsiccates also can be deposited in a herbarium, to form a database of host plants and to give the possibility of comparing and checking with another samples (Fig. 16.5b). It is also recommended that samples of the galls, which are the expression of the interaction between galler and host plant (Weis et al. 1988; Pan et al. 2015), are present in the exsiccates or placed in individualized boxes and labeled for deposit as vouchers in specific collections. The largest collection of galls in the world is deposited in the

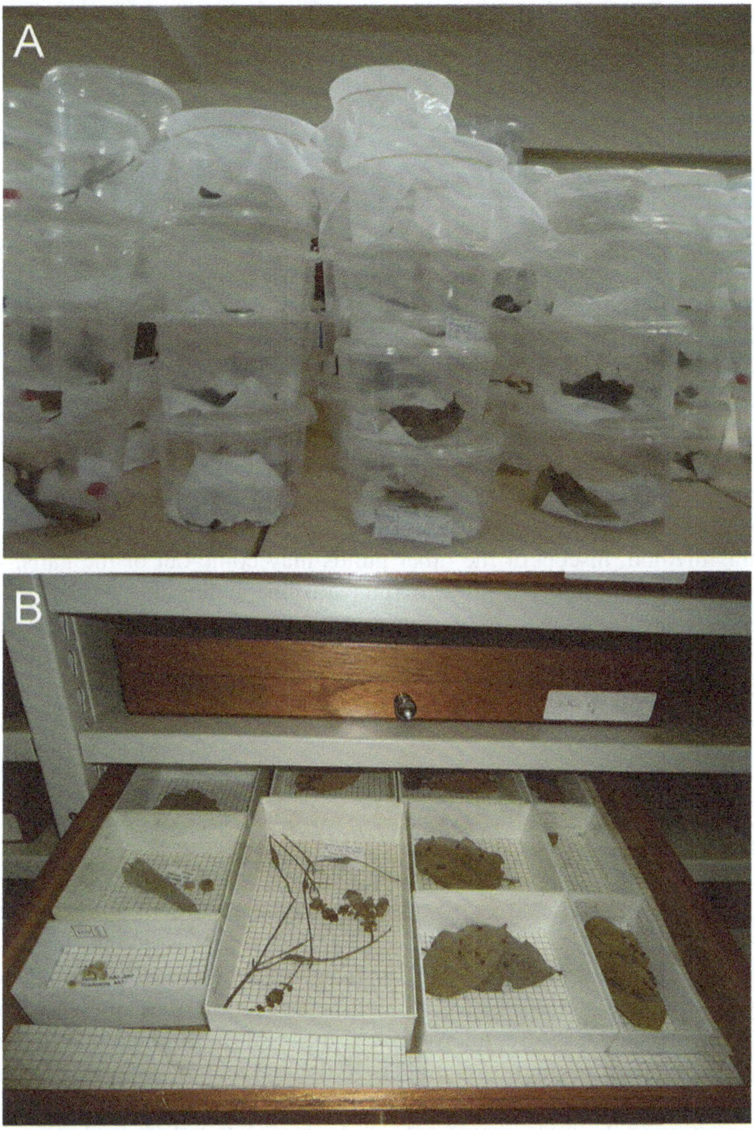

Fig. 16.5 Treatment of gall morphotypes in the laboratory. (**a**) Plastic pots containing separate morphotypes for obtaining adults. (**b**) Reference collection of gall morphotypes registered in the Collection of Galls of the National Museum of Rio de Janeiro, Brazil. (Photos: VC Maia)

United States National Museum of Natural History, Smithsonian Institute and the largest collection of the Neotropical galls was in the Museu Nacional do Rio de Janeiro (MNRJ), but it was burned in the big fire that destroyed the Museum in 2018. In the laboratory, each gall must be characterized morphologically (morphotyped), photographed (see Sect. 16.4.4 for further details), and separated into two lots, one for dissection to describe internal structure and obtain immature insects and the other for rearing adults of gallers and associated fauna.

16.4.2 Obtaining Adults

To obtain adults from galling insects, galls can be packed in plastic pots or bags (Fig. 16.5a). It is important to emphasize that when there is more than one gall morphotype simultaneously in the same organ (which is quite common in the case of leaves), the morphotypes must be separated so that there is no mixing of material from different galls. In the case of pots, the lids can be replaced by a screen to prevent the escape of adults and facilitate air circulation. It is recommended to put a piece of cotton moistened with water to maintain the humidity of the environment. It is advisable to substitute cotton for soil in the case of species with larvae that abandon the gall and bury themselves in the soil (Gagné 1994), in order for the galler to complete its development for obtaining adults. It is important to note that using this method, insects of the other guilds may come out of the galls, such as cecidophages, inquilines, parasites, kleptoparasites, symbionts, predators, or successors (Luz and Mendonça 2019). Alternatively, the researchers can also bag the branches with galls directly on the plants in the field. In this case, daily observation is necessary to prevent weather conditions such as winds and rains from damaging insects that emerge from galls.

16.4.3 Processing Zoological Material

After the morphological characterization and documentation of the galls (see Sect. 16.4.4 for further details), two lots are separated to be processed, one for dissection of some galls to obtain immatures and characterize structure and the other for rearing adults of gallers and associated fauna. The dissection of the galls can be made with sharp blades under stereomicroscope (Fig. 16.6). In this step it is important to annotate the number of internal chambers in the gall and the number of immatures found in each chamber. The immatures found in the process (full-grown larvae and pupae or exuviae) should be preserved in anhydrous ethanol for further preparation in slides and identification. The galls used for rearing adults should preferably be kept in the same conditions of humidity and temperature to their natural environment. Vials should be checked daily for adults, with an annotation on the emergency date of each adult. The exuvia of the pupae must be actively searched for observation together with the adult emerging from it. Adults and specimens found should be removed from the pots with soft bristle brushes moistened in anhydrous ethanol. It is recommended to place the breeding pot or bag in the freezer for a few minutes to cause lethargy in the insect and prevent it from flying out of the pot or bag. Specimens

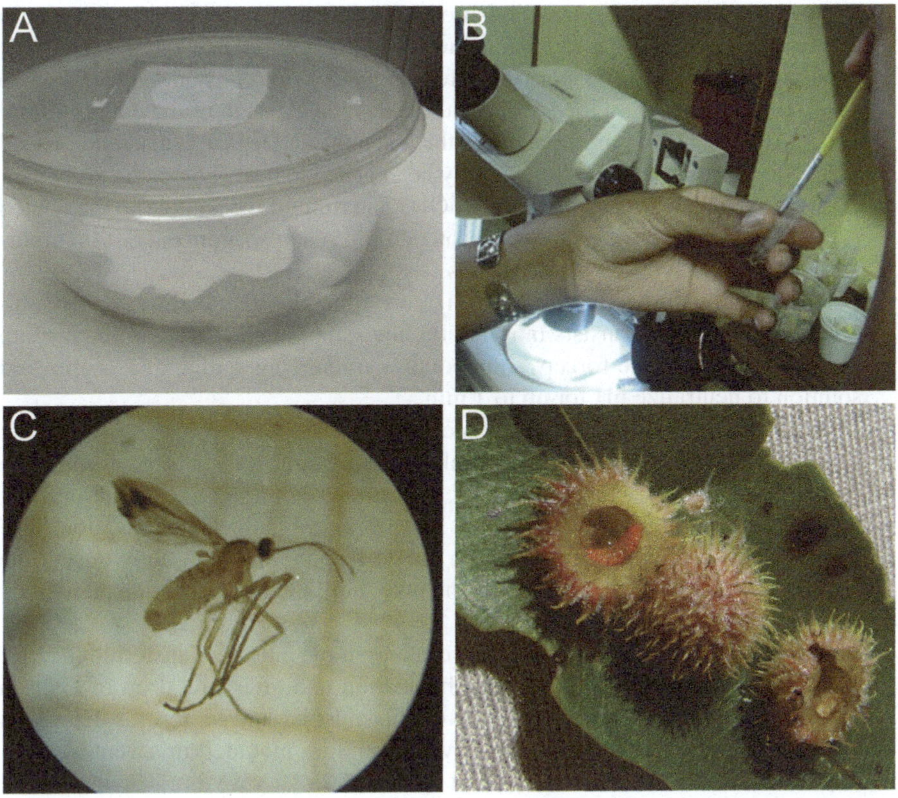

Fig. 16.6 Procedure for obtaining gall inducers. (**a**) Pot for gall insect rearing. (**b**) Removing adults with a brush for 70% alcohol storage. (**c**) Adult of galling insect (gall midge) emerged from a gall. (**d**) Gall dissection to obtain galling insect larvae. (Photos: WS Araújo)

should be stored and labeled in vials with anhydrous ethanol for later preparation and identification.

The permanent slide mounting technique is outlined in Gagné (1994), which consists of the following procedures:

1. *Diaphanization* Procedure used to lighten the cuticle and remove internal tissues from specimens, especially the adipose tissue. The procedure consists in the immersion of the sample in 10% NaOH solution (two or three pellets of NaOH in 10 ml of water). Before the immersion of the sample in the solution, two steps must be observed: it is important to prick the abdomen of the sample once or twice with a fine insect pin to ensure the entry of the NaOH solution into the body and the dissolution of the mentioned tissues, and for adults, it is necessary to remove the wings with the aid of tweezers and razor blades. The wings remain in anhydrous ethanol for later inclusion in permanent glass slides in conjunction with the corresponding insect body. After the perforation of the bodies and the removal of the wings, the specimens are placed in the NaOH solution at room

temperature for approximately 1.5 h or until the cuticle is clear. A variation to increase the speed of the diaphanization is the heating of the sample in NaOH solution in a hot plate to boil for a few minutes with constant supervision to verify the dissolution of the tissues and the desired clearing of the cuticle. If not done carefully, this method can cause the antennae and legs to break, and instead of boiling the material, they can simply be exposed to heat. This same procedure can be done using KOH instead of NaOH. The time of the sample in the heated solution varies according to the sclerotization of each specimen, so it is important to carefully supervise this process.

2. *Neutralization* A procedure to neutralize the effects of the alkaline NaOH solution. The procedure consists in removing the sample from the solution at room temperature using fine forceps, immediately immersing it in a 10% acetic acid solution to neutralize pH for up to 1 min.
3. *Dehydration* Successive transference of the sample through an alcoholic sequence of 70%, 80%, 95%, and anhydrous ethanol for about 5 min in each liquid for the gradual dehydration of the sample. If the passage is abrupt, the cuticle may become wrinkled, with loss of morphological information. The wings should be placed together with the corresponding adult body during the anhydrous ethanol phase and follow the body until the mounting in a permanent glass slide.
4. *Inclusion in Canada Balsam* The dehydrated sample must be transferred to clove oil for 5 minutes to finalize the diaphanization and after to a drop of Canada balsam on a permanent slide, and cover slip must be laid on top for deposit in zoological collection (Fig. 16.7). Larvae should be mounted in a ventral view, pupae in ventral and lateral view, and the adults in lateral view. The male head and terminalia should be separated from the body and mounted, respectively, in frontal and dorsal view, near the body location from which they were removed. Antennas must be aligned so that the details of their morphology can be seen. If more than one male is available, some must have the terminalia left on the body in lateral and dorsal views. The wings should be positioned one above the other on the right side of the body under a different cover slip. Legs should be aligned and positioned so that the anterior, middle, and posterior pairs are next to each other in lateral view, especially in the region of the tarsal claws.
5. *Labeling*Attach a label to each end of the slide, one to transfer the information of the original label that consists in place and date of collection, collector, and identification of the host plant and another for specimen identification and identifier name followed by "det" (Fig. 16.8).

16.4.4 Method Using Digital Photography

Before the popularization of digital photography, researchers used scientific illustrations in their studies on gall inventories. Gall morphology was portrayed through drawings. This was obviously costly and expensive. It was necessary to collect the

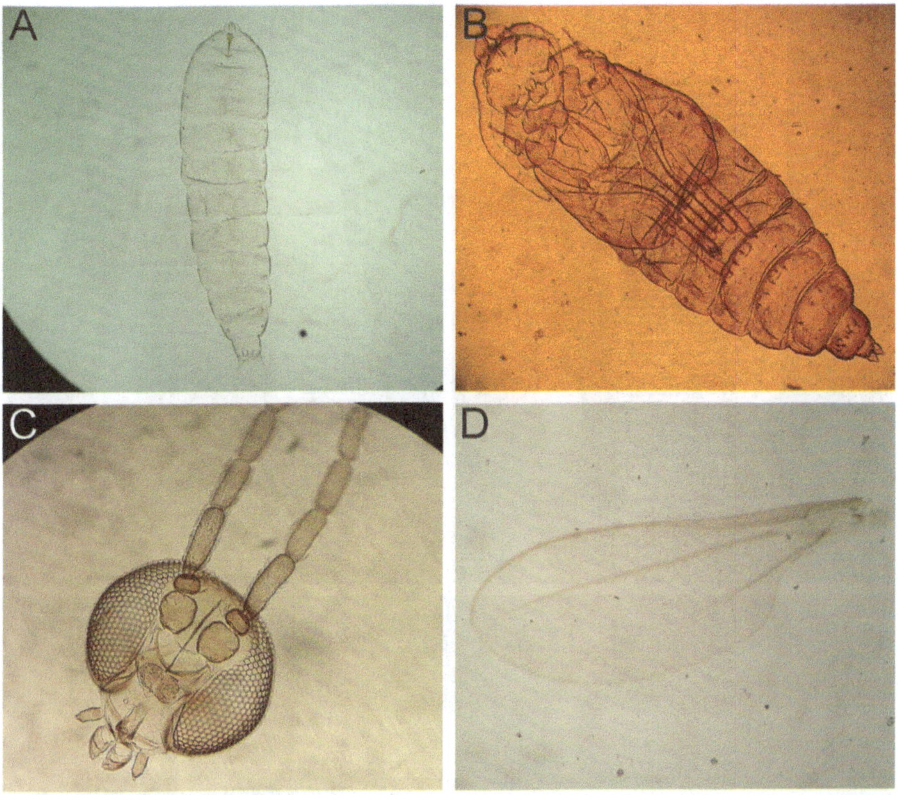

Fig. 16.7 Different diagnostic features emphasized in the slides mounted for gall midges (Cecidomyiidae, Diptera). (**a**) Larvae. (**b**) Pupae. (**c**) Adult—head. (**d**) Adult—wings. (Photos: B Proença (**a**, **d**) and VC Maia (**b**, **c**))

material with galls and preserve it using the same techniques for preparing botanical exsiccates. Afterward, all of this material was sent to a draftsman, who took a certain amount of time to draw each gall. Even with good-quality drawings, gall morphology was usually not well represented (Fig. 16.9a–c).

After the popularization of digital photography, digital cameras became better and cheaper, and this facilitated the photographic records of galls in the field and/or in the laboratory. Photographic sampling is an effective, low-cost, and low-impact method for galling insect populations. In the last decade, the use of digital photography for studies of galling insect inventories has become essential, and it is seen in virtually all studies. Therefore, digital photography represents an excellent tool to register gall morphology in a faithful way. Some advantages are evident: (a) galls can be portrayed with precision in the field; (b) photo samples do not deteriorate over time, compared to plant material used to obtain illustrations; and (c) there is virtually no limit to the number of photos that can be taken from the galls, and the storage of the files is relatively simple. However, pictures taken without proper

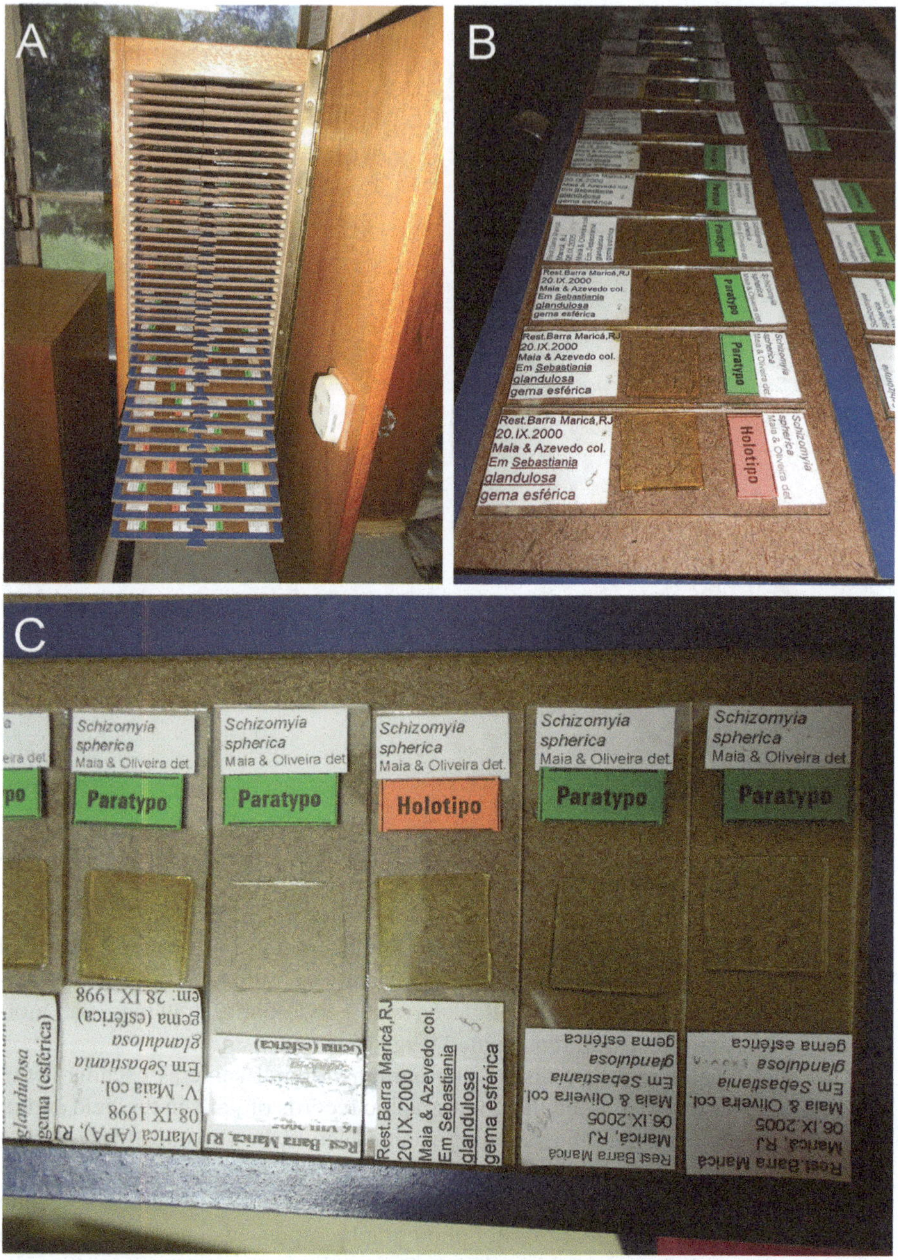

Fig. 16.8 Reference collection of gall midges (Cecidomyiidae, Diptera) from the National Museum of Rio de Janeiro, Brazil. (**a**) Laminar cabinet containing holotypes and paratypes. (**b**) Blades assembly process. (**c**) Blades containing the specimens of gall midges. (Photos: VC Maia)

Fig. 16.9 (**a–c**) Recurring examples of inadequate photos found in articles on galling insect inventories. (**a**) The framing of the photo is poor. The gall is not centralized, although plant leaf is. In addition, focal length to the gall is too long, and the detailed morphology of the gall cannot be properly seen. (**b**) The gall is not in clear focus. In addition, the photo contains elements that are not necessary for the composition of the photo. (**c**) The photo was taken in the shade, so it is dark and the gall cannot be seen clearly. (**d–f**) Some examples of useful photos for gall inventories: Cylindrical leaf galls of *Lopesia mataybae* Garcia and Urso-Guimarães (Diptera: Cecidomyiidae) on *Matayba guianensis* Aubl. (Sapindaceae). (**d**) The galls are centered, in focus, and occupying maximum space in the composition of the photo. This is important because the photos are organized on plates with numerous other photos. If the gall is evident in the photo, there will be no problem in identifying gall morphology. (**e**) A change in perspective may reveal gall morphology. In this photo, we have the same gall of figure **a** but in another perspective. However, there is usually a limitation on the number of figures in journals. One suggestion is to add additional photos as supplementary material. (**f**) This image shows how complex galls can be. It illustrates the same galls of figures (**d**) and (**e**). However, the image shows the projection of the gall on the adaxial part of the leaf (on the left) and the projection of the other gall on the abaxial part of the leaf (on the right). (Photos: JC Santos)

photographic techniques can result in poor photos, with low-quality images, which do not contribute to the study of gall inventories (see Fig. 16.9a–c). Here, we pass on what we believe are fundamental tips for high-quality photos (Fig. 16.9d–f):

1. *Equipment* The type of photographic equipment certainly makes a difference. The vast majority of galls have small dimensions, and, normally, many conventional equipment will not capture morphological details of these galls with quality. Thus, using DSLR (digital single-lens reflex) cameras and a macro lens is preferable. These cameras have great photo quality performance. The price of the equipment can be high, but there are affordable options in most markets. However, compact digital cameras or even high-resolution cell phone cameras can be useful, as long as some technical details are followed.
2. *Framework* It is of fundamental importance to place the galls in the center of the photo.
3. *Focus* It is essential that galls are in focus. For this, it is a need to check if the galls appear sharp in photos and whether autofocus and manual focus functions are available on the used equipment.
4. *Focal Distance* Researchers can experiment with different focal lengths depending on the lens-type variations. Try to get as close as possible to the galls to get clearer photos.
5. *Macro* Among the camera pre-programming modes, there should be a macro (close-up) mode that can be used for small objects. Using a macro mode will give small apertures to ensure maximum depth of field. The advantage is that it has a single focal point.
6. *Perspectives* Typically, galls are depicted from only one perspective. It is recommended that researcher shoot multiple perspectives. Galls can have different shapes in the abaxial and adaxial regions of the leaves, for example, so capturing different perspectives would help in characterizing shape.
7. *Light* Light is fundamental in photography. Inside forests or on cloudy days, photos can end up dark. In this case, it is preferable to use a flash. If the researcher does not have a flash, the use of natural light can be necessary, collecting the gall and placing it in a brighter spot, for example.
8. *Scale* A scale is not normally used in photos taken in the field, but it is highly recommended so as to give an idea of gall size.
9. *In the Laboratory* Galls may also be collected in the field and taken to a laboratory for photo shoot. There, there are higher chances of better control of camera and lighting conditions. It is possible to use a mini photo studio. They are portable and some have LED lighting.
10. *Post-production* It is possible to use various photography software and post-production techniques to improve the quality of photos, for example, correct framing and light. However, it is advisable to check the effect this has on the final pictures, as color and even shapes could be rendered unnatural when using too much retouching.

16.5 Identification of Galling Groups and Associated Fauna

The habit of inducing galls in plants clearly has evolved independently several times among the phytophagous insects (Gullan and Cranston 2014) occurring in at least 53 families distributed in the orders Coleoptera, Diptera, Hemiptera, Hymenoptera, Lepidoptera, and Thysanoptera (Table 16.2) (Fig. 16.10) (Shorthouse and Rohfritsch 1992; Raman et al. 2005; Fernandes et al. 2012). The insect galls are common in several families of vascular plants, especially in the association with angiosperms, which is estimated to have at least two morphotypes per gall inducer for each flowering plant species (Brown 2009). Even though the six orders are found in all biogeographic regions, some specific taxa are more species-rich in some regions than in others. Some species of mites (order Trombidiformes) are also known to induce galls or erineum. The orders of gall inducers are presented bellow as well as the most relevant families in different regions.

16.5.1 Order Coleoptera

Coleoptera is a megadiverse order with about 350,000 described species (Lawrence and Newton Jr 1995) and varied habits such as phytophagous, predatory, fungivorous, detritivores, parasites, and the gall inducers. The coleopterous gall makers belong to five superfamilies, Buprestoidea, Chrysomeloidea, Cucujoidea, Curculionoidea, and Tenebrionoidea (Triplehorn and Johnson 2011). Galls induced by coleopterans (Fig. 16.10a) can be recognized by the presence of large internal chambers, uni- or multichamber, and generally hosting only one larva per chamber. Stem is the preferred plant organ of the coleopteran galls (Korotyaev et al. 2005). Coleopterans induce galls in various plant families, including Asteraceae, Solanaceae, Brassicaceae, and Fabaceae. Grandez-Rios et al. (2016) related 20 species of Coleoptera to Latin America, mostly in Fabaceae and Asteraceae. Due to the great diversity of gall-inducing coleopteran families, several taxonomic keys may be necessary to identify these taxa.

16.5.2 Order Diptera

Diptera is a megadiverse order with over 150,000 species and varied habits including phytophagous, predatory, fungivorous, detritivores, parasites, and the gall inducers. The families of Diptera that include gall inducer species are the Agromyzidae, Anthomyiidae, Cecidomyiidae, Chloropidae, Clythiidae,

Table 16.2 Main characteristic of galls induced by different arthropod orders

Order	Main galling families worldwide	Main host plant families worldwide	Gall characteristics	References
Coleoptera	Buprestidae Brentidae Cerambycidae Curculionidae Chrysomelidae Nitidulidae	Asteraceae Brassicaceae Fabaceae Solanaceae	Stem galls, usually large and uni- or multichambered	Korotyaev et al. (2005)
Diptera	Agromyzidae Anthomyiidae Cecidomyiidae Chloropidae Clythiidae Lauxaniidae Lonchaeidae Tephritidae	Asteraceae Bignoniaceae Cannabaceae Cyperaceae Fabaceae Fagaceae Myrtaceae Pinaceae Salicaceae	Leaf galls, usually closed and unichambered	Gagné (1994)
Hemiptera	Adelgidae Aleyrodidae Aphididae Asterolecaniidae Beesoniidae Cicadellidae Coccidae Calophyidae Diaspididae Eriococcidae Kermisidae Pemphigidae Psyllidae Phylloxeridae Phacopteronidae Tingidae Triozidae	Asteraceae Fabaceae Myrtaceae	Leaf galls, usually open type including pit, curling, folding, convex, or amorphous	Maia (2006) Hanson et al. (2014)
Hymenoptera	Agaonidae Cynipidae Eulophidae Eurytomidae Figitidae Pteromalidae Scelionidae Tanaostigmatidae Tenthredinidae	Fabaceae Fagaceae Myrtaceae Rosaceae	Stem and bud galls, closed type	Maia (2012)
Lepidoptera	Cecidosidae Cosmopterigidae Gelechiidae Momphidae Tineidae Tortricidae	Anacardiaceae Asteraceae Fabaceae Salicaceae	Closed stem and bud galls	Maia (2006)

(continued)

Table 16.2 (continued)

Order	Main galling families worldwide	Main host plant families worldwide	Gall characteristics	References
Thysanoptera	Phlaeothripidae	Moraceae Myrtaceae	Leaf folding	Maia (2006)
Trombidiformes	Eriophyidae	Lauraceae Melastomataceae Rubiaceae	Leaf closed galls or erinea (open galls)	Nasareen and Ramani (2014)

Lauxaniidae, Lonchaeidae, and Tephritidae (Raman et al. 2005; Gullan and Cranston 2014; Gagné and Jaschhof 2017). The highest number of gall-inducing species among the Diptera belongs to one family, Cecidomyiidae (Bibionomorpha, Sciaroidea), with over 6500 described species, 80% of them gall inducers. Most of the cecidomyiid species are gall inducers, and some of them are considered pests of grains, fruits, and vegetables (Table 16.3) (Gagné and Jaschhof 2017). Cecidomyiids can induce galls in any vegetal organ, stem, flower, fruit, bud, and root, although the leaf is the most attacked organ worldwide (Mani 1964). The galls induced by cecidomyiids (Fig. 16.10b) can be recognized by the organization of the modified tissue of plant, mostly in closed galls, unichambered and hosting only one larva per chamber, although development of larvae in open flowers and a few species with gregarious larvae (Gagné 1994). In the Nearctic region, 1300 species of gall midges are known (Gagné and Jaschhof 2017), and the most common families of host plants associated with the gall midges are Salicaceae, Pinaceae, Fagaceae, Cyperaceae, and Cannabaceae. In the Palaearctic region, there are known 3113 species in 344 genera, with about 1800 species in 270 genera in Europe (Skuhravá and Skuhravý 2009). In Neotropical region, around 600 species in 170 genera are known, and the most common host plant families are Fabaceae, Asteraceae, Myrtaceae, and Bignoniaceae, but thousands of undescribed species are related in surveys in several biomes of South America (Araújo et al. 2019a). The main key to identify the cecidomyiid genera are Gagné (1994, 2018).

16.5.3 Order Hemiptera

The order Hemiptera is a megadiverse order with 89,000 species worldwide, much diversified morphologically and in feeding habits, although most species are phytophagous. The gall inducer habit is found in 16 families of hemipterans (Triplehorn and Johnson 2011): 1 in suborder Heteroptera (family Tingidae) and 16 in the suborder Sternorrhyncha (families Adelgidae, Aleyrodidae, Aphididae, Asterolecaniidae,

Fig. 16.10 Examples of galls induced by different arthropod groups. (**a**) Coleopteran gall induced by *Apion* sp. (Brentidae) on the stem of *Diospyros hispida* DC. (Ebenaceae). (**b**) Dipteran gall induced by *Myrciariamyia admirabilis* Maia and Fernandes (Cecidomyiidae) on the leaf of *Erythroxylum suberosum* A. St.-Hil. (Erythroxylaceae). (**c**) Hemipteran galls induced on the leaf of *Pouteria ramiflora* (Mart.) Radlk. (Sapotaceae). (**d**) Hymenopteran galls induced on the leaf of *Caryocar brasiliense* Cambess. (Caryocaraceae). (**e**) Lepidopteran gall induced by *Palaeomystella oligophaga* Becker and Adamski (Momphidae) on the branch of *Macairea radula* (Bonpl.) DC. (Melastomataceae). (**f**) Mite galls induced by Eriophyidae (Trombidiformes) on the leaf of *Miconia ligustroides* (DC.) Naudin (Melastomataceae). (Photos: WS Araújo (**a–e**) and LL Bergamini (**f**))

Beesoniidae, Cicadellidae, Coccidae, Calophyidae, Diaspididae, Eriococcidae, Kermisidae, Pemphigidae, Psyllidae, Phylloxeridae, Phacopteronidae, Triozidae) (Grazia et al. 2012; Fernandes et al. 2012). In Neotropical region, around 135 species of Psyllidae, Coccidae, Aphididae, and Brachyscelidae are recorded; the most common host plant families are Myrtaceae, Asteraceae, and Fabaceae; the preferred organ is the leaf, usually of open type (Fig. 16.10c), curling, folding, convex, or amorphous and with gregarious nymphs in a single gall (Maia 2006; Hanson et al. 2014).

Table 16.3 Cecidomyiid pests in species of plant of economic importance in the world

Common name	Plant species	Cecidomyiid species	Geographical distribution
Apple	*Pyrus communis*	*Cecidomyia pyricola* Nordlinger	Germany, Italy
		Cecidomyia pyri Bouche	Palearctic, Nearctic, New Zealand
		Contarinia piri Tavares	Spain
		Moreschiella contariniperda Del Guercio	Italy
	Pyrus malus	*Cecidomyia pyrivora* Riley	Palearctic, USA
		Contarinia mali Barnes	Japan
	Pyrus malus, P. sylvestris	*Dasineura mali* (Kieffer)	Palearctic, USA, Canada, Argentina, New Zealand
	Pyrus salicifolia	*Resseliela pyricola* Skuhrava	Greece
Avocado	*Persea americana*	*Asphondylia websteri* Felt	USA, Mexico, Honduras, Guatemala, Dominican Republic, Guadeloupe FWI
		Bruggmanniella perseae Gagne	Costa Rica, Colombia

(continued)

Table 16.3 (continued)

Common name	Plant species	Cecidomyiid species	Geographical distribution
Mango	*Mangifera indica*	*Gephyraulus indica* (Grover and Prasad)	India
		Gephyraulus mangiferae (Felt)	India, Hawaii, Puerto Rico, Guadeloupe FWI
		Procontarinia allahabadensis (Grover)	India
		Procontarinia amraeomyia (Rao)	India
		Procontarinia biharana Gagné	India
		Procontarinia brunneigallicola (Rao)	India
		Procontarinia echinogalliperda (Mani)	India
		Procontarinia frugivora Gagné	Philippines
		Procontarinia keshopurensis (Rao)	India
		Procontarinia mangicola (Shi)	China, Japan, India, Réunion; Iran, Guadeloupe FWI, St. Lucia, St. Vincent, Trinidad, Brazil, St. Vincent
		Procontarinia mangifoliae (Grover)	India
		Procontarinia matteiana Kieffer and Cecconi	Kenya, South Africa, Mauritius, Réunion, India, Java, Italy
		Procontarinia pustulata Kolesik	Australia, Papua New Guinea
		Procontarinia robusta Li, Bu and Zhang	China, East Timor
		Procontarinia tenuispatha (Kieffer)	India
		Procontarinia viridigallicola (Rao)	India
Manioc	*Manihot utilissima*	*Iatrophobia brasiliensis* Rübsaamen	Brazil
Rice	*Oryza sativa, O. perennis*	*Orseolia oryzae* Wood-Mason	India, Southern China, Thailand, and Indonesia
	Oryza glaberrima	*Orseolia oryzivora* Harris and Gagné	Senegal, Ivory Coast, Burkina Faso, Nigeria, Sudan

(continued)

Table 16.3 (continued)

Common name	Plant species	Cecidomyiid species	Geographical distribution
Sorghum	*Sorghum vulgare*	*Thomasiella sorghivora* Harris	Nigeria
	Sorghum arundinaceum, S. guineense, S. verticilliflorum	*Stenodiplosis sorghi* Harris	Senegal, Nigeria, Uganda
	Sorghum bicolor, Sorghum spp.	*Contarinia sorghicola* Coquillett	Africa, India, Southeastern China, Japan, Philippines, USA, Mexico, Dominican Republic, Argentina
	Sorghum intrans	*Contarinia intrans* Harris	Australia
	Sorghum plumosum	*Contarinia roperi* Harris	Australia

16.5.4 Order Hymenoptera

Hymenoptera is a megadiverse order of insects and includes bees, ants, and wasps, with 153,088 species in 132 families (Aguiar et al. 2013) and varied habits such as phytophagous, pollinators, predatory, parasites, and gall inducers. Hymenopteran galls are more frequent in Europe and North America, occurring specially on three superfamilies of wasps: Cynipoidea contains hundreds of species of gall wasps (Cynipidae, at least 1300 species), especially on oaks (Fagaceae) and roses (Rosaceae); in Tenthredinoidea several species of galling species are present in Tenthredinidae (sawflies); and Chalcidoidea includes several families of gall inducers, such as Agaonidae, Eurytomidae, and Pteromalidae (Gullan and Cranston 2014). In South America, hymenopteran gall inducers (Fig. 16.10d) belong to different families, such as Agaonidae, Cynipidae, Eulophidae, Eurytomidae, Figitidae, Pteromalidae, Scelionidae, and Tanaostigmatidae, inducing galls more frequently in plant families Myrtaceae and Fabaceae, forming closed galls especially in stems and buds (Maia 2012). Due to the great diversity of gall-inducing hymenopteran families, several taxonomic keys may be necessary to identify these taxa.

16.5.5 Order Lepidoptera

Lepidoptera is a megadiverse order composed of moths and butterflies. The predominant habit is phytophagy but with relatively few lepidopteran species inducing galls. In the Nearctic and Palaearctic regions, of 352 examples reported worldwide, 173 were not identified further than order; the 179 identified examples are distributed among 20 families in 11 superfamilies (Miller et al. 2005). The most common lepidopteran families in Europe are Tortricidae and Tineidae; in South America are Gelechiidae and Tortricidae and with few representatives in Cosmopterigidae and Momphidae families (Maia 2012). In the Neotropical region, lepidopteran species induce galls in at least 41 families of host plants, especially Asteraceae, Salicaceae, and Fabaceae (Fernandes et al. 2012); in South America the lepidopterous closed stems galls in Melastomataceae are predominant (Fig. 16.10e) (Maia 2012). Due to the great diversity of gall-inducing lepidopteran families, several taxonomic keys may be necessary to identify these taxa.

16.5.6 Order Thysanoptera

Thysanoptera is a small order of insects with 6288 species in over 782 genera (ThripsWiki 2020). The suborder Terebrantia has less than five gall inducer species; the suborder Tubulifera includes most of the gall-inducing species, the majority from Phlaeothripidae (Mound and Morris 2005). In South America, the plant families most attacked are Myrtaceae and Moraceae; the majority of the galls occurred on leaves, which modifies the morphology to partial or complete leaf folding (Maia 2006). The keys to identify the main genus are in ThripsWiki (2020).

16.5.7 Order Trombidiformes

The family Eriophyidae is the main family of arachnids known as gall inducers (Araújo et al. 2019b). Mites inducing galls are recorded on angiosperms, conifers, and ferns throughout the world (Oldfield 1996). The galls of eriophyids usually are amorphous (Fig. 16.10f), closed, or opened (also called erinea), and dozens of mites can be registered in a single gall (Nasareen and Ramani 2014).

16.6 Implications for Conservation of Galls and Galling Arthropods

We discussed throughout the previous topics how the diversity of arthropod galls can be diverse, despite the taxonomic knowledge still incipient for a gigantic fraction of this fauna. For example, for Brazilian Cerrado, one of the main global hotspots of arthropod gall diversity (Araújo et al. 2014b), only 49 species of Cecidomyiidae of the more than 9000 estimated for the biome have already been described (Cintra et al. 2020), which represents less than 1%. This brings us to the reflection on the need to inventory, characterize, and describe this galling arthropod fauna, in face of the strong and fast anthropic changes that are occurring in the planet. This is an emergency in order to reduce the fraction of gallers species that have been affected by the silent extinction that affects the little-known taxonomic groups of insects (Fonseca 2009). Given this scenario, it is essential that researchers carry out research with a sample design that maximizes the success of the sampling and identifies a collection of material that has studies with different approaches.

Acknowledgments The authors thank Marco Antônio Alves Carneiro and Valeria Cid Maia for important suggestions to the text; Barbara Proença, Leonardo Lima Bergamini, and Valeria Cid Maia for providing the photos used in some figures (credits in the captions); and National Council for Scientific and Technological Development (CNPq) (grant processes #312752/2018-0 for JCS, #311298/2019-2 for MMJ).

References

Abrahamson WG, McCrea KD (1986) The impacts of galls and gall makers on plants. Proc Entomol Soc Wash 88:364–367

Abrahamson WG, Weis AE (1987) Nutritional ecology of arthropod gall makers. In: Slansky-Junior F, Rodriquez JG (eds) Nutritional ecology of insects, mites, spiders, and related invertebrates. Wiley, New York, pp 235–258

Abrahamson WG, Weis AE (1997) Evolutionary ecology across three trophic levels: goldenrods, gallmakers, and natural enemies. Princeton University Press, Princeton

Aguiar A, Deans A, Engel M et al (2013) Animal biodiversity: an outline of higher-level classification and survey of taxonomic richness (Addenda 2013). Zootaxa 3703:51–62. https://doi.org/10.11646/zootaxa.3703.1.12

Araújo WS (2017) Plant species richness mediates the effects of vegetation structure, but not soil fertility, on insect gall richness in a savanna in Brazil. J Trop Ecol 33:197–204. https://doi.org/10.1017/S0266467417000086

Araújo WS (2018) 30 years of research on insect galls in Brazil: a scientometric review. Pap Avulsos Zool 58:1–11. https://doi.org/10.11606/1807-0205/2018.58.34

Araújo WS, Espírito-Santo Filho K (2012) Edge effect benefits galling insects in the Brazilian Amazon. Biodivers Conserv 21:2991–2997. https://doi.org/10.1007/s10531-012-0333-z

Araújo WS, Scareli-Santos C, Guilherme FAG et al (2013) Comparing galling insect richness among Neotropical savannas: effects of plant richness, vegetation structure and super-host presence. Biodivers Conserv 22:1083–1094

Araújo WS, Sobral FL, Maracahipes L (2014a) Insect galls of the Parque Nacional das Emas (Mineiros, GO, Brazil). Check List 10:1445–1451. https://doi.org/10.15560/10.6.1445

Araújo WS, Santos BB, Guilherme FAG et al (2014b) Galling insects in the Brazilian Cerrado: ecological patterns and perspectives. In: Fernandes GW, Santos JC (eds) Neotropical insect galls. Springer, Dordrecht, pp 257–272. https://doi.org/10.1007/978-94-017-8783-3_15

Araújo WS, Fernandes GW, Santos JC (2019a) An overview of inventories of gall-inducing insects in Brazil: looking for patterns and identifying knowledge gaps. An Acad Bras Ciênc 91:e20180162. https://doi.org/10.1590/0001-3765201920180162

Araújo WS, Freitas ÉVD, Kollár J et al (2019b) Host specialization in plant-galling interactions: contrasting mites and insects. Diversity 11:1–11. https://doi.org/10.3390/d11100180

Blanche KR (2000) Diversity of insect-induced galls along a temperature–rainfall gradient in the tropical savannah region of the Northern Territory, Australia. Austral Ecol 25:311–318. https://doi.org/10.1046/j.1442-9993.2000.01040.x

Blanche KR, Westoby M (1995) Gall-forming insect diversity is linked to soil fertility via host plant taxon. Ecology 76:2334–2337. https://doi.org/10.2307/1941706

Brown BV (2009) Manual of Central American Diptera. NRC Research Press, Ottawa

Carneiro MAA, Branco CSA, Braga CED et al (2009) Are gall midge species (Diptera, Cecidomyiidae) host-plant specialists? Rev Bras Entomol 53:365–378. https://doi.org/10.1590/S0085-56262009000300010

Carvalho-Fernandes SP, Ascendino S, Maia VC et al (2016) Diversity of insect galls associated with coastal shrub vegetation in Rio de Janeiro, Brazil. An Acad Bras Ciênc 88:1407–1418. https://doi.org/10.1590/0001-3765201620150658

Cintra CF, Araújo WS, Maia VC et al (2020) Plant-galling insect interactions: a dataset of host plants and their gall-inducing insects for the Cerrado. Ecology: e03149. https://doi.org/10.1002/ecy.3149

Cuevas-Reyes P, Quesada M, Hanson P et al (2004) Diversity of gall-inducing insects in a Mexican tropical dry forest: the importance of plant species richness, life-forms, host plant age and plant density. J Ecol 92:707–716. https://doi.org/10.1111/j.0022-0477.2004.00896.x

Dalbem RV, Mendonça MS Jr (2006) Diversity of galling arthropods and host plants in a subtropical forest of Porto Alegre, Southern Brazil. Neotrop Entomol 35:616–624. https://doi.org/10.1590/S1519-566X2006000500007

Espírito-Santo MM, Fernandes GW (2007) How many species of gall-inducing insects are there on earth, and where are there? Ann Entomol Soc Am 100:95–99. https://doi.org/10.1603/0013-8746(2007)100[95,HMSOGI]2.0.CO;2

Fernandes GW, Price PW (1988) Biogeographical gradients in galling species richness. Oecologia 76:161–167. https://doi.org/10.1007/BF00379948

Fernandes GW, Júnior PDM, Schönrogge K (2008) Plant organ abscission and the green island effect caused by gallmidges (Cecidomyiidae) on tropical trees. Arthropod Plant Interact 2:93–99. https://doi.org/10.1007/s11829-008-9031-x

Fernandes GW, Carneiro MAA, Isaias R (2012) Gall-inducing insects: from anatomy to biodiversity. In: Panizzi AR, Parra JRP (eds) Insect bioecology and nutrition for integrated pest management. CRC Press, Boca Raton, pp 369–395

Fernandes GW, Coelho MS, Santos JC (2014) Neotropical insect galls: status of knowledge and perspectives. In: Fernandes GW, Santos JC (eds) Neotropical insect galls. Springer, Dordrecht, pp 1–14. https://doi.org/10.1007/978-94-017-8783-3_1

Fonseca CR (2009) The silent mass extinction of insect herbivores in biodiversity hotspots. Conserv Biol 23:1507–1515. https://doi.org/10.1111/j.1523-1739.2009.01327.x

Gagné RJ (1994) The gall midges of the region Neotropical. Cornell University Press, New York

Gagné RJ (2018) Key to adults of North American genera of the Subfamily Cecidomyiinae (Diptera: Cecidomyiidae). Zootaxa 4392:401–457. https://doi.org/10.11646/zootaxa.4392.3.1

Gagné RJ, Jaschhof M (2017) A catalog of Cecidomyiidae of the world. https://www.ars.usda.gov/ARSUserFiles/80420580/Gagne_2017_World_Cat_4th_ed.pdf. Accessed 20 May 2019

Gonçalves-Alvim SJ, Fernandes GW (2001) Biodiversity of galling insects: historical, community and habitat effects in four Neotropical savannas. Biodivers Conserv 10:79–98. https://doi.org/10.1023/A:1016602213305

Grandez-Rios JM, García-Villacorta R, Cuevas-Reyes P, Araújo WS (2016) Insectos inductores de Agallas en América Latina: Ecología, Importancia y Nuevas Perspectivas. Rev Biol Neotrop 12:92–103. https://doi.org/10.5216/rbn.v12i2.32619

Grazia J, Cavichioli RR, Wolff VRS et al (2012) Hemiptera Linnaeus, 1758. In: Rafael JA, GAR M, CJB C et al (eds) Insetos do Brasil: diversidade e taxonomia. Holos Editora, Ribeirão Preto, pp 347–406

Gullan PJ, Cranston PS (2014) The insects: an outline of entomology. Wiley-Blackwell, Hoboken

Hanson P, Nishida K, Gómez-Laurito J (2014) Insect galls of Costa Rica and their parasitoids. In: Fernandes GW, Santos JC (eds) Neotropical insect galls. Springer, Dordrecht, pp 497–517. https://doi.org/10.1007/978-94-017-8783-3_23

Harris KM (1966) Gall midge genera of economic importance (Diptera: Cecidomyiidae) Part 1: introduction and subfamily Cecidomyiinae; supertribe Cecidomyiidi. Trans R Entomol Soc Lond 118:313–358. https://doi.org/10.1111/j.1365-2311.1966.tb03156.x

Hebert PD, Ratnasingham S, Zakharov EV et al (2016) Counting animal species with DNA barcodes: Canadian insects. Philos T R Soc B 371: 20150333. https://doi.org/10.1098/rstb.2015.0333

Inbar M, Izhaki I, Lupo I et al (2010) Why do many galls have conspicuous colors? A new hypothesis. Arthropod Plant Interact 4:1–6. https://doi.org/10.1007/s11829-009-9082-7

Isaias RMS, Carneiro RGS, Oliveira DC et al (2013) Illustrated and annotated checklist of Brazilian gall morphotypes. Neotrop Entomol 42:230–239. https://doi.org/10.1007/s13744-013-0115-7

Isaias RMS, Oliveira DC, Carneiro RGS et al (2014) Developmental anatomy of galls in the Neotropics: arthropods stimuli versus host plant constraints. In: Fernandes GW, Santos JC (eds) Neotropical insect galls. Springer, Dordrecht, pp 15–34. https://doi.org/10.1007/978-94-017-8783-3_2

Julião GR, Venticinque EM, Fernandes GW (2008) Richness and abundance of gall-forming insects in the Mamirauá Várzea, a flooded Amazonian forest. Uakari 1:59–64. https://doi.org/10.31420/uakari.v1i1.5

Julião GR, Venticinque EM, Fernandes GW et al (2014) Unexpected high diversity of galling insects in the Amazonian upper canopy: the savanna out there. PLoS One 9:1–20. https://doi.org/10.1371/journal.pone.0114986

Korotyaev BA, Konstantinov AS, Lingafelter SW et al (2005) Gall-inducing Coleoptera. In: Raman A, Schaefer CW, Withers TM (eds) Biology, ecology, and evolution of gall-inducing arthropods. Science Publishers, New Hampshire, pp 239–271

Krassilov V, Shuklina S (2008) Arthropod trace diversity on fossil leaves from the mid-Cretaceous of Negev, Israel. Alavesia 2:239–245

Labandeira CC, Wilf P, Johnson KR et al. (2007) Guide to insect (and other) damage types on compressed plant fossils (Version 3.0). Smithsonian Institution, Washington, D.C. http://paleobiology.si.edu/pdfs/insectDamageGuide3.01.pdf. Accessed 20 April 2020

Lawrence JF, Newton AF Jr (1995) Families and subfamilies of Coleoptera (with selected genera, notes, references and data on family-group names). In: Pakaluk JF, Slipinski SA (eds) Biology, phylogeny and classification of Coleoptera: papers celebrating the 80th birthday of Roy A. Crowson. Muzeum i Instytut Zoologii PAN, Warszawa, pp 779–1006

Lillo E, Pozzebon A, Valenzano D et al (2018) An intimate relationship between Eriophyoid mites and their host plants - a review. Front Plant Sci 9:1786. https://www.frontiersin.org/articles/10.3389/fpls.2018.01786/full

Luz FA, Mendonça MS Jr (2019) Guilds in insect galls: who is who. Fla Entomol 102:207–210. https://doi.org/10.1653/024.102.0133

Maia VC (2006) Galls of Hemiptera, Lepidoptera and Thysanoptera from Central and South America. Pub Avul Museu Nac 110:1–24

Maia VC (2012) Coleopterous galls from the Neotropical region. Pap Avul Zool 52:175–184. https://doi.org/10.1590/S0031-10492012001500001

Maia VC (2018) Gall-inducing insects of restinga areas (Atlantic Forest) in Brazil: economic importance. Pap Avul Zool 58:1–13. http://orcid.org/0000-0001-9396-5618

Mani MS (1964) Ecology of plant galls. Junk, The Hague

Mendonça MS Jr, Romanowski HP (2002) Life history of the gall-maker *Eugeniamyia dispar* Maia, Mendonça-Jr. and Romanowski, 1996 (Diptera, Cecidomyiidae). Braz J Biol 62:277–283. https://doi.org/10.1590/S1519-69842002000200012

Mendonça MS Jr, Stiling P (2018) Subtropical interactions: comparing galling insect and host plant diversity in Southern Brazil and Florida. Neotrop Entomol 47:628–633. https://doi.org/10.1007/s13744-017-0563-6

Mendonça MS Jr, Piccardi HM, Jahnke SM et al (2010) Galling arthropod diversity in adjacent swamp forests and restinga vegetation in Rio Grande do Sul, Brazil. Neotrop Entomol 39:513–518. https://doi.org/10.1590/S1519-566X2010000400008

Mendonça MS Jr, Toma T, Silva JS (2014) Galls and galling arthropods of Southern Brazil. In: Fernandes GW, Santos JC (eds) Neotropical insect galls. Springer, Dordrecht, pp 221–256. https://doi.org/10.1007/978-94-017-8783-3_14

Miller DG III, Raman A (2019) Host-plant relations of gall-inducing insects. Ann Entomol Soc Am 112:1–19. https://doi.org/10.1093/aesa/say034

Miller WE, Raman A, Schaefer CW et al (2005) Gall-inducing Lepidoptera. In: Raman A, Schaefer CW, Withers TM (eds) Biology, ecology, and evolution of gall-inducing arthropods. Science Publishers, New Hampshire, pp 431–465

Mound LA, Morris DC (2005) Gall-inducing thrips: an evolutionary perspective. In: Raman A, Schaefer CW, Withers TM (eds) Biology, ecology, and evolution of gall-inducing arthropods. Science Publishers, New Hampshire, pp 59–72

Nasareen PNM, Ramani N (2014) Survey on gall forming Eriophyid Mites (Acari: Eriophyidae) associated with economically important plants of North Kerala. J Entomol Zool Stud 2:126–133

Oldfield GN (1996) 1.4.3 Diversity and host plant specificity. In: Lindquist EE, Sabelis MW, Bruin J (eds) World crop pests. Elsevier, Amsterdam, pp 199–216

Oyama K, Pérez-Pérez MA, Cuevas-Reyes P et al (2003) Regional and local species richness of gall-inducing insects in two tropical rain forests in Mexico. J Trop Ecol 19:595–598. https://doi.org/10.1017/S0266467403003651

Pan LY, Chen WN, Chiu ST et al (2015) Is a gall an extended phenotype of the inducing insect? A comparative study of selected morphological and physiological traits of leaf and stem galls on *Machilus thunbergii* (Lauraceae) induced by five species of *Daphnephila* (Diptera: Cecidomyiidae) in Northeastern Taiwan. Zool Sci 32:314–321. https://doi.org/10.2108/zs140244

Paniagua MR, Medianero E, Lewis OT (2009) Structure and vertical stratification of plant galler–parasitoid food webs in two tropical forests. Ecol Entomol 34:310–320. https://doi.org/10.1111/j.1365-2311.2008.01079.x

Price PW, Fernandes GW, Lara ACF et al (1998) Global patterns in local number of insect galling species. J Biogeogr 25:581–591. https://doi.org/10.1046/j.1365-2699.1998.2530581.x

Raman A (2011) Morphogenesis of insect-induced plant galls: facts and questions. Flora 206:517–533. https://doi.org/10.1016/j.flora.2010.08.004

Raman A, Schaefer CW, Withers TM (eds) (2005) Biology, ecology, and evolution of gall-inducing arthropods. Science Publishers, New Hampshire

Ribeiro SP, Basset Y (2007) Gall-forming and free-feeding herbivory along vertical gradients in a lowland tropical rainforest: The importance of leaf sclerophylly. Ecography 30:663–672. https://doi.org/10.1111/j.2007.0906-7590.05083.x

Ribeiro SP, Basset Y, Kitching R (2014) Density of insect galls in the forest understorey and canopy: Neotropical, Gondwana or global patterns? In: Fernandes GW, Santos JC (eds) Neotropical insect galls. Springer, Dordrecht, pp 129–141. https://doi.org/10.1007/978-94-017-8783-3_8

Santana AP, Isaias RMS (2014) Galling insects are bioindicators of environmental quality in a Conservation Unit. Acta Bot Bras 28:594–608. https://doi.org/10.1590/0102-33062014abb3510

Shorthouse JD, Rohfritsch O (1992) Biology of Insect induced Galls. Oxford University Press, Oxford

Shorthouse JD, Wool D, Raman A (2005) Gall-inducing insects – Nature's most sophisticated herbivores. Basic Appl Ecol 6:407–411. https://doi.org/10.1016/j.baae.2005.07.001

Skuhravá M, Skuhravý V (2009) Species richness of gall midges (Diptera: Cecidomyiidae) in Europe (West Palaearctic): biogeography and coevolution with host plants. Acta Soc Zool Bohem 73:87–156

Stone GN, Schönrogge K (2003) The adaptive significance of insect gall morphology. Trend Ecol Evol 18:512–522. https://doi.org/10.1016/S0169-5347(03)00247-7

ThripsWiki (2020) ThripsWiki - Providing Information on the World's Thrips. http://thrips.info/wiki/. Accessed 22 Mar 2020

Toma TS, Fernandes GW, Souza DG et al (2014) Galling insects as indicators of habitat quality. In: Fernandes GW, Santos JC (eds) Neotropical insect galls. Springer, Dordrecht, pp 143–150. https://doi.org/10.1007/978-94-017-8783-3_9

Triplehorn CA, Johnson N (2011) Borror and Delong's Introduction to the Study of Insects. Cengage Learning, Boston

Uechi NAMI, Yukawa J, Yamaguchi D (2004) Host alternation by gall midges of the genus *Asphondylia* (Diptera: Cecidomyiidae). Bishop Museum Bulletin Entomol 12:53–66

Veldtman R, McGeoch MA (2003) Gall-forming insect species richness along a non-scleromorphic vegetation rainfall gradient in South Africa: The importance of plant community composition. Austral Ecol 28:1–13. https://doi.org/10.1046/j.1442-9993.2003.01234.x

Weis AE, Walton R, Crego CL (1988) Reactive plant tissue sites and the population biology of gall makers. Annu Rev Entomol 33:467–486. https://doi.org/10.1146/annurev.en.33.010188.002343

Wright MG, Samways MJ (1998) Insect species richness tracking plant species richness in a diverse flora: gall-insects in the Cape Floristic Region, South Africa. Oecologia 115:427–433. https://doi.org/10.1007/s004420050537

Yukawa J, Tokuda M, Uechi N et al (2001) Species richness of galling arthropods in Manaus, Amazon and the surroundings of the Iguassu Falls. Esakia 41:11–15

Chapter 17
Collecting, Rearing, and Preserving Leaf-Mining Insects

Carlos Lopez-Vaamonde, Natalia Kirichenko, and Issei Ohshima

17.1 Introduction

Leaf mines are characteristic cavities of various shapes made by larvae of some moths, sawflies, flies, and beetles that feed on the leaf tissues between the lower and upper epidermis (Hering 1951) (Fig. 17.1a–h). The mine provides a microenvironment that is supposed to protect the larva from adverse conditions, although the temperature inside mines can be higher than ambient air temperature by more than 10 °C (Pincebourde and Casas 2015); also to shield the larva from predators and parasitoids, although relatively high parasitism rates are often recorded (Kato 1985a, 1994; Hawkins 1994; Aoyama and Ohshima 2019a, b); or to enhance the nutritional quality of plant tissues (Kaiser et al. 2010). Indeed, the remarkable diversity of leaf-mining patterns is thought to have resulted in part from adaptations against parasitoids that track the mine to locate the larvae (Kato 1984, 1985b; Salvo and Valladares 2004; Ayabe et al. 2008; Aoyama and Ohshima 2019a).

Leaf miners are not restricted to leaves. Some leaf-mining species of moths and beetles mine leaf petioles and larger veins before feeding in tissues inside the leaf

C. Lopez-Vaamonde (✉)
INRAE, UR0633 Zoologie Forestière, Orléans, France

Institut de Recherche sur la Biologie de l'Insecte UMR 7261, CNRS – Université de Tours, Tours, France
e-mail: carlos.lopezvaamonde@inrae.fr

N. Kirichenko
Sukachev Institute of Forest, Siberian Branch of Russian Academy of Sciences, Krasnoyarsk, Russia
e-mail: carlos.lopezvaamonde@inrae.fr

Siberian Federal University, Krasnoyarsk, Russia

I. Ohshima
Department of Life and Environmental Sciences, Kyoto Prefectural University, Kyoto, Japan
e-mail: issei@kpu.ac.jp

J. C. Santos, G. W. Fernandes (eds.), *Measuring Arthropod Biodiversity*,
https://doi.org/10.1007/978-3-030-53226-0_17

Fig. 17.1 Leaf mines and larvae of leaf-mining insects from four orders found on poplars, *Populus* spp., in Siberia. **Lepidoptera**, (**a**, **b**) *Phyllocnistis labyrinthella* (Bjerkander, 1790) (Gracillariidae); **Diptera**, (**c**, **d**) *Agromyza albitarsis* Meigen, 1830 (Agromyzidae); **Coleoptera**, (**e**, **f**) *Zeugophora scutellaris* Suffrian, 1840 (Megalopodidae); **Hymenoptera**, (**g**, **h**) *Heterarthrus ochropoda* (Klug, 1818) (Tenthredinidae). Sampling localities: (**a–h**) Novosibirsk, Central Siberian botanical garden, 20.VI.2009 (**a**), ibidem, 7.VIII.2011 (**b–h**). (Photos: N. Kirichenko)

lamina (Hering 1951). Some species of moths and flies are known to mine soft stems or young shoots (Hering 1951; Pitkin et al. 2019). Other species of moths can even mine inflorescences and flower petals and the surface of young fruits and seeds (Hering 1951) or switch their feeding mode from normal leaf miner to gall inducer (Guiguet et al. 2018, 2019).

Most leaf-mining insects are host-specific, and physical characteristics of leaf mines (i.e., the oviposition site where the mine begins, shape, position within a leaf, frass (excrements) pattern, etc.) can be used to identify leaf miners even down to species level (Eiseman 2019; Ellis 2020) (Figs. 17.2, 17.3, 17.4, and 17.5). The type of mine largely depends on the feeding behavior of larvae (Hering 1951). Those larvae that move only in one direction and thus eat their way through tissues only going forward create linear mines that can vary in length, width, and depth (Figs. 17.1a and 17.2). Linear mines can be long and found not only on the leaf lamina but also on stems, shoots, fruits, and seeds. There are larvae that feed on tissue in several directions producing so-called blotch mines which can be of different sizes and depth and can be often situated between secondary veins or cover big parts of leaf lamina across secondary veins. Some moths and flies make branched or star-like mines, where larvae are based in a central patch by making short galleries in the leaf lamina in various directions, and larvae immediately return to the central site when they sense danger (Fig. 17.3). In some species, mines are represented by the combination of a linear superficial mine (in early stages) and a deep blotch (in later stages) (Fig. 17.4a–h). In other cases, blotch mines are three-dimensional, so-called tentiform mines, whereby larvae gather the epidermis of mines by spinning silk, shaping the blotch mine as if a tent (Fig. 17.5b).

Frass from larvae can accumulate inside the leaf mine, and its arrangement varies among leaf miner groups (Figs. 17.1 and 17.2). Some species eject frass outside of the leaf mine, in particular larvae of some moths and sawflies (Hering 1951; Eiseman 2019). In dipteran mines, frass particles containing large amount of moisture can be distributed either irregularly (in blotch mines) (Fig. 17.1c) or in one or both sides of the mine (Fig. 17.2h).

Lepidopteran larvae have a cylindrical body (larvae of a few species and early-stage larvae are flattened), with well-developed head capsule, three pairs of thoracic legs, and five pairs of abdominal prolegs (rarely without prolegs). In contrast, larvae of Hymenoptera have six or more pairs of prolegs (Fig. 17.1h); rarely prolegs are reduced or completely absent (i.e., *Profenusa*). Coleopteran leaf-mining larvae have no prolegs. Larvae of dipteran miners are easily distinguished by the absence of legs and well-defined sclerotized head capsule and by the presence of clawlike mandibles (except in the occasional sciarid or chironomid leaf miner) (Fig. 17.1d).

17.2 Collecting Leaf Mines

Leaf mines can be found on many flowering plants (angiosperms) (Hering 1951). As an exception some species of moths and beetles make mines on needles of gymnosperms (Falkovich 1964; Kumata 1964). Due to a high degree of trophic specialization, mines of a given insect species are usually found only on one host plant species or on closely related species of the same genus or family (Lopez-Vaamonde et al. 2003). Leaf miners seem to prefer shaded leaves (Faeth 1991, but see Potter 1992), and therefore leaf mines can often be found in the lower branches, but sometimes a long-reach telescopic pruner is needed to cut the branches with mined leaves in tree crowns.

Fig. 17.2 Tunnel mines of **Lepidoptera**: (**a**) *Phyllocnistis labyrinthella* (Bjerkander, 1790) (Gracillariidae) on *Populus tremula*, (**b**) *Ph. unipunctella* (Stephens, 1834) (Gracillariidae) on *Populus laurifolia*, (**c**) *Stigmella anomalella* (Goeze, 1783) (Nepticulidae) on *Rosa* sp., (**d**) *S. lemniscella* (Zeller, 1839) (Nepticulidae) on *Ulmus laevis*, (**e**) unknown species (Gracillariidae, Ornixolinae) on *Bauhinia* sp.; **Coleoptera**: (**f**) *Orchestes rusci* (Herbst, 1795) (Coleoptera, Curculionidae) on *Betula pendula*; **Diptera**: (**g**) *Agromyza aristata* Malloch, 1915 (Agromyzidae) on *Ulmus pumila*, (**h**) *Agromyza alnibetulae* Hendel, 1931 (Agromyzidae) on *Betula pendula*. Localities in Russia: (**a, g**) Krasnoyarsk, forest 15.VI.2015 (**a**), ornamental plantation, 18.VI.2013 (**g**); (**b–d, f**) Novosibirsk, Central Siberian botanical garden SB RAS, arboretum, 16.VIII.2013 (**b–c**), ibidem, 11.VII.2012 (**d, f**); (**h**) Omsk, Victory park, 29.VI.2013. In France: (**e**) French Guiana: Parare, Nouragues Field Station, 2.XII.2017. Arrows indicate: *p*, pupation place; *lar*, larva inside the mine; *e*, exit site of larva. (Photos: N. Kirichenko (**a–d, f–h**) and I. Ohshima (**e**). Zoom up: (**d–f**) fragment of a mine with frass)

Fig. 17.3 Coexistence of a starlike mine of *Liocrobyla lobata* Kuroko, 1960 (Lepidoptera, Gracillariidae, Ornixolinae) (**a**) and tunnel-blotch mine of *Spulerina dissotoma* (Meyrick, 1931) (Lepidoptera, Gracillariidae, Acrocercopinae) (**b**) on the same leaflet of *Pueraria montana var. lobata*, Japan: Niimi, Okayama, 30.IX.2016. Arrows indicate: *s*, starlike mine; *t*, tunnel part of the mine; *b*, blotch part of the mine. (Photo: I. Ohshima)

To verify whether a leaf mine has living larvae/pupae inside, one can use a pocket size magnifying lens holding the leaf against the sky (or light) and use the natural backlight to see the contents better, if not a stereomicroscope (Fig. 17.6a–e). Healthy larvae will typically be actively eating or moving when disturbed. However, immobile larvae may be molting and/or have already been killed or paralyzed by parasitoids.

Active leaf mines (i.e., with feeding larvae) are normally found on young, soft leaves but can also be observed in fallen leaves on the ground, in particular in temperate areas during the autumn when leaves are shed, but also in the tropics. Some leaf mines can be very easily spotted because in aging leaves the mined part of the leaf stays green, creating a green patch on a brown or yellow leaf, an effect named "green island" (Hering 1951). This phenomenon represents a textbook example of how insects can manipulate the physiology of the host plant to conserve photosynthetic activity in the mined area of the senescent leaf so it can continue feeding (Kaiser et al. 2010; Gutzwiller et al. 2015) (Fig. 17.7) a phenomenon also reported in gall-inducing insects (Fernandes et al. 2008). On the other hand, exceptionally,

Fig. 17.4 Tunnel-blotch mines of **Lepidoptera**: (**a**) *Caloptilia betulicola* (Hering, 1928) (Gracillariidae) on *Betula pendula*, (**b**) *Callisto denticulella* (Thunberg, 1794) (Gracillariidae) on *Malus* sp., (**c**) *Parornix* sp. (Gracillariidae) on *Amelanchier* sp., (**d**) *Phyllonorycter platani* Staudinger, 1870 (Gracillariidae) on *Platanus* sp., (**e**) *Phyllonorycter issikii* (Kumata, 1963) on *Tilia cordata*, (**f**) *Ectoedemia rosae* van Nieukerken & Berggren, 2011 (Nepticulidae) on *Rosa* sp., (**g**) *Eriocrania sparrmannella* (Bosc, 1791) (Eriocraniidae) on *Betula pendula*; **Coleoptera**: (**h**) *Orchestes steppensis* Korotyaev 2016 (Curculionidae). Localities in Russia: (**a**, **c**, **h**) Krasnoyarsk, forest, 19.VI.2013 (**a**), Trostentsovo, private garden, 19.07.2011 (**c**), Krasnoyarsk, 18.VI.2013; (**b**) Tyumen, Zagorodny park, 30.VI.2013; (**e–g**) Novosibirsk, Central Siberian botanical garden SB RAS, 10.VI.2009 (**e**), ibidem, 03.VIII.2011 (**f**), ibidem, 10.VIII.2010 (**g**). In France: (**d**) Chambord, garden, 21.XI.2009. Arrows indicate: *t*, tunnel part of a leaf mine; *r*, rolled leaf tip; and *b*, folded leaf margin (inside these constructions larva continued feeding after vacating the mine). (Photos: N. Kirichenko; (**a**) and (**g**) are reproduced from Kirichenko et al. (2017))

Fig. 17.5 Blotch mines of irregular (**a**, **d–f**, **h**) and oval (**b**, **c**, **g**) shape in **Lepidoptera**: (**a**) *Gracillaria syringella* (Fabricius, 1794) (Gracillariidae), (**b**) tentiform mine of *Phyllonorycter agilella* (Zeller, 1846) (Lepidoptera, Gracillariidae) on *Ulmus laevis*, (**c**) *Incurvaria pectinea* Haworth, 1828 (Incurvariidae) on *Betula pendula*; **Diptera**: (**d**) *Phytomyza ilicis* (Curtis, 1846) on *Ilex* sp.; **Hymenoptera**: (**e**) *Profenusa* sp. (Tenthredinidae) on *Betula pendula*, (**f**) *Fenusa ewaldi* Smith, 2011 (Tenthredinidae) on *Ulmus pumila*; **Coleoptera**: (**g**) *Isochnus sequensi* (Stierlin, 1894) (Curculionidae) on *Populus balsamifera*, (**h**) *Zeugophora scutellaris* Suffrian, 1840 (Megalopodidae) on *Populus alba*. Localities in Russia: (**a–c**, **e–h**) Krasnoyarsk, Akademgorodok, ornamental plantation, 20.VI.2011 (**a**), ibidem, 19.VI.2011 (**h**), ibidem, along the River Yenisei, 19.VI.2013 (**c**), ibidem, Trostentsovo, forest, 11.VII.2011 (**e**), ibidem, Red Square, ornamental plantation, 18.VI.2013 (**f**), ibidem, Svobodniy Prospect, ornamental plantation, 19.VII.2011 (**g**); (**b**) Omsk, Victory park, 29.VI.2013. In France: (**d**) Orleans, University of Orleans, campus, plantation along the road, 22.XI.2009. Zoom up: (**c**) an individual mine with young larva and frass. (Photos: N. Kirichenko)

Fig. 17.6 Examining leaf mines and arranging rearing tests in laboratory: (**a**) checking for the presence of living immature stages inside leaf mines using stereomicroscope; (**b**) feeding leaf-mining larvae of a beetle under magnification; (**c**) photographing larvae and the mine using a digital camera attached to the stereomicroscope; (**d**, **e**) preparing leaves with mines for rearing in plastic boxes, cutting tissue around the petiole, wrapping the petioles with a strip of wiping tissue; (**f**) moisturizing petioles with sucrose solution by punching rearing box lid with syringe. (Photos: N. Kirichenko (**a**, **c–f**), C. Lopez-Vaamonde (**b**))

there are species whose larvae feed on the dead part of the leaf, as in *Artifodina japonica* Kumata, 1985 (Gracillariidae, Acrocercopinae) (Fig. 17.8).

Leaf miners are frequently parasitized, so to have a chance of obtaining adult leaf miners, leaf mines of the target species should be collected from several plant

Fig. 17.7 Manipulation of leaf physiology by gracillariid moth: green island in the lower side blotch leaf mine of *Phyllonorycter platani* Staudinger, 1870 on *Platanus* sp., France: Chambord, garden, 21.XI.2009. (**a**) A plane leaf with a mine, upper side view; (**b**, **c**) zoom up of the mine region—upper side view (**b**), lower side view (**c**). (Photos: N. Kirichenko)

individuals if possible. Host plant vouchers should be collected with reproductive parts if possible (Fig. 17.9a, b) and labeled in the field (Fig. 17.9d,f). Every collecting event should be illustrated with photographs of the collected leaves and the reproductive parts of the plants while alive (Fig. 17.9c). This field label containing a code is attached to a twig or branch of the sample and should be linked to the leaf-mining insect that emerged from the preserved mines (Fig. 17.9f). The field code can be composed of the initials of the collector, followed by collection year (two or four digits as preferred) and then a sequential three-digit code, if you do not do more than 999 collection events in a year, for instance, CLV20001. It is of paramount

Fig. 17.8 Manipulation of leaf by gracillariid moth: the distal part of the leaf damaged by *Artifodina japonica* (Gracillariidae, Acrocercopinae), (**a**, **b**, **d**) Japan, Hirai, Kozagawa, Wakayama, 21.IV.2013; (**c**) Japan, Setouchi, Amami Island, Kagoshima, 16.V.2015. (**a–c**) Early-stage mine: young instar larvae transverse the distal part of the leaf making long narrow tunnel mine, weakening the damaged area. (**d**) Late-stage mine causing browning and drying of the damaged leaf area. Arrows indicate: *m*, mined area of the distal part of the leaf; *t*, narrow long tunnel. (Photos: I. Ohshima; (**d**), reproduced from Ohshima 2013)

importance to assign a unique rearing code for each sampling event and retain it across all related specimens of the larvae and pupae extracted from those mines, as well as adults and/or parasitoids reared from those mines. This allows revealing the full complexity of trophic interactions (Valladares et al. 2001; Lopez-Vaamonde et al. 2005; Ohshima et al. 2015; Leppänen et al. 2013; Kirichenko et al. 2019a, b) and host race formation in leaf miners (Ohshima 2008).

17.3 Rearing Leaf-Mining Insects

Immature stages (larvae and pupae), if kept in the right conditions (i.e., together with their mines), can be reared in the laboratory to adulthood. It is important to keep the mined leaves as fresh as possible until the larvae finish their development. Various methods exist to rear leaf miners in laboratory conditions and prolong the life of mined leaves so as to provide food for larvae feeding in mines (Hering 1951;

Fig. 17.9 Collecting leaf mines: (**a**) sampling mined leaves high in the tree crown using long telescopic pruner; (**b**) searching for leaf mines in the lower part of the tree crown; (**c**) photographing leaf mines in the field; (**d**, **f**) collecting and labeling host plant voucher; (**e**) preparing host plant vouchers in the field. (Photos: C. Lopez-Vaamonde)

Johansson et al. 1990; Elliott 2007; Eiseman 2019; Pitkin et al. 2019). Here we describe a rearing protocol that focuses on gracillariid leaf-mining moths (Ohshima 2005, 2012, 2013), but it can also be easily adapted to rearing leaf-mining insects belonging to other families and orders.

Once the mined leaves are collected, samples must be kept in a plastic bag (e.g., Ziploc©), which should be labeled with a field code and georeferenced data. At the same time, a sample of the identical host plant from which the leaves were obtained should be collected and labeled with the same field code as the mined leaf and preserved as a herbarium voucher. This is especially critical in the tropics, where identification of host plants is very difficult and requires assistance of botanical experts and/or DNA barcoding. It is also important to photograph the host plant as well as leaf mines in the field and to document the habitat in which the leaf mine was found. The mine should be photographed from both the upper and lower sides of the leaf. Care should be taken to make sure the sample of the plant voucher corresponds to the same plant individual from which the leaf mine was collected, again, in particular in the tropics where the density of different tree species is very high and often plants are entangled.

The larvae of many species pupate outside the mines so sampling plastic bags should be tightly closed because larvae excel at escaping through any tiny hole. We have found that it is important to inflate the plastic bags, to increase the CO_2 content inside the bag, which plays a potential role in keeping the mined leaves fresh, especially at night. Although leaf mines are generally resistant to heat, leaves should be kept in a shaded area without direct sunlight after collection, even during the field work, because condensation has undesired effects, including promoting mortality among the leaf miners. Especially, herbaceous leaves wilt fast in the heat.

In the laboratory, mined leaves are to be transferred to rearing boxes (Fig. 17.10). Ideally, leaf mines should be kept individually in separate rearing boxes. In some cases, one leaf can contain several species of leaf-mining insects (Fig. 17.3). Thus, care should be taken not to keep several individuals of different species in the same rearing boxes.

Cut the lower part of the leaf, but leave the main vein (Fig. 17.10a, b, c). Wrap a stripe of tissue paper or wiping paper tightly around the main vein (Fig. 17.10d, e), and immediately add a 2% sucrose solution to the rolled-up paper using a pipette or syringe (Fig. 17.10f). The sucrose concentration of the solution added should be more concentrated than in the plants so the sugar flows into the plant (a method commonly used in the ornamental flower trade) (Arrom and Munné-Bosch 2012). When it is difficult to wrap the tissue paper around the main vein of a leaf, cut at a sharp angle the base of the petiole, and wrap the tissue paper around the petiole (Fig. 17.10g, h).

Fig. 17.10 (continued) the left main vein by tissue paper, transferring the leaf to the rearing box and adding sucrose solution to the rolled main vein immediately after cutting each leaf; (**g**, **h**) the cases rolling up petioles [(**g**) leaf miner, *Deoptilia heptadeta* (Meyrick, 1936) (Gracillariidae); host plant, *Mallotus japonicus*, Euphorbiaceae; (**h**) leaf miner, *Psydrocercops wisteriae* (Kuroko, 1982) (Gracillariidae); host plant, *Wisteria floribunda*, Fabaceae]. (Photos: I. Ohshima, reproduced from Ohshima 2012)

Fig. 17.10 Preparation of mined leaves in the laboratory for rearing. (**a**) A mined leaf (leaf miner, *Lyonia* race of *Acrocercops transecta* Meyrick, 1931 (Gracillariidae); host plant, *Lyonia ovalifolia*, Ericaceae); (**b**, **c**) cutting the basal part of the leaf body but leaving the main vein; (**d–f**) rolling up

The larvae of some species (e.g., Momphidae and some Gracillariidae in Lepidoptera, and some Galerucinae, Chrysomelidae in Coleoptera) have the ability to exit the initial mines and start secondary mining in new leaves (Hering 1951; Inoue 1990; Pappers et al. 2002; Kirichenko et al. 2016; Pitkin et al. 2019), so adding extra new fresh leaves may be required. Also, some species make a leaf roll using a different leaf/leaflet in their later larval stages (e.g., in Gracillariidae, *Caloptilia* moths), so adding host leaves is essential for the successful completion of life cycle. As mentioned above, other species mine part of the stem and complete development on neighboring leaves (Langmaid and Corley 2007), so mined leaves will be sampled with a twig/branch. For the species beginning gall induction from a particular instar, collection should be taken place after completion of gall development; otherwise larvae collected at miner or immature galler stages will fail to survive until pupation due to the lack of gall tissues (Guiguet et al. 2018, 2019).

When rearing species in a climate that differs from where they were collected, rearing boxes should be kept in climatic chambers. Rearing boxes should be checked daily and few drops of the sucrose solution added regularly to keep moisture levels high, but not many, to avoid mold. Species that pupate inside leaf mines require to keep a bit higher moisture avoiding the leaves to dry because the cuticle can become hard and it will be difficult for the adult leaf miner to emerge.

For species pupating outside the mine on leaf remains, (e.g., some gracillariid genera like *Acrocercops*, *Caloptilia*, and *Micrurapteryx*: Hering 1951; Kirichenko et al. 2016), or on the walls of the rearing box, one has to be careful when adding the drops of sucrose solution because pupating larvae can get stuck in a drop and drown. Some *Caloptilia* larvae make cocoons whose upper surface is well tensioned (Fig. 17.11a), folding the leaf (Fig. 17.11b). However, these highly tensioned cocoons are often detached during the cocooning stage, especially when the cocoons are fixed to the wall of the plastic rearing boxes, and then the larvae are very likely to fail to pupate. To prevent this, place a tissue or wiping paper at the bottom of the rearing box before the cocoon is spun (Fig. 17.11c).

Some larvae will pupate on or in the ground, notably Nepticulidae and Eriocraniidae, the majority of leaf-mining sawflies and flies, and few species of leaf-mining beetles (Hering 1951). In that case add moss or soil or sand collected preferably at the base or around the host plant to the rearing box. Moss/soil should ideally be sterile, as there could be harmful bacteria, fungi, predatory mites, etc.

Leaf mines sampled at high latitudes can be kept in wooden boxes in the snow for overwintering. Otherwise, they can be placed in a refrigerator with a temperature just above 0 °C. In this case, leaves with mines should be checked at least once a week to make sure there is no problem with humidity and fungal infection.

Fig. 17.11 Preparation for *Caloptilia* pupation. (**a**) Cocoon on the underside of host leaflet (cocoon, *Caloptilia* sp.; host plant, *Toxicodendron trichocarpum*); (**b**) sharply folded upper surface of the same leaflet; (**c**) rearing boxes lined with tissue paper. Larvae make their cocoon on the surface of host leaves or on the tissue paper. (Photos: I. Ohshima)

17.4 Mounting and Preserving Adults

Here we focus on how to double mount micromoths. Once adults emerge from the leaf mines, they can be collected with the help of a handheld aspirator (Fig. 17.12a). A drop of ammonia solution (ca. 30%) is placed inside a small glass vial (Fig. 17.12b, c), and after a few minutes, the adult micromoth will die. Alternatively,

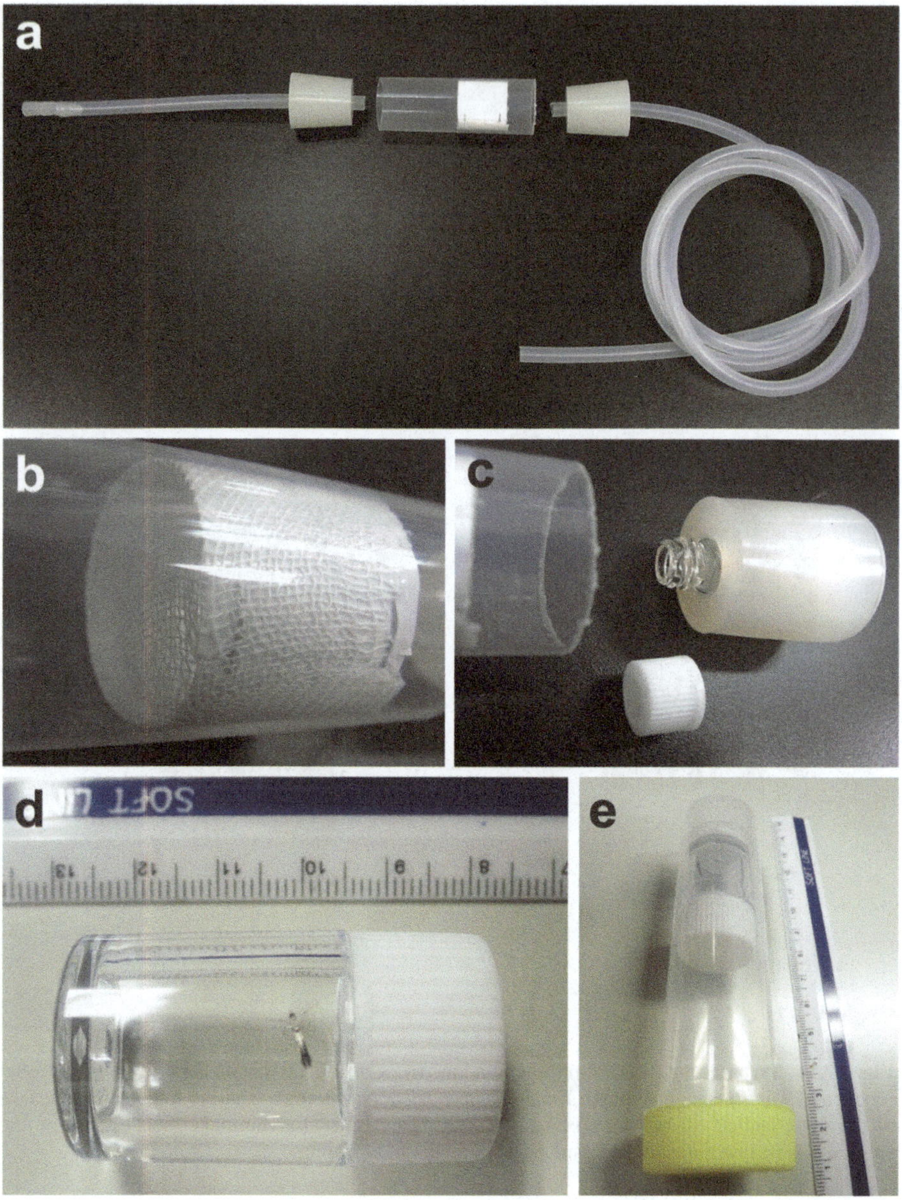

Fig. 17.12 Aspirator for killing emerged leaf miner adults. (**a**) Whole parts of the aspirator except for an ammonium-vial silicon stopper. Insects pass through the left tube and remain in the left part of the center tube when aspirated from the right tube by mouth; (**b**) filtering mesh in the center tube; (**c**) after aspirating, the right tube is replaced by the ammonium-vial silicon stopper to kill stored insects. There is an absorbent cotton soaked by ammonium solution in the vial, and the lid of the tube should be removed just before setting in the center tube; (**d**) a plastic vial including an adult; (**e**) the plastic vial is inserted in the centrifuge tube (50 ml falcon tube) whose bottom has an absorbent cotton soaked by the ammonium solution. The lid of the vial should be slightly opened before inserted in the centrifuge tube. (Photos: I. Ohshima; (**a–c**), reproduced from Ohshima 2012)

a micromoth can be collected by allowing it to walk into a small plastic/glass tube (Fig. 17.12d). Then the tube containing the micromoth will be placed inside a large falcon tube (Fig. 17.12e) with some ethyl acetate or ammonia solution. It is safer to use glass containers with ethyl acetate since it dissolves many plastics. The small tube with the micromoth should have the lid slightly unscrewed so the ammonia/ethyl acetate fumes can penetrate the tube and kill the moth.

Unless they are placed in the deep freezer, micromoths will soon become rigid after killing, so immediate preparation after death is crucial (Fig. 17.13a,b). Nepticulidae are very difficult to spread once dead, so it is better to sedate them with ethyl acetate first, then mount and spread, and then place them in a killing jar. Due to the tiny size of microlepidoptera, a good stereomicroscope will be required to pin

Fig. 17.13 Arranging dry collections in field. (**a**, **b**) Spreading micromoth specimens in the field using a stereomicroscope; (**c**) spreading board containing micromoths; (**d**) close-up of spread specimen. (Photos: C. Lopez-Vaamonde, French Guiana)

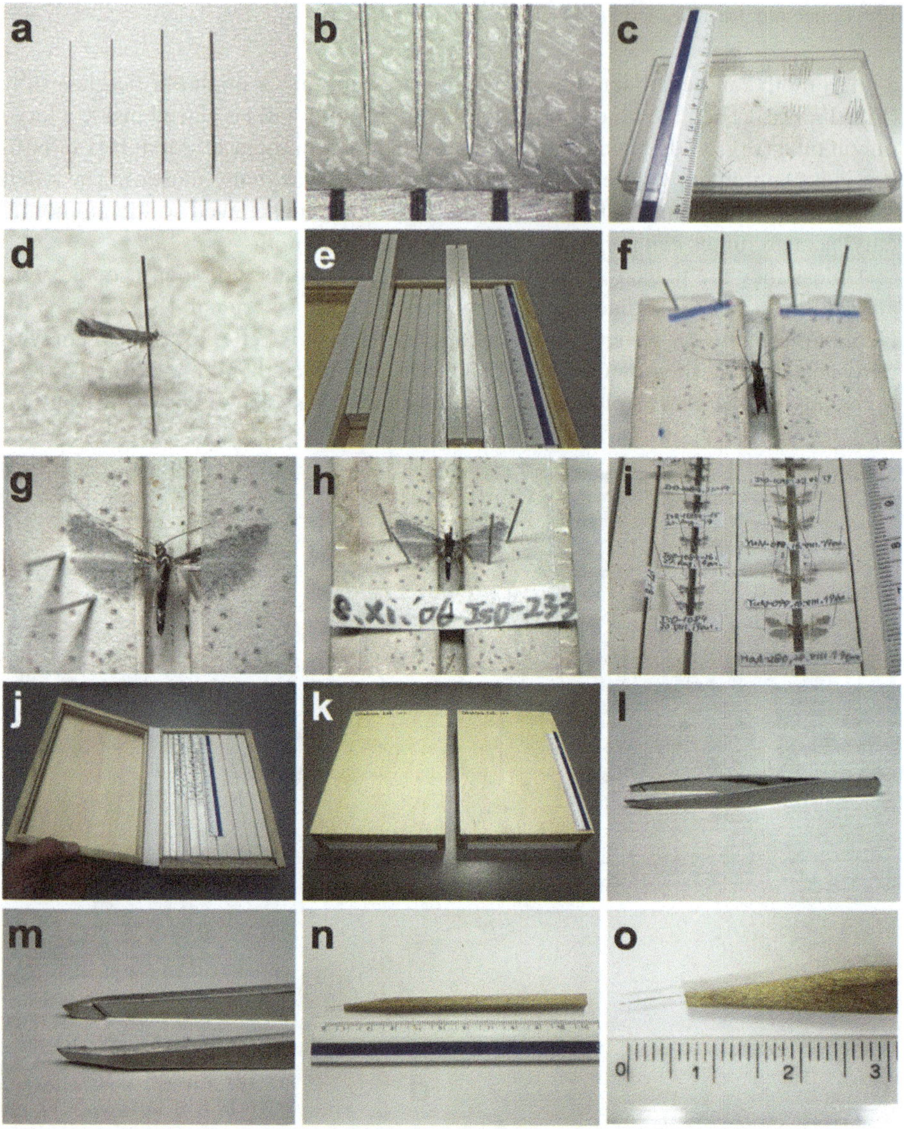

Fig. 17.14 Pinning and spreading emerged micromoths. (**a**, **b**) Micropins for minute insects, MINUCIE, produced by Ento Sphinx, Czech Republic, No. 10, 15, 20, 25 from the left—10, for *Phyllocnistis*, *Porphyrosela* (Gracillariidae), and Nepticulidae; 15, for *Phyllonorycter*, smaller Acrocercopinae (Gracillariidae); 20, larger Acrocercopinae and Gracillariinae (Gracillariidae); 25, larger Gracillariinae—(**c**) small plastic box laid with polyurethane foam for pinning; (**d**) pinned micromoths; (**e**) spreading boards. Various sizes of width and slit; (**f**) a pinned micromoth is set in the slit of the spreading board; (**g**) for a right-handed person, start spreading from the left wings; (**h**) spread micromoth with a tentative label with emerged date and field/rearing ID; (**i**) sequentially spread micromoths on the boards; (**j**) Kumata spreading boards. This system has been designed by late Dr. Tosio Kumata. All boards are fixed in the box and the box can easily be carried when the lid is closed (**k**); (**l**) tweezers for cosmetics; tweezers are very useful for handling micropins but strong forceps may also be needed; (**m**) tips of the tweezers; (**n**) microneedle with timber stick handle made from micropins and chopsticks; (**o**) tip of the microneedle. (**d**, **f–h**) *Acrocercops transecta*. (**f–h**) 20 mm width with 2 mm slit spreading board. (Photos: I. Ohshima; (**d**, **f–h**, **l**, **m**), reproduced from Ohshima 2012)

them using minuten pins (Fig. 17.14a,b). The size of the minuten pin will depend on the size of the micromoth (Fig. 17.13c,d).

Under a stereomicroscope, killed adults should be placed on polyethylene/polyurethane foam (Fig. 17.14c). Care should be taken to pin vertically through the mesothorax (Fig. 17.14d). The size (width of the board and the slit) of spreading boards also depends on the insect size (wingspan and width of the body) (Fig. 17.14e). For example, a board width of 15 mm with 0.6 mm slit size is useful for Nepticulidae and tiny Gracillariidae, such as *Phyllocnistis*, and a 25 mm width with 2 mm slit is appropriate for *Caloptilia* (Gracillariidae). A paraffin paper strip precisely cut to the width and length of one side of the board is used for pressing the wings. Tips of the paraffin tapes should be folded upward (Fig. 17.14f). Pin the minuten pin with insect body at the center of the slit (Fig. 17.14f). It is very important that the wing base is level with the board surface (Fig. 17.14g). Wings are spread with the help of a micropin mounted on a wooden support (Fig. 17.14n, o) between the board surface and paraffin paper. Adjust the angle of the fore and hind wings as shown (Fig. 17.14g). Once the wings are correctly arranged, pull the paraffin tape slightly to press and fix the angle of the wings, and immediately place pins near and around the wings to better apply pressure to the paper onto the wings (Fig. 17.14g, h). After fixing both left and right wings, a label including at least the date of emergence and ID code should be pinned below the spread specimen (Fig. 17.14h, i). Spreading boards with pinned specimens can be kept inside wooden boxes (Fig. 17.14j, k). The spreading boards are fixed tightly inside the wooden boxes so pinned specimens can be hand carried in the field without the risk of being damaged. Spread specimens should be dried for about 2 weeks (4 weeks if they are killed with ammonia). Once dried the spread specimens can be removed from the board (Fig. 17.15a). Because the antennae and wings have been pressed, they are often a bit attached to the paraffin strip and/or the surface of the board. Thus, the strip should be removed very carefully. Pinned specimens with their labels are then transferred from the spreading boards to a polyethylene/polyurethane foam sheet (Fig. 17.15b, c). Mounting stages (Fig. 17.15d) are prepared using hard polyethylene/polyurethane foam and cut into uniform sizes (e.g., 3 mm × 10 mm for Nepticulidae and tiny Gracillariidae such as *Phyllocnistis* and 5 mm × 13 mm for larger micromoths). These stages are pinned with standard entomological pins (size 2), at an equal height with the help of a pinning block (Fig. 17.15e). Each specimen is mounted on the upper side of the stage (Fig. 17.15f). Formal labels are printed and cut into appropriate sizes (Fig. 17.15g). For reared specimens from immature stages, we need at least two labels: sampling label and rearing label (Fig. 17.15g). The sampling label is pinned first and will contain country, state or province, exact locality, georeference (latitude/longitude), the dates that the leaf mine was collected and of adult emergence, and collector's name. The second label will contain rearing code, the data of adult emergence, and the species name of the host plant (Fig. 17.15h, i). If the specimens have been identified, dissected, and/or DNA barcoded, a third label will be added with the species name, name of identifier, genitalia preparation number, and the process ID (barcode) number. Final specimens should look neat and uniform (Fig. 17.15j, k) and are arranged in the specimen box (Fig. 17.15l).

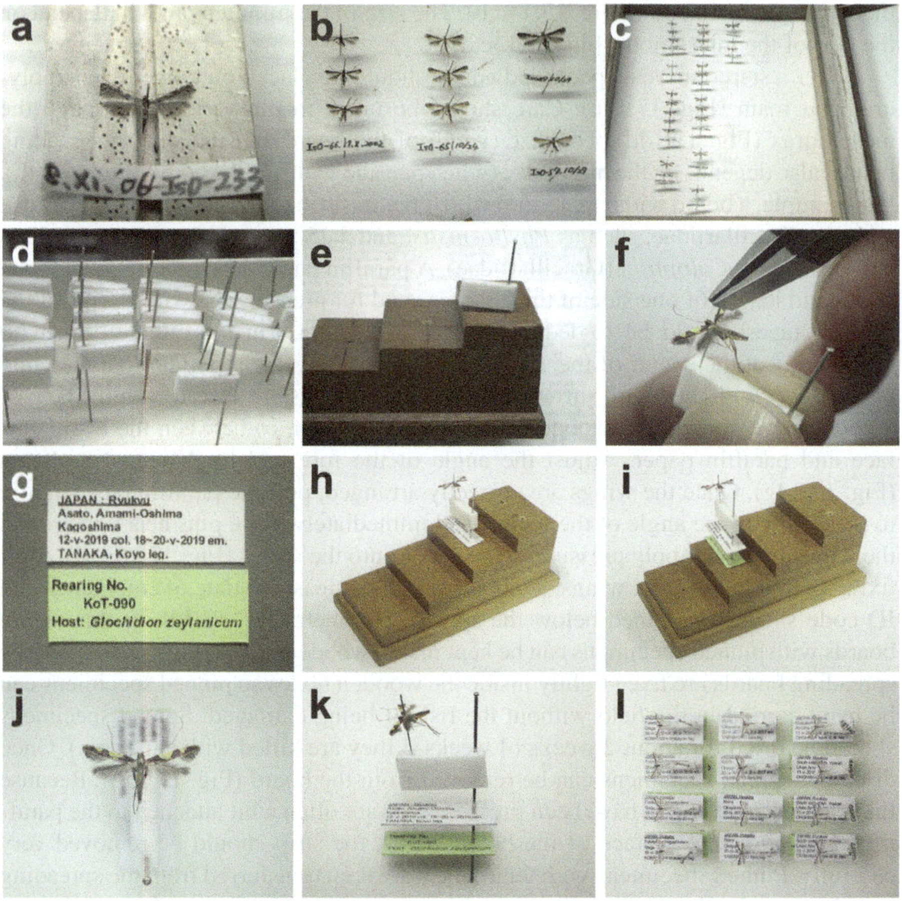

Fig. 17.15 Mounting the spread specimens and labeling. (**a**) Shifting the press tape sideways; (**b**, **c**) plastazote box for pinning before mounting; (**d**) a series of plastazote strips pinned for double mounting (**e**) stair-step wooden pinning block to set pinned specimens and labels at proper heights on the pin; (**f**) double mounting on plastazote strip; (**g**) labels (upper, collection label; bottom, rearing label with field/rearing ID and host plant name); (**h**, **i**) pinning block helps to standardise the collection by setting all specimens and labels mounted at appropriate heights ; (**j**) dorsal view and (**k**) lateral view of the finished specimen; (**L**) finished specimens. (**a**) *Acrocercops transecta*; (**f**, **h–k**) *Caloptilia ryukyuensis* Kumata, 1966; (**l**) *Psydrocercops* sp. (Photos: I. Ohshima; **a–e**, reproduced from Ohshima 2012)

17.5 Fixing and Preserving Immature Stages

It is important to preserve some samples of larvae and/or pupae in ethanol for subsequent morphological and molecular analysis, especially when rearing is difficult to set up in the field (Lopez-Vaamonde et al. 2012; Kirichenko et al. 2017, 2019a, b). Indeed, DNA barcoding has been successfully applied to reveal new trophic interactions of leaf miners by DNA sequencing living larvae/pupae or even the remains (molts, head capsules) sampled inside leaf mines (Lees et al. 2011, 2013; Kirichenko et al. 2019a, b).

To dissect immature stages out of leaf mines, a dissecting needle can be used to open the epidermis covering the mines. Syringe needles are recommended for dissecting mines; they have a long-sharpened edge and can be easily inserted into the mine, allowing accurate cutting of the epidermis along the edge of the mine. The mines are to be opened under a stereomicroscope. Flexible tweezers, a fine brush, or a dissecting needle can be used to transfer larvae and pupae from mines to tubes with ethanol ≥96%. Samples are stored in a freezer for further morphological and molecular research.

Insects recovered from mines should be stored in hermetically sealed tubes (Fig. 17.16a). Plastic microcentrifuge tubes with a volume of 1.5–2 ml and screw caps with a rubber ring which close tightly, preventing evaporation, are ideal for storing tiny insects. Snap cap tubes are not recommended since ethanol easily leaks and evaporates if the caps become loose.

Each tube must be properly labeled. Writing with a felt pen on the plastic must be avoided since the annotation can easily be rubbed off by ethanol. Labels can be handwritten with pigmented ink or printed with a laser or an inkjet printer (use alcohol proof ink). The label should be positioned so that data can be read without pulling out the label from the tube (note that it is easy to lose the tiny specimens when taking the label out of the tube for reading) (Fig. 17.16a). The tubes should be placed in storage boxes in vertical position (to ensure that all specimens in a tube are covered by ethanol) (Fig. 17.16b, c) and stored in a freezer at −20 °C (Fig. 17.16d). Storing ethanol-preserved specimens at low temperatures slows down DNA degradation (Doorenweerd and Beentjes 2012).

17.6 Pressing and Storing Mined Leaves

Leaves with mines, especially those from which larvae and pupae have been sampled, represent an important source of data and should therefore be preserved in a herbarium.

A botanical press can be used to preserve the samples with twigs; leaves without twigs are better to dry under the press of a pile of books to get well-flattened leaf specimen. Mined leaves from which adults and/or immature stages have been sampled should also be pressed and preserved as vouchers. If books are used to press leaves, it is important that the area of the lowest book is large enough to cover fully and exert even pressure on the herbarium in order to properly flatten leaves.

For drying, leaves are placed between newspaper sheets or blotting paper (Fig. 17.9e). Place leaf vouchers between newspaper pages; two to three pages should be left on either side of the samples to absorb moisture. Alternatively, pieces of corrugated cardboard could be used on either side of leaf sample—cardboard will absorb more moisture and allow some air circulation, leading to faster drying. Thus filled newspapers should be left overnight. If conditions are very humid, particularly in the tropics, drying is best accelerated with the help of an oven (Fig. 17.17a). Newspapers in which leaves are pressed should be replaced with a frequency

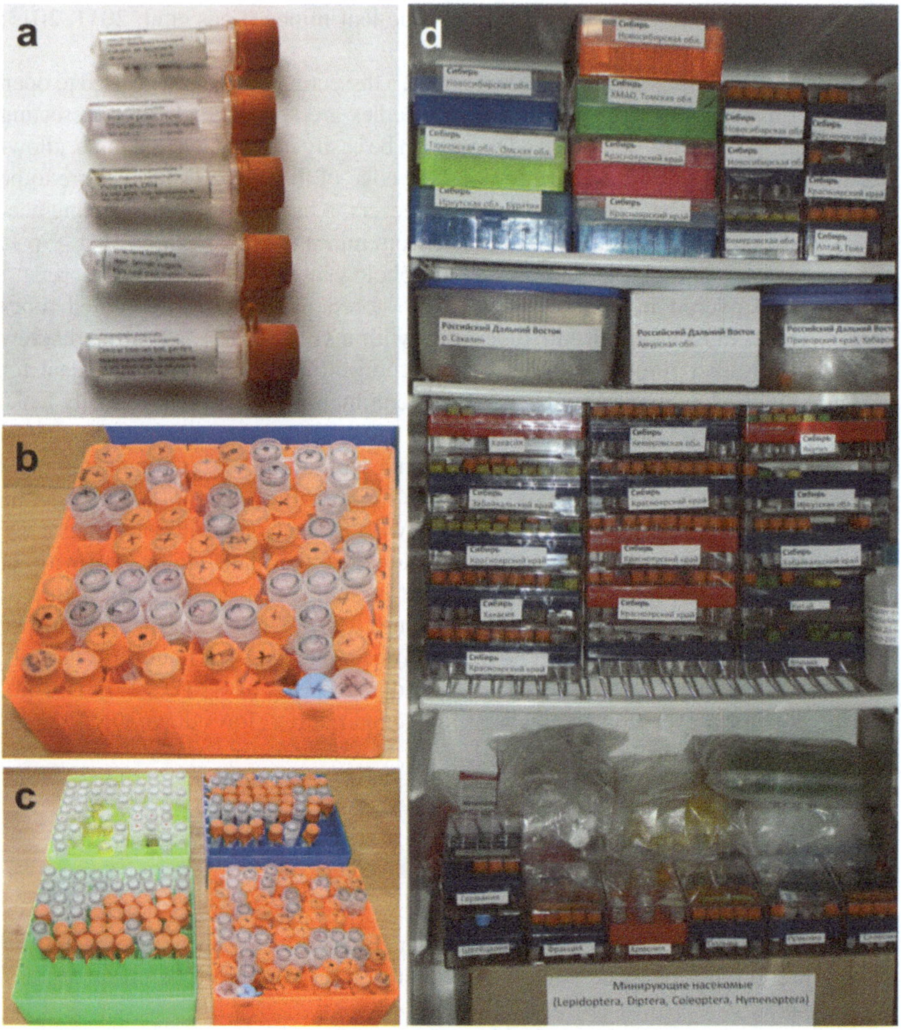

Fig. 17.16 Preserving and storing immature stages. (**a**) Microtube with larvae and pupae of leaf miners and the label placed tightly to the tube wall from inside the tube; (**b**, **c**) boxes storing tubes with specimens in vertical position; (**d**) storing the specimens in the freezer. (Photos: N. Kirichenko)

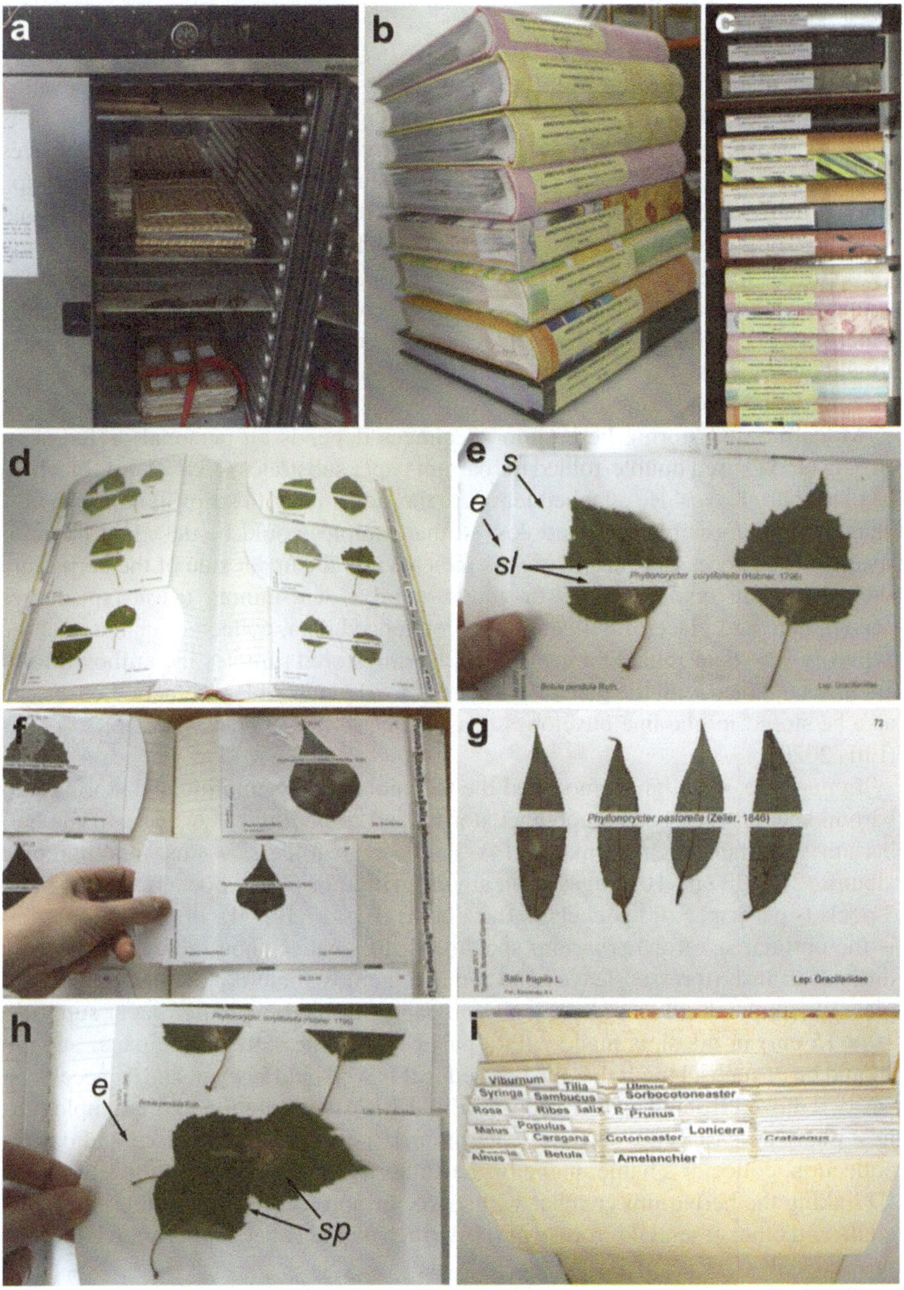

Fig. 17.17 Preparation of herbarium collection (pressed leaves with mines). (**a**) Botanical press containing herbarium specimens being dried out in oven (especially important in tropical conditions); (**b**, **c**) annotated herbarium albums; (**d**) album pages each with three plastic pockets filled in with pressed leaves; (**e**) arrangement of pressed leaves on the labeled paper support; (**f**) pulling paper support with the "demonstration" herbarized leaf and the envelope with the duplicates (situated below the paper support) out of plastic bag; (**g**) arrangement of several herbarized leaves on the paper support; (**h**) paper envelope containing extra specimens (duplicates); (**i**) side labels with the genus name of plants. Arrows indicate: *s* labeled paper support with pressed leaves; *e*, envelope with extra specimens; *sl*, double cut made in paper support to accommodate herbarized leaf; *sp*, specimens being stored in the envelope. (Photos: (**a**) C. Lopez-Vaamonde, (**b**–**i**) N. Kirichenko)

depending on humidity and leaf size and thickness; replacing paper will achieve the best drying results and minimize the color change of samples. Usually a week is enough for drying (depending on room humidity and size and thickness of leaves and twigs). The leaves of some plants (in particular, leaves of poplars, *Populus* spp., and willows, *Salix* spp.) can darken when dried. To speed up the drying process and avoid darkening of leaves and mines, freshly picked leaves can be ironed (Hering 1951). Insert leaves between two layers of filter paper and flatten with an iron set at temperature around 100 °C in several steps. After each ironing, the top layer of filter paper should be lifted up to allow the leaves to "breathe." The cooled dried samples should then be left between newspaper sheets under pressure for a day or two before placing them in the herbarium.

Mounting and storing herbarium specimens depends on personal preferences. Hering (1951) used double-folded herbarium paper substrates (33 × 21 cm) in which he placed dry leaves. He attached leaves to the substrate with strips of adhesive tape placed, where possible over petioles, so that samples could be detached for reinspection. The necessary information is indicated on the upper side of the herbarium substrate: plant taxon (species or genus, family), leaf-mining taxon (species or genus), date and place of collection, georeferenced data, collector's name, and the authority who determined the leaf miner. Hering stored samples in cardboard folders in accordance with the genera/families of food plants of leaf miners. Leaves can also be stored in glassine envelopes (Caspers et al. 2019) or plastic Ziploc© bags (Ellis 2020).

In our case, we slightly modified the technique for assembling and storing herbarium specimens assuring compact storing and providing easy access to the herbarium specimens (Kirichenko 2014). For storage folders, we use photographic albums (21 × 33 cm; 100 pages in an album) with transparent pockets (10 × 15 cm; 3 pockets per page; 300 pockets in the album) (Fig. 17.17b–d). In our opinion, this is the optimal size for herbarium specimens. In smaller photo albums, it is more difficult to insert pressed leaves and remove them without damaging fragile samples. We fix the dried leaves vertically on paper sheets (hereinafter paper supports; 10 × 15 cm) in the slots made with a paper knife. The slots on the paper support should be of such a length that the pressed leaf is well held in them and, if necessary, easily removed from the paper support (Fig. 17.17e). Annotation to the sample (i.e., host plant, at least genus or family, leaf miner species or genus, place and date of collection, collector's name, name of the authority who determined the leaf miner, ID linking the herbarium voucher with larvae or pupae taken from mines or reared adult, and other essential information) is written on the paper support into which the pressed leaves are inserted.

The labeled paper supports with leaves are placed into the transparent plastic pockets of the albums (Fig. 17.17d). If necessary, the paper supports can be easily removed from the plastic pocket (Fig. 17.17f). On the paper support, one can position a pressed leaf with its characteristic mine, and if the leaves are small, then a few more if they will fit (Fig. 17.17g). Additional (duplicate) samples of leaves that are not included on the paper support can be placed in an appropriately sized paper envelope with an oval protrusion on the side (Fig. 17.17h), allowing the envelope to be easily removed from the album pocket. Also, if herbarized leaf is too big to fit the paper support, we suggest to cut the leaf into two or three fragments (avoid cutting the mines) and insert the leaf fragment carrying mine/mines into the paper support and the remaining parts store in the paper envelope in the same plastic pocket. Thus, in one plastic pocket, we store a paper support with a "demonstration" mined leaf (or leaves) and under it a paper envelope with duplicates or leaf fragments (Fig. 17.17e, h). Leaves from one album pocket represent a collection exclusively from one individual plant and mines of one leaf miner species. Sampling more leaves and storing leaf samples in paper envelopes will allow assessment of the variability of the characteristics of mines, increasing the chances of finding insects in mines (in case they are needed for morphological study or DNA barcoding). Finally, even in the eventuality that one of the leaves has crumbled upon microscopic examination, the bulk of the collection would not lose its value.

Herbarium samples are placed in the album according to the genus of the host plant. To facilitate searching for samples, we place side labels with the generic name of host plants arranged in alphabetical order on the sides of the album pages (Fig. 17.17i). In our case, each album represents a collection for a particular year. On the front and the spine of the album, we attach the label stating the year, place of collection, and the names of the compilers. The albums are stored in horizontal position (Fig. 17.17c) in a cabinet in dark and dry conditions.

Whatever method of herbarium arrangement is used, the most important points that must be kept in mind when mounting and storing pressed leaves with mines are:

1. Herbarium specimens must be mounted and stored in institutes that provide free access to them, ideally using methods that allow easy detachment from the herbarium sheets or paper backing for investigation, photographing, or scanning. This allows for a detailed examination of mines and leaves on both sides.
2. Herbarium samples must be provided with a label containing all the essential biological and geographical information and other important compiler-specific data. Any numbers or codes linking the herbarium specimens with related insect specimens in dry or ethanol-preserved collections must be included on the label.
3. It is recommended that herbarium samples are properly systematized to provide fast search of the needed specimen. This can be done in accordance with the genus/family of host plant or following the taxonomy of the leaf miners.
4. Herbarium specimens must be stored in folders or albums in a horizontal position in a dark and dry place, to avoid loss of parts or damage by light and water. Glassine envelopes or plastic Ziploc© bags can also be placed vertically in small boxes as used for butterflies (Caspers et al. 2019; Ellis 2020).

Acknowledgments We are grateful to Camiel Doorenweerd and Erik Van Nieukerken (the Netherlands), Charley Eiseman and John Tooker (USA), Atsushi Kawakita (Japan), David Lees (UK), Sylvain Pincebourde (France), and Ayco Tack (Sweden) for insightful comments on the manuscript. CLV was funded by Centre d'Etude de la Biodiversité Amazonienne (CEBA) (project LEAFMINE) and a Fellowship of the Japan Society for the Promotion of Science and visiting researcher of Kyoto Prefectural University, Japan (2014). NK contribution was partially supported by the Russian Foundation for Basic Research (RFBR) (project no. 19-04-01029-a). IO was funded in part by JSPS KAKENHI grant numbers 04J09250, 08J05555, 26840120, and 17H06260 and Bilateral Joint Research Project between CNRS and JSPS.

References

Aoyama H, Ohshima I (2019a) Changing leaf geometry provides a refuge for a leaf miner from a parasitoid. Zool Sci 36:31–37. https://doi.org/10.2108/zs180062

Aoyama H, Ohshima I (2019b) Observation of *Aneurobracon philippinensis* (Hymenoptera: Braconidae) immatures shows how koinobiont offspring flexibly adjust their development to host growth. Ann Entomol Soc Am 112:490–496. https://doi.org/10.1093/aesa/saz036

Arrom L, Munné-Bosch S (2012) Sucrose accelerates flower opening and delays senescence through a hormonal effect in cut lily flowers. Plant Sci 188–189:41–47. https://doi.org/10.1016/j.plantsci.2012.02.012

Ayabe Y, Tuda M, Mochizuki A (2008) Benefits of repeated mine trackings by a parasitoid when the host leafminer has a tortuous feeding pattern. Anim Behav 76:1795–1803. https://doi.org/10.1016/j.anbehav.2008.07.026

Caspers M, Willemse L, Gasso L et al (2019) Butterflies in bags: permanent storage of Lepidoptera in glassine envelopes. Nota Lepidopterol 42:1–16. https://doi.org/10.3897/nl.42.28654

Doorenweerd C, Beentjes K (2012) Extensive guidelines for preserving specimen or tissue for later DNA work. Science. Naturalis. https://www.yumpu.com/en/document/-read/51104633/extensive-guidelines-for-preserving-specimen-or-tissue-for-later-dna-. Accessed 15 Feb 2020

Eiseman C (2019) Leaf miners of North America. 1st edn. June version, USA. http://charleyeiseman.com/leafminers/

Elliott B (2007) Rearing leaf miners. British leafminers. Newsletter. February 2007, Issue 2. http://www.leafmines.co.uk/pdfs/news2.pdf. Accessed 15 Feb 2020

Ellis W (2020) Plant parasites of Europe: leafminers, galls and fungi. The Netherlands. https://bladmineerders.nl/introduction/mines/practical/. Accessed 23 Jan 2020

Faeth SH (1991) Novel aspects of host tree resistance to leafminers. In: Baranchikov YN, Matison WJ, Hain FP et al. (eds) Forest insect guilds: patterns of interaction with host trees, US Dep Agric For Serv Gen Tech Rep NE-153, pp 219–239

Falkovich MI (1964) Casebarer moths (Lepidoptera, Coleophoridae) damaging larch in the USSR, their distribution and historical links with the host plants. Zool J 43(6):851–858

Fernandes GW, De Marco JP, Schönrogge K (2008) Plant organ abscission and the green island effect caused by gallmidges (Cecidomyiidae) on tropical trees. Arthropod-Plant Inte 2:93–99. https://doi.org/10.1007/s11829-008-9031-x

Guiguet A, Hamatani A, Amano T et al (2018) Inside the horn of plenty: leaf-mining micromoth manipulates its host plant to obtain unending food provisioning. PLoS One 13(12):e0209485. https://doi.org/10.1371/journal.pone.0209485

Guiguet A, Ohshima I, Takeda S et al (2019) Origin of gall-inducing from leaf-mining in a micromoth showing a mixed feeding strategy. Sci Rep 9:6794. https://doi.org/10.1038/s41598-019-43213-7

Gutzwiller F, Dedeine F, Kaiser W et al (2015) Correlation between the green-island phenotype and *Wolbachia* infections during the evolutionary diversification of Gracillariidae leaf-mining moths. Ecol Evol 18:4049–4062. https://doi.org/10.1002/ece3.1580

Hawkins BA (1994) Pattern and process in host-parasitoid interactions. Cambridge University Press, Cambridge

Hering EM (1951) Biology of the leaf miners. Junk's – Gravenhage

Inoue T (1990) Habitat selection of two fleabeetle species of the genus *Argopistes* (Coleoptera: Chrysomelidae) on their host trees. Jpn J Appl Entomol Z 34:217–226. https://doi.org/10.1303/jjaez.34.217. (in Japanese with English abstract)

Johansson R, Nielsen ES, van Nieukerken EJ et al (1990) The Nepticulidae and Opostegidae (Lepidoptera) of North West Europe. Fauna Entomologica Scandinavica, vol 23. Brill Academic Pub, Leiden

Kaiser W, Huguet E, Casas J et al (2010) Plant green-island phenotype induced by leaf-miners is mediated by bacterial symbionts. Proc Biol Sci 277:2311–2319. https://doi.org/10.1098/rspb.2010.0214

Kato M (1984) Mining pattern of the honeysuckle leaf-miner *Phytomyza lonicerae*. Res Popul Ecol 26:84–96

Kato M (1985a) Life history and mortality pattern of the honeysuckle leaf-miner *Phytomyza lonicerae* Robineau-Desboidy (Diptera, Agromyzidae). Jpn J Entomol 53:604–615

Kato M (1985b) The adaptive significance of leaf-mining pattern as an anti-parasitoid strategy: a theoretical study. Res Popul Ecol 27:265–275

Kato M (1994) Structure, organization, and response of a species-rich parasitoid community to host leafminer population dynamics. Oecoligia 97:17–25

Kirichenko NI (2014) Methodological approaches to the study of insects mining leaves of woody plants. Trans SPb State For Tech Acad 207:235–246

Kirichenko N, Triberti P, Mutanen M et al (2016) Systematics and biology of some species of *Micrurapteryx* Spuler (Lepidoptera, Gracillariidae) from the Holarctic region, with re-description of *M. caraganella* (Hering) from Siberia. Zookeys 579:99–156. https://doi.org/10.3897/zookeys.579.7166

Kirichenko NI, Petko VM, Magnoux E, Lopez-Vaamonde C (2017) Diversity and distribution of leaf-mining insects on Birches (*Betula* spp.) in Siberia. Entomol Rev 97:183–198. https://doi.org/10.1134/S0013873817020051

Kirichenko N, Triberti P, Akulov E et al (2019a) Exploring species diversity and host plant associations of leaf-mining micromoths (Lepidoptera: Gracillariidae) in the Russian Far East using DNA barcoding. Zootaxa 4652:1–55. https://doi.org/10.11646/zootaxa.4652.1.1

Kirichenko NI, Triberti P, Akulov EN et al (2019b) Novel data on the taxonomic diversity, distribution and host plants of leafmining moths of the family Gracillariidae (Lepidoptera) in Siberia based on DNA-barcoding. Entomol Rev 99(6):796–819. https://doi.org/10.1134/S0013873819060083

Kumata T (1964) Description of a new stem-miner of coniferous trees from Japan (Lepidoptera: Gracillariidae). Insecta Matsumurana 27(1):31–34

Langmaid JR, Corley MFV (2007) *Phyllocnistis ramulicola* sp. nov. (Lepidoptera: Gracillariidae) in England and Portugal. Entomologist's Gazette 58:227–237

Lees DC, Lack HW, Rougerie R et al (2011) Tracking origins of invasive herbivores using herbaria and archival DNA: the case of the horse-chestnut leaf miner. Front Ecol Environ 9:322–328. https://doi.org/10.1890/100098

Lees DC, Kawahara AY, Bouteleux O et al (2013) DNA barcoding reveals a largely unknown fauna of Gracillariidae leaf-mining moths in the Neotropics. Mol Ecol Resour 14:286–296. https://doi.org/10.1111/1755-0998.12178

Leppänen SA, Altenhofer E, Liston AD et al (2013) Ecological versus phylogenetic determinants of trophic associations in a plant-leafminer-parasitoid food web. Evolution 67:1493–1502. https://doi.org/10.1111/evo.12028

Lopez-Vaamonde C, Godfray HCJ, Cook JM (2003) Evolutionary dynamics of host plant use in a genus of leafmining moths. Evolution 57:1804–1821. https://doi.org/10.1554/02-470

Lopez-Vaamonde C, Godfray HCJ, West S et al (2005) The evolution of host use and unusual reproductive strategies in *Achrysocharoides* parasitoid wasps. J Evolution Biol 18:1029–1041. https://doi.org/10.1111/j.1420-9101.2005.00900.x

Lopez-Vaamonde C, Breman FC, Lees DC et al (2012) Analysis of tissue dependent DNA yield for optimal sampling of micro-moths in large-scale biodiversity surveys. Eur. J Entomol 109:1–6. https://doi.org/10.14411/eje.2012.001

Ohshima I (2005) Techniques for continuous rearing and assessing host preference of a multivoltine leaf-mining moth, *Acrocercops transecta* (Lepidoptera: Gracillariidae). Entomol Sci 8:227–228. https://doi.org/10.1111/j.1479-8298.2005.00120.x

Ohshima I (2008) Host race formation in the leaf-mining moth *Acrocercops transecta* (Lepidoptera: Gracillariidae). Biol J Linn Soc 93:135–145. https://doi.org/10.1111/j.1095-8312.2007.00949.x

Ohshima I (2012) Natural history of leaf miners, and the methods for collecting, rearing and preparing specimens of them. In: Kawakita A, Okuyama Y (eds) Biology of species interactions. Bun-ichi Sogo Shuppan Co., Tokyo, pp 331–356. (in Japanese)

Ohshima I (2013) Methods for studying micromoths and their observation. In: Nasu Y, Hirowatari T, Kishida Y (eds) The standard of moths in Japan IV. Gakken Education Publishing, Tokyo, pp 10–13. (in Japanese)

Ohshima I, Watanabe K, Kawamura T (2015) Distinct parasitoid communities associated with host races of the leaf-mining moth Acrocercops transecta on distantly related host plants (Juglandaceae and Ericaceae). J Nat Hist 49:815–828. https://doi.org/10.1080/00222933.2014.953613

Pappers SM, van der Velde G, Ouborg NJ (2002) Host preference and larval performance suggest host race formation in Galerucella nymphaeae. Oecologia 130:433–440. https://doi.org/10.1007/s00442-001-0822-3

Pincebourde S, Casas J (2015) Warming tolerance across insect ontogeny: influence of joint shifts in microclimates and thermal limits. Ecology 96:986–997. https://doi.org/10.1890/14-0744.1

Pitkin B, Ellis W, Plant C et al. (2019) The leaf and stem mines of British flies and other insects. http://www.ukflymines.co.uk/index.php. Accessed 23 Jan 2020

Potter DA (1992) Abundance and mortality of a specialist leafminer in response to experimental shading and fertilization of American holly. Oecologia 91:14–22. https://doi.org/10.1007/BF00317235

Salvo A, Valladares G (2004) Looks are important: parasitic assemblages of agromyzid leaf miners (Diptera) in relation to mine shape and contrast. J Anim Ecol 73:494–505. https://doi.org/10.1111/j.0021-8790.2004.00824.x

Valladares GR, Salvo A, Godfray HCJ (2001) Quantitative food webs of dipteran leafminers and their parasitoids in Argentina. Ecol Res 16:925–939. https://doi.org/10.1046/j.1440-1703.2001.00453.x

Chapter 18
Canopy Insect Sampling

Timothy Schowalter and Jung-Tai Chao

18.1 Introduction

Forest canopies may harbor half of the world's arthropod species (Erwin 1982; Ødegaard 2000; Stork 1987), currently estimated at seven million species globally (Stork 2018), as a result of complex three-dimensional structure and the variety of canopy food and habitat resources. The difficulty of canopy access restricted research on arthropod diversity until Beebe's pioneering work in Ecuador during the 1940s. Canopy access continues to be a limitation for study in many forests, and canopy access techniques will be addressed in this chapter. A variety of sampling methods are available for measuring insect diversity in forest canopies and will be described and compared in this chapter.

18.2 Canopy Access

Until the 1940s, forest canopies could only be observed from the ground, using binoculars, severely restricting measurement of arthropod diversity. Ground-based methods, such as shooting branches down with a shotgun, using long-handled nets and pruning poles to enclose branches (Parker III et al. 1993; Schowalter et al. 1981), and fogging the canopy with knockdown insecticides (Erwin 1982; Lowman 1984), have been widely used. However, a number of methods are now available for researchers to access the canopy directly.

T. Schowalter (✉)
Entomology Department, Louisiana State University, Baton Rouge, LA, USA
e-mail: tschowalter@agcenter.lsu.edu

J.-T. Chao
Taiwan Forestry Research Institute, Taipei, Taiwan, ROC

J. C. Santos, G. W. Fernandes (eds.), *Measuring Arthropod Biodiversity*,
https://doi.org/10.1007/978-3-030-53226-0_18

18.2.1 Ladders and Spurs

Beebe (1949) used rope ladders, hoisted into the canopy on lines shot over branches, to access tropical canopies in Ecuador, and was the first to introduce the general public to tropical canopy biodiversity with his popular book, *High Jungle*. A variety of rope, wooden, and aluminum ladders, some with interlocking sections to extend height, are available to reach forest canopies (Gunatilleke and Gunatilleke 1996; Mehra and Bawa 1968, 1969). Climbing spurs represent a similar means for accessing forest canopies (Mori 1984). Spurs are no longer recommended due to their bark penetration that can lead to inoculation of tree pathogens. Tree bicycles have seen limited use (Gentry 1991). All of these methods largely restrict sampling to areas reachable from the bole or to weight-bearing branches near the bole. In addition, the weight of materials and difficulty maneuvering rigid ladders or tree bicycles through dense understory typically restrict the number of trees sampled and their distance from roads or trails.

18.2.2 Rope Climbing

Rope-climbing techniques, originally designed for caving and mountaineering, have been adapted for tree climbing, making these techniques relatively safe and inexpensive (Perry 1986; Lowman 1999; Lowman et al. 2012). Rope climbing requires throwing or shooting a weighted monofilament line over a selected, or accessible, weight-bearing branch (Fig. 18.1). This step can be facilitated by using a crossbow with an attached fishing reel and line (Fig. 18.1a). A climbing rope is tied to the end of the monofilament and pulled over the branch. This technique is more physically demanding than most other canopy access techniques. For safety purposes, the ends of the climbing lines should be anchored, often around the base of the tree (Fig. 18.1b), and the condition of ropes, carabiners, and other climbing equipment checked regularly. The portability of necessary equipment for direct canopy access has allowed researchers with modest budgets to sample biodiversity throughout the canopy and to replicate sampling between and among different tree canopies (Fig. 18.1c). However, the perimeters of tree crowns remain inaccessible, because ropes must be anchored to weight-bearing branches. The double rope technique (DRT) is preferred over the single rope technique (SRT) because of improved safety of a climber (Lowman et al. 2012).

Fig. 18.1 Climbing techniques. (**a**) Using a crossbow to shoot a guide line over a canopy branch in preparation for rope climbing. (**b**) Initiation of climb, showing the climbing rope anchored around the base of the tree. (**c**) Climber near top of tree (in circle) preparing to collect a sample

18.2.3 Towers and Walkways

Towers (Fig. 18.2a) and canopy walkways (Fig. 18.2b) can be constructed in forests where construction equipment can be moved. Towers and walkways have now been constructed in forests around the world, largely for studying canopy-atmosphere interactions as part of global research networks such as FLUXNET and NEON or for ecotourism (Lowman et al. 2012). Walkways are constructed by connecting multiple towers and can vary in length and height in the canopy. In contrast to rope climbing and ladders, towers and walkways largely restrict sampling to branch ends of the relatively limited number of trees surrounding the tower or walkway. Furthermore, biodiversity in this zone may be affected by the tower "footprint" (altered environmental conditions in the open cylinder resulting from tower construction). Towers and walkways also provide access to forest canopies for monitoring disease vectors and other insect groups (Haddow et al. 1961; Schowalter et al. 2017). Towers and walkways also allow more researchers to work from the platform and to hoist heavy equipment into the canopy.

18.2.4 Balloons and Cranes

The following two canopy access techniques provide for sampling throughout the canopy volume of forests. However, these platforms are expensive and require collaborative research and funding.

Fig. 18.2 (**a**) Example of a tower for canopy sampling. (**b**) A canopy walkway

Hot-air balloons were introduced for canopy access by Hallé (1990) in French Guiana. An inflatable raft 27 m in diameter formed a platform that could be placed on top of the forest canopy and utilized as a base camp for research on the uppermost canopy (Basset et al. 2007). The hot-air balloon can be used to move the raft to new positions throughout the canopy and give researchers a camp for temporary residence while sampling. Researchers can rappel down to sample anywhere in the canopy volume. This technique facilitates the rapid collection of canopy arthropods or other organisms (Lowman et al. 1993, 2012).

Canopy construction cranes were pioneered at the Smithsonian Tropical Research Institute in Panama (Basset et al. 2007; Parker et al. 1992; Lowman et al. 2012). Cranes clearly must be located where road or river access permit construction of the base and erection of the tower that must extend above the canopy (Fig. 18.3). Cranes have now been constructed in a number of forests around the world. Construction cranes consist of a tower and a boom, with a trolley that moves a hook block holding the gondola, in which researchers can ride (Fig. 18.3b). An operator sitting in a cab at the top of the tower can rotate the boom 360° above the canopy and lower the gondola to any depth between the top of the canopy and the ground. Construction cranes provide access to the uppermost canopy as well as to any section of the understory that is within the circumference of the crane boom and provide access to trees beyond the tower footprint. Multiple researchers can work collaboratively and comparatively, both within and between crane sites (Lowman et al. 2012; Schowalter and Ganio 1998; Stork 2007).

Fig. 18.3 (**a**) Canopy crane at Wind River Canopy Research Facility in Washington. (**b**) Bagging canopy foliage for arthropod diversity from crane gondola. (Photo A by J.F. Franklin. Reprinted from Forest Canopies, 2nd Edition, Chapter 4, Vertical Organization of Canopy Biota, D. Shaw, Fig. 4-1, page 75, copyright 2004, with permission from Elsevier)

18.3 Sampling Methods

A variety of methods have been developed to sample canopy invertebrate diversity, but techniques vary in their level of accuracy (Bar-Ness et al. 2012; Basset et al. 2007, 2012; Lowman et al. 1996, 2012; Ozanne 2005) and representation of invertebrate taxa (Basset et al. 2012; Blanton 1990; Leather 2005; Majer and Recher 1988; Southwood 1978). Therefore, a method appropriate for groups of interest, or a combination of techniques to represent a broader diversity of groups, should be selected. Sampling techniques for invertebrates also differ in their ability to provide qualitative or quantitative data. Qualitative methods provide primarily diversity and relative abundance data. They do not represent specific units of area that can be used to calculate population density. Common methods include branch beating and interception traps. Quantitative sampling methods represent units of area that permit calculation of population density for comparison among experimental treatments or forest types. Quantitative methods include branch bagging, epiphyte mat sampling, and canopy fogging.

18.3.1 Branch Beating

Branch beating involves beating a foliated branch, using a heavy stick or bat, over a collecting sheet or pan (Fig. 18.4). Dislodged insects are collected from the sheet (Lowman 1985; Southwood 1978). An aspirator is useful for collecting small or fast-moving arthropods. Branch beating is most practical where the canopy access method leaves both hands free to hold the bat and sheet. This method can provide a somewhat quantitative estimate of population density if the beaten foliage area or branch length is measured.

18.3.2 Interception Traps

Various types of interception traps have been developed for sampling invertebrates. Malaise traps, flight interception traps, window traps, and light traps are the most commonly used. Bar-Ness et al. (2012) compared arthropod assemblages collected with sticky traps, Malaise traps, and hanging traps. The trap designs collected different assemblages. Sticky traps caught primarily Diptera, Hymenoptera, and Coleoptera; funnel traps caught Diptera, Hemiptera, and Coleoptera; and hanging traps caught Diptera, Coleoptera, and Lepidoptera. They rated sticky traps best and funnels worst for collecting canopy arthropods, based on relative cost, transport, durability, construction, placement, retrieval, sorting, and arthropod condition.

Interception traps can be suspended from ropes or attached to bole or branches at different heights to represent vertical distribution of arthropods (Hirao et al. 2007).

Fig. 18.4 (**a**) Branch beating using a beating stick and sheet. (**b**) Sample material in the sheet

An advantage of all these interception traps is that they collect passively over a designated time period (usually 5–15 days), so they represent a substantial portion of the diversity of insects flying both day and night. Furthermore, separate collecting jars on each side can provide information on direction from which collected insects arrived. A disadvantage is that these traps do not represent nonflying arthropods, nor sessile and internal (galling or mining) arthropods, which often are the most abundant and functionally important members of the canopy community (e.g., Schowalter and Ganio 1998, 2003; Schowalter et al. 2017).

18.3.2.1 Malaise Traps

Malaise traps are large mesh tents that funnel flying insects into a collecting jar at the peak of the tent (Fig. 18.5). This trap is particularly effective for collecting Diptera and Hymenoptera (Grimbacher and Stork 2007; Southwood 1978). The trap is named after Swedish entomologist, Dr. René Malaise, who invented the trap (Malaise 1937). Improved designs of the Malaise trap were later published by Townes (1962) and Gressitt and Gressitt (1962). The Townes design is the most commonly used (van Achterberg 2009). van Achterberg (2009) also introduced several modifications of the Townes-style Malaise trap.

Fig. 18.5 Setting up a Malaise trap

The Townes-style Malaise trap is open at two sides and consists of two short end mesh walls, one central mesh wall, a tilted fabric roof which is peaked on one end, and a collecting bottle attached to the outlet of the roof. For most commercially manufactured Malaise traps, the mesh walls are black and the roof is white. This design is based on the fact that when flying insects, particularly tiny ones, encounter an obstacle, such as the central mesh wall of the trap, they tend to climb up toward the white-colored roof to escape. On reaching the apex of the tent, the only way out is into the collecting bottle, which is filled with a killing agent such as ethanol. Collecting bottles should be emptied or changed every day, every week, or even up to a month, depending on the purpose of the study and the type of killing agent.

Malaise traps set for canopy studies are placed through the high branches under the canopy using two guide ropes attached to opposite ends of the frame within which the trap is erected (Fig. 18.6a). The size of Malaise traps makes positioning in the canopy problematic. It may take some time to set a canopy Malaise trap in an ideal location and position. The trap needs to be as close to the canopy as possible (Fig. 18.6b). In addition, canopy foliage can easily tear the fabric of the traps, especially during high winds. Raising or lowering the trap must be handled with care.

Proper placement of a Malaise trap is extremely important for optimal collecting efficiency. Improper placement may lower captures significantly in the same microhabitat (Malaise 1937; van Achterberg 2009). The aerial Malaise trap, like that set on the ground, should be placed perpendicular to an insect flyway. Modifications of the Townes-style Malaise trap, e.g., the Sea, Land, and Air Malaise (SLAM) trap (Fig. 18.7a) or the four-headed (i.e., four collecting bottles) SLAM trap (Fig. 18.7b), collect flying insects from four sides, making placement, wind direction, and exposure to sunlight less of a concern although they may catch fewer insects (see Matthews and Matthews 1983).

18.3.2.2 Flight Interception Traps

Flight interception traps are simpler in designs but similar in function to Malaise traps. These traps are constructed of a fabric, glass, or clear plastic panel that is either coated with an adhesive to trap insects flying into the panel or suspended over collecting pans filled with preservative (Fig. 18.8) (Southwood 1978). Solid panels exploit the tendency for many flying Coleoptera and some other insect groups to fall when they hit an obstacle. Falling insects are collected in the pans. The diversity of insects collected can be increased by coating the panel with adhesive to retain all insects landing on the surface. However, many small, soft-bodied insects, especially flies and moths, are damaged by the adhesive, reducing the quality of specimens for identification. Attention must be paid to the weight of the preservative-filled pans, to the rate of evaporation, and to the toxicity of the preservative, which may attract thirsty vertebrates. Propylene glycol (animal-safe antifreeze) is preferred as a preservative. Traps generally are checked every 5–15 days, depending on evaporation rate.

Fig. 18.6 Malaise trap suspended in canopy. (**a**) Viewed from ground. (**b**) Viewed laterally

Fig. 18.7 A Sea Land and Air Malaise (SLAM) trap collects flying insects from four directions. (**a**) Trap ready to be raised into the canopy. (**b**) Detail of four bottles which collect flying insects from four directions. (Source: MegaView Science Co., Ltd., Taichung, Taiwan. Photo courtesy of Dr. S.-C. Kang)

Fig. 18.8 Flight interception traps attached on a frame to represent insects flying at different heights

18.3.2.3 Window Traps

Window traps are flight interception traps that consist of a roof, two transparent vanes that intersect vertically, and a yellow plastic collecting tray (Fig. 18.9a). Window traps are designed to intercept a wide range of flying insects that collide with transparent vanes (the "windows") or attracted by the color of the tray or the reflection of the vanes and subsequently drop into the collecting tray. The collecting tray is filled with water and a few drops of detergent so that the falling insects drown. The roof of the trap covers the whole trap and prevents water intrusion. The overflow hole of the plastic tray prevents it from overflowing and potentially flushing out the collected insect samples. Window traps are suspended by rope from a canopy pulley (Fig. 18.9b). Trapping intervals depend on the purpose of the study. Traps may be emptied at daily intervals during each sampling event. In each sampling event, the water containing the sample is poured through a sieve, lined with a piece of very fine nylon gauze. The collected sample is then wrapped in nylon gauze and stored in a pre-labeled plastic bottle filled with 80% ethanol.

18.3.2.4 Light Traps

Light traps consist of a UV or fluorescent light that can be suspended over a collecting pan (Fig. 18.10a) or placed against a white sheet or panel that funnels attracted insects into a collecting jar (Lowman 1985; Southwood 1978). Nocturnal flying insects attracted to the light source (Fig. 18.10b) can be collected either from the sheet or the jar (Hirao et al. 2007). Lights can be powered by portable generator or battery, requiring attention to support of these heavy items and/or electric cabling in the canopy.

18.3.3 Branch Bagging

Branch bagging (Fig. 18.11) involves slipping a closable plastic bag over a foliage-bearing branch to collect invertebrates (Blanton 1990; Majer and Recher 1988; Schowalter and Ganio 2003). Accessible branches can be bagged by climbing into the canopy (Fig. 18.2), from a tower, walkway, or crane gondola (Fig. 18.3) or from the ground using a long-handled net (Fig. 18.11). Samples can be collected at measured heights, most easily from towers or cranes, to evaluate the effect of crown height and environmental conditions on invertebrates (Schowalter and Ganio 1998). Long-handled nets can reach branch ends at heights of 10–15 m, depending on slope angle. To collect samples using a long-handled net, a drawstring is inserted around the rim of a 50-liter plastic bag (Fig. 18.12a, b), the bag is attached to the rim of a net (Fig. 18.12c), using binder clips, at the end of a telescoping or extendable pole, the net is slipped quickly over a foliage-bearing branch (Fig. 18.11), the bag is

Fig. 18.9 Window trap. (**a**) Construction of window trap. (**b**) Window trap suspended in lower canopy

Fig. 18.10 (**a**) Light trap construction. (**b**) Light trap turned on at night for collection

closed with the drawstring, and the branch is clipped for sorting of invertebrates in the lab.

This technique represents abundances of nonflying scale insects, gall formers, and leaf miners that are missed by flight interception techniques, and it provides quantitative estimates (number of insects per unit leaf area or mass) of abundances of invertebrate species most closely associated with the particular tree species sampled. This technique has maximum mobility among trees, but elevating a long pole through dense understories may limit use in some forests, and only accessible branch ends can be sampled. A 2-m pole that telescopes to 8–10 m facilitates use in dense forests. Furthermore, this technique provides only a snapshot of invertebrates in the sample at the time of sampling and yields underestimates of the abundances of highly mobile (especially flying) invertebrates. Nocturnal sampling is necessary to represent nocturnal invertebrates (Schowalter and Ganio 2003). Foliage in samples also can be used to estimate the effect of herbivory, i.e., from measurement of leaf area missing.

18.3.4 Canopy Knockdown

Canopy knockdown is a way to collect arthropods by infusing the canopy with pyrethrins (botanical insecticides derived from chrysanthemum flowers) or pyrethroids (synthetic compounds derived from pyrethrins) that immobilize and knock down unattached arthropods inhabiting the canopy (Blanton 1990; Dial et al. 2006; Erwin 1982, 1995; Floren 2010; Floren and Schmidl 2008; Gering et al. 2007; Majer and Recher 1988; Paarmann and Stork 1987; Roberts 1973; Stork et al. 1997). Fogging

Fig. 18.11 Branch bagging using long-handled insect net

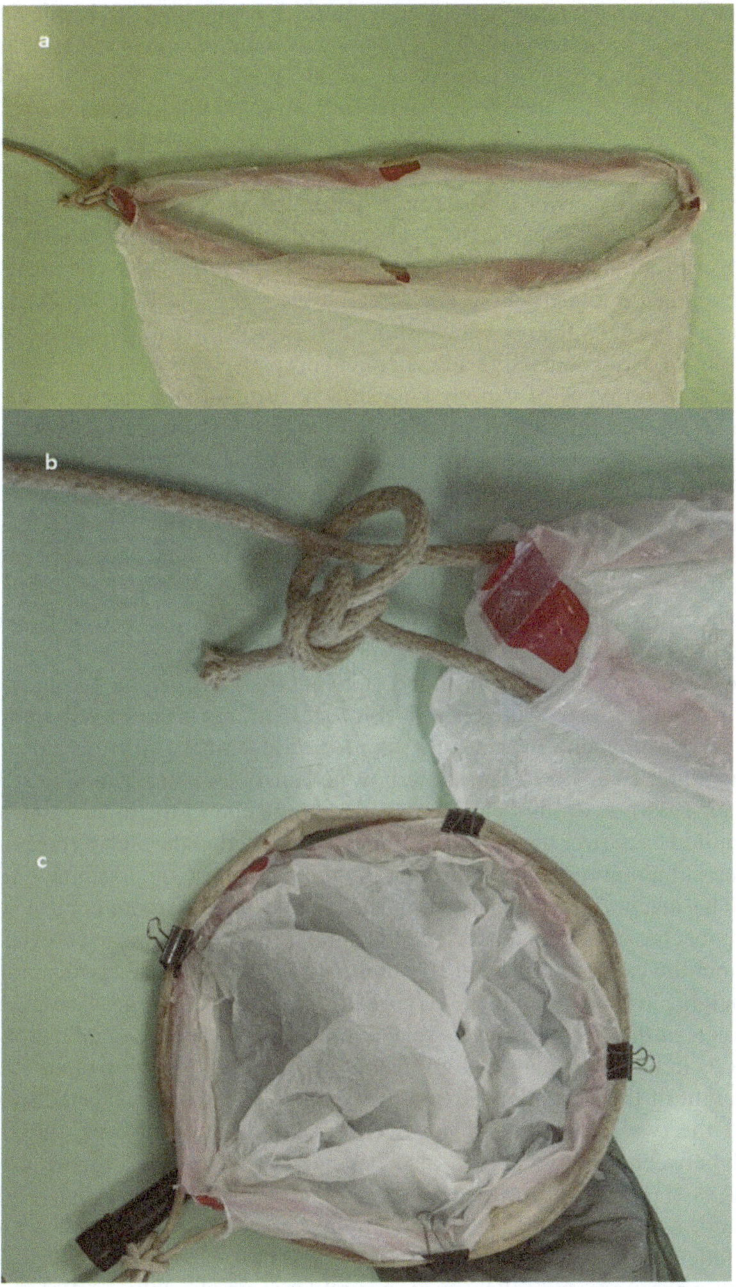

Fig. 18.12 (**a**) Plastic bag with drawstring inserted. (**b**) Detail of slip knot to close bag. (**c**) Attachment of bag to net rim using binder clips

machines or backpack misters are used to produce a hot cloud of chemical droplets which remain suspended until they evaporate. Smaller droplets vaporize first, and larger droplets disperse in a column of still air, gradually becoming smaller until they vaporize. Fogging, therefore, can reach the tops of tall trees, such as 70-m-tall trees in the tropics. In some cases, synergistic piperonyl butoxide is added to enhance the effectiveness of the knockdown compound. Arthropods knocked down by pyrethrins may recover later through detoxification mechanisms (Erwin 1990). In addition, these chemicals break down quickly in direct sunlight and are relatively safe to use in a natural environment. Nevertheless, use of these chemicals should follow manufacturer instructions and research site protocols. Researchers and assistants engaged in canopy knockdown should wear protective clothing and take other precautions to avoid contacting any chemical droplets.

There are two types of canopy knockdown: fogging and misting (Stork 1988, Ozanne 2005). Fogging releases particles averaging less than 50 microns in diameter, whereas misting releases particles averaging 50–100 microns in diameter. The smaller particle size of fog increases the rate of evaporation (especially under conditions of low humidity) and of dispersal of the fog away from a desired spatial column (especially under windy conditions). Therefore, fogging should be conducted under windless conditions, generally at night. In contrast, the larger and heavier mist particles take much longer to evaporate and are less likely to drift away from the target tree or air column (Chey et al. 1998).

Ropes and pulleys must be installed on target trees to hoist the fogging machine or backpack mister to a desired height (Fig. 18.13). This is done by shooting a lead weight and fishing line over a high, stout central branch with the aid of a line-throwing gun (Erwin 1983), crossbow, bow and arrow, or sling (Kitching et al. 1993, Adis et al. 1998). A strong nylon rope is tied to the end of the fishing line so that it can be pulled back over the branch. A pulley is attached to the end of the nylon rope, and a second nylon rope is run through the pulley to raise the machine (Adis et al. 1998). The machine can be started on the ground or by remote control when the machine reaches a desired height. The machine spins on the rope and creates a cloud throughout the canopy (Toda and Kitching 1999). Two separate applications at different heights in each plot may be required for adequate coverage. The first application targets arthropods in lower vegetation, and the second application targets arthropods in the high canopy after raising the machine (Kitching et al. 1993). The exact height of the spray depends on the purpose of the study, but the spatial volumes of the two sprays should not overlap. The duration of each spray and the interval between the two sprays also depend on the purpose of the study. The machine is shut off and lowered to the ground after spraying. After the spray is completed, workers should leave the sampling area for 30 min and then return to collect fallen arthropods.

Fallen arthropods are collected at ground level on sheets (e.g., Roberts 1973), cloth funnels (e.g., Gagné 1979), cloth trays (e.g., Southwood et al. 1982), suspended funnel trays (e.g., Erwin 1983), or suspended funnels, with a collecting bottle or vial containing 75–80% ethanol attached to each funnel (Fig. 18.14a) (e.g., Erwin 1990; Kitching et al. 1993; Toda and Kitching 1999). A 10 × 10 m plot can

Fig. 18.13 Insecticidal mister for canopy fogging

be used to define a "column" of canopy to be sampled (Kitching et al. 1993). A "spider web" (Adis et al. 1998) or cat's cradle (Kitching et al. 1993, Toda and Kitching 1999) of nylon ropes is erected at head height over the area of the plot (Fig. 18.14b), and collecting sheets or funnels covering at least 75% of the canopy area are hung throughout the plot (Adis et al. 1998; Kitching et al. 1993; Toda and Kitching 1999). This allows arthropod abundance to be recorded as density (number/m^2). Arthropods on the sheets or funnels must be brushed into the vials within 3–4 h after each spraying event, before they recover from the knockdown (Kitching et al. 1993, Toda and Kitching 1999). Samples are then collected for sorting and identification. Useful suggestions and cautions for applying canopy misting in the field are provided by Adis et al. (1998) and Toda and Kitching (1999).

18.3.5 Bait Stations

More focused sampling of target insect groups may involve use of attractants to lure relatively dispersed individuals to bait stations. For example, Roubik (1993) used traps baited with cineole to compare diversity and abundances of euglossine bees at different canopy heights in tropical rainforest surrounding a canopy crane. Heatwole (1989) used various attractants to measure diversity and distribution of ant species

Fig. 18.14 (**a**) Collecting funnels suspended above ground. (**b**) Initiation of canopy fogging

Fig. 18.15 Banana trap suspended in canopy for attraction of butterflies and flies

along a forest gradient. Malaise traps baited with ripe or rotting fruit can be used to attract butterflies or flies (Fig. 18.15) (e.g., Bossart et al. 2006).

18.3.6 Specialized Habitats

Ellwood and Foster (2004) and Richardson et al. (2000) reported that the arthropods associated with canopy ferns, bromeliads, and phytotelmata (Figs. 18.16 and 18.17), respectively, represented unique subcommunities in tropical forest canopies. Similarly, perched soils provide habitat for unique arboreal specialists representing

Fig. 18.16 Canopy bromeliads and perched soils can be sampled for associated arthropods

soil/litter fauna (Lindo and Winchester 2007). Arthropods associated with these specialized canopy habitats can be extracted from these habitats for diversity measurement using a variety of methods.

Ferns and bromeliads can be turned upside down and shaken or fumigated over a container to collect dislodged arthropods. Foliage may be flushed with water to extract remaining specimens. Plants also can be dismantled leaf by leaf and washed to ensure that all arthropods are collected (Richardson et al. 2000).

Soil/litter species are commonly extracted from samples using Berlese funnels. This method involves placing a sample in a funnel suspended under a light/heat source and collecting invertebrates moving downward from the adverse conditions into a collector placed under the funnel (Lindo and Winchester 2007).

18.4 Comparison of Methods

Sampling methods differ in their representation of various invertebrate taxa, as noted above, and ability to represent or replicate canopies over large areas. Several studies have compared sampling methods in terms of their representation of canopy invertebrate diversity and abundance.

Fig. 18.17 Canopy phytotelmata can be sampled for associated arthropods

Blanton (1990), Majer and Recher (1988), and Ozanne (2005) compared branch bagging and canopy knockdown methods and concluded that branch bagging provided better representation of resident arthropod biomass (including scale insects, gall formers, and leaf miners), but canopy knockdown provided better representation of arthropod species richness. Depending on the degree of complementarity (overlap of species representation) of sampling methods, combinations of methods can improve efficiency of canopy invertebrate sampling (Basset et al. 2007, 2012). For example, Longino and Colwell (1997) reported that Berlese funnels, Malaise traps, and canopy fogging individually were equally efficient in capturing ants but differed in the composition of ant species collected. Combining Berlese samples with either Malaise or fogging methods improved the efficiency of representation of total ant diversity, compared to individual methods, because the species sampled were highly complementary. Combining Malaise and fogging methods did not improve efficiency because of high overlap (low complementarity) of the faunas collected. Basset et al. (2007) used 14 different collecting methods at 12 sites in a tropical forest in Panama to collect about 4000 species.

Independent replication for representation of canopies over large areas or to compare treated and untreated canopies is easier with some methods than with others. Multiple samples within trees do not represent independent replicates and should be pooled for statistical analyses (Hurlbert 1984). Sampling from towers or walkways provides for limited replication of crowns accessible from these platforms, unless multiple towers or walkways can be sampled. Similarly, canopy knockdown and light trapping typically are limited to a few trees near access roads. Greater mobility and replication are provided by branch bagging (using a telescoping, long-handled net), Malaise traps, flight interception traps, and window traps.

A few studies have addressed effects of insect outbreaks, disturbances of forest harvest, fragmentation, or other management practices on canopy insect diversity (e.g., Schowalter 1995; Schowalter et al. 2017). Some studies that have compared multiple samples within a single pair of treated and untreated plots suffer from pseudoreplication (Hurlbert 1984). Such studies provide no degrees of freedom for treatment comparison. Evaluation of treatment effects on biodiversity requires independent replication, i.e., multiple pairs of treated and untreated plots or geographically intermixed plots representing all treatments.

Acknowledgments Use of canopy cranes was provided by the Wind River Canopy Crane Research Facility, operated by the US Forest Service and University of Washington, and the Smithsonian Tropical Research Institute in Panama. The malaise trap, window trap, and canopy knockdown equipment used for canopy study by JT Chao were kindly provided by the Center for Ecological Research, Kyoto University, Secretariat of DIVERSITAS in the Western Pacific and Asia (DIWPA).

References

Adis J, Basset Y, Floren A et al (1998) Canopy fogging of an overstory tree – recommendations for standardization. Ecotropica 4:93–97

Bar-Ness YD, McQuillan PB, Whitman M et al (2012) Sampling forest canopy arthropod biodiversity with three novel minimal-cost trap designs. Aust J Entomol 51:12–21. https://doi.org/10.1111/j.1440-6055.2011.00836.x

Basset Y, Corbara B, Barrios H et al (2007) IBISCA-Panama, a large-scale study of arthropod beta-diversity and vertical stratification in a lowland rainforest: rationale, study sites and field protocols. Bull Inst R Sci Nat Belg Entomol 77:39–69

Basset Y, Cizek L, Cuénoud P (2012) Arthropod diversity in a tropical forest. Science 338:1481–1484. https://doi.org/10.1126/science.1226727

Beebe W (1949) High jungle. Duell, Sloan and Pearce Publishers, New York

Blanton CM (1990) Canopy arthropod sampling: a comparison of collapsible bag and fogging methods. J Agric Entomol 7:41–50

Bossart JL Opuni-Frimpong E, Kuudaar S, Nkrumah S (2006) Richness, abundance and complementarity of fruit-feeding butterfly species in relict sacred forests and forest reserves of Ghana. Biodiv Cons 15:333–359

Chey VK, Holloway JD, Hambler C et al (1998) Canopy knockdown of arthropods in exotic plantations and natural forest in Sabah, north-east Borneo, using insecticidal mist-blowing. Bull Entomol Res 88:15–24. https://doi.org/10.1017/S0007485300041511

Dial R, Ellwood MDF, Turner EC, Foster WA (2006) Arthropod abundance, canopy structure, and microclimate in a Bornean lowland tropical rain forest. Biotropica 38:643–652. https://doi.org/10.1111/j.1744-7429.2006.00181.x

Ellwood MDF, Foster WA (2004) Doubling the estimate of invertebrate biomass in a rainforest canopy. Nature 429:549–551

Erwin TL (1982) Tropical forests: their richness in Coleoptera and other arthropod species. Coleop Bull 36:74–75. https://www.jstor.org/stable/4007977

Erwin TL (1983) Tropical forest canopies, the last biotic frontier. Bull Entomol Soc Am 29(1):14–19. https://doi.org/10.1093/besa/29.1.14

Erwin TL (1990) Canopy arthropod biodiversity: a chronology of sampling techniques and results. Rev Peru Entomol 2:71–77

Erwin TL (1995) Measuring arthropod biodiversity in the topical forest canopy. In: Lowman MD, Nadkarni NM (eds) Forest canopies. Academic Press, San Diego, pp 109–127

Floren A (2010) Sampling arthropods from the canopy by insecticidal knockdown. In: Eymann J, Degref J, Häuser C et al (eds) Manual on field recording techniques and protocols for all taxa biodiversity inventories, vol 8, part 1. ABC Taxa, pp 158–172

Floren A, Schmidl J (eds) (2008) Canopy arthropod research in Europe: basic and applied studies from the high frontier. Bioform Entomology & Equipment, Nűrnberg, DE

Gagné WC (1979) Canopy-associated arthropods in *Acacia kea* and *Metrosideros* tree communities along an altitudinal transect on Hawaii Island. Pac Insects 21:56–82

Gentry AH (1991) The distribution and evolution of climbing plants. In: Putz FE, Mooney HA (eds) The biology of vines. Cambridge University Press, Cambridge, pp 3–42. https://doi.org/10.1017/CBO9780511897658.003

Gering JC, DeRennaux KA, Crist TO (2007) Scale dependence of effective specialization: its analysis and implications for estimates of global insect species richness. Divers Distrib 13:115–125. https://doi.org/10.1111/j.1366-9516.2006.00301.x

Gressitt JL, Gressitt MK (1962) An improved Malaise trap. Pac Insects 4:87–90

Grimbacher PS, Stork NE (2007) Vertical stratification of feeding guilds and body size in beetle assemblages from an Australian tropical rainforest. Aust Ecol 32:77–85. https://doi.org/10.1111/j.1442-9993.2007.01735.x

Gunatilleke N, Gunatilleke S (1996) Sinharaja: world heritage site, Sri Lanka/Sri Lanka. Natural Resources, Energy & Science Authority, NARESA, Colombo, Sri Lanka

Haddow AJ, Corbet PS, Gilett JD (1961) Studies from a high tower in Mpanga Forest, Uganda. Trans R Entomol Soc London 113:249–368. https://doi.org/10.1111/j.1365-2311.1961.tb02295.x

Hallé F (1990) A raft atop the rain forest. Natl Geogr 178:129–138

Heatwole H (1989) Changes in ant assemblages across an arctic treeline. Rev d'Entomol du Quebec 34:10–22

Hirao T, Murakami M, Kashizaki A, Tanabe S-I (2007) Additive apportioning of lepidopteran and coleopteran species diversity across spatial and temporal scales in a cool-temperate deciduous forest in Japan. Ecol Entomol 32:627–636. https://doi.org/10.1111/j.1365-2311.2007.00913.x

Hurlbert SH (1984) Pseudoreplication and the design of ecological field experiments. Ecol Monogr 54:187–211

Kitching RL, Bergelson JM, Lowman MD et al (1993) The biodiversity of arthropods from Australian rainforest canopies: general introduction, methods, sites and ordinal results. Aust J Ecol 18:181–191. https://doi.org/10.1111/j.1442-9993.1993.tb00442.x

Leather SR (ed) (2005) Insect sampling in forest ecosystems. Blackwell Science, Malden, MA

Lindo Z, Winchester NN (2007) Oribatid mite communities and foliar litter decomposition in canopy suspended soils and forest floor habitats of western red cedar forests, Vancouver Island, Canada. Soil Biol Biochem 39:2957–2966. https://doi.org/10.1016/j.soilbio.2007.06.009

Longino JT, Colwell RK (1997) Biodiversity assessment using structured inventory: capturing the ant fauna of a tropical rain forest. Ecol Appl 7:1263–1277. https://doi.org/10.1890/1051-0761(1997)007[1263:BAUSIC]2.0.CO;2

Lowman MD (1984) An assessment of techniques for measuring herbivory: is rainforest defoliation more intense than we thought? Biotropica 16:264–268. https://doi.org/10.2307/2387934
Lowman MD (1985) Spatial and temporal variability in herbivory of Australian rain forest canopies. Aust J Ecol 10:7–14
Lowman MD (1999) Life in the treetops. Yale University Press, New Haven
Lowman M, Moffett M, Rinker HB (1993) A new technique for taxonomic and ecological sampling in rain forest canopies. Selbyana 14:75–79. https://www.jstor.org/stable/41760420
Lowman MD, Kitching RL, Carruthers G (1996) Arthropod sampling in Australian subtropical rain forest: how accurate are some of the more common techniques? Selbyana 17:36–42. https://www.jstor.org/stable/41759922
Lowman MD, Schowalter TD, Franklin JF (2012) Methods in forest canopy research. University of California Press, Berkeley, CA
Majer JD, Recher HF (1988) Invertebrate communities on Western Australian eucalypts - a comparison of branch clipping and chemical knockdown procedures. Aust J Ecol 13:269–278. https://doi.org/10.1111/j.1442-9993.1988.tb00974.x
Malaise R (1937) A new insect-trap. Entomol Tidskr 58:148–160
Matthews RW, Matthews JR (1983) Malaise traps: the Townes model catches more insects. Contrib Am Entomol Inst 20:428–443
Mehra PN, Bawa KS (1968) B-chromosomes in some Himalayan hardwoods. Chromosoma 25:90–95. https://doi.org/10.1007/BF00338235
Mehra PN, Bawa KS (1969) Chromosomal evolution in tropical hardwoods. Evolution 23:466–481. https://doi.org/10.1111/j.1558-5646.1969.tb03529.x
Mori SA (1984) Use of Swiss tree grippers for making botanical collections of tropical trees. Biotropica 16:79–80. https://doi.org/10.2307/2387899
Ødegaard F (2000) The relative importance of trees versus lianas as hosts for phytophagous beetles (Coleoptera) in tropical forests. J Biogeogr 27:283–296. https://doi.org/10.1046/j.1365-2699.2000.00404.x
Ozanne CMP (2005) Techniques and methods for sampling canopy insects. In: Leather SR (ed) Insect sampling in forest ecosystems. Blackwell, Oxford, pp 146–165
Paarmann W, Stork NE (1987) Canopy fogging, a method of collecting living insects for investigation of life history strategies. J Nat Hist 21:563–566. https://doi.org/10.1080/00222938700770341
Parker TA III, Holst BK, Emmons LH et al (1993) A biological assessment of the Columbia River forest reserve, Toledo District, Belize. Conserv. Intl. Report 381
Parker GG, Smith AP, Hogan KP (1992) Access to the upper forest canopy with a large tower crane. Bioscience 42:664–670. https://doi.org/10.2307/1312172
Perry DR (1986) Life above the jungle floor. Simon & Schuster, New York
Richardson BA, Rogers C, Richardson MJ (2000). Nutrients, diversity, and community structure of two phytotelm systems in a lower montane forest, Puerto Rico. Ecol Entomol 25:348–356
Roberts HR (1973) Arboreal Orthoptera in the rain forest of Costa Rica collected with insecticide: a report on the grasshoppers (Acrididae) including new species. Proc Acad Nat Sci Phila 125:46–66. https://www.jstor.org/stable/4064682
Roubik DW (1993) Tropical pollinators in the canopy and understory: field data and theory for stratum "preferences". J Ins Behav 6:659–673. https://doi.org/10.1007/BF01201668
Schowalter TD, Ganio LM (1998) Vertical and seasonal variation in canopy arthropod communities in an old-growth conifer forest in southwestern Washington, USA. Bull Entomol Res 88:633–640. https://doi.org/10.1017/S0007485300054304
Schowalter TD, Ganio LM (2003) Diel, seasonal and disturbance-induced variation in invertebrate assemblages. In: Basset Y, Novotny V, Miller S et al (eds) Arthropods of tropical forests. Cambridge University Press, Cambridge, pp 315–328
Schowalter TD, Webb JW, Crossley DA Jr (1981) Community structure and nutrient content of canopy arthropods in clearcut and uncut forest ecosystems. Ecology 62:1010–1019. https://doi.org/10.2307/1937000

Schowalter TD (1995) Canopy arthropod communities in relation to forest age and alternative harvest practices in western Oregon. For Ecol Manage 78:115–125
Schowalter TD, Willig MR, Pressley SJ (2017) Post-hurricane successional dynamics in abundance and diversity of canopy arthropods in a tropical rainforest. Environ Entomol 46:11–20. https://doi.org/10.1093/ee/nvw155
Southwood TRE (1978) Ecological methods with particular reference to the study of insect populations. Chapman and Hall, London
Southwood TRE, Moran VC, Kennedy CEJ (1982) The assessment of arboreal insect fauna: comparisons of knockdown sampling and faunal lists. Ecol Entomol 7:331–340. https://doi.org/10.1111/j.1365-2311.1982.tb00674.x
Stork NE (1987) Guild structure of arthropods from Bornean rain forest trees. Ecol Entomol 12:69–80. https://doi.org/10.1111/j.1365-2311.1987.tb00986.x
Stork NE (1988) Insect diversity: facts, fiction and speculation. Biol J Linn Soc 35:321–337. https://doi.org/10.1111/j.1095-8312.1988.tb00474.x
Stork NE (2007) Australian tropical forest canopy crane: new tools for new frontiers. Aust Ecol 32:4–9. https://doi.org/10.1111/j.1442-9993.2007.01740.x
Stork NE (2018) How many species of insects and other terrestrial arthropods are there on Earth? Annu Rev Entomol 63:31–45
Stork NE, Adis J, Didham RK (eds) (1997) Canopy arthropods. Chapman & Hall, London
Toda MJ, Kitching RL (eds) (1999) Forest ecosystems: the assessment of plant and animal biodiversity in forest ecosystems. The manual. DIVERSITAS International Biodiversity Observation Year
Townes HK (1962) Design for a Malaise trap. Proc Entomol Soc Wash 64:253–262
van Achterberg K (2009) Can Townes type Malaise traps be improved? Some recent developments. Entomol Berichten 69(4):129–135

Chapter 19
Sampling Methods for Soil and Litter Fauna

Grizelle González, Maria Fernanda Barberena-Arias, Wei Huang, and Claudia M. Ospina-Sánchez

19.1 Importance and Challenges in the Study of Soil and Litter Fauna

Shaping aboveground biodiversity, regulating the functioning of terrestrial ecosystems, and influencing the ecological and evolutionary response of ecosystems to environmental change are some of the major roles of belowground communities revealed by recent advances in soil biodiversity research (Bardgett et al. 2005; Bardgett and Wardle 2010; Bardgett and van der Putten 2014). Soils host high biodiversity, with significant control over aboveground ecosystems. Belowground decomposers drive essential ecosystem functions, such as organic matter turnover and nutrient cycling (Wardle et al. 2004; Bardgett and Wardle 2010), and are therefore key determinants of soil fertility and nutrient uptake by plants (Coleman et al. 2004; Wardle et al. 2004; Bardgett and Wardle 2010; Culliney 2013; Menta and Remelli 2020). The soil is a key resource for several ecosystem and biosphere processes. These processes include plant production, cycling of organic matter and nutrients, storage of carbon and water, and release of nitrous oxides, carbon dioxide, and methane (Swift et al. 1979; Lavelle and Spain 2001).

Soil biology is challenged to understand the specific role of organisms, how they are organized in niches along soil resources, and how they respond to changing environments. Knowledge of soil species is difficult because of their small size and

G. González (✉) · C. M. Ospina-Sánchez
USDA-FS, International Institute of Tropical Forestry, Río Piedras, Puerto Rico
e-mail: grizelle.gonzalez@usda.gov

M. F. Barberena-Arias
School of Natural Sciences and Technology, Universidad Ana G. Méndez, Gurabo, Puerto Rico

W. Huang
Department of Environmental Sciences, Nanjing Forestry University, College of Biology and the Environment and University of Puerto Rico, Río Piedras, Puerto Rico

J. C. Santos, G. W. Fernandes (eds.), *Measuring Arthropod Biodiversity*,
https://doi.org/10.1007/978-3-030-53226-0_19

large biodiversity (Anderson 1975; André et al. 1994). Populations of small invertebrate animals spending most of their lifecycle in the soil are extremely large, reaching numbers as high as a quarter of a million per square meter (Fletcher 1976). It is estimated there are more than seven million unique terrestrial arthropods species living on Earth (Stork 2018). Yet this number could be extrapolated to ten million when animals of less than 200 μ are included (May 1990). This great range in species estimates can be attributed to the lack of knowledge of small arthropods (May 1988, 1994; André et al. 1994; Eisenhauer et al. 2017).

High numbers of individuals and species belonging to all terrestrial phyla can be found in soil and litter environments (e.g., Rotifera, Annelida, Mollusca, Tardigrada, Nematoda, and Arthropoda). Usually Arthropoda and Nematoda have the highest diversity. Indeed, in terms of species richness, arthropods may represent as much as 85% of the soil fauna (Culliney 2013). Representatives of all the arthropodan subphyla and of all their terrestrial classes can be found in the soil: Cheliceromorpha (scorpions, pseudoscorpions, spiders, harvestmen, and mites), Crustacea (amphipods and woodlice), Myriapoda (centipedes and millipedes), and Hexapoda (which include Collembola, Protura, and Diplura) and 20 out of the 26 orders of ectognathous hexapoda (true insects) (Greenslade 1985). Thus, the term microarthropod refers to an artificial grouping of the small arthropods that range in size from about 100 μm to a few millimeters and include the mites (Acari), Collembola, Symphyla, Protura, Diplura, Pauropoda, small centipedes and millipedes, and small insects from several orders (Coleman et al. 1999). Microarthropods include representatives of most trophic groups within the belowground food web (Walter et al. 1987; Moore et al. 1988).

Given the difficulty in the identification of all the species present in soil, many studies use functional groups divisions and abundance or species richness parameters to understand their role in ecosystems (Bengtsson 1998; Wolters 2001). Groups of soil fauna separated by size are also useful to understand their role in soil ecosystems. Classification by body size has widespread repercussions on sampling and study of the different groups. Most authors (e.g., Wallwork 1970) differentiate three size classes: micro-, meso-, and macrofauna based on body width with microfauna (<100 μm) and mesofauna (100 μm–2 mm) that participate as regulators in microorganisms' activities and macrofauna (>2 mm), which create microhabitats for other soil biota by reworking the soil (Wallwork 1970; Brussaard 1998). Other classifications include their role in ecosystem processes as they interact with microorganisms and function as micro-predators, litter transformers, and ecosystem engineers (Lavelle 1996). Microarthropods (definition from Coleman et al. 1999) can be identified to species level in addition to their characterization in functional groups (Potapov et al. 2016). Identification to species level makes comparisons among species within functional groups possible and can help determine how different species respond to environmental factors (Ponge and Salmon 2013; Potapov et al. 2016; Coyle et al. 2017).

Diversity estimates of soil and litter fauna that can be generalized are difficult to achieve because their belowground habitats are complex and vary in physicochemical characteristics that give rise to methodological challenges (Moldenke 1994). Results are affected by the mechanisms used to recover organisms from environmental samples. Thus, the type of sampling system used for any ecological investigation must be carefully planned as an integral part of the whole study (Macfadyen 1962); aas no one method can extract all soil and litter arthropods with equal efficiency from all types of soil and, it is most unlikely that any of the extraction methods currently available will extract any group or species of arthropod from soil and litter with complete efficiency (Fletcher 1976).

Rhea-Fournier and González (2017) discuss the plethora of challenges faced in soil research and argue that no singular method or tool is a panacea to the difficulties that may arise. Similarly, in this book chapter, we review numerous methodologies in the faunal toolbox of techniques and approaches to evaluate and quantify the influences of soil and litter fauna although qualitative techniques are also mentioned. In the following sections, we review passive and dynamic methods in the study of soil and litter fauna and provide methodological considerations for ecological studies of soil fauna with emphasis on the study of decomposition processes.

19.2 Methods in the Study of Soil and Litter Fauna

It is nearly impossible to count all the animals per unit area of soil in a site, so populations must be estimated from the numbers recovered from small samples of soil (Fletcher 1976). These samples can be extracted from samples of litter, cores, or cubes of soil of known surface area or volume. The numbers of animals recovered from the samples can then be converted into populations per unit area of soil at a known depth (usually 25 or 30 cm) providing the number and size of samples taken is adequate. Soil fauna can be extracted from a soil and litter sample either through passive (also known as mechanical) methods or via the use of dynamic (behavior) modification techniques. Mechanical methods are generally labor-intensive and work best on compact clay soils with low organic matter content, where meso- and macro-arthropods are most efficiently extracted (Moldenke 1994). Inducing the biota to leave the soil on their own is labor-frugal and works best on soils with high organic matter content; microarthropods are efficiently extracted as well as larger taxa (Moldenke 1994). Most authors consider that dynamic methods for extracting arthropods from soil are more efficient on loose-textured soils with large air spaces that contain appreciable amounts of decaying plant material (Fletcher 1976). Dynamic methods have the advantage that animals are recovered undamaged and are suitable for taxonomic studies (see Fig. 19.1), whereas mechanical methods, although tending to damage specimens, yield data on resting stages of arthropods, particularly eggs and pupae (Fletcher 1976; Moldenke 1994).

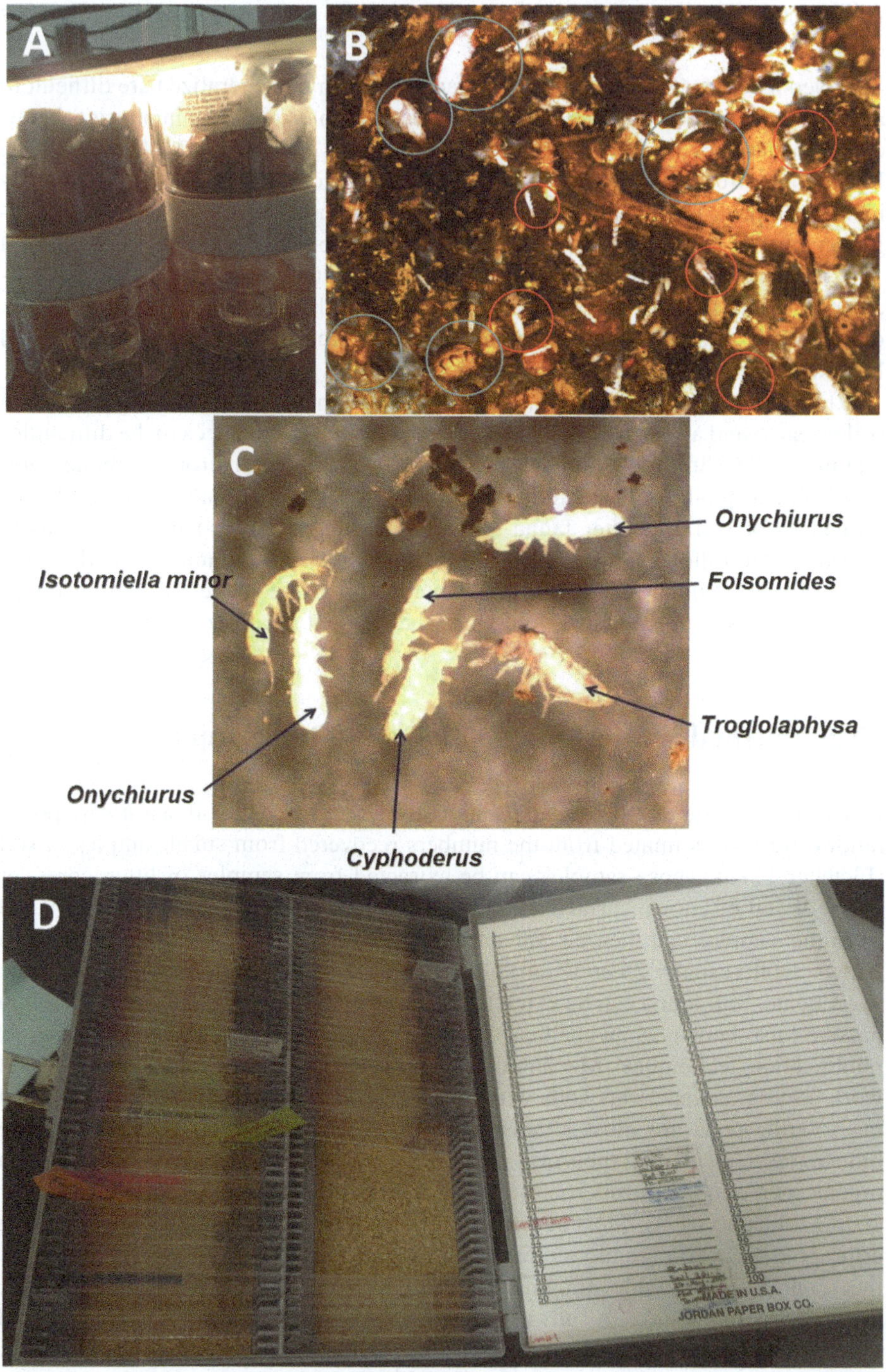

Fig. 19.1 Collembola species identification: separation. (**a**) After field collections, the sample is exposed to a light source in the Berlese-Tullgren funnels; after 7–10 days the sample is dry, and the arthropods are in 70% alcohol inside the collecting vial. (**b**) The collembolans (red circles) are separated from the other arthropods (blue circles). (**c**) The morphospecies of Collembola are separated under the stereoscopy. (**d**) Some Collembola specimens are prepared in slides for further identification. Photos by Claudia M. Ospina-Sánchez

19.2.1 Passive (and Mechanical) Methods on the Study of Soil and Litter Fauna

Observational Studies The ability to individually observe and identify organisms is an important research component for effective conservation and management programs (González et al. 2006). Methods of marking or tagging individual organisms are especially important in genetics and selective breeding studies, where individual identification is essential (Arce et al. 2003). Yet visual identification is often difficult. Paints, dyes, and stains are widely used in studies of insect population dynamics (e.g., Coppedge et al. 1979; Haagsma and Rust 1993). Fluorescent markers have several advantages over liquid paints, and they are becoming the best technique for color-marking insects and aquatic fauna (e.g., Doupé et al. 2003; Welker et al. 1975), but are not as often used in the study of soil invertebrates (González et al. 2006). Fluorescent markers can be a useful tool for carrying out short-term mark and recount studies of earthworms so we can better understand the worm's basic biology, demography, and ethology (González et al. 2006). The technique allows for rapid and inexpensive marking of large numbers of individuals (Turchin 1998). It is also environmentally safe, cost-effective, and easy to use.

There are other observational studies of earthworms that focused on their activities or incidence of casts. Rhea-Fournier and González (2017) described how passive methods for soil within the sphere of earthworm influence (also known as drilosphere) exist. The burrows of soil-dwelling earthworms open to the soil surface, providing visual evidence allowing investigators to sample burrow soil (Parkin and Berry 1999; Gorres et al. 2001). Alternatively, the presence of surface-casting species provides investigators with a visual cue to compare drilosphere and non-drilosphere soil by sampling the casts themselves (Aira et al. 2003; Hong et al. 2011) or the underlying soil (Bohlen et al. 2004). Other visual indicators can be exploited in northern temperate forests where patches of soil invaded with non-native earthworms contrast greatly with earthworm-free patches where thick organic horizons remain (Burtelow et al. 1998; Suárez et al. 2006). A quite different passive approach employed by Lavelle et al. (1992) is utilizing the difference in size of soil aggregates and the casts of geophagous earthworms by passing soil through a 2 mm mesh sieve to exclude casts. This technique works by looking specifically at casts; however, the effect of earthworms extends beyond the casts themselves. For direct study, fresh casts can be obtained by lightly squeezing on the posterior end of a collected or raised worm (Barois and Lavelle 1986; Lavelle et al. 1992).

There are other techniques that allow for the repetitive, nondestructive observation of the soil biota. Root boxes can facilitate photographic documentation of soil arthropod activities (Rygiewicz et al. 1988; Unestam and Stenstrom 1989). Density estimates of microarthropods observed with minirhizotrons can be underestimated depending on time efforts of the observation periods, levels of lights, and optical resolution (Lussenhop and Fogel 1993). Yet, time-lapse video can be used to sample microarthropod density in rhizotrons, where better visualization of all the biota can be achieved by using long working length microscope objectives and stains (Lussenhop and Fogel 1993).

Nicholas and Parkinson (1967) have reviewed in detail several techniques that have been developed to quantify the structure of soils on a scale relevant to arthropods based on the preparation of undisturbed soil sections. These methods seek to accomplish three objectives: preserve the soil micro-architecture in situ, transport the sample to the laboratory unaltered, and section it for examination under the microscope (Moldenke 1994). In the field, soils can either be quickly frozen with water followed by liquid nitrogen (Froehlich and Miles 1986) or embedded in agar, gelatin, or epoxy (Haarlov and Weis-Fogh 1953; Anderson and Healey 1970; Rusek 1985). However, all these techniques suffer from one or another serious drawback and, thus, their application has been very restricted. For example, the success of the agar-gelatin technique depends on obtaining thin sections of good quality that could be inspected under intense illumination with a binocular microscope. However, this is possible only if the soil is rich in organic matter, for gelatin is unable to bind mineral particles effectively (Pande 1975).

Observation using dissection stereoscopes and microscopy remains an essential tool for the separation and identification of individual animal specimens in the study of soil biology (see Box 19.1 for case on the study of collembolans and Figs. 19.1 and 19.2).

Pitfall Traps Pitfall traps rely on animal movement and are used to collect ground active organisms that fall into the collecting pan or trap. This trap can consist of a vial buried into the ground and carefully aligned with the ground surface to prevent any protruding edge to become a potential barrier for the free movement of organisms. Pitfall traps usually have an inserted cone-shaped device to prevent the escape

Box 19.1 Species Identification, the Case of Collembola Separation and Identification in the Tropics

1. *Separation.* After field collections, the samples are installed in Berlese-Tullgren funnels and exposed to a light source for 7–10 days until the sample is dried (Fig. 19.1a). Samples extracted via Berlese-Tullgren funnels produce a rich number of animals. Thus, it is necessary to separate the Collembola from the other arthropods (Fig. 19.1b). Although a few conspicuous species can be identified under the dissection stereoscope, most of the specimens would need further preparation to be identified using microscopy. This process begins with the separation of specimens of Collembola into morphospecies using a dissecting microscope (Fig. 19.1c). Two or three specimens of each morphospecies are cleared using Nesbitt solution and fixed in slides using Mac André II solution (Mari Mutt 1976). To harden the solution, the slides are dried in a slide warmer at 45–50 °C for at least 7 days. At the final step, each specimen in the slide is labeled with its collecting data (Fig. 19.1d).

2. *Description.* Collembolans are not well known in the tropics; therefore, it is common to find new species. The first characteristic to note is the coloration pattern that must be seen under the dissecting microscope (Fig. 19.2a). For the specimens that cannot be identified using primary literature, it is necessary to look for revisions of the group and the original descriptions of the species and genera. Then one can determine if the specimens belong to a new species. The principal criterion used for this determination is morphological differentiation (Gisin 1967; Soto-Adames 2002; Yoshii 1989), in combination with differences in chaetotaxy (Carapelli et al. 1995; Jordana and Baquero 2005; Soto-Adames 2002). These characteristics must be seen under the microscope (Fig. 19.2b). For new species description, it is necessary to draw all the morphological characteristics, made with the aid of a drawing tube (Fig. 19.2c–d). As the final step of description, all drawings are digitized and, with species descriptions, submitted for review as a new species in a specialized peer review journal (Fig. 19.2e–f).

of fallen organisms and a covering ceiling to prevent the trap to be flooded with rainfall (see Fig. 19.3). In addition, traps can be partially filled with a killing solution. For collembolans, for example, the use of pitfall tramps in 1.5 ml microtubes (more commonly used for molecular biology laboratory work) is common. The traps are filled with 1–1.2 ml of ethanol and baited with a small amount of limburger cheese and then placed wherever there was suitable substrate over 2 days (Soto-Adames and Taylor 2010).

Pitfall trap efficiency, as any other ecological method, varies because of several factors such as location of the traps and the intensity, pattern, and timing of sampling (Southwood and Henderson 2000). In addition, there has been an enormous amount of information about the efficiency and variation of pitfall traps (Brown and Matthews 2016) including reports of variation in pitfall trap size and in the materials from which they are made, such as glass, plastic, or metallic (Luff 1975). Luff (1975) reports glass vials to be more effective than plastic and metallic because there was almost no escape in the glass vials. In addition, the size of organisms can affect their escape as larger organisms escape more from small traps and small organisms escape more from large traps. There have been also reports of limitations in accurately reflecting the species composition and sex ratio that come from more extensive and intensive samplings (Topping and Sunderland 1992). As a consequence, there has been a call to standardize the use of these traps by taking into account potential sources of variation such as level of activity of organisms (e.g., due to mate or oviposition site searching) (Brown and Matthews 2016), the choice of materials, and size of the pitfall traps, among others.

Physical Removal Passive methods of soil and litter fauna sampling include hand sorting, washing and sifting, and flotation techniques (Coleman et al. 1999). Coleman et al. (1999) give an overview of sorting soils by hand for earthworm

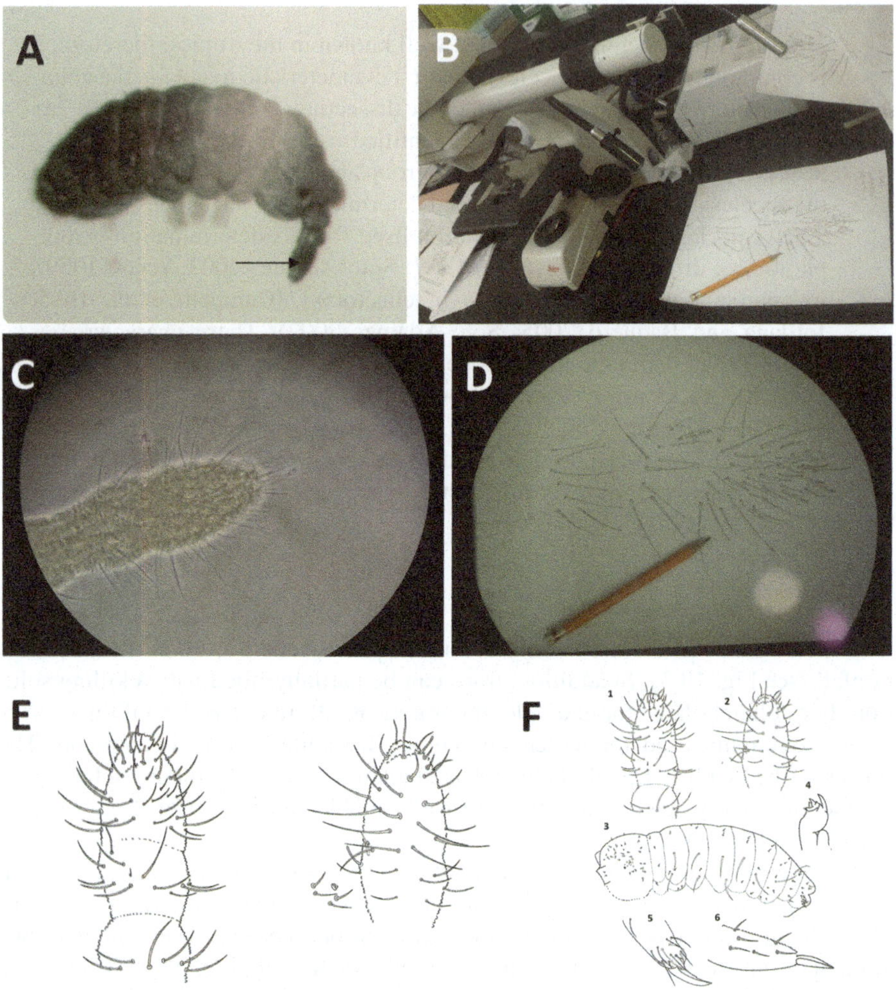

Fig. 19.2 Collembola species identification: description. (**a**) Specimen of *Brachystomella cyrillae* sp. n. under stereoscopy; the arrow indicates the antenna. (**b**) The new species needs a full description including morphological schemes under the microscopy using a lucid camera. (**c**) Detail of the antenna of *B. cyrillae* sp. n. at 100×. (**d**) The same antenna drawing under the lucid camera. (**e**) The figures are inked and digitalized. (**f**) All the figures are included in the new species description. (Photos and drawings by Claudia M. Ospina-Sánchez)

sampling. Hand sorting typically samples a 25 × 25 cm area to a known depth. While small sample areas increase the fraction of fragmented worms and can be inefficient where population densities are low, larger areas can decrease efficiency purely due to the time required to process the amount of soil. Wet sieving or washing can be applied in addition to detect smaller species and cocoons or instead of hand sorting in grassland systems where fibrous roots are very dense (Coleman

Fig. 19.3 Pitfall traps installed in a tropical rain forest consisting of a funnel connected to a vial buried into the ground and carefully aligned with the ground surface and a covering ceiling to prevent the trap to be flooded with rainfall. (Photo by María M. Rivera)

et al. 1999). Hand sorting or sieving can be undesirable in field experiments as it destroys the pedology and soil texture of the site (thus porosity and hydrology) and likely disturbs many other classes of soil biota (Rhea-Fournier and González 2017). While time- and labor-consuming, control (or ambient) earthworm treatments could also be sorted without earthworm removal to duplicate the disturbance; however, the resulting soil would not reflect any natural system, even rarely agroecosystems (Rhea-Fournier and González 2017). Zhang et al. (2010) applied a 1 cm sieve to sort out native earthworm cocoons, stating this mesh size was too large to disrupt soil aggregates, replacing the soil into field-installed mesocosms in a rubber plantation and then inoculating the mesocosm with worms.

There are mechanical methods which separate the arthropods from soil by physical means or depend on the oleophilic and hydrophobic properties of the arthropod cuticle. Physical extraction techniques are usually based on dispersing soil samples in a solvent with the subsequent flotation of soil invertebrates based on density differences (salt or sugar solutions) or on the affinity of the arthropod cuticle for organic solvents (Moldenke 1994). Detailed descriptions of the heptane flotation method are provided by Walter et al. (1987) and Geurs et al. (1991). The heptane flotation technique is based on the principle that the waxy cuticle of the arthropods has chemical affinity for the heptane and not the water. Walter et al. (1987) prove this method is an efficient extractant of arthropods for dryland soils and soils with little organic material and for studies where both dead and live animals in all life stages are to be collected. However, the heptane flotation technique has multiple disadvantages because soft-bodied animals could get damaged during the processing of the sample and it is time-consuming (limiting the number of replicate samples that can be processed in a given time) and expensive (requires the availability of a fume hood and proper organic waste disposal facilities) (Moldenke 1994).

An elutriation technique using upward currents of water to separate soil and animals was used by Törne (1962) to separate Pauropoda and small Collembola from light mineral soils with low contents of organic matter. Müller and Naglitsch (1957) and Müller (1962) put soil samples in saturated sodium chloride solution and centrifuged them to collect the arthropods. The supernatant liquid was then decanted and the residual animals counted and identified and the procedure repeated until no more animals were recovered (Fletcher 1976).

19.2.2 Dynamic/Behavior Methods in the Study of Soil and Litter Fauna

Those methods where animals leave the sample in response to either physical or chemical repellents or attractants are referred to as "behavior methods" (Macfadyen 1957) or "dynamic methods" (Murphy 1962). Behavior or dynamic methods of extracting arthropods from soil are divided into two main types, dry funnel methods and wet funnel methods (Fletcher 1976). Simple dry and wet funnel extractions are usually not labor-intensive. They require no expensive equipment and can yield a wide variety of fauna (Moldenke 1994). The efficiency of both dry and wet funnel extractions is dependent upon the volume of the sample (usually 5–20 g); the greater the surface-to-volume ratio, the more efficient is the extraction. Since the extraction is dependent upon behavior induced in the invertebrate, extraction efficiency is species-specific as well as site-specific (Moldenke 1994). The stimuli responsible for the removal of microarthropods from samples during the extraction process are usually associated with the moisture content of the sample in relation to its water-retaining capacity (Fletcher 1976).

Dry Funnels Funnels can be used to collect smaller and more slowly moving organisms from soil and litter samples when compared with pitfall traps. Funnels have a mesh where the sample is placed and left to dry out and a bottom collecting vial partially filled with killing solution. As the sample dries, an environmental gradient is created within the sample to which organisms differentially respond. The dry funnel apparatus described by Berlese (1905) was one of the first behavior or dynamic methods. The Berlese extractor employed the direct application of heat to a littler sample via a heated water jacket. The surrounding flow of warm water established a temperature gradient throughout the sample, driving the arthropods into a collection vial. Later, Tullgren (1918) introduced a modification to the Berlese via an electric light bulb as a source of heat to dry out sample material and collect arthropods from it. Macfadyen (1953) further modified the funnel setup to create a high gradient to maximize extraction. In this high-gradient design, samples are placed in bulb-heated funnels that are immersed in a water bath to which ice is periodically added to cool the water, therefore creating a large temperature difference that stimulates the downward movement of organisms to fall in the collecting vial (Moldenke 1994). Around the same time as the development of the Berlese funnel, the Winkler-Moczarski eclector was manufactured (Holdhaus 1910). The Winkler funnels were lightweight and did not rely on electricity, which allowed for the extraction of arthropods from substrate, using a basic setup that comprised a series of thin mesh bags hanging vertically within a funnel casing. The Winkler funnels are collapsible and versatile to take to remote sites, and the sample drying accounts partially for recovery of organisms, as samples need to be periodically mixed to promote the movement of organisms and their fall into the collecting vial (Besuchet et al. 1987).

Berlese-Tullgren funnels (as modernly known) use light to intensify the environmental gradient as the sample dries out and thus accelerate the recovery of organisms (see Fig. 19.1). Heat, coming from a light bulb that is placed on top of the sample avoiding direct contact, is often applied to samples to force the outward movement of fauna (Coleman et al. 2004), but it is expensive and logistically laborious in remote field conditions in comparison to extractions without light. Heat increases sample temperature and speeds drying (Coleman et al. 2004), but it may also burn organisms before their collection, decreasing estimates of their abundance (Walter et al. 1987). While extraction with light is based on photophobia and dehydration, extraction without light is based only on photophobia. It is useful for microarthropods to be extracted from soils using cores placed on a high-gradient-type extractor (Crossley Jr and Blair 1991). Although grab litter or soil samples from different habitat types are useful for initial qualitative surveys, intact soil cores of known diameter and depth are important to report population numbers on a per unit area or unit volume basis. Coleman et al. (2004) recommend a core diameter of 5 cm, preferably collected with a split corer containing an inner sleeve to keep the cores intact (see Fig. 19.4).

Tuf (2015) compared pitfalls and Berlese-Tullgren funnels efficiency to assess Chilopoda diversity and found more chilopods recovered from Berlese-Tullgren than from pitfalls, but some ecological groups such as large lithobiomorphs were unique to pitfalls. Several authors have compared the efficiency of Winkler and Berlese funnels, for example, Owens and Carlton (2015) compared the diversity of Coleoptera and found similar species richness, including similar richness of rare species, although the total richness using Berlese was higher than using Winkler in all but one collection site (see Fig. 19.1).

In assessing the efficiency of Winkler and Berlese-Tullgren funnels to extract arthropods from litter samples, Sakchoowong et al. (2007) compared data at different taxonomic resolution levels, i.e., at the broad arthropod orders/classes level, at the Coleoptera family level, and at the Staphylinidae (Coleoptera), subfamily Pselaphinae genera level. Across all resolution levels, they reported higher total

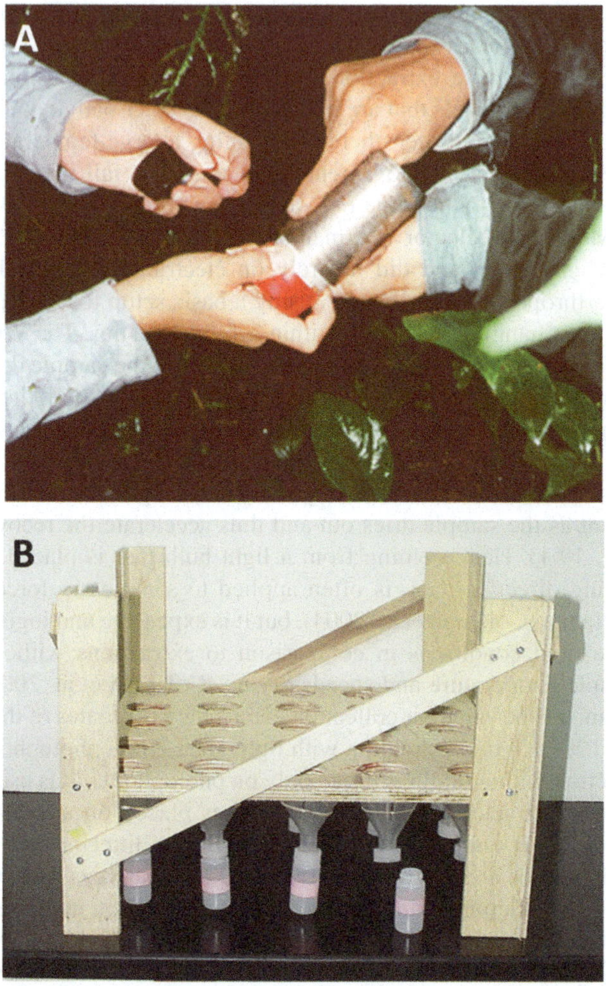

Fig. 19.4 (**a**) Soil cores taken with a double-cylinder hammer sampler are efficient at sampling arthropods due to minimal compaction during sampling. The inner sleeve of the corer is usually aluminum, and placing plastic caps at the ends of inner corer after collection facilitates transportation of the sample from the field into the laboratory, where they can be placed surface side down on a (**b**) high-gradient extractor with light bulbs on top of the samples

arthropod abundance and richness in Tullgren extractions than in Winkler extractions, including the collection of some groups only in Tullgren extractions. For example, almost all Chilopoda, eight Coleoptera families (e.g., Salpingidae, Nitidulidae), and eight Pselaphinae genera (e.g., *Leptoplectus* sp., *Euplectodina* sp.) were only collected using Berlese-Tullgren. Nevertheless, there were similar patterns of dominance and proportions of arthropods common to both methodologies, at all resolution levels. For example, with both funnels Hymenoptera-Formicidae was the dominant group followed by Coleoptera, and arachnids and myriapods were among the least abundant groups. Also, Staphylinidae (Coleoptera) and *Plagiophorus* sp1. (Staphylinidae, Pselaphinae) were consistently the most abundant groups at the family and genera resolution levels.

These authors compared Winkler with no use of light and Berlese-Tullgren with a 60 W light bulb, therefore modifying the environmental conditions within the samples by allowing Winkler samples to naturally dry out, while Berlese-Tullgren samples were artificially dried out. Over a week, Winkler samples' humidity decreased from 35.2% to 31.6%, while Berlese-Tullgren samples' humidity decreased from 37.5% to 3.6%. These data show that the use of light in Berlese-Tullgren results in a steeper environmental gradient within the sample that produced higher arthropod abundance and richness and the collection of some arthropod groups unique to extraction with light. Nevertheless, both extraction methodologies showed similar efficiency in mirroring the abundance and dominance pattern of arthropods.

Barberena-Arias et al. (2012) compared arthropod extraction efficiencies using light and with no light for litter and soil samples coming from habitats with contrasting temperature and rainfall, i.e., dry and wet forests. For litter arthropods, they used Berlese-Tullgren funnels and found that extraction with light resulted in higher litter arthropod abundance than extraction with no light. On the other hand, for soil arthropods, they used modified metallic funnels covered with wooden ceilings to place samples collected with a soil corer or manually with a shovel. They found that extraction with no light resulted in higher soil arthropod abundance than extractions with light for both manual and corer samples. These data show different patterns of extraction as litter arthropods were better extracted with the use of light, while soil arthropods were better extracted with no use of light.

Furthermore, litter samples that used light consistently recovered more arthropod groups than litter samples with no light (Table 19.1, data from Barberena-Arias et al. 2012). Soil samples collected with corers and that used light during extraction consistently returned fewer arthropod groups than corers with no light and shovel samples. In addition, some arthropod groups were unique to some methodologies, for example, Chilopoda and Ricinulei, both predators, were exclusively recovered with the use of light, while others such as Blattodea were recovered with no light. These data suggest that litter organisms are more resistant to dehydration, therefore requiring a steeper environmental gradient within the sample to exit, such as the one created with the use of artificial light. Soil samples collected manually returned higher number of arthropod groups including some unique groups (e.g., Opiliones) than corer samples suggesting that the use of corers may compact the sample making it difficult for organisms to exit.

Varying results come not only from variations in collection and extraction techniques but also in the time that samples are left to allow arthropods to respond to the gradient within the sample and exit. Delsinne and Arias-Penna (2012) found that increasing extraction time doubled the abundance of recovered arthropods as the sample moisture affected total numbers of recovered arthropods, but the proportion and dominance of species were correctly mirrored. Barberena-Arias et al. (2012) found that the time it took arthropods to exit the sample was higher for samples

Table 19.1 Average abundance (ind/m^2) for arthropod groups collected from litter and soil samples in dry and wet forests in Puerto Rico

Habitat type	Litter				Soil							
Forest type	Dry forest		Moist forest		Dry forest				Wet forest			
Extraction type	With light	No light	With light	No light	Corer with light	Corer no light	Manual with light	Manual no light	Corer with light	Corer no light	Manual with light	Manual no light
O. Acari	1565	453	550	231	270	224	269	301	150	189	100	129
O. Araneae	133	200	225	135			300	100	100	130	113	100
O. Pseudoscorpiones	310	263	180	282	167	114	100	300	100	100	100	100
O. Coleoptera	138	163	160	133	100	100	167	130	100	100	100	100
O. Collembola	1000	455	376	384	180	165	203	212	140	132	157	248
O. Diplura	100	186	100	175		100	100		100	114	100	133
O. Diptera	338	246	294	207	675	100	100	100	100	100	182	122
O. Homoptera	200	150	188	325		125	100	100	100	120	100	100
O. Hymenoptera	1771	300	325	543	117	700	120	263	125	171	167	360
O. Protura	100	100	157	240		114	100	117	100	174	100	100
O. Thysanoptera	217	100	333	100	100		150					
O. Isopoda	100	150	120	173	100	100				117		100
O. Schizomida	100		100	100								
O. Embioptera	100		100									
O. Orthoptera	100			100								
O. Hemiptera	308	155	100			100		150		100		
O. Solifugae	300			100			100				100	
O. Psocoptera	120	100			100	100				100		
O. Opiliones	114	100					100					100
C. Chilopoda	200											
C. Diplopoda		200					100					
O. Blattodea		100										
O. Lepidoptera		100										

O. Ricinulei			100									
O. Isoptera			200						200		200	
O. Neuroptera				100								
O. Palpigradi							100					
O. Thysanura							100					
Average abundance	366	196	212	208	201	170	138	177	120	127	127	141
Number of groups	20	18	17	16	9	12	16	10	11	13	12	12
Total abundance	7313	3520	3609	3328	1809	2043	2208	1773	1315	1647	1518	1692

Arthropod extraction efficiency was compared with and without light and for soil samples collected with a corer or shovel. Letters in front of animal group denominate order (O.) or class (C.) levels of taxonomic classification

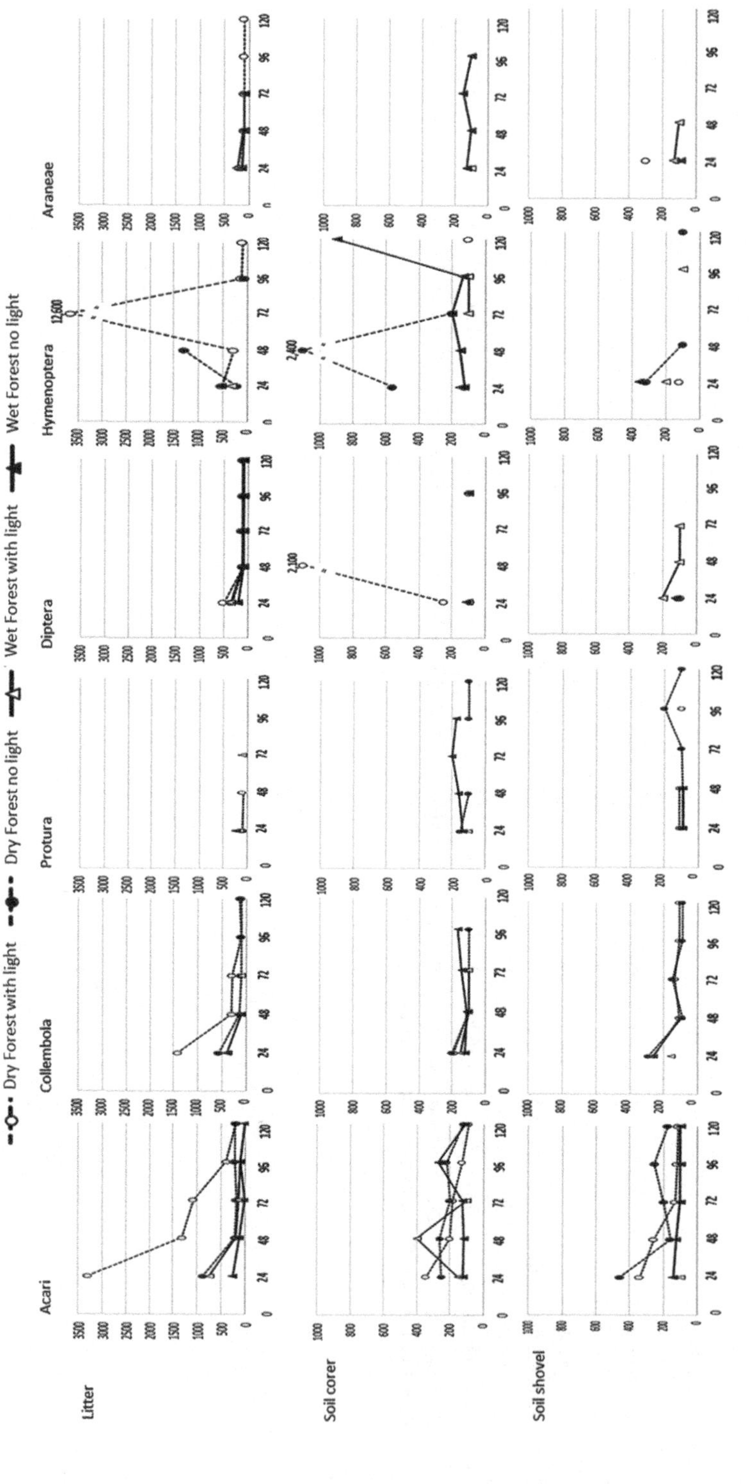

Fig. 19.5 Examples of extraction variation through time. Figures show average abundance (ind/m^2) through extraction time (24, 48, 72, 96, 120 h) for arthropods recovered from litter and soil samples from dry and wet forests in Puerto Rico. Please note that Y-axis scales vary among collection methodologies, also broken lines show outlier abundances, and single points show arthropods recovered only at that time

coming from dry forests than for samples coming from wet forests as 90% of wet forest arthropods exited the samples in the first 48 h, while it took 144 h for 90% of dry forest arthropods to exit the sample. Furthermore, different groups of arthropods showed peak extractions at different times (Fig. 19.5, data from Barberena-Arias et al. 2012). For example, in litter samples, Acari and Collembola progressively exited the sample within the first 24–48 h, but the majority of Hymenoptera left the sample after 72 h. In soil corers, the majority of Diptera and Hymenoptera left the sample at 48 h, while Protura peakextraction occurred in manual no light samples at 96 h.

*Wet Funnels*Wet funnels are used to separate organisms from samples and are particularly effective to recover hydrophilic organisms (Moldenke 1994). This technique is recommended for the extraction of animals extremely sensible to desiccation, works well for the separation of nematodes and rotifers from soil, and can be used to separate nematode larvae from soil samples. Wet funnels are based on the mechanisms originally devised by Baermann (1917). The Baermann apparatus consists of a funnel connected to a clipped rubber pipe. Soil samples are placed inside cloth bags, securely tied up and fastened to an upper supporting rod. Water is added to the funnel ensuring that the sample is completely immersed in the water, and it is left to stand for at least 24 h, and the average extraction is run for 2 to 4 days. Nematode larvae will actively migrate from the sample to the water and then sink to the bottom. The bottom aliquot is then recovered to count and identify organisms (Demelash et al. 2016). Dinaburg (1942) reports that Cort et al. (1922) suggested water to be warmer than the soil to maximize the larval migration from the sample to the water. Warmer water uses the thermotropism of nematodes to separate larvae from the sample and maximize larval recovery (Demelash et al. 2016) for more accurate density estimates.

There are several variations to the wet funnel method that include cold-water extractions, hot-water extractions, sample heating extraction, and duration of the extraction (Didden et al. 1995). The cold-water extraction relies on hydrophilic behavior of organisms but fails to recover inactive life stages such as eggs or pupae (Edwards 1991). The wet funnel method was modified by applying heat with a light bulb to the surface, which speeds the separation process (O'Connor 1955). Milne et al. (1958) (in Moldenke 1994) reversed the temperature gradient by heating the soil sample and accelerating the migration of other invertebrates to the water including copepods, enchytraeids, and tardigrades. Didden et al. (1995) report longer extraction periods, and extraction without heating the sample nor the water to render the highest extraction efficiencies, for samples collected in Germany during the fall. In addition, variations in efficiency can come from different size of organisms and behavior (Edwards 1991) and different soils and environmental conditions to which organisms are adapted.

Chemical Repellents The use of chemical repellents or attractants are dynamic or behavioral methodologies to study soil fauna. One of the most common of such behavior methods involves liquid earthworm irritants applied to soils for extraction at the soil surface. When effective, these liquids can be considered vermifuges (Rhea-Fournier and González 2017). There are toxic and nontoxic options to expel worms from the soil by means of irritation, and the efficiency of extraction of these vermifuges might depend on the climatic conditions, habitat characteristics, and the animal functional type. The application of dilute formalin to a known area of soil to expel earthworms has been widely used. Reviewing this method, Coleman et al. (1999) concluded that formalin is better for vertical burrowing (anecic) species, less useful for horizontal burrowing species, and ineffective for megascolecid species. In addition, climate restricts the formalin efficacy in cold (below 8 °C) or very wet or dry soils. Furthermore, as the flow path of formalin cannot be determined, it is difficult if not impossible to determine the volume of soil sampled with this method. A nontoxic alternative to formalin that acts largely in the same way is "mustard flour" or "hot mustard" with the active ingredient allyl isothiocyanate. Gunn (1992) concluded that mustard was an effective vermifuge, with better extracting efficiencies than formalin, potassium permanganate, and household detergent. Like mustard, cultivars of the genus *Allium* (onions) produce natural sulfur compounds in high densities that act as irritants to many animals including humans and earthworms. Steffen et al. (2013) tested the application of an onion solution as a vermifuge in both a sandy Ultisol and a clayey Oxisol compared to formalin. Results indicated that 175 g onion extract L^{-1} was the ideal concentration, with higher concentrations yielding less earthworms and lower concentrations being less efficient than formalin.

Other Environmental Stimuli The use of grunting and electroshocking can provide for environmental stimuli that would propel the extraction of invertebrates from the soil based on their reactive behavior. Grunting is a qualitative technique of worm collection used commonly by locals for generations in the Florida's Apalachicola National Forest. It involves driving a wooden stake into the ground and then rubbing a metal bar across the top, sending vibrations down the stake and into the ground. Earthworms emerge up to 12 m away and thousands can be collected in hours (Rhea-Fournier and González 2017). Another earthworm extraction method that shows great potential in limiting both physical soil disturbance and nontarget species effects is that of electroshock extraction. The electroshocking technique often involves stainless steel cylinders driven into the earth attached in a series by copper wires where a known electrical current and timing is applied. Findings by Blair et al. (1995) and Staddon et al. (2003) suggest that any nontarget effects of this method are limited or undetectable, making it ideal for earthworm exclusion experiments that do not aim to reduce other soil fauna. Electrical extraction is the least destructive and thus more desirable among other methods with no reliance on hazardous materials (Rhea-Fournier and González 2017).

19.3 Methodological Considerations for Ecological Studies of Soil and Litter Fauna

Soil organisms are important to many ecosystem processes, such as leaf-litter breakdown, soil formation, and nutrient cycling (D'Haese 2013). The chemical composition of plant residues and the nature of the decomposer community play an important role in decomposition and nutrient availability to plants (Tian and Brussaard 1993).

Faunal influences are strongest in the tropics (Heneghan et al. 1998; González and Seastedt 2001). The amount and quality of the litter layer may control the diversity and action of important soil organisms (Crossley Jr et al. 1992; Wall and Moore 1999; Wardle et al. 1999; González and Seastedt 2001). There are three main levels of litter decomposition control, which operate in the following order: climate> litter chemistry> soil organisms (Swift et al. 1979; Seastedt 1984; McClaugherty et al. 1985; Zak et al. 1990; Lavelle et al. 1993; Aerts 1997; Bengtsson 1998). Decomposition can be considered as a two-stage process. First, litter is broken down by detritivores to small pieces which can be chemically reduced. Second, through the activities of microorganisms (bacteria and fungi), these small pieces of organic matter are further reduced and mineralized into basic inorganic molecules, such as ammonium, phosphate, carbon dioxide, and water (González 2002). These can be taken up by plants or microorganisms, leached out of the system, or, in the case of gaseous breakdown products, released to the atmosphere (Swift et al. 1979; Golley 1983).

Chemical Use for Exclusion Experiments: Use of Naphthalene The addition of naphthalene, a polycyclic aromatic hydrocarbon ($C_{10}H_8$), to suppress soil fauna has been used for decades in the field (Williams and Wiegert 1971; Seastedt 1984). It is generally considered to have several advantages over other chemical biocides. It is volatile; it is water insoluble, which contributes to its immobility and persistence in the soil or litter; and it contains only carbon and hydrogen; therefore, it does not contribute exogenous nutrients (Blair et al. 1989).

Naphthalene additions significantly reduced soil invertebrate abundance and richness at global scale (Wall et al. 2008). A reduction in the total number of litter arthropods ranging between 86% and 99% by naphthalene was found in tropical and subalpine forests with contrasting soil moisture regimes (González and Seastedt 2001). Naphthalene treatment significantly inhibited densities of litter fauna 73–85% and the species richness of litter fauna 15–28% in a subalpine forest of center China (Liu et al. 2019). Naphthalene application suppressed arthropod abundances within litterbags by 60–65% in evergreen broadleaf forest, coniferous forest, dwarf forest, and alpine meadow of southeastern China (Wang et al. 2009). Naphthalene addition significantly reduced the abundance of microarthropod 45–52% in a tall grass prairie ecosystem (Cotrufo et al. 2014). The difference in the efficacy of naphthalene suppression may be due to the different soil fauna densities among these sites. The efficacy of naphthalene additions did not appear to be altered by time (Cotrufo et al. 2014; Liu et al. 2019), implying that soil microarthropods do not adapt to the presence of naphthalene. Surface additions of naphthalene are sufficient to reduce soil microarthropods down to a 20 cm depth (Cotrufo et al. 2014). While nematodes were not affected by the naphthalene additions (Coleman et al. 1994; Xiong et al. 2008; Cotrufo et al. 2014). It is likely an effect of the inability of naphthalene to be dissolved in water minimizing the exposure of nematodes to this biocide (Cotrufo et al. 2014).

Naphthalene represents a C source for soil microorganisms (Blair et al. 1989). Naphthalene may indirectly influence litter and soil processes through possible non-target effects on the microbial community (Newell et al. 1987). There is controversy about the nontarget effects of naphthalene. Early microcosm studies suggested that naphthalene might directly affect microbial populations and activity and alter nitrogen dynamics (Newell et al. 1987; Blair et al. 1989). By contrast, a field incubation in tall grass prairie ecosystem reported that naphthalene addition has negligible direct effects on microbial abundance, community structure or activity, and C dynamics, although bacteria utilized naphthalene-derived C (Cotrufo et al. 2014). The discrepancy between these results could have emerged from fundamental differences between the two methods. The use of microcosms for studying the effects of naphthalene is not recommended as a model for understanding field experiments because the small volume of the microcosm exposes the soil biota to higher concentrations than under field conditions. During field studies it is not possible to differentiate whether the naphthalene directly affects microbial communities or indirectly through its effects on soil fauna. Nontarget effects of naphthalene may vary substantially in the field with variations in climatic conditions and soil types. A recent field experiment in the subalpine forests of western China showed that the naphthalene treatment had nontarget effects on the active bacterial community abundance (Lan et al. 2019).

*Mechanical Exclosures: Litterbags and Mesh of Different Sizes*Litterbags are often used to determine the effects of soil animals on litter decomposition rates using varied mesh sizes that exclude soil animals based on body size (Kampichler and Bruckner 2009). By choosing adequate mesh size, researchers can restrict or permit access of the three large groups of soil animals, as defined by their body width: microfauna (<0.1 mm), mesofauna (0.1–2 mm), and macrofauna (>2 mm) (Swift et al. 1979). There is general agreement that soil fauna abundance, biomass, and/or composition were significantly affected by mesh size (Brennan et al. 2009; Richardson et al. 2010), with lower abundance (Brennan et al. 2009; Yang et al. 2017) or biomass (Richardson et al. 2010) in the small mesh litterbags. But a study reported that the densities of major groups of litter arthropods did not significantly differ between small and large mesh size litterbags (González et al. 2003).

However, there are two main drawbacks to this technique. First, the faunal community surrounding the litterbags has not been excluded despite potentially large indirect effects of their activity on biotic and abiotic processes that control litter decomposition at the habitat scale (Bradford et al. 2002). It has been reported that increased soil-faunal functional complexity surrounding the litterbags inhibited litter decomposition by microfauna, bacteria, and fungi by ca 11.5% (Bradford et al. 2002). To predict reliably the effects of changed faunal community composition on litter decomposition requires experiments that exclude faunal groups from both the experimental litter and the habitat where the litter is placed. Second, larger mesh sizes will allow proportionally more litter fragments to exit the bags than smaller mesh sizes (Frouz et al. 2015). Loss of undecomposed fragmented litter was 10% higher from the coarse mesh (2 mm) than the fine mesh (50 μm) bags (Bradford et al. 2002). Additional approaches are required to determine the relative contribution of undecomposed fragments being lost from coarse mesh bags to litter mass loss from litterbags.

Concerns have been raised regarding microclimatic and leaching losses that can co-vary with mesh size and that might affect mass loss rates from litterbags independent of soil animal effects (Kampichler and Bruckner 2009). Enclosing the litter in bags may change the microclimatic conditions, which may then affect the activity of the soil biota (Bradford et al. 2002). However, Bokhorst and Wardle (2013) indicated that microclimatic differences between litterbags of different mesh sizes were minimal, and the effect of mesh size on litter mass because of UV radiation loss is likely to be minor in most cases. Fine mesh size physically protects confined litter against abrasive forces (e.g., wind, precipitations, water flow, etc.) (Lecerf 2017). Hence, leaching of soluble compounds could be different between litterbags of different mesh sizes. But previous study suggested that there was a small difference in leaching losses among different mesh sizes (Bokhorst and Wardle 2013). The effect of litter mass loss in the absence of soil animals is similar between mesh sizes, so method of excluding soil fauna through different mesh size litterbags is proven to be reliable methods for studying the soil fauna (Bokhorst and Wardle 2013).

19.4 Concluding Remarks

Now more than ever, understanding the role of soils and soil organisms in cycling carbon and other nutrients that shape our terrestrial ecosystems is important. Climate strongly affects the composition and activity of soil organisms, and they in turn affect the rates of turnover among carbon pools. The consequences of changing climate and land use affect natural processes as well as global society. Within this context, the study of soil ecology and health remains a relevant field of research. The study of soil biology as all other ecological sciences faces many challenges due to current acceleration of changing environments. One of the critical needs in many ecological fields is the identification of species (Anderson 1975; Eisenhauer et al. 2017). For soils, the relative importance of small organisms means that the number of species is very high, and thus the soil is the natural habitat for millions of species of bacteria, fungi, arachnids, arthropods, and worms that make up the broad functional groups that are widely studied in soil ecology (Andrén et al. 1995; Wardle 2006). A challenge faced in soil research is that no singular method or tool can be used to describe the abundance and diversity of life and functions that inhabit the litter and soil environments. We hope this review of methodologies for the study of soil and litter fauna will help elucidate the most appropriate technique to use based on resources at hand, organism of interest, and habitat characteristics.

Acknowledgments Support was provided by grants DEB 1239764 and 1546686 from the US National Science Foundation to the Institute for Tropical Ecosystem Studies, University of Puerto Rico, and to the International Institute of Tropical Forestry (IITF) USDA Forest Service, as part of the Luquillo Long-Term Ecological Research Program. González was additionally supported by the Luquillo Critical Zone Observatory (EAR-1331841). The US Forest Service (Department of Agriculture) Research and Development Unit and the University of Puerto Rico provided additional support. We thank Drs. William A. Gould, Ariel E. Lugo, Geraldo Wilson Fernandes, and Jean Carlos Santos for kindly reviewing an earlier version of this manuscript.

References

Aerts R (1997) Climate, leaf litter chemistry and leaf litter decomposition in terrestrial ecosystems: a triangular relationship. Oikos 79:439–449. https://doi.org/10.2307/3546886

Aira M, Monroy F, Dominguez J (2003) Effects of two species of earthworms (*Allolobophora* spp.) on soil systems: a microfaunal and biochemical analysis. Pedobiologia 47:877–881. https://doi.org/10.1078/0031-4056-00274

Anderson J (1975) The enigma of soil animal species diversity. In: Vaněk J (ed) Progress in soil zoology. Czechoslovak Academy of Sciences. Springer, Dordrecht, pp 51–58

Anderson JM, Healey IN (1970) Improvements in the gelatine embedding technique for woodland soil and litter samples. Pedobiologia 10:108–120

André HM, Noti M-I, Lebrun P (1994) The soil fauna: the other last biotic frontier. Biodivers Conserv 3:45–56. https://doi.org/10.1007/BF00115332

Andrén O, Bengtsson J, Clarholm M (1995) Biodiversity and species redundancy among litter decomposers. In: Collins HP, Robertson GP, Klug MJ (eds) The significance and regulation of soil biodiversity, Developments in plant and soil sciences, vol 63. Springer, Dordrecht, pp 141–151

Arce SM, Argue BJ, Thompson DA et al (2003) Evaluation of a fluorescent, alphanumeric tagging system for penaeid shrimp and its application in selective breeding programs. Aquaculture 228:267–278. https://doi.org/10.1016/S0044-8486(03)00267-9

Baermann G (1917) Eine einfache methode zur affindung von ankylostomum (Nematoden) Larven in erdproben. Meded. geneesk. Lab. Weltevr, pp 41–47

Barberena-Arias MF, González G, Cuevas E (2012) Quantifying variation of soil Arthropods using different sampling protocols: is diversity affected? Tropical Forests. p. 51–70. Sudarshana, P. Ed. ISBN: 978-953-51-0255-7. InTech

Bardgett RD, van der Putten WH (2014) Belowground biodiversity and ecosystem functioning. Nature 515:505–511. https://doi.org/10.1038/nature13855

Bardgett RD, Wardle DA (2010) Aboveground-belowground linkages: biotic interactions, ecosystem processes, and global change. Oxford University Press, Oxford

Bardgett RD, Bowman WD, Kaufmann R et al (2005) A temporal approach to linking aboveground and belowground ecology. Trends Ecol Evol 20:634–641. https://doi.org/10.1016/j.tree.2005.08.005

Barois I, Lavelle P (1986) Changes in respiration rate and some physicochemical properties of a tropical soil during transit through *Pontoscolex corethrurus* (Glossoscolecidae, Oligochaeta). Soil Biol Biochem 18:539–541. https://doi.org/10.1016/0038-0717(86)90012-X

Bengtsson J (1998) Which species? What kind of diversity? Which ecosystem function? Some problems in studies of relations between biodiversity and ecosystem function. Appl Soil Ecol 10:191–199. https://doi.org/10.1016/S0929-1393(98)00120-6

Berlese A (1905) Apparecchio per raccogliere presto ed in gran numero piccoli artropodi. Redia 2:35–89

Besuchet C, Burckhardt DH, Löbl I (1987) The "Winkler/Moczarski" eclector as an efficient extractor for fungus and litter Coleoptera. Coleopt Bull 41:392–394. https://www.jstor.org/stable/4008527

Blair JM, Crossley DA Jr, Rider S (1989) Effects of naphthalene on microbial activity and nitrogen pools in soil-litter microcosms. Soil Biol Biochem 21:507–510. https://doi.org/10.1016/0038-0717(89)90122-3

Blair JM, Bohlen PJ, Edwards CA et al (1995) Manipulations of earthworm populations in field experiments in agroecosystems. Acta Zool Fenn 196:48–51. https://doi.org/10.1016/0038-0717(95)00025-A

Bohlen PJ, Pelletier DM, Groffman PM et al (2004) Influence of earthworm invasion on redistribution and retention of soil carbon and nitrogen in northern temperate forests. Ecosystems 7:13–27. https://doi.org/10.1007/s10021-003-0127-y

Bokhorst S, Wardle DA (2013) Microclimate within litter bags of different mesh size: implications for the 'arthropod effect' on litter decomposition. Soil Biol Biochem 58:147–152. https://doi.org/10.1016/j.soilbio.2012.12.001

Bradford MA, Tordoff GM, Eggers T et al (2002) Microbiota, fauna, and mesh size interactions in litter decomposition. Oikos 99:317–323. https://doi.org/10.1034/j.1600-0706.2002.990212.x

Brennan KEC, Christie FJ, York A (2009) Global climate change and litter decomposition: more frequent fire slows decomposition and increases the functional importance of invertebrates. Glob Chang Biol 15:2958–2971. https://doi.org/10.1111/j.1365-2486.2009.02011.x

Brown GR, Matthews IM (2016) A review of extensive variation in the design of pitfall traps and a proposal for a standard pitfall trap design for monitoring ground-active arthropod biodiversity. Ecol Evol 6:3953–3964. https://doi.org/10.1002/ece3.2176

Brussaard L (1998) Soil fauna, guilds, functional groups and ecosystem processes. Appl Soil Ecol 9:123–135. https://doi.org/10.1016/S0929-1393(98)00066-3

Burtelow AE, Bohlen PJ, Groffman PM (1998) Influence of exotic earthworm invasion on soil organic matter, microbial biomass and denitrification potential in forest soils of the northeastern United States. Appl Soil Ecol 9:197–202. https://doi.org/10.1016/S0929-1393(98)00075-4

Carapelli A, Fanciulli PP, Frati F et al (1995) The use of genetic markers for the diagnosis of sibling species in the genus *Isotomurus* (Insecta, Collembola). Ital J Zool 62:71–76. https://doi.org/10.1080/11250009509356053

Coleman DC, Crossley DA, Ingham ER (1994) Use of sulfamethoxazole-penicillin, oxytetracycline, carbofuran, carbaryl, naphthalene and temik to remove key organism groups in soil in a corn agroecosystem. J Sustain Agr 4:7–30. https://doi.org/10.1300/J064v04n03_03

Coleman DC, Blair JM, Elliott ET et al (1999) Soil invertebrates. In: Robertson GP, Coleman DC, Bledsoe CS et al (eds) Standard soil methods for long-term ecological research. Network series. Oxford University Press, New York, pp 349–377

Coleman DC, Crossley DA Jr, Hendrix PF (2004) Fundamentals of soil ecology, 2nd edn. Elsevier Academic Press, Burlington, San Diego, London

Coppedge JR, Spencer JP, Brown HE et al (1979) A new dye marking technique for screwworm. J Econ Entomol 72:40–42. https://doi.org/10.1093/jee/72.1.40

Cort WW, Ackert JE, Augustine DL, Payne FK (1922) Investigations on the control of hookworm disease. II. The description of an apparatus for isolating infective hookworm larvae from soil. Am. J. Hyg 2:1–16

Cotrufo MF, Soong J, Vandegehuchte ML et al (2014) Naphthalene addition to soil surfaces: a feasible method to reduce soil micro-arthropods with negligible direct effects on soil C dynamics. Appl Soil Ecol 74:21–29. https://doi.org/10.1016/j.apsoil.2013.09.008

Coyle DR, Nagendra UJ, Taylor MK et al (2017) Soil fauna responses to natural disturbances, invasive species, and global climate change: current state of the science and a call to action. Soil Biol Biochem 110:116–133. https://doi.org/10.1016/j.soilbio.2017.03.008

Crossley DA Jr, Blair JM (1991) A high-efficiency, low-technology Tullgren-type extractor for soil microarthropods. Agric Ecosyst Environ 34:187–192. https://doi.org/10.1016/0167-8809(91)90104-6

Crossley DA Jr, Mueller BR, Perdue JC (1992) Biodiversity of microarthropods in agricultural soils: relations to processes. Agric Ecosyst Environ 40:37–46. https://doi.org/10.1016/0167-8809(92)90082-M

Culliney TW (2013) Role of arthropods in maintaining soil fertility. Agriculture 3:629–659. https://doi.org/10.3390/agriculture3040629

Delsinne TD, Arias-Penna TM (2012) Influence of leaf litter moisture on the efficiency of the Winkler method for extracting ants. J Insect Sci 12:57. https://doi.org/10.1673/031.012.5701

Demelash K, Abebaw M, Negash A et al (2016) A review on diagnostic techniques in veterinary helminthology. Nat Sci 14:109–118. https://doi.org/10.7537/marsnsj140716.15

D'Haese CA (2013) Homology and morphology in Poduromorpha (Hexapoda, Collembola). Eur J Entomol 100:385–407. https://doi.org/10.14411/eje.2003.060

Didden W, Born H, Domm H et al (1995) The relative efficiency of wet funnel techniques for the extraction of Enchytraeidae. Pedobiologia 39:52–57

Dinaburg AG (1942) The efficiency of the Baermann apparatus in the recovery of larvae of *Haemonchus contortus*. J Parasitol 28:433–440. https://doi.org/10.2307/3272903

Doupé RG, Partridge GJ, Lymbery AJ (2003) Visible implant fluorescent elastomer tags as pedigree markers for applied aquaculture: an evaluation using black bream *Acanthopagrus butcheri*. Aquac Res 34:681–683. https://doi.org/10.1046/j.1365-2109.2003.00860.x

Edwards CA (1991) The assessment of populations of soil-inhabiting invertebrates. Agric Ecosyst Environ 34:145–176. https://doi.org/10.1016/0167-8809(91)90102-4

Eisenhauer N, Antunes PM, Bennett AE et al (2017) Priorities for research in soil ecology. Pedobiologia 63:1–7. https://doi.org/10.1016/j.pedobi.2017.05.003

Fletcher KE (1976) Studies on soil arthropods particularly in relation to population assessments. Dissertation, University of London

Froehlich HA, Miles DWR (1986) A freezing technique for sampling skeletal, structureless forest soils. Soil Sci Soc Am J 50:1640–1642.

Frouz J, Roubíčková A, Heděnec P et al (2015) Do soil fauna really hasten litter decomposition? A meta-analysis of enclosure studies. Eur J Soil Biol 68:18–24. https://doi.org/10.1016/j.ejsobi.2015.03.002

Geurs M, Bongers J, Brussaard L (1991) Improvements of the heptane flotation method for collecting microarthropods from silt loam soil. Agric Ecosyst Environ 34:213–221. https://doi.org/10.1016/0167-8809(91)90108-A

Gisin VH (1967) La systématique idéale. J Zool Syst Evol Res 5:111–128. https://doi.org/10.1111/j.1439-0469.1967.tb00903.x

Golley FB (ed) (1983) Ecosystems of the world, Tropical rain forest ecosystems: structure and functions, Part A, vol 14. Elsevier, Amsterdam

González G (2002) Soil organisms and litter decomposition. In: Ambasht RS, Ambasht NK (eds) Modern trends in applied terrestrial ecology. Springer, Boston, pp 315–329

González G, Seastedt TR (2001) Soil fauna and plant litter decomposition in tropical and subalpine forests. Ecology 82:955–964. https://doi.org/10.2307/2679895

González G, Seastedt TR, Donato Z (2003) Earthworms, arthropods and plant litter decomposition in aspen (Populus tremuloides) and lodgepole pine (*Pinus contorta*) forests in Colorado, USA. Pedobiologia 47:863–869. https://doi.org/10.1078/0031-4056-00272

González G, Espinoza E, Liu Z et al (2006) A fluorescent marking and re-count technique using the invasive earthworm, *Pontoscolex corethrurus* (Annelida: Oligochaeta). Caribb J Sci 42:371–379

Gorres JH, Savin MC, Amador JA (2001) Soil micropore structure and carbon mineralization in burrows and casts of an anecic earthworm (*Lumbricus terrestris*). Soil Biol Biochem 33:1881–1887. https://doi.org/10.1016/S0038-0717(01)00068-2

Greenslade PJN (1985) Pterygote insects and the soil: their diversity, their effects on soils and the problem of species identification. Quaest Entomol 21:571–585

Gunn A (1992) The use of mustard to estimate earthworm populations. Pedobiologia 36:65–67

Haagsma KA, Rust MK (1993) Two marking dyes useful for monitoring field populations of *Reticulitermes hesperus* (Isoptera: Rhinotermitidae). Sociobiology 23:155–164

Haarlov N, Weis-Fogh T (1953) A microscopical technique for studying the undisturbed texture of soils. Oikos 4:44–57. https://doi.org/10.2307/3565074

Heneghan L, Coleman D, Zou X et al (1998) Soil microarthropod community structure and litter decomposition dynamics: a study of tropical and temperate sites. Appl Soil Ecol 9:33–38. https://doi.org/10.1016/S0929-1393(98)00050-X

Holdhaus K (1910) Die Siebetechnik zum Aufsammeln der Terricolfauna. Zeitschrift für Wissenschaftliche Insektenbiologie 6:44–57

Hong NH, Rumpel C, Henry des Tureaux T et al (2011) How do earthworms influence organic matter quantity and quality in tropical soils? Soil Biol Biochem 43:223–230. https://doi.org/10.1016/j.soilbio.2010.09.033

Jordana R, Baquero E (2005) A proposal of characters for taxonomic identification of *Entomobrya* species (Collembola, Entomobryomorpha), with description of a new species. Abhandlungen und Berichte des Naturkundemuseums Görlitz 76:117–134

Kampichler C, Bruckner A (2009) The role of microarthropods in terrestrial decomposition: a meta-analysis of 40 years of litterbag studies. Biol Rev 84:375–389. https://doi.org/10.1111/j.1469-185X.2009.00078.x

Lan L, Yang F, Zhang L et al (2019) Non-target effects of naphthalene on the soil microbial biomass and bacterial communities in the subalpine forests of Western China. Sci Rep 9:9811. https://doi.org/10.1038/s41598-019-46394-3

Lavelle P (1996) Diversity of soil fauna and ecosystem function. Biol Int 33

Lavelle P, Spain A (2001) Soil ecology. Springer Science & Business Media

Lavelle P, Melendez G, Pashanasi B et al (1992) Nitrogen mineralization and reorganization in casts of the geophagous tropical earthworm *Pontoscolex corethrurus* (Glossoscolecidae). Biol Fertil Soils 14:49–53. https://doi.org/10.1007/BF00336302

Lavelle P, Blanchart E, Martin A et al (1993) A hierarchical model for decomposition in terrestrial ecosystems: application to soils of the humid tropics. Biotropica 25:130–150. https://doi.org/10.2307/2389178

Lecerf A (2017) Methods for estimating the effect of litterbag mesh size on decomposition. Ecol Model 362:65–68. https://doi.org/10.1016/j.ecolmodel.2017.08.011

Liu Y, Yang F, Yang W et al (2019) Effects of naphthalene on soil fauna abundance and enzyme activity in the subalpine forest of western Sichuan, China. Sci Rep 9:2849. https://doi.org/10.1038/s41598-019-39603-6

Luff ML (1975) Some features influencing the efficiency of pitfall traps. Oecologia 19:345–357. https://doi.org/10.1007/BF00348110

Lussenhop J, Fogel R (1993) Observing soil biota in situ. Geoderma 56:25–36. https://doi.org/10.1016/0016-7061(93)90098-6

Macfadyen A (1953) Notes on methods for the extraction of small soil arthropods. J Anim Ecol 22:65–77. https://doi.org/10.2307/1691

Macfadyen A (1957) Animal ecology. Aims and methods. Isaac Pitman and Sons, London

Macfadyen A (1962) Soil arthropod sampling. In: Cragg JB (ed) Advances in ecological research, vol 1. Academic Press, London, pp 1–34

Mari Mutt JA (1976) Genera of Collembola (Insecta) in Puerto Rico: keys, diagnoses, and general comments. J Agr U Puerto Rico 60:113–128

May RM (1988) How many species are there on earth? Science 241:1441–1449. https://doi.org/10.1126/science.241.4872.1441

May RM (1990) How many species? Philos Trans R Soc B 330:293–304. https://doi.org/10.1098/rstb.1990.0200

May RM (1994) Past efforts and future prospects towards understanding how many species there are. In: Solbrig OT, van Emden HM, van Oordt PGWJ (eds) Biodiversity and global change. CAB International in association with the International Union of Biological Sciences, Wallingford, pp 71–84

McClaugherty CA, Pastor J, Aber JD et al (1985) Forest litter decomposition in relation to soil nitrogen dynamics and litter quality. Ecology 66:266–275. https://doi.org/10.2307/1941327

Menta C, Remelli S (2020) Soil health and arthropods: from complex system to worthwhile investigation. Insects 11:54. https://doi.org/10.3390/insects11010054

Milne A, Coggins RE, Laughlin R (1958) The determination of numbers of leatherjackets in sample turves. J Anim Ecol 27:125–145. https://doi.org/10.2307/2178

Moldenke AR (1994) Arthropods. In: Bottomley PJ, Angle JS, Weaver RW (eds) Methods of soil analysis, part 2: microbiological and biochemical properties, vol 5. Soil Science Society of America, Madison, WI, pp 517–542

Moore JC, Walter DE, Hunt HW (1988) Arthropod regulation of micro–and mesobiota in belowground detrital food webs. Annu Rev Entomol 33:419–439. https://doi.org/10.1146/annurev.en.33.010188.002223

Müller G (1962) A centrifugal-flotation extraction technique and its comparison with two funnel extractors. In: Murphy PW (ed) Progress in soil zoology. Butterworth's, London, pp 207–211
Müller G, Naglitsch F (1957) Vergleichende prtifung bodenzoologischen auslesemethoden fur kleinarthropoden. Zoologische Jahrbücher. Abteilung für Systematik 85:177–210
Murphy PW (1962) Extraction methods for soil animals. I. Dynamic methods with particular reference to funnel processes. In: Murphy PW (ed) Progress in soil zoology. Butterworth's, London, pp 75–114
Newell K, Frankland JC, Whittaker JB (1987) Effects on microflora of using naphthalene or X-rays to reduce arthropod populations in the field. Biol Fertil Soils 3:11–13. https://doi.org/10.1007/BF00260572
Nicholas DP, Parkinson D (1967) A comparison of methods for assessing the amount of fungal mycelium in soil samples. Pedobiologia 7:23–41
O'Connor FB (1955) Extraction of enchytraeid worms from a coniferous forest soil. Nature 175:815–816. https://doi.org/10.1038/175815b0
Owens BE, Carlton CE (2015) "Berlese vs. Winkler": comparison of two forest litter coleoptera extraction methods and the ECOLI (extraction of Coleoptera in litter) protocol. Coleopt Bull 69:645–661. https://doi.org/10.1649/0010-065X-69.4.645
Pande YD (1975) Scope of the gelatin-embedding technique for studying soil arthropod fauna, with particular reference to Oribatei. In: Vaněk J (ed) Progress in soil zoology. Czechoslovak Academy of Sciences. Springer, Dordrecht
Parkin T, Berry EC (1999) Microbial nitrogen transformations in earthworm burrows. Soil Biol Biochem 31:1765–1771. https://doi.org/10.1016/S0038-0717(99)00085-1
Ponge JF, Salmon S (2013) Spatial and taxonomic correlates of species and species trait assemblages in soil invertebrate communities. Pedobiologia 56:129–136. https://doi.org/10.1016/j.pedobi.2013.02.001
Potapov AA, Semenina EE, Korotkevich AY et al (2016) Connecting taxonomy and ecology: trophic niches of collembolans as related to taxonomic identity and life forms. Soil Biol Biochem 101:20–31. https://doi.org/10.1016/j.soilbio.2016.07.002
Rhea-Fournier D, González G (2017) Methodological considerations in the study of earthworms in forest ecosystems. In: Chakravarty S, Shukla G (eds) Forest ecology and conservation. InTech, Rijeka
Richardson BA, Richardson MJ, González G et al (2010) A canopy trimming experiment in Puerto Rico: the response of litter invertebrate communities to canopy loss and debris deposition in a tropical forest subject to hurricanes. Ecosystems 13:286–301. https://doi.org/10.1007/s10021-010-9317-6
Rusek J (1985) Soil microstructures-contributions on specific soil organisms. In: J. Spence (ed) Faunal influences on soil structure. Quaest Entomol 21:497–514
Rygiewicz PT, Miller SL, Durall DM (1988) A root-mycocosm for growing ectomycorrhizal hyphae apart from host roots while maintaining symbiotic integrity. Plant Soil 109:281–284. https://doi.org/10.1007/BF02202096
Sakchoowong W, Nomura S, Ogata K et al (2007) Comparison of extraction efficiency between Winkler and Tullgren extractors for tropical leaf litter macroarthropods. Thai J Agric Sci 40:97–105
Seastedt T (1984) The role of microarthropods in decomposition and mineralization processes. Annu Rev Entomol 29:25–46. https://doi.org/10.1146/annurev.en.29.010184.000325
Soto-Adames FN (2002) Molecular phylogeny of the Puerto Rican *Lepidocyrtus* and *Pseudosinella* (Hexapoda: Collembola), a validation of Yoshii's "color pattern species". Mol Phylogenet Evol 25:27–42. https://doi.org/10.1016/S1055-7903(02)00250-6
Soto-Adames FN, Taylor SJ (2010) Status assessment survey for springtails (Collembola) in Illinois caves: the Salem Plateau. Illinois Natural History Survey Technical Report 13:1–76
Southwood TRE, Henderson PA (2000) Ecological methods, 3rd edn. Blackwell Science, Oxford
Staddon PL, Ostle N, Fitter AH (2003) Earthworm extraction by electroshocking does not affect canopy CO_2 exchange, root respiration, mycorrhizal fungal abundance or mycorrhizal fungal vitality. Soil Biol Biochem 35:421–426. https://doi.org/10.1016/S0038-0717(02)00293-6

Steffen GPK, Antoniolli ZI, Steffen RB et al (2013) Earthworm extraction with onion solution. Appl Soil Ecol 69:28–31. https://doi.org/10.1016/j.apsoil.2012.12.013
Stork NE (2018) How many species of insects and other terrestrial arthropods are there on earth? Annu Rev Entomol 63:31–45. https://doi.org/10.1146/annurev-ento-020117-043348
Suárez E, Fahey T, Yavitt J et al (2006) Patterns of litter disappearance in a northern hardwood forest invaded by exotic earthworms. Ecol Appl 16:154–165. https://doi.org/10.1890/04-0788
Swift MJ, Heal OW, Anderson JM (1979) Decomposition in terrestrial ecosystems. Studies in ecology, vol 5. University of California Press, Berkeley, CA
Tian G, Brussaard L (1993) Biological effects of plant residues with contrasting chemical compositions under humid tropical conditions: effects on soil fauna. Soil Biol Biochem 25:731–737. https://doi.org/10.1016/0038-0717(93)90114-Q
Topping CJ, Sunderland KD (1992) Limitations to the use of pitfall traps in ecological studies exemplified by a study of spiders in a field of winter wheat. J Appl Ecol 29:485–491. https://doi.org/10.2307/2404516
Törne EV (1962) An elutriation and sieving apparatus, for extracting micro–arthropods from soil. In: Murphy PW (ed) Progress in soil zoology. Butterworth's, London, pp 204–206
Tuf IH (2015) Different collecting methods reveal different ecological groups of centipedes (Chilopoda). Fortschr Zool 32:345–350. https://doi.org/10.1590/S1984-46702015000500003
Tullgren A (1918) Ein sehr einfacher Ausleseapparat für terricole Tierformen. Z Angew Entomol 4:149–150
Turchin P (1998) Quantitative analysis of movement: measuring and modeling population redistribution in animals and plants. Sinauer Associates, Sunderland
Unestam T, Stenstrom E (1989) A method for observing and manipulating roots and root-associated fungi on plants growing in nonsterile substrates. Scand J For Res 4:51–58. https://doi.org/10.1080/02827588909382545
Wall DH, Moore JC (1999) Interactions underground: soil biodiversity, mutualism, and ecosystem processes. Bioscience 49:109–117. https://doi.org/10.2307/1313536
Wall DH, Bradford MA, John MGS et al (2008) Global decomposition experiment shows soil animal impacts on decomposition are climate-dependent. Glob Chang Biol 14:2661–2677. https://doi.org/10.1111/j.1365-2486.2008.01672.x
Wallwork JA (1970) Ecology of soil animals. McGraw-Hill, London
Walter DE, Kethley J, Moore JC (1987) A heptane flotation method for microarthropods from semi-arid soils, with comparison to the Merchant-Crossley high-gradient extraction method and estimates of microarthropod biomass. Pedobiologia 30:221–232
Wang S, Ruan H, Wang B (2009) Effects of soil microarthropods on plant litter decomposition across an elevation gradient in the Wuyi Mountains. Soil Biol Biochem 41:891–897. https://doi.org/10.1016/j.soilbio.2008.12.016
Wardle DA (2006) The influence of biotic interactions on soil biodiversity. Ecol Lett 9:870–886. https://doi.org/10.1111/j.1461-0248.2006.00931.x
Wardle DA, Nicholson KS, Bonner KI et al (1999) Effects of agricultural intensification on soil-associated arthropod population dynamics, community structure, diversity and temporal variability over a seven-year period. Soil Biol Biochem 31:1691–1706. https://doi.org/10.1016/S0038-0717(99)00089-9
Wardle DA, Bardgett RD, Klironomos JN et al (2004) Ecological linkages between aboveground and belowground biota. Science 304:1629–1633. https://doi.org/10.1126/science.1094875
Welker BD, Clark SH, Fontaine CT et al (1975) A comparison of Petersen tags and biological stains used with internal tags as marks for shrimp. Gulf Res Rep 5:1–5. https://doi.org/10.18785/grr.0501.02
Williams JE, Wiegert RG (1971) Effects of naphthalene application on a coastal plain broomsedge (Andropogon) community. Pedobiologia 11:58–65
Wolters V (2001) Biodiversity of soil animals and its function. Eur J Soil Biol 37:221–227. https://doi.org/10.1016/S1164-5563(01)01088-3
Xiong Y, Shao Y, Xia H et al (2008) Selection of selective biocides on soil microarthropods. Soil Biol Biochem 40:2706–2709. https://doi.org/10.1016/j.soilbio.2008.07.014

Yang B, Zhang W, Xu H et al (2017) Effects of soil fauna on leaf litter decomposition under different land uses in eastern coast of China. J For Res 29:973–982. https://doi.org/10.1007/s11676-017-0521-5

Yoshii R (1989) On some Collembola of New Caledonia, with notes on the "colour pattern species". Contributions from the Biological Laboratory, Kyoto University 27:233–259

Zak DR, Grigal DF, Gleeson S et al (1990) Carbon and nitrogen cycling during old-field succession: constraints on plant and microbial biomass. Biogeochemistry 11:111–129. https://doi.org/10.1007/BF00002062

Zhang M, Zou X, Schaefer DA (2010) Alteration of soil labile organic carbon by invasive earthworms (*Pontoscolex corethrurus*) in tropical rubber plantations. Eur J Soil Biol 46:74–79. https://doi.org/10.1016/j.ejsobi.2009.11.004

Chapter 20
Sampling Methods for Aquatic Insects

Marcos Callisto, Riccardo Mugnai, Diego M. P. Castro, and Marden S. Linares

20.1 Introduction

Aquatic insects are important elements in the structure and function of freshwater ecosystems because they contribute to organic matter processing, nutrient cycling, and energy flux and serve as food for vertebrate and invertebrate predators in aquatic food webs. They comprise a highly varied group of animals that are dependent on water habitats at some stage in their life cycles (Table 20.1). Aquatic insects are ubiquitous in the lakes, streams and rivers, springs, and wetlands of the world. The habitats where they can be found are varied, including hyporheic zones, bottom sediments, air-water interfaces, temporary systems, and brackish, salty, and hypersaline waters.

Aquatic insect collection methods vary depending on the freshwater habitat type (lentic or lotic) and substrate type (e.g., sand, stones, leaves, and vegetation). Moreover, sampling aquatic insects can be qualitative or quantitative, depending on the type of study performed. Depending on the study objectives, researchers can collect individual specimens from single habitats a single time or entire assemblages across river basins, ecoregions, or continents through use of long-term systematic surveys (Oliveira and Pes 2014; USEPA 2016; Ligeiro et al. 2020). Additional concerns must include sample site location, sampling frequency (e.g., daily, monthly, yearly), and numbers of samples, true replicates, pseudo-replicates, and subsamples (Hughes and Peck 2008).

M. Callisto (✉) · D. M. P. Castro · M. S. Linares
Laboratório de Ecologia de Bentos, Instituto de Ciências Biológicas, Universidade Federal de Minas Gerais, Belo Horizonte, Minas Gerais, Brazil
e-mail: callistom@ufmg.br; dcastro@ufmg.br; mslinares@ufmg.br

R. Mugnai
Laboratório BIOCICLOS, Centro de Ciências Agrárias e Ambientais, Universidade Federal do Maranhão, Chapadinha, Maranhão, Brazil
e-mail: mugnai.riccardo@ufma.br

J. C. Santos, G. W. Fernandes (eds.), *Measuring Arthropod Biodiversity*,
https://doi.org/10.1007/978-3-030-53226-0_20

Table 20.1 Insect orders with aquatic forms

Order	Common names	Aquatic development phase (excluding egg)
Ephemeroptera	Mayflies	Nymphs
Odonata	Damselflies and dragonflies	Nymphs
Orthoptera	Grasshoppers and crickets	All
Plecoptera	Stoneflies	Nymphs
Hemiptera	True bugs	All
Mecoptera	Scorpionflies and hangingflies	Larvae
Neuroptera	Lacewings	Larvae
Megaloptera	Dobsonflies and alderflies	Larvae
Coleoptera	Beetles	All
Trichoptera	Caddisflies	Larvae and pupae
Lepidoptera	Moths	Larvae and pupae
Diptera	True flies	Larvae and pupae
Hymenoptera	Wasps	Larvae

The collection of aquatic insects can be done with a wide variety of equipment and methods, from simply hand-lifting rocks at the bottom of a river or examining aquatic vegetation. Sampling efficiency of aquatic insects is significantly increased when hand nets, kick nets, dredges, and other inexpensive apparatus is used. After a sample has been collected, labeled, stored, and safely transported to a laboratory, samples should be washed over standard sieves (e.g., 250 μm, 500 μm) to remove organic debris and leaf detritus, algae, and sediment particles, which facilitates the screening of aquatic insects (Fig. 20.1). In general, aquatic insect samples are fixed with 70% alcohol and preserved for later taxonomic identification using stereomicroscopes. Eventually, slides are prepared for the identification of small-size morphological structures and observation of gills and lamellae and other structures not visible under stereomicroscopes (Fig. 20.2). Lastly, DNA barcoding has been used to identify insects, but the necessary DNA barcode libraries are just beginning to be developed for the Neotropics (Nichols et al. 2019).

20.2 Freshwater Habitats and Sampling Devices

20.2.1 Lotic

Shallow lotic ecosystems can show high substrate diversity and are generally rich in aquatic insect species (Fig. 20.3). Kick nets (D-shape net with 0.3–0.5 m aperture and 1 m x 1 m square of 500 μm coupled to a 1.5 m wooden handle) are used to obtain semiquantitative data (Fig. 20.4a–d), whereas Surber samplers (a 0.3 × 0.3 m quadrat linked to a 1 m × 1 m square of 500 μm net) offer quantitative samples (Fig. 20.4e–h). Both methods require in disturbing bottom sediments (e.g., cobbles,

Fig. 20.1 Field and laboratory sediment wash over standard sieves

gravel, sand, leaf deposits) and catching organisms in a net downstream. Besides, a researcher can collect stones and pebbles by hand to visually inspect and collect the aquatic insects with tweezers, transfer them to a flask, and take them to the laboratory for further taxonomic identification (Fig. 20.9c–d). When the collected material

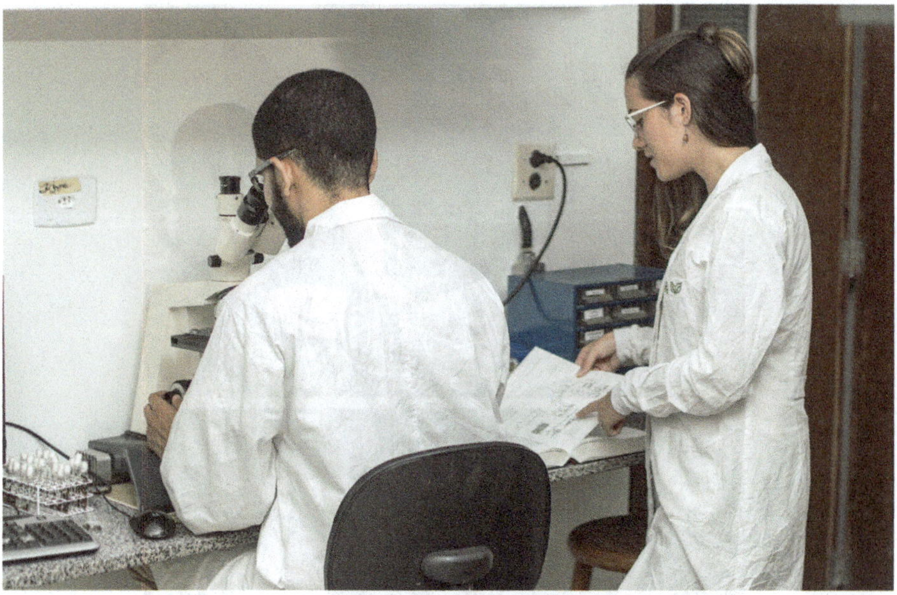

Fig. 20.2 Slide preparation for taxonomic identification of morphological structures using a microscope in laboratory

has sediment, leaf debris, or organic matter left in the sample, it should be washed (Fig. 20.9a) transferred to a white tray and manually sorted live invertebrates are transferred to vials in the laboratory (Fig. 20.9b). Ideally, sediment samples have to be collected from downstream to upstream, to avoid drift addition.

20.2.2 Lentic

Lentic ecosystems can be roughly divided between littoral and limnetic zones (Esteves 2011). Littoral habitats form a transition zone between aquatic and terrestrial ecosystems, concentrating much of the biodiversity of a lake (Kutyła 2014; Gownaris et al. 2018). To sample aquatic insects in the littoral zone, D-shape nets are used for semiquantitative sampling. This methodology is similar to the one used in lotic ecosystems, consisting of disturbing bottom sediments, with the difference being that due to the lack of a strong downstream current, it is usually necessary to push the sediment to the net or sweep the net through the sediment plume. Corer samplers (8–20 cm diameter acrylic cylinder tube) are used for quantitative sampling, gathering sediment samples of a known area (Fig. 20.5a–d).

Pelagic habitats usually need the assistance of a boat to be sampled, due to the depth. Eckman-Birge and Petersen dredges (a clamshell bucket with two opposing jaws for the collection of a sediment sample) are the most common samplers in these conditions, being able to acquire sediment samples in deepwater ecosystems,

Fig. 20.3 Examples of running waters in reference conditions

but can be also used in the littoral zone (Fig. 20.6a–c). Additional samples using D-net can be collected in littoral habitats (Figs. 20.4b and 20.6g–h).

20.2.3 Hyporheic Zone

The hyporheic zone (Greek *hypo* and *rheos* = under and river) is the connecting ecotone between surface and groundwater and is functionally part of both fluvial and groundwater ecosystems (Mugnai et al. 2015). Some aquatic insects can live in

Fig. 20.4 Pictures of field samplings using D-nets (**a–d**), drift net (**e**), and Surber sampler (**f–h**)

the hyporheic zone to avoid competition or predation, as a nursery, or to escape seasonal flooding and droughts (Dole-Olivier et al. 1997; James and Suren 2009). Larvae of some Coleoptera, Diptera, Ephemeroptera, and Plecoptera and adult forms like elmid and staphylinid coleopterans are frequently observed in hyporheic zones. The larvae are generally confined to the upper layers, but some species penetrate deeper and are important components of the hyporheic fauna (Mugnai et al. 2019).

To sample aquatic insects in hyporheic zones, several techniques can be used (Fig. 20.7) including the Karaman-Chappuis technique for gravel bed streams, where a hole is dug in the bank of a river and the water is filtered; Bou-Rouch pump modified, where a perforated bar is inserted directly into the riverbed and water is pumped through a membrane; standpipe cores, where perforated bars are inserted semipermanently into a riverbed and the water is pumped periodically using a hand pump; freeze-cores, where a freezing medium is used to freeze and extract part of a riverbed; artificial substrates and baited; and baited and un-baited traps buried in the riverbed allow the collection of specimens. All these methods substantially differ in efficiency, financial cost, and field effort (Hahn 2005; Malard et al. 2002).

20.2.4 Springs

Springs are natural points of water resurgence and are threatened and poorly known freshwater environments (Malard et al. 2002). These environments are subject to deforestation, trampling by livestock, and human recreational activities. These habitats can be very heterogeneous from a geomorphological point of view, amount of water provided, aquatic vegetation, and water chemistry.

From biotic point of view, in general, the benthic and nektonic insect fauna is relatively scarce due to the characteristics of the water (usually low pH, nutrient, and dissolved oxygen), but some are specialists for crenon habitats such as some Diptera (Chironomidae), Ephemeroptera (Baetidae), Plecoptera (*Leuctra*), and Trichoptera (Lepidostomatidae) (Dobrin and Giberson 2003; Ilmonen and Paasivirta 2005).

To sample aquatic insects in springs, it is advisable to use a 10 × 10 cm D-shape net (Rosati et al. 2016), artificial substrate for benthic and nektonic fauna, and modified Hess sampler (Dobrin and Giberson 2003). For interstitial organisms, the gear and techniques described for hyporheic habitats can be used (Malard et al. 2002).

20.2.5 Marine, Brackish, and Inland Saltwaters

Aquatic insects occasionally live in waters with greater than 0.5‰ salinity. These environments are represented by brackish waters, which include mixtures of coastal freshwater with marine water as estuary, mangrove, coastal lagoons, etc., with a variable salinity between 0.5 and 32‰. Saline inland waters consist of saline lakes occurring typically in arid or semiarid regions with a variable salinity between 0.5 and 300 ‰. The marine water is constituted by intertidal areas and neritic or oceanic environments with salinity between 34 and 37‰.

Species that have occupied marine and brackish environments are currently constituted by a few hundred species belonging to the Hemiptera, Dermaptera, Hymenoptera, Coleoptera, and Diptera. Some few species of Lepidoptera, Odonata, and Trichoptera may live in environments with salinity up to 15‰ (Ward 1992). Intertidal or benthos specimens may be collected by hand or using samplers depending on the nature of substrata (kick samplers, Surber sampler, grab samplers, corer, etc.). Neustonic (small organisms that inhabit below the surface of a body of water), and nektonic (actively swimming aquatic organisms in a body of water) species are best sampled with a handheld plankton net screwed to an extendable handle and

Fig. 20.5 Field sediment sampling using a corer sampler

obtained by snorkling or SCUBA diving (Thorp and Covich 2010; Merritt and Cummings 1996).

20.2.6 Habitats Outside Water Bodies

20.2.6.1 Phytotelmata

The word phytotelmata is derived from the Greek *phyton* and *telm* = plant and pond, consisting in small waterbodies of hollow trees or inflorescences such as bromeliads. Thienemann (1954), Kitching (1971), and Greeney (2001) divided this habitat into six major types: tree holes, leaf axils, bowers, modified leaves, fallen vegetative

Fig. 20.6 Field sediment sampling using Eckman-Birge dredge (**a**, **b**, **c**), Petersen dredge (**d**, **e**, **f**), and D-net (**g**, **h**) at the littoral region of a lentic habitat

Fig. 20.7 Example of field equipment used to sample in hyporheic zones

parts (such as leaves or bracts), and fallen fruit husks. Stem rots, traditionally not included, are considered as transition zones between phytotelmata and other similar habitats (Greeney 2001). This habitat is heterogeneous and is generally considered as temporary but often being available all year round. From a dimensional standpoint, they are usually less than 200 ml, even those of 30 ml or less being quite common, to at least 45 l.

Phytotelmata occur in a wide range of ecosystems from subarctic to tropics in tree holes of deciduous woodlands, in Sarraceniaceae and Nepenthaceae plants, and in tropical plants that can retain water such as bromeliaceans and bamboos. These habitats are most common in tropical and subtropical areas and are utilized by a wide range of arthropods. Besides insects, mites, entomostracods (Copepoda, Ostracoda, Cladocera), brachiurans, crabs, and tartigrads are often found in phytotelmata habitats. Seventy families from 11 orders of insects have been reported living in phytotelmata habitats. That includes truly aquatic insects such as Diptera, Coleoptera, and Odonata (as Pseudostigmatidae and Megapodagrionidae) that are best represented, but Hemiptera, Trichoptera, and Plecoptera are also common. Besides that, semi-aquatic species and terrestrial species foraging in or around phytotelmata habitats are included (Greeney 2001; Frank and Lounibos 2009).

The interest in the phytotelmata habitat, such as in some plants as Sarraceniaceae, Nepenthaceae, and bromeliaceans, can be destructive, due to invasive methods that result in the dissection and removal of the plant. Water-sucking devices of various types can be used such as laboratory pipettes, cooking basters, syringes or syringes with tubes can be used to collect in this environment with good results (Derraik 2009; Jocque et al. 2010).

20.2.6.2 Hygropetric and Madicolous Habitats

Hygropetric and madicolous are two habitats characterized by a tiny layer of water. The word hygropetric derived from the Greek *hydro* and Latin *pedra* = water and stone. It is constituted by the rock environment always covered by a subtle veil of water usually due to the splash of water from waterfalls or turbulent waters. Madicolous habitats are characterized by a flowing, often permanently, thin layer of water (<2 mm), over many types of surfaces, usually limited to mountain areas and restricted in size. In these environments, rich in organic muck and mosses, can be found specialized species in the Coleoptera (Torridincolidae, Hydrophilidae, Dytiscidae), Diptera (Ceratopogonidae, Chironomidae, Dolichopodidae, Psychodidae, Stratiomyidae, Thaumaleidae, Tipulidae), Ephemeroptera (Heptageniidae), Odonata (Megapodagrionidae), and Trichoptera (Hydroptilidae), besides semiterrestrial organisms (Ferrington et al. 1995; Hájek et al. 2019; Miyairi and Tojo 2007; Tennessen 2010). Aquatic specimens can be collected by hand, or with small hand nets (Thorp and Covich 2010), flying adults are collected by sticky traps or emergence traps (Shimabukuro et al. 2015).

20.3 Drift Samples

Insects and other aquatic invertebrates are transported downstream in a phenomenon known as drift. The entry of invertebrates into the water column can be active or passive and may be the result of several factors, such as (a) changes in water flow and velocity, (b) presence of predators, (c) changes in the water physical and chemical characteristics, and (iv) redistribution of invertebrate populations as a function of competitive pressures (Brittain and Eikeland 1988). The transport of invertebrates downstream is not constant and varies with the season, day to day, and at different times of the day. Differences in the densities of drift organisms may also vary for each species and their propensity to drift at different stages of the life cycle or when insects are emerging. Invertebrate drift is of great importance to the functioning of aquatic ecosystems as it is a primary mechanism for invertebrate redistribution and colonization (Brittain and Eikeland 1988; Naman et al. 2016).

Techniques and methods for sampling insect drift are generally simple, and different equipment can be used. The most widespread method has been the use of a frame (square or rectangular) (Fig. 20.4e) or tube mouth fitted with a long net (~0.5–2.0 m) which has a removable container at the end. Drift nets must be placed in the stream with the net face perpendicular to the direction of flow and anchored with iron bars driven into the substrate (Fig. 20.8a–c). The top edge of the frame must be above the water surface, and the bottom edge close to the stream bottom, but clear of the bottom to ensure that only drifting insects enter the net, avoiding insects crawling directly into the net. The mesh size of the nets will depend on the objectives of the study, but typical mesh size used is between 200 and 500 μm. Smaller mesh size nets can clog rapidly and can only be deployed for short amounts of time to avoid backflow.

It is important to measure the volume of water passing through a drift sampler. The net filters a column of water with a known cross-sectional area. Using a flow meter near the mouth of the sampler and the cross-sectional area, it is possible to calculate the volume of water flowing through a drift sampler. With the volume filtered and the number of invertebrates collected, it is possible to calculate the drift density (i.e., the number of invertebrates per unit volume of water) and the drift rate (the number of invertebrates passing a sampling point in unit time) (Allan and Russek 1985).

The sampling period can vary from a few minutes to several hours depending on the sampling station characteristics and the aim of the research. It is important to monitor the net during the sampling because the net may become clogged with organic detritus and thus produce a backflow that causes a decrease in sampling efficiency.

20.4 Sampling Aquatic Insect Adults

Flying adults of aquatic insects can be sampled using ultraviolet (UV)-light traps, mercury vapor lights, Malaise traps, sweep netting riparian vegetation, and rearing from immature stages. These sampling methodologies are unselective and provide big samples including mostly terrestrial nocturnal insects.

Specific techniques to sample emerging aquatic insect adults include enclosed channels (Wartinbee and Coffman 1976), floating emergence traps (LeSage and Harrison 1979), hand screen collectors (Usinger 1956), light trap (Southwood 1978), madicolous trap (Shimabukuro et al. 2015), Mundie pyramid trap (Mundie 1956), pan trap (Grigarick 1959), stationary screen trap (Hamilton 1969), subaquatic light traps (Aiken 1979), surface film samplers (Coffman 1973), and window trap (Chapman and Kinghorn 1995). All these sample methods differ in construction effort, cost, and microhabitat utilization.

The sampling of emerging adults is mostly influenced by some environmental factors such as seasonality, temperature, and moon phases (Ivković et al. 2013). Aquatic insect adults can also be obtained by direct association between larval and adult forms (Edmunds et al. 1976; Merritt and Cummings 1996).

20.5 List of Supplies for Field Sampling

The organization of the necessary material is an essential phase for any field sampling. It is essential to anticipate the needs of the sampling, as usually forgotten equipment can cost a day's worth of work or even more: (1) plastic buckets; (2) 250 or 500 μm sieves (filters 60–90 micron pore for hyporheic); (3) plastic bags, wash bottles, and 70% ethanol; (4) current meter; (5) Surber (Fig. 20.4f–h), D-net (Figs. 20.4b and 20.6g–h), drift nets (Fig. 20.8a–c), Eckman-Birge (Fig. 20.6a–c), or Petersen (Fig. 20.6d–f) dredges; (6) metal holding rods; (7) permanent marker pens and labeling paper; and (8) forceps.

20.6 Laboratory Procedures

After field sampling, the collected samples are taken to the laboratory. The first procedure is washing (Fig. 20.9a) and sorting that involves the separation of the aquatic invertebrates from the inorganic and organic matter collected with the sample. This is a time-consuming procedure but essential. Normally this procedure is done using lighted tables, plastic trays, and forceps (Fig. 20.9b). Flotation using salt or sugar is commonly used to remove small larvae (but not for mollusks) in sandy sediments. After samples have been sorted, the following step is to identify and

Fig. 20.8 Pictures of the use of a drift net in a headwater stream

Fig. 20.9 Laboratory processing activities of sediment samples: (**a**) washing, (**b**) sorting, (**c, d**) identifying aquatic insects

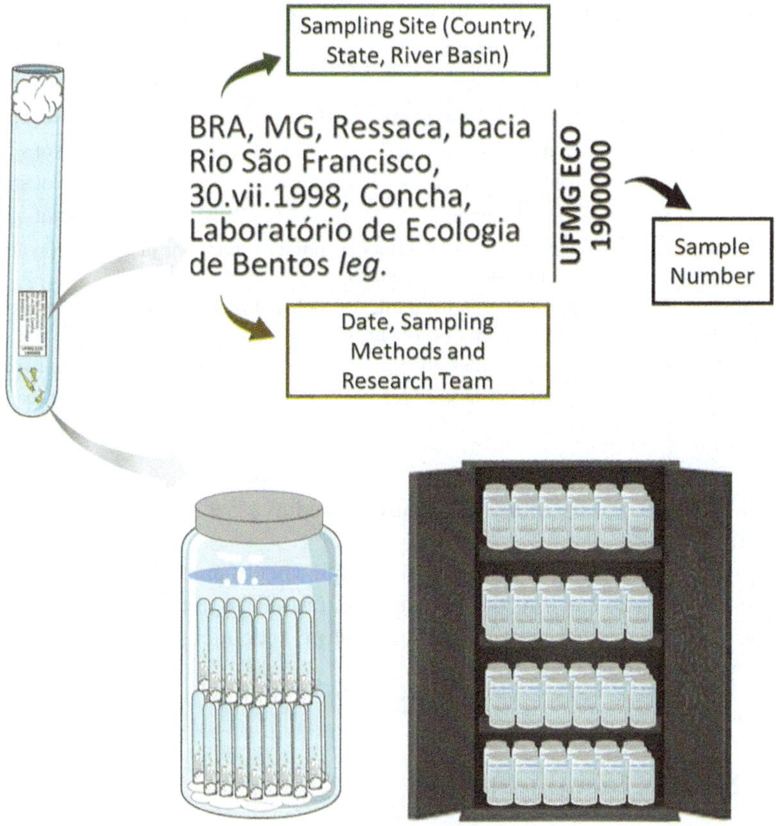

Fig. 20.10 Scheme of a reference collection of aquatic insects

count the organisms through the use of taxonomic keys (Fig. 20.9c, d). The results are then analyzed by different procedures depending on the research questions, aim, and experimental design. All aquatic insect forms must be preserved in a Reference Collection, with field sampling method, number of individuals, and taxonomic identification and registered in a public institution (Fig. 20.10).

20.7 Specimen Preparation

To prepare aquatic insects, most specimens consist of juvenile phases but can also often include pupae in the case of Holometabola forms. To conserve larvae and pupae, several authors suggest adding to the 70–80% alcohol a few drops of glycerol; to conserve adult phases, specimens can be stored in a dry way, properly pinned or glued to cards depending on size (Lincoln and Sheals 1979). To pin speci-

Table 20.2 Pinning scheme for different aquatic insect taxa groups

Group	Position of pin	Wing
Large Coleoptera	Right helitrae	Not spread
Large Ephemeroptera	Center of mesothorax	Not spread
Large Heteroptera	Center of scutellum	Not spread
Slender Heteroptera	Center of prothorax	Not spread
Large Hymenoptera	Right forewings	Not spread
Large Lepidoptera	Through middle at thickest point of thorax or just behind base of forewings	Spread
Large Odonata	Through middle at thickest point of thorax or just behind base of forewings	Spread
Large Plecoptera	Center of mesothorax	Spread
Large Trichoptera	Center of mesothorax	Spread

mens, it is necessary to follow standard procedures according to the taxonomic group (Table 20.2).

Despite the possibility of storing dry adult specimens, several aquatic insect research laboratories have in the last decades chosen to conserve these specimens wet, due to the necessary physical space to store specimens and the skill and time necessary to pinning.

20.8 Species Identification

Species identification in biodiversity surveys has two major aspects. The first is the identification of species not known from the taxonomic point of view with the purpose to describing and naming them, defined as taxonomy. This activity should follow the precepts illustrated by Winston (1999) in the book *Describing Species: Practical Taxonomic Procedure for Biologists* and the rules dictated by the International Code of Zoological Nomenclature (ICZN).

The second activity aimed at quantifying biological diversity and to develop ecological works aimed at identifying species already known taxonomically known as parataxonomy. The term parataxonomy was introduced by Jansen et al. in 1993 coming from the term parataxonomist (or "biological diversity technicians" in some parts of the world) the specialist involved in the sampling collection of natural history or inventory data and in the activities of specimen sampling, sorting, and identification. For this activity, unlike taxonomy, identification at the species level is not strictly necessary, and often in ecological works, different levels are used such as genus or family, technically defined as operational taxonomic units (OTUs), recognizable taxonomic units (RTUs), or parataxonomic units (Pus) or morphospecies (Krell 2004; Oliver and Beattie 1993, 1996a, b).

To aid the parataxonomy identification activity in the last decades, many specialized publications have been produced by condensing taxonomic work on identification keys, often in the form of juxtaposed dichotomous keys, but also as grouped, identified, graphical, or combined dichotomous keys. More recently, multiple-entry interactive identification electronic keys have been introduced. Such keys illustrate orders or families or brings together various taxonomic groups (e.g., insects, macroinvertebrates, etc.) in the form of identification manuals.

Regarding geographical area, identification keys and identification manuals are heterogeneous, ranging from local (state, microregion, river basin) to large extent publications (North America, South America, Europe). From the point of view of the amount of published works for area, there are notable differences. In the northern hemisphere areas, such as the United States or Europe, there is a large amount of available bibliographic material. On the other hand, in other areas, as Southern Hemisphere (Africa and South America or some places in Asia), bibliographic specific works are scarce or even nonexistent. The absence of specific bibliographic material may lead the researcher to use manuals and keys from other geographic regions to try to identify the specimens found, which is not always possible and generally inappropriate.

Another problem of identification using dichotomous keys, applied to aquatic insects and the most commonly used in biomonitoring programs of water bodies, is related to the continual updating of scientific information as new species and/or cases of synonyms and homonyms are discovered and new identification keys are published. In this case, it is important to point out that the use of the most up-to-date bibliographic material in research may create a problem of identification standardization precluding the possibility to compare the work with previous studies or result in distortions of results between researchers from different teams or even within the same team. Therefore it is important to maintain a library of taxa names to facilitte cross-referencing of data.

20.9 Conservation and Curatory

Biological samples naturally tend to degrade with time. Curatory is the set of actions that involve physical and/or chemical actions, routinely implemented, to minimize or avoid physical or biological damage of stored specimens in the collection and preserve the associated information (Herholdt 1990; Horie 1986).

Ultraviolet light, produced either by lamps or direct sunlight, can damage dry specimens (Rose and Torres 2009; Simmons and Munhoz-Saba 2005), and these sources can be avoided using specific adhesive filters. In addition, damage can be caused by some biological factors as the beetle *Anthrenus* spp. (Dermestidae) and some fungi. These pests can be controlled by temperature, humidity, and chemical preservation with paradichlorobenzene. The biggest problem with wet specimen conservation is the loss of preservative fluid (e.g., 70% alcohol) that must be periodically added and replaced to ensure correct concentration.

20.10 Conservation Implications of Aquatic Insects

As insect biodiversity is threatened in all parts of the world (Sánchez-Bayo and Wyckhuys 2019), it is of paramount importance to train students, researchers, and the general public about the importance of aquatic insects. Specific sampling methods for different types of aquatic ecosystems are available at reduced prices and enable qualitative and quantitative sampling of species and estimations of their abundance. Many of these aquatic insects are bioindicators of water quality (Bonada et al. 2006) and have been widely used in environmental monitoring programs and development of multimetric indices (Silva et al. 2017) and as environmental education tools in citizen science activities (França et al. 2019).

The conservation of freshwater ecosystems is necessary to ensure water supply for multiple human uses and ecosystem services but we have a fundamental duty to ensure adequate habitat and conditions for the maintenance of aquatic insect species. These organisms, by their participation in food chains and detritus decomposition, linking different compartments in aquatic ecosystems through energy flow and nutrient cycling in the riparian meta-ecosystems (Callisto et al. 2019) reflect environmental quality in entire river basins, ecoregions and nations (USEPA 2016).

Acknowledgments This chapter is part of the PROECOS Project, continued supported by the Pesquisa & Desenvolvimento of Agência Nacional de Energia Elétrica and Companhia Energética de Minas Gerais (P&Ds ANEEL/CEMIG GT-599 and GT-611). We appreciate the support of Coordenação de Aperfeiçoamento de Pessoal de Nível Superior – Brasil (CAPES) – Finance Code 001, Conselho Nacional de Desenvolvimento Científico e Tecnológico (CNPq) via research fellowship to MC (303380/2015-2), and the Fundação de Apoio à Pesquisa do Estado de Minas Gerais (FAPEMIG) via grants to MC (PPM 00104-18). DMPC and MSL are postdoc fellows by the Pesquisa & Desenvolvimento of Agência Nacional de Energia Elétrica and Companhia Energética de Minas Gerais (P&D Aneel-Cemig GT-611 and GT-599, respectively). We thank Ricardo Solar who kindly took photos of all field sampling devices, Anderson Santos prepared the collection figure, and Robert M. Hughes reviewed the English language and contributed with critical inputs to an early version and the page proofs of this chapter.

References

Aiken R (1979) A size selective underwater light trap. Hydrobiologia 65:65–68. https://doi.org/10.1007/BF00032720

Allan J, Russek E (1985) The quantification of stream drift. Can J Fish Aquat Sci 42:210–215. https://doi.org/10.1139/f85-028

Bonada N, Prat N, Resh VH et al (2006) Developments in aquatic insect biomonitoring: a comparative analysis of recent approaches. Annu Rev Entomol 51:495–523. https://doi.org/10.1146/annurev.ento.51.110104.151124

Brittain JE, Eikeland TJ (1988) Invertebrate drift - a review. Hydrobiologia 166:77–93. https://doi.org/10.1007/BF00017485

Callisto M, Solar R, Silveira F et al (2019) A Humboldtian approach to mountain conservation and freshwater ecosystem services. Front Environ Sci 7:195. https://doi.org/10.3389/fenvs.2019.00195

Chapman J, Kinghorn J (1995) Window-trap for flying insects. Can Entomol 82:46–47

Coffman W (1973) Energy flow in a woodland stream ecosystem: II. The taxonomic composition and phenologgy of the Chironomidae as determined by the collection of pupal exuviae. Arch fur Hydrobiol 71:281–322

Derraik J (2009) A tool for sampling mosquito larvae from phytotelmata. J Vector Ecol 34:155–156. https://doi.org/10.1111/j.1948-7134.2009.00019.x

Dobrin M, Giberson DJ (2003) Life history and production of mayflies, stoneflies, and caddisflies (Ephemeroptera, Plecoptera, and Trichoptera) in a spring-fed stream in Prince Edward Island, Canada: evidence for population asynchrony in spring habitats? Can J Zool 81:1083–1095. https://doi.org/10.1139/z03-091

Dole-Olivier M, Marmonier P, Beffy J (1997) Response of invertebrates to lotic disturbance: is the hyporheic zone a patchy refugium? Freshw Biol 37:257–276. https://doi.org/10.1046/j.1365-2427.1997.00140.x

Edmunds JGFE, Jensen SL, Berner L (1976) The Mayflies of North and Central America. University of Minnesota Press, Minneapolis

Esteves F (2011) Fundamentos de Limnologia, 3rd edn. Editora Interciência, Rio de Janeiro

Ferrington LCJ, Kavanaugh R, Schmidt F et al (1995) Habitat separation among Chironomidae (Diptera) in big springs. J Kansas Entomol Soc 68:152–165

França JS, Solar R, Hughes RM et al (2019) Student monitoring of the ecological quality of neotropical urban streams. Ambio 48:867–878. https://doi.org/10.1007/s13280-018-1122-z

Frank J, Lounibos L (2009) Insects and allies associated with bromeliads: a review. Terr Arthropod Rev 1:125–153. https://doi.org/10.1163/187498308X414742

Gownaris N, Rountos K, Kaufman L et al (2018) Water level fluctuations and the ecosystem functioning of lakes. J Great Lakes Res 44:1154–1163. https://doi.org/10.1016/j.jglr.2018.08.005

Greeney H (2001) The insects of plant-held waters: a review and bibliography. J Trop Ecol 17:241–260. https://doi.org/10.1017/S026646740100116X

Grigarick A (1959) A floating pan trap for insects associated with the water surface. J Econ Entomol 52:348–349. https://doi.org/10.1093/jee/52.2.348a

Hahn H (2005) Unbaited phreatic traps: a new method of sampling stygofauna. Limnologica 35:248–261. https://doi.org/10.1016/j.limno.2005.04.004

Hájek J, Alarie Y, Stastny J et al (2019) The first hygropetric Platynectes and its larva from eastern China (Coleoptera: Dytiscidae). Acta Entomol Musei Natl Pragae 59:217–228. https://doi.org/10.2478/aemnp-2019-0019

Hamilton A (1969) A new type of emergence trap for collecting stream insects. J Fish Res Board Canada 26:1685–1689. https://doi.org/10.1139/f69-155

Herholdt E (1990) Vertebrate collections management: a personal perspective with special reference to Southern Africa. In: Natural history collections: their management and value. Transvaal Museum, Pretoria, pp 1–11

Horie C (1986) Who is a curator? Int J Museum Manag Curatorsh 5:267–272. https://doi.org/10.1080/09647778609515029

Hughes RM, Peck DV (2008) Acquiring data for large aquatic resource surveys: the art of compromise among science, logistics, and reality. J N Am Benthol Soc 27:837–859

Ilmonen J, Paasivirta L (2005) Benthic macrocrustacean and insect assemblages in relation to spring habitat characteristics: patterns in abundance and diversity. Hydrobiologia 533:99–113. https://doi.org/10.1007/s10750-004-2399-4

Ivković M, Miliša M, Previšić A et al (2013) Environmental control of emergence patterns: case study of changes in hourly and daily emergence of aquatic insects at constant and variable water temperatures. Int Rev Hydrobiol 98:104–115. https://doi.org/10.1002/iroh.201301483

James A, Suren A (2009) The response of invertebrates to a gradient of flow reduction - an instream channel study in a New Zealand lowland river. Freshw Biol 54:2225–2242. https://doi.org/10.1111/j.1365-2427.2009.02254.x

Jocque M, Kernahan A, Nobes A et al (2010) How effective are non-destructive sampling methods to assess aquatic invertebrate diversity in bromeliads? Hydrobiologia 649:293–300. https://doi.org/10.1007/s10750-010-0272-1

Kitching R (1971) An ecological study of water-filled tree-holes and their position in the woodland ecosystem. J Anim Ecol 40:281. https://doi.org/10.2307/3247

Krell F (2004) Parataxonomy vs. taxonomy in biodiversity studies–pitfalls and applicability of 'morphospecies' sorting. Biodivers Conserv 13:795–812

Kutyła S (2014) Characteristics of water level fluctuations in Polish lakes – a review of the literature/Charakterystyka wahań poziomu wody w jeziorach polskich – przegląd piśmiennictwa. Ochr Sr i Zasobów Nat 25:27–34. https://doi.org/10.2478/oszn-2014-0011

LeSage L, Harrison A (1979) Improved traps and techniques for the study of emerging aquatic insects. Entomol News 90:65–78

Ligeiro R, Hughes RM, Kaufmann PR et al (2020) Choice of field and laboratory methods affects the detection of anthropogenic disturbances using stream macroinvertebrate assemblages. Ecol Indic 115:106382. https://doi.org/10.1016/j.ecolind.2020.106382

Lincoln R, Sheals J (1979) Invertebrate animals - collection and preservation. Cambridge University Press, London

Malard F, Dole-Olivier M, Mathieu J (2002) Sampling manual for the assessment of regional groundwater biodiversity. Technical Report. European Project Pascalis

Merritt R, Cummings K (1996) Aquatic insects of North America, 2nd edn. Kendall Hunt Pub Co, New York

Miyairi K, Tojo K (2007) Biology of the mayfly *Bleptus fasciatus* Eaton (Insecta: Ephemeroptera, Heptageniidae), with special reference to the distribution, habitat environment, life cycle, and nuptial behavior. Limnology 8:85–93. https://doi.org/10.1007/s10201-007-0203-7

Mugnai R, Messana G, Di Lorenzo T (2015) The hyporheic zone and its functions: revision and research status in Neotropical regions. Braz J Biol 75:524–534. https://doi.org/10.1590/1519-6984.15413

Mugnai R, Serpa-Filho A, Nessimian J et al (2019) Morphological traits and vertical distribution of hyporheic chironomid larvae in Atlantic Forest streams. Trop Zool 32:119–134. https://doi.org/10.1080/03946975.2019.1639034

Mundie J (1956) Emergence traps for aquatic insects. SIL Commun 1953-1996 7:1–13. https://doi.org/10.1080/05384680.1956.11904089

Naman S, Rosenfeld J, Richardson J (2016) Causes and consequences of invertebrate drift in running waters: from individuals to populations and trophic fluxes. Can J Fish Aquat Sci 73:1292–1305. https://doi.org/10.1139/cjfas-2015-0363

Nichols SJ, Kefford BJ, Campbell CD et al (2019) Towards routine DNA metabarcoding of macroinvertebrates using bulk samples for freshwater bioassessment: effects of debris and storage conditions on the recovery of target taxa. Fresh Biol 65:607–620. https://doi.org/10.1111/fwb.13443

Oliveira V, Pes A (2014) Inventário da fauna de insetos aquáticos: coleta, preservação e criação. In: Hamada N, Nessimian J, Querino R (eds) Insetos Aquáticos na Amazônia Brasileira: Taxonomia. Biologia e Ecologia. Editora do INPA, Manaus, pp 151–167

Oliver I, Beattie A (1993) A possible method for the rapid assessment of biodiversity. Conserv Biol 7:562–568. https://doi.org/10.1046/j.1523-1739.1993.07030562.x

Oliver I, Beattie A (1996a) Invertebrate morphospecies as surrogates for species: a case study. Conserv Biol 10:99–109. https://doi.org/10.1046/j.1523-1739.1996.10010099.x

Oliver I, Beattie A (1996b) Designing a cost-effective invertebrate survey: a test of methods for rapid assessment of biodiversity. Ecol Appl 6:594–607. https://doi.org/10.2307/2269394

Rosati M, Cantonati M, Fenoglio S et al (2016) Is there an ideal protocol for sampling macroinvertebrates in springs? J Freshw Ecol 31:199–209. https://doi.org/10.1080/02705060.2016.1149892

Rose C, Torres A (2009) Storage of natural history collections: ideas and practical solutions. Washington, DC

Sánchez-Bayo F, Wyckhuys KAG (2019) Worldwide decline of the entomofauna: a review of its drivers. Biol Conserv 232:8–27. https://doi.org/10.1016/j.biocon.2019.01.020

Shimabukuro E, Pepinelli M, Perbiche-Neves G et al (2015) A new trap for collecting aquatic and semi-aquatic insects from madicolous habitats. Insect Conserv Divers 8:578–583. https://doi.org/10.1111/icad.12135

Silva DRO, Herlihy AT, Hughes RM et al (2017) An improved macroinvertebrate multimetric index for the assessment of wadeable streams in the neotropical savanna. Ecol Indic 81:514–525. https://doi.org/10.1016/j.ecolind.2017.06.017

Simmons J, Munhoz-Saba Y (2005) Conservación preventiva y causas del deterioro de las colecciones. In: Cuidado Manejo y Conservación de Colecciones Biológicas. Universidad Nacional de Colombia, Bogotá, pp 54–71

Southwood T (1978) Ecological methods with particular reference to the study of insect populations

Tennessen K (2010) The madicolous nymph of Heteropodagrion sanguinipes Selys (Odonata: Megapodagrionidae). Zootaxa 2531:29–38. https://doi.org/10.11646/zootaxa.2531.1.3

Thienemann A (1954) *Chironomus*. Leben, Verbreitung und wirtschaftliche Bedeutung der Chironomiden. Die Binnengewässer, XX. Schweizerbart'sche Verlagbuchhandlung, Stuttgart

Thorp J, Covich A (2010) Ecology and classification of North American freshwater invertebrates. Academic Press, New York

USEPA (U. S. Environmental Protection Agency) (2016) National rivers and streams assessment 2008–2009: a collaborative survey. EPA/841/R-16/007. Office of Water/Office of Research and Development. Washington, DC. http://www.epa.gov/national-aquatic-resource-surveys/nrsa

Usinger R (1956) Aquatic insects of California, with keys to North American genera and California species. University of California Press, Berkeley

Ward J (1992) Aquatic insect ecology: biology and habitat. Wiley, New York

Wartinbee D, Coffman W (1976) Quantitative determination of chironomid emergence from enclosed channels in a small lotic ecosystem. Am Midl Nat 95:479–484. https://doi.org/10.2307/2424416

Winston J (1999) Describing species: practical taxonomic procedure for biologists. Columbia University Press, New York

Chapter 21
Sampling Methods for Blood-Feeding Insects Diversity

Álvaro Eduardo Eiras, Elis Paula de Almeida Batista, and Marcelo Carvalho de Resende

21.1 Introduction

Among all groups of animals, the phylum Arthropoda is considered the most abundant. They represent more than seven million species, of which 70% are insects and can be found in all environments and regions of the world (Foley et al. 2007; Stork 2018).

Class Insecta deserves to be highlighted as many species of insects live in close association with humans and domestic and wild animals. They can affect humans and animals directly, causing allergic reactions and poisoning, and by blood feeding, they act as vectors or pathogen hosts for organisms such as viruses, fungi, bacteria, protozoa, and helminths (Khamesipour et al. 2018).

Insects belonging to the orders Hemiptera (kissing bugs), Siphonaptera (fleas), and Diptera (mosquitoes) are those most involved in the transmission of pathogens to humans and animals. Some families of the order Diptera are important due to their blood-feeding behavior and the transmission of diseases, such as the Simuliidae (onchocerciasis), Psychodidae (leishmaniasis and viral infections), and Ceratopogonidae (bluetongue virus and epizootic hemorrhagic disease virus) that affect domestic and wild ruminants (Shelley 1991; Houghton et al. 2010; McGregor et al. 2019). Other families such as Tabanidae that consist of blood-sucking flies known as tabanids are pests of livestock. However, the transmission of diseases by

Á. E. Eiras (✉) · E. P. de Almeida Batista
Laboratory of Innovation Technologies in Vector Control, Department of Parasitology, Biological Sciences Institute, Federal University of Minas Gerais, Belo Horizonte, Brazil

M. C. de Resende
Laboratory of Innovation Technologies in Vector Control, Department of Parasitology, Biological Sciences Institute, Federal University of Minas Gerais, Belo Horizonte, Brazil

Department of Entomology, Ministry of Health of Brazil, National Health Foundation, Belo Horizonte, Brazil

J. C. Santos, G. W. Fernandes (eds.), *Measuring Arthropod Biodiversity*,
https://doi.org/10.1007/978-3-030-53226-0_21

tabanids is highly controversial, and only nematode infection is unquestionable (Martins-Neto 2003). Other families of flies, such as Chloropidae, represented by the genera *Liohippelates* and *Hippelates*, are attracted to hosts by sebaceous secretions, wound, and blood which accelerate the entry of pathogenic organisms into their hosts (Mulla and Chansang 2007). Also the flies of the families Muscidae (housefly), Glossinidae (tsetse flies), Calliphoridae (flies of metallic color), and Sarcophagidae (flies of scavenging habits) are important in public and veterinary health in the transmission of organisms that cause bacterial infections, protozoa intestinal diseases, and trypanosomiasis, among others (Khamesipour et al. 2018).

Among all Diptera, Culicidae (mosquitoes) are the most numerous and important because of their blood feeding on vertebrates, which can lead to decreased productivity and weight loss in livestock and mainly act as the only vectors of diseases such as malaria, arboviruses such as dengue, yellow fever, Chikungunya and Zika virus, filariasis, and viral encephalitis to man (Clements 1992; Forattini 2002; Vasconcelos and Calisher 2016; Tandina et al. 2018).

The objective of this chapter is to provide an overview of the main sampling methods used in studies of blood-feeding insects due to their relevant role in transmission of several diseases to humans and animals, so that these sampling tools can be used for surveillance. Laboratory studies that used sampling tools were not included, and therefore this chapter describes only sampling methods used in the field. Because there have been many studies for developing new sampling tools in the past, this chapter is addressed to those that are more recent and used by the scientific community. The chapter also describes the sampling methods, technologies, and capture tools available for collecting the blood-sucking insects of medical and veterinary importance. Several techniques available for data analysis and the appropriate techniques for preserving samples obtained by each collection method are also described.

21.2 Sampling Tools for Blood-Feeding Insects

When selecting an appropriate vector sampling method, some essential attributes must be thoroughly considered, such as the design of the sampling tool, costs, transport, handling, and the ecological characteristics and environmental conditions under which the target insects are found. Efficient trap deployment is a compromise between a high trap density to maximize the likelihood of insect collection and a low trap density, to minimize the costs. In this context, specific sampling methods are indeed necessary to sample different vectors based upon their biology, developmental stage, behavior, and habitat.

Commonly, sampling methodologies consist of active searching or attraction strategies, known as "active sampling" and "passive sampling," respectively. Active sampling involves the collection of insects in their habitat or surround human or animal habitats, whereas passive sampling approaches the attraction of the specimens by stimuli known as "baits."

The number of methods available to sample vector species, including many new tools developed and evaluated, is significantly high, which makes it unrealistic to describe them all in this review. In this chapter, we considered only the traps that were evaluated under the vector's natural environment conditions and which correspond to the resources most used by researchers and professionals in the field, which are summarized in Table 21.1.

21.2.1 Active Sampling

Active sampling techniques are applied to remove insects from their habitat or resting places, using forceps, aspirators, or dippers, or even during flight by sweep nets, as described below.

Forceps This is the main tool to collect kissing bugs (Hemiptera, Reduviidae, Triatominae) nymphs and adult stages and blackfly (Diptera, Simuliidae) larvae. Usually, kissing bugs colonize domestic environments, and it is usually active manual searches and subsequent collection with forceps that are done. These insects are found in cracks, fissures, and holes in walls, ceilings, and other places that are their harborages indoors, while outdoors they are found in animal shelters (chicken coop, pigsty, kennel, etc.) (Costa et al. 2003; Curtis-Robles et al. 2018). Sampling the larvae of blackflies is carried out in running freshwater by using forceps to remove them from the substrate (e.g., leaves and stems, twigs, plant roots, and rocks) to which they are fixed (Ya'cob et al. 2016; Hadi et al. 2019).

Aspirators Aspirators are particularly used for sampling small dipterans such as mosquitoes (Diptera, Culicidae) (Silver 2008), sandflies (Diptera, Psychodidae, Phlebotominae) (Vilela et al. 2018), and blackflies (Diptera, Simuliidae) (Walsh et al. 1978). The suction of insects into the aspirator can be done orally or by battery, and aspirators are recommended for sampling vectors that remain in shelters resting after hatching, copulation, or blood meal. This can take place outdoors (vegetation, holes in the ground, animal dens, among others) or indoors (wall cracks or other sheltered locations). Mouth aspirators are one of the oldest tools for sampling vectors and have been used to sample indoor-resting insects or in association with other sample methods, such as Shannon's tent, cone trap, and human landing catch (Rubio-Palis and Curtis 1992; Odiere et al. 2007). This tool has several advantages in its use, such as low cost and can be used at anytime and anywhere, including in remote areas. It is a selective tool, which allows the collection to be directed at the target species only.

The mouth aspirator is the most used of its kind due to its simplicity, economy, and easiness to produce. The simplest model and also the most common consists of a flexible rubber tube (~60 cm long) connected to an acrylic tube or other transparent materials (~30 cm long) and a screen such as a piece of fine fabric located between the tubes to block small insects from moving from one tube to another. In this method, sampling is carried out based on mouth suction at the end of the flexible tube, which generates negative pressure and sucks insects from the end of the

Table 21.1 Summary of sampling methods for insect vectors diversity

Sampling methods		Specificity	Attractants	Behavior of interest	Advantages	Limitation
Forceps	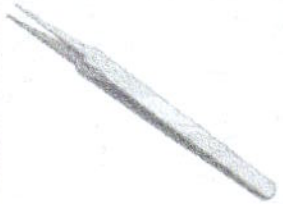	Triatominae (kissing bugs) and immature stages of Simuliidae (blackflies)	No	Resting	Light and easy to handle	Labor-intensive and requires collector patience
Mouth aspirator	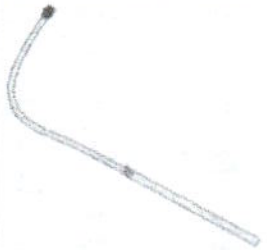	Blackflies, Phlebotominae (sandflies) and Culicidae (mosquitoes)			Light, easy to make and handle	Requires collector practice and patience
Prokopack aspirator				Resting and flying		
Dippers		Specific for mosquito immature stages		Swimming larvae		
Sweep nets	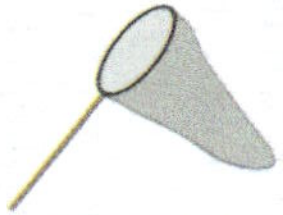	Blackflies, sandflies, and mosquitoes		Flying		
Human landing catch	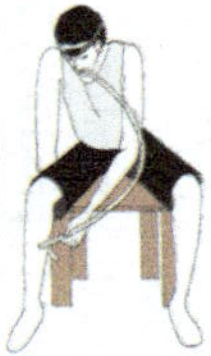		Human as bait	Host-seeking	Uses the collector itself as attractive	Requires the presence of the collector throughout the capture period; exposure to potentially infectious bites; difficulty in standardization due to differences in attractiveness
Bednet traps	See Table 21.2	Specific for mosquitoes (*Anopheles* spp.)				Requires the presence of the collector throughout the capture period; difficulty in standardization due to differences in attractiveness; its robustness makes transport and use in remote areas difficult

Shannon trap	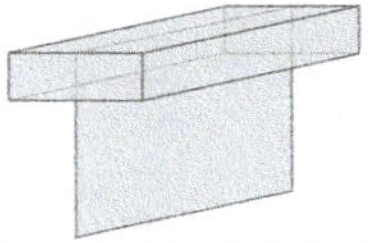	Blackflies, sandflies, and mosquitoes	Light, human, and animal as bait	Nocturnal activity	Easy to make, light and foldable, which makes it easy to transport and install	It needs the presence of the collector during the entire capture period
Biconical trap		Specific for Glossinidae (tsetse flies)	Color, shape, and size	Flying	Light and easy to handle	Its robustness makes use in remote areas difficult
Esperanza window trap		Specific for blackflies	Color and host odor			
Malaise trap	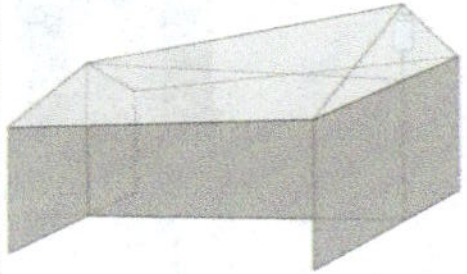	Tabanidae (horse flies), sandflies, Calliphoridae (blow flies), and tsetse flies				
Pitfall trap	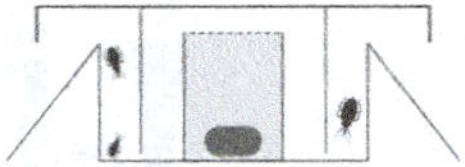	Kissing bugs and Cimicidae (bedbugs)	CO_2	Walking	Light and easy to make	Build-up of dust and debris inside the trap and regular maintenance
Disney trap	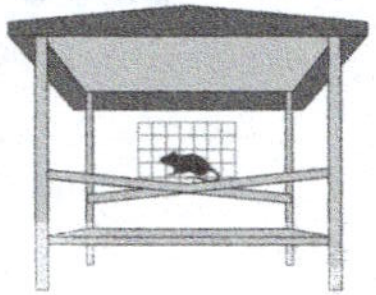	Sandflies	Small animal (rodent) as bait	Host-seeking	Doesn't require electricity	High cost, robustness makes transport and use in remote areas difficult

(continued)

Table 21.1 (continued)

Sampling methods		Specificity	Attractants	Behavior of interest	Advantages	Limitation
MosquiTRAP		Mosquitoes (*Aedes aegypti*, *Aedes albopictus*, and *Culex* spp.)	Synthetic oviposition attractant	Oviposition site-seeking	Light and compact; the mosquitoes can be identified during trap inspection in the field and can be used in large scale	Low sensitivity trap and requires synthetic attractants
Autocidal gravid Ovitrap	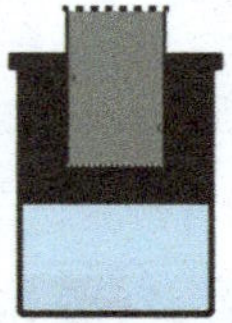		Hay infusion		Low-cost device that requires no electricity and can be used in large scale	Identification of mosquitoes requires lab facilities
Gravid *Aedes* trap			Synthetic oviposition attractant and host odors		Light and compact; the mosquitoes can be identified during trap inspection in the field, highly sensitive trap and can be used in large scale for *Aedes* surveillance and control	Use of insecticide can interfere in the trapping because of the odor
CDC light traps	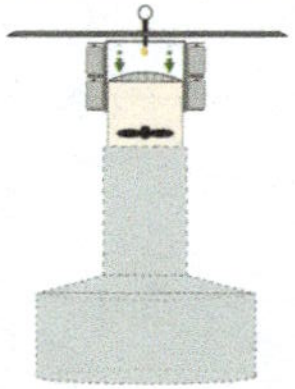	Sandflies and mosquitoes	Light, CO_2, and host odor	Host-seeking	Light, easy to handle and transport	Need batteries and high cost

Mosquito magnetic-X	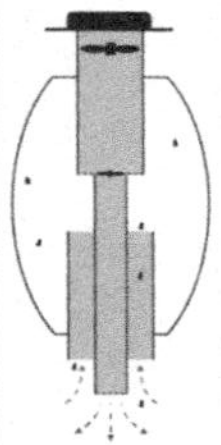	Blackflies, Ceratopogonidae (culicoids), sandflies, and mosquitoes	CO_2 and synthetic human odors		Light and easy to handle	High cost, robustness makes transport and use in remote areas difficult
BG-sentinel	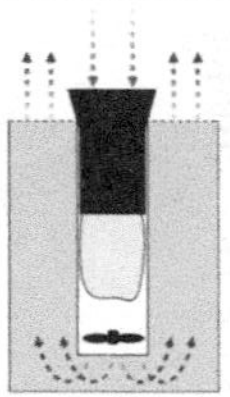	Blackflies, culicoids, sandflies, and mosquitoes			Light and foldable, which makes it easy to transport and install	High cost, the need for power supply makes it difficult to use in remote areas
BG-malaria	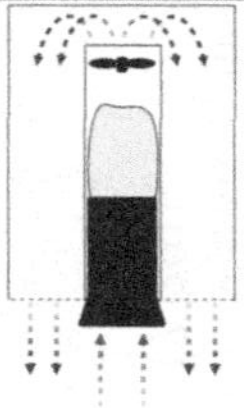	Mosquitoes (*Anopheles* spp.)				
Suna trap	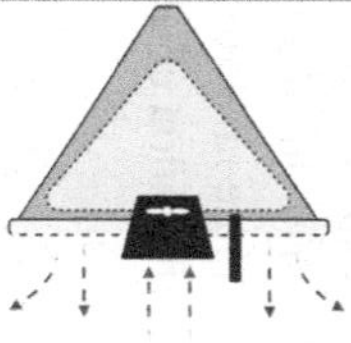				Light and easy to handle and install	

other tube into the collecting chamber. Depending on the purpose of the collection, the captured insects are removed to carriage cages or directly to a killing jar.

Battery-powered aspirators do not require much effort on the part of the collector, as the suction is generated by an electric fan, so that the sampling area is greater and provides collections of insects in large quantities and in a shorter time. There are several types of battery-powered aspirators (Nelson and Chamberlain 1955; Husbands and Holten 1967; Trpis 1968). Currently, the CDC Backpack aspirator (CDC-BP) (Clark et al. 1994) is one of the most used to capture mosquitoes (Maia et al. 2011) and sandflies (Oca-Aguilar et al. 2013). However, the high cost and weight (12 kg) are its main limitations, as well as the limited reach to higher or narrower places, due to its robust size. The Prokopack aspirator (Vazquez-Prokopec et al. 2009), which has the same operating characteristics as the CDC-BP but more compact and lightweight (900 g), has gained prominence in mosquito collection since its development. In comparison studies, Prokopack performed the same (Maia et al. 2011) or better than CDC-BP (Vazquez-Prokopec et al. 2009) for catching mosquitoes, and Prokopack has now become the standard method for collecting resting mosquitoes (Onyango et al. 2013; Charlwood et al. 2018; Koyoc-Cardeña et al. 2019).

Entomological Dippers This tool is regularly used to sample immature stages of mosquitoes in large water collections (Silver 2008). Dippers can vary in the volume of the sampling cup, but the standard model has 350 mL, and according to the size of the breeding sites, it can be used with short or long handles. In addition, for larvae and pupae of mosquitoes found in small reservoirs, such as plant with tightly overlapping leaf bases that form a cup for catching rainwater (e.g., bromeliads) or small containers, the sampling is conducted with a laboratory dropper or a Pasteur pipette.

Sweep Nets They are also known as entomological nets, and they are used to collect fast-flying vectors such as mosquitoes (Chen et al. 2011) and blackflies (Hempolchom et al. 2019). Like the manual mouth aspirator, sweep nets can be easily made, requiring only a metal ring (~50 cm in diameter); a 1 m long handle, which can be made with different materials (e.g., wood, PVC tube, metal); and a conical bag made of fine fabric, the length of which is twice the diameter of the ring. The handling of this tool requires a little practice, because after catching the insects in a fast and precise movement, the net must be twisted by rotating the handle, to prevent the insects from leaving. In order to empty the net, the captured specimens can be removed by a mouth aspirator or dumped in a cooled tray after being properly anesthetized or killed. Recently, this tool has been widely used for sampling male mosquitoes in swarms formed to attract females for mating (Dabiré et al. 2013; Kaindoa et al. 2017, 2019).

21.2.2 Passive Sampling

Although vectors spend most of their time in shelters, they are usually more attracted than sought. In the passive sampling strategy, the insects are attracted to the collection place through visual (shape, color, and contrast), physical (heat, humidity, convection currents), and/or olfactory (host odors, pheromones, oviposition attractants) cues (Clements 1992; Takken and Knols 2010). Among the various methods used to attract and catch vector, we can first classify them according to the bait used, which can be natural (e.g., humans or other animals) or artificial (e.g., light and synthetic odors) and are as follows.

21.2.2.1 Methods with Natural Baits

Human Landing Catch (HLC) The HLC relies on human attraction to attract mostly anthropophilic blood-feeding insects, and it is considered the gold standard for mosquitoes and blackflies sampling (WHO 1975). Human or animal as a bait is considered the most attractive physical and olfactory stimuli (CO_2, lactic acid, carboxylic acid, ammonia, octenol) (Eiras and Jepson 1991, 1994; Bosch et al. 2000; Pinto et al. 2001; Williams et al. 2006; Takken and Knols 2010) than any available sampling tools to date for anthropophilic blood-feeding insects. It is also the most accurate method for determining essential entomological parameters, such as the rate of bites/man/hour and entomological inoculation rate (Kelly-Hope and McKenzie 2009). In this method, the attraction is carried out by the collecting person him-/herself, who remains seated with legs exposed during the collection and equipped with a handheld mouth or Prokopack aspirator and a flashlight to capture the approaching mosquitoes; once an insect makes contact with human legs, it is caught, ideally before the collecting person is bitten.

Despite the advantages described above, the use of HLC has been widely debated due to ethical issues involving the exposure of the collector and the risk of contracting pathogens. Currently, its use is strictly limited, subject to approval by an ethics committee, and recommended only when there is no another equivalent method (Achee et al. 2015), as is the case with malaria and onchocerciasis vectors, for which there is no other more efficient collection method (Lima et al. 2014; Hendy et al. 2017).

Bednet Traps To minimize the limitations to the use of HLC, several attempts have been made to improve this technique by increasing the protection of volunteers. Bednet traps also consist of attraction to humans, but without direct contact with blood-feeding insects. They all have the same basic design, two chambers separated by screens, one with openings that allow insects to enter the trap and the other closed so that the volunteer remains protected by the nets (Table 21.2).

Shannon Trap This is another method commonly used for sampling nocturnal mosquitoes and sandflies, among other insects, in forest areas (Huemer et al. 2017; Demari-Silva et al. 2020). Made of white fabric, it was originally designed as a box with flaps to house a source of natural attraction (large animal or human) and/or light (Shannon 1939). However, the shape and dimensions of the trap have been improved over time, to a greater or lesser extent, and a simple version of the trap has the shape of a wall with flaps (Vexanet et al. 1986; Hashiguchi et al. 2014). Another remarkable change was made by Galati et al. (2001) that evaluated the modification of the trap color, in which more sandflies were attracted to the new black Shannon

Table 21.2 Examples of bednet traps that use humans as bait

Trap	Design	Characteristics	References
Mbita trap	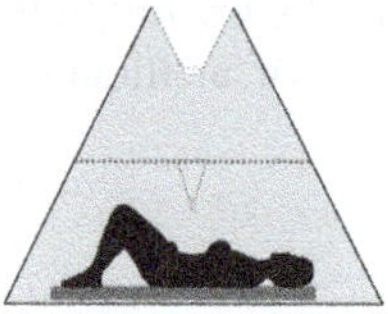	Made of white cotton fabric in a conical shape with the inverted top, creating a funnel shape with an opening for the mosquitoes to enter. The lower chamber has an elastic band on the lower edge to fix the trap to a mattress	Mathenge et al. (2002)
Ifakara tent trap (ITT)	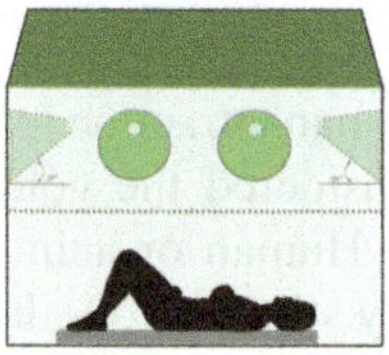	The trap has a rectangular shape, covered externally with green canvas. The attracted mosquitoes enter the interior through six funnels with the opening facing upward at a 37° angle. The collecting agent is protected by a Teflon-coated screen room equipped with a zipper that allows the collecting agent to catch mosquitoes from inside the trap	Govella et al. (2009)
Human-baited double net (HDN)	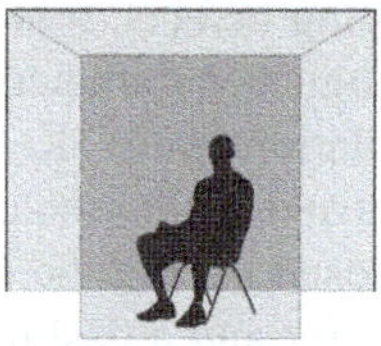	The collecting agent remains in an internal rectangular chamber of polyester fabric. A second larger external rectangular chamber raised 30 cm above the ground allows mosquitoes to enter the space between the two compartments. Subsequently, mosquitoes can be collected between the two cameras	Tangena et al. (2015)
MosqTent	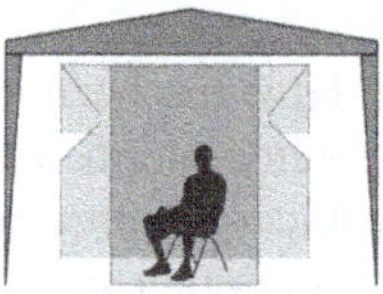	The human collector remains in the internal screened chamber. A second chamber, raised 10 cm from the floor and with upper flaps on the wall of the external chamber, allows mosquitoes to enter the compartment. A gazebo-like tent protects the human collector in light rainy conditions the sun and rain	Lima et al. (2017)

rather than the regular white Shannon. Essentially, nocturnal insects are attracted by the bait (light source or human/animal) inside the box, and they land on the fabric and/or are trapped in the flaps. The collector person can use a manual or mouth aspirator to collect the insects.

21.2.2.2 Methods with Artificial Baits (Traps)

Generally, traps are divided into passive and active traps. This classification refers to the use of mechanisms to capture and retain the insects once within the trap, which may require electricity for their operation (active traps) or not (passive traps).

Passive Traps (*No Electricity Is Required*)

Intercepting Traps These traps consist of a barrier to intercept the flight path of insects and rely on color contrast, light source, and host odors to enhance sampling. They are commonly hung in wild areas to catch insects such as blackflies (Rodríguez-Pérez et al. 2013; Hendy et al. 2018), horse flies (Ježek et al. 2019; Lucas et al. 2020), sandflies (Shimabukuro et al. 2016), blow flies (Castro et al. 2019), and tsetse flies (Kuzoe and Schofield 2004; Lindh et al. 2009). These types of traps exploit the responses of insects to visual stimuli, using spectral reflectance and polarization of reflected light fabric material as key determinants (Kriska et al. 2009). The collection with this method is performed by a dish positioned underneath and filled with preservative liquid (e.g., ethanol), where the intercepted insects fall.

Biconical Traps Interception traps are the main method used to sample tsetse flies because of their high-speed flight, which is about 6 m/s reaching up to speeds of 10 m/s (Gibson and Brady 1988). Among the several traps developed over 100 years of studies with different species of tsetse flies, the biconical trap, developed by Challier et al. (1977) for sampling *Glossina morsitans*, is the standard device currently used (Malele et al. 2016). As the name implies, this trap consists of two conical-shape parts, attached by their largest end. The upper cone is made of mosquito netting, and tsetse flies are caught in the top part. The lower cone is made of electric-blue cotton or synthetic cloth. The inner part is divided into four compartments by four segments of black cloth, which has four openings that allow the flies to enter the blue cone. The biconical traps attract tsetse flies through its color, shape, and size and exploit their behavior of flying over short periods with a relatively low capacity for active dispersal (Lindh et al. 2009). Additionally, they must be impregnated with biodegradable insecticides, which is a key component of tsetse control itself.

Esperanza Window Traps This is another type of trap that uses flight interception as the sampling mechanism. This device uses visual cues, such as vertical blue stripes on a black background, and host odor to attract host-seeking blackflies (Rodríguez-Pérez et al. 2013). This trap consists of a $1m^2$ piece of black satin fabric sandwiched between two sheets of $1m^2$ glass (3 mm thickness) that were each coated on their outer faces with a sticky coating. The Esperanza window trap, when baited with CO_2 and olfactory lures, collected substantial numbers of *Simulium ochraceum* s.l., one of the main vectors of *Onchocerca volvulus* in Latin America, and the addition of human scent (a worn shirt or BG-Lure) increased the efficacy of the trap, which approached the attractiveness of a HLC sampling (Hendy et al. 2018).

Malaise Traps This is also an intercepting trap that is used for the collection of flying insects, particularly dipteran. This trap is a tentlike structure that intercepts flying insects through a transverse screen barrier, which directs them to the highest point where a collection jar is positioned, usually filled with preservative liquid like the method described previously (Gressitt and Gressitt 1962). Malaise traps have been successfully used for sampling horse flies (Ježek et al. 2019; Lucas et al. 2020), sandflies (Shimabukuro et al. 2016), blow flies (Castro et al. 2019), and tsetse flies (Vale et al. 1988; Kappmeier 2000).

*Pitfall and Harborage Traps*Pitfall traps are specially designed to catch crawling insects, including kissing bugs (Guerenstein et al. 1995) and bedbugs (Hemiptera, Cimicidae) (Wang et al. 2016). These traps attempt to exploit the thigmotactic (organism's response to the stimulus of contact or touch) and refuge behaviors (Schofield 1979; Suchy and Lewis 2011).

The triatomine active vector surveillance consists of manual surveillance techniques, which are labor-intensive for searching triatomine nymphs and adult stages (Curtis-Robles et al. 2018) or by using trained dogs to find triatomines (Rolón et al. 2011). The cardboard type like harborage trap developed and used for detecting triatomine was compared with active surveillance technique, and the trap caught very few specimens, but nevertheless provided evidence that triatomines can be found in heavily populated urban environments (Wisnivesky-Colli et al. 1987; Dye-Braumuller et al. 2019). Also, pitfall traps utilizing yeast cultures to produce CO_2 as an attractant (Pimenta et al. 2007) and live-baited traps plus a shelter for the triatomines have been evaluated and successfully increased the triatomine samples (Angulo et al. 2012). To date, no traps have been validated for collecting triatomine species, neither in the field nor indoors.

Bedbugs (Hemiptera, Cimicidae) are very difficult to locate during visual inspections due to their small size and cryptic behavior. A complete inspection of an occupied room usually demands two experienced technicians and intensive labor. The visual inspections are unreliable and can underestimate the number of bedbugs in dwellings. Monitoring bedbugs by traps is effective for detecting low-level bedbug activity, which is often missed during visual inspections (Vaidyanathan and Feldlaufer 2013). Pitfall traps have been shown effective for early detection of bedbugs (nymphs and adults of both sexes), and when deployed under the legs of beds and upholstered furniture for 1 or 2 weeks, they were found to be the most effective for detection of low-level infections (Wang et al. 2016). There are several bedbugs traps commercially available, but not for kissing bugs.

Sticky Traps A simple sheet of card coated with odorless glue constitutes another simple method of sampling vector by different means. The sticky card can be deployed as an insect trap itself or within an insect trap as a retention method. The sticky card has been successfully used on the surface of water for sampling egg-laying malaria mosquitoes (Harris et al. 2011) and sandflies (Molyneux and Ashford 1983; Beresford and Sutcliffe 2006; Kaufman et al. 2005; Doggett et al. 2011). The sticky cards are generally inexpensive and easy to manufacture in large numbers. They can be prepared in large numbers and stored until required.

Disney Trap This trap was developed in the early 1960s, and it has commonly used a live rodent to lure and trap insects, especially sandflies (Disney 1966). Basically, it consists of a metal tray with a cage, holding an animal as bait, in the middle. Castor oil is sprayed onto the tray for sampling the insects attracted to the animal bait. The insects are stuck on the oil when they land on the surface of the tray, before or after contact with the bait. This trap relies on sandflies' behavior, in which the insect usually does not land on a host directly; they alight on a close-by surface and approach the animal in a series of short flight hops. Newly engorged females also tend to fly away from the host in a sequence of hops, and they also become stuck on the sticky surface of the tray (Brilhante et al. 2019).

The Disney trap was compared with Malaise, sticky traps, CDC light trap, CO_2 light trap, Shannon traps, and animal-baited trap in an endemic focus of cutaneous leishmaniasis in the Isfahan Province, Iran (Hesam-Mohammadi et al. 2014). Disney trap demonstrated the greatest efficiency compared to other traps which are currently used in routine sandfly surveillance programs. Some modification aiming to simplify handling and protection of the animal bait and durability of the trap in the field of Disney trap has also been shown to increase the trap performance when compared with Shannon and automatic light traps (Dorval et al. 2007).

MosquiTRAP This is a passive trap that was developed based on the behavior of gravid *Aedes aegypti* mosquito inside the breeding site (Eiras 2002). MosquiTRAP can also collect *Ae. albopictus* and *Culex* spp. The trap lures the gravid mosquitoes by olfactory (synthetic oviposition attractants) (Eiras et al. 2010) and visual stimuli (black matte color). The trap consists of a matte black container (33 cm high by 15 cm wide) filled with approximately 300 ml of tap water and an odorless sticky card that holds the captured mosquitoes and is attached from the waterline (Eiras and Resende 2009). The trap uses synthetic oviposition attractants identified in grass infusions (Eiras et al. 2010). When compared with the conventional method of larval search, it is more sensitive and captures the *Ae. aegypti* throughout the year, whereas larval research does not detect them during the dry season (Gama et al. 2007). The mosquitoes can be identified in the field during the trap inspection, thus reducing time, labor, and laboratory infrastructure (Fávaro et al. 2006; Eiras and Resende 2009). In Brazil, this sticky trap has been used in over 200 cities for *Ae. aegypti* surveillance in real time using quantitative reverse transcription polymerase chain reaction technique (qRT-PCR) for detection of infected mosquitoes by arboviroses (dengue, Chikungunya and Zika virus) (Sedda et al. 2018) and showed to be cost-effective (Pepin et al. 2013).

CDC Autocidal Gravid Ovitrap (CDC-AGO) The CDC-AGO trap is also a trap for sampling gravid *Ae. aegypti* and *Ae. albopictus* and a sticky card to retain the mosquitoes. The trap is shaped like a black matte bucket with an opening at the top, through which the insect enters the trap. Hay infusion is used as a natural attractant at the bottom of the trap (Mackay et al. 2013). A fine mesh screen prevents mosquitoes from reaching the water at the bottom of the trap. This trap has been used successfully in Puerto Rico for the control of *Ae. aegypti* (Sharp et al. 2019). Interestingly this gravid trap also catches host-seeking *Ae. aegypti* when baited with artificial human odor (Liu et al. 2019).

Gravid Aedes Trap (GAT) Like the previous two traps described above, the GAT is also a trap designed to catch gravid *Ae. aegypti* passively but also captures *Ae. albopictus*, *Culex quinquefasciatus*, and *Ae. polynesiensis* (Eiras et al. 2014; Ritchie et al. 2014; Cilek et al. 2017a; Harwood et al. 2018). Gravid mosquitoes are attracted to a black matte bucket base containing oviposition attractant (infusion or synthetic) (Eiras et al. 2018a) and are trapped in a translucent chamber containing a sticky card, or alternatively insecticide, canola oil, sprayed insecticide, or long-lasting insecticide-impregnated nets (LLINs) (Heringer et al. 2016). The sampled mosquitoes can be identified also during the trap inspection and subsequently submitted to qRT-PCR for detection of infected mosquitoes by arboviruses (Eiras et al. 2018a, b). The sensitivity of GAT was compared with other sticky traps, and it was shown that its success was approximately 4 times greater than MosquiTRAP and 3 to 6 times greater than CDC-AGO, in field studies in Australia and Florida, respectively (Ritchie et al. 2014; Cilek et al. 2017b). In addition, adaptations in GAT have been made for sampling male *Ae. aegypti* by the addition of a device that emitted the sound of the female wing beats at a frequency of 484 Hz to 65 dB (Johnson and Ritchie 2015; Johnson et al. 2018).

Active Traps (*Need Electricity*)

Light Traps Light-baited suction traps have been one of the most widely used tools for nocturnal insects sampling for long time. It has been used for over 130 years to sample a variety of nocturnal and crepuscular insets of both sexes. The range of attraction of light traps may vary from 2 m to 30 m (Rigot and Gilbert 2012). These traps can sample a large variety of insects of both sexes such as mosquitoes (Service 1970), biting midges (Diptera, Ceratopogonidae) (Bakhoum et al. 2016), sandflies, blackflies (Service 1979), horn flies (*Haematobia irritans*) (Diptera, Muscidae), fleas (Siphonaptera, Pulicidae) (Dryden and Broce 1993; Dryden et al. 1999; Müller et al. 2011), and kissing bugs (Vazquez-Prokopec et al. 2004; Rebollar-Téllez et al. 2009; Updyke and Allan 2018). The source of light that can be used varies from gas lamps, incandescent, UV (ultraviolet) light tubes, and more recently LEDs (light-emitting diodes).

Centers for Disease Control and Prevention Light Trap (CDC-LT) The CDC light trap was developed in the 1960s (Sudia and Chamberlain 1962) and has been considered as the standard trap for adult mosquito (except for *Ae. aegypti* and anophelines species) and sandflies sampling and variety of small nocturnal insects. The light source attracts the nocturnal insects that are then sucked into the trap by the downward current produced by the electrical fan and kept in a fabric mesh compartment. CDC-LT should be deployed before dusk and collected in the morning after dawn, because it can operate overnight to collect nocturnal insects. The specimens collected can sometimes become damaged, however, by the electrical fan. The addition of attractants to CDC-LT, such as CO_2 and other synthetic odors, is done for sampling mosquitoes (Newhouse et al. 1966; Ponlawat et al. 2017; Xue et al. 2008) and sandflies (Pinto et al. 2001; Ballart et al. 2016; Silva et al. 2019).

Mosquito Magnet-X® (MMX) Developed to capture anophelines (malaria vectors), it collects a variety of mosquito species in different environments (Brown et al. 2008). This trap uses the concept of counterflow, which is produced by two electric fans that generate, at the same time, downward (forms the odor plums of attractants) and upward air currents (sucks in the mosquitoes) (Kline 1999). Initially, MMX was designed to use CO_2 as attractant, but the addition of other synthetic odors increased the attraction of mosquitoes (Qiu et al. 2007). In comparison with CDC-LT, MMX captured more mosquitoes in general and also the largest number of species (Heidi et al. 2008). Sandflies can be also collected by the addition of colored LEDs (red, blue, and green) and attractants to MMX trap (Mann et al. 2009).

BG-Sentinel® (BGS) The original BGS trap was developed to capture host-seeking female *Ae. aegypti* (Kröckel et al. 2006) but can also catch a variety of mosquito species (Mercer et al. 2012; Cilek et al. 2017a). The traps consist of cylindrical shape base and one single electric fan which can produce the upward airflow to release odor plumes for attracting mosquitoes and a central downward airflow for sucking the insects into a collection bag. The BGS trap uses synthetic human odors (Kröckel et al. 2006; Englbrecht et al. 2015; Akaratovic et al. 2017), and the addition of CO_2 has a synergist effect in mosquito sampling (Ázara et al. 2013; Roiz et al. 2016; Jerry et al. 2017). Natural host odors can also be used as an attractant, such as socks and t-shirts already worn by people (Owino et al. 2014), or even mice to attract male and female *Aedes* mosquitoes (Le Goff et al. 2016). BGS has been also considered the gold standard trap for mosquito collection because of its ability to catch more mosquitoes when compared with several commercial traps (Williams et al. 2006; Farajollahi et al. 2009; Verhulst et al. 2015). BGS showed significantly reduced biting pressure by mass trapping control of *Ae. aegypti* in Brazil (Degener et al. 2014) and *Ae. albopictus* population in Italy (Englbrecht et al. 2015).

*BG-Malaria (BGM)*BGM uses the BGS airflow concept, but the trap deployment is different. BGS is deployed vertically and on the ground, whereas BGM hangs upside down 40 cm from the ground (Gama et al. 2013; Rodrigues et al. 2014; Batista et al. 2017). BGM was effective in attracting malaria mosquitoes by adding synthetic human odors (Batista et al. 2017, 2018). In field studies conducted in Brazil (Gama et al. 2013) and Tanzania (Batista et al. 2018), BGM and HLC have collected similar numbers of *Nyssorhynchus darlingi* and *Anopheles funestus.*

BG-Suna The BG-Suna trap was also developed by BioGents (Regensburg, Germany) and uses the BGM airflow concept, but it has a rigid external structure in a conical shape (Hiscox et al. 2014). This trap was also developed for sampling anopheline mosquitoes. In field studies in Kenya, BG-Suna trap captured similar numbers of *An. gambiae s.l.* and *An. funestus* as the MMX trap (Hiscox et al. 2014) and the HLC in the field in Malawi (Mburu et al. 2019). The Suna trap was the first trap evaluated for the mass capture of malaria vectors in a study carried out in Rusinga Island, Kenya, whose reduction in the populations of *An. funestus* demonstrated the possibility of using traps based on synthetic odors for malaria control (Homan et al. 2016).

21.3 Preparation, Identification, and Conservation

After sampling the insect vectors, they can be transported alive to the laboratory or killed at the capture site. The transport of live insects can be done when laboratory bioassays are intended, such as insecticide susceptibility tests, dissection to investigate parity (ovaries) or infection (intestine and salivary glands), or even to start a new colony.

The transport of live insects to the laboratory needs caution to ensure that the samples arrive as intact as possible. Containers, insect-collecting traps, or cages should be used, with enough space for the insects to be able to move without damaging their structures (e.g., legs and wings). The packaging must be rigid and with a lid, preferably of insulating material (e.g., Styrofoam box), and at the bottom should lined with some quilting material (towel, cotton, etc.) to avoid, as much as possible, movements and vibrations. A cotton pad soaked in water or a sugar solution for feeding is placed on the screen of each cage.

For immature forms (larvae and pupa) that develop in water, it should be placed in partially submerged or with high humidity container. The transport must be carried out in bottles containing water or substrate in which they were collected, demanding a space with air in the bottle.

When the objective of sampling is for identification and/or taxonomic studies, the insects can be killed using the techniques described below:

70% alcohol: Should be used to kill the insects only the larvae, pupae, and nymph stages and adult insects with soft or delicate bodies, which are placed in flasks containing only 70% alcohol.

Toxic gases: Toxic gases in a deadly flask is the ideal method for killing adult insects in general. A wide-mouthed glass jar with a lid labeled "POISON" should be used, containing a plaster layer of 2 or 3 cm at the bottom, containing a deadly agent (ethyl acetate is the safest and easy to acquire). During the procedure, the bottle must be closed, and the insects remain there until they die.

Freezing: This is an alternative to toxic gases and can be done in the field using dry ice. However, if it is not possible due to the logistics in the field, you can transport the live insects and use the freezer (−20 °C) to kill them. The insects should be placed in tightly closed containers and with absorbent paper at the bottom, to avoid drying out, and remain there for up to 24 h or long enough for them to die. The handling of insects that have been killed in this way must be done with care, and defrosting must be carried out before starting the preparation techniques to preserve them. As it takes longer than the deadly flask to kill the insects, a label with location, date of collection, and name of the collector must be inserted inside the container for freezing.

21.3.1 Preparation

Most adult insects are conserved dry, mounted on pins, and vectors are no different. However, adults of some species and immature soft-bodied forms are preserved in preservative liquid or mounted on microscope slides, due to the wrinkling and deterioration normally caused by drying. The preparation methods for the three methods mentioned are detailed below, and the most suitable for each family of vectors are summarized in Table 21.3.

Table 21.3 Preparation methods according to families of insect vector

Order	Family	Pinning		Microscope slides	Liquid
		Direct	Double mount		
Diptera	Ceratopogonidae	No	No	Yes	Yes
	Culicidae	No	Yes	Yes	Yes
	Glossinidae	Yes	No	No	Yes
	Muscidae	Yes	Yes	No	Yes
	Psychodidae	No	No	Yes	Yes
	Simuliidae	No	Yes	Yes	Yes
	Tabanidae	Yes	No	No	Yes
Hemiptera (Heteroptera)	Reduviidae (Triatominae)	Yes	No	No	Yes
Siphonaptera	All	No	No	Yes	Yes

21.3.1.1 Mounting on Entomological Pins

The best time to assemble the insects is just after their killing, to prevent them from drying out and hardening, becoming breakable, which would make handling difficult. If the insects are dry for assembly, it is recommended to place them in a relaxing chamber so that their appendages and other parts of the body can be placed in the correct position. Relaxing chamber can be made of a container, preferably of glass, flat and wide and with a lid. A damp sponge or a layer of damp sand should be placed on the bottom of this container and filter paper above to separate the insects from the wet substrate. To avoid the proliferation of fungi in the container, small pieces of naphthalene should be mixed with the substrate. The time required for the insects to become soft depends on their size and the ambient temperature, which can vary from a few hours to days, but only until they ensure that their legs and wings can be moved freely.

The pins used in the assemblies are flexible and have 38 mm in length and vary in thickness and vary in size from 000 to 7. The assembly in pins is very simple and needs few materials, only entomological pins, fine-tipped forceps (type watchmaker), and a pinning block, which consists of a ladder with heights of 7, 15, and 25 mm used to standardize the height of the insects on the pin (Fig. 21.1).

Direct Pinning: The pin is inserted directly into the insect's body at a specific location, in the pronotum in triatomines and mesothorax in dipteran (Fig. 21.2a, b). The inserting of an entomological pin should be carried out slightly to the right, because if any damage occurs during the process, the structures on the left side will remain preserved, due to the bilateral symmetry presented by the insects. The correct position of the assembled samples is horizontal perpendicular to the pin, with its appendages arranged in order to facilitate future identification (Fig. 21.2c, d).

Double Mount: This type of assembly is performed using small triangles made of rigid white paper (2 mm at the base × 6 mm long) or micro-pins, which are produced with the same material as standard entomological pins (10–15 mm long).

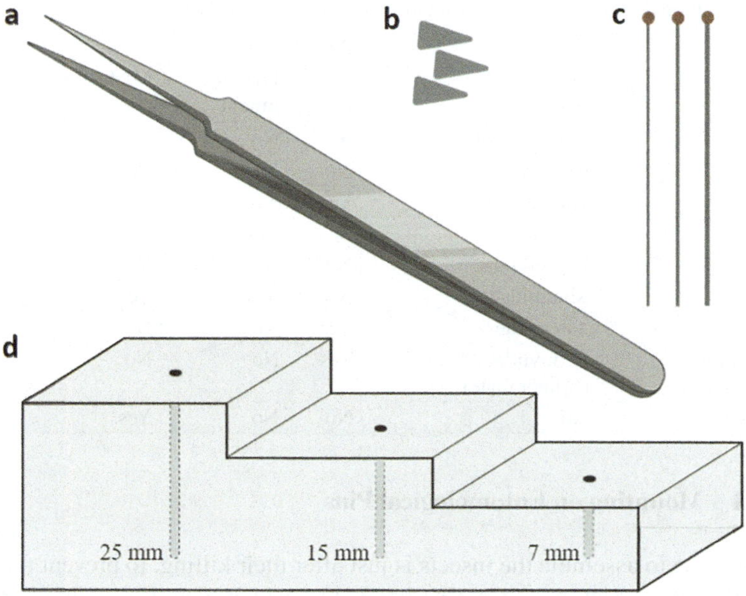

Fig. 21.1 Tools used for mounting specimens on entomological pins: (**a**) forceps, (**b**) points, (**c**) pins, and (**d**) pinning block

The triangular assembly is suitable for mosquito adults, in which the procedure involves inserting an entomological pin centrally at the base of the triangle. Then, the specimen should be attached to the tip, using a small amount of clear glue or colorless nail polish. In mosquitoes, the attachment must be performed on the thoracic pleura, below the wings, so that the legs are facing the pin and the head is positioned to the right base of the triangle (Fig. 21.2e, f). Micro-pin assembly is recommended also to preserve simulids and small species of flies. The procedure is performed by sticking the sample with the micro-pin, which, in turn, can be inserted in a platform (small squares or rectangular strips) or in a small cube, both made of firm and porous material (e.g., Styrofoam or cork) and pierced with a normal entomological pin.

21.3.1.2 Liquid Preservation

Ethanol (70–80%) is the most commonly used liquid preservative for preserving immature and soft adult forms in the long term. Mosquitoes and other adult insects that have hairs and scales should not be preserved in alcohol, regardless of concentration. For immature forms, it is recommended to use a fixative fluid before being placed in alcohol, of which the most used to fix vector larvae and pupae is KAA (two parts of glacial acetic acid, ten parts of 95% ethanol, and one part kerosene).

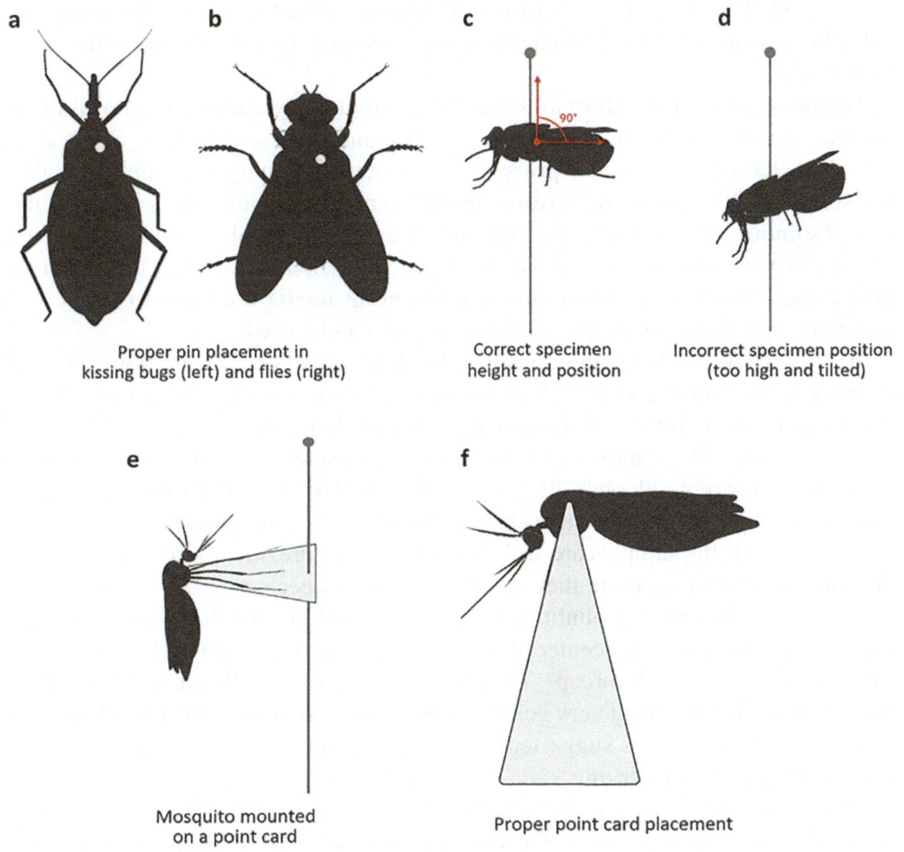

Fig. 21.2 Mounting on entomological pins: direct pinning (**a**, **b**, **c**, and **d**) and double mount (**e** and **f**)

21.3.1.3 Mounting Insects on Microscope Slides

Very small and delicate insects, immature forms, or certain larger insect structures that require detailed study are mounted temporarily or permanently on slides to be examined microscopically. Although there are several ways of preparing the slides, which vary according to the specimen to be assembled, the standard procedure is basically the same for everyone (Consoli and Lourenço-de-Oliveira 1994; Linardi and Guimarães 2000; Calvão-Brito and Maia-Herzog 2003; Vilela et al. 2018).

The mounting media can be of the aqueous type, based on gum arabic and chloral hydrate (e.g., Hoyer medium and Berlese liquid) or based on resins, such as Canada balsam and Euparal®. The choice of mounting media depends on the purpose of the preparation, with aqueous media indicated for temporary assembly (for identification purposes), as they deteriorate over time, and resin media, although the preparation process is longer, are recommended for specimens to be stored for a long period. The entire process of assembling the samples follows the steps provided below, and the times shown are average and may vary according to the specimen studied.

Diaphanization: Procedure to "clean" the samples, by chemical removal of the internal contents of the body (organs, muscles, and other soft tissues) and leaving only the sclerotized parts. The process takes place by dipping the samples in alkaline solutions, 10% potassium hydroxide (KOH) or 10% sodium hydroxide (NaOH), or acid solutions (lactic acid or lactophenol), leaving them soaking for several hours, which can vary between 3 h and 24 h, because the more chitinized the insect the longer the time needed. Most aqueous mounting media continue to clean after assembly, in which case weaker cleaning agents can be used.

Washing: The washing process stops the diaphanization and the removal of all cleaning agent from the sample. Samples must be soaked for approximately 15 minutes in acetic acid (10%) and washed again in distilled water.

Dehydration: The removal of water is done gradually, usually carried out in an alcoholic series (e.g., alcohol 70%, 80%, 90%, and 100%). The residence time in each stage depends on the samples but can be 10 to 20 minutes.

Bleaching: If the samples are still dark after all the previous steps that prevent to the microscope examination, they can be clarified in beech creosote for 24 h.

Mounting: The insect-mounting process consists of placing the insect in an aqueous or resin medium at the center of the glass slide and then gently placing a coverslip over the sample with forceps. The coverslip must be slightly angled by touching first one side and releasing very gently, in order to prevent bubbles. For a long-term storage of the slides, it is suggested to use coatings at the ends of the coverslip to seal it and thus prevent drying.

Drying: The final stage in the preparation of the slides is the drying in an oven set at 45 °C at horizontal position. The required time for drying differs from few days using aqueous up to several weeks for resin-based mounting process.

21.3.1.4 Labeling

For the correct storage of the insect specimens, tags must be always placed to be easily identified. Generally, two tags are produced, one with information on the sampling origin and the other of the identification, which must be made on white paper and written with ink, pencil or printed. Although the size of the tags may vary, the dimensions of 2.0 × 1.0 cm are recommended. The first tag should include the origin of the sampling location (city, state, and country), date (with month in Roman numerals), and collector's name. The second identification tag must contain as much information as possible about the sample identification (order, family, genus, or species).

Belo Horizonte, MG, Brazil 20-IV-2020 Batista, E.P.A. col.	Diptera Culicidae *Aedes aegypti*

In pinned specimens, the tags must be positioned parallel to the insect's body, at a uniform height on the pin; the mounting block can be used for this. In liquid preservation, insects must be complemented by only tag label that includes all information, which must be written in pencil on stiff white paper, such as index card stock.

Insects preserved between the slide and coverslip should also be properly labeled. All information related to the sample can be placed on just one label or on two, in the latter case positioned at each end of the slide.

21.3.2 Identification

The identification of insect vectors is carried out mainly through a comprehensive examination of morphological characters (traditional taxonomy). However, identifications based on DNA (molecular taxonomy) is also widely used. The use of both methods provides additional information to the identification of specimens belonging to a complex of cryptic species (Li and Wilkerson 2005; Erlank et al. 2018; Laurito et al. 2018) or proposes changes in the classification of groups (Foster et al. 2017).

21.3.2.1 Traditional Taxonomy

Morphological identification is carried out by using identification keys or with the help of a taxonomist. There are numerous publications to identify each group of vectors, among which we highlight Ronderos et al. (2018) for biting midges; Consoli and Lourenço-de-Oliveira (1994), Forattini (2002), Harbach (2007), and Coetzee (2020) for mosquitoes; Zumpt (1973) and Carvalho et al. (2002) for flies; Galati (2018) for sandflies; Shelley et al. (2004) and Hamada and Nascimento (2018) for blackflies; Lent and Wygodzinsky (1979) for kissing bugs; and Linardi and Guimarães (2000) for fleas.

21.3.2.2 Molecular Taxonomy

In mitochondria, structures present in practically all cells in large numbers, present genes are considered sources of information very useful for several molecular studies in recent decades. Mitochondrial DNA has been vastly used to identify complexes of species, as well as to analyze phylogenetic relationships among insect populations, historical patterns of genetic diversity and migration, species invasion, and biogeography (Wilkinson and Fleming 1996; Wahlberg et al. 2005; Leonard 2008; Zink and Barrowclough 2008; De Barro and Ahmed 2011; Erlank et al. 2018).

In DNA-based identification, sequences of base pairs of one or more genes are used to distinguish species, a procedure denominated DNA barcoding. A 648 base pair fragment at the 5′ end of the mitochondrial cytochrome C oxidase subunits I (COI) (the Folmer region) is the most widely used barcode in molecular analysis and is currently the standard genetic marker for most animals, including insects (Hebert et al. 2003). Moreover, the sequence of the COI gene presents a relatively high degree of genetic variation in this region but differs slightly in length, and depending on which group of organisms is examined, a highly conserved segment of this gene is used (Marshall 2005; Hajibabaei et al. 2006; Ashfaq et al. 2018). For presenting a genetic organization considered simple and uniform, besides allowing the retrieval of genetic information from dimensions in inadequate states of conservation, this molecular marker has been widely used (Zhang and Hewitt 1997).

For DNA extraction, amplification, and sequencing procedures, there are several protocols that have been developed in the last 30 years, but classic protocols, such as the one described by Livak (1984) DNA extraction, still remain among the most used by researchers (Kamgang et al. 2017; Singh et al. 2020). Regarding DNA amplification by polymerase chain reaction (PCR), each vectors group follows protocols with taxon-specific primers, rather than the universal primers established by Folmer et al. (1994). The standard protocol for PCR assays to identify members of the *Anopheles gambiae* complex, for example, is the one described by Scott et al. (1993). In addition, procedures like the PCR-restriction fragment length polymorphism (PCR-RFLP) are widely used for species identification of many organisms (Ratcliffe et al. 2003; Dokianakis et al. 2016; Singh et al. 2020). Furthermore, the DNA of unknown samples can be identified by comparing their barcode sequences with a reference library, such as Barcode of Life Data Systems (BOLD) (Ratnasingham and Hebert 2007).

Nevertheless, the success of species identification through DNA barcoding approaches is highly dependent on the adequate collection and preservation of the samples, which must be carried out considering the DNA conservation. Therefore, after collection it is recommended to store fresh specimens in a freezer, preferably at −80 °C, or in refrigerated 95% ethanol, due to its potential to prevent DNA degradation (Flournoy et al. 1996; Nagy 2010).

21.3.3 Conservation

The pinned insects are stored in entomological strongboxes or drawers that must be made of wood and with a standard size of 50 × 40 × 5 cm. The boxes or drawers have a glass lid and the bottom covered with Styrofoam plates or another firm and porous material that allows fixing the pins and covered with a white paper, and thus, it improves the contrast.

As a precaution against the attack of other insects and the proliferation of fungi, it is recommended to add naphthalene or paraformaldehyde (powder or tablets), which must be kept in order not to destroy the insects. However, if the specimens still mold, they can be cleaned with a brush or cotton swab moistened with ether. Insect grease can occur due to the dissolution of fatty tissues, in which case cleaning is carried out by immersing them in the ether for 2 days.

For specimens preserved in 70% alcohol, evaporation is the biggest threat, so replacing alcohol periodically is highly recommended to keep it at levels suitable for conservation. Packing smaller vials into larger vials, also containing alcohol, facilitates the replacement of the liquid, since it is carried out only in the largest containers. In addition, large collections must be kept in separate fireproof areas.

The slides can be stored horizontally or vertically, usually in wooden or plastic boxes. The boxes have different sizes, and the most used are those that contain 50 to 100 slides, whose internal side walls have devices to keep them in a vertical position separated from each other. It is important to highlight that the slides storage in the vertical should only be carried out for those that are completely dry, as the coverslips can slip, especially those that were assembled with Hoyer's medium, which can soften in very humid environments.

In general, collections tend to deteriorate quickly, and in order to keep them for long periods, basic precautions like maintaining low levels of humidity and temperature also prevent mold and infestation of other insects that may damage the samples. Conservation protected from direct light preserves the color of the specimens longer.

21.4 Data Collection and Management

The details of the sampling methods used and the habitat in which the collections are made must be recorded. The data can be obtained in portable units and the global positioning satellite (GPS) systems for the sampling site coordinates together with information on ecology, epidemiology, and mapping (Thomson and Connor 2000; Eiras et al. 2018a, b; Rutto et al. 2018; Vijayan and Anbalagan 2018; Salah et al. 2020).

Throughout studies targeting a single species of medical or veterinary importance, it should also identify other species present. Thus, this information of associated species that can coexist in a habitat provides useful indicators of the characteristic of the environment (Dorville 1996; Ferreira et al. 2017; Rutto et al. 2018). Also, by identifying the species in their corresponding genera and families, the number of individuals registered in the databases can be retrieved for further analysis, including diversity studies.

Studies of vector population fluctuations must follow a rigorous sampling regime that allows replication without suppressing population density. These studies can provide reliable information on the abundance of vector that can cover several years of sampling, in order to assist in the implementation of control measures (Sanchis-Marin et al. 1989; Pepin et al. 2013; Barrera et al. 2014; Eiras et al. 2018a, b; Sedda et al. 2019).

Many studies of blood-feeding population density are directly related to the bite rate and the increase in cases of a disease, since the more vectors, the greater the probability of them becoming infected and transmitting the pathogen to a healthy individual (Phaijoo and Gurung 2015; Suparit et al. 2018). This reflects in a greater vectoral capacity, which is the ability of a vector to transmit a given pathogen, which represents the daily rate of potentially infective contacts between vectors and susceptible humans (Garret-Jones 1964; Mitchell et al. 2015; Lounibos and Kramer 2016).

To estimate the size of the blood-feeding population in an area, mark-release-recapture studies allow investigating dispersion or measuring survival over time for mosquitoes (Russell et al. 2005; Maciel-de-Freitas et al. 2008; Marini et al. 2010), kissing bugs (Schofield et al. 1992), bedbugs (Cooper et al. 2015), sandflies (Silva et al. 2013), and blackflies (Rodríguez-Pérez et al. 2001). The markers used are varied, such as rubidium, incorporated into blood sources (Kimsey and Kimsey 1984), for vegetable sources of sugar (Alexander and Usma 1994), and detected by flame ionization spectrometry. Other simple techniques like the use of colored sugars sprayed on natural substrates in the environment (Schlein and Yuval 1987) and fluorescent powders sprayed on insects retained in a container, which are released, and the recaptured insects are illuminated with ultraviolet light (Maciel-de-Freitas et al. 2007). Radio telemetry was also successfully used to demonstrate as a potential tool for studying fine-scale non-flight movement of triatomines (kissing bug) and the discovery of cryptic resting habitats in savanna and urban areas (Hamer et al. 2018).

The study of population density can provide valuable information on the ecology and behavior of the vector with the help of traps; the efficiency of a control intervention can be evaluated, estimating the density of mosquitoes before and after treatment (Fávaro et al. 2006; Resende et al. 2013). The estimation of the size of the mosquito population, for example, can be made by means of adult traps in studies with different density of traps and with the aid of the Box-Lucas function (Box and Lucas 1959) to estimate the size of the Culicidae population (Resende et al. 2012).

The entomological indices for blood-feeding insects indicate the infestation rates, which can be estimated in the egg, larva, pupa, and adult stages of the vector insects. The absolute values of entomological samplings are useful for planning and management of the vector control. The majority of entomological indices studies have been used on mosquitoes, and the weekly surveillance of *Ae. aegypti* is a good example of how important the temporal and geographical comparisons are for triggering vector control measures (Eiras and Resende 2009; Steffler et al. 2011; De Melo et al. 2012; Pepin et al. 2013, 2015; Sedda et al. 2018).

In Brazil, an innovative mosquito surveillance technology (MI-Aedes) has been successfully implemented over 200 cities (Ecovec—Rentokil, Brazil) (Eiras et al. 2018a, b). MI-Aedes consists of GPS-located sticky traps for *Ae. aegypti*females citywide at fine spatial and temporal scales for vector surveillance. The data is entered by cell phones and directly sent to a database, which automatically generates maps of mosquito density and hot spot and clusters (Eiras and Resende 2009; Pepin et al. 2013). All trapped mosquitoes are submitted to molecular biology technique (reverse transcription polymerase chain reaction—RT-PCR) for arbovirus detection (dengue, Chikungunya and Zika virus) (Eiras et al. 2018a, b). The information of infected mosquitoes is crucial for early warning system for epidemics since viral detection in mosquitoes can precede detection in humans. The main advantage of the MI-Aedes platform over traditional mosquito surveillance is the integration of continuous vector monitoring coupled with an information technology platform for near real-time data collection, analysis, and decision-making (Pepin et al. 2013; Eiras et al. 2018a, b; Sedda et al. 2018). Additional environmental data (temperature, precipitation, and humidity) and the mosquito data from MI-Aedes have been used for prediction modeling for prediction of *Ae. aegypti* population (Barsante et al. 2014; Ferreira et al. 2017), population dynamics (Lana et al. 2018), and vector control (Pepin et al. 2015).

Data from sampling blood-feeding insects can exhibit as simple as transforming them into relative values and are represented by indices or proportions (Clark et al. 1994; Gomes 1998; Dorval et al. 2007; Resende et al. 2013; Barrera et al. 2014; Lima et al. 2014; Cilek et al. 2017b; Hendy et al. 2017). It can be analyzed using generalized linear models (continuous and discrete data) (Zeger and Liang 1986; Sedda et al. 2018), survival analysis models (Ohm et al. 2016), and mixed effects models (longitudinal data) (Breslow and Clayton 1993; Kaldun and Otti 2016; Mantilla et al. 2018). These models can be used to assess complex ecological processes, such as competition (Connor and Simberloff 1986; Lourenço et al. 2018; Moon et al. 2019).

Information on climatic factors, such as temperature and precipitation, can be useful in predicting the probability of the occurrence of diseases carried by blood-feeding insects, through mathematical modeling (Barsante et al. 2017; Ferreira et al. 2017; Mallya et al. 2018; Sedda et al. 2018, 2019). These seasonal factors modulate the activities of vectors, pathogens, and reservoir animals and are widely used in epidemiological studies of diseases transmitted by blood-feeding insects (Eldridge 2000).

The methods of sampling arthropods related to the human health and other animals, based on an adequate planning of collection techniques, as well as the duration, frequency, spatial distribution, and analysis, should contribute so that entomological surveillance can constitute a research tool (De Melo et al. 2012; Mallya et al. 2018; Sedda et al. 2019). In this way, entomological surveillance based on the observation and evaluation of the collected data must determine changes in the geographic distribution of the infestation areas, obtain relative measurements of the vector population over time, and facilitate appropriate and timely decisions regarding interventions in the prevention programs control (Cohnstaedt et al. 2012; Eiras et al. 2018a, b; Sedda et al. 2019).

In summary, the selection of an appropriate method depends on the objectives and type of study to be carried out. The species, sex or physiological state of the insects, restrictions on the preservation, and transport of the samples have also to be considered. This information can contribute to knowledge of the biology of most blood-feeding insect species in nature.

Acknowledgments We are grateful to Prof. Owen Jones for English revision of this chapter and to Dr. Juliana Maria Trindade Bezerra for the critical review. AEE acknowledges the research grants from Conselho Nacional de Desenvolvimento Científico e Tecnológico, CNPq (312335/2018-0 and 440358/2016-7); Coordenação de Aperfeiçoamento de Pessoal de Nível Superior, CAPES (88881.130779/2016-01); Departamento de Ciência e Tecnologia, DECIT; and Secretaria de Ciência, Tecnologia e Insumos Estratégicos, SCTIE. EPAB acknowledges the post-doctoral fellowship from CAPES (Grant 88887.467092/2019-00). We would like to extend our sincere gratitude to Ms. Sarah Carvalho Batista for her wonderful work creating the human landing catch illustration. The other illustrations were created with BioRender.com.

References

Achee NL, Youngblood L, Bangs MJ et al (2015) Considerations for the use of human participants in vector biology research: a tool for investigators and regulators. Vector Borne Zoonotic Dis 15:89–102. https://doi.org/10.1089/vbz.2014.1628

Akaratovic KI, Kiser JP, Gordon S et al (2017) Evaluation of the trapping performance of four biogents AG traps and two lures for the surveillance of *Aedes albopictus* and other host-seeking mosquitoes. J Am Mosq Control Assoc 33:108–115. https://doi.org/10.2987/16-6596.1

Alexander B, Usma MC (1994) Potential sugar sources of *Lutzomyia youngi* (Diptera: Psychodidae) in a Colombian coffee plantation. Annals Trop Med Parasitol 88:543–549. https://doi.org/10.1080/00034983.1994.11812902

Angulo VM, Esteban L, Luna KP (2012) *Attalea butyracea* palms adjacent to housing as a source of infestation by *Rhodnius prolixus* (Hemiptera: Reduviidae). Biomedica 32:277–285

Ashfaq M, Sabir JSM, El-Ansary HO et al (2018) Insect diversity in the Saharo-Arabian region: revealing a little-studied fauna by DNA barcoding. PLoS One 13(7):e0199965. https://doi.org/10.1371/journal.pone.0199965

Ázara TMFD, Degener CM, Roque RA et al (2013) The impact of CO_2 on collection of *Aedes aegypti* (Linnaeus) and *Culex quinquefasciatus* say by BG-sentinel traps in Manaus, Brazil. Mem Inst Oswaldo Cruz 108:229–232. https://doi.org/10.1590/0074-0276108022013016

Bakhoum MT, Fall M, Seck MT et al (2016) Foraging range of arthropods with veterinary interest: new insights for Afrotropical Culicoides biting midges (Diptera: Ceratopogonidae) using the ring method. Acta Trop 157:59–67. https://doi.org/10.1016/j.actatropica.2016.01.023

Ballart C, Vidal G, Picado A et al (2016) Intradomiciliary and peridomiciliary captures of sandflies (Diptera: Psychodidae) in the leishmaniasis endemic area of Chapare province, tropic of Cochabamba, Bolivia. Acta Trop 154:121–124. https://doi.org/10.1016/j.actatropica.2015.11.007

Barrera R, Amador M, Acevedo V et al (2014) Use of the CDC autocidal gravid ovitrap to control and prevent outbreaks of *Aedes aegypti* (Diptera: Culicidae). J Med Entomol 51:145–154. https://doi.org/10.1603/ME13096

Barsante LS, Paixão KS, Laass KH et al. (2014) A model to predict the population size of the dengue fever vector based on rainfall data. arXiv preprint arXiv:1409.7942

Barsante L, Cardoso R, Acebal J et al (2017) Modelo entomológico determinístico sob efeito da pluviosidade para o *Aedes aegypti* e o *Aedes albopictus*. Proc Ser Braz Soc Comput Math 5:1–7. https://doi.org/10.5540/03.2017.005.01.0066

Batista EPA, Ngowo HS, Opiyo M et al (2017) Semi-field assessment of the BG-malaria trap for monitoring the African malaria vector, *Anopheles arabiensis*. PLoS One 12(10):e0186696. https://doi.org/10.1371/journal.pone.0186696

Batista EPA, Ngowo H, Opiyo M et al (2018) Field evaluation of the BG-malaria trap for monitoring malaria vectors in rural Tanzanian villages. PLoS One 13(10):e0205358. https://doi.org/10.1371/journal.pone.0205358

Beresford D, Sutcliffe JF (2006) Studies on the effectiveness of Coroplast sticky traps for sampling stable flies (Diptera: Muscidae), including a comparison to Alsynite. J Econ Entomol 99:1025–1035. https://doi.org/10.1093/jee/99.3.1025

Bosch OJ, Geier M, Boeckh J (2000) Contribution of fatty acids to olfactory host finding of female *Aedes aegypti*. Chem Senses 25:323–330. https://doi.org/10.1093/oxfordjournals.chemse.a014042

Box GEP, Lucas HL (1959) Design of experiments for non-linear situations. Biometrika 46:77–90. https://doi.org/10.2307/2332810

Breslow NE, Clayton DG (1993) Approximate inference in generalized linear mixed models. J Am Stat Assoc 88:9–25. https://doi.org/10.1080/01621459.1993.10594284

Brilhante AF, Dorval MEC, Cristaldo G et al (2019) Disney trap attractiveness for the cortelezzii series (Diptera: Psychodidae, Phlebotominae) in a fishing tourism area, Bonito Municipality, Mato Grosso do Sul State, Brazil. J Med Entomol 56:271–275. https://doi.org/10.1093/jme/tjy156

Brown HE, Paladini M, Cook RA et al (2008) Effectiveness of mosquito traps in measuring species abundance and composition. J Med Entomol 45:517–521. https://doi.org/10.1093/jmedent/45.3.517

Calvão-Brito RHS, Maia-Herzog M (2003) Modificação na técnica para montagem de simulídeos (Insecta, Diptera) em lâmina/lamínula. Rev Bras Zool 20:773–774. https://doi.org/10.1590/S0101-81752003000400034

Carvalho CJBD, Moura MO, Ribeiro PB (2002) Chave para adultos de dípteros (Muscidae, Fanniidae, Anthomyiidae) associados ao ambiente humano no Brasil. Rev Bras Entomol 46(2):107–144. https://doi.org/10.1590/S0085-56262002000200001

Castro M, Centeno N, González-Vainer P (2019) An initial study of insect succession on pig carcasses in open pastures in the northwest of Uruguay. For Sci Intern 302:109837. https://doi.org/10.1016/j.forsciint.2019.05.053

Challier A, Eyraud M, Lafaye A et al (1977) Amelioration du rendement du piege biconique pour glossines (Diptera, Glossinidae) par l'emploi d'un cone inferieur bleu [improving the performance of the biconical trap for flies (Diptera, Glossinidae) by using a lower blue cone]. Serie Entomologie Medicale et Parasitologie 15:283–286

Charlwood JD, Kessy E, Yohannes K et al (2018) Studies on the resting behaviour and host choice of *Anopheles gambiae* and *An. arabiensis* from Muleba, Tanzania. Med Vet Entomol 32:263–270. https://doi.org/10.1111/mve.12299

Chen YC, Wang CY, Teng HJ et al (2011) Comparison of the efficacy of CO_2-baited and unbaited light traps, gravid traps, backpack aspirators, and sweep net collections for sampling mosquitoes infected with Japanese encephalitis virus. J Vector Ecol 36:68–74. https://doi.org/10.1111/j.1948-7134.2011.00142.x

Cilek JE, Weston JR, Richardson AG (2017a) Comparison of adult mosquito abundance from Biogents-2 sentinel and biogents gravid Aedes traps in northeastern Florida. J Am Mosq Control Assoc 33:358–360. https://doi.org/10.2987/17-6678.1

Cilek JE, Knapp JA, Richardson AG (2017b) Comparative efficiency of biogents gravid Aedes trap, CDC autocidal gravid Ovitrap, and CDC gravid trap in northeastern Florida. J Am Mosq Control Assoc 33:103–107. https://doi.org/10.2987/16-6628.1

Clark GG, Seda H, Gubler DJ (1994) Use of the "CDC backpack aspirator" for surveillance of *Aedes aegypti* in San Juan, Puerto Rico. J Am Mosq Control Assoc 10:119–124

Clements AN (1992) The biology of mosquitoes. Development, nutrition and reproduction, vol 1. Chapman & Hall, London

Coetzee M (2020) Key to the females of Afrotropical *Anopheles* mosquitoes (Diptera: Culicidae). Mal J 19:1–20. https://doi.org/10.1186/s12936-020-3144-9

Cohnstaedt LW, Rochon K, Duehl AJ et al (2012) Arthropod surveillance programs: basic components, strategies, and analysis. An Entomol Soc Am 2:135–149. https://doi.org/10.1603/AN11127

Connor E, Simberloff D (1986) Competition, scientific method, and null models in ecology: because field experiments are difficult to perform, ecologists often rely on evidence that is nonexperimental and that therefore needs to be rigorously evaluated. Am Scientist 74:155–162. www.jstor.org/stable/27854031

Consoli RA, Lourenço-de-Oliveira R (1994) Principais mosquitos de importância sanitária no Brasil. Editora Fiocruz, Rio de Janeiro

Cooper R, Wang C, Singh N (2015) Mark-release-recapture reveals extensive movement of bed bugs (*Cimex lectularius* L.) within and between apartments. PLoS One 10(9):e0136462. https://doi.org/10.1371/journal.pone.0136462

Costa J, Almeida CE, Dotson EM et al (2003) The epidemiologic importance of *Triatoma brasiliensis* as a Chagas disease vector in Brazil: a revision of domiciliary captures during 1993-1999. Mem Inst Oswaldo Cruz 98:443–449. https://doi.org/10.1590/S0074-02762003000400002

Curtis-Robles R, Auckland LD, Snowden KF et al (2018) Analysis of over 1500 triatomine vectors from across the US, predominantly Texas, for *Trypanosoma cruzi* infection and discrete typing units. Infect Genet Evol 58:171–180. https://doi.org/10.1016/j.meegid.2017.12.016

Dabiré KR, Sawadogo PS, Hien DF et al (2013) Occurrence of natural *Anopheles arabiensis* swarms in an urban area of Bobo-Dioulasso city, Burkina Faso, West Africa. Acta Trop 130:44–50. https://doi.org/10.1016/j.actatropica.2013.12.012

De Barro P, Ahmed MZ (2011) Genetic networking of the *Bemisia tabaci* cryptic species complex reveals pattern of biological invasions. PLoS One 6:e25579. https://doi.org/10.1371/journal.pone.0025579

De Melo DPO, Scherrer LR, Eiras AE (2012) Dengue fever occurrence and vector detection by larval survey, ovitrap and MosquiTRAP: a space-time clusters analysis. PLoS One 7(7). https://doi.org/10.1371/journal.pone.0042125

Degener CM, Eiras AE, Azara TMF et al (2014) Evaluation of the effectiveness of mass trapping with BG-sentinel traps for dengue vector control: a cluster randomized controlled trial in Manaus, Brazil. J Med Entomol 51:408–420. https://doi.org/10.1603/ME13107

Demari-Silva B, Laporta GZ, Oliveira TMP et al (2020) *Plasmodium* infection in *Kerteszia cruzii* (Diptera: Culicidae) in the Atlantic tropical rain forest, southeastern Brazil. Infect Genet Evol 78:104061. https://doi.org/10.1016/j.meegid.2019.104061

Disney RH (1966) A trap for phlebotomine sandflies attracted to rats. Bull Entomol Res 56:445–451. https://doi.org/10.1017/S0007485300056510

Doggett S, Lilly D, Russell RC (2011) Battling bed bugs; the latest in weaponry. Prof Pest Manag 15:25–26

Dokianakis E, Tsirigotakis N, Christodoulou V et al (2016) DNA sequencing confirms PCR-RFLP identification of wild caught Larroussius sandflies from Crete and Cyprus. Acta Trop 164:314–320. https://doi.org/10.1016/j.actatropica.2016.09.003

Dorval MEC, Alves TP, Oliveira AG et al (2007) Modification of Disney trap for capture of sandflies (Diptera: Psychodidae: Phlebotominae). Mem Inst Osw Cruz 102:877–878. https://doi.org/10.1590/S0074-02762007005000111

Dorville FM (1996) Mosquitoes as bioindicators of forest degradation in southeastern Brazil. A statistical evaluation of published data in the literature. Stud Neotrop Fauna Environ 31:68–78. https://doi.org/10.1076/snfe.31.2.68.13331

Dryden MW, Broce AB (1993) Development of a trap for collecting newly emerged *Ctenocephalides felis* (Siphonaptera: Pulicidae) in homes. J Med Entomol 30:901–906. https://doi.org/10.1093/jmedent/30.5.901

Dryden MW, Perez HR, Ulitchny DM (1999) Control of fleas on pets and in homes by use of imidacloprid or lufenuron and a pyrethrin spray. J Am Vet Med Assoc 215:36–39

Dye-Braumuller KC, Gorchakov R, Gunter SM et al (2019) Identification of Triatomines and their habitats in a highly developed urban environment. Vector Borne Zoonotic Dis 19:265–273. https://doi.org/10.1089/vbz.2018.2352

Eiras AE (2002) Armadilha para captura de mosquitos. Brazil Patent PI0203907, 5 Sept 2002

Eiras AE, Jepson PC (1991) Host location by *Aedes aegypti* (Diptera: Culicidae): a wind tunnel study of chemical cues. Bull Entomol Res 81:151–160. https://doi.org/10.1017/S0007485300051221

Eiras AE, Jepson PC (1994) Responses of female *Aedes aegypti* (Diptera: Culicidae) to host odours and convection currents using an olfactometer bioassay. Bull Entomol Res 84:207–211. https://doi.org/10.1017/S0007485300039705

Eiras AE, Resende MC (2009) Preliminary evaluation of the 'Dengue-MI' technology for *Aedes aegypti* monitoring and control. Cad Saude Publica 25:45–58. https://doi.org/10.1590/S0102-311X2009001300005

Eiras AE, Geier M, Rose A et al (2010) Practical application of olfactory cues for monitoring and control of *Aedes aegypti* in Brazil: a case study. In: Takken W, Knols BGJ (eds) Olfaction in vector-host interactions, Ecology and control of vector-borne diseases, vol 2. Wageningen, Wageningen Academic Publishers, pp 365–398

Eiras AE, Buhagiar TS, Ritchie SA (2014) Development of the gravid Aedes trap for the capture of adult female container-exploiting mosquitoes (Diptera: Culicidae). J Med Entomol 51:200–209. https://doi.org/10.1603/ME13104

Eiras AE, Pires SF, Staunton KM et al (2018a) A high-risk Zika and dengue transmission hub: virus detections in mosquitoes at a Brazilian university campus. Parasit Vectors 11:359. https://doi.org/10.1186/s13071-018-2883-8

Eiras AE, Resende MC, Acebal JL et al (2018b) New cost-benefit of Brazilian technology for vector surveillance using trapping system. In: Kasenga FH (ed) Malaria. IntechOpen. https://doi.org/10.5772/intechopen.78781

Eldridge BF (2000) Surveillance for arthropodborne diseases. In: Eldridge BC, Edman JD (eds) Medical entomology. Springer, Dordrecht, pp 515–538. https://doi.org/10.1007/978-94-007-1009-2_13

Englbrecht C, Gordon S, Venturelli C et al (2015) Evaluation of BG-sentinel trap as a management tool to reduce *Aedes albopictus* nuisance in an urban environment in Italy. J Am Mosq Control Assoc 31:16–25. https://doi.org/10.2987/14-6444.1

Erlank E, Koekemoer LL, Coetzee M (2018) The importance of morphological identification of African anopheline mosquitoes (Diptera: Culicidae) for malaria control programmes. Mal J 17:43. https://doi.org/10.1186/s12936-018-2189-5

Farajollahi A, Kesavaraju B, Price DC et al (2009) Field efficacy of BG-sentinel and industry-standard traps for *Aedes albopictus* (Diptera: Culicidae) and West Nile virus surveillance. J Med Entomol 46:919–925. https://doi.org/10.1603/033.046.0426

Fávaro EA, Dibo MR, Mondini A et al (2006) Physiological state of *Aedes* (*Stegomyia*) *aegypti* mosquitoes captured with MosquiTRAPs in Mirassol, São Paulo, Brazil. J Vector Ecol 31:285–291. https://doi.org/10.3376/1081-1710(2006)31[285:PSOASA]2.0.CO;2

Ferreira DA, Degener CM, Almeida MTC et al (2017) Meteorological variables and mosquito monitoring are good predictors for infestation trends of *Aedes aegypti*, the vector of dengue, chikungunya and Zika. Parasit Vectors 10:78. https://doi.org/10.1186/s13071-017-2025-8

Flournoy LE, Adams RP, Pandy RN (1996) Interim and archival preservation of plant specimens in alcohols for DNA studies. BioTechniques 20:657–660

Foley DH, Rueda LM, Wikerson RC (2007) Insight into global mosquito biogeography from country species records. J Med Entomol 44:554–567

Folmer O, Black M, Hoeh W et al (1994) DNA primers for amplification of mitochondrial cytochrome c oxidase subunit I from diverse metazoan invertebrates. Mol Mar Biol Biotechnol 3:294–299

Forattini OP (2002) Culicidologia Médica. Identificação, Biologia, Epidemiologia, vol 2. Edusp, São Paulo

Foster PG, Oliveira TMP, Bergo ES et al (2017) Phylogeny of Anophelinae using mitochondrial protein coding genes. R Soc Open Sci 4:170758. https://doi.org/10.1098/rsos.170758

Galati EA (2018) Phlebotominae (Diptera, Psychodidae): classification, morphology and terminology of adults and identification of American taxa. In: Rangel EF, Shaw JJ (eds) Brazilian sandflies. Springer, Cham, pp 9–212. https://doi.org/10.1007/978-3-319-75544-1_2

Galati EAB, Nunes VLB, Dorval MEC et al (2001) Attractiveness of black Shannon trap for phlebotomines. Mem Inst Osw Cruz 96:641–647. https://doi.org/10.1590/S0074-02762001000500008

Gama RA, Silva EM, Silva IM et al (2007) Evaluation of the sticky MosquiTRAP™ for detecting *Aedes* (*Stegomyia*) *aegypti* (L.) (Diptera: Culicidae) during the dry season in Belo

Horizonte, Minas Gerais, Brazil. Neotrop Entomol 36:294–302. https://doi.org/10.1590/S1519-566X2007000200018

Gama RA, Silva IM, Geier M et al (2013) Development of the BG-malaria trap as an alternative to human-landing catches for the capture of *Anopheles darlingi*. Mem Inst Osw Cruz 108:763–771. https://doi.org/10.1590/0074-0276108062013013

Garret-Jones C (1964) The human blood index of malaria vectors in relation to epidemiological assessment. Bull World Health Organ 30:241–261

Gibson G, Brady J (1988) Flight response of tsetse flies on host odour plumes: the initial response to entering or leaving a plume. Physiol Entomol 13:29–42

Gomes AC (1998) Medidas dos níveis de infestação urbana para *Aedes* (*Stegomyia*) *aegypti* e *Aedes* (*Stegomyia*) *albopictus* em programas de Vigilância Entomológica. Informe Epidemiológico do SUS 7:49–57. https://doi.org/10.5123/S0104-16731998000300006

Govella NJ, Chaki PP, Geissbuhler Y et al (2009) A new tent trap for sampling exophagic and endophagic members of the *Anopheles gambiae* complex. Mal J 8:157. https://doi.org/10.1186/1475-2875-8-157

Gressitt JL, Gressitt MK (1962) An improved malaise trap. Pacific insects 4:87–90

Guerenstein PG, Lorenzo MG, Núñez J et al (1995) Baker's yeast, an attractant for baiting traps for Chagas' disease vectors. Experientia 51:834–837. https://doi.org/10.1007/BF01922439

Hadi UK, Soviana S, Rohmah IL (2019) Diversity and ecology of black flies in tea plantation area of Puncak Bogor, Indonesia. Acta Trop 199:104986. https://doi.org/10.1016/j.actatropica.2019.04.010

Hajibabaei M, Janzen DH, Burns JM et al (2006) DNA barcodes distinguish species of tropical Lepidoptera. Proc Natl Acad Sci U S A 103:968–971. https://doi.org/10.1073/pnas.0510466103

Hamada N, Nascimento JMC (2018) Family Simuliidae. In: Hamada N, Thorp JH, Rogers DC (eds) Thorp and Covich's freshwater invertebrates. Keys to Neotropical Hexapoda, vol 3. Academic Press, Amsterdam, pp 701–718. https://doi.org/10.1016/B978-0-12-804223-6.00032-9

Hamer GL, Bejcek JR, Valdez EA et al (2018) A pilot radio telemetry field study of triatomine vectors (Hemiptera: Reduviidae) of the Chagas disease parasite. J Med Entomol 55:1380–1385. https://doi.org/10.1093/jme/tjy094

Harbach RE (2007) The Culicidae (Diptera): a review of taxonomy, classification and phylogeny. Zootaxa 1668:591–638. https://doi.org/10.11646/zootaxa.1668.1.28

Harris C, Kihonda J, Lwetoijera D et al (2011) A simple and efficient tool for trapping gravid *Anopheles* at breeding sites. Parasit Vectors 4:125. https://doi.org/10.1186/1756-3305-4-125

Harwood JF, Rama V, Hash JM et al (2018) The attractiveness of the gravid Aedes trap to dengue vectors in Fiji. J Med Entomol 55:481–484. https://doi.org/10.1093/jme/tjx221

Hashiguchi K, Kato H, Hashiguchi Y (2014) Sand fly fauna (Diptera, Psychodidae, Phlebotominae) in different leishmaniasis-endemic areas of Ecuador, surveyed by using a newly named mini-Shannon trap. Trop Med Health 4:163–170. https://doi.org/10.2149/tmh.2014-20

Hebert PD, Cywinska A, Ball SL et al (2003) Biological identifications through DNA barcodes. Proc Biol Sci 270:313–321. https://doi.org/10.1098/rspb.2002.2218

Heidi E, Paladini M, Cook RA et al (2008) Effectiveness of mosquito traps in measuring species abundance and composition. J Med Entomol 1:517–521. https://doi.org/10.1093/jmedent/45.3.517

Hempolchom C, Reamtong O, Sookrung N et al (2019) Proteomes of the female salivary glands of *Simulium nigrogilvum* and *Simulium nodosum*, the main human-biting black flies in Thailand. Acta Trop 194:82–88. https://doi.org/10.1016/j.actatropica.2019.03.026

Hendy A, Sluydts V, Tushar T et al (2017) Esperanza window traps for the collection of anthropophilic blackflies (Diptera: Simuliidae) in Uganda and Tanzania. PLoS Negl Trop Dis 11(6):e0005688. https://doi.org/10.1371/journal.pntd.0005688

Hendy A, Krüger A, Pfarr K et al (2018) The blackfly vectors and transmission of *Onchocerca volvulus* in Mahenge, south eastern Tanzania. Acta Trop 181:50–59. https://doi.org/10.1016/j.actatropica.2018.01.009

Heringer L, Johnson BJ, Fikrig K et al (2016) Evaluation of alternative killing agents for *Aedes aegypti* (Diptera: Culicidae) in the gravid Aedes trap (GAT). J Med Entomol 53:873–879. https://doi.org/10.1093/jme/tjw051

Hesam-Mohammadi M, Abai MR et al (2014) Efficacy of different sampling methods of sandflies (Diptera: Psychodidae) in endemic focus of cutaneous leishmaniasis in Kashan district, Isfahan province, Iran. J Arthropod Borne Dis 8:156–162

Hiscox A, Otieno B, Kibet A et al (2014) Development and optimization of the Suna trap as a tool for mosquito monitoring and control. Mal J 13:257. https://doi.org/10.1186/1475-2875-13-257

Homan T, Hiscox A, Mweresa CK et al (2016) The effect of mass mosquito trapping on malaria transmission and disease burden (SolarMal): a stepped-wedge cluster-randomised trial. Lancet 388:1193–1201. https://doi.org/10.1016/S0140-6736(16)30445-7

Houghton R, Morkowski S, Stevens Y et al (2010) Monoclonal and polyclonal antibodies for development of rapid immunoassays for detection of sand fly fever viruses. Am Soc Trop Med Hyg 83:345–346

Huemer H, Prudhomme J, Amaro F et al (2017) Practical guidelines for studies on sandfly-borne phleboviruses: part ii: important points to consider for fieldwork and subsequent virological screening. Vector-Borne and Zoo Dis 17:81–90. https://doi.org/10.1089/vbz.2016.1965

Husbands R, Holten J (1967) An improved mechanical method of aspirating insects. Calif Vec Views 14:78–80

Jerry DC, Mohammed T, Mohammed A (2017) Yeast-generated CO_2: a convenient source of carbon dioxide for mosquito trapping using the BG-sentinel® traps. Asian Pacific J Trop Biomed 7:896–900. https://doi.org/10.1016/j.apjtb.2017.09.014

Ježek J, Votýpka J, Brzoňová J et al (2019) Horse flies (Diptera: Tabanidae) collected in Central African Republic, Gabon and Liberia with comments on their updated distribution. Acta Musei Silesiae, Scientiae Naturales 68:263–274. https://doi.org/10.2478/cszma-2019-0025

Johnson BJ, Ritchie SA (2015) The siren's song: exploitation of female flight tones to passively capture male *Aedes aegypti* (Diptera: Culicidae). J Med Entomol 53:245–248. https://doi.org/10.1093/jme/tjv165

Johnson BJ, Rohde BB, Zeak N et al (2018) A low-cost, battery-powered acoustic trap for surveilling male *Aedes aegypti* during rear-and-release operations. PLoS One 13(8):e0201709. https://doi.org/10.1371/journal.pone.0201709

Kaindoa EW, Ngowo HS, Limwagu A et al (2017) New evidence of mating swarms of the malaria vector, *Anopheles arabiensis* in Tanzania. Wellcome Open Res 2:88. https://doi.org/10.12688/wellcomeopenres.12458.1

Kaindoa EW, Ngowo HS, Limwagu A et al (2019) Swarms of the malaria vector *Anopheles funestus* in Tanzania. Mal J 18:29. https://doi.org/10.1186/s12936-019-2660-y

Kaldun B, Otti O (2016) Condition-dependent ejaculate production affects male mating behavior in the common bedbug *Cimex lectularius*. Ecol Evol 6:2548–2558. https://doi.org/10.1002/ece3.2073

Kamgang B, Yougang AP, Tchoupo M et al (2017) Temporal distribution and insecticide resistance profile of two major arbovirus vectors *Aedes aegypti* and *Aedes albopictus* in Yaoundé, the capital city of Cameroon. Parasit Vectors 10:469. https://doi.org/10.1186/s13071-017-2408-x

Kappmeier K (2000) A newly developed odour-baited "H trap" for the live collection of *Glossina brevipalpis* and *Glossina austeni* (Diptera: Glossinidae) in South Africa. Onderstepoort J Vet Res 67:15–26

Kaufman PE, Rutz DA, Frisch S (2005) Large sticky traps for capturing house flies and stable flies in dairy calf greenhouse facilities. J Dairy Sci 88:176–181. https://doi.org/10.3168/jds.S0022-0302(05)72676-X

Kelly-Hope LA, McKenzie FE (2009) The multiplicity of malaria transmission: a review of entomological inoculation rate measurements and methods across sub-Saharan Africa. Mal J 8:19. https://doi.org/10.1186/1475-2875-8-19

Khamesipour F, Lankarani KB, Honarvar B et al (2018) A systematic review of human pathogens carried by the housefly (*Musca domestica* L.). Public Health 18:1049. https://doi.org/10.1186/s12889-018-5934-3

Kimsey RB, Kimsey PB (1984) Identification of arthropod blood meals using rubidium as a marker: a preliminary report. J Med Entomol 21:714–719. https://doi.org/10.1093/jmedent/21.6.714

Kline DL (1999) Comparison of two American biophysics mosquito traps: the professional and a new counterflow geometry trap. J Am Mosq Control Assoc 15:276–282

Koyoc-Cardeña E, Medina-Barreiro A, Cohuo-Rodríguez A et al (2019) Estimating absolute indoor density of *Aedes aegypti* using removal sampling. Parasit Vectors 12:250. https://doi.org/10.1186/s13071-019-3503-y

Kriska G, Bernáth B, Farkas R et al (2009) Degrees of polarization of reflected light eliciting polarotaxis in dragonflies (Odonata), mayflies (Ephemeroptera) and tabanid flies (Tabanidae). J Insect Physiol 55:1167–1173. https://doi.org/10.1016/j.jinsphys.2009.08.013

Kröckel U, Rose A, Eiras AE et al (2006) New tools for surveillance of adult yellow fever mosquitoes: comparison of trap catches with human landing rates in an urban environment. J Am Mosq Control Assoc 22:229–238. https://doi.org/10.2987/8756-971X(2006)22[229:NTFSOA]2.0.CO;2

Kuzoe F, Schofield C (2004) Strategic review of traps and targets for tsetse and African trypanosomiasis control (No TDR/IDE/TRY/05.1). World Health Organization, Geneva

Lana RM, Morais MM, Lima TFM et al (2018) Assessment of a trap based *Aedes aegypti* surveillance program using mathematical modeling. PLoS One 13:e0190673. https://doi.org/10.1371/journal.pone.0190673

Laurito M, Briscoe AG, Almiron WR et al (2018) Systematics of the *Culex coronator* complex (Diptera: Culicidae): morphological and molecular assessment. Zoo J Linnean Soc 182:735–757. https://doi.org/10.1093/zoolinnean/zlx053

Le Goff G, Damiens D, Payet L et al (2016) Enhancement of the BG-sentinel trap with varying number of mice for field sampling of male and female *Aedes albopictus* mosquitoes. Parasit Vectors 9:514. https://doi.org/10.1186/s13071-016-1801-1

Lent H, Wygodzinsky P (1979) Revision of the Triatominae (Hemiptera, Reduviidae), and their significance as vectors of Chagas' disease. Bull Am Museum Nat Hist 163:123–520

Leonard JA (2008) Ancient DNA applications for wildlife conservation. Mol Ecol 17:4186–4196. https://doi.org/10.1111/j.1365-294X.2008.03891.x

Li C, Wilkerson RC (2005) Identification of *Anopheles* (*Nyssorhynchus*) *albitarsis* complex species (Diptera: Culicidae) using rDNA internal transcribed spacer 2-based polymerase chain reaction primes. Mem Inst Osw Cruz 100:495–500. https://doi.org/10.1590/S0074-02762005000500009

Lima JBP, Rosa-Freitas MG, Rodovalho CM et al (2014) Is there an efficient trap or collection method for sampling *Anopheles darlingi* and other malaria vectors that can describe the essential parameters affecting transmission dynamics as effectively as human landing catches? A review. Mem Inst Osw Cruz 109:685–705. https://doi.org/10.1590/0074-0276140134

Lima JBP, Galardo AKR, Bastos LS et al (2017) MosqTent: an individual portable protective double-chamber mosquito trap for anthropophilic mosquitoes. PLoS Negl Trop Dis 11:e0005245. https://doi.org/10.1371/journal.pntd.0005245

Linardi PM, Guimarães LR (2000) Sifonápteros do Brasil. Museu de Zoologia da USP/FAPESP, São Paulo

Lindh JM, Torr SJ, Vale GA et al (2009) Improving the cost- effectiveness of artificial visual baits for controlling the tsetse fly *Glossina fuscipes fuscipes*. PLoS Negl Trop Dis 3:e474. https://doi.org/10.1371/journal.pntd.0000474

Liu H, Dixon D, Bibbs CS et al (2019) Autocidal gravid ovitrap incorporation with attractants for control of gravid and host-seeking *Aedes aegypti* (Diptera: Culicidae). J Med Entomol 56:576–578. https://doi.org/10.1093/jme/tjy207

Livak KJ (1984) Organization and mapping of a sequence on the *Drosophila melanogaster* X and Y chromosomes that is transcribed during spermatogenesis. Genetics 107:611–634

Lounibos LP, Kramer LD (2016) Invasiveness of *Aedes aegypti* and *Aedes albopictus* and vectorial capacity for chikungunya virus. J Infec Dis 214:S453–S458. https://doi.org/10.1093/infdis/jiw285

Lourenço J, Tennant W, Faria NR et al (2018) Challenges in dengue research: a computational perspective. Evol Appl 11:516–533. https://doi.org/10.1111/eva.12554

Lucas M, Krolow TK, Riet-Correa F et al (2020) Diversity and seasonality of horse flies (Diptera: Tabanidae) in Uruguay. Sci Rep 10:1–9. https://doi.org/10.1038/s41598-019-57356-0

Maciel-de-Freitas R, Codeço CT, Lourenço-de-Oliveira R (2007) Daily survival rates and dispersal of *Aedes aegypti* females in Rio de Janeiro, Brazil. Am J Trop Med Hyg 76:659–665. https://doi.org/10.4269/ajtmh.2007.76.659

Maciel-de-Freitas R, Eiras ÁE, Lourenço-de-Oliveira R (2008) Calculating the survival rate and estimated population density of gravid *Aedes aegypti* (Diptera, Culicidae) in Rio de Janeiro, Brazil. Cad Saude Publica 24:2747–2754. https://doi.org/10.1590/S0102-311X2008001200003

Mackay AJ, Amador M, Barrera R (2013) An improved autocidal gravid ovitrap for the control and surveillance of *Aedes aegypti*. Parasit Vectors 6:225. https://doi.org/10.1186/1756-3305-6-225

Maia MF, Robinson A, John A et al (2011) Comparison of the CDC backpack aspirator and the Prokopack aspirator for sampling indoor- and outdoor-resting mosquitoes in southern Tanzania. Parasit Vectors 4:124. https://doi.org/10.1186/1756-3305-4-124

Malele II, Ouma JO, Nyingilili HS et al (2016) Comparative performance of traps in catching tsetse flies (Diptera: Glossinidae) in Tanzania. Onderstepoort J Vet Res 83:1–7. https://doi.org/10.4102/ojvr.v83i1.1057

Mallya S, Sander B, Roy-Gagnon MH et al (2018) Factors associated with human West Nile virus infection in Ontario: a generalized linear mixed modelling approach. BMC Infect Dis 18:141. https://doi.org/10.1186/s12879-018-3052-6

Mann RS, Kaufman PE, Butler JF (2009) *Lutzomyia* spp. (Diptera: Psychodidae) response to olfactory attractant- and light emitting diode-modified mosquito magnet X (MM-X) traps. J Med Entomol 46:1052–1061. https://doi.org/10.1603/033.046.0512

Mantilla JS, Moncada LI, Matta NE et al (2018) Distribution of black flies (Diptera: Simuliidae) along an elevational gradient in the Andes Mountains of Colombia during the El Niño Southern Oscillation. Acta Trop 183:162–172. https://doi.org/10.1016/j.actatropica.2018.04.001

Marini F, Caputo B, Pombi M et al (2010) Study of *Aedes albopictus* dispersal in Rome, Italy, using sticky traps in mark–release–recapture experiments. Med Vet Entomol 24:361–368. https://doi.org/10.1111/j.1365-2915.2010.00898.x

Marshall E (2005) Will DNA bar codes breathe life into classification? Science 307:1037. https://doi.org/10.1126/science.307.5712.1037

Martins-Neto RG (2003) The fossil tabanids (Diptera Tabanidae): when they began to appreciate warm blood and when they began transmit diseases? Mem Inst Osw Cruz 98:29–34. https://doi.org/10.1590/S0074-02762003000900006

Mathenge EM, Killeen GF, Oulo DO et al (2002) Development of an exposure-free bednet trap for sampling Afrotropical malaria vectors. Med Vet Entomol 16:67–74. https://doi.org/10.1046/j.0269-283x.2002.00350.x

Mburu MM, Mzilahowa T, Amoah B et al (2019) Biting patterns of malaria vectors of the lower Shire valley, southern Malawi. Acta Trop 197:105059. https://doi.org/10.1016/j.actatropica.2019.105059

McGregor BL, Sloyer KE, Sayler KA et al (2019) Field data implicating *Culicoides stellifer* and *Culicoides venustus* (Diptera: Ceratopogonidae) as vectors of epizootic hemorrhagic disease virus. Parasit Vectors 12:258. https://doi.org/10.1186/s13071-019-3514-8

Mercer DR, Bossin H, Sang MC et al (2012) Monitoring temporal abundance and spatial distribution of *Aedes polynesiensis* using BG-sentinel traps in neighboring habitats on Raiatea, Society Archipelago, French Polynesia. J Med Entomol 49:51–60. https://doi.org/10.1603/ME11087

Mitchell SN, Kakani EG, South A et al (2015) Evolution of sexual traits influencing vectorial capacity in anopheline mosquitoes. Science 347:985–988. https://doi.org/10.1126/science.1259435

Molyneux DH, Ashford RW (1983) The biology of *Trypanosoma* and *Leishmania*, parasites of man and domestic animals. Taylor & Francis, London

Moon SA, Cohnstaedt LW, McVey DS et al (2019) A spatio-temporal individual-based network framework for West Nile virus in the USA: spreading pattern of West Nile virus. PLoS Comput Biol 15(3):e1006875. https://doi.org/10.1371/journal.pcbi.1006875

Mulla MS, Chansang U (2007) Pestiferous nature, resting sites, aggregation, and host-seeking behavior of the eye fly *Siphunculina funicola* (Diptera: Chloropidae) in Thailand. J Vector Ecol 32:292–301. https://doi.org/10.3376/1081-1710(2007)32[292:PNRSAA]2.0.CO;2

Müller GC, Dryden MW, Revay EE et al (2011) Understanding attraction stimuli of the cat flea, *Ctenocephalides felis*, in non-chemical control methods. Med Vet Entomol 25:413–420. https://doi.org/10.1111/j.1365-2915.2011.00960.x

Nagy ZT (2010) A hands-on overview of tissue preservation methods for molecular genetic analyses. Org Divers Evol 10:91–105. https://doi.org/10.1007/s13127-010-0012-4

Nelson DB, Chamberlain RW (1955) A light trap and mechanical aspirator operating on dry cell batteries. Mosq News 15:28–32

Newhouse VF, Chamberlain RW, Johnston JG et al (1966) Use of dry ice to increase mosquito catches of the CDC miniature light trap. Mosq News 26:30–35

Oca-Aguilar ACM, Moo-Llanes D, Rebollar-Téllez EA (2013) Adult sand fly species from diurnal resting sites on the peninsula of Yucatan, Mexico. Southwestern Entomol 38:241–250. https://doi.org/10.3958/059.038.0209

Odiere M, Bayoh MN, Gimnig J et al (2007) Sampling outdoor, resting *Anopheles gambiae* and other mosquitoes (Diptera: Culicidae) in western Kenya with clay pots. J Med Entomol 44:14–22. https://doi.org/10.1093/jmedent/41.5.14

Ohm JR, Teeple J, Nelson WA et al (2016) Fitness consequences of altered feeding behavior in immune-challenged mosquitoes. Parasit Vectors 9:113. https://doi.org/10.1186/s13071-016-1392-x

Onyango SA, Kitron U, Mungai P et al (2013) Monitoring malaria vector control interventions: effectiveness of five different adult mosquito sampling methods. J Med Entomol 50:1140–1151. https://doi.org/10.1603/ME12206

Owino EA, Sang R, Sole CL et al (2014) Field evaluation of natural human odours and the biogent-synthetic lure in trapping *Aedes aegypti*, vector of dengue and chikungunya viruses in Kenya. Parasit Vectors 7:451. https://doi.org/10.1186/1756-3305-7-451

Pepin KM, Marques-Toledo C, Scherer L et al (2013) Cost-effectiveness of novel system of mosquito surveillance and control, Brazil. Emerg Infec Dis 19:542. https://doi.org/10.3201/eid1904.120117

Pepin KM, Leach CB, Marques-Toledo C et al (2015) Utility of mosquito surveillance data for spatial prioritization of vector control against dengue viruses in three Brazilian cities. Parasit Vectors 8:98. https://doi.org/10.1186/s13071-015-0659-y

Phaijoo GR, Gurung DB (2015) Mathematical study of biting rates of mosquitoes in transmission of dengue disease. J Sci Eng Technol 11:25–33

Pimenta FV, Diotaiuti L, Lustosa ACL et al (2007) Evaluation of cultures of *Saccharomyces cerevisiae* as baits for *Triatoma dimidiata* and *Triatoma pallidipennis*. Mem Inst Oswaldo Cruz 102:229–231. https://doi.org/10.1590/S0074-02762007005000014

Pinto MC, Campbell-Lendrum DH, Lozovei AL et al (2001) Phlebotomine sandfly responses to carbon dioxide and human odour in the field. Med Vet Entomol 15:132–139. https://doi.org/10.1046/j.1365-2915.2001.00294.x

Ponlawat A, Khongtak P, Jaichapor B et al (2017) Field evaluation of two commercial mosquito traps baited with different attractants and colored lights for malaria vector surveillance in Thailand. Parasit Vectors 10:378. https://doi.org/10.1186/s13071-017-2315-1

Qiu YT, Smallegange RC, Braak CJ et al (2007) Attractiveness of MM-X traps baited with human or synthetic odor to mosquitoes (Diptera: Culicidae) in the Gambia. J Med Entomol 44:970–983. https://doi.org/10.1093/jmedent/44.6.970

Ratcliffe ST, Webb DW, Weinzievl RA et al (2003) PCR-RFLP identification of Diptera (Calliphoridae, Muscidae and Sarcophagidae)-a generally applicable method. J For Sci 48:783–785. https://doi.org/10.1520/JFS2002136

Ratnasingham S, Hebert PDN (2007) BOLD: the barcode of life data system (www.barcodinglife.org). Mol Ecol Notes 7:355–364. https://doi.org/10.1111/j.1471-8286.2007.01678.x

Rebollar-Téllez EA, Reyes-Villanueva F, Escobedo-Ortegón J et al (2009) Abundance and nightly activity behavior of a sylvan population of *Triatoma dimidiata* (Hemiptera: Reduviidae: Triatominae) from the Yucatan, México. J Vector Ecol 34:304–310. https://doi.org/10.1111/j.1948-7134.2009.00038.x

Resende MC, Ázara TMF, Costa IO et al (2012) Field optimisation of MosquiTRAP sampling for monitoring *Aedes aegypti* Linnaeus (Diptera: Culicidae). Mem Inst Osw Cruz 107:294–302. https://doi.org/10.1590/S0074-02762012000300002

Resende MC, Silva IM, Ellis BR et al (2013) A comparison of larval, ovitrap and MosquiTRAP surveillance for *Aedes* (*Stegomyia*) *aegypti*. Mem Inst Osw Cruz 108:1024–1030. https://doi.org/10.1590/0074-0276130128

Rigot T, Gilbert M (2012) Quantifying the spatial dependence of *Culicoides* midge samples collected by Onderstepoort-type blacklight traps: an experimental approach to infer the range of attraction of light traps. Med Vet Entomol 26:152–161. https://doi.org/10.1111/j.1365-2915.2011.00988.x

Ritchie SA, Buhagiar TS, Townsend M et al (2014) Field validation of the gravid Aedes trap (GAT) for collection of *Aedes aegypti* (Diptera: Culicidae). J Med Entomol 51:210–219. https://doi.org/10.1603/ME13105

Rodrigues MS, Silva IM, Leal LB et al (2014) Development of a new mosquito retention system for the BG-malaria trap to reduce the damage to mosquitoes. J Am Mosq Control Assoc 30:184–190. https://doi.org/10.2987/14-6423R.1

Rodríguez-Pérez MA, Vazquez-Sanchez R, Reyes-Villanueva F (2001) Estimation of survival and gonotrophic cycle length of *Simulium Metallicum* (Diptera: Simuliidae) in southern Mexico. J Am Mosq Control Assoc 17:254–259

Rodríguez-Pérez MA, Adeleke MA, Burkett-Cadena ND et al (2013) Development of a novel trap for the collection of black flies of the *Simulium ochraceum* complex. PLoS One 8(10):e76814. https://doi.org/10.1371/journal.pone.0076814

Rodríguez-Rojas JJ, Arque-Chunga W, Fernández-Salas I et al (2016) Comparative field evaluation of different traps for collecting adult phlebotomine sandflies (Diptera: Psychodidae) in an endemic area of cutaneous leishmaniasis in Quintana Roo, Mexico. J Am Mosq Control Assoc 32:103–116. https://doi.org/10.2987/moco-32-02-103-116.1

Roiz D, Duperier S, Roussel M (2016) Trapping the tiger: efficacy of the novel BG-sentinel 2 with several attractants and carbon dioxide for collecting *Aedes albopictus* (Diptera: Culicidae) in southern France. J Med Entomol 53:460–465. https://doi.org/10.1093/jme/tjv184

Rolón M, Vega MC, Román F et al (2011) First report of colonies of sylvatic *Triatoma infestans* (Hemiptera: Reduviidae) in the Paraguayan Chaco, using a trained dog. Plos Neg Trop Dis 5:e1026. https://doi.org/10.1371/journal.pntd.0001026

Ronderos MM, Díaz F, Marino PI et al (2018) Family Ceratopogonidae. In: Hamada N, Thorp JH, Rogers DC (eds) Thorp and Covich's freshwater invertebrates. Keys to Neotropical Hexapoda, vol 3. Academic Press, Amsterdam, pp 625–659. https://doi.org/10.1016/B978-0-12-804223-6.00030-5

Rubio-Palis Y, Curtis CF (1992) Biting and resting behaviour of anophelines in western Venezuela and implications for control of malaria transmission. Med Vet Entomol 6:325–334. https://doi.org/10.1111/j.1365-2915.1992.tb00628.x

Russell RC, Webb CE, Williams CR et al (2005) Mark–release–recapture study to measure dispersal of the mosquito *Aedes aegypti* in Cairns, Queensland, Australia. Med Vet Entomol 19:451–457. https://doi.org/10.1111/j.1365-2915.2005.00589.x

Rutto JJ, Osano O, Odenyo V (2018) Influence of human, wildlife and livestock husbandry on epidemiology of human African trypanosomiasis at the transboundary of Western Kenya and Southeast Uganda. J Infect Dis Ther 6:364. https://doi.org/10.4172/2332-0877.1000364

Salah I, Abbasi I, Warburg A et al (2020) The spatial and temporal distribution, species composition, and host preference of phlebotomine sandflies in the Bethlehem District of Palestine. Acta Trop 203:105327. https://doi.org/10.1016/j.actatropica.2019.10532

Sanchis-Marin MC, Morillas-Marquez J, Gonzalez-Castro J et al (1989) Seasonal dynamics of phlebotomine sandflies (Diptera: Phlebotomidae) in the province of Almeria, Spain. Revista Ibérica de Parasitologia 46:284–292

Schlein Y, Yuval B (1987) Leishmaniasis in the Jordan Valley. IV. Attraction of Phlebotomus papatasi to plants in the field. J Med Entomol 24:87–90. https://doi.org/10.1093/jmedent/24.1.87

Schofield CJ (1979) The behaviour of Triatominae (Hemiptera: Reduviidae): a review. Bull Entomol Res 69:363–379. https://doi.org/10.1017/S0007485300018897

Schofield CJ, Lehane MJ, McEwen P et al (1992) Dispersive flight by *Triatoma infestans* under natural climatic conditions in Argentina. Med Vet Entomol 6:51–56. https://doi.org/10.1111/j.1365-2915.1992.tb00035.x

Scott JA, Brogdon WG, Collins FH (1993) Identification of single specimens of the *Anopheles gambiae* complex by the polymerase chain reaction. Am J Trop Med Hyg 49:520–529. https://doi.org/10.4269/ajtmh.1993.49.520

Sedda L, Vilela APP, Aguiar ERGR et al (2018) The spatial and temporal scales of local dengue virus transmission in natural settings: a retrospective analysis. Parasit Vectors 11:79. https://doi.org/10.1186/s13071-018-2662-6

Sedda L, Lucas ER, Djogbénou LS et al (2019) Improved spatial ecological sampling using open data and standardization: an example from malaria mosquito surveillance. J R Soc Interface 16:20180941. https://doi.org/10.1098/rsif.2018.0941

Service MW (1970) A battery-operated light-trap for sampling mosquito populations. Bull World Health Organ 43:635

Service MW (1979) Light trap collections of ovipositing *Simulium squamosum* in Ghana. Ann Trop Med Parasitol 73:487–490. https://doi.org/10.1080/00034983.1979.11687288

Shannon RC (1939) Methods for collecting and feeding mosquitoes in jungle yellow fever studies. Am J Trop Med Hyg 1:131–140. https://doi.org/10.4269/ajtmh.1939.s1-19.131

Sharp TM, Lorenzi O, Torres-Velasquez B et al (2019) Autocidal gravid ovitraps protect humans from chikungunya virus infection by reducing *Aedes aegypti* mosquito populations. Plos Neg Trop Dis 13:e0007538. https://doi.org/10.1371/journal.pntd.0007538

Shelley AJ (1991) Simuliidae and the transmission and control of human Onchocerciasis in Latin America. Cad Saude Publica 7:310–327. https://doi.org/10.1590/S0102-311X1991000300003

Shelley AJ, Hernández LM, Davies JB (2004) Blackflies (Diptera: Simuliidae) of southern Guyana with keys for the identification of adults and pupae: a review. Mem Inst Osw Cruz 99:443–470. https://doi.org/10.1590/S0074-02762004000500001

Shimabukuro PHF, Moreira JAC, Costa TSD (2016) Sandflies (Diptera: Psychodidae) from the Brazilian Atlantic Forest domain collected with malaise traps. J Med Entomol 53:1488–1491. https://doi.org/10.1093/jme/tjw107

Silva NMMG, De Melo SCCS, Massafera R et al (2013) Dispersal and memory of sandflies in an endemic area of cutaneous leishmaniasis, southern Brazil. J Med Entomol 50:986–993. https://doi.org/10.1603/ME12065

Silva JAO, Silva FJD, Macedo LOD et al (2019) Sandflies in an endemic area for visceral Leishmaniasis in northeastern Brazil. Braz J Vet Parasitol 28:569–573. https://doi.org/10.1590/s1984-29612019073

Silver JB (2008) Mosquito ecology: field sampling methods. Springer, Dordrecht

Singh OP, Nanda N, Chandra D et al (2020) Modified PCR-based assay for the differentiation of members of *Anopheles fluviatilis* complex in consequence of the discovery of a new cryptic species (species V). Mal J 19:96. https://doi.org/10.1186/s12936-020-03172-1

Steffler LM, Marteis LS, Dolabella SS et al (2011) Risk of dengue occurrence based on the capture of gravid *Aedes aegypti* females using MosquiTRAP. Mem Inst Osw Cruz 106:365–367. https://doi.org/10.1590/S0074-02762011000300018

Stork NE (2018) How many species of insects and other terrestrial arthropods are there on earth? Annu Rev Entomol 63:31–45. https://doi.org/10.1146/annurev-ento-020117-043348

Suchy JT, Lewis VR (2011) Host-seeking behavior in the bed bug, *Cimex lectularius*. Insects 2:22–35. https://doi.org/10.3390/insects2010022

Sudia WD, Chamberlain RW (1962) Battery-operated light trap, an improved model. Mosq News 22:126–129

Suparit P, Wiratsudakul A, Modchang C (2018) A mathematical model for Zika virus transmission dynamics with a time-dependent mosquito biting rate. Theor Biol Med Modelling 15:11. https://doi.org/10.1186/s12976-018-0083-z

Takken W, Knols BG (eds) (2010) Olfaction in vector-host interactions, Ecology and control of vector-borne diseases, vol 2. Wageningen Academic Publishers, Wageningen

Tandina F, Doumbo O, Yaro AS et al (2018) Mosquitoes (Diptera: Culicidae) and mosquito-borne diseases in Mali, West Africa. Parasit Vectors 11:467. https://doi.org/10.1186/s13071-018-3045-8

Tangena J-AA, Thammavong P, Hiscox A et al (2015) The human-baited double net trap: an alternative to human landing catches for collecting outdoor biting mosquitoes in Lao PDR. PLoS One 10:e0138735. https://doi.org/10.1371/journal.pone.0138735

Thomson M, Connor SJ (2000) Environmental information systems for the control of arthropod vectors of disease. Med Vet Entomol 14:227–244. https://doi.org/10.1046/j.1365-2915.2000.00250.x

Trpis M (1968) A suction apparatus for collecting mosquitoes and other insects. Mosq News 28:647

Updyke EA, Allan BF (2018) An experimental evaluation of cross-vane panel traps for the collection of sylvatic triatomines (Hemiptera: Reduviidae). J Med Entomol 55:485–489. https://doi.org/10.1093/jme/tjx224

Vaidyanathan R, Feldlaufer MF (2013) Bed bug detection: current technologies and future directions. Am J Trop Med Hyg 88:619–625. https://doi.org/10.4269/ajtmh.12-0493

Vale GA, Lovemore DF, Flint S et al (1988) Odour-baited targets to control tsetse flies *Glossina* spp. (Diptera: Glossinidae), in Zimbabwe. Bull Entomol Res 78:31–49. https://doi.org/10.1017/S0007485300016059

Vasconcelos PF, Calisher CH (2016) Emergence of human arboviral diseases in the Americas, 2000–2016. Vector Borne Zoonotic Dis 16:295–301. https://doi.org/10.1089/vbz.2016.1952

Vazquez-Prokopec GM, Ceballos LA, Kitron U et al (2004) Active dispersal of natural populations of *Triatoma infestans* (Hemiptera: Reduviidae) in rural northwestern Argentina. J Med Entomol 41:614–621. https://doi.org/10.1603/0022-2585-41.4.614

Vazquez-Prokopec GM, Galvin WA, Kelly R et al (2009) A new, cost-effective, battery-powered aspirator for adult mosquito collections. J Med Entomol 46:1256–1259. https://doi.org/10.1603/033.046.0602

Verhulst NO, Bakker JW, Hiscox A (2015) Modification of the Suna trap for improved survival and quality of mosquitoes in support of epidemiological studies. J Am Mosq Control Assoc 31:223–232. https://doi.org/10.2987/moco-31-03-223-232.1

Vexanet JA, Barreto AC, Cuba CC et al (1986) Características epidemiológicas da leishmaniose tegumentar americana em uma região endêmica do estado da Bahia. III. Fauna flebotomínica. Mem Inst Osw Cruz 81:293–301. https://doi.org/10.1590/S0074-02761986000300005

Vijayan S, Anbalagan S (2018) Assemblage pattern and seasonality of larval black flies (Simuliidae: Diptera) in a stream of southern eastern Ghats. Intern J Curr Trends Sci Technol 8:20187–20196. https://doi.org/10.15520/ctst.v8i03.383.pdf

Vilela ML, Zwetsch A, Silva JDS (2018) Methods for capturing, processing and preserving Phlebotominae. In: Rangel EF, Shaw JJ (eds) Brazilian sandflies. Springer, Cham, pp 443–466. https://doi.org/10.1007/978-3-319-75544-1_10

Wahlberg N, Brower AVZ, Nylin S (2005) Phylogenetic relationships and historical biogeography of tribes and genera in the subfamily Nymphalinae (Lepidoptera: Nymphalidae). Biol J Linn Soc 86:227–251. https://doi.org/10.1111/j.1095-8312.2005.00531.x

Walsh JF, Davies JB, Le Berre R et al (1978) Standardization of criteria for assessing the effect of *Simulium* control in onchocerciasis control programmes. Trans R Soc Trop Med Hyg 72:675–676. https://doi.org/10.1016/0035-9203(78)90039-1

Wang C, Singh N, Zha C et al (2016) Bed bugs: prevalence in low-income communities, resident's reactions, and implementation of a low-cost inspection protocol. J Med Entomol 53:639–646. https://doi.org/10.1093/jme/tjw018

Wilkinson GS, Fleming TH (1996) Migration and evolution of lesser long-nosed bats *Leptonycteris curasoae*, inferred from mitochondrial DNA. Mol Ecol 5:329–339. https://doi.org/10.1046/j.1365-294X.1996.00081.x

Williams CR, Long SA, Russell RC et al (2006) Field efficacy of the BG-sentinel compared with CDC backpack aspirators and CO_2-baited EVS traps for collection of adult *Aedes aegypti* in Cairns, Queensland, Australia. J Am Mosq Control Assoc 22:296–300. https://doi.org/10.2987/8756-971X(2006)22[296:FEOTBC]2.0.CO;2

Wisnivesky-Colli C, Paulone I, Perez A et al (1987) A new tool for continuous detection of the presence of triatomine bugs, vectors of Chagas disease, in rural households. Medicina (Buenos Aires) 47:45–50. https://doi.org/10.2987/8756-971X(2006)22[296:FEOTBC]2.0.CO;2

World Health Organization (1975) Manual on practical entomology in Malaria. Methods and techniques, Part II, vol 131. WHO Offset Publication, p 47

Xue RD, Doyle MA, Kline DL (2008) Field evaluation of CDC and mosquito magnet® X traps baited with dry ice, CO_2 sachet, and octenol against mosquitoes. J Am Mosq Control Assoc 24:249–252. https://doi.org/10.2987/5701.1

Ya'cob Z, Takaoka H, Pramual P et al (2016) Distribution pattern of black fly (Diptera: Simuliidae) assemblages along an altitudinal gradient in peninsular Malaysia. Parasit Vectors 9:219. https://doi.org/10.1186/s13071-016-1492-7

Yoshioka M, Couret J, Kim F et al (2012) Diet and density dependent competition affect larval performance and oviposition site selection in the mosquito species *Aedes albopictus* (Diptera: Culicidae). Parasit Vectors 5:225. https://doi.org/10.1186/1756-3305-5-225

Zeger S, Liang K (1986) Longitudinal data analysis for discrete and continuous outcomes. Biometrics 42:121–130. https://doi.org/10.2307/2531248

Zhang DX, Hewitt GM (1997) Insect mitochondrial control region: a review of its structure, evolution and usefulness in evolutionary studies. Biochem Syst Ecol 25:99–120. https://doi.org/10.1016/S0305-1978(96)00042-7

Zink RM, Barrowclough GF (2008) Mitochondrial DNA under siege in avian phylogeography. Mol Ecol 17:2107–2121. https://doi.org/10.1111/j.1365-294X.2008.03737.x

Zumpt F (1973) The Stomoxyinae biting flies of the world. Taxonomy, biology, economic importance and control measures. Gustav Fischer Verlag, Stuttgart

Index

J. C. Santos, G. W. Fernandes (eds.), *Measuring Arthropod Biodiversity*,
https://doi.org/10.1007/978-3-030-53226-0

M

N

Q

R

The manufacturer's authorised representative in the EU is Springer Nature Customer Service Centre GmbH, Europaplatz 3, 69115 Heidelberg, Germany. If you have any concerns regarding our products, please contact ProductSafety@springernature.com

Printed and bound by CPI Group (UK) Ltd, Croydon, CR0 4YY
30/09/2026
02225495-0017